Classes of Finite Groups

Mathematics and Its Applications

Managing Editor:

M. HAZEWINKEL

Centre for Mathematics and Computer Science, Amsterdam, The Netherlands

Classes of Finite Groups

by

Adolfo Ballester-Bolinches

Universitat de València,
València, Spain

and

Luis M. Ezquerro

Universidad Pública de Navarra,
Pamplona, Spain

 Springer

A C.I.P. Catalogue record for this book is available from the Library of Congress.

ISBN-10 1-4020-4718-5 (HB)
ISBN-13 978-1-4020-4718-3 (HB)
ISBN-10 1-4020-4719-3 (e-book)
ISBN-13 978-1-4020-4719-0 (e-book)

Published by Springer,
P.O. Box 17, 3300 AA Dordrecht, The Netherlands.

www.springer.com

Printed on acid-free paper

For the ones we love:
Fran, Isabel, Eneko

Contents

Preface

In the sixties and seventies of the last century, in parallel to the tremendous effort to classify the simple groups, a large number a papers created a beautiful and comprehensive view of finite soluble groups. In 1980, when the classification was almost completed, Helmut Wielandt proposed giving priority after the classification to the extension of these brilliant results of the theory of finite soluble groups to the more ambitious universe of all finite groups.

Almost at the same time Klaus Doerk and Trevor Hawkes started to write a volume gathering, ordering, and systematising the rich stuff of soluble groups. This encyclopedic work took more than ten years to accomplish. The publication of *Finite soluble groups* (De Gruyter, 1992) is a crucial milestone in the history of the development of the theory of classes of finite soluble groups. In fact lots of separate pieces of the manuscript, generously distributed by the authors to all interested specialists, had a strong influence on the research of the area even before the publication of the volume.

In the last decade, the Doerk-Hawkes' book has been one of the most powerful tools for undertaking Wielandt's task. The consequence is an impressive flourishing of ideas, methods and results illuminating the structure of finite groups. Furthermore, this process has produced a new arithmetic-free approach to understand some aspects of the soluble case.

We believe that there is already a lot of work published in this area and consequently there is a need for a detailed account of the theory of classes of groups in the general finite universe. The present book represents an attempt to meet this need.

Our main objective in this book is to present the latest achievements and investigations continuing the Doerk-Hawkes book to enlarge and adapt the methods of the soluble case to classes of finite non-necessarily soluble, according to Wielandt's proposal.

The contents of the book are organised in seven chapters. Chapter 1 begins with primitive groups and crowns. These concepts are central to our approach. It continues with the study of solid sets and systems of maximal subgroups. They are, together with the generalised Jordan-Hölder theorem, the ingredients combined to introduce the prefrattini subgroups in Chapter 4. Chapter 2 contains definitions, and elementary and basic results on classes of groups. Chapter 3 deals with partially saturated formations. A unified extension of the theorems of Gaschütz-Lubeseder-Schmid and Baer on the local character of the saturated and solubly saturated formations is presented there. Normalisers associated with Schunck classes $\mathfrak{H}$ of the form $\mathrm{E}_\Phi\,\mathfrak{F}$ for some formation $\mathfrak{F}$ and prefrattini subgroups associated to arbitrary Schunck classes are studied in Chapter 4, whereas Chapter 5 is devoted to presenting an alternative approach to a theory of projectors and covering subgroups in arbitrary finite groups resembling the corresponding theory in finite soluble groups. It is based on Salomon's Dissertation *Strukturerhaltende Untergruppen, Schunckklassen und extreme Klassen endlicher Gruppen*, Johannes Gutenberg-Universität, Mainz, 1987. Subnormal subgroups associated to formations are the main theme of Chapter 6. This concept was introduced by Hawkes in 1969 in the soluble universe and it turns out to be very useful in the study of the structure of finite groups. The last chapter contains some of the recent developments of the theory of Fitting classes, focusing our attention on injective Fitting classes and supersoluble Fitting classes. In particular, a detailed account of Salomon's unpublished example of a non-injective Fitting class is included.

To end this preface, we would like to pay a tribute to the figure of Professor Klaus Doerk, recently deceased. Without Doerk and his research team's collaboration, this book would have never ever come to be.

Acknowledgements

We would like to conclude by expressing our deepest gratitude to Ramón Esteban-Romero for his patient work with our manuscripts. His knowledge on this book's issues as well as his master skills on the use of TeX made him the best helper for the meticulous task of editing this book.

We would also like to thank Homer Bechtell, John Cossey, Arny Feldman, María Jesús Iranzo, Paz Jiménez-Seral, Carmen Lacasa-Esteban, Julio Lafuente, Inmaculada Lizasoain, María del Carmen Pedraza-Aguilera, Tatiana Pedraza and Francisco Pérez-Monasor for their valious collaborations, as well as to the *Ministerio de Educación y Ciencia* (Spanish Government) and *FEDER* (European Union) for their financial support via the grants MTM2004-08219-C02-01 and MTM2004-08219-C02-02.

To conclude, we must thank Springer for converting this project into a reality and for their continuous patience and help while writing this book.

Torres-Torres,   *A. Ballester-Bolinches*
Pamplona,   *Luis M. Ezquerro*
January, 2006

Maximal subgroups and chief factors

1.1 Primitive groups

This book, devoted to classes of finite groups, begins with the study of a class, the class of primitive groups, with no hereditary properties, the usual requirement for a class of groups, but whose importance is overwhelming to understand the remainder. We shall present the classification of primitive groups made by R. Baer and the refinement of this classification known as the O'Nan-Scott Theorem. The book of H. Kurzweil and B. Stellmacher [KS04], recently appeared, presents an elegant proof of this theorem. Our approach includes the results of F. Gross and L. G. Kovács on induced extensions ([GK84]) which are essential in some parts of this book.

We will assume our reader to be familiar with the basic concepts of permutation representations: G-sets, orbits, faithful representation, stabilisers, transitivity, the Orbit-Stabiliser Theorem, ... (see [DH92, A, 5]). In particular we recall that the stabilisers of the elements of a transitive G-set are conjugate subgroups of G and any transitive G-set Ω is isomorphic to the G-set of right cosets of the stabiliser of an element of Ω in G.

Definition 1.1.1. *Let G be a group and Ω a transitive G-set. A subset $\Phi \subseteq \Omega$ is said to be a* block *if, for every $g \in G$, we have that $\Phi^g = \Phi$ or $\Phi^g \cap \Phi = \emptyset$.*

Given a G-set Ω, trivial examples of blocks are $\emptyset$, Ω and any subset with a single element $\{\omega\}$, for any $\omega \in \Omega$. In fact, these are called *trivial blocks*.

Proposition 1.1.2. *Let G be a group which acts transitively on a set Ω and $\omega \in \Omega$. There exists a bijection*

$$\{block\ \Phi\ of\ \Omega : \omega \in \Phi\} \longrightarrow \{H \leq G : G_\omega \leq H\}$$

which preserves the containments.

Proof. Given a block Φ in Ω such that $\omega \in \Phi$, then $G_\Phi = \{g \in G : \Phi^g = \Phi\}$ is a subgroup of G and the stabiliser G_ω is a subgroup of G_Φ. Conversely, if H is a subgroup of G containing G_ω, then the set $\Phi = \{\omega^h : h \in H\}$ is a block and $\omega \in \Phi$. These are the mutually inverse bijections required. $\qquad \square$

The following result is well-known and its proof appears, for instance, in Huppert's book [Hup67, II, 1.2].

Theorem 1.1.3. *Let G be a group which acts transitively on a set Ω and assume that Φ is a non-trivial block of the action of G on Ω. Set $H = \{g \in G : \Phi^g = \Phi\}$. Then H is a subgroup of G.*

Let $\mathcal{T}$ be a right transversal of H in G. Then

1. *$\{\Phi^t : t \in \mathcal{T}\}$ is a partition of Ω.*
2. *We have that $|\Omega| = |\mathcal{T}||\Phi|$. In particular $|\Phi|$ divides $|\Omega|$.*
3. *The subgroup H acts transitively on Φ.*

Notation 1.1.4. If H is a subgroup of a group G, the *core of H in G* is the subgroup

$$\mathrm{Core}_G(H) = \bigcap_{g \in G} H^g.$$

Along this chapter, in order to make the notation more compact, the core of a subgroup H in a group G will often be denoted by H_G instead of $\mathrm{Core}_G(H)$.

Theorem 1.1.5. *Let G be a group. The following conditions are equivalent:*

1. *G possesses a faithful transitive permutation representation with no non-trivial blocks;*
2. *there exists a core-free maximal subgroup of G.*

Proof. 1 implies 2. Suppose that there exists a transitive G-set Ω with no non-trivial blocks and consider any $\omega \in \Omega$. The action of G on Ω is equivalent to the action of G on the set of right cosets of G_ω in G. The kernel of this action is $\mathrm{Core}_G(G_\omega)$ and, by hypothesis, is trivial. By Proposition 1.1.2, if H is a subgroup containing G_ω, there exists a block $\Phi = \{\omega^h : h \in H\}$ of Ω such that $\omega \in \Phi$ and $H = G_\Phi = \{g \in G : \Phi^g = \Phi\}$. Since G has no non-trivial blocks, either $\Phi = \{\omega\}$ or $\Phi = \Omega$. If $\Phi = \{\omega\}$, then $G_\omega = H$ and if $\Phi = \Omega$, then $H = G_\Omega = G$. Hence the stabiliser G_ω is a core-free maximal subgroup of G.

2 implies 1. If U is a core-free maximal subgroup of G, then the action of G on the set of right cosets of U in G is faithful and transitive. By maximality of U, this action has no non-trivial blocks by Proposition 1.1.2. $\qquad \square$

Definitions 1.1.6. *A a faithful transitive permutation representation of a group is said to be* primitive *if it does not have non-trivial blocks.*

A primitive group *is a group which possesses a primitive permutation representation. Equivalently, a group is* primitive *if it possesses a core-free maximal subgroup.*

A primitive pair is a pair (G, U), where G is a primitive group and U a core-free maximal subgroup of G,

Each conjugacy class of core-free maximal subgroups affords a faithful transitive and primitive permutation representation of the group. Thus, in general, it is more precise to speak of primitive pairs. Consider, for instance, the alternating group of degree 5, $G = \mathrm{Alt}(5)$. There exist three conjugacy classes of maximal subgroups, namely the normalisers of each type of Sylow subgroup. Obviously all of them are core-free. This gives three non-equivalent primitive representations of degrees 5 (for the normalisers of the Sylow 2-subgroups), 10 (for the normalisers of the Sylow 3-subgroups) and 6 (for the normalisers of the Sylow 5-subgroups).

The remarkable result that follows, due to R. Baer, classifies all primitive groups (a property defined in terms of maximal subgroups) according to the structure of the *socle*, i.e. the product of all minimal normal subgroups.

Theorem 1.1.7 ([Bae57]).

1. *A group G is primitive if and only if there exists a subgroup M of G such that $G = MN$ for all minimal normal subgroups N of G.*

2. *Let G be a primitive group. Assume that U is a core-free maximal subgroup of G and that N is a non-trivial normal subgroup of G. Write $C = C_G(N)$. Then $C \cap U = 1$. Moreover, either $C = 1$ or C is a minimal normal subgroup of G.*

3. *If G is a primitive group and U is a core-free maximal subgroup of G, then exactly one of the following statements holds:*

   a) *$\mathrm{Soc}(G) = S$ is a self-centralising abelian minimal normal subgroup of G which is complemented by U: $G = US$ and $U \cap S = 1$.*

   b) *$\mathrm{Soc}(G) = S$ is a non-abelian minimal normal subgroup of G which is supplemented by U: $G = US$. In this case $C_G(S) = 1$.*

   c) *$\mathrm{Soc}(G) = A \times B$, where A and B are the two unique minimal normal subgroups of G and both are complemented by U: $G = AU = BU$ and $A \cap U = B \cap U = A \cap B = 1$. In this case $A = C_G(B)$, $B = C_G(A)$, and A, B and $AB \cap U$ are non-abelian isomorphic groups.*

Proof.   1. If G is a primitive group, and U is a core-free maximal subgroup of G, then it is clear that $G = UN$ for every minimal normal subgroup N of G. Conversely, if there exists a subgroup M of G, such that $G = MN$ for every minimal normal subgroup N of G and U is a maximal subgroup of G such that $M \leq U$, then U cannot contain any minimal normal subgroup of G, and therefore U is a core-free maximal subgroup of G.

2. Since U is core-free in G, we have that $G = UN$. Since N is normal, then C is normal in G and then $C \cap U$ is normal in U. Since $C \cap U$ centralises N, then $C \cap U$ is in fact normal in G. Therefore $C \cap U = 1$.

If $C \neq 1$, consider a minimal normal subgroup X of G such that $X \leq C$. Since X is not contained in U, then $G = XU$. Then $C = C \cap XU = X(C \cap U) = X$.

3. Let us assume that N_1, N_2, and N_3 are three pairwise distinct minimal normal subgroups. Since $N_1 \cap N_2 = N_1 \cap N_3 = N_2 \cap N_3 = 1$, we have that $N_2 \times N_3 \leq C_G(N_1)$. But then $C_G(N_1)$ is not a minimal normal subgroup of G, and this contradicts 2. Hence, in a primitive group there exist at most two distinct minimal normal subgroups.

Suppose that N is a non-trivial abelian normal subgroup of G. Then $N \leq C_G(N)$. Since by 2, $C_G(N)$ is a minimal normal subgroup of G, we have that N is self-centralising. Thus, in a primitive group G there exists at most one abelian minimal normal subgroup N of G. Moreover, $G = NU$ and N is self-centralising. Then $N \cap U = C_G(N) \cap U = 1$.

If there exists a unique minimal non-abelian normal subgroup N, then $G = NU$ and $C_G(N) = 1$.

If there exist two minimal normal subgroups A and B, then $A \cap B = 1$ and then $B \leq C_G(A)$ and $A \leq C_G(B)$. Since $C_G(A)$ and $C_G(B)$ are minimal normal subgroups, we have that $B = C_G(A)$ and $A = C_G(B)$. Now $A \cap U = C_G(B) \cap U = 1$ and $B \cap U = C_G(A) \cap U = 1$. Hence $G = AU = BU$.

Since $A = C_G(B)$, it follows that B is non-abelian. Analogously we have that A is non-abelian.

By the Dedekind law [DH92, I, 1.3], we have $A(AB \cap U) = AB = B(AB \cap U)$. Hence $A \cong A/(A \cap B) \cong AB/B \cong B(AB \cap U)/B = AB \cap U$. Analogously $B \cong AB \cap U$. □

Baer's theorem enables us to classify the primitive groups as three different types.

Definition 1.1.8. *A primitive group G is said to be*

1. *a primitive group of type 1 if G has an abelian minimal normal subgroup,*
2. *a primitive group of type 2 if G has a unique non-abelian minimal normal subgroup,*
3. *a primitive group of type 3 if G has two distinct non-abelian minimal normal subgroups.*

We say that G is a monolithic *primitive group if G is a primitive group of type 1 or 2.*

Definition 1.1.9. *Let U be a maximal subgroup of a group G. Then U/U_G is a core-free maximal subgroup of the quotient group G/U_G. Then U is said to be*

1. *a maximal subgroup of type 1 if G/U_G is a primitive group of type 1,*
2. *a maximal subgroup of type 2 if G/U_G is a primitive group of type 2,*
3. *a maximal subgroup of type 3 if G/U_G is a primitive group of type 3.*

We say that U is a monolithic *maximal subgroup if G/U_G is a monolithic primitive group.*

Obviously all primitive soluble groups are of type 1. For these groups, there exists a well-known description called Galois' theorem. The proof appears in Huppert's book [Hup67, II, 3.2 and 3.3].

Theorem 1.1.10. *1. (Galois) If G is a soluble primitive group, then all core-free maximal subgroups are conjugate.*
2. If N is a self-centralising minimal normal subgroup of a soluble group G, then G is primitive, N is complemented in G, and all complements are conjugate.

Remarks 1.1.11. 1. The statement of Theorem 1.1.10 (1) is also valid if G is p-soluble for all primes dividing the order of $\mathrm{Soc}(G)$.

2. If G is a primitive group of type 1, then its minimal normal subgroup N is an elementary abelian p-subgroup for some prime p. Hence, N is a vector space over the field $\mathrm{GF}(p)$. Put $\dim N = n$, i.e. $|N| = p^n$. If M is a core-free subgroup of G, then M is isomorphic to a subgroup of $\mathrm{Aut}(N) = \mathrm{GL}(n, p)$. Therefore G can be embedded in the affine group $\mathrm{AGL}(n, p) = [C_p^n]\,\mathrm{GL}(n, p)$ in such a way that N is the translation group and $G \cap \mathrm{GL}(n, p)$ acts irreducibly on N. Thus, clearly, primitive groups of type 1 are not always soluble.

3. In his book B. Huppert shows that the affine group $\mathrm{AGL}(3, 2) = [C_2 \times C_2 \times C_2]\,\mathrm{GL}(3, 2)$ is an example of a primitive group of type 1 with non-conjugate core-free maximal subgroups (see [Hup67, page 161]).

4. Let G be a primitive group of type 2. If N is the minimal normal subgroup of G, then N is a direct product of copies of some non-abelian simple group and, in particular, the order of N has more than two prime divisors. If p is a prime dividing the order of N and $P \in \mathrm{Syl}_p(N)$, then $G = \mathrm{N}_G(P)N$ by the Frattini argument. Since P is a proper subgroup of N, then $\mathrm{N}_G(P)$ is a proper subgroup of G. If U is a maximal subgroup of G such that $\mathrm{N}_G(P) \leq U$, then necessarily U is core-free. Observe that if $P_0 \in \mathrm{Syl}_p(G)$ such that $P \leq P_0$, then $P = P_0 \cap N$ is normal in P_0 and so $P_0 \leq U$. In other words, U has p'-index in G. This argument can be done for each prime dividing $|N|$. Hence, the set of all core-free maximal subgroups of a primitive group of type 2 is not a conjugacy class.

5. In non-soluble groups, part 2 of Theorem 1.1.10 does not hold in general. Let G be a non-abelian simple group, p a prime dividing $|G|$ and $P \in \mathrm{Syl}_p(G)$. Suppose that P is cyclic. Let $G_{\Phi, p}$ be the maximal Frattini extension of G with p-elementary abelian kernel $A = \mathrm{A}_p(G)$ (see [DH 92; Appendix β] for details of this construction). Write $J = \mathrm{J}(KG)$ for the Jacobson radical of the group algebra KG of G, over the field $K = \mathrm{GF}(p)$. Then the section $N = A/AJ$ is irreducible and $\mathrm{C}_G(N) = \mathrm{O}_{p',p}(G) = 1$. Consequently $G_{\Phi, p}/AJ$ is a group with a unique minimal normal subgroup, isomorphic to N, self-centralising and non-supplemented.

In primitive groups of type 1 or 3, the core-free maximal subgroups complement each minimal subgroup. This characterises these types of primitive groups. In case of primitive groups of type 2 we will see later that the minimal normal subgroup could be complemented by some core-free maximal subgroup in some cases; but even then, there are always core-free maximal subgroups supplementing and not complementing the socle.

Proposition 1.1.12 ([Laf84a]). *For a group G, the following are pairwise equivalent:*

1. *G is a primitive group of type 1 or 3;*
2. *there exists a minimal normal subgroup N of G complemented by a subgroup M which also complements $C_G(N)$;*
3. *there exists a minimal normal subgroup N of G such that G is isomorphic to the semidirect product $X = [N](G/C_G(N))$.*

Proof. Clearly 1 implies 2. For 2 implies 1 observe that, since $N \cap M_G = 1$, then $M_G \leq C_G(N)$. But, since also $M_G \cap C_G(N) = 1$, we have that $M_G = 1$. Suppose that S is a proper subgroup of G such that $M \leq S$. Then the subgroup $S \cap N$ is normal in S and is centralised by $C_G(N)$. Hence $S \cap N$ is normal in $S C_G(N) = G$. By minimality of N, we have that $S \cap N = 1$ and then $S = M$. Then M is a core-free maximal subgroup of G and the group G is primitive. Observe that the minimal normal subgroup of a primitive group of type 2 has trivial centraliser.

2 implies 3. Observe that $G = NM$, with $N \cap M = 1$, and $M \cong G/C_G(N)$. The map $\alpha\colon G \longrightarrow [N](G/C_G(N))$ given by $(nm)^\alpha = (n, m\, C_G(N))$ is the desired isomorphism.

3 implies 2. Write $C = C_G(N)$. Assume that there exists an isomorphism

$$\alpha\colon [N](G/C) \longrightarrow G$$

and consider the following subgroups $N^* = (\{(n, C) : n \in N\})^\alpha$, $M^* = (\{(1, gC) : g \in G\})^\alpha$, and $C^* = (\{(n, gC) : ng \in C\})^\alpha$. For each $n \in N$, the element $(n^{-1}, nC)^\alpha$ is a non-trivial element of C^*. Hence $C^* \neq 1$. It is an easy calculation to show that N^* is a minimal normal subgroup of G, $C^* = C_G(N^*)$ and M^* complements N^* and C^*. □

Corollary 1.1.13. *The following conditions for a group G are equivalent:*

1. *G is a primitive group of type 3.*
2. *The group G possesses two distinct minimal normal subgroups N_1, N_2, such that*
   a) *N_1 and N_2 have a common complement in G;*
   b) *the quotient groups G/N_i, for $i = 1, 2$, are primitive groups of type 2.*

Proof. 1 implies 2. By Theorem 1, if G is a primitive group of type 3, then G possesses two distinct minimal normal subgroups N_1, N_2 which have a common complement M in G. Observe that $M \cong G/N_1$ and N_2N_1/N_1 is a minimal normal subgroup of G/N_1. If $gN_1 \in C_{G/N_1}(N_2N_1/N_1)$, then $[n, g] \in N_1$, for all $n \in N_2$. But then $[n, g] \in N_1 \cap N_2 = 1$, and therefore $g \in C_G(N_2) = N_1$. Hence

$C_{G/N_1}(N_2N_1/N_1) = 1$. Consequently G/N_1 is a primitive group of type 2 and therefore so are M and G/N_2.

2 implies 1. Let M be a common complement of N_1 and N_2. Then $G/N_i \cong M$ is a primitive group of type 2 such that $\operatorname{Soc}(G/N_i) = N_1N_2/N_i$ and $C_G(N_1N_2/N_i) = N_i$. Therefore $C_G(N_2) = N_1$ and $C_G(N_1) = N_2$. By Proposition 1.1.12, this means that G is a primitive group of type 3. $\qquad \Box$

Proposition 1.1.14 ([Laf84a]). *For a group G, the following statements are pairwise equivalent.*

1. *G is a primitive group of type 2.*
2. *G possesses a minimal normal subgroup N such that $C_G(N) = 1$.*
3. *There exists a primitive group X of type 3 such that $G \cong X/A$ for a minimal normal subgroup A of X.*

Proof. 3 implies 2 is Corollary 1.1.13 and 2 implies 1 is the characterisation of primitive groups of type 2 in Theorem 1. Thus it only remains to prove that 1 implies 3. If G is a primitive group of type 2 and N is the unique minimal normal subgroup of G, then N is non-abelian and $C_G(N) = 1$. By Proposition 1.1.12, the semidirect product $X = [N]G$ is a primitive group of type 3. Clearly if $A = \{(n, 1) : n \in N\}$, then $X/A \cong G$. $\qquad \Box$

Consequently, if M is a core-free maximal subgroup of a primitive group G of type 3, then M is a primitive group of type 2 and $\operatorname{Soc}(M)$ is isomorphic to a minimal normal subgroup of G.

According to Baer's Theorem, the socle of a primitive group of type 2 is a non-abelian minimal normal subgroup and therefore is a direct product of copies of a non-abelian simple group (see [Hup67, I, 9.12]). Obviously, the simplest examples of primitive groups of type 2 are the non-abelian simple groups. Observe that if S is a non-abelian simple group, then $\operatorname{Z}(S) = 1$ and we can identify S and the group of inner automorphisms $\operatorname{Inn}(S)$ and write $S \leq \operatorname{Aut}(S)$. Since $C_{\operatorname{Aut}(S)}(S) = 1$, any group G such that $S \leq G \leq \operatorname{Aut}(S)$ is a primitive group of type 2 such that $\operatorname{Soc}(G)$ is a non-abelian simple group. Conversely, if G is a primitive group of type 2 and $S = \operatorname{Soc}(G)$ is a simple group, then, since $C_G(S) = 1$, we can embed G in $\operatorname{Aut}(S)$.

Definition 1.1.15. *An* almost simple *group G is a subgroup of $\operatorname{Aut}(S)$ for some simple group S, such that $S \leq G$.*

If G is an almost simple group and $S \leq G \leq \operatorname{Aut}(S)$, for a non-abelian simple group S, then $C_G(S) = 1$. Hence G possesses a unique minimal normal subgroup S and every maximal subgroup U of G such that $S \not\leq U$ is core-free in G.

Proposition 1.1.16. *Suppose that S is a non-abelian simple group and let G be an almost simple group such that $S \leq G \leq \operatorname{Aut}(S)$. If U is a core-free maximal subgroup of G, then $U \cap S \neq 1$.*

Proof. Recall Schreier's conjecture ([KS04, page 151]) which states that the group of outer automorphisms $\mathrm{Out}(S) = \mathrm{Aut}(S)/\mathrm{Inn}(S)$ of a non-abelian simple group S is always soluble. The classification of simple groups has allowed us to check that this conjecture is true.

Suppose that $U \cap S = 1$. We know that $U \cong US/S \leq \mathrm{Aut}(S)/\mathrm{Inn}(S)$ and, by Schreier's conjecture ([KS04, page 151]) we deduce that U is soluble. Let Q be a minimal normal subgroup of U. Then Q is an elementary abelian q-group for some prime q. Observe that $\mathrm{C}_G(Q)$ is normalised by U. Therefore $\mathrm{C}_S(Q)$ is normalised by U and then $U\,\mathrm{C}_S(Q)$ is a subgroup of G. Since U is maximal in G and $\mathrm{C}_G(S) = 1$, then $\mathrm{C}_S(Q) = 1$. The q-group Q acts fixed-point-freely on S and then S is a q'-group. By the Odd Order Theorem ([FT63]), we have that $q \neq 2$. Now Q acts by conjugation on the elements of the set $\mathrm{Syl}_2(S)$ and by the Orbit-Stabiliser Theorem ([DH92, A, 5.2]) we deduce that Q normalises some $P \in \mathrm{Syl}_2(S)$. If P and $P^{x^{-1}}$, for $x \in S$, are two Sylow 2-subgroups of S which are normalised by Q, then Q, $Q^x \in \mathrm{Syl}_q(\mathrm{N}_{QS}(P))$ and there exists an element $g \in \mathrm{N}_{QS}(P)$, such that $Q^g = Q^x$. Write $g = yz$, with $y \in Q$ and $z \in S$. Then $Q^x = Q^z$ with $z \in \mathrm{N}_S(P)$. Hence $[Q, xz^{-1}] \leq Q \cap S = 1$ and $xz^{-1} \in \mathrm{C}_S(Q) = 1$. Therefore $x = z \in \mathrm{N}_S(P)$ and we conclude that Q normalises exactly one Sylow 2-subgroup P of S. Hence $\mathrm{N}_G(Q) \leq \mathrm{N}_G(P)$. But $U = \mathrm{N}_G(Q)$, by maximality of U. The subgroup UP is a proper subgroup of G which contains properly the maximal subgroup U. This is a contradiction. Hence $U \cap S \neq 1$.   □

For our purposes, it will be necessary to embed the primitive group G in a larger group. Suppose that $\mathrm{Soc}(G) = S_1 \times \cdots \times S_n$, where the S_i are copies of a non-abelian simple group S, i.e. $\mathrm{Soc}(G) \cong S^n$, the direct product of n copies of S. Since $\mathrm{C}_G\big(\mathrm{Soc}(G)\big) = 1$, the group G can be embedded in $\mathrm{Aut}(S^n)$. The automorphism group of a direct product of copies of a non-abelian simple group has a well-known structure: it is a *wreath product*.

Thus, the study of some relevant types of subgroups of groups which are wreath products and the analysis of some special types of subgroups of a direct product of isomorphic non-abelian simple groups will be essential.

Definition 1.1.17. *Let X and H be two groups and suppose that H has a permutation representation φ on a finite set $\mathcal{I} = \{1, \ldots, n\}$ of n elements. The wreath product $X \wr_\varphi H$ (or simply $X \wr H$ if the action is well-known) is the semidirect product $[X^\natural]H$, where $X^\natural$ is the direct product of n copies of X: $X^\natural = X_1 \times \cdots \times X_n$, with $X_i = X$ for all $i \in \mathcal{I}$, and the action is*

$$(x_1, \ldots, x_n)^h = (x_{1(h^{-1})\varphi}, \ldots, x_{n(h^{-1})\varphi}) \tag{1.1}$$

for $h \in H$ and $x_i \in X$, for all $i \in \mathcal{I}$.

The subgroup $X^\natural$ is called the base group *of $X \wr H$.*

Remarks 1.1.18. Consider a wreath product $G = X \wr_\varphi H$.
   1. If φ is faithful, then $\mathrm{C}_G(X^\natural) \leq X^\natural$.

2. For any $g \in G$, then $g = xh$, with $x \in X^\natural$ and $h \in H$. For each $i = 1, \ldots, n$, we have that $X_i^g = X_i^h = X_{i^{h^\varphi}}$.

3. Thus, the group G acts on $\mathcal{I}$ by the following rule: if $i \in \mathcal{I}$, for any $g = xh \in G$, with $x \in X^\natural$ and $h \in H$, then $i^g = i^{h^\varphi}$. In particular $i^h = i^{h^\varphi}$, if $h \in H$.

4. If $\mathcal{S} \subseteq \mathcal{I}$, then write

$$\pi_{\mathcal{S}} : X^\natural \longrightarrow \prod_{j \in \mathcal{S}} X_j$$

for the projection of $X^\natural$ onto $\prod_{j \in \mathcal{S}} X_j$. Then for any $y \in X^\natural$ and any $g \in G$, we have that

$$(y^g)^{\pi_{\mathcal{S}^g}} = (y^{\pi_{\mathcal{S}}})^g.$$

Proposition 1.1.19. *Let S be a non-abelian simple group and write $S^n = S_1 \times \cdots \times S_n$ for the direct product of n copies $S_1, \ldots, S_n$ of S, for some positive integer n. Then the minimal normal subgroups of S^n are exactly the S_i, for any $i = 1, \ldots, n$,*

Proof. Let N be a minimal normal subgroup of S^n. Suppose that $N \cap S_i = 1$ for all $i = 1, \ldots, n$. Then N centralises all S_i and hence $N \leq Z(S^n) = 1$. This is a contradiction. Therefore $N \cap S_i = N$ for some index i. Then $N = S_i$. $\quad\square$

Proposition 1.1.20. *Let S be a non-abelian simple group and write $S^n = S_1 \times \cdots \times S_n$ for the direct product of n copies $S_1, \ldots, S_n$ of S, for some positive integer n. Then $\mathrm{Aut}(S^n) \cong \mathrm{Aut}(S) \wr \mathrm{Sym}(n)$, where $\mathrm{Sym}(n)$ is the symmetric group of degree n.*

Proof. If σ is a permutation in $\mathrm{Sym}(n)$, the map α_σ defined by

$$(x_1, \ldots, x_n)^{\alpha_\sigma} = (x_{1^{\sigma^{-1}}}, \ldots, x_{n^{\sigma^{-1}}})$$

is an element of $\mathrm{Aut}(S^n)$ associated with σ. Now $H = \{\alpha_\sigma \in \mathrm{Aut}(S^n) : \sigma \in \mathrm{Sym}(n)\}$ is a subgroup of $\mathrm{Aut}(S^n)$ and $\sigma \longmapsto \alpha_\sigma$ defines an isomorphism between $\mathrm{Sym}(n)$ and H. By Proposition 1.1.19, the minimal normal subgroups of the direct product $S_1 \times \cdots \times S_n$ are exactly the $S_1, \ldots, S_n$. Therefore, if $\gamma \in \mathrm{Aut}(S^n)$, then there exists a $\sigma \in \mathrm{Sym}(n)$ such that $S_i^\gamma = S_{i^\sigma} = S_i^{\alpha_\sigma}$, for all $i = 1, \ldots, n$.

Let D be the subgroup of all elements β in $\mathrm{Aut}(S^n)$ such that $S_i^\beta = S_i$ for all i. The maps $\beta_1, \ldots, \beta_n$ defined by $(x_1, \ldots, x_n)^\beta = (x_1^{\beta_1}, \ldots, x_n^{\beta_n})$ are automorphisms of S and the map $\beta \mapsto (\beta_1, \ldots, \beta_n)$ defines an isomorphism between D and $\mathrm{Aut}(S)^n$. Moreover, by Proposition 1.1.19 again, if $\beta \in D$ and $\gamma \in \mathrm{Aut}(S^n)$, then $(S_i^\gamma)^\beta = S_i^\gamma$. This means that D is a normal subgroup of $\mathrm{Aut}(S^n)$.

Observe that $\alpha_\sigma \in D$ if and only if $\sigma = 1$, or, in other words, $D \cap H = 1$. Moreover for all $\gamma \in \mathrm{Aut}(S^n)$, we have that $\gamma \alpha_\sigma^{-1} \in D$. Therefore $\mathrm{Aut}(S^n) = [D]H$. This allows us to define a bijective map between $\mathrm{Aut}(S^n)$ and $\mathrm{Aut}(S) \wr \mathrm{Sym}(n)$ which is an isomorphism. $\quad\square$

F. Gross and L. G. Kovács published in [GK84] a construction of groups, the so-called *induced extensions*, which is crucial to understand the structure of a, non-necessarily finite, group that possesses a normal subgroup which is a direct product of copies of a group. It is clear that primitive groups of type 2 are examples of this situation. We present in the sequel an adaptation of this construction to finite groups.

Proposition 1.1.21. *Consider the following diagram of groups and group homomorphisms:*

$$
\begin{array}{ccc}
 & & Z \\
 & & \downarrow g \\
X & \xrightarrow{\ f\ } & Y
\end{array}
\tag{1.2}
$$

where g is a monomorphism. Let G be the following subset of X:

$$G = \{x \in X : x^f = z^g \text{ for some } z \in Z\},$$

and the following mapping

$$h \colon G \longrightarrow Z \quad x^h = x^{fg^{-1}} \text{ for every } x \in G.$$

Then G is a subgroup of X and h is a well-defined group homomorphism such that the following diagram of groups and group homomorphisms is commutative:

$$
\begin{array}{ccc}
G & \xrightarrow{\ h\ } & Z \\
\downarrow \iota & & \downarrow g \\
X & \xrightarrow{\ f\ } & Y
\end{array}
$$

(where ι is the canonical inclusion of G in X). Moreover $\mathrm{Ker}(h^\iota) = \mathrm{Ker}(f)$.

Further, if (G_0, ι_0, h_0) is a triple, with G_0 a group, $\iota_0 \colon G_0 \longrightarrow X$ a monomorphism and $h_0 \colon G_0 \longrightarrow Z$ is a group homomorphism, such that the diagram

$$
\begin{array}{ccc}
G_0 & \xrightarrow{\ h_0\ } & Z \\
\downarrow \iota_0 & & \downarrow g \\
X & \xrightarrow{\ f\ } & Y
\end{array}
$$

is commutative, then there exists a monomorphism $\Phi \colon G_0 \longrightarrow G$, such that $\Phi h = h_0$, $\Phi \iota = \iota_0$ and $\left(\mathrm{Ker}(h_0)\right)^\Phi \leq \left(\mathrm{Ker}(h)\right)^\iota = \mathrm{Ker}(f)$.

Proof. It is an easy exercise to prove that G is a subgroup of X and, since g is a monomorphism, the mapping h is a well-defined group homomorphism. It is not difficult to see that $\mathrm{Ker}(h)^\iota = \mathrm{Ker}(f)$.

For the second statement, let $x \in G_0$ and observe that x^{h_0} is an element of Z such that $(x^{h_0})^g = (x^{\iota_0})^f$, and then $x^{\iota_0} \in G$ and $(x^{\iota_0})^h = x^{h_0}$. Write $\Phi \colon G_0 \longrightarrow G$ such that $x^\Phi = x^{\iota_0}$. $\qquad\square$

Definition 1.1.22. *The triple* (G, ι, h) *introduced in Proposition 1.1.21 is said to be the* pull-back *of the diagram* (1.2).

Proposition 1.1.23. *Consider the following extension of groups:*

$$1 \longrightarrow K \longrightarrow X \overset{f}{\longrightarrow} Y \longrightarrow 1$$

and a monomorphism $g \colon Z \longrightarrow Y$. *Consider the triple* (G, ι, h), *the pull-back of the diagram* (1.2).

1. *There exists an extension*

$$Eg \colon 1 \longrightarrow K \longrightarrow G \overset{h}{\longrightarrow} Z \longrightarrow 1$$

   such that the following diagram of groups and group homomorphisms is commutative:

$$
\begin{array}{ccccccccc}
Eg \colon 1 & \longrightarrow & K & \longrightarrow & G & \overset{h}{\longrightarrow} & Z & \longrightarrow & 1 \\
& & \downarrow{\scriptstyle \mathrm{id}} & & \downarrow{\scriptstyle \iota} & & \downarrow{\scriptstyle g} & & \\
E \colon 1 & \longrightarrow & K & \longrightarrow & X & \overset{f}{\longrightarrow} & Y & \longrightarrow & 1
\end{array}
$$

2. *Moreover, if*

$$E_0 \colon 1 \longrightarrow K \longrightarrow G_0 \overset{h_0}{\longrightarrow} Z \longrightarrow 1$$

   is another extension such that the diagram

$$
\begin{array}{ccccccccc}
E_0 \colon 1 & \longrightarrow & K & \longrightarrow & G_0 & \overset{h_0}{\longrightarrow} & Z & \longrightarrow & 1 \\
& & \downarrow{\scriptstyle \mathrm{id}} & & \downarrow{\scriptstyle \iota_0} & & \downarrow{\scriptstyle g} & & \\
E \colon 1 & \longrightarrow & K & \longrightarrow & X & \overset{f}{\longrightarrow} & Y & \longrightarrow & 1
\end{array}
$$

   is commutative, there exists a group isomorphism $\Phi \colon G_0 \longrightarrow G$ *such that* $\Phi h = h_0$, $\Phi \iota = \iota_0$ *and* $\Phi|_K = \mathrm{id}_K$.

Proof. The proof of 1 is a direct exercise. To see 2, first notice that, by the Short Five Lemma ([Hun80, IV, 1.17]), the homomorphism ι_0 is a monomorphism. By Proposition 1.1.21, there exists a group monomorphism $\Phi \colon G_0 \longrightarrow G$ such that $\Phi h = h_0$, $\Phi \iota = \iota_0$ and $\Phi|_K = \mathrm{id}_K$. Furthermore, since $|G| = |Z|/|K| = |G_0|$, we have that Φ is an isomorphism. $\qquad \square$

Definition 1.1.24. *The extension* Eg *is said to be the* pull-back extension *of the extension* E *and the monomorphism* g.

Hypotheses 1.1.25. *Let* B *be a group. Assume that* C *a subgroup of a group* B *such that* $|B : C| = n$ *and let* $\mathcal{T} = \{t_1 = 1, \ldots, t_n\}$ *be a right transversal* *of C in B. Then B, acting by right multiplication on the set of right cosets of C in B, induces a transitive action $\rho\colon B \longrightarrow \mathrm{Sym}(n)$ on the set of indices $\mathcal{I} = \{1, \ldots, n\}$ in the following way. For each $i \in \mathcal{I}$ and each $h \in B$, the element $t_i h$ belongs to some coset $C t_j$, i.e. $t_i h = c_{i,h} t_j$, for some $c_{i,h} \in C$. Then $i^{h^\rho} = j$. Write $P = B^\rho \leq \mathrm{Sym}(n)$.*

Let $\alpha\colon A \longrightarrow B$ be a group homomorphism and write $C = A^\alpha$ and $S = \mathrm{Ker}(\alpha)$. Write $W = A \wr_\rho P$. There exists an induced epimorphism $\bar{\alpha}\colon A \wr_\rho P \longrightarrow C \wr_\rho P$ defined by $\big((a_1, \ldots, a_n)x\big)^{\bar{\alpha}} = (a_1^\alpha, \ldots, a_n^\alpha)x$, for $a_1, \ldots, a_n \in A$ and $x \in P$. Write $M = \mathrm{Ker}(\bar{\alpha})$. Observe that $(a_1, \ldots, a_n)x \in M$ if and only if $a_j^\alpha = 1$, for all $j \in \mathcal{I}$ and $x = 1$. This is to say that $M = \mathrm{Ker}(\bar{\alpha}) = \mathrm{Ker}(\alpha) \times \ldots \times \mathrm{Ker}(\alpha) = S_1 \times \ldots \times S_n$. We have the exact sequence:

$$E\colon 1 \longrightarrow M \longrightarrow A \wr_\rho P \overset{\bar{\alpha}}{\longrightarrow} C \wr_\rho P \longrightarrow 1$$

Lemma 1.1.26. *Assume the hypotheses and notation of Hypotheses 1.1.25.*

1. *The mapping $\lambda = \lambda_{\mathcal{T}}\colon B \longrightarrow C \wr_\rho P$ such that $h^\lambda = (c_{1,h}, \ldots, c_{n,h})h^\rho$, for any $h \in B$, is a group monomorphism.*
2. *Consider the pull-back exact sequence $E\lambda$:*

$$
\begin{array}{ccccccccc}
E\lambda\colon 1 & \longrightarrow & M & \longrightarrow & G & \overset{\sigma}{\longrightarrow} & B & \longrightarrow & 1 \\
& & \downarrow{\scriptstyle \mathrm{id}} & & \downarrow & & \downarrow{\scriptstyle \lambda} & & \\
E\colon 1 & \longrightarrow & M & \longrightarrow & A \wr_\rho P & \overset{\bar{\alpha}}{\longrightarrow} & C \wr_\rho P & \longrightarrow & 1
\end{array}
$$

Then, the isomorphism class of the group G is independent from the choice of transversal of C in B.

Proof. 1. Let $h, h' \in B$. Observe that

$$c_{i,hh'} t_{i^{(hh')^\rho}} = t_i hh' = c_{i,h} t_{i^{h^\rho}} h' = c_{i,h} c_{i^{h^\rho},h'} t_{i^{(hh')^\rho}}.$$

Hence, by (1.1) in Definition 1.1.17, we have that

$$
\begin{aligned}
h^\lambda h'^\lambda &= (c_{1,h}, \ldots, c_{n,h}) h^\rho (c_{1,h'}, \ldots, c_{n,h'}) h'^\rho \\
&= (c_{1,h}, \ldots, c_{n,h}) (c_{1,h'}, \ldots, c_{n,h'})^{(h^\rho)^{-1}} (hh')^\rho \\
&= (c_{1,h}, \ldots, c_{n,h}) (c_{1^{h\rho},h'}, \ldots, c_{n^{h\rho},h'}) (hh')^\rho \\
&= (c_{1,hh'}, \ldots, c_{n,hh'}) (hh')^\rho = (hh')^\lambda
\end{aligned}
$$

and λ is a group homomorphism.

Suppose that $h^\lambda = h'^\lambda$. Then $(c_{1,h}, \ldots, c_{n,h}) h^\rho = (c_{1,h'}, \ldots, c_{n,h'}) h'^\rho$ and therefore, since $C \wr_\rho P_n = [C^n] P_n$ is a semidirect product, we have that

$$c_{j,h} = c_{j,h'} = c_j, \quad j \in \mathcal{I}; \qquad h^\rho = h'^\rho = \tau.$$

Therefore, for any index $j \in \mathcal{I}$, we have that $t_j h = c_j t_{j^\tau} = t_j h'$ and then $h = t_j^{-1} c_j t_{j^\tau} = h'$. Hence λ is a group monomorphism.

2. Let $\mathcal{T}' = \{t_1', \ldots, t_n'\}$ be some other right transversal of C in B such that $Ct_i' = Ct_i$, for each $i \in \mathcal{I}$: there exist elements $b_1, \ldots, b_n \in C$ such that $t_i' = b_i t_i$, for $i = 1, \ldots, n$. For each $i \in \mathcal{I}$ and each $h \in B$, the element $t_i' h$ belongs to the coset $Ct_j' = Ct_j$, for $i^{h^\rho} = j$, and $t_i' h = c_{i,h}' t_j'$, for some $c_{i,h}' \in C$. Then

$$t_i' h = b_i t_i h = b_i c_{i,h} t_j = c_{i,h}' t_j' = c_{i,h}' b_j t_j \quad \text{and} \quad c_{i,h} = b_i^{-1} c_{i,h}' b_j$$

and it appears the element $(b_1, \ldots, b_n) \in C^\natural$ associated with $\mathcal{T}'$. Then, for $\lambda' = \lambda_{\mathcal{T}'}$, we have that

$$h^\lambda = (c_{1,h}, \ldots, c_{n,h}) h^\rho = \left((b_1, \ldots, b_n)^{-1} (c_{1,h}', \ldots, c_{n,h}')(b_{1^{h^\rho}}, \ldots, b_{n^{h^\rho}})\right) h^\rho$$
$$= \left((b_1, \ldots, b_n)^{-1} (c_{1,h}', \ldots, c_{n,h}')(b_1, \ldots, b_n)^{(h^{-1})^\rho}\right) h^\rho$$
$$= (b_1, \ldots, b_n)^{-1} (c_{1,h}', \ldots, c_{n,h}') h^\rho (b_1, \ldots, b_n)$$
$$= \left((c_{1,h}', \ldots, c_{n,h}') h^\rho\right)^{(b_1, \ldots, b_n)} = (h^{\lambda'})^{(b_1, \ldots, b_n)},$$

for any $h \in B$, and then $\left(\mathrm{Im}(\lambda')\right)^{(b_1, \ldots, b_n)} = \mathrm{Im}(\lambda)$. For each $i \in \mathcal{I}$, let a_i be an element of A such that $a_i^\alpha = b_i$. This is to say that $(a_1, \ldots, a_n)^{\bar{\alpha}} = (b_1, \ldots, b_n)$. If $x \in G$, then

$$\left(x^{(a_1, \ldots, a_n)}\right)^{\bar{\alpha}} = (x^{\bar{\alpha}})^{(b_1, \ldots, b_n)} = (h^\lambda)^{(b_1, \ldots, b_n)} = h^{\lambda'}$$

and then $x^{(a_1, \ldots, a_n)} \in G^* = \{w \in W : w^{\bar{\alpha}} = h^{\lambda'} \text{ for some } h \in B\}$, which is the pull-back defined with the monomorphism λ':

$$
\begin{array}{ccccccccc}
E\lambda' : 1 & \longrightarrow & M & \longrightarrow & G^* & \overset{\sigma'}{\longrightarrow} & B & \longrightarrow & 1 \\
& & \downarrow{\scriptstyle \mathrm{id}} & & \downarrow & & \downarrow{\scriptstyle \lambda'} & & \\
E : 1 & \longrightarrow & M & \longrightarrow & A \wr_\rho P & \overset{\bar{\alpha}}{\longrightarrow} & C \wr_\rho P & \longrightarrow & 1
\end{array}
$$

Thus, $G^* = G^a$ for some $a \in A^\natural$ associated with the transversals $\mathcal{T}$ and $\mathcal{T}'$, i.e. the pull-back groups constructed from two different transversals are conjugate in W. In other words, the isomorphism class of the group G is independent from the choice of transversal. $\qquad\square$

Definition 1.1.27 ([GK84]). *In the above situation and with that notation, we will say that $E\lambda$ is the* induced extension *defined by $\alpha\colon A \longrightarrow B$.*

Recall that G is a subgroup of $W = A \wr_\rho P$ defined by:

$$G = \{x \in W : x^{\bar{\alpha}} = h^\lambda, \text{ for some } h \in B\}$$

and σ is defined by $\sigma = \bar{\alpha}|_G \lambda^{-1}$.

Proposition 1.1.28. *With the notation introduced above, we have the following.*

1. $N_G(A_1) = N_G(S_1) = N_G(S_2 \times \cdots \times S_n) = N = \{x \in W : x^{\bar{\alpha}} = h^\lambda,$ *for some* $h \in C\}$.
2. $N/(S_2 \times \cdots \times S_n) \cong A$. *Moreover, the image of* $M/(S_2 \times \cdots \times S_n)$ *under this isomorphism is* $S = \operatorname{Ker}(\alpha)$.
3. *In particular* $N^\sigma = C$ *and* $|G : N| = |B : C| = n$. *Thus, if* $\rho' : G \longrightarrow$ *Sym(n) is the action of G on the right cosets of N in G by multiplication, then* $\rho' = \sigma\rho$.
4. *The set* $\{S_1, \ldots, S_n\}$ *is the conjugacy class of the subgroup* S_1 *in* G.

Proof.   1. We can consider the subgroup

$$N = \{w \in W : w^{\bar{\alpha}} = h^\lambda, \text{ for some } h \in C\}.$$

Observe that if $(a_1, \ldots, a_n)x \in N$, for $a_i \in A$ and $x \in P$, then there exists $h \in C$, such that

$$h^\lambda = (c_{1,h}, \ldots, c_{n,h})h^\rho = (a_1^\alpha, \ldots, a_n^\alpha)x.$$

Since $h \in C$, it is clear that $c_{1,h} = h$ and h^ρ belongs to the stabiliser P_1 of 1. In other words

$$N \leq A_1 \times (A_2 \times \cdots \times A_n)P_1 = N_W(A_1) = N_W(S_1) = N_W(S_2 \times \cdots \times S_n)$$

and hence $N \leq N_G(A_1)$. Conversely, if $(a_1, \ldots, a_n)x \in N_G(A_1)$, then $x \in P_1$ and there exists $h \in B$ such that $a_i^\alpha = c_{i,h}$ and $x = h^\rho \in P_1$, i.e. $1^{h^\rho} = 1$. Hence $h = t_1 h = c_{1,h}t_1 = c_{1,h} = a_1^\alpha \in C$. Then $N_G(A_1) \leq N$. Hence $N = N_G(A_1) = N_G(S_1)$.

2. Consider the projection $e_1 : A_1 \times (A_2 \times \cdots \times A_n)P_1 = N_W(A_1) \longrightarrow A$ on the first component. Obviously, $\operatorname{Ker}(e_1) = (A_2 \times \cdots \times A_n)P_1$.

Let e be the restriction to N of the projection e_1:

$$e = e_1|_N : N \longrightarrow A.$$

Observe that if $x \in N$, then $x^{\bar{\alpha}} = c^\lambda$ for some $c \in C$. We can characterise this $c = x^\sigma$ in the following way. Assume that $x = (a_1, \ldots, a_n)y$. Then $x^{\bar{\alpha}} = (a_1^\alpha, \ldots, a_n^\alpha)y = c^\lambda = (c, c_{2,c}, \ldots, c_{n,c})c^\rho$. Hence $c = a_1^\alpha = x^{e\alpha}$.

We have that $\operatorname{Ker}(e) = \operatorname{Ker}(e_1) \cap N$. If $x \in \operatorname{Ker}(e)$, then $x^{\bar{\alpha}} = (x^{e\alpha})^\lambda = 1$. Thus $x \in \operatorname{Ker}(\bar{\alpha}) = M$ and then $\operatorname{Ker}(e) \leq M$. Therefore $\operatorname{Ker}(e) = \operatorname{Ker}(e_1) \cap M = (A_2 \times \cdots \times A_n)P_1 \cap M = S_2 \times \cdots \times S_n$.

For any $a \in A$, consider the element $c = a^\alpha \in C$. Then $c^\rho \in P_1$ and $c_{i,c} = t_i c t_j^{-1} \in C$, where $j = i^{c^\rho}$, for $i = 2, \ldots, n$. Since $C = A^\alpha$, there exist elements $a_2, \ldots, a_n$ in A such that $a_j^\alpha = c_{j,c}$, for $j = 2, \ldots, n$. The element $x = (a, a_2, \ldots, a_n)c^\rho \in N$, since $x^{\bar{\alpha}} = (a^\alpha, a_2^\alpha, \ldots, a_n^\alpha)c^\rho = (c, c_{2,c}, \ldots, c_{n,c})c^\rho = c^\lambda$. Now $x^e = a$, and then e is an epimorphism. Hence

$$N/\operatorname{Ker}(e) = N/(S_2 \times \cdots \times S_n) \cong A.$$

Finally observe that $M^e \cong M/\operatorname{Ker}(e|_M) = M/(S_2 \times \cdots \times S_n) \cong S$. Since $M^e \leq S = \operatorname{Ker}(\alpha)$ and these two subgroups have the same order, equality holds.

3. Choose a right transversal of N in G, $\{g_1 = 1, \ldots, g_n\}$ such that $g_i^\sigma = t_i$. Then for each $g \in G$, we have that $g_i g = x_{i,g} g_{i^{g^{\rho'}}}$, for some $x_{i,g} \in N$. Then

$$c_{i,g^\sigma} t_{i^{g^{\sigma\rho}}} = t_i g^\sigma = g_i^\sigma g^\sigma = x_{i,g}^\sigma g_{i^{g^{\rho'}}}^\sigma = x_{i,g}^\sigma t_{i^{g^{\rho'}}}$$

and then $i^{g^{\sigma\rho}} = i^{g^{\rho'}}$, for every $i \in \mathcal{I}$. Therefore $g^{\sigma\rho} = g^{\rho'}$ for each $g \in G$, and then $\sigma\rho = \rho'$.

4. Observe that for each $i \in \mathcal{I}$, the permutation t_i^ρ moves 1 to i. Therefore, having in mind (1.1) of Definition 1.1.17, we see that $S_1^{g_i} = S_i$, and then $\{S_1, \ldots, S_n\}$ is the conjugacy class of the subgroup S_1 in G.   □

We prove next that in fact the structure of the group G analysed in Proposition 1.1.28 characterises the induced extensions.

Theorem 1.1.29. *Let G be a group. Suppose that we have in G the following situation: there exist a normal subgroup M of G and a normal subgroup S of M such that $\{S_1, \ldots, S_n\}$ is the set of all conjugate subgroups of S in G and $M = S_1 \times \cdots \times S_n$. Write $N = \mathrm{N}_G(S_1)$ and $K = S_2 \times \cdots \times S_n$.*

Let $\alpha \colon N/K \longrightarrow G/M$ be defined by $(Kx)^\alpha = Mx$. Then G is the induced extension defined by α.

Proof. Let $\sigma \colon G \longrightarrow G/M$ and $e \colon N \longrightarrow N/K$ be the natural epimorphisms. If $\mathcal{T} = \{t_1 = 1, \ldots, t_n\}$ is a right transversal of N in G, then $\mathcal{T}^\sigma$ is a right transversal of N/M in G/M. Consider $\rho \colon G/M \longrightarrow \mathrm{Sym}(n)$ the permutation representation of G/M on the right cosets of N/M in G/M. Then $\bar{\rho} = \sigma\rho$ is the permutation representation of G on the right cosets of N in G. Write $P = G^{\bar{\rho}} = (G/M)^\rho$. Let

$$\bar{\lambda} = \lambda_{\mathcal{T}} \colon G \longrightarrow N \wr_{\bar{\rho}} P$$

be the embedding of G into $N \wr_{\bar{\rho}} P$ defined in Lemma 1.1.26 and

$$\lambda = \lambda_{\mathcal{T}^\sigma} \colon G/M \longrightarrow (N/M) \wr_\rho P$$

be the embedding of G/M into $(N/M) \wr_\rho P$. As usual, for each $x \in G$, write $t_i x = c_{i,x} t_j$, for some $c_{i,x} \in N$, and $i^{x^{\bar{\rho}}} = j$. Observe that $c_{i,g^\sigma} = (c_{i,g})^\sigma$. Write $S_i = S^{t_i}$. For each $i \in \mathcal{I} = \{1, \ldots, n\}$, write also $K_i = \prod_{j \in \mathcal{I} \setminus \{i\}} S_j$. Then $K = K_1$ and $K_i = K^{t_i}$.

If we write $\bar{\sigma} \colon N \wr_{\bar{\rho}} P \longrightarrow (N/M) \wr_\rho P$ for the epimorphism induced by σ, then $\sigma\lambda = \bar{\lambda}\bar{\sigma}$. Consider

$$\bar{e} \colon N \wr P \longrightarrow (N/K) \wr P, \quad \text{induced by } e$$

and

$$\bar{\alpha} \colon (N/K) \wr P \longrightarrow (N/M) \wr P, \quad \text{induced by } \alpha.$$

Since $e\alpha = \sigma|_N$, we find that $\bar{e}\bar{\alpha} = \bar{\sigma}$. Therefore $\bar{\lambda}\bar{e}\bar{\alpha} = \bar{\lambda}\bar{\sigma} = \sigma\lambda$ and the following diagram is commutative:

$$G \xrightarrow{\;\sigma\;} G/M$$

$$\Big\downarrow{\bar\lambda\bar e} \qquad\qquad \Big\downarrow{\lambda}$$

$$(N/K) \wr P \xrightarrow{\;\bar\alpha\;} (N/M) \wr P$$

The commutativity of the diagram shows that $M^{\bar\lambda\bar e\bar\alpha} = M^{\sigma\lambda} = 1$ and then $M^{\bar\lambda\bar e} \le \mathrm{Ker}(\bar\alpha)$.

Consider an element $x \in G$ such that $x^{\bar\lambda} = (c_{1,x}, \ldots, c_{n,x})x^{\bar\rho} \in G^{\bar\lambda} \cap \mathrm{Ker}(\bar e)$. Then we have $1 = (Kc_{1,x}, \ldots, Kc_{n,x})x^\rho$. This means that $x^\rho = \mathrm{id}$ and $c_{i,x} \in K$, for $i \in \mathcal{I}$. Therefore, $c_{i,x} = t_i x t_i^{-1}$, for $i \in \mathcal{I}$. Hence, $x \in \bigcap_{i=1}^n K^{t_i} = \bigcap_{i=1}^n K_i = 1$. Therefore $G^{\bar\lambda} \cap \mathrm{Ker}(\bar e) = 1$ and then $\bar\lambda\bar e$ is a monomorphism. Observe that $\mathrm{Ker}(\bar\alpha) = (M/K)^\natural = (M/K)_1 \times \cdots \times (M/K)_n$ and then $|\mathrm{Ker}(\bar\alpha)| = |M|$. Thus, the restriction $\bar\lambda\bar e|_M : M \longrightarrow \mathrm{Ker}(\bar\alpha)$ is an isomorphism. Therefore, the following diagram is commutative:

$$1 \longrightarrow M \longrightarrow G \xrightarrow{\;\sigma\;} G/M \longrightarrow 1$$

$$\Big\downarrow \qquad\quad \Big\downarrow{\bar\lambda\bar e} \qquad\quad \Big\downarrow{\lambda}$$

$$1 \longrightarrow \mathrm{Ker}(\bar\alpha) \longrightarrow (N/K) \wr P \xrightarrow{\;\bar\alpha\;} (N/M) \wr P \longrightarrow 1$$

Therefore G is the induced extension defined by α. $\qquad\qquad\square$

Remark 1.1.30. We are interested in the action of the group G on the normal subgroup $M = S_1 \times \cdots \times S_n$, when G is an induced extension. We keep the notation of Theorem 1.1.29. The action of the group N on S, $\psi \colon N \longrightarrow \mathrm{Aut}(S)$, is defined by conjugation: if $x \in N$, then x^ψ is the automorphism of S given by the conjugation in N by the element x: for every $s \in S$, we have $s^{x^\psi} = s^x$.

The induced extension G can be considered as a subgroup of the wreath product $W = N \wr_\rho P$, via the embedding

$$\bar\lambda = \lambda_{\mathcal{T}} \colon G \longrightarrow N \wr_{\bar\rho} P \quad \text{given by } x^{\bar\lambda} = (c_{1,x}, \ldots, c_{n,x})g^{\bar\rho}, \text{ for all } x \in G.$$

If $(x_1, \ldots, x_n) \in M = S_1 \times \cdots \times S_n$ and $x \in G$, then, by Definition 1.1.17,

$$(x_1, \ldots, x_n)^x = \left(x_1^{c_{1,x}^\psi}, \ldots, x_n^{c_{n,x}^\psi} \right)^{x^{\bar\rho}} = (y_1, \ldots, y_n),$$

where $x_i^{c_{i,x}^\psi} = y_{i^{x^{\bar\rho}}}$, for $i \in \{1, \ldots, n\}$.

Proposition 1.1.31. *In the hypotheses 1.1.25, assume that S is a group and C acts on S by a group homomorphism $\psi \colon C \longrightarrow \mathrm{Aut}(S)$. Then the group B acts on the direct product $S^n = S_1 \times \cdots \times S_n$ by a group homomorphism*

$$\psi^B \colon B \longrightarrow C^\psi \wr_\rho P \le \mathrm{Aut}(S^n)$$

such that for $(x_1, \ldots, x_n) \in S^n$ *and* $h \in B$, *then*

$$(x_1, \ldots, x_n)^{h^{\psi B}} = (y_1, \ldots, y_n), \text{ where } x_i^{c_{i,h}^{\psi}} = y_{i^{h\bar{\rho}}}, \text{ for } i \in \{1, \ldots, n\}.$$
(1.3)

Moreover, $\mathrm{Ker}(\psi^B) = \mathrm{Core}_B(\mathrm{Ker}(\psi))$.

Proof. If $\bar{\psi} \colon C \wr_\rho P \longrightarrow C^\psi \wr_\rho P$ is induced by ψ and λ is the monomorphism of Lemma 1.1.26, then $\psi^B = \lambda \bar{\psi}$. Clearly ψ^B is a group homomorphism. Observe that $h \in \mathrm{Ker}(\psi^B)$ if and only if h^ρ is the identity permutation and $c_{i,h} \in \mathrm{Ker}(\psi)$, for all $i \in \mathcal{I}$. This means that $t_i h t_i^{-1} = c_{i,h} \in \mathrm{Ker}(\psi)$, for all $i \in \mathcal{I}$. And this is equivalent to saying that $h \in \mathrm{Core}_B(\mathrm{Ker}(\psi))$. In other words, $\mathrm{Ker}(\psi^B) = \mathrm{Core}_B(\mathrm{Ker}(\psi))$. □

These observations motivate the following definition.

Definition 1.1.32. *With the notation of Proposition 1.1.31, the action* ψ^B *is called* the induced B-action from ψ, *and the* B-group (S^n, ψ^B) *is the* induced B-group.
    The semidirect product $[S^n]_{\psi^B} B = [S_1 \times \cdots \times S_n]B$ *is called the* twisted wreath product *of* S *by* B; *it is denoted by* $S \wr_{(C,\psi)} B$.

Thus, if G is the induced extension defined by the map $\alpha \colon N/K \longrightarrow G/M$ as in Theorem 1.1.29, then the conjugacy action of G on the normal subgroup $M = S_1 \times \cdots \times S_n$ is the induced G-action from the conjugacy action of $N = \mathrm{N}_G(S_1)$ on S_1.

Remarks 1.1.33.   1. The structure of induced B-group does not depend, up to equivalence of B-groups, on the chosen transversal of C in B.
    2. The construction of induced actions is motivated by the classical construction of induced modules. If S is a C-module, the induced B-action gives to S^n the well-known structure of induced B-module: $S^n \cong S^B$. This explains the name and the notation.

Proposition 1.1.34. *Let* S *and* B *be groups and* C *a subgroup of* B. *Suppose that* (S, ψ) *is a* C-group *and consider the twisted wreath product* $G = S \wr_{(C,\psi)} B$. *Then*

1. $\mathrm{N}_B(S_1) = C$ *and* $\mathrm{C}_B(S_1) = \mathrm{Ker}(\psi)$.
2. $\mathrm{C}_B(S^\natural) = \mathrm{Core}_B(\mathrm{Ker}(\psi))$. *Moreover if* $\mathrm{Core}_B(C) = 1$, *then* $\mathrm{C}_G(S^\natural) = \mathrm{Z}(S^\natural)$.

Proof.   1. If $h \in \mathrm{N}_B(S_1)$, then, by (1.3), $1^{h^\rho} = 1$ and $h = c_{1,h} \in C$. Conversely, if $c \in C$, then $c = c_{1,c}$ and $1^{c^\rho} = 1$; moreover $(x, 1, \ldots, 1)^{c^\rho} = (x^{c^\psi}, 1, \ldots, 1)$. Hence $C \leq \mathrm{N}_B(S_1)$.
    Observe that the elements of $\mathrm{C}_B(S_1)$ are elements $c \in C$ such that $c^\psi = \mathrm{id}_{S_1}$. Hence $\mathrm{C}_B(S_1) = \mathrm{Ker}(\psi)$.

2. Observe that $S_1^{t_i} = S_i$, for all $i \in \mathcal{I}$. Therefore $\mathrm{C}_B(S^{\natural}) = \bigcap_{i=1}^{n} \mathrm{C}_B(S_i) = \bigcap_{i=1}^{n} \mathrm{C}_B(S_1^{t_i}) = \mathrm{Core}_B\big(\mathrm{C}_B(S_1)\big) = \mathrm{Core}_B\big(\mathrm{Ker}(\psi)\big)$.

If $(x_1, \dots, x_n)h \in \mathrm{C}_G(S^{\natural})$, then $h \in \bigcap_{i=1}^{n} \mathrm{N}_B(S_i) = \mathrm{Core}_B(C) = 1$. Therefore $(x_1, \dots, x_n) \in \mathrm{Z}(S^{\natural})$. $\square$

If $1 \longrightarrow M \longrightarrow G \longrightarrow B \longrightarrow 1$ is the induced extension defined by a group homomorphism $\alpha \colon A \longrightarrow B$, then G splits over M if and only if G is isomorphic to the twisted wreath product $S\wr_{(C,\psi)} B$. F. Gross and L. G. Kovács characterise when the induced extension splits. This characterisation, which will be crucial in Chapter 7, is just a consequence of a deep analysis of the supplements of M in G.

Theorem 1.1.35 (([GK84])). *Let G be a group in which there exists a normal subgroup M of G such that $M = S_1 \times \cdots \times S_n$, where $\{S_1, \dots, S_n\}$ is the set of all conjugate subgroups of a normal subgroup S_1 of M. Write $N = \mathrm{N}_G(S_1)$ and $K = S_2 \times \cdots \times S_n$.*

1. *Let L/K be a supplement of M/K in N/K. Then, there exists a supplement H of M in G satisfying the following:*
   a) *$L = (H \cap N)K$ and $H \cap M = (H \cap S_1) \times \cdots \times (H \cap S_n)$. Further, $\{H \cap S_1, \dots, H \cap S_n\}$ is a conjugacy class in H, and $H \cap S_1 = L \cap S_1$.*
   b) *Suppose that H_0 is a supplement of M in G such that $H_0 \cap N \le L$. Then there is an element $k \in K$ such that $H_0^k \le H$. Moreover, $H_0^k = H$ if and only if $L = (H_0 \cap N)K$ and $H_0 \cap M = (H_0 \cap S_1) \times \cdots \times (H_0 \cap S_n)$.*
   c) *In particular, H is unique up to conjugacy under K.*
2. *Suppose that H is a supplement M in G such that $H \cap M = (H \cap S_1) \times \cdots \times (H \cap S_n)$. Write $L = (H \cap N)K$. Assume further that R is a subgroup of G such that $G = RM$. Then the following are true:*
   a) *R is conjugate in G to a subgroup of H if and only if $R \cap N$ is conjugate in N to a subgroup of L.*
   b) *R is conjugate to H in G if and only if $(R \cap N)K$ is conjugate to L in N and also $R \cap M = (R \cap S_1) \times \cdots \times (R \cap S_n)$.*
3. *There is a bijection between, on the one hand, the conjugacy classes in G of supplements H of M in G such that $H \cap M = (H \cap S_1) \times \cdots \times (H \cap S_n)$, and, on the other hand, the conjugacy classes in N/K of supplements L/K of M/K in N/K, Moreover, under this bijection, we have the following:*
   a) *the conjugacy classes in G of supplements U of M which are maximal subgroups of G such that $U \cap M = (U \cap S_1) \times \cdots \times (U \cap S_n)$ are in one-to-one correspondence with the conjugacy classes in N/K of supplements of M/K which are maximal subgroups of N/K.*
   b) *the conjugacy classes in G of complements of M, if any, are in one-to-one correspondence with the conjugacy classes in N/K of complements of M/K.*

Proof. By Theorem 1.1.29, the group G is the induced extension defined by $\alpha \colon N/K \longrightarrow G/M$ given by $(Kx)^{\alpha} = Mx$, for all $x \in G$. Let $\mathcal{T} = \{t_1 = $

$1, \ldots, t_n\}$ be a right transversal of N in G and write $\rho\colon G \longrightarrow \mathrm{Sym}(n)$ the permutation representation of G on the right cosets of N in G. As usual, for each $x \in G$, write $t_i x = c_{i,x} t_j$, for some $c_{i,x} \in N$, and $i^{x^\rho} = j$. Write $S_i = S^{t_i}$. For each $i \in \mathcal{I} = \{1, \ldots, n\}$, write also $K_i = \prod_{j \in \mathcal{I} \setminus \{i\}} S_j$. Then $K = K_1$ and $K_i = K^{t_i}$. For $P = G^\rho$, let λ be the embedding of G into $(N/K) \wr_\rho P$ defined by $\lambda\colon G \longrightarrow (N/K) \wr_\rho P$ such that $x^\lambda = (Kc_{1,x}, \ldots, Kc_{n,x})x^\rho$, for any $x \in G$.

1a. Define

$$H = \big((L/K) \wr_\rho P\big)^{\lambda^{-1}} = \{x \in G : c_{i,x} \in L, \text{ for all } i \in \mathcal{I}\}.$$

This subgroup H satisfies the required properties.

Fix an element $g \in G$. Then, for each $i \in \mathcal{I}$, we have that $c_{i,g} \in N = ML$ and there exists $m_{i,g} \in M$ such that $m_{i,g}^{-1} c_{i,g} \in L$.

Observe that, if $m \in M$, then $c_{i,m} = m^{t_i^{-1}}$. Then

$$m^\lambda = (Km^{t_1^{-1}}, \ldots, Km^{t_n^{-1}}) = (Km, Km^{t_2^{-1}}, \ldots, Km^{t_n^{-1}}).$$

Write $m = (s_1, \ldots, s_n)$. Then, for any $i \in \mathcal{I}$, using (1.1) in Definition 1.1.17, $(m^{t_i^{-1}})^{\pi_1} = s_i$, since $1^{t_i^\rho} = i$. Therefore $(s_1, \ldots, s_n)^\lambda = (Ks_1, \ldots, Ks_n)$.

Since the restriction of λ to M is an isomorphism onto $(M/K)^\natural$, i.e. $M^\lambda = (M/K)^\natural$, there exists a unique $m_g \in M$ such that $m_g^\lambda = (Km_{1,g}, \ldots, Km_{n,g})$. Hence

$$(m_g^{-1}g)^\lambda = (m_g^\lambda)^{-1} g^\lambda = (Km_{1,g}^{-1}, \ldots, Km_{n,g}^{-1})(Kc_{1,g}, \ldots, Kc_{n,g})g^\rho =$$
$$= (Km_{1,g}^{-1}c_{1,g}, \ldots, Km_{n,g}^{-1}c_{n,g})g^\rho \in (L/K) \wr_\rho P,$$

and then $m_g^{-1}g \in H$. Hence, $G = HM$.

Observe that $Km_{i,g} = Kc_{i,m_g} = Km_g^{t_i^{-1}}$. If $g \in L$, then we can choose $m_{1,g} = 1$, and then $m_g \in K$. Thus $m_g^{-1}g \in H \cap N$. Then $L \leq K(H \cap N)$. On the other hand, if $h \in H \cap N$, then $h = c_{1,h} \in L$. Hence $L = K(H \cap N)$.

If $m = (s_1, \ldots, s_n) \in M \cap H$, then $Ks_i \in L/K$, for all $i \in \mathcal{I}$. Observe that, for any $i \in \mathcal{I}$, we have that $(1, \ldots, s_i, \ldots, 1)^\lambda = (K, \ldots, Ks_i, \ldots, K) \in \big((L \cap M)/K\big)^\natural$ and then $(1, \ldots, s_i, \ldots, 1) \in H \cap S_i$. Hence, $H \cap M = (H \cap S_1) \times \cdots \times (H \cap S_n)$.

Since $G = HM$, we can choose the transversal $\mathcal{T} \subseteq H$. Hence, for all $i \in \mathcal{I}$, we have that $H \cap S_i = (H \cap S_1)^{t_i}$. Therefore $\{H \cap S_1, \ldots, H \cap S_n\}$ is a conjugacy class in H. Moreover $(L \cap M)/K = \big((H \cap N)K \cap M\big)/K = (H \cap M)K/K = (H \cap S_1)K/K \cong H \cap S_1$ and also $(L \cap M)/K = (L \cap S_1)K/K \cong L \cap S_1$. Hence $|H \cap S_1| = |L \cap S_1|$. Since $H \cap S_1 = H \cap N \cap S_1 \leq L \cap S_1$, we have the equality $H \cap S_1 = L \cap S_1$.

1b. Assume now that H_0 is a subgroup of G such that $G = MH_0$ and $H_0 \cap N \leq L$. For each $i \in \mathcal{I}$, there must be an element $m_i \in M$ such that $t_i \in m_i^{-1}H_0$, i.e. $m_i t_i \in H_0$. We may choose $m_1 = 1$. Now, there exists a unique $k \in M$ such that

$$(K, Km_2, \ldots, Km_n) = k^\lambda = (Kk^{t_1^{-1}}, \ldots, Kk^{t_n^{-1}}).$$

This implies that $k \in K$ and $t_i k t_i^{-1} m_i^{-1} \in K$, for all $i \in \mathcal{I}$. We show that $H_0^{k^{-1}} \leq H$.

Let $x \in H_0$ and consider $y = x^{k^{-1}}$. Observe that, for all $i \in \mathcal{I}$, $Mt_i x = Mt_i x^{k^{-1}} = Mt_i y$ and then $i^{x^\rho} = i^{y^\rho}$. Now

$$
\begin{aligned}
c_{i,y} &= t_i y t_{i^{y^\rho}}^{-1} = t_i y t_{i^{x^\rho}}^{-1} = t_i k x k^{-1} t_{i^{x^\rho}}^{-1} = \\
&= t_i k (t_i^{-1} m_i^{-1} m_i t_i) x (t_{i^{x^\rho}}^{-1} m_{i^{x^\rho}}^{-1} m_{i^{x^\rho}} t_{i^{x^\rho}}) k^{-1} t_{i^{x^\rho}}^{-1} = \\
&= (t_i k t_i^{-1} m_i^{-1})(m_i t_i x t_{i^{x^\rho}}^{-1} m_{i^{x^\rho}}^{-1})(m_{i^{x^\rho}} t_{i^{x^\rho}} k^{-1} t_{i^{x^\rho}}^{-1}).
\end{aligned}
$$

Now observe that $m_i t_i$ and $t_{i^{x^\rho}}^{-1} m_{i^{x^\rho}}^{-1}$ are in H_0 and then, $m_i t_i x t_{i^{x^\rho}}^{-1} m_{i^{x^\rho}}^{-1} \in H_0$. On the other hand, $t_i x t_{i^{x^\rho}}^{-1} = c_{i,x} \in N$, and then $m_i t_i x t_{i^{x^\rho}}^{-1} m_{i^{x^\rho}}^{-1} \in N$. Since $t_i k t_i^{-1} m_i^{-1} \in K$ and also $m_{i^{x^\rho}} t_{i^{x^\rho}} k^{-1} t_{i^{x^\rho}}^{-1} \in K$, we have that

$$c_{i,y} = (t_i k t_i^{-1} m_i^{-1})(m_i t_i x t_{i^{x^\rho}}^{-1} m_{i^{x^\rho}}^{-1})(m_{i^{x^\rho}} t_{i^{x^\rho}} k^{-1} t_{i^{x^\rho}}^{-1}) \in K(H_0 \cap N)K \leq L$$

for all $i \in \mathcal{I}$. This means that $y \in H$.

Assume that $H_0^k \leq H$, for $k \in K$. Clearly, if $H_0^k = H$, then $L = (H_0 \cap N)K$ and $H_0 \cap M = (H_0 \cap S_1) \times \cdots \times (H_0 \cap S_n)$. Conversely, suppose that $L = (H_0 \cap N)K$ and $H_0 \cap M = (H_0 \cap S_1) \times \cdots \times (H_0 \cap S_n)$. Observe that H_0^k satisfies the same properties. Thus, we can assume that $H_0 \leq H$.

As in 1a, since $G = H_0 M$, we have that $\{H_0 \cap S_1, \ldots, H_0 \cap S_n\}$ is a conjugacy class in H_0, and $H_0 \cap S_1 = L \cap S_1$. Hence, $|H \cap S_1| = |H_0 \cap S_1|$, and then $H \cap S_1 = H_0 \cap S_1$. Therefore, $H \cap M = H_0 \cap M$. Then, from $G = H_0 M = HM$, we deduce that $|G : H_0| = |M : M \cap H_0| = |M : M \cap H| = |G : H|$. Hence, $|H| = |H_0|$ and then, $H = H_0$.

Part 1c is a direct consequence of 1b.

2a. Clearly L/K is a supplement of M/K in N/K. By 1c, the subgroup H is determined, up to conjugacy in K, by L.

Suppose that $G = RM$ and $R \cap N$ is conjugate to a subgroup of L in N. Since $N = RM \cap N = (R \cap N)M$, there is an element $m \in M$ such that $(R \cap N)^m \leq L$. Write $H_0 = R^m$. Then $G = H_0 M$ and $H_0 \cap N \leq L$. It follows, from 1b, that H_0 is conjugate to a subgroup of H. Hence R is conjugate to a subgroup of H. Conversely, if R is conjugate to a subgroup of H, then, since $G = RM$, we have that $R^m \leq H$, for some $m \in M$. Then $(R \cap N)^m = R^m \cap N \leq H \cap N \leq L$.

2b. If $G = RM$ and L is conjugate to $(R \cap N)K$ in N, there is an element $m \in M$ such that $L = ((R \cap N)K)^m = (R \cap N)^m K = (R^m \cap N)K$. If $R \cap M = (R \cap S_1) \times \cdots \times (R \cap S_n)$, by 1b, we deduce that $H_0 = R^m$ is conjugate to H. The rest of 2b follows easily.

3. The bijection follows easily from 1 and 2.

3a. Let L be a maximal subgroup of N such that $K \leq L$ and $N = LM$ and consider one of the supplements U of M in G determined by the conjugacy class of L in N under the bijection. Suppose that $U \leq H < G$. Then $N = (H \cap N)M$. Set $L_0 = (H \cap N)K$. Then L_0/K is a supplement of M/K in N/K. Clearly $L = (U \cap N)K \leq L_0$. By maximality of L, we have that $L = L_0$. But then $H \cap N \leq L$ and, by 1b, $H^k \leq U$, for some $k \in K$. Clearly, this implies that $U = H$. Hence U is maximal in G.

Conversely, let U be a maximal subgroup of G which supplements M in G such that $U \cap M = (U \cap S_1) \times \cdots \times (U \cap S_n)$. Write $L = (U \cap N)K$. Suppose that $L \leq L_0 < N$. Consider a supplement R of M in G determined by L_0 under the bijection. Then $L_0 = (R \cap N)K$. Since $U \cap N \leq L_0$, then $U^k \leq R$, for some $k \in K$. By maximality of U, we have that $R = U^k$. This implies that L and L_0 are conjugate in N and, since $L \leq L_0$, equality holds.

3b. Observe that if L/K is a complement of M/K in N/K, then $L \cap S_1 = 1$. Hence $H \cap S_1 = 1$ and therefore $H \cap M = 1$. This is to say that H is a complement of M in G. Conversely, if H is a complement of M in G, then $(H \cap N)K \cap M = (H \cap N \cap M)K = K$. □

The following result, due also to F. Gross and L. G. Kovács, is an application of the induced extension procedure to the construction of groups which are not semidirect products. We will use it in Chapter 5.

Theorem 1.1.36 ([GK84]). *Let B be any finite simple group. Then there exists a finite group G with a minimal normal subgroup M such that M is a direct product of copies of* $\mathrm{Alt}(6)$, *the alternating group of degree 6, the quotient group G/M is isomorphic to B and G does not split over M.*

Proof. Consider the group $A = \mathrm{Aut}\big(\mathrm{Alt}(6)\big)$. Let D denote the normal subgroup of inner automorphisms, $D \cong \mathrm{Alt}(6)$, of A. It is well-known that the quotient group A/D is isomorphic to an elementary abelian 2-group of order 4 and A does not split over D, i.e. there is no complement of D in A (see [Suz82]).

By the Odd Order Theorem ([FT63]), the Sylow 2-subgroups of B are non-trivial. By the Burnside Transfer Theorem (see [Suz86, 5.2.10, Corollary 2]), a Sylow 2-subgroup of B cannot by cyclic. By a theorem of R. Brauer and M. Suzuki (see [Suz86, page 306]), the Sylow 2-subgroups of G cannot by isomorphic to a quaternion group. Hence a Sylow 2-subgroup of B has two transpositions generating a dihedral 2-group (see [KS04, 5.3.7 and 1.6.9]). Therefore B must contain a subgroup G which is elementary abelian of order 2. Then there is a homomorphism α of A into B such that $A^\alpha = C$ and Ker $(\alpha) = D$

Now let G be the induced extension defined by $\alpha \colon A \longrightarrow B$. Since A does not split over D, the group G has the required properties. □

Let G be a group which is an induced extension of a normal subgroup $M = S_1 \times \cdots \times S_n$. We have presented above a complete description of those supplements of M in G whose intersection with M is a direct product of the projections in each component $H \cap M = (H \cap S_1) \times \cdots \times (H \cap S_n)$. But nothing is said about those supplements H whose projections $\pi_i \colon H \cap M \longrightarrow S_i$ are surjective. Subgroups D of a direct product M such that all projections $\pi_i \colon D \longrightarrow S_i$ are surjective are fully described by M. Aschbacher and L. Scott in [AS85]. In the sequel we present here an adaptation of their results suitable for our purposes.

Definition 1.1.37. *Let $G = \prod_{i=1}^{n} S_i$ be a direct product of groups. A subgroup H of G is said to be* diagonal *if each projection $\pi_i \colon H \longrightarrow S_i$, $i = 1, \dots, n$, is injective.*

If each projection $\pi_i \colon H \longrightarrow S_i$ is an isomorphism, then the subgroup H is said to be a full diagonal subgroup.

Obviously if H is a full diagonal subgroup of $G = \prod_{i=1}^{n} S_i$, then all the S_i are isomorphic. Observe that if $x = (x_1, \dots, x_n) \in H$, then $x_i = x^{\pi_i}$, for all $i = 1, \dots, n$, and then $x = (x_1, x_1^{\pi_1^{-1}\pi_2}, \dots, x_1^{\pi_1^{-1}\pi_n})$. All $\varphi_i = \pi_1^{-1}\pi_i$ are isomorphisms of S_1 and then $\varphi = (\varphi_1 = 1, \varphi_2, \dots, \varphi_n) \in \operatorname{Aut}(S_1)^n$. Conversely, given a group S and $\varphi = (\varphi_1, \varphi_2, \dots, \varphi_n) \in \operatorname{Aut}(S)^n$, it is clear that $\{x^\varphi = (x^{\varphi_1}, x^{\varphi_2}, \dots, x^{\varphi_n}) : x \in S\}$ is a full diagonal subgroup of S^n.

More generally, given a direct product of groups $G = \prod_{i=1}^{n} S_i$ such that all S_i are isomorphic copies of a group S, to each pair (Δ, φ), where $\Delta = \{\mathcal{I}_1, \dots, \mathcal{I}_l\}$ is a partition of the set $\mathcal{I} = \{1, \dots, n\}$ and $\varphi = (\varphi_1, \dots, \varphi_n) \in \operatorname{Aut}(S)^n$, we associate a direct product $D_{(\Delta, \varphi)} = D_1 \times \cdots \times D_l$, where each D_j is a full diagonal subgroup of the direct product $\prod_{i \in \mathcal{I}_j} S_i$ defined by the automorphisms $\{\varphi_i : i \in \mathcal{I}_j\}$. It is easy to see that if Γ is a partition of $\mathcal{I}$ refining Δ, then $D_{(\Delta, \varphi)} \leq D_{(\Gamma, \varphi)}$. In particular, the trivial partition $\Omega = \{\{1\}, \dots \{n\}\}$ of $\mathcal{I}$ gives $D_{(\Omega, \varphi)} = G$, for any $\varphi \in \operatorname{Aut}(S)^n$.

For groups S with trivial centre, the group G can be embedded in the wreath product $W = \operatorname{Aut}(S) \wr \operatorname{Sym}(n)$. In particular, if S is a non-abelian simple group, then $G \leq \operatorname{Aut}(S^n)$. In the group W the conjugacy by the element $\varphi \in W$ makes sense and $D_{(\Delta, \varphi)} = D_{(\Delta, \operatorname{id})}^{\varphi}$, where id denotes the n-tuple composed by all identity isomorphisms.

Lemma 1.1.38. *Let H be a full diagonal subgroup of the direct product $G = \prod_{i=1}^{n} S_i$, where the S_i are copies of a non-abelian simple group S. Then H is self-normalising in G.*

Proof. Since H is a full diagonal subgroup of G, all π_i are isomorphisms of H onto S_i. Observe that $(x_1, \dots, x_n) \in H$ if and only if $x_j = x_1^{\pi_1^{-1}\pi_j}$, for $j = 2, \dots, n$ and for all $x_1 \in S$. Write $\varphi_j = \pi_1^{-1}\pi_j$, for $j = 2, \dots, n$.

If $g = (g_1, \dots, g_n) \in \mathrm{N}_G(H)$, then for all $x \in S$ we have that

$$(x, x^{\varphi_2}, \dots, x^{\varphi_n})^g = \left(x^{g_1}, (x^{\varphi_2})^{g_2}, \dots, (x^{\varphi_n})^{g_n}\right) \in H.$$

Hence, for $j = 2, \dots, n$, $(x^{\varphi_j})^{g_j} = (x^{g_1})^{\varphi_j} = (x^{\varphi_j})^{g_1^{\varphi_j}}$ and the automorphism $g_1^{\varphi_j} g_j^{-1}$ is the trivial automorphism of S_j. Hence $g_1^{\varphi_j} = g_j$ and $g \in H$. This is to say that H is self-normalising in G. $\qquad\square$

Proposition 1.1.39. *Suppose that H is a subgroup of the direct product $G = \prod_{i=1}^{n} S_i$, where the S_i are non-abelian simple groups for all $i \in \mathcal{I} = \{1, \ldots, n\}$. Assume that all projections $\pi_i \colon H \longrightarrow S_i$, $i \in \mathcal{I}$, are surjective.*

1. *There exists a partition Δ of $\mathcal{I}$ such that the subgroup H is the direct product*

$$H = \prod_{\mathcal{D} \in \Delta} H^{\pi_{\mathcal{D}}},$$

   where

   a) *each $H^{\pi_{\mathcal{D}}}$ is a full diagonal subgroup of $\prod_{i \in \mathcal{D}} S_i$,*
   b) *the partition Δ is uniquely determined by H in the sense that if $H = \prod_{\mathcal{D} \in \Delta} H^{\pi_{\mathcal{D}}} = \prod_{\mathcal{G} \in \Gamma} H^{\pi_{\mathcal{G}}}$, for Δ and Γ partitions of $\mathcal{I}$, then $\Delta = \Gamma$, and*
   c) *if $H \leq K \leq G$, then $K = \prod_{\mathcal{G} \in \Gamma} H^{\pi_{\mathcal{G}}}$, where Γ is a partition of $\mathcal{I}$ which refines Δ.*

2. *Suppose that the S_i are isomorphic copies of a non-abelian simple group S, for all $i \in \mathcal{I}$, i.e. $G \cong S^n$. Let U be a subgroup of $\mathrm{Aut}(G)$. Then U, acting by conjugation on the simple components S_i of $\mathrm{Soc}\big(\mathrm{Aut}(G)\big)$, is a permutation group on the set $\{S_1, \ldots, S_n\}$ (and therefore on $\mathcal{I}$).*
   Observe that the action of U on $\mathcal{I}$ induces an action on the set of all partitions of $\mathcal{I}$. We can say that a partition Δ of $\mathcal{I}$ is U-invariant if $\Delta^x = \Delta$ for all $x \in U$.
   If H is U-invariant, i.e. $U \leq \mathrm{N}_{\mathrm{Aut}(G)}(H)$, then the partition Δ is a U-invariant set of blocks of the action of U on $\mathcal{I}$.

3. *In the situation of 2, if Γ is a U-invariant partition of $\mathcal{I}$ which refines Δ and every member of Γ is again a block for the action of U on $\mathcal{I}$, then the subgroup $K = \prod_{\mathcal{G} \in \Gamma} H^{\pi_{\mathcal{G}}}$ is also U-invariant.*

Proof. 1a. Let $\mathcal{D}$ be a subset of $\mathcal{I}$ minimal such that the subgroup $D = H \cap \big(\prod_{i \in \mathcal{D}} S_i\big)$ is non-trivial. It is clear that D is a normal subgroup of H and then every projection of D is a normal subgroup of the corresponding projection of H. Since, by minimality of $\mathcal{D}$, D^{π_j} is non-trivial, for each $j \in \mathcal{D}$, we have that $D^{\pi_j} = S_j$. Moreover, for each $j \in \mathcal{D}$, we have that $\mathrm{Ker}(\pi_j) \cap D = H \cap \big(\prod_{i \in \mathcal{D}, i \neq j} S_i\big) = 1$, by minimality of $\mathcal{D}$. Therefore D is a full diagonal subgroup of $\prod_{i \in \mathcal{D}} S_i$. Let $E = H^{\pi_{\mathcal{D}}}$ be the image of the projection of H in $\prod_{i \in \mathcal{D}} S_i$. Then $D = D^{\pi_{\mathcal{D}}}$ is normal in E. By Lemma 1.1.38, $D = E$. Write $F = H \cap \prod_{i \notin \mathcal{D}} S_i$. Clearly $D \times F \leq H$. For each $x \in H$, we can write $x = x_1 x_2$, where x_1 is the projection of x onto $\prod_{i \in \mathcal{D}} S_i$ and x_2 is the projection of x onto $\prod_{i \notin \mathcal{D}} S_i$. Observe that $x_1 \in D \leq H$ and then $x_2 \in F$. This is to say that $H = D \times F$. Now the result follows by induction on the cardinality of $\mathcal{I}$.

To prove 1b suppose that $H = \prod_{\mathcal{D} \in \Delta} H^{\pi_{\mathcal{D}}} = \prod_{\mathcal{G} \in \Gamma} H^{\pi_{\mathcal{G}}}$, for Δ and Γ partitions of $\mathcal{I}$. Observe that for each $\mathcal{D} \in \Delta$, since $H^{\pi_{\mathcal{D}}}$ is a full diagonal subgroup of $\prod_{i \in \mathcal{D}} S_i$, we have that the following statements are equivalent for a non-trivial element $h \in H$:

1. $h \in H^{\pi_{\mathcal{D}}}$;
2. $h^{\pi_i} \neq 1$ if and only if $i \in \mathcal{D}$;
3. there exists an $i \in \mathcal{D}$ such that $h^{\pi_i} \neq 1$ and for each $\mathcal{D}' \in \Delta$, with $\mathcal{D}' \neq \mathcal{D}$, there exists a $j \in \mathcal{D}'$ such that $h^{\pi_j} = 1$.

Suppose that $h \in H^{\pi_{\mathcal{D}}}$. Then $h^{\pi_i} \neq 1$, for all $i \in \mathcal{D}$, and $h^{\pi_j} = 1$, for all $j \notin \mathcal{D}$. If $i \in \mathcal{D}$, there exists $\mathcal{G} \in \Gamma$ such that $i \in \mathcal{G}$. Thus $h \in H^{\pi_{\mathcal{G}}}$ and in fact $\mathcal{D} = \mathcal{G}$. Hence $\Delta = \Gamma$.

1c. Suppose finally that K is a subgroup of G containing H. Obviously, the projections $\pi_i \colon K \longrightarrow S_i$ are surjective. Then, by the above arguments, we have that $K = \prod_{\mathcal{G} \in \Gamma} K^{\pi_{\mathcal{G}}}$, where Γ is a partition of $\mathcal{I}$, and, for each $\mathcal{G} \in \Gamma$, $K^{\pi_{\mathcal{G}}}$ is a full diagonal subgroup of $\prod_{j \in \mathcal{G}} S_j$. In particular, for all $i \in \mathcal{G}$, the S_i are isomorphic to a non-abelian simple group $S_{\mathcal{G}}$. Since $H = \prod_{\mathcal{D} \in \Delta} H^{\pi_{\mathcal{D}}}$, we have $H^{\pi_{\mathcal{G}}} = \prod_{\mathcal{D} \cap \mathcal{G}, \mathcal{D} \in \Delta} H^{\pi_{\mathcal{D} \cap \mathcal{G}}}$. If $\mathcal{G} \cap \mathcal{D} \neq \emptyset$, then $H^{\pi_{\mathcal{D} \cap \mathcal{G}}} \cong S_{\mathcal{G}}$. Observe that $H^{\pi_{\mathcal{G}}}$ is a direct product contained in $K^{\pi_{\mathcal{G}}} \cong S_{\mathcal{G}}$. This implies that the direct product has a unique component which is equal to $K^{\pi_{\mathcal{G}}}$. Hence, for each $\mathcal{G} \in \Gamma$, $H^{\pi_{\mathcal{G}}} = K^{\pi_{\mathcal{G}}}$, and $\mathcal{G} \subseteq \mathcal{D}$, for some $\mathcal{D} \in \Delta$, i.e. Γ is a partition of $\mathcal{I}$ which refines Δ.

2. By Proposition 1.1.20, we can consider that U is a subgroup of the wreath product $A \wr \mathrm{Sym}(n)$, for $A = \mathrm{Aut}(S)$ and S a non-abelian simple group such that $S \cong S_i$, for all $i \in \mathcal{I}$. We see in Remark 1.1.18 (2) of that U acts by conjugation on the set $\{A_1, \dots, A_n\}$ of factors of the base group. Since S is the unique minimal normal subgroup of A, the group U acts by conjugation on $\{S_1, \dots, S_n\}$.

Suppose that H is U-invariant. Then, for any $x \in U$, by Remark 1.1.18 (4), we have

$$H = H^x = \prod_{\mathcal{D} \in \Delta} (H^{\pi_{\mathcal{D}}})^x = \prod_{\mathcal{D} \in \Delta} (H^x)^{\pi_{\mathcal{D}^x}} = \prod_{\mathcal{D}^x \in \Delta^x} H^{\pi_{\mathcal{D}^x}}$$

and then $\Delta = \Delta^x$, by 1b. Hence Δ is U-invariant. Moreover $\mathcal{D}^x$ is an element of the partition Δ. Therefore either $\mathcal{D} = \mathcal{D}^x$ or $\mathcal{D} \cap \mathcal{D}^x = \emptyset$. Hence the elements of Δ are blocks for the action of U on $\mathcal{I}$.

3. This follows immediately from Remark 1.1.18 (4): for any $x \in U$, we have

$$K^x = \prod_{\mathcal{G} \in \Gamma} (H^{\pi_{\mathcal{G}}})^x = \prod_{\mathcal{G} \in \Gamma} (H^x)^{\pi_{\mathcal{G}^x}} = \prod_{\mathcal{G} \in \Gamma} H^{\pi_{\mathcal{G}}} = K,$$

and therefore K is U-invariant.   □

The purpose of the following is to present a proof of the Theorem of O'Nan and Scott classifying all primitive groups of type 2. The first version of this theorem, stated by Michael O'Nan and Leonard Scott at the symposium on Finite Simple Groups at Santa Cruz in 1979, appeared in the proceedings in [Sco80] but one of the cases, the primitive groups whose socle is complemented by a maximal subgroup, is omitted. In [Cam81], P. J. Cameron presented an outline of the proof of the O'Nan-Scott Theorem again with the same omission. Finally, in [AS85] a corrected and expanded version of the theorem appears. Independently, L. G. Kovács presented in [Kov86] a completely different approach to the same result.

We are indebted to P. Jiménez-Seral for her kind contributions in [JS96]. These personal notes, written for a doctoral course at the Universidad de Zaragoza and adapted for her students, are motivated mainly by the self-contained version of the O'Nan-Scott Theorem which appears in [LPS88].

To study the structure of a primitive group G of type 2 whose socle $\text{Soc}(G)$ is non-simple, we will follow the following strategy. We observe that in general, for any supplement M of $\text{Soc}(G)$ in G, we have that M is a maximal subgroup of G if and only if $M \cap \text{Soc}(G)$ is a maximal M-invariant subgroup of $\text{Soc}(G)$. We will focus our attention in the structure of the intersection $U \cap \text{Soc}(G)$ of a core-free maximal subgroup U of G with the socle.

General remarks and notation 1.1.40. We fix here the main notation which is used in our study of primitive groups of type 2 in the sequel. We also review some previously known facts and make some remarks. All these observations give rise to the first steps of the classification theorem of O'Nan and Scott.

Let G be a primitive group of type 2.

1. Write $\text{Soc}(G) = S_1 \times \cdots \times S_n$ where the S_i are copies of a non-abelian simple group S, for $i \in \mathcal{I} = \{1, \ldots, n\}$. Write also $K_j = \prod_{i \in \mathcal{I}, i \neq j} S_i$, for each $j \in \mathcal{I}$.

2. Write $N = \text{N}_G(S_1)$ and $C = \text{C}_G(S_1)$. Let $\psi \colon N \longrightarrow \text{Aut}(S_1)$ denote the conjugacy action of $N = \text{N}_G(S_1)$ on S_1. Sometimes we will make the identification $S_1^{\psi} = \text{Inn}(S_1) = S_1$.

3. The quotient group $X = N/C$ is an almost simple group with $\text{Soc}(X) = S_1 C/C$.

4. Suppose that U is a core-free maximal subgroup of G.

5. The subgroup $U \cap \text{Soc}(G)$ is maximal with respect to being a proper U-invariant subgroup of $\text{Soc}(G)$.

6. By Proposition 1.1.19, the group G, acting by conjugation on the elements of the set $\{S_1, \ldots, S_n\}$, induces the structure of a G-set on $\mathcal{I}$. Write $\rho \colon G \longrightarrow \text{Sym}(n)$ for this action. The kernel of this action is $\text{Ker}(\rho) = \bigcap_{i=1}^{n} \text{N}_G(S_i) = Y$. Therefore G/Y is isomorphic to a subgroup $G^{\rho} = P_n$ of $\text{Sym}(n)$. For any $g \in G$, we write g^{ρ} for the image of g in P_n.

Moreover, since $\text{Soc}(G)$ is a minimal normal subgroup, the conjugacy action of G on $\{S_1, \ldots, S_n\}$, and on $\mathcal{I}$, is transitive. Observe that $S_{i^{x^{\rho}}} = S_i^x$ and $K_{i^{x^{\rho}}} = K_i^x$, for $x \in G$ and $i \in \mathcal{I}$.

It is worth remarking here that the action of $\text{Soc}(G)$ on $\mathcal{I}$ is trivial. Therefore if H is a supplement of $\text{Soc}(G)$ in G and Δ is a partition of $\mathcal{I}$, then Δ is H-invariant if and only if Δ is G-invariant. Also, a subset $\mathcal{D} \subseteq \mathcal{I}$ is block for the action of H if and only if $\mathcal{D}$ is a block for the action of G.

Since the set $\{S_1, \ldots, S_n\}$ is a conjugacy class of subgroups of G, we have that $Y = \text{Core}_G(N)$. In particular $\text{Soc}(G) \leq Y$.

Now U is core-free and maximal in G and therefore $G = UY$. This means that if τ is a permutation of $\mathcal{I}$ in P_n, there exists an element $x \in U$ such that the conjugation by x permutes the S_i in the same way τ does: $S_{i^\tau} = S_i^x$, for all $i \in \mathcal{I}$. In other words, $x^\rho = \tau$. This is to say that the projection of U onto P_n is surjective.

7. The stabiliser of the element 1 for the action of G on $\mathcal{I}$ is $N = \mathrm{N}_G(S_1)$. Therefore $|G : N| = n$. Observe that $N = \mathrm{N}_G(S_1) = \mathrm{N}_G(K_1)$. Let $\mathcal{T} = \{1 = t_1, t_2, \ldots, t_n\}$ be a right transversal of N in G such that $S_1^{t_i} = S_i$, for $i \in \mathcal{I}$.

8. Observe that $\mathrm{Soc}(G) \leq N$ and then $G = UN$. For this reason the transversal $\mathcal{T}$ can be chosen such that $\mathcal{T} \subseteq U$.

9. Write $V = U \cap N = \mathrm{N}_U(S_1)$. Then $\mathcal{T}$ is a right transversal of V in U. Observe that $N = N \cap U \operatorname{Soc}(G) = (N \cap U) \operatorname{Soc}(G) = V \operatorname{Soc}(G) = V \mathrm{C} S_1$.

10. The conjugation in S_1 by the elements of V induces a group homomorphism $\varphi \colon V \longrightarrow \mathrm{Aut}(S_1)$. It is clear that $\mathrm{Ker}(\varphi) = \mathrm{C}_U(S_1)$.

11. For any $i \in \mathcal{I}$, we have

$$t_i g = a_{i,g} t_j, \text{ with } a_{i,g} \in N \text{ and } i^{g^\rho} = j.$$

Moreover, since $\mathcal{T} \subseteq U$, if $g \in U$, then $a_{i,g} \in V$.

12. Denote with a star ($*$) the projection of N in X: if $a \in N$, then $a^* = aC \in X$.

13. The group G is the induced extension defined by $\alpha \colon N/K_1 \longrightarrow G/\operatorname{Soc}(G)$. Hence, the action of G on $\operatorname{Soc}(G)$ is the induced G-action from ψ:

$$\psi^G \colon G \longrightarrow X \wr_\rho P_n \leq \mathrm{Aut}(S^n),$$

given by $g^{\psi^G} = (a_{1,g}^*, \ldots, a_{n,g}^*) g^\rho$, for any $g \in G$. Observe that $\mathrm{Ker}(\psi^G) = \mathrm{Core}_G\big(\mathrm{Ker}(\psi)\big) = \mathrm{Core}_G\big(\mathrm{C}_G(S_1)\big) = 1$. Hence ψ^G is injective. In other words, ψ^G is an embedding of G into the wreath product $X \wr_\rho P_n$, and then into $\mathrm{Aut}(S^n)$. We identify G and G^{ψ^G}.

With this identification, $\mathrm{N}_G(S_1) = G \cap (X_1 \times [X_2 \times \cdots \times X_n] P_{n-1})$, where P_{n-1} is the stabiliser of 1. If $g \in \mathrm{N}_G(S_1)$, then g^ρ fixes 1, i.e. $g^\rho \in P_{n-1}$. Moreover $a_{1,g} = g$. Hence $g^{\psi^G} = (g^*, a_{2,g}^*, \ldots, a_{n,g}^*) g^\rho \in (X_1 \times [X_2 \times \cdots \times X_n] P_{n-1})$. Hence the projection of $\mathrm{N}_G(S_1)$ on X_1 is surjective.

14. Observe that, for each $i \in \mathcal{I}$, any element x_i of S_i can be written as $x_i = e_i^{t_i}$, for certain $e_i \in S_1$. For any $j \neq i$, we have that $x_i^{t_j^{-1}} \in S_k$, for some $k \neq 1$ and therefore $x_i^{t_j^{-1}} \in \mathrm{C}_G(S_1)$. This implies that $a_{j,x_i}^* = 1$ for any $j \neq i$. Moreover $a_{i,x_i} = e_i$. Also it is clear that x_i normalises all the S_j, for $j = 1, \ldots, n$ and then $x_i^\rho = 1$. Hence $x_i^{\psi^G} = e_i^*$. This is to say that, with the identification of 2, $S_i^{\psi^G} = S_i$, for all $i \in \mathcal{I}$, and then $\operatorname{Soc}(G)^{\psi^G} = S^\natural$.

15. For each $i \in \mathcal{I}$, the quotient group $Y \mathrm{C}_G(S_i)/\mathrm{C}_G(S_i)$ is isomorphic to a subgroup of $\mathrm{Aut}(S_i)$ and then $Y/\bigcap_{i=1}^n \mathrm{C}_Y(S_i) \cong Y$ is embedded in $\mathrm{Aut}(S_1) \times \cdots \times \mathrm{Aut}(S_n)$. Observe that the kernel of the homomorphism which assigns to each n-tuple of $\mathrm{Aut}(S_1) \times \cdots \times \mathrm{Aut}(S_n)$ the n-tuple of the corresponding projections of $\mathrm{Out}(S_1) \times \cdots \times \mathrm{Out}(S_n)$ is $\operatorname{Soc}(G)$. Hence the quotient group $Y/\operatorname{Soc}(G)$ is isomorphic to a subgroup of $\mathrm{Out}(S_1) \times \cdots \times \mathrm{Out}(S_n)$. Hence,

by the Schreier's conjecture ([KS04, page 151]), the group $Y/\operatorname{Soc}(G) = \big(Y \cap U \operatorname{Soc}(G)\big)/\operatorname{Soc}(G) = (Y \cap U)\operatorname{Soc}(G)/\operatorname{Soc}(G) \cong (U \cap Y)/\big(U \cap \operatorname{Soc}(G)\big)$ is soluble.

16. As in Remarks 1.1.18, if $\mathcal{S} \subseteq \mathcal{I}$, then we write

$$\pi_{\mathcal{S}} \colon \operatorname{Soc}(G) \longrightarrow \prod_{j \in \mathcal{S}} S_j$$

for the projection of $\operatorname{Soc}(G)$ onto $\prod_{j \in \mathcal{S}} S_j$. If $\mathcal{S} = \{j\}$, then the projection onto S_j is denoted simply π_j.

17. Write $R_j = \big(U \cap \operatorname{Soc}(G)\big)^{\pi_j}$. Since the action of G on $\mathcal{I}$ is transitive and $G = U \operatorname{Soc}(G)$, then all projections R_j, $j = 1, \ldots, n$ are conjugate by elements of U. Hence $U \cap \operatorname{Soc}(G) \le R_1 \times \cdots \times R_n = R_1 \times R_1^{t_2} \times \cdots \times R_1^{t_n}$.

18. By Remark 1.1.18 (4), if $y \in U \cap \operatorname{Soc}(G)$ and $g \in V$, then $(y^g)^{\pi_1} = (y^{\pi_1})^g$. This is to say that R_1 is a V-invariant subgroup of S_1.

Therefore $R_1 \times \cdots \times R_n = R_1 \times R_1^{t_2} \times \cdots \times R_1^{t_n}$ is a V-invariant subgroup of $\operatorname{Soc}(G)$.

19. By 5 and 18, we have two possibilities for each R_i:
a) either R_i is a proper subgroup of S_i; in this case,

$$U \cap \operatorname{Soc}(G) = R_1 \times \cdots \times R_n = (U \cap S_1) \times \cdots \times (U \cap S_n),$$

b) or $R_i = S_i$, i.e. the projections of $U \cap \operatorname{Soc}(G)$ on each S_i are surjective.

20. Let us deal first with the Case 19a: suppose that R_1 is a proper subgroup of S_1. Suppose that $R_1 \le T_1 < S_1$ and T_1 is a V-invariant subgroup of S_1. Then $T_1 \times T_1^{t_2} \cdots \times T_1^{t_n}$ is U-invariant in $\operatorname{Soc}(G)$ and, by 5, we have that $T_1 \times T_1^{t_2} \cdots \times T_1^{t_n} = U \cap \operatorname{Soc}(G) = R_1 \times \cdots \times R_n$. Hence $R_1 = T_1$.

This means that if R_1 is a proper subgroup of S_1, then R_1 is maximal with respect to being a proper V-invariant subgroup of S_1.

21. If the projection π_1 of $U \cap \operatorname{Soc}(G)$ on S_1 is not surjective, then two possibilities arise:
a) either $R_1 = 1$, i.e. $U \cap \operatorname{Soc}(G) = 1$: the core-free maximal subgroup U complements $\operatorname{Soc}(G)$;
b) or $1 \ne R_1 < S_1$.

22. Suppose that $1 \ne R_1 < S_1$. Then, by 19a, $R_1 = U \cap S_1$ and then $R_1 \le V$. Hence $R_1 = V \cap S_1$.

Moreover, if we suppose that there exists a proper subgroup M of N such that $VC \le M < N$, then $M \cap S_1$ is a V-invariant subgroup of S_1 and $R_1 \le M \cap S_1$. Observe that if $S_1 \le M$, then $N = VCS_1 \le M$ and $N = M$, against our assumption. Hence, $R_1 \le M \cap S_1 \ne S_1$. By maximality of R_1, we have that $R_1 = M \cap S_1$ and then $M = M \cap CVS_1 = CV(M \cap S_1) = CVR_1 = CV$. Therefore VC is a maximal subgroup of N.

23. Now we consider the Case 19b. Assume that H is a supplement of $\operatorname{Soc}(G)$ in G and we suppose that the projection of $H \cap \operatorname{Soc}(G)$ on each component S_i of $\operatorname{Soc}(G)$ is surjective. Then, by Proposition 1.1.39, there exists an H-invariant partition Δ of $\mathcal{I}$ into blocks for the action of H on $\mathcal{I}$ such that

$$H \cap \mathrm{Soc}(G) = \prod_{\mathcal{D} \in \Delta} \left(H \cap \mathrm{Soc}(G) \right)^{\pi_{\mathcal{D}}},$$

and, for each $\mathcal{D} \in \Delta$, the projection $\left(H \cap \mathrm{Soc}(G) \right)^{\pi_{\mathcal{D}}}$ is a full diagonal subgroup of the direct product $\prod_{i \in \mathcal{D}} S_i$.

Now we prove that H is maximal in G if and only if Δ is a minimal non-trivial G-invariant partition of $\mathcal{I}$ in blocks for the action of G on $\mathcal{I}$.

Suppose $1 < \Gamma < \Delta$, where all are H-invariant partitions of $\mathcal{I}$ into blocks for the action of H on $\mathcal{I}$. Then by Proposition 1.1.39 (3), the product of projections of $H \cap \mathrm{Soc}(G)$ obtained from Γ is an H-invariant subgroup J of $\mathrm{Soc}(G)$. By Proposition 1.1.39 (1b), $H \cap \mathrm{Soc}(G) < J < \mathrm{Soc}(G)$. But if H is maximal in G, then $H \cap \mathrm{Soc}(G)$ is maximal as an H-invariant subgroup of $\mathrm{Soc}(G)$ as in 5. Hence H is not maximal in G.

Now suppose $H < L < G$. Then $H \mathrm{Soc}(G) = L \mathrm{Soc}(G) = G$ implies $H \cap \mathrm{Soc}(G) < L \cap \mathrm{Soc}(G) < \mathrm{Soc}(G)$. Then, by Proposition 1.1.39 (1c), $L \cap \mathrm{Soc}(G)$ is the product of projections of $H \cap \mathrm{Soc}(G)$ (which are the same as the projections of $L \cap \mathrm{Soc}(G)$) obtained from a non-trivial proper refinement Γ of Δ. Then by Proposition 1.1.39 (2), Γ is L-invariant so, like Δ, it is an H-invariant set of blocks for the action of H on $\mathcal{I}$. Thus if Δ is a minimal such partition of $\mathcal{I}$, then H is maximal in G.

Finally, any H-invariant block is G-invariant, by 6.

24. If the projection of $U \cap \mathrm{Soc}(G)$ on each component S_i of $\mathrm{Soc}(G)$ is surjective, then $U \cap \mathrm{Soc}(G) = D_1 \times \cdots \times D_l$, with $1 \leq l < n$, and each D_i is isomorphic to S. Hence $\mathrm{Soc}(G) = \left(U \cap \mathrm{Soc}(G) \right) K_1$ and then $G = U K_1$.

25. In this study we have observed three different types of core-free maximal subgroups U of a primitive group G of type 2 according to the image of the projection $\pi_1 \colon U \cap \mathrm{Soc}(G) \longrightarrow S_1$.

a) $\left(U \cap \mathrm{Soc}(G) \right)^{\pi_1} = S_1$, i.e. the projection π_1 of $U \cap \mathrm{Soc}(G)$ on S_1 is surjective.

b) $1 \neq R_1 = \left(U \cap \mathrm{Soc}(G) \right)^{\pi_1} < S_1$, i.e. the image of the projection π_1 of $U \cap \mathrm{Soc}(G)$ on S_1 is a non-trivial proper subgroup of S_1. In this case

$$1 \neq U \cap \mathrm{Soc}(G) = R_1 \times \cdots \times R_n = (U \cap S_1) \times \cdots \times (U \cap S_n).$$

c) $\left(U \cap \mathrm{Soc}(G) \right)^{\pi_1} = 1$, i.e. U is a complement of $\mathrm{Soc}(G)$ in G.

26. With all the above remarks, we have a first approach to the O'Nan-Scott classification of primitive groups of type 2. We have the following five situations:

a) $\mathrm{Soc}(G)$ is a simple group, i.e. $n = 1$: the group G is almost simple;

b) $n > 1$ and $U \cap \mathrm{Soc}(G) = D$ is a full diagonal subgroup of $\mathrm{Soc}(G)$;

c) $n > 1$ and $U \cap \mathrm{Soc}(G) = D_1 \times \cdots \times D_l$, a direct product of l subgroups, with $1 < l < n$, such that, for each $j = 1, \ldots, l$, the subgroup D_j is a full diagonal subgroup of a direct product $\prod_{i \in \mathcal{I}_j} S_i$, and $\{\mathcal{I}_1, \ldots, \mathcal{I}_l\}$ is a minimal non-trivial G-invariant partition of $\mathcal{I}$ in blocks for the action of U on $\mathcal{I}$;

d) $n > 1$ and the projection $R_1 = \left(U \cap \mathrm{Soc}(G)\right)^{\pi_1}$ is a non-trivial proper subgroup of S_1; here, $R_1 = VC \cap S_1$ and VC/C is a maximal subgroup of X.

e) $U \cap \mathrm{Soc}(G) = 1$.

This enables us to describe all configurations of primitive groups of type 2.

Proposition 1.1.41. *Let S be a non-abelian simple group and consider an almost simple group X such that $S \leq X \leq \mathrm{Aut}(S)$. Let P_n be a primitive group of permutations of degree n. Construct the wreath product $W = X \wr P_n$ and consider the subgroups $D_X = \{(x,\ldots,x) : x \in X\} \leq X^\natural$ and $D_S = \{(s,\ldots,s) : s \in S\} \leq S^\natural$. Clearly $P_n \leq \mathrm{C}_W(D_X)$. Suppose that U is a subgroup of W such that $D_S \leq U \leq D_X \times P_n$, and the projection of U on P_n is surjective.*

Then the group $G = S^\natural U$ is a primitive group of type 2 and U is a core-free maximal subgroup of G.

Proof. It is clear that $S^\natural$ is a minimal normal subgroup of G and $\mathrm{C}_G(S^\natural) = 1$. Hence G is a primitive group of type 2 and $\mathrm{Soc}(G) = S^\natural$.

Observe that $D_S = \mathrm{Soc}(G) \cap D_X = \mathrm{Soc}(G) \cap U$. Since P_n is a primitive group, the action of U on the elements of the set $\{S_1, \ldots, S_n\}$ is primitive and there are no non-trivial blocks. By 1.1.40 (23), U is a maximal subgroup of G. □

Definition 1.1.42. *A primitive pair (G, U) constructed as in Proposition 1.1.41 is called a primitive pair with simple diagonal action.*

A detailed and complete study of these primitive groups of simple diagonal type appears in [Kov88].

Remarks 1.1.43. In a primitive pair (G, U) with simple diagonal action, we have the following.

1. $U \cap \mathrm{Soc}(G) = D_S \neq 1$: this is the case 26b in 1.1.40
2. $D_S \cap (S_2 \times \cdots \times S_n) = 1$ and $\mathrm{Soc}(G) = D_S(S_2 \times \cdots \times S_n)$. Hence $\mathrm{N}_G(S_1) = \mathrm{N}_U(S_1)\mathrm{Soc}(G) = \mathrm{N}_U(S_1)(S_2 \times \cdots \times S_n)$, and analogously for the centraliser. Hence

$$\mathrm{N}_G(S_1)/\mathrm{C}_G(S_1) \cong \mathrm{N}_U(S_1)/\mathrm{C}_U(S_1).$$

Proposition 1.1.44. *Let (Z, H) be a primitive pair such that either Z is an almost simple group or (Z, H) is a primitive pair with simple diagonal action. Write $T = \mathrm{Soc}(Z)$. Given a positive integer $k > 1$, let P_k be a transitive group of degree k and construct the wreath product $W = Z \wr P_k$. Write P_{k-1} for the stabiliser of 1.*

Consider a subgroup $G \leq W$ such that

1. $\mathrm{Soc}(W) = T^\natural = T_1 \times \cdots \times T_k \leq G$,
2. *the projection of G onto P_k is surjective,*

  3. the projection of $N_G(T_1) = N_W(T_1) \cap G = (Z_1 \times [Z_2 \times \cdots \times Z_k]P_{k-1}) \cap G$
  onto Z_1 *is surjective.*

Put $U = G \cap (H \wr P_k)$. *Then* G *is a primitive group of type 2 and* U *is a
core-free maximal subgroup of* G.

Proof. Set $M = H \cap T$; clearly $N_Z(M) = H$. With the obvious notation,
write $M^\natural = M_1 \times \cdots \times M_k$. Then clearly $H \wr P_k \le N_W(M^\natural)$. Moreover if
$(z_1, \ldots, z_k)x \in N_W(M^\natural)$, then $z_i \in N_{Z_i}(M_i) = H_i$ for any $i = 1, \ldots, k$. Hence
$H \wr P_k = N_W(M^\natural)$ and therefore $U = N_G(M^\natural)$.

  Notice that $T_1 \times \cdots \times T_k$ is a minimal normal subgroup of G and $C_G(T_1 \times
\cdots \times T_k) = 1$. Hence G is a primitive group of type 2 and $\operatorname{Soc}(G) = T_1 \times \cdots \times T_k$.

  Clearly $G = U \operatorname{Soc}(G)$. Since W is a semidirect product, every element of
W can be written uniquely as a product of an element of $Z^\natural$ and an element
of P_k. Hence, if $(h_1, \ldots, h_k)x \in T^\natural$, for $x \in P_k$ and $h_i \in H_i$, $i = 1, \ldots, k$, then
$x = 1$ and $h_i \in T_i \cap H_i = M_i$. Hence $U \cap \operatorname{Soc}(G) = M^\natural$. In particular, U is
core-free in G. Let us see that U is a maximal subgroup of G.

  Observe that $N_G(T_1) = N_W(T_1) \cap U \operatorname{Soc}(G) = N_U(T_1) \operatorname{Soc}(G)$. Let V_1
be the projection of $N_U(T_1)$ on Z_1. It is clear that V_1 is contained in the
projection of U on Z_1, i.e. $V_1 \le H_1$. Since the projection of $N_G(T_1)$ onto Z_1
is surjective and the projection of $\operatorname{Soc}(G)$ on Z_1 is T_1, then $Z_1 = V_1 T_1$. Since
clearly $M_1 \le N_G(T_1)$, then $M_1 \le V_1 \le H_1$, so $V_1 \cap T_1 = M_1$ and by easy
order calculations, $V_1 = H_1$.

  Let L be an intermediate subgroup $U \le L < G$. By the above arguments,
the projection of $N_L(T_1)$ on Z_1 is an intermediate subgroup between H_1 and
Z_1. By maximality of H in Z, we have that this projection is either H_1 or Z_1.

  Write Q_i for the projection of $L \cap \operatorname{Soc}(G)$ on T_i, for $i = 1, \ldots, k$. Since L
acts transitively by conjugation on the elements of the set $\{T_1, \ldots, T_k\}$, we
have that all Q_i are isomorphic to a subgroup Q such that $M \le Q \le T$ and
$L \cap \operatorname{Soc}(G) \le Q_1 \times \cdots \times Q_k$. The subgroup $L \cap \operatorname{Soc}(G)$ is normal in L and
then in $N_L(T_1)$. Hence Q_1 is normal in the projection of $N_L(T_1)$ on Z_1. If this
projection is H_1, then Q is normal in H and then $M \le Q \le H \cap T = M$, i.e.
$Q = M$. In this case $L = L \cap U \operatorname{Soc}(G) = U(L \cap \operatorname{Soc}(G)) = U$.

  Suppose that the projection of $N_L(T_1)$ on Z_1 is the whole of Z_1. Then Q is
a normal subgroup of Z and therefore $Q = T$. If for each $i = 1, \ldots, k$ we write
$T_i = S_{i1} \times \cdots \times S_{ir}$, where all the S_{ij} are isomorphic copies of a non-abelian
simple group S, then we can put

$$\operatorname{Soc}(G) = (S_{11} \times \cdots \times S_{1r}) \times \cdots \times (S_{k1} \times \cdots \times S_{kr}).$$

The projection of $L \cap \operatorname{Soc}(G)$ on each simple component is surjective. By Re-
mark 1.1.40 (23), $L \cap \operatorname{Soc}(G) = \prod_{\mathcal{D} \in \Delta} (L \cap \operatorname{Soc}(G))^{\pi_{\mathcal{D}}}$ is a direct product of full
diagonal subgroups and the partition Δ of the set $\{11, \ldots, 1r, \ldots, k1, \ldots, kr\}$
associated with $L \cap \operatorname{Soc}(G)$ is a set of blocks for the action of L. Observe that
$M_1 \times 1 \times \cdots \times 1 \le L \cap \operatorname{Soc}(G)$. If Z is an almost simple group, then $r = 1$
and $\mathcal{D} = \{1\}$ is a block of Δ. Hence, in this case, Δ is the trivial partition of

$\{1, \ldots, k\}$. If (Z, H) is a primitive pair of simple diagonal action, then M is a full diagonal subgroup of T. Hence the set $\{11, \ldots, 1r\}$ is the union set of some members $\mathcal{D}_1, \ldots, \mathcal{D}_l$ of the partition Δ. Since the projection of $L \cap \mathrm{Soc}(G)$ on T_1 is surjective, then $T_1 = \prod_{i=1}^{l} \left(L \cap \mathrm{Soc}(G) \right)^{\pi \mathcal{D}_i} \cong S_1 \times \cdots \times S_l$ (here the S_i's are simply the names of the projections). Hence $l = r$. Since L is transitive on the T_i's, so that because the blocks corresponding to T_1 have one element, all the blocks do. In other words, $L \cap \mathrm{Soc}(G) = T_1 \times \cdots \times T_k$. Hence $L = G$. □

Definition 1.1.45. *A primitive pair* (G, U) *constructed as in Proposition 1.1.44 is called a* primitive pair with product action.

A detailed and complete study of these primitive groups in product action appears in [Kov89].

Remarks 1.1.46. 1. If (Z, H) is a primitive pair, then Z is a permutation group on the set of right cosets of H in Z and the cardinality of Ω is $|Z : H|$ (the degree of the permutation group Z). Now, if (G, U) is a primitive pair with product action, as in Proposition 1.1.44, then the degree of the permutation group G is

$$|G : U| = |G : G \cap (H \wr P_k)| = |W : H \wr P_k| = |Z : H|^k.$$

2. Observe that we have two different types of primitive pairs with product action:

a) If Z is an almost simple group, $T = \mathrm{Soc}(Z)$, and $R = H \cap T$, then $1 \neq R < T$ and the projection $R_1 = \left(U \cap \mathrm{Soc}(G) \right)^{\pi_1}$ is a non-trivial proper subgroup of T_1, by Proposition 1.1.16; this is Case 26d in 1.1.40.

b) If (Z, H) is a primitive pair with simple diagonal action, then $U \cap \mathrm{Soc}(G) = D_1 \times \cdots \times D_k$ a direct product of k full diagonal subgroups, with $1 < k < n$; here we are in Case 26c of 1.1.40.

Examples 1.1.47. 1. Let S be a non-abelian simple group and H a maximal subgroup of S. If C is a cyclic group of order 2, construct the wreath product $G = S \wr C$ with respect to the regular action. The group G is a primitive group of type 2 and $\mathrm{Soc}(G) = S^\natural = S_1 \times S_2$.

Consider the diagonal subgroup $D = \{(x, x) : x \in S\}$. Then $U = D \times C$ is a core-free maximal subgroup of G and (G, U) is a primitive pair with diagonal action.

Consider now the subgroup $U^* = H \wr C = [H_1 \times H_2]C$. Then U^* is also a core-free maximal subgroup of G and the pair (G, U^*) is a primitive pair with product action.

2. Let G be the primitive group of Example 1 and construct the wreath product $W = G \wr Z$ with respect to the regular action of the cyclic group Z of order 2. Then, the socle of W is isomorphic to the direct product of four copies of S: $\mathrm{Soc}(W) = S_1 \times S_2 \times S_3 \times S_4$. Moreover $\mathrm{Soc}(W)$ is complemented by a 2-subgroup P isomorphic to the wreath product $C_2 \wr C_2$, that is, isomorphic to the dihedral group of order 8. The group W is a primitive group of type 2.

If we consider the maximal subgroup U of G and construct $M = U \wr Z$, we obtain a core-free maximal subgroup of index $|W : M| = |S|^2$ such that $M \cap \mathrm{Soc}(W) = D_1 \times D_2$. Taking now the maximal subgroup U^* of G, then the subgroup $M^* = U^* \wr Z$ is another core-free maximal subgroup of W of index $|S : H|^4$ such that $M^* \cap \mathrm{Soc}(G) = H_1 \times H_2 \times H_3 \times H_4$.

Therefore the pairs (W, M) and (W, M^*) are non-equivalent primitive pairs of type 2 with product action.

Write $D_S = \{(s, s, s, s) : s \in S\}$, the full diagonal subgroup of $\mathrm{Soc}(W)$. Observe that M contains properly the subgroup $M_0 = D_S \times P$ and therefore M_0 is non-maximal in W.

According to Remark 1.1.40 (26), there still remains another structure of primitive group of type 2 to describe: those primitive groups of type 2 with the special property that the core-free maximal subgroup is a complement of the socle. This new configuration is in fact a twisted wreath product.

Theorem 1.1.48.   *1. If (G, U) is a primitive pair of type 2 and $U \cap \mathrm{Soc}(G) = 1$, then, with the notation of Definition 1.1.32, $G \cong S \wr_{(V, \varphi)} U$.*

   2. Conversely, let S be a non-abelian simple group and a group U with a subgroup V such that there exists a group homomorphism $\varphi \colon V \longrightarrow \mathrm{Aut}(S)$. Construct the twisted wreath product $G = S \wr_{(V, \varphi)} U$. If $\mathrm{Core}_U(V) = 1$ then G is a primitive group of type 2. Moreover, if U is maximal in G, then (G, U) is a primitive pair of type 2. By construction, $U \cap \mathrm{Soc}(G) = 1$.

Proof.   1. Recall that G is the induced extension defined by $\alpha \colon N/K_1 \longrightarrow G/\mathrm{Soc}(G)$. Hence $\mathrm{Soc}(G)$ is the induced U-group from the action φ of V on S (see Remark 1.1.40 (10)). Since G splits on $\mathrm{Soc}(G)$, then G is isomorphic to the twisted wreath product $G \cong S \wr_{(V, \varphi)} U$.

2. To prove the converse, it is enough to recall that in the twisted wreath product $G = S \wr_{(V, \varphi)} U$, we have that $\mathrm{C}_G(Z^\natural) = \mathrm{Z}(S^\natural) = 1$, by Proposition 1.1.34, and the conclusion follows.   □

Definition 1.1.49. *A primitive pair (G, U) constructed as in Theorem 1.1.48 is called* a primitive pair with twisted wreath product action.

Maximal subgroups of a primitive group G of type 2 complementing $\mathrm{Soc}(G)$ are called by some authors *small maximal subgroups*.

Obviously one can wonder about the existence of primitive groups of type 2 with small maximal subgroups. P. Förster, in [För84a], gives sufficient conditions for U, V, and S to obtain a primitive group with small maximal subgroups.

Theorem 1.1.50 ([För84a]). *Let U be a group with a non-abelian simple non-normal subgroup S such that whenever A is a non-trivial subgroup of U such that $S \leq \mathrm{N}_U(A)$, then $S \leq A$. Write $V = \mathrm{N}_U(S)$ and $\varphi \colon V \longrightarrow \mathrm{Aut}(S)$ for the obvious group homomorphism induced by the conjugation. Construct the twisted wreath product $G = S \wr_{(V, \varphi)} U$.*

Then G is a primitive group of type 2 such that $\mathrm{Soc}(G) = S^\natural$, the base group, is complemented by a maximal subgroup of G isomorphic to U.

Proof. First we see that if $\mathrm{C}_U(S) \neq 1$, then, by hypothesis, we have that $S \leq \mathrm{C}_U(S)$ and this contradicts the fact that S is a non-abelian simple group. Hence $\mathrm{C}_U(S) = 1$ and φ is in fact a monomorphism of V into $\mathrm{Aut}(S)$ and V is an almost simple group such that $\mathrm{Soc}(V) = S$.

Write $n = |U : V|$ and $S^\natural = S_1 \times \cdots \times S_n$. Since U acts a transitive permutation group by right multiplication on the set of right cosets of V in U, and then on the set $\mathcal{I} = \{1, \ldots, n\}$, $S^\natural$ is a minimal non-abelian subgroup of G. Moreover, if $C = \mathrm{Core}_U(V) \neq 1$, then $S \leq \mathrm{N}_U(C) = U$. Now C is an almost simple group with $\mathrm{Soc}(C) = S$. Hence S is normal in U, giving a contradiction. Hence $C = \mathrm{Core}_U(V) = 1$. Therefore, to prove that (G, U) is a primitive pair of type 2 with twisted wreath product action by Theorem 1.1.48, it only remains to prove U is a maximal subgroup of G. To do this, let M be a maximal subgroup of G such that $U \leq M$. Observe that $M = M \cap G = M \cap U \mathrm{Soc}(G) = U(M \cap \mathrm{Soc}(G))$. All projections $R_j = (M \cap \mathrm{Soc}(G))^{\pi_j}$, for $j \in \mathcal{I}$, are conjugate by elements of M, that is, all R_i are isomorphic to the subgroup R_1 and $S_1 \cap U \leq R_1 \leq S_1$ and $M \cap \mathrm{Soc}(G) \leq R_1 \times \cdots \times R_k$. Observe that $V \leq \mathrm{N}_G(S_1)$ by (1.3) in Proposition 1.1.31, since $v = v_{1,v}$, for all $v \in V$, and $1^v = 1$. By 1.1.18 (4), $(y^v)^{\pi_1} = (y^{\pi_1})^v$, for all $y \in M \cap \mathrm{Soc}(G)$. Since the subgroup S normalises $M \cap \mathrm{Soc}(G)$, then S normalises $R_1 = (M \cap \mathrm{Soc}(G))^{\pi_1}$. The automorphisms induced in S_1 by S are the inner automorphisms. Hence R_1 is a normal subgroup of S_1, and, since S_1 is a simple group, we have that $R_1 = 1$ or $R_1 = S_1$. In the first case, we have that $M \cap \mathrm{Soc}(G) = 1$ and then $M = U$. Thus, assume that the projections π_j are surjective, for all $j \in \mathcal{I}$.

By 1.1.40 (23), there exists a minimal non-trivial M-invariant partition Δ of $\mathcal{I}$ in blocks for the action of M on $\mathcal{I}$ such that

$$M \cap \mathrm{Soc}(G) = \prod_{\mathcal{D} \in \Delta} (M \cap \mathrm{Soc}(G))^{\pi_{\mathcal{D}}},$$

and, for each $\mathcal{D} \in \Delta$, the projection $(M \cap \mathrm{Soc}(G))^{\pi_{\mathcal{D}}}$ is a full diagonal subgroup of the direct product $\prod_{i \in \mathcal{D}} S_i$.

For each $y \in M \cap \mathrm{Soc}(G)$ ad $x \in M$, we have that $(y^x)^{\pi_{\mathcal{D}^x}} = (y^{\pi_{\mathcal{D}}})^x$ for any $\mathcal{D} \in \Delta$. Suppose that Δ_0 is an orbit of the action of M on Δ. Then the subgroup

$$T = \prod_{\mathcal{D} \in \Delta_0} (M \cap \mathrm{Soc}(G))^{\pi_{\mathcal{D}}}$$

is normal in M. If Δ_0 is a proper subset of Δ, then there exists some j which is not in a member of Δ_0. Then S_j centralises T and then T is normal in $\langle M, S_j \rangle$. Since T is a proper subgroup of $\mathrm{Soc}(G)$, we have that $S_j \leq M$, by maximality of M. But this implies that $\mathrm{Soc}(G) \leq M$, and this is not true. Hence, M acts transitively on Δ. And so does U, since $M = U(M \cap \mathrm{Soc}(G))$.

Assume that each member $\mathcal{D}$ of Δ has m elements of $\mathcal{I}$ and $|\Delta| = l$, i.e. $n = lm$. Since Δ is a non-trivial partition, then $m > 1$.

Suppose that $l = 1$. This means that $M \cap \mathrm{Soc}(G)$ is a full diagonal subgroup of $\mathrm{Soc}(G)$. Hence $M = [M \cap \mathrm{Soc}(G)]U$ and $M \cap \mathrm{Soc}(G)$ is a normal subgroup of M which is isomorphic to S (π_1 is an isomorphism between $M \cap \mathrm{Soc}(G)$ and S_1). This gives a homomorphism $\psi \colon U \longrightarrow \mathrm{Aut}(S)$ whose restriction to V is the monomorphism φ. Notice that $\mathrm{Ker}(\psi)$ is a normal subgroup of U and, by hypothesis, if $\mathrm{Ker}(\psi) \neq 1$, then $S \leq \mathrm{Ker}(\psi)$. This contradicts the fact that φ is a monomorphism. Therefore $\mathrm{Ker}(\psi) = 1$ and ψ is a monomorphism. Since $S^\psi = \mathrm{Inn}(S)$ is normal in $U^\psi \leq \mathrm{Aut}(S)$, then S is normal in U. But this contradicts the fact that $\mathrm{Core}_U(S) = 1$. Hence $l > 1$.

The partition Δ has l members which are blocks for the action of M (or U) on $\mathcal{I}$. Write $\Delta = \{\mathcal{D}_1, \ldots, \mathcal{D}_l\}$. The subgroup U acts transitively on Δ. We can assume without loss of generality that $1 \in \mathcal{D}_1$. Let U_1 denote the stabiliser of $\mathcal{D}_1$ by the action of U on Δ. Clearly $|U : U_1| = l$.

For any $x \in V$, since $V \leq \mathrm{N}_G(S_1)$, then $1^x = 1$ and $1 \in \mathcal{D}_1 \cap \mathcal{D}_1^x$. Hence $\mathcal{D}_1 = \mathcal{D}_1^x$ and $x \in U_1$. Therefore $V \leq U_1$. Since $\mathcal{D}_1 = \left(M \cap \mathrm{Soc}(G)\right)^{\pi_{\mathcal{D}_1}} \cong S$, there exists a group homomorphism $\psi \colon U_1 \longrightarrow \mathrm{Aut}(\mathcal{D}_1) \cong \mathrm{Aut}(S)$ whose restriction to V is the monomorphism φ. Repeating the arguments of the above paragraph, we obtain that S^ψ is normal in U_1^ψ and then $U_1 \leq \mathrm{N}_U(S) = V$. Therefore $V = U_1$.

But now we have that $l = |U : U_1| = |U : V| = n$, and then $m = 1$. This is the final contradiction. Thus we deduce that U is a maximal subgroup of G. $\qquad\square$

Remarks 1.1.51. 1. Examples of pairs U, S satisfying the conditions of the hypothesis of Theorem 1.1.48 are $S = \mathrm{Alt}(n)$ and $U = \mathrm{Alt}(n+1)$, for $n \geq 5$. In this case S is maximal in U. Also $S = \mathrm{PSL}(2, p^n)$ and $U = \mathrm{PSL}(2, p^{2n})$, for $p^n \geq 3$ satisfies the hypothesis. Here $\mathrm{N}_U(S) \cong \mathrm{PGL}(2, p^n)$ is maximal in U.

2. In [Laf84b], J. Lafuente proved that if G is a primitive group of type 2 and U is a small maximal subgroup of G, then U is also a primitive group of type 2 and each simple component of $\mathrm{Soc}(U)$ is isomorphic to a section of the simple component of $\mathrm{Soc}(G)$.

The O'Nan-Scott Theorem proves that these are all possible configurations of primitive groups of type 2.

Theorem 1.1.52 (M. O'Nan and L. Scott). *Let G be a primitive group of type 2 and U a core-free maximal subgroup of G. Then one of the following holds:*

1. *G is an almost simple group;*
2. *(G, U) is equivalent to a primitive pair with simple diagonal action; in this case $U \cap \mathrm{Soc}(G)$ is a full diagonal subgroup of $\mathrm{Soc}(G)$;*
3. *(G, U) is equivalent to a primitive pair with product action such that $U \cap \mathrm{Soc}(G) = D_1 \times \cdots \times D_l$, a direct product of $l > 1$ subgroups such that, for each $j = 1, \ldots, l$, the subgroup D_j is a full diagonal subgroup of a direct product $\coprod_{i \in \mathcal{I}_j} S_i$, and $\{\mathcal{I}_1, \ldots, \mathcal{I}_l\}$ is a minimal non-trivial G-invariant partition of $\mathcal{I}$ in blocks for the action of U on $\mathcal{I}$.*

4. (G, U) *is equivalent to a primitive pair with product action such that the projection* $R_1 = (U \cap \mathrm{Soc}(G))^{\pi_1}$ *is a non-trivial proper subgroup of* S_1; *in this case* $R_1 = VC \cap S_1$ *and* VC/C *is a maximal subgroup of* X;

5. (G, U) *is equivalent to a primitive pair with twisted wreath product action; in this case* $U \cap \mathrm{Soc}(G) = 1$.

Proof. Recall that by 1.1.40 we can distinguish five different cases.

Case 1. If $n = 1$, then G is an almost simple group. Thus we suppose that $n > 1$.

Case 2. Assume that $n > 1$ and $U \cap \mathrm{Soc}(G) = D$ is a full diagonal subgroup Then there exist automorphisms $\varphi_i \in \mathrm{Aut}(S)$, $i \in \mathcal{I}$, such that $D = U \cap \mathrm{Soc}(G) = \{(x^{\varphi_1}, x^{\varphi_2}, \ldots, x^{\varphi_n}) : x \in S\}$. Since D is normal in U and U is maximal in G, we have that $U = \mathrm{N}_G(D)$. Let P_n be the permutation group induced by the conjugacy action of G on the simple components of $\mathrm{Soc}(G)$: $P_n = G/Y$ (see 1.1.40 (13)). By 1.1.40 (23), the group P_n is transitive and primitive. We embed G in $X \wr P_n$ as in 1.1.40 (13) and then in $\mathrm{Aut}(S) \wr P_n$. Consider $\varphi = (\varphi_1^{-1}, \ldots, \varphi_n^{-1}) \in \mathrm{Aut}(S)^n \le \mathrm{Aut}(S) \wr P_n$. By conjugation by φ in $\mathrm{Aut}(S) \wr P_n$, we have that $D^\varphi = D_S = \{(x, \ldots, x) : x \in S\}$ and $U^\varphi = \mathrm{N}_{G^\varphi}(D_S) = G^\varphi \cap (D_X \times P_n)$, where $D_X = \{(x, \ldots, x) : x \in X\}$. Then $G^\varphi = U^\varphi S^\natural$ and, since $S_i^\varphi = S_i$, for all $i \in \mathcal{I}$, the action of U^φ and of U on $\mathcal{I}$ are the same. Hence, the projection of U^φ onto P_n is surjective. By Proposition 1.1.41, we have that (G^φ, U^φ) is a primitive pair with simple diagonal action and is equivalent to (G, U).

Case 3. Assume that $n > 1$ and $U \cap \mathrm{Soc}(G) = D_1 \times \cdots \times D_l$, a direct product of $l > 1$ subgroups such that, for each $j = 1, \ldots, l$, the subgroup D_j is a full diagonal subgroup of a direct product $\prod_{i \in \mathcal{I}_j} S_i$, and $\{\mathcal{I}_1, \ldots, \mathcal{I}_l\}$ is a minimal non-trivial U-invariant partition of $\mathcal{I}$ in blocks for the action of U on $\mathcal{I}$.

Suppose that the S_i are ordered in such a way that $\mathcal{I}_1 = \{1, \ldots, m\}$. Write $K = S_1 \times \cdots \times S_m$, $N^* = \mathrm{N}_G(K)$, $C^* = \mathrm{C}_G(K)$. Observe that $\mathcal{I}_1$ is a minimal block for the action of G on $\mathcal{I}$. Then N^* acts transitively and primitively on $\mathcal{I}_1$. Hence, $X^* = N^*/C^*$ is a primitive group whose socle is $\mathrm{Soc}(X^*) = KC^*/C^*$. Put $V^* = U \cap N^*$. Since $\mathrm{Soc}(G) \le N^*$, then $N^* = N^* \cap U \mathrm{Soc}(G) = V^* \mathrm{Soc}(G) = V^* C^* K$. Moreover $K \cap V^* = K \cap N^* \cap U = K \cap U = D_1$. Let $\{g_1 = 1, \ldots, g_l\}$ be a right transversal of V^* in U (and of N^* in G). We can assume that this transversal is ordered in such a way that $D_1^{g_i} = D_i$, for $i = 1, \ldots, l$, and put $K_i = K^{g_i}$, for $i = 1, \ldots, l$. Then G acts transitively, by conjugation of the K_i's, on the set $\{K_1, \ldots, K_l\}$.

Clearly D_1 is a V^*-invariant subgroup of K. Suppose that $D_1 \le T_1 < K_1$ and T_1 is a V^*-invariant subgroup of K_1. Then $T_1 \times T_1^{g_2} \cdots \times T_1^{g_l}$ is U-invariant in $\mathrm{Soc}(G)$ and, by maximality of U, we have that $T_1 \times T_1^{g_2} \cdots \times T_1^{g_l} = U \cap \mathrm{Soc}(G) = D_1 \times \cdots \times D_l$. Hence $D_1 = T_1$. In other words, D_1 is maximal as V^*-invariant subgroup of K and then a maximal $V^* C^*$-invariant subgroup of K. Suppose that $s \in S_1 \cap V^* C^*$. There exist $v \in V^*$ and $c \in C^*$, such that $s = vc$. Now $v = sc^{-1} \in \mathrm{C}_G(S_i)$, for $i = 2, \ldots, m$ and $v \in S_1 \mathrm{C}_G(S_1) \le \mathrm{N}_G(S_1)$.

Consider the element $(t, t^{\varphi_2} \ldots, t^{\varphi_m}) \in D_1$ associated with some $t \in S_1$; then $(t, t^{\varphi_2} \ldots, t^{\varphi_m})^v = (t^v, t^{\varphi_2} \ldots, t^{\varphi_m}) \in D_1$, since D_1 is normal in V^*. Hence $t^v = t$. This happens for any $t \in S_1$ and therefore $v \in \mathrm{C}_G(S_1)$. Hence $s \in \mathrm{C}_{S_1}(S_1) = 1$. Therefore $S_1 \cap V^*C^* = 1$ and then $K \neq K \cap V^*C^*$. Since $D_1 \leq V^*C^* \cap K \leq K$ and D_1 is maximal as V^*C^*-subgroup of K, we have that $D_1 = V^*C^* \cap K$. And, finally, if M is a maximal subgroup of N^* such that $V^*C^* \leq M$, then $M \cap K$ is a V^*C^*-invariant subgroup of K containing D_1. Hence $D_1 = V^*C^* \cap K = M \cap K$. Now $M = M \cap N^* = M \cap V^*C^*K = V^*C^*(M \cap K) = V^*C^*$. Therefore V^*C^*/C^* is a core-free maximal subgroup of X^*.

Observe that $(V^*C^*/C^*) \cap \mathrm{Soc}(X^*) = D_1C^*/C^*$ is a full diagonal subgroup of $\mathrm{Soc}(X^*)$. Thus X^* is a group of Case 2. Hence $(X^*, V^*C^*/C^*)$ is a primitive pair with simple diagonal action.

Write $P_l = G/(\bigcap_{i=1}^{l} \mathrm{N}_G(K_i))$ for the permutation group induced by the action of G by conjugation of the K_i's. For any $g \in G$, we write g^ρ for the projection of g in P_l. On the other hand, for each $g \in G$ and each $i \in \{1, \ldots, l\}$, let $a_{i,g}$ be the element of N^* such that $g_i g = a_{i,g} g_j$, for some j. For any $a \in N^*$, write $\bar{a} = aC^*$ for the projection of a on X^*. Consider the conjugacy action $\psi \colon N^* \longrightarrow \mathrm{Aut}(K)$ and the induced G-action on $(X^*)^\natural$:

$$\psi^G \colon G \longrightarrow X^* \wr P_l \quad \text{given by} \quad g^{\psi^G} = (\bar{a}_{1,g}, \ldots, \bar{a}_{l,g})g^\rho, \quad \text{for any } g \in G.$$

Arguing as in 1.1.40 (13–14), we have that

1. the map ψ^G is a group homomorphism and is injective; the projection of G^{ψ^G} on P_l is surjective;
2. $\mathrm{N}_G(K_1)^{\psi^G} = G^{\psi^G} \cap (X_1^* \times [X_2^* \times \cdots \times X_l^*]P_{l-1})$, where P_{l-1} is the stabiliser of 1. The image of $\mathrm{N}_G(K_1)$ by the projection on the first component of $(X^*)_1^\natural$ is the whole of X_1^*;
3. the elements of $\mathrm{Soc}(G)$ can be written as $(e_1, e_2^{g_2}, \ldots, e_l^{g_l})$, for certain $e_1, \ldots, e_l \in K_1$. The image by ψ^G of the elements of the socle is

$$(e_1, e_2^{g_2}, \ldots, e_l^{g_l})^{\psi^G} = (\bar{e}_1, \bar{e}_2, \ldots, \bar{e}_l),$$

   and then $(KC^*/C^*)^\natural = \mathrm{Soc}(X^* \wr P_l) \leq G^{\psi^G}$.

Now, for any $g \in U$, since the $g_i \in U$, we have that $a_{i,g} \in N^* \cap U = V^*$. Hence $U^{\psi^G} \leq G^{\psi^G} \cap ((V^*C^*/C^*) \wr P_l)$. Since V^*C^*/C^* is maximal in X^* and U^{ψ^G} is maximal in G^{ψ^G}, we have that $U^{\psi^G} = G^{\psi^G} \cap ((V^*C^*/C^* \wr P_l))$. By Proposition 1.1.44, this means that (G, U) is equivalent to (G^{ψ^G}, U^{ψ^G}) which is a primitive pair with product action.

Case 4. Suppose now $n > 1$ and the projection $R_1 = (U \cap \mathrm{Soc}(G))^{\pi_1}$ is a non-trivial proper subgroup of S_1.

Moreover, $R_1 = VC \cap S_1$ and VC is a maximal subgroup of N.

Consider the embedding $\psi^G \colon G \longrightarrow X \wr P_n$ of 1.1.40 (13). Then X is almost simple and G is isomorphic to a subgroup G^{ψ^G} of $X \wr P_n$ satisfying all conditions of Proposition 1.1.44. Hence $\overline{U}^{\psi^G} \leq G^{\psi^G} \cap \left((VC/C) \wr P_n\right)$. Since VC/C is maximal in X and U^{ψ^G} is maximal in G^{ψ^G}, we have that $U^\psi = G^\psi \cap \left((VC/C) \wr P_n\right)$. Therefore (G, U) is equivalent to a primitive pair with product action.

Case 5. Assume finally that $U \cap \mathrm{Soc}(G) = 1$. Then, by Theorem 1.1.48, $G \cong S \wr_{(V,\varphi)} U$ and the pair (G, U) is equivalent to a primitive pair with twisted wreath product action. $\square$

If U is a core-free maximal subgroup of a primitive group G of type 2, then there are exactly three different possibilities as we saw in 1.1.40 (25):

1. $\left(U \cap \mathrm{Soc}(G)\right)^{\pi_1} = S_1$, i.e. the projection π_1 of $U \cap \mathrm{Soc}(G)$ on S_1 is surjective.
2. $1 \neq R_1 = \left(U \cap \mathrm{Soc}(G)\right)^{\pi_1} < S_1$, i.e. the image of the projection π_1 of $U \cap \mathrm{Soc}(G)$ on S_1 is a non-trivial proper subgroup of S_1.

$$1 \neq U \cap \mathrm{Soc}(G) = R_1 \times \cdots \times R_n = (U \cap S_1) \times \cdots \times (U \cap S_n).$$

3. $\left(U \cap \mathrm{Soc}(G)\right)^{\pi_1} = 1$, i.e. U is a complement of $\mathrm{Soc}(G)$ in G.

As we saw in 1.1.35, in a primitive group G of type 2, there exists a bijection between

1. the set of all conjugacy classes of maximal subgroups U of G such that the projection $\left(U \cap \mathrm{Soc}(G)\right)^{\pi_1}$ is a proper subgroup of S_1,
2. the set of all conjugacy classes of maximal subgroups of $N/(S_2 \times \cdots \times S_n)$ supplementing $\mathrm{Soc}(G)/(S_2 \times \cdots \times S_n)$.

Under this bijection, the complements, if any, of $\mathrm{Soc}(G)$ in G are in correspondence with the complements of $\mathrm{Soc}(G)/K_1$ in N/K_1. Thus, this bijection works in Cases 2 and 3. Since core-free maximal subgroups of Case 2 occur in every primitive group of type 2, these are called *frequent maximal subgroups* by some authors. We complete this study in the following way.

Proposition 1.1.53. *Let G be a primitive group of type 2. There exist bijections between the following sets:*

1. *the set of all conjugacy classes of maximal subgroups U of G such that the projection $\left(U \cap \mathrm{Soc}(G)\right)^{\pi_1}$ is a non-trivial proper subgroup of S_1,*
2. *the set of all conjugacy classes of maximal subgroups of $N/(S_2 \times \cdots \times S_n)$ supplementing but not complementing $\mathrm{Soc}(G)/(S_2 \times \cdots \times S_n)$, and*
3. *the set of all conjugacy classes of core-free maximal subgroups of X.*

Proof. We only have to see the bijection between the sets in 2 and 3. Write $K = S_2 \times \cdots \times S_n$ and observe that if L/C is core-free maximal subgroup of X, then obviously L/K is a maximal subgroup of N/K and $N = L \mathrm{Soc}(G)$. If L/K complements $\mathrm{Soc}(G)/K$ in N/K, then $K = L \cap \mathrm{Soc}(G)$; in particular $L \cap S_1 = 1$. But $L \cap S_1 C = C(L \cap S_1) = C$ and this contradicts the fact that $(L/C) \cap (S_1 C/C)$ is non-trivial by Proposition 1.1.16. Thus L does not complement $\mathrm{Soc}(G)/K$ in N.

Conversely, let L/K be a maximal subgroup of N/K such that $N = L\,\mathrm{Soc}(G)$ and $K < \mathrm{Soc}(G) \cap L$. Let us see that $C \leq L$. Consider $L_0/K = \mathrm{Core}_{N/K}(L/K)$. Since $\mathrm{Soc}(G)/K$ is a minimal normal subgroup of N/K, then $L_0/K \leq \mathrm{C}_{N/K}(\mathrm{Soc}(G)/K) = C/K$ and $L_0 \leq C$. If $L_0 = C$, then $C \leq L$ and we are done. Suppose that C/L_0 is nontrivial. Since L/L_0 is a core-free maximal subgroup of N/L_0, it is clear that N/L_0 is a primitive group. Observe that $\mathrm{Soc}(G)L_0/L_0$ is a minimal normal subgroup of N/L_0 and $\mathrm{C}_{N/L_0}(\mathrm{Soc}(G)L_0/L_0) = C/L_0$. Since we are assuming that C/L_0 is nontrivial, the primitive group N/L_0 is of type 3. Hence L/L_0 complements $\mathrm{Soc}(G)L_0/L_0$. This is to say that $L \cap \mathrm{Soc}(G) \leq L_0$, i.e. $L \cap \mathrm{Soc}(G) = L_0 \cap \mathrm{Soc}(G)$. Therefore $L \cap \mathrm{Soc}(G)$ is a normal subgroup of N between K and $\mathrm{Soc}(G)$. Since $\mathrm{Soc}(G)/K \cong S$, a non-abelian simple group, and L supplements $\mathrm{Soc}(G)$ in N, we have that $K = \mathrm{Soc}(G) \cap L$. This is not possible.   □

As we saw in 1.1.35, the existence of complements of the socle in a primitive group G of type 2 is characterised by the existence of complements of $\mathrm{Soc}(G)/(S_2 \times \cdots \times S_n)$ in $\mathrm{N}_G(S_1)/(S_2 \times \cdots \times S_n)$. We wonder whether it is possible to obtain a characterisation of the existence of complements of $\mathrm{Soc}(G)$ in G in terms of complements of $\mathrm{Soc}(X)$ in X as we saw in 1.1.53 for supplements. The answer is partially affirmative.

Corollary 1.1.54. *With the notation of 1.1.40, let G be a primitive group of type 2 such that $\mathrm{Soc}(X)$ is complemented in X. Then $\mathrm{Soc}(G)$ is complemented in G.*

The converse does not hold in general.

Proof. Suppose that there exists a subgroup $Y \leq N$ such that $C \leq Y$ and $N = YS_1$ and $Y \cap S_1 C = C$. Then it is clear that

$$S_2 \times \cdots \times S_n \leq Y \cap \mathrm{Soc}(G) \leq Y \cap S_1 C \cap \mathrm{Soc}(G) = C \cap \mathrm{Soc}(G) = S_2 \times \cdots \times S_n$$

and therefore Y is a complement of $\mathrm{Soc}(G)/(S_2 \times \cdots \times S_n)$ in $N/(S_2 \times \cdots \times S_n)$. The conclusion follows by Theorem 1.1.35.

It is well-known that if $S = \mathrm{Alt}(6)$, the alternating group of degree 6, the automorphism group $A = \mathrm{Aut}(S)$ is an almost simple group whose socle is non-complemented. With the cyclic group $C \cong C_2$ we consider the regular wreath product $H = A \wr C$. In H we consider the diagonal subgroups $D_S = \{(x,x) : x \in S)\}$ and $D_A = \{(x,x) : x \in A\}$. Then $\mathrm{N}_H(D_S) = D_A C$. Since $D_S \cong S$, the conjugacy action of $\mathrm{N}_H(D_S)$ on D_S gives a group homomorphism $\varphi \colon \mathrm{N}_H(D_S) \longrightarrow \mathrm{Aut}(S)$. We construct the twisted wreath product $G = S \wr_{(\mathrm{N}_H(D_S),\varphi)} H$. Then $\mathrm{Soc}(G) = S_1 \times \cdots \times S_n$ is a minimal normal subgroup of G and it is the direct product of $n = |H : \mathrm{N}_H(D)|$ copies of S. Moreover since $\mathrm{Core}_H(\mathrm{N}_H(D_S)) = 1$, then $\mathrm{C}_G(\mathrm{Soc}(G)) = 1$ by Proposition 1.1.34 (2). Hence G is a primitive group of type 2. Clearly $\mathrm{Soc}(G)$ is complemented in G. $N_H(S_1) = N_H(D_S) = D_A C$ and $C_H(S_1) = \text{Ker}(\varphi) = C_H(D_S) = C$. Hence, $X \cong D_A \cong A$ and $\text{Soc}(X)$ is not complemented in X.

$\square$

Primitive pairs (G, U) of diagonal type, i.e. core-free maximal subgroups U of primitive groups G of type 2 such that the projection π_1 of $U \cap \text{Soc}(G)$ on S_1 is surjective, appear in Cases (2) and (3) of the O'Nan-Scott Theorem. In this case $U \cap \text{Soc}(G)$ is a direct product of l full diagonal subgroups, with $1 \leq l < n$, and $U = N_G(D)$.

Proposition 1.1.55. *Let G be a primitive group of type 2. Given a minimal non-trivial partition $\Delta = \{\mathcal{I}_1, \dots, \mathcal{I}_l\}$ of $\mathcal{I}$ in blocks for the action of G on $\mathcal{I}$ and a subgroup $D = D_1 \times \cdots \times D_l$, where D_j is a full diagonal subgroup of $\prod_{i \in \mathcal{I}_j} S_i$, for each $j = 1, \dots, l$, associated with Δ. The following statements are pairwise equivalent:*

1. *there exists a maximal subgroup U of G such that $U \cap \text{Soc}(G) = D$;*
2. *$N_G(D)$ is a maximal subgroup of G;*
3. *$G = N_G(D) \text{Soc}(G)$.*

Proof. 1 implies 2. Suppose that there exists a maximal subgroup U of G such that $U \cap \text{Soc}(G) = D$. Then $U \leq N_G(D)$ and, by maximality of U in G, we have that $U = N_G(D)$.

2 implies 3. Observe that $N_G(D) \cap \text{Soc}(G) = N_{\text{Soc}(G)}(D) = D$, by Lemma 1.1.38, and then $\text{Soc}(G) \not\leq N_G(D)$. Therefore $G = N_G(D) \text{Soc}(G)$.

3 implies 1. Let H be a maximal subgroup of G such that $N_G(D) \leq H$. Then $D = N_{\text{Soc}(G)}(D) = \text{Soc}(G) \cap N_G(D) \leq \text{Soc}(G) \cap H$. Then $H \cap \text{Soc}(G)$ is a direct product of full diagonal subgroups associated with a partition of $\mathcal{I}$ which refines $\{\mathcal{I}_1, \dots, \mathcal{I}_l\}$, by Proposition 1.1.39. By minimality of the blocks, we have that $H \cap \text{Soc}(G) = D$ and therefore $H = N_G(D)$. $\square$

Example 1.1.56. We construct a primitive group G of type 2 with no maximal subgroup of diagonal type. Consider the symmetric group of degree 5, $H \cong \text{Sym}(5)$ and denote with S the alternating group of degree 5. If C is a cyclic group of order 2, let G be the regular wreath product $G = H \wr C$. Then $\text{Soc}(G) = S_1 \times S_2 \cong \text{Alt}(5) \times \text{Alt}(5)$. Any full diagonal subgroup of $\text{Soc}(G)$ is isomorphic to $\text{Alt}(5)$ and its normaliser N is isomorphic to $\text{Sym}(5) \times C_2$. Observe that $|G/\text{Soc}(G)| = 8 > 4 = |N \text{Soc}(G)/\text{Soc}(G)|$. Hence N does not satisfy 3. Clearly $N \text{Soc}(G)$ is a normal maximal subgroup of G containing N.

Proposition 1.1.57. *Let G be a primitive group of type 2. Two maximal subgroups U, U^* of G, such that $U \cap \text{Soc}(G)$ and $U^* \cap \text{Soc}(G)$ are direct products of full diagonal subgroups, are conjugate in G if and only if $U \cap \text{Soc}(G)$ and $U^* \cap \text{Soc}(G)$ are conjugate in $\text{Soc}(G)$.*

Proof. Suppose that $U^g = U^*$ for some $g \in G$. Then $g = xh$, with $x \in N_G(U \cap \text{Soc}(G))$ and $h \in \text{Soc}(G)$. Hence $U^* \cap \text{Soc}(G) = (U \cap \text{Soc}(G))^g = (U \cap$

$\mathrm{Soc}(G))^h$. Conversely, if $U^* \cap \mathrm{Soc}(G) = (U \cap \mathrm{Soc}(G))^h$ for some $h \in \mathrm{Soc}(G)$, then $U^* = \mathrm{N}_G(U^* \cap \mathrm{Soc}(G)) = \mathrm{N}_G((U \cap \mathrm{Soc}(G))^h) = \mathrm{N}_G(U \cap \mathrm{Soc}(G))^h = U^h$. □

1.2 A generalisation of the Jordan-Hölder theorem

In the first book dedicated to Group Theory, the celebrated *Traité des substitutions et des équations algébriques* ([Jor70]), published in Paris in 1870, the author, C. Jordan, presents the first version of a theorem known as the Jordan-Hölder Theorem: *The length of all composition series of a finite group is an invariant of the group and the orders of the composition factors are uniquely determined by the group.* Nineteen years later, in 1889, O. Hölder ([Höl89]) completed his contribution to the theorem proving that not only the orders but even the composition factors are uniquely determined by the group.

In recent years a number of generalisations of the classic Jordan-Hölder Theorem have been done. For example it has been proved that *given two chief series of a finite group G, there is a one-to-one correspondence between the chief factors of the series, corresponding factors being G-isomorphic, such that the Frattini chief factors of one series correspond to the Frattini chief factors of the other* (see [DH92, A, 9.13]). This result was first published by R. W. Carter, B. Fischer, and T. O. Hawkes (see [CFH68]) for soluble groups, and for finite groups in general by J. Lafuente (see [Laf78]). A further contribution is given by D. W. Barnes (see [Bar72]), for soluble groups, and again by J. Lafuente [Laf89] for finite groups in general, describing the bijection in terms of common supplements.

But if we restrict our arguments to a proper subset of the set of all maximal subgroups, we find that this is no longer true. For instance, in the elementary abelian group G of order 4, there are three maximal subgroups, say A, B, and C. If we consider the set $\mathbf{X} = \{A, B\}$, the maximal subgroup B is a common complement in $\mathbf{X}$ for the chief factors A and C. Also G/A is complemented by $A \in \mathbf{X}$. However G/C has no complement in $\mathbf{X}$.

In general, the key of the proof of these Jordan-Hölder-type theorems is to prove the result in the particular case of two pieces of chief series of a group G of the form

$$1 < N_1 < N_1 \times N_2 \qquad 1 < N_2 < N_1 \times N_2$$

where N_1 and N_2 are minimal normal subgroups of G. It is not difficult to prove that if $N_1 N_2 / N_1$ is supplemented by a maximal subgroup M, then M also supplements N_2 (see Lemma 1.2.16), but the converse is not true. The particular case in which N_1 and N_2 are supplemented and either $N_1 N_2 / N_1$ or $N_1 N_2 / N_2$ is a Frattini chief factor is the hardest one (see [DH92, A, 9.12]) and, in fact, proving the generalised Jordan-Hölder Theorem is reduced to proving that, in the above situation, $N_1 N_2 / N_1$ and $N_1 N_2 / N_2$ are simultaneously Frattini chief factors of G.

For this reason J. Lafuente, in [Laf89], wonders about the precise condition on a set $\mathbf{X}$ of maximal subgroups of a group G which allows a proof that, in the above situation, if N_1 and N_2 have supplements in $\mathbf{X}$, then $N_1 N_2 / N_1$ and $N_1 N_2 / N_1$ possess simultaneously supplements in $\mathbf{X}$, or, in other words, which is the precise condition on $\mathbf{X}$ to prove a Jordan-Hölder-type Theorem. In this section we present, among other related results, an answer to this question.

Definition 1.2.1. *Given a group G and two normal subgroups K, H of G such that $K \leq H$, we say that the section H/K is a* chief factor *of G if there is no normal subgroup of G between K and H, i.e. if N is a normal subgroup of G and $K \leq N \leq H$, then either $H = N$ or $K = N$.*

Equivalently, H/K is a chief factor of G if H/K is a minimal normal subgroup of G/K.

Hence H/K is a direct product of copies of a simple group and we have two possibilities:

1. either H/K is abelian, and there exists a prime p such that H/K is an elementary abelian p-group, or
2. H/K is non-abelian, and there exists a non-abelian simple group S such that $H/K \cong S_1 \times \cdots \times S_n$, where $S_i \cong S$ for all $i = 1, \ldots, n$.

Given a group G and two normal subgroups K, H of G such that $K \leq H$, the group G acts by conjugation on the cosets of the section H/K: for $h \in H$ and $g \in G$, then $(hK)^g = h^g K$. This action of G on H/K defines a group homomorphism $\varphi \colon G \longrightarrow \operatorname{Aut}(H/K)$ such that

$$\operatorname{Ker}(\varphi) = \mathrm{C}_G(H/K) = \{g \in G : h^g K = hK \text{ for all } h \in H\}.$$

We say that $\mathrm{C}_G(H/K)$ is the *centraliser* of H/K in G. We write $\operatorname{Aut}_G(H/K) = \operatorname{Im}(\varphi) \cong G / \mathrm{C}_G(H/K)$ for the group of automorphisms of H/K induced by the conjugation of the elements of G. The set of G composed of all elements which induce inner automorphisms on H/K is the subset $\mathrm{C}_G^*(H/K) = H\,\mathrm{C}_G(H/K)$.

Definition 1.2.2. *Given a chief factor H/K of a group G, the* inneriser *of H/K in G is the subgroup*

$$\mathrm{C}_G^*(H/K) = H\,\mathrm{C}_G(H/K).$$

It is clear that if H/K is abelian, then $\mathrm{C}_G^*(H/K) = \mathrm{C}_G(H/K)$

Definition 1.2.3. *Let G be a group and let F_1 and F_2 two chief factors of G. A map $\gamma \colon F_1 \longrightarrow F_2$ is a* G-isomorphism *if γ is a group isomorphism and $(x^g)^\gamma = (x^\gamma)^g$, for any $x \in F_1$ and any $g \in G$.*

Two chief factors F_1, F_2 of G are G-isomorphic *if there exists a G-isomorphism $\gamma \colon F_1 \longrightarrow F_2$.*

If two chief factors F_1, F_2 of G are G-isomorphic, then write $F_1 \cong_G F_2$.

Proposition 1.2.4. *Let G be a group and let H_1/K_1 and H_2/K_2 be two chief factors of G.*

1. *If H_1/K_1 and H_2/K_2 are G-isomorphic, then $\mathrm{C}_G(H_1/K_1) = \mathrm{C}_G(H_2/K_2)$.*
2. *In general, the converse of 1 is not true.*
3. *Suppose that H_1/K_1 and H_2/K_2 are non-abelian. Then H_1/K_1 and H_2/K_2 are G-isomorphic if and only if $\mathrm{C}_G(H_1/K_1) = \mathrm{C}_G(H_2/K_2)$.*

Proof. Since clearly 1 is true, we prove 3 and give a counterexample to prove 2.

Suppose that H_1/K_1 and H_2/K_2 are non-abelian chief factors of G such that $C = \mathrm{C}_G(H_1/K_1) = \mathrm{C}_G(H_2/K_2)$. We have that $K_i \leq C \cap H_i \leq H_i$, for $i = 1$, 2. Since the chief factors are non-abelian, H_i is not contained in C. Therefore $K_i = C \cap H_i$, for $i = 1$, 2. Hence, $H_i/K_i \cong_G H_i C/C$, for $i = 1$, 2. Observe that $H_1 C/C$ is a minimal normal subgroup of the group G/C with trivial centraliser. This means that G/C is a primitive group of type 2, by Proposition 1.1.14. Since $H_2 C/C$ is also a minimal normal subgroup of G/C, then $H_1 C = H_2 C$. Hence H_1/K_1 and H_2/K_2 are G-isomorphic.

To see that this does not hold when the chief factors are abelian, let P be an extraspecial p-group, p an odd prime, of order p^3. Let F be a field of characteristic q, with $q \neq p$, such that F contains a primitive p-th root of unity. Then there exist $p - 1$ non-equivalent irreducible and faithful P-modules over F of dimension p (see [DH92, B, 9.16]). Since $p - 1 > 1$, we can consider two non-isomorphic such P-modules, V_1, V_2. If V is the direct sum $V = V_1 \oplus V_2$, construct the semidirect product $G = [V]P$. The group G has two isomorphic minimal normal subgroups V_1, V_2 such that $\mathrm{C}_G(V_i) = V$, for $i = 1$, 2. But V_1 and V_2 are not G-isomorphic. □

Observe that in a primitive group G of type 3, the two minimal normal subgroups are not G-isomorphic. In other words, G-isomorphism is an equivalence relation in the set of all chief factors of G which is too "narrow" to include the case of the relation between the two minimal normal subgroups of a primitive group of type 3. J. Lafuente and P. Förster [För83] propose two equivalent "enlargements" of G-isomorphism. Here we follow Lafuente's definition.

Definition 1.2.5. *Let G be a group. We say that two given chief factors of G are G-connected if either they are G-isomorphic or there exists a normal subgroup N of G such that G/N is a primitive group of type 3 whose minimal normal subgroups are G-isomorphic to the given chief factors.*

Obviously, in a group G, two abelian chief factors are G-connected if and only if they are G-isomorphic.

Proposition 1.2.6 ([Laf84a]). *In a group G, the relation of being G-connected is an equivalence relation on the set of all chief factors of G.*

Proof. The only non-obvious property to prove is transitivity. Let F_1, F_2, F_3 be chief factors of G such that F_1 is G-connected to F_2 and F_2 is G-connected to F_3. We may suppose that no two are G-isomorphic. Therefore

1. there exists a normal subgroup N of G such that G/N is a primitive group of type 3 whose minimal normal subgroups are $A/N \cong_G F_1$ and $B/N \cong_G F_2$, and
2. there exists a normal subgroup M of G such that G/M is a primitive group of type 3 whose minimal normal subgroups are $C/M \cong_G F_2$ and $D/M \cong_G F_3$.

Observe that $C_G(F_2) = C_G(B/N) = A$ and also $C_G(F_2) = C_G(C/M) = D$. Hence $A = D$. Moreover $N \leq NM \leq A$ and A/N is a chief factor. If $N = NM$, then $M \leq N \leq A$ and $M = N$. This implies that $F_1 \cong_G F_3$ and, in particular, F_1 and F_3 are G-connected.

Now suppose that $A = MN$. Then the group G/A is isomorphic to $(G/N)/(A/N)$, which is the quotient group of a primitive group of type 3 over one of its minimal normal subgroups. Therefore G/A is a primitive group of type 2 by Corollary 1.1.13. On the other hand $BA/A \cong_G B/(B \cap A) = B/N \cong_G F_2$ and, since $M = A \cap C$, we have that $CA/A \cong_G C/(C \cap A) = C/M \cong_G F_2$, so BA/A and CA/A are minimal normal subgroups of G/A. Hence $AC = AB$. Analogously, working with G/B, we obtain that $AB = BC$.

Note that if C is contained in B, then $AB = B$ and then $A = B$, giving a contradiction. If B is contained in C, then $AB = C$. Since $M < A \leq AB = C$ and C/M is a chief factor of G, we have that $A = C$ and then $A = B$, which gives again a contradiction. Hence the subgroup $E = B \cap C$ is a proper subgroup of B and of C. Consider the group G/E. We have that

$$B/E \cong_G BC/C = AC/C \cong_G A/(A \cap C) = A/M \cong_G F_3$$

and then

$$C_G(B/E) = C_G(F_3) = C_G(A/M) = C.$$

Also

$$C/E \cong_G BC/B = AB/B \cong_G A/(A \cap B) = A/N \cong_G F_1$$

and then

$$C_G(C/E) = C_G(F_1) = C_G(A/N) = B.$$

On the other hand, let U, V be maximal subgroups of G such that $N \leq U$ and U is a common complement of A/N and B/N and $M \leq V$ and V is a common complement of A/M and C/M. Consider the subgroup $X = (U \cap V)E$.

If $X = G$, then $U = U \cap X = (U \cap V)(U \cap E) = (U \cap V)(N \cap C) = (U \cap V)(N \cap M) = U \cap V$. This contradicts the fact that $U \neq V$. Hence X is a proper subgroup of G. Now we have:

$$XB = (U \cap V)B = (U \cap VN)B = UB = G$$

and

$$XC = (U \cap V)C = (UM \cap V)C = VC = G.$$

Moreover $B \cap X$ is a normal subgroup of X and $(B \cap X)/E$ is centralised by $C_G(B/E) = C$. Hence $B \cap X$ is a normal subgroup of $XC = G$. Since B/E is a chief factor of G and X is a proper subgroup of G, then $B \cap X = E$. Analogously $C \cap X = E$. In other words, the subgroup X is a common complement of B/E and C/E. By Corollary 1.1.13, the group $G/(B \cap C)$ is a primitive group of type 3. Consequently, F_1 is G-connected to F_3. □

Definition 1.2.7. *Let H/K be a chief factor of a group G.*

1. *We say that H/K is a* Frattini *chief factor of G if $H/K \leq \Phi(G/K)$.*
2. *If there exists a proper subgroup M of G such that $G = MH$ and $K \leq H \cap M$, we say that H/K is a* supplemented *chief factor of G and M is a* supplement *of H/K in G. If H/K in a non-Frattini chief factor of G, then H/K is supplemented in G by a maximal subgroup of G.*
3. *If H/K is a chief factor of G supplemented by a subgroup M of G and $K = H \cap M$, then we say that H/K is a* complemented *chief factor of G and M is a* complement *of H/K in G.*

Remarks 1.2.8. Let G be a group and H/K a supplemented chief factor of G. Consider a maximal subgroup M of G supplementing H/K in G. Clearly, in the quotient group G/M_G, the maximal subgroup M/M_G is core-free. Therefore G/M_G is a primitive group. We get $K = H \cap M_G$ and then note that if $M_G < X < HM_G$ and X is normal in G, then $X = M_G(X \cap H)$, where $K \leq X \cap H \leq H$. Hence $X \cap H = K$ or H. In both cases we have a contradiction. Thus HM_G/M_G is a minimal normal subgroup of the primitive group G/M_G.

1. Note that if M is a maximal subgroup of type 1 or 3 of a group G, then each chief factor of G supplemented by M is in fact complemented by M. In these cases, HM_G/M_G is a minimal normal subgroup of the primitive group G/M_G, which is of type 1 or 3, and then $M \cap HM_G = M_G$. Therefore $M \cap H = M_G \cap H = K$, as claimed.

2. Observe that $HM_G/M_G \cong_G H/K$. Write

$$C = C_G(H/K) = C_G(HM_G/M_G).$$

a) If H/K is abelian, then the primitive group G/M_G is of type 1; in this case $C = HM_G$ and $M/M_G \cong G/C$; therefore G/M_G is isomorphic to the semidirect product $[H/K](G/C)$.

b) if H/K is non-abelian, then two cases arise:
   i. If $C = M_G$, then G/M_G is a primitive group of type 2; clearly $\mathrm{Soc}(G/C) = HC/C \cong_G H/K$.
   ii. If M_G is contained in C, then G/M_G is a primitive group of type 3 whose minimal normal subgroups are HM_G/M_G and C/M_G; in this case G/C is a primitive group of type 2 and $\mathrm{Soc}(G/C) = HC/C \cong_G$

H/K. If S is a maximal subgroup supplementing HC/C in G, then $G = HS$ and $K = H \cap C = H \cap S_G$. Hence S is also a supplement of H/K in G and $S_G = C$ as in 2(b)i.

Hence for any supplemented chief factor H/K of G, there exists a maximal subgroup M of G supplementing H/K in G such that G/M_G is a monolithic primitive group. We say then that M is a *monolithic supplement* of H/K in G. This observation leads us to two definitions.

Definition 1.2.9. *For any chief factor H/K of a group G, we define the primitive group associated with H/K in G to be*

1. *the semidirect product $[H/K](G/\operatorname{C}_G(H/K))$, if H/K is abelian, or*
2. *the quotient group $G/\operatorname{C}_G(H/K)$, if H/K is non-abelian.*

Notation 1.2.10. The primitive group associated with H/K is denoted by $[H/K] * G$.

It is easy to see that if H/K is a supplemented chief factor of a group G, and M is a monolithic supplement of H/K in G, then $[H/K] * G \cong G/M_G$.

Definition 1.2.11. *Let H/K be a supplemented chief factor of the group G. Assume that M is a maximal subgroup G supplementing H/K in G such that G/M_G is a monolithic primitive group. We say that the chief factor $\operatorname{Soc}(G/M_G) = HM_G/M_G$ is the precrown of G associated with M and H/K, or simply, a precrown of G associated with H/K.*

Remarks 1.2.12.   1. If H/K is a non-abelian chief factor of the group G, then for each maximal subgroup M of G supplementing H/K in G such that G/M_G is a monolithic primitive group, we have that $M_G = \operatorname{C}_G(H/K)$. Therefore the unique precrown of G associated with H/K is

$$\operatorname{Soc}(G/M_G) = HM_G/M_G$$
$$= H\operatorname{C}_G(H/K)/\operatorname{C}_G(H/K) = \operatorname{C}_G^*(H/K)/\operatorname{C}_G(H/K).$$

2. If H/K is a complemented abelian chief factor of G and M is a complement of H/K in G, then the precrown of G associated with M and H/K is

$$\operatorname{Soc}(G/M_G) = HM_G/M_G = \operatorname{C}_{G/M_G}(HM_G/M_G) = \operatorname{C}_G(H/K)/M_G.$$

For this reason it is interesting to know how many different precrowns are associated with a particular abelian chief factor. The answer, in a soluble group, is particularly elegant.

Proposition 1.2.13. *Let H/K be a complemented chief factor of a soluble group G. Then the function which assigns to each conjugacy class of complements of H/K in G, $\{M^g : g \in G\}$ say, the common core M_G of its elements*

induces a bijection between the set of all conjugacy classes of complements of
H/K *in* G *and the set of all normal subgroups of* G *which complement* H/K
in $C_G(H/K)$.

Therefore there exists a bijection between the set of all precrowns of G
associated with H/K and the set of all conjugacy classes of complements of
H/K in G.

Proof. Write $C = C_G(H/K)$. Let N be a normal subgroup of G such that
$C = HN$ and $H \cap N = K$. Then $HN/N \cong_G H/K$ and HN/N is a self-
centralising minimal normal subgroup of the group G/N. By Theorem 1.1.10,
HN/N is complemented in G/N and all complements are conjugate. If M/N
is one of these complements, then $N = M_G$. Hence the correspondence is
surjective.

Let M and S be two complements of H/K in G such that $N = M_G =
S_G$. Then G/N is a soluble primitive group such that and S/N, M/N are
complements of $\mathrm{Soc}(G/N) = HN/N$. By Theorem 1.1.10, there exists an
element $g \in G$ such that $S^g = M$. Hence the correspondence is injective.

Finally observe that, since H/K is abelian, the precrowns of G associated
with H/K have a common numerator $C_G(H/K)$ and different denominators
M_G, one for each conjugacy class of complements of H/K in G. $\square$

Our next goal is to give a characterisation of the property of being G-
connected. Observe that in a primitive group G of type 3, if A and B are
the minimal normal subgroups, then $C_G^*(A) = C_G^*(B) = AB = \mathrm{Soc}(G)$. This
means that two G-connected chief factors have the same inneriser. But this
cannot be a characterisation as we can see from the example in Proposition 1.2.4 (2). To characterise the property of being G-connected in terms of
the inneriser we have to be more precise.

But before that we have to include here a technical lemma, which will be
crucial in our presentation.

Lemma 1.2.14 (see [För88]).

1. *Let* $N_1, \ldots, N_n$ *be normal subgroups of a group* G *(*$n \geq 2$*), and consider*
   $N = \prod_{i=1}^{n} N_i$. *Suppose that* $\bigcap_{i=1}^{n} N_i = 1$ *and that* $|N| = \prod_{i=1}^{n} |N/N_i|$.
   For $i = 1, \ldots, n$, *write* $p_i \colon G/N_i \longrightarrow G/N$ *for the natural projection:*
   $(gN_i)^{p_i} = gN$, *for all* $g \in G$. *Then the following statements are equivalent:*
   a) *There exists a subgroup* U *of* G *which complements all the* N_i*'s in* G.
   b) *There exist group isomorphisms* $\varphi_i \colon G/N_1 \longrightarrow G/N_i$, *for* $i = 2, \ldots, n$,
      such that $\varphi_i p_i = p_1$, *for all* $i = 2, \ldots, n$.
2. *Let* N_1 *and* N_2 *be two normal subgroups of a group* G *such that* $N_1 \cap
   N_2 = 1$. *Write* $N = N_1 N_2$. *Suppose that, for* $i = 1, 2$, *there exist group
   isomorphisms* γ_i *between* G/N_i *and a semidirect product* $X = [Z]Y$, *where*
   Z *is a normal subgroup of* X, *such that* $(N/N_i)^{\gamma_i} = Z$.
   Then there exists a subgroup H *of* G *such that* $G = HN$ *and* $H \cap N = 1$.
   For such H *the following statements are equivalent:*

a) there exists a subgroup U of G such that $H \leq U$ and U is a common complement of N_1 and N_2 in G, and

b) $N_1 \cong_H N_2$.

If, moreover, the N_i, $i = 1, 2$, are abelian, then each of the previous statements is equivalent to

c) $N_1 \cong_G N_2$.

Proof. 1. Define $\varphi \colon N \longrightarrow N/N_1 \times \cdots \times N/N_n$, by $x^\varphi = (xN_1, \ldots, xN_n)$, for every $x \in N$. It is clear that φ is a group homomorphism. If $x \in \mathrm{Ker}(\varphi)$, then $x \in \bigcap_{i=1}^n N_i = 1$. Moreover, since $|N| = \prod_{i=1}^n |N/N_i|$, we have that φ is an isomorphism.

Suppose that there exist group isomorphisms $\varphi_i \colon G/N_1 \longrightarrow G/N_i$, for $i = 2, \ldots, n$, such that $\varphi_i p_i = p_1$, for all $i = 2, \ldots, n$. Given $g_1 N_1 \in G/N_1$, we consider $g_i N_i = (g_1 N_1)^{\varphi_i}$, for $i = 2, \ldots, n$. Then $(g_1 N_1)^{\varphi_i p_i} = g_i N$ and $(g_1 N_1)^{p_1} = g_1 N$. Hence $g_1^{-1} g_i \in N$, for all $i = 1, \ldots, n$.

Since φ is an isomorphism, there exists a unique element $x_0 \in N$ such that

$$(N_1, g_1^{-1} g_2 N_2, \ldots, g_1^{-1} g_n N_n) = (x_0 N_1)^\varphi = (x_0 N_1, \ldots, x_0 N_n)$$

and then $x_0 \in N_1$ and $x_0^{-1} g_1^{-1} g_i \in N_i$, for $i = 2, \ldots, n$. Therefore $g_i N_i = g_1 x_0 N_i$, for all $i = 2, \ldots, n$. Then, $(g_1 x_0 N_1)^{\varphi_i} = (g_1 N_1)^{\varphi_i} = g_i N_i = g_1 x_0 N_i$. For the element $g = g_1 x_0 \in g_1 N_1 \cap g_2 N_2 \cap \cdots \cap g_n N_n$, we have that $(gN_1)^{\varphi_i} = gN_i$, for $i = 2, \ldots, n$.

For each $i = 1, \ldots, n$, we choose a system of coset representatives $\mathcal{U}_i = \{x_{1i}, \ldots, x_{ri}\}$ of N_i in G, such that $(x_{k1} N_1)^{\varphi_i} = x_{ki} N_i$ for all $i = 2, \ldots, n$ and all $k = 1, \ldots, r$. The above arguments show that there exist $z_k \in x_{k1} N_1 \cap x_{k2} N_2 \cap \cdots \cap x_{kn} N_n$ such that $(z_k N_1)^{\varphi_i} = z_k N_i$, for all $i = 2, \ldots, n$. Thus we obtain a common system of coset representatives $U = \{z_1, \ldots, z_k\}$ of all the N_i's in G.

Let us prove that U is a subgroup of G. If we suppose that $x_{11} N_1 = N_1$, which forces $z_1 N_1 = N_1$, we obtain $N_i = N_1^{\varphi_i} = (z_1 N_1)^{\varphi_i} = z_1 N_i$, for all $i = 2, \ldots, n$. Hence $z_1 \in \bigcap_{i=1}^n N_i = 1$ and $1 \in U$.

Suppose that $(z_k N_1)^{-1} = z_t N_1$ for some t. Then $z_k z_t \in N_1$. Hence $z_k z_t \in \bigcap_{i=1}^n N_i = 1$. Therefore $z_k^{-1} = z_t \in U$.

For $z_k, z_j \in U$, we have that $z_k z_j N_1 = z_t N_1$ for some t. Then $z_t^{-1} z_k z_j \in N_1$. As above this implies that $z_t^{-1} z_k z_j \in \bigcap_{i=1}^n N_i = 1$ and $z_k z_j = z^t \in U$.

Therefore U is a subgroup of G and is the required common complement of all the N_i's in G.

To prove the converse, let U be a common complement of the N_i's in G and define $\varphi_i \colon G/N_1 \longrightarrow G/N_i$ by $(gN_1)^{\varphi_i} = uN_i$, where $g = un$, $u \in U$, and $n \in N_i$. This is a well-defined homomorphism and it is injective. Since all the N_i have a common complement, they have, in particular, the same order and $|G/N_1| = |G/N_i|$, for all $i = 2, \ldots, n$. Then the φ_i are group isomorphisms. Finally note that, for all $i = 2, \ldots, n$ and all $g \in G$, $(gN_1)^{\varphi_i p_i} = uN = gN = (gN_1)^{p_1}$, i.e. $\varphi_i p_i = p_1$.

2. Since Y is a complement of Z in X and γ_i is a group isomorphism, then $H_i/N_i = Y^{\gamma_i^{-1}}$ is a complement of $Z^{\gamma_i^{-1}} = N/N_i$ in G/N_i, for each $i = 1, 2$. Consider the subgroup $H = H_1 \cap H_2$. Observe that $G = H_1 N = H_1 N_2 = H_2 N_1$. Then $HN = (H_1 \cap H_2)N = (H_1 \cap H_2 N_1)N_2 = H_1 N_2 = G$, and $H \cap N = H_1 \cap H_2 \cap N = N_1 \cap N_2 = 1$.

Suppose that there exists a subgroup U of G such that $H \le U$ and U is a common complement of N_1 and N_2 in G. Consider the isomorphisms φ_i between $G/N_i = U N_i/N_i$ and U defined by $(uN_i)^{\varphi_i} = u$. Then $(N/N_i)^{\varphi_i} = U \cap N$. Write τ_i for the restriction of φ_i to N/N_i.

Consider also the isomorphisms $\rho_i \colon N/N_i \longrightarrow N_{3-i}$, $i = 1, 2$, given by $(nN_i)^{\rho_i} = n_{3-i}$, for all $n \in N$, where $n = n_1 n_2$, $n_1 \in N_1$ and $n_2 \in N_2$.

Consider the isomorphism $\psi = \rho_1^{-1}\tau_1\tau_2^{-1}\rho_2$ between N_2 and N_1. It is not difficult to see that if $n_2 \in N_2$, then $n_2^{\psi} = n_1^{-1}$, where $n_2 = u n_1$ for $u \in U$ and $n_1 \in N_1$. The fact that ψ is H-invariant is an easy consequence of the fact that U is H-invariant. Therefore 2a implies 2b.

Conversely, if φ is an H-isomorphism between N_1 and N_2, then $T = \{aa^{\varphi} : a \in N_1\}$ is a subgroup of $N = N_1 N_2$, and $H \le N_G(T)$. Consider $U = HT$. Since $N = TN_i$, then $G = UN_i$, for $i = 1, 2$. Moreover, $U \cap N_i \le HT \cap N = T(H \cap N) = T$, and then $U \cap N_i \le T \cap N_i = 1$, for $i = 1, 2$. Hence U is a common complement of N_1 and N_2 in G. Therefore 2b implies 2a.

If, moreover, the N_i, $i = 1, 2$, are abelian and 2a is true, then it is easy to see that any H-isomorphism between N_1 and N_2 is a G-isomorphism. $\square$

Proposition 1.2.15. *Let G be a group and H_i/K_i, $i = 1, 2$, two supplemented chief factors of G. Then the following are equivalent.*

1. *H_1/K_1 and H_2/K_2 are G-connected;*
2. *for each $i = 1, 2$, there exists a precrown C_i/R_i associated with H_i/K_i, such that*
   a) *$C_1 = C_2$, and*
   b) *there exists a common complement U of the factors $R_i/(R_1 \cap R_2)$ in G, $i = 1, 2$.*

Proof. 1 implies 2. If the H_i/K_i, $i = 1, 2$, are abelian, then $H_1/K_1 \cong_G H_2/K_2$. In this case $C_1 = C_G(H_1/K_1) = C_2 = C_G(H_2/K_2) = C$. Hence the numerators of the precrowns coincide. For each $i = 1, 2$, let M_i be a complement of H_i/K_i in G. Then $C = H_1(M_1)_G = H_2(M_2)_G$. If $R = (M_1)_G = (M_2)_G$, then both chief factors have the same precrown C/R and we can take $U = G$. Otherwise $R_1 = (M_1)_G \ne (M_2)_G = R_2$. We can assume without loss of generality that $R_1 \cap R_2 = 1$. In particular, $C = R_1 \times R_2$ and $R_1 \cong_G R_2 \cong_G H_1/K_1$.

Note that $G/R_1 \cong G/R_2 \cong [H_1/K_1](G/C)$ and the isomorphisms map the C/R_i onto H_1/K_1. By the previous lemma, there exists a common complement to R_1 and R_2 in G.

Suppose now that the H_i/K_i, $i = 1$, 2, are non-abelian and $H_1/K_1 \cong_G$ H_2/K_2. Then they have the same precrown and we can take G as complement of the trivial factor.

Assume finally that H_i/K_i, $i = 1$, 2, are non-abelian and there exists a normal subgroup N of G such that G/N is a primitive group of type 3 with minimal normal subgroups A_1/N and A_2/N such that $A_1/N \cong_G H_1/K_1$ and $A_2/N \cong_G H_2/K_2$. Clearly $C_G(A_1/N) = A_2$ and $C_G(A_2/N) = A_1$. Hence the precrown of G associated with H_1/K_1 and with A_1/N is A_1A_2/A_2 and the precrown of G associated with H_2/K_2 and with A_2/N is A_1A_2/A_1. Since $A_1 \cap A_2 = N$ and G/N is a primitive group of type 3, the conclusion follows easily from Theorem 1 (3c).

2 implies 1. Suppose that there exist normal subgroups C, R_1, R_2 of G such that C/R_i is a precrown associated with H_i/K_i and there exists a common complement U of the factors $R_i/(R_1 \cap R_2)$ in G, $i = 1$, 2.

If H_1/K_1 and H_2/K_2 are non-abelian, then $R_i = C_G(H_i/K_i)$ and G/R_i is a primitive group of type 2, $i = 1$, 2. If $R_1 = R_2$, then H_1/K_1 and H_2/K_2 are G-isomorphic and then G-connected. If $R_1 \neq R_2$, we apply Corollary 1.1.13 to conclude that $G/(R_1 \cap R_2)$ is a primitive group of type 3 whose minimal normal subgroups are $R_i/(R_1 \cap R_2) \cong_G H_i/K_i$, $i = 1$, 2. Therefore H_1/K_1 and H_2/K_2 are G-connected.

Assume that H_1/K_1 and H_2/K_2 are abelian. If $R_1 = R_2$, then H_1/K_1 and H_2/K_2 are G-isomorphic and if $R_1 \neq R_2$, then both factors are G-isomorphic to $\operatorname{Soc}(G/U_G)$. In both cases, they are G-connected. $\qquad\square$

Lemma 1.2.16 ([Bra88]). *Let G be a group and suppose that Z, Y, X, W are normal subgroups of G such that $Z = XY$ and $X \cap Y = W$.*

1. *If Z/X is complemented in G by M, then Y/W is complemented in G by M.*

2. *Moreover, if M complements Z/X and S complements X/W, then $(M \cap S)Y$ complements Z/Y; in this case $M \cap S$ complements Z/W in G.*

3. *Parts 1 and 2 hold in terms of supplements.*

   When Y/W is a non-abelian chief factor of G, we can say even more:

4. *the set of monolithic supplements of Y/W in G coincides with the set of monolithic supplements of Z/X in G;*

5. *moreover, if X/W is an abelian chief factor of G then the (possibly empty) set of complements of X/W in G coincides with the set of complements of Z/Y in G.*

Proof. 1, 3. If $G = MZ$ and $X \leq Z \cap M$, then $G = MY$. Moreover $W = X \cap Y \leq M \cap Z \cap Y = M \cap Y$. Then M is a supplement of Y/W in G.

2, 3. If $G = MZ$ with $X \leq Z \cap M$ and $G = SX$ with $W \leq S \cap X$, then $\big((M \cap S)Y\big)Z = (M \cap S)Z = (M \cap S)XY = (M \cap SX)Y = MY = M(XY) = MZ = G$. Moreover $\big((M \cap S)Y\big) \cap Z = (M \cap S \cap Z)Y$ contains $(X \cap S)Y$ and $Y = WY \leq (X \cap S)Y$. Hence $(M \cap S)Y$ is a supplement of

Z/Y in G. Moreover, in this case, $G = (M \cap S)Z$ and W is contained in $S \cap X = S \cap X \cap Z \leq M \cap S \cap Z$. This is to say that $M \cap S$ supplements Z/W in G.

A substitution of the above inequalities by equalities gives the result in terms of complements.

For the remainder of the proof we can suppose without loss of generality that $W = 1$ and then Y is a non-abelian minimal normal subgroup of G centralising X.

4. If M is a monolithic supplement of Y in G then $G = MY$ and the group G/M_G is a monolithic primitive group of non-abelian socle. Then we have $\operatorname{Soc}(G/M_G) = M_G Y/M_G$ and $\mathrm{C}_G(Y) = \mathrm{C}_G(M_G Y/M_G) = M_G$. Hence $X \leq \mathrm{C}_G(Y) = M_G \leq M$. Then $G = MZ$ with $X \leq Z \cap M$ and M is a monolithic supplement of Z/X in G. Conversely, if M is a monolithic supplement of Z/X in G, then, by Statement 3, M supplements Y in G.

5. Suppose that X is an abelian minimal normal subgroup of G complemented by M. Then $\mathrm{C}_G(X) = XM_G$ and then $Z = X \times (Z \cap M_G)$. Since Y is non-abelian, this implies that $Y = Z'$ is contained in $Z \cap M_G$. Then Y is contained in M and M complements Z/Y. Note that the roles of X and Y in the original hypothesis can be interchanged without loss. Hence, by Statement 1, the (possibly empty) set of complements in G of X coincides with the set of complements of Z/Y in G.   □

Lemma 1.2.17 (see [Haw67]). *Let U and S be two maximal subgroups of a group G such that $U_G \neq S_G$. Suppose that U and S supplement the same chief factor H/K of G. Then $M = (U \cap S)H$ is a maximal subgroup of G such that $M_G = H(U_G \cap S_G)$.*

1. *Assume that H/K is abelian. Then M is a maximal subgroup of type 1 and complements the chief factors $U_G/(U_G \cap S_G)$ and $S_G/(U_G \cap S_G)$. Moreover $M \cap U = M \cap S = U \cap S$.*
2. *Assume that H/K is non-abelian. Then either U or S is of type 3. Suppose that U is of type 3 and S is monolithic. Then $U_G < S_G = \mathrm{C}_G(H/K)$. Moreover M is a maximal subgroup of type 2 of G such that M supplements the chief factor S_G/U_G.*
3. *Assume that U and S are of type 3. Then M is a maximal subgroup of type 3 of G such that M complements the chief factors HS_G/M_G and HU_G/M_G. Moreover $M \cap U = M \cap S = U \cap S$.*

Proof. 1. Assume that H/K is abelian and denote $C = \mathrm{C}_G(H/K)$. First observe that $M \cap U = H(U \cap S) \cap U = (H \cap U)(U \cap S) = K(U \cap S) = U \cap S$, since $H \cap U = K$, by the abelian nature of H/K. Analogously $M \cap S = U \cap S$. Hence M is a proper subgroup of G. Note also that $C = U_G H = S_G H = U_G S_G$ and $U_G/(U_G \cap S_G)$ is a G-chief factor which is G-isomorphic to the precrown C/S_G. Hence $U_G/(U_G \cap S_G)$ is G-isomorphic to H/K. Now, $MU_G = (U \cap S)HU_G = (U \cap SU_G)H = UH = G$ and $U_G \cap S_G \leq M \cap U_G$ and then we deduce that M is a maximal subgroup of G which complements $U_G/(U_G \cap S_G)$. The same argument holds for the chief factor $S_G/(U_G \cap S_G)$. Since M also complements the chief factor $C/(U_G \cap S_G)H$, we have that $M_G = H(U_G \cap S_G)$.

2. Assume that H/K is non-abelian. If U and S were both monolithic, of type 2, then $U_G = S_G = C_G(H/K)$. This is not true by hypothesis and then either U or S is of type 3.

Assume that U is of type 3 and S is monolithic. It is clear that $S_G = C_G(H/K)$. Observe that HU_G/U_G is a chief factor of G which is G-isomorphic to H/K. Then HU_G/U_G and S_G/U_G are the two minimal normal subgroups of the primitive group G/U_G of type 3. Both are complemented by U; in particular, $G = US_G$.

Observe that $MS_G = H(U \cap S)S_G = H(US_G \cap S) = HS = G$ and $M \cap S_G = (U \cap S)H \cap S_G$ contains $U_G H \cap S_G = U_G(H \cap S_G) = U_G K = U_G$ and then M supplements the chief factor S_G/U_G.

Now the group $G/U_G H = (M/U_G H)(S_G H/U_G H)$ is primitive of type 2. If the normal subgroup $M_G/U_G H$ were non-trivial, then $S_G H$ would be contained in M_G and so $S_G \leq M$. This is not possible. Hence $M_G = U_G H$.

Consider a subgroup T such that $U \cap S \leq T \leq U$. Then $S = (U \cap S)S_G \leq TS_G \leq US_G = G$. By maximality of S in G we have that either $S = TS_G$ or $G = TS_G$. Observe that $T \cap S_G = U \cap S_G = U_G$, and then, $U \cap TS_G = T(U \cap S_G) = T(T \cap S_G) = T$, so $U \cap S = T$ or $U = U \cap G = T$. This means that $U \cap S$ is a maximal subgroup of U. In the isomorphism $U/(U \cap H) \cong G/H$, the image of $(U \cap S)/(U \cap H)$ is M/H. Hence M is a maximal subgroup of G of type 2.

3. Assume now that U and S are maximal subgroups of type 3: the quotient groups G/U_G and G/S_G are primitive groups of type 3.

If $C = C_G(H/K)$, then U complements the chief factors HU_G/U_G and C/U_G. Analogously, S complements the chief factors HS_G/S_G and C/S_G. In particular, $U_G \not\leq S_G$ and $S_G \not\leq U_G$. Therefore $G = US_G = SU_G$. Now, by an analogous argument to that presented at the end of 2, we have that $M = (U \cap S)H$ is a maximal subgroup of G.

On the other hand, since C/S_G and C/U_G are chief factors of G and $U_G \neq S_G$, then $C = U_G S_G$. Write $L = U_G \cap S_G$. Observe that $HU_G/HL \cong_G U_G/(U_G \cap HL) = U_G/L \cong_G C/S_G$ and then HU_G/HL is a chief factor of G and $C_G(HU_G/HL) = C_G(C/S_G) = HS_G$. Similarly HS_G/HL is a chief factor of G and $C_G(HS_G/HL) = HU_G$. Hence the quotient group $G^* = G/HL$ has two minimal normal subgroups, namely $N = HS_G/HL$ and $C_{G^*}(N) = HU_G/HL$. Observe that $M(S_G H) = (U \cap S)S_G H = (US_G \cap S)H = SH = G$. Because U complements C/U_G, we have that $U \cap U_G S_G = U_G$, so $U \cap S_G = U_G \cap S_G = L$ and $M \cap HS_G = (U \cap S \cap HS_G)H = (U \cap S_G)H = HL$. Analogously $G = M(U_G H)$ and $M \cap HU_G = HL$. Therefore, the maximal subgroup $M^* = M/HL$ of G^* complements N and $C_{G^*}(N)$. By Proposition 1.1.12, the group G^* is a primitive group of type 3. Hence $M_G = HL$.

Finally observe that $M \cap U = H(U \cap S) \cap U = (H \cap U)(U \cap S) = K(U \cap S) = U \cap S$. Analogously $M \cap S = U \cap S$. □

Definitions 1.2.18. *Let* **X** *be a set of maximal subgroups of a group G.*

1. *If* **X** *is non-empty, then the* **X***-Frattini subgroup of G is defined to be the intersection of the cores of all members of* **X**. *It is denoted by* $\Phi_{\mathbf{X}}(G)$. *If* **X** $= \emptyset$, *we define* $\Phi_{\mathbf{X}}(G) = G$.
2. *Let* H/K *be a chief factor of G. We say that* H/K *is an* **X***-supplemented (respectively,* **X***-complemented) chief factor if it has a supplement (respectively, complement) in* **X**; *otherwise* H/K *is said to be an* **X***-Frattini chief factor.*
3. *If* C^*/N *is a precrown of G associated with an* **X***-supplemented chief factor* H/K *of G, we shall say that* C^*/N *is an* **X***-precrown of G associated with* H/K.

Notation 1.2.19. Let N be a normal subgroup of a group G and let **X** be a set of maximal subgroups of G. We write

$$\mathbf{X}/N = \{Z/N : Z \in \mathbf{X} \text{ and } N \text{ is contained in } Z\}$$

and if $\varphi \colon G \longrightarrow H$ is a group homomorphism, we write

$$\mathbf{X}^{\varphi} = \{S^{\varphi} : S \in \mathbf{X}\}.$$

The following lemma will be used frequently in the sequel.

Lemma 1.2.20. *Let* **X** *be a set of maximal subgroups of a group G. Let* H/K *be a chief factor of a group G.*

1. H/K *is an* **X***-Frattini chief factor of G if and only if* $H/K \leq \Phi_{\mathbf{X}/K}(G/K)$.
2. *If A is a normal subgroup of G contained in K, then* H/K *is* **X***-Frattini in* G/K *if and only if* $(H/A)/(K/A)$ *is* **X**/A*-Frattini in* G/A. *Furthermore, if* H/K *is* **X***-supplemented in G, then a maximal subgroup* $U \in \mathbf{X}$ *is a supplement of* H/K *in G if and only if* U/A *is an* **X**/A*-supplement of* $(H/A)/(K/A)$ *in* G/A.

Definition 1.2.21. *A set* **X** *of maximal subgroups of a group G is said to be* solid for the Jordan-Hölder theorem, *or simply* JH-solid, *if it satisfies the following condition:*

(JH) If U, S $\in \mathbf{X}$ *with* $U_G \neq S_G$ *and both supplement a chief factor* H/K *of G, then there exists* $M \in \mathbf{X}$ *such that* $M_G = (U_G \cap S_G)H$.

Applying Lemma 1.2.17, the set of all maximal subgroups of a group G that supplement a single chief factor, the set $\mathrm{Max}(G)$ of all maximal subgroups of a group G, and the set $\mathrm{Max}^*(G)$ of all monolithic maximal subgroups of a group G are JH-solid.

Note that

$$\Phi(G) = \bigcap\{M \in \mathrm{Max}(G)\} = \bigcap\{M \in \mathrm{Max}^*(G)\}.$$

We will use the following results in inductive arguments.

Theorem 1.2.22. *Let G be a group factorised as $G = MN$, where M is a subgroup of G and N is a normal subgroup of G. Then $G/N \cong M/(N \cap M)$, and we have the following.*

1. If

$$N = H_n < \cdots < H_0 = G \qquad (1.4)$$

is a piece of chief series of G, then

$$M \cap N = M \cap H_n < \cdots < M \cap H_0 = M \qquad (1.5)$$

is a piece of chief series of M. If S is a maximal subgroup of G which supplements a chief factor H_i/H_{i+1} in (1.4), then $M \cap S$ is a maximal subgroup of M which supplements the chief factor $(H_i \cap M)/(H_{i+1} \cap M)$ in (1.5). Moreover, the core of $M \cap S$ in M is $(M \cap S)_M = M \cap S_G$.

2. Conversely, if

$$M \cap N = M_n < \cdots < M_0 = M \qquad (1.6)$$

is a piece of chief series of M, then

$$N = M_n N < \cdots < M_0 N = MN = G \qquad (1.7)$$

is a piece of chief series of G. If U is a maximal subgroup of M which supplements a chief factor M_i/M_{i+1} in (1.6), then UN is a maximal subgroup of G which supplements the chief factor $M_i N/M_{i+1} N$ in (1.7). Moreover, the core of UN in G is $(UN)_G = U_M N$.

Lemma 1.2.23. *Let $\mathbf{X}$ be a JH-solid set of maximal subgroups of a group G and N a normal subgroup of G.*

1. The set $\mathbf{X}/N$ is a JH-solid set of maximal subgroups of G/N.
2. Suppose that the subgroup M supplements N in G: $G = MN$. Then the set

$$(\mathbf{X} \cap M)/(N \cap M) = \{(S \cap M)/(N \cap M) : N \le S \in \mathbf{X}\}$$

is a JH-solid set of maximal subgroups of $M/(N \cap M)$.
Moreover, if φ is the isomorphism between G/N and $M/(N \cap M)$ then we have that $(\mathbf{X}/N)^\varphi = (\mathbf{X} \cap M)/(M \cap N)$.

Now we can prove the announced strengthened form of the Jordan-Hölder theorem for chief series of finite groups and give an answer to Lafuente's question. To do this we proceed following Lafuente's arguments in [Laf89]. It must be observed that these arguments deal with the modular lattice of all normal subgroups of a group in which we can use the Duality Principle (see [Bir69, Chapter 1, Theorem 2]).

Notation 1.2.24. If A/B and C/D are sections of a group G, then we write $A/B \ll C/D$ (or $C/D \gg A/B$) if $C = AD$ and $B = A \cap D$.

Observe that if $A/B \ll C/D$, then $A/B \cong_G C/D$. In particular, A/B is a chief factor of G if and only if C/D is a chief factor of G.

Lemma 1.2.25. *Let K and H be normal subgroups of a group G and let*

$$K = Y_0 < Y_1 < \cdots < Y_{m-1} < Y_m = H$$

be a piece of chief series between K and H. Suppose that X^*/X is a chief factor of G between H and K.

1. *If $X^*Y_j = XY_j$, then $X^*Y_k = XY_k$ for $j \le k \le m$.*
2. *If $X^* \cap Y_{j-1} = X \cap Y_{j-1}$, then $X^* \cap Y_{k-1} = X \cap Y_{k-1}$, for $1 \le k \le j$.*
3. *If $X^*Y_{j-1} > XY_{j-1}$, then $X^*Y_{k-1} > XY_{k-1}$, for $1 \le k \le j$ and $X^* \cap Y_{j-1} = X \cap Y_{j-1}$. In this case,*

$$X^*Y_{j-1}/XY_{j-1} \gg X^*Y_{k-1}/XY_{k-1} \gg X^*/X.$$

4. *If $X^* \cap Y_j > X \cap Y_j$, then $X^* \cap Y_k > X \cap Y_k$, for $j \le k \le m$ and $X^*Y_j = XY_j$. Moreover*

$$X^*/X \gg (X^* \cap Y_k)/(X \cap Y_k) \gg (X^* \cap Y_j)/(X \cap Y_j).$$

Proof. Note that Statement 1 and its dual, which is Statement 2, are obvious.

3. By Statement 1, if $X^*Y_{j-1} > XY_{j-1}$, then $X^*Y_{k-1} > XY_{k-1}$, for $1 \le k \le j$. On the other hand, we have

$$(X^*Y_{k-1})(XY_{j-1}) = X^*Y_{j-1} \qquad X^*Y_{k-1} = X^*(XY_{k-1}).$$

Moreover $X \le X(X^* \cap Y_{j-1}) = X^* \cap XY_{j-1} \le X^*$. Since X^*/X is a chief factor of G, then either $X = X(X^* \cap Y_{j-1}) = X^* \cap XY_{j-1}$ or $X^* \cap XY_{j-1} = X^*$. In the last case $X^* \le XY_{j-1}$ and then $X^*Y_{j-1} = XY_{j-1}$, contrary to our supposition. Hence $X^* \cap Y_{j-1} \le X$ and then $X^* \cap Y_{j-1} = X \cap Y_{j-1}$. By Statement 2, $X^* \cap Y_{k-1} = X \cap Y_{k-1}$. Hence

$$X^* \cap XY_{k-1} = X(X^* \cap Y_{k-1}) = X(X \cap Y_{k-1}) = X$$

and

$$\begin{aligned} XY_{j-1} \cap X^*Y_{k-1} &= (XY_{j-1} \cap X^*)Y_{k-1} \\ &= X(Y_{j-1} \cap X^*)Y_{k-1} = X(X \cap Y_{j-1})Y_{k-1} = XY_{k-1}. \end{aligned}$$

Statement 4 is dual of Statement 3.   □

Definition 1.2.26. *Let A/B, A/C and C/D be chief factors of a group G such that $A/B \gg C/D$. If $\mathbf{X}$ is a set of maximal subgroups of G, such that A/B is $\mathbf{X}$-Frattini and C/D is $\mathbf{X}$-supplemented, we will say that the situation $A/B \gg C/D$ is an $\mathbf{X}$-crossing. We write $[A/B \gg C/D]$ to denote an $\mathbf{X}$-crossing.*

Remark 1.2.27.   1. If $A/B \gg C/D$ and A/B is **X**-supplemented, then C/D is **X**-supplemented, by Lemma 1.2.16.

2. If $[A/B \gg C/D]$ is an **X**-crossing, then C/D is abelian. If C/D is a non-abelian **X**-supplemented chief factor, then A/B is also **X**-supplemented, by Lemma 1.2.16 (4), against our supposition.

Next we see a characterisation of JH-solid sets of monolithic maximal subgroups in terms of **X**-crossing situations.

Theorem 1.2.28. *Let* **X** *be a set of maximal subgroups of a group G.*

1. *Assume that* **X** *is JH-solid. Let Z/Y, Y/W and X/W be chief factors of G. If $[Z/X \gg Y/W]$ is an* **X***-crossing, then $[Z/Y \gg X/W]$ is an* **X***-crossing. Moreover, in this case, a maximal subgroup $U \in$ **X** supplements Y/W if and only if U supplements X/W.*

2. *Conversely, assume that* **X** *is a monolithic set of maximal subgroups of G such that whenever we have chief factors Z/Y, Y/W and X/W of G such that $[Z/X \gg Y/W]$ is an* **X***-crossing, then $[Z/Y \gg X/W]$ is an* **X***-crossing. Then* **X** *is JH-solid.*

Proof.   1. We can assume that $W = 1$. We have to prove that if X and Y are minimal normal subgroups of G, Z/X is **X**-Frattini chief factor and Y is **X**-suplemented, then Z/Y is **X**-Frattini and X is **X**-supplemented.

Assume that U is an **X**-supplement of Y. If $X \leq U$, then $G = UZ$ and $X \leq U \cap Z$, so U supplements Z/X. This contradiction yields that X is not contained in U and then U supplements X. Suppose that, in this case, there exists $S \in$ **X** supplementing Z/Y. Then S also supplements X. Since $Y \nleq U_G$ and $Y \leq S_G$, by the property (JH), there exists $M \in$ **X** such that $M_G = (U_G \cap S_G)X$. If $Z \leq M$, then $Z = Z \cap M_G = X(U_G \cap S_G \cap Z) = X(U_G \cap Y) = X$, which is a contradiction. Hence M supplements Z/X, which we have supposed to be **X**-Frattini. We deduce that Z/Y must be an **X**-Frattini chief factor of G.

2. Suppose that we have $U, S \in$ **X**, both supplementing the same chief factor H/K of G and $U_G \neq S_G$. Since U and S are monolithic, the chief factor H/K must be abelian, by Lemma 1.2.17 (2). Therefore $K = U \cap H = U_G \cap H = S_G \cap H = S \cap H$.

Observe that $C = \mathrm{C}_G(H) = HS_G = HU_G = U_G S_G$. Write $A = U_G \cap S_G$. Then

$$C/HA = HU_G/HA \cong_G U_G/(U_G \cap HA) = U_G/A \cong_G C/S_G$$

and then C/HA is a chief factor of G and $C/HA \gg U_G/A$. Observe that U_G/A is **X**-complemented by S. Suppose that C/HA is **X**-Frattini. Then $[C/HA \gg U_G/A]$ is an **X**-crossing. By hypothesis, $[C/U_G \gg HA/A]$ is an **X**-crossing. But C/U_G is obviously **X**-complemented by U. This contradiction yields that C/HA is **X**-complemented in G, i. e. there exists $M \in$ **X** such that $G = MC$ and $HA = M_G$. Therefore **X** is JH-solid.   □

Proposition 1.2.29. *With the hypotheses of Lemma 1.2.25, assume that X^*/X is an* **X**-*supplemented chief factor of G. Let*

$$j' = \max\{j : X^*Y_{j-1}/XY_{j-1} \text{ is } \mathbf{X}\text{-supplemented chief factor of } G\}$$

and set $Y^* = Y_{j'}$ and $Y = Y_{j'-1}$. Then Y^*/Y is **X**-*supplemented.*
Furthermore the following conditions are satisfied:

1. *If $X^*Y^* = XY^*$, then $X^*Y^* = XY^* = X^*Y$. Write $R^* = X^*Y^*$ and $R = XY$. Then $X^*/X \ll R^*/R \gg Y^*/Y$. Moreover $X^* \cap Y = X \cap Y = X \cap Y^*$. Write $S = X \cap Y$ and $S^* = X^* \cap Y^*$, then $X^*/X \gg S^*/S \ll Y^*/Y$.*
2. *If $X^*Y^* \neq XY^*$, then $[X^*Y^*/XY^* \gg X^*Y/XY]$ is an* **X**-*crossing and $X^*/X \ll X^*Y/XY$ and $XY^*/XY \gg Y^*/Y$.*

In particular, in both cases X^*Y/XY and XY^*/XY are **X**-*supplemented chief factors of G.*

Proof. Observe that $X^*Y_0/XY_0 = X^*/X$ is **X**-supplemented. Hence j' is well-defined.

Assume that $XY^* = XY$. Then $X^*Y^* = X^*Y$. So $X^*Y^*/XY^* = X^*Y/XY$ is **X**-supplemented, giving a contradiction to the election of j'. Therefore $XY^*/XY \gg Y^*/Y$ and XY^*/XY is a chief factor.

1. Assume that $X^*Y^* = XY^*$. Then $XY \le X^*Y \le X^*Y^* = XY^*$. Therefore $X^*Y^* = X^*Y$ because $X^*Y > XY$ by hypothesis. From part 3 of Lemma 1.2.25, it follows that $X^*/X \ll R^*/R \gg Y^*/Y$. On the other hand, $X^* = XY^* \cap X^* = (X^* \cap Y^*)X$. Hence $X^*/X \gg (X^* \cap Y^*)/(X \cap Y^*)$. Now, from part 3 of Lemma 1.2.25, $X^* \cap Y = X \cap Y = X \cap Y^*$. Thus, $X^*/X \gg S^*/S \ll Y^*/Y$.

In this case $R^*/R = X^*Y^*/XY = XY^*/XY = X^*Y/XY$ is **X**-supplemented, by definition of j'.

2. Now consider $X^*Y^* \neq XY^*$. From the choice of j', it follows that X^*Y^*/XY^* is an **X**-Frattini chief factor of G. Then $XY \le XY^* \cap X^*Y \le X^*Y$. If $XY^* \cap X^*Y = X^*Y$, it follows that $X^*Y^* = XY^*$ contrary to our assumption. Hence $XY = XY^* \cap X^*Y$ and $[X^*Y^*/XY^* \gg X^*Y/XY]$ is an **X**-crossing. Moreover $X^*/X \ll X^*Y/XY$ and $XY^*/XY \gg Y^*/Y$.

Since $[X^*Y^*/XY^* \gg X^*Y/XY]$ is an **X**-crossing, we have that X^*Y/XY and XY^*/XY are **X**-supplemented chief factors of G.   □

Proposition 1.2.30. *With the hypotheses of Lemma 1.2.25, assume that X^*/X is an* **X**-*Frattini chief factor of G. Let*

$$j' = \min\{j : (X^* \cap Y_j)/(X \cap Y_j) \text{ is an } \mathbf{X}\text{-Frattini chief factor of } G\}$$

and set $Y^* = Y_{j'}$ and $Y = Y_{j'-1}$. Then Y^*/Y is **X**-*Frattini.*
Furthermore the following conditions are satisfied:

1. *If $X^* \cap Y = X \cap Y$, then $X \cap Y = X \cap Y^* = X^* \cap Y^*$. Write $S^* = X^* \cap Y^*$ and $S = X \cap Y$. Then $X^*/X \gg S^*/S \ll Y^*/Y$. Moreover $X^*Y = X^*Y^* = XY^*$. Write $R = XY$ and $R^* = X^*Y^*$, then $X^*/X \ll R^*/R \gg Y^*/Y$.*

2. If $X^* \cap Y \neq X \cap Y$, then $[(X^* \cap Y^*)/(X \cap Y^*) \gg (X^* \cap Y)/(X \cap Y)]$ is an **X**-crossing and $X^*/X \gg (X^* \cap Y^*)/(X \cap Y^*)$ and $(X^* \cap Y^*)/(X^* \cap Y) \ll Y^*/Y$.

In particular, in both cases $(X^* \cap Y^*)/(X^* \cap Y)$ and $(X^* \cap Y^*)/(X \cap Y^*)$ are **X**-Frattini chief factors of G.

Proof. This is the dual statement of Proposition 1.2.29.   □

Definition 1.2.31. *Given a set* **X** *of maximal subgroups of a group* G, *we say that two chief factors of* G, *say* X^*/X *and* Y^*/Y, *are* **X**-related *if one of these properties is satisfied:*

1. *There exists an* **X**-supplemented *chief factor* R^*/R *such that* $X^*/X \ll R^*/R \gg Y^*/Y$,
2. *There exists an* **X**-crossing $[A/Z \gg T/B]$ *such that* $X^*/X \ll Z/B$ *and* $T/B \gg Y^*/Y$.
3. *There exists an* **X**-Frattini *chief factor* S^*/S *such that* $X^*/X \gg S^*/S \ll Y^*/Y$,
4. *There exists an* **X**-crossing $[A/Z \gg T/B]$ *such that* $X^*/X \gg A/Z$ *and* $A/T \ll Y^*/Y$.

The importance of the **X**-relation becomes clear in the following theorem.

Theorem 1.2.32. *Let* **X** *be a JH-solid set of maximal subgroups of a group* G. *If the chief factors* X^*/X *and* Y^*/Y *are* **X**-related, *then*

1. X^*/X *and* Y^*/Y *are* G-connected, *and*
2. X^*/X *is* **X**-Frattini *if and only if* Y^*/Y *is* **X**-Frattini.
3. *If* X^*/X *and* Y^*/Y *are* **X**-supplemented, *there exists a common* **X**-supplement *to both.*

Furthermore, if **X** *is composed of monolithic maximal subgroups of* G *then any two* **X**-related *chief factors are* G-isomorphic.

Proof.   1. Observe that in Cases 1 and 3 of the definition of **X**-relation, we have that X^*/X is G-isomorphic to Y^*/Y. Suppose that there exists an **X**-crossing $[A/Z \gg T/B]$ such that $X^*/X \ll Z/B$ and $T/B \gg Y^*/Y$. Since **X** is JH-solid, there exists a common **X**-supplement U of Z/B and T/B, by Theorem 1.2.28. Then TU_G/U_G and ZU_G/U_G are minimal normal subgroups of the primitive group G/U_G. If $ZU_G = TU_G$, then $Z/B \cong_G T/B$; in this case $X^*/X \cong_G Y^*/Y$. Otherwise G/U_G is a primitive group of type 3 whose minimal normal subgroups are TU_G/U_G and ZU_G/U_G. Since $X^*/X \cong_G Z/B$ and $Y^*/Y \cong_G T/B$, then X^*/X and Y^*/Y are G-connected. The analysis of Case 4 is analogous.

Observe that if all elements of **X** are monolithic maximal subgroups of G, then necessarily $ZU_G = TU_G$ in the above analysis. Therefore $X^*/X \cong_G Y^*/Y$.

2. If X^*/X is **X**-Frattini, then we are not in Case 1 of the definition of **X**-relation. Suppose that there exists an **X**-crossing $[A/Z \gg T/B]$ such that $X^*/X \ll Z/B$ and $T/B \gg Y^*/Y$. Then $[A/T \gg Z/B]$ is an **X**-crossing by Theorem 1.2.28. Then Z/B is **X**-complemented. This implies that X^*/X is **X**-supplemented by Lemma 1.2.16. Therefore we are not in Case 2 of Definition 1.2.31 either. If we are in Case 3, then Y^*/Y is **X**-Frattini by Lemma 1.2.16. In Case 4, $[A/T \gg Z/B]$ is an **X**-crossing by Theorem 1.2.28 and again Y^*/Y is **X**-Frattini by Lemma 1.2.16.

3. If X^*/X and Y^*/Y are **X**-supplemented, we are either in Case 1 or in Case 2 of Definition 1.2.31. In Case 1, if U is an **X**-supplement of R^*/R, then U supplements X^*/X and Y^*/Y, In Case 2, there exists an **X**-crossing $[A/Z \gg T/B]$ such that $X^*/X \ll Z/B$ and $T/B \gg Y^*/Y$. By Theorem 1.2.28, we know that there exists a common **X**-supplement U to Z/B and T/B. By Lemma 1.2.16, U also **X**-supplements X^*/X and Y^*/Y, □

Lemma 1.2.33. *Under the hypotheses of Lemma 1.2.25, assume that X^*/X and Y_j/Y_{j-1} are* **X**-*related.*

1. X^*/X *and* Y_j/Y_{j-1} *are* **X**-*supplemented in G if and only if* X^*Y_{j-1}/XY_{j-1} *is* **X**-*supplemented in G.*
2. X^*/X *and* Y_j/Y_{j-1} *are* **X**-*Frattini if and only if* $(X^* \cap Y_j)/(X \cap Y_j)$ *are* **X**-*Frattini.*

Proof. 1. Set $Y^* = Y_j$, $Y = Y_{j-1}$ and assume that there exists an **X**-supplemented chief factor R^*/R such that $X^*/X \ll R^*/R \gg Y^*/Y$. Since $(X^*Y)R = R^*$ and $XY \leq R$, then $XY < X^*Y$. By part 3 of Lemma 1.2.25, $X^*Y/XY \gg X^*/X$ and in particular, X^*Y/XY is a chief factor. On the other hand, $XY \leq X^*Y \cap R \leq X^*Y$. As X^*Y is not contained in R, then $R^*/R \gg X^*Y/XY$. Therefore X^*Y_{j-1}/XY_{j-1} is **X**-supplemented in G. Now suppose that there exists an **X**-crossing $[A/Z \gg T/B]$ such that $Z/B \gg X^*/X$y $T/B \gg Y^*/Y$. Since $XY \leq B$ and $(X^*Y)B = Z$, we have that $X^*Y > XY$ and, as above, $X^*/X \ll X^*Y/XY$. Now $Z = (X^*Y)B$ and $X^*Y \cap B = Y(X^* \cap B) = XY$. Hence $Z/B \gg X^*Y/XY$. Therefore X^*Y_{j-1}/XY_{j-1} is **X**-supplemented in G.

The converse follows from part 3 of Lemma 1.2.25.

2. This is the dual statement of 1. □

Theorem 1.2.34. *Let G be a group and* **X** *a JH-solid set of maximal subgroups of G. For any pair K, H of normal subgroups of G such that $K < H$ and two pieces of chief series of G between K and H*

$$K = X_0 \leq X_1 \leq \cdots \leq X_n = H$$

and

$$K = Y_0 \leq Y_1 \leq \cdots \leq Y_m = H,$$

then $n = m$ and there exists a unique permutation $\sigma \in \mathrm{Sym}(n)$ such that X_i/X_{i-1} and $Y_{i^\sigma}/Y_{i^\sigma-1}$ are **X**-*related, for $1 \leq i \leq n$. Furthermore*

$$i^\sigma = \max\{j : X_i Y_{j-1}/X_{i-1} Y_{j-1} \text{ is } \mathbf{X}\text{-supplemented}\}$$

if X_i/X_{i-1} is $\mathbf{X}$-supplemented, and

$$i^\sigma = \min\{j : (X_i \cap Y_j)/(X_{i-1} \cap Y_j) \text{ is } \mathbf{X}\text{-Frattini}\}$$

if X_i/X_{i-1} is $\mathbf{X}$-Frattini.

Proof. We can assume without loss of generality that $m \leq n$. Write $X^* = X_i$, $X = X_{i-1}$, $Y^* = Y_{i^\sigma}$ and $Y = Y_{i^\sigma-1}$.

By Proposition 1.2.29, if X^*/X is $\mathbf{X}$-supplemented, then so is Y^*/Y. Furthermore, if $X^*Y^* = XY^*$, then $X^*/X \ll R^*/R \gg Y^*/Y$, where $R^* = X^*Y^* = X^*Y$ and $R = XY$, by part 1 of Proposition 1.2.29. Hence R^*/R is $\mathbf{X}$-supplemented by definition of i^σ. So, this is Case 1 of the definition of $\mathbf{X}$-relation. And if $X^*Y^* \neq XY^*$, then we are in Case 2 of Definition 1.2.31 by part 2 of Proposition 1.2.29.

Dually, by Proposition 1.2.30, if X^*/X is $\mathbf{X}$-Frattini, then so is Y^*/Y. Furthermore, if $X^* \cap Y^* = X \cap Y^*$, then $X^*/X \gg S^*/S \ll Y^*/Y$, where $S^* = X^* \cap Y^*$ and $S = X \cap Y$, by part 1 of Proposition 1.2.30. Hence S^*/S is $\mathbf{X}$-Fratttini by definition of i^σ. So, this is Case 3 of the definition of $\mathbf{X}$-relation. and if $X^* \cap Y^* \neq X \cap Y$, then we are in Case 4 of Definition 1.2.31.

Therefore, in any case, X_i/X_{i-1} and $Y_{i^\sigma}/Y_{i^\sigma-1}$ are $\mathbf{X}$-related, for $1 \leq i \leq n$.

Now we prove that the map $\sigma : \{1, \ldots, n\} \rightarrow \{1, \ldots, m\}$ defined above is injective. Write $Z^* = X_k$ and $Z = X_{k-1}$, where $i < k$ and $i^\sigma = k^\sigma$.

Suppose that X^*/X is $\mathbf{X}$-supplemented; then so are Y^*/Y and Z^*/Z. Assume that $X^*Y^* = XY^*$. From $X^* \leq Z$ we get that $ZY^* = ZY$. Since Z^*/Z is $\mathbf{X}$-supplemented and $k^\sigma = j$, Z^*Y/ZY is a chief factor of G and then $ZY = ZY^* < Z^*Y = Z^*Y^*$. By part 2 of Proposition 1.2.29, $ZY^*/ZY \gg Y^*/Y$. In particular $ZY^* > ZY$ and yields a contradiction. Hence $X^*Y^* > XY^*$. Then $[X^*Y^*/XY^* \gg X^*Y/XY]$ is an $\mathbf{X}$-crossing by part 2 of Proposition 1.2.29. The chief factor X^*Y^*/X^*Y is $\mathbf{X}$-Frattini. Since $k^\sigma = j$, then Z^*Y/ZY and ZY^*/ZY are $\mathbf{X}$-supplemented chief factors of G. As $X^* \leq Z$ gives $X^*Y \leq ZY$ and $X^*Y^* \leq ZY^*$. Observe that $ZY^* = (ZY)(X^*Y^*)$. Moreover $ZY \cap X^*Y^* = X^*(Z \cap Y^*)Y$ In the situation $Y \leq (Z \cap Y^*)Y \leq Y^*$ and Y^*/Y chief factor of G, we cannot have $ZY \cap Y^* = Y^*$, since this would imply $Y^* \leq ZY$ and then $ZY^* = ZY$ and this contradicts the fact that ZY^*/ZY is a chief factor. Hence $ZY \cap X^*Y^* = X^*Y$. In other words, $ZY^*/ZY \gg X^*Y^*/X^*Y$ and we deduce that ZY^*/ZY is $\mathbf{X}$-Frattini by Lemma 1.2.16. This is a contradiction.

We have shown that the restriction to σ to the subset $\mathcal{I}$ of $\{1, \ldots, n\}$ composed of all indices i corresponding to $\mathbf{X}$-supplemented chief factors X_i/X_{i-1}, is injective. Applying dual arguments we show that the restriction of σ to the subset of $\{1, \ldots, n\} \setminus \mathcal{I}$ composed of all indices i corresponding to $\mathbf{X}$-Frattini chief factors X_i/X_{i-1}, is injective. By the arguments at the beginning of the proof, σ is injective. Therefore $n = m$ and σ is a permutation of the set $\{1, \ldots, n\}$.

Finally if τ is any permutation with the above properties, then the definition of σ requires that $i^\tau \leq i^\sigma$ for all $i \in \mathcal{I}$ and $i^\tau \geq i^\sigma$ for all $i \in \{1, \ldots, n\} \setminus \mathcal{I}$ by Lemma 1.2.33. Consequently $\sigma = \tau$. $\qquad \square$

Remark 1.2.35. By Theorem 1.2.32, the bijection constructed in Theorem 1.2.34 satisfies that if X_i/X_{i-1} and $Y_{i^\sigma}/Y_{i^\sigma-1}$ are **X**-supplemented, there exists a common **X**-supplement to both. Clearly when X_i/X_{i-1}, and $Y_{i^\sigma}/Y_{i^\sigma-1}$, is abelian we can change the **X**-supplementation by **X**-complementation. But we can go further and say the same even for non-abelian **X**-complemented chief factors. We know, by Theorem 1.1.48, the existence of non-abelian chief factors complemented by maximal subgroups. Observe that if X_i/X_{i-1} and $Y_{i^\sigma}/Y_{i^\sigma-1}$ are **X**-complemented non-abelian chief factors, then we are in Case 1 of Definition 1.2.31, since Case 2 is not possible by Remark 2 of 1.2.27. If U is an **X**-complement of the non-abelian chief factor X_i/X_{i-1} and $Y_{i^\sigma}/Y_{i^\sigma-1} \ll R^*/R \gg X_i/X_{i-1}$, then U also supplements R^*/R, by of Lemma 1.2.16 (4), and the same for $Y_{i^\sigma}/Y_{i^\sigma-1}$. Observe that U/U_G is a small maximal subgroup of the primitive group G/U_G of type 2. Then, $\mathrm{Soc}(G/U_G) = X_i U_G/U_G = R^* U_G/U_G = Y_{i^\sigma} U_G/U_G$ and $U_G = U \cap Y_{i^\sigma} U_G$. Thus, $U \cap Y_{i^\sigma} = U_G \cap Y_{i^\sigma} = Y_{i^\sigma-1}$ and U complements $Y_{i^\sigma}/Y_{i^\sigma-1}$.

Theorem 1.2.36. *Let G be a group and **X** a set of monolithic maximal subgroups of G. Then the following conditions are equivalent:*

1. ***X** is a JH-solid set.*
2. *For any pair K, H of normal subgroups of G such that $K < H$ and two pieces of chief series of G between K and H*

$$K = X_0 \leq X_1 \leq \cdots \leq X_n = H \quad and \quad K = Y_0 \leq Y_1 \leq \cdots \leq Y_m = H,$$

then $n = m$ and there exists $\sigma \in \mathrm{Sym}(n)$ such that
   a) $X_i/X_{i-1} \cong_G Y_{i^\sigma}/Y_{i^\sigma-1}$;
   *b) X_i/X_{i-1} is **X**-Frattini if and only if $Y_{i^\sigma}/Y^{i^\sigma-1}$ is **X**-Frattini;*
   *c) if X_i/X_{i-1} is **X**-supplemented (respectively, complemented) in G, there exists a maximal subgroup $U \in **X**$ of G such that G supplements (respectively, complements) both X_i/X_{i-1} and $Y_{i^\sigma}/Y_{i^\sigma-1}$.*

Proof. After Theorem 1.2.34 we have only to see that 2 implies 1.

Suppose that we have $U, S \in **X**$, both supplementing the same chief factor H/K of G and $U_G \neq S_G$. Since U and S are monolithic, the chief factor H/K must be abelian, by Lemma 1.2.17 (2). Therefore $K = U \cap H = U_G \cap H = S_G \cap H = S \cap H$.

Observe that $C = \mathrm{C}_G(H) = HS_G = HU_G = U_G S_G$. Write $A = U_G \cap S_G$. Then

$$C/HA = HU_G/HA \cong_G U_G/(U_G \cap HA) = U_G/A \cong_G C/S_G$$

and then C/HA is a chief factor of G and $C/HA \gg U_G/A$. Observe that U_G/A is **X**-complemented by S and C/U_G is obviously **X**-complemented by U. By

Statement 2, all chief factors of G between C and A are $\mathbf{X}$-complemented. In particular C/HA is $\mathbf{X}$-complemented in G, i. e. there exists a maximal subgroup $M \in \mathbf{X}$ such that $G = MC$ and $HA = M \cap C$. This implies that $M_G = HA$. Therefore $\mathbf{X}$ is JH-solid.   □

Corollary 1.2.37. *If $\mathbf{X}$ is a JH-solid set of maximal subgroups of a group G and H is a normal subgroup of G such that all chief factors H/K_i, $i = 1, \ldots, n$, of G are $\mathbf{X}$-supplemented, and $\bigcap_{i=1}^{n} K_i = K$, then every chief factor between K and H is $\mathbf{X}$-supplemented.*

Proof. Denote $K^j = \bigcap_{i=1}^{j} K_i$ and $K^0 = H$. Then

$$K = K^n \leq K^{n-1} \leq \cdots \leq K^0 = H$$

is a piece of a chief series of G. Assume that $K^i \neq K^{i+1}$. Then $H = K^i K_{i+1}$, K^i/K^{i+1} is a chief factor of G and $K^i/K^{i+1} \cong_G H/K_{i+1}$. If M is an $\mathbf{X}$-supplement of H/K_{i+1} in G, then M is an $\mathbf{X}$-supplement of K^i/K^{i+1} in G by Lemma 1.2.16 (1). We deduce that all chief factors in the above series are $\mathbf{X}$-supplemented. Now apply Theorem 1.2.36 to conclude the proof.   □

Corollary 1.2.38. *Let $\mathbf{X}$ be a JH-solid set of monolithic maximal subgroups of a group G and write $R = \Phi_{\mathbf{X}}(G)$. Suppose that N is a normal subgroup of G such that $N = N_1 \times \cdots \times N_n$, where N_i is a minimal normal subgroups of G, $1 \leq i \leq n$. If $R \cap N = 1$, then every chief factor of G below N is $\mathbf{X}$-supplemented in G.*

Proof. We use induction on n. If $n = 1$, the result is obvious. Thus we assume that $n \geq 2$.

If N_1 is $\mathbf{X}$-Frattini, then $N_1 \leq R \cap N = 1$, giving a contradiction. Hence there exists $M \in \mathbf{X}$ such that $G = MN_1$. The quotient group G/M_G is a monolithic primitive group and then $NM_G/M_G = N_1 M_G/M_G = \operatorname{Soc}(G/M_G)$. Then $N = N_1 \times (N \cap M_G)$. By Theorem 1.2.36, every piece of chief series of G between N_1 and N has exactly $n - 1$ chief factors and so every piece of chief series of G below $N_0 = N \cap M_G$ has exactly $n - 1$ chief factors. Since the normal subgroup N_0 is contained in $\operatorname{Soc}(G)$, we have that N_0 can be written as a direct product of $n-1$ minimal normal subgroups of G. Since $R \cap N_0 = 1$, it follows that every chief factor of G below N_0 is $\mathbf{X}$-supplemented by induction. Since clearly M supplements N/N_0, we have that all chief factors of G below N are $\mathbf{X}$-supplemented, by Theorem 1.2.36.   □

Observe that in a primitive group G of type 3 with minimal normal subgroups N_1 and N_2, if M is a core-free maximal subgroup, then $\mathbf{X} = \{M\}$ is a JH-solid set of maximal subgroups of G, $R = M_G = 1$, and $N = \operatorname{Soc}(G)$ satisfies that $R \cap N = 1$. However neither N/N_1 nor N/N_2 are $\mathbf{X}$-supplemented.

Remarks 1.2.39.   1. Given a modular lattice $\mathcal{L}$, J. Lafuente in [Laf89] introduced the concept of *M-set in $\mathcal{L}$* and he proved a general Jordan-Hölder theorem in modular lattices with an M-set.

In fact, Theorem 1.2.28 shows that, for a set of maixmal subgroups $\mathbf{X}$ of a group G, the set $\mathcal{M}_{\mathbf{X}}$ of all $\mathbf{X}$-supplemented chief factors of G is an M-set in the modular lattice $\mathcal{N}$ of all normal subgroups of G if and only if $\mathbf{X}$ is JH-solid.

2. For JH-solid sets containing some maximal subgroups of type 3, a converse of Theorem 1.2.34, giving an equivalence analogous to Theorem 1.2.36, does not hold.

Let T be a non-abelian simple group and consider the group G which is the direct product of three copies of T: $G = T_1 \times T_2 \times T_3$. Suppose that $\mathbf{X}$ is the set whose elements are three monolithic maximal subgroups M_1, M_2, and M_3, such that $(M_i)_G = T_j \times T_k$, where $\{i, j, k\} = \{1, 2, 3\}$. Consider the subgroups $U_1 = \Delta_{23} \times T_1 = \{(x, y, y) : x, y \in T\}$, which is a maximal subgroup of type 3 of G such that $(U_1)_G = T_1$, and $U_2 = \Delta_{13} \times T_2 = \{(x, y, x) : x, y \in T\}$, a maximal subgroup of type 3 of G such that $(U_2)_G = T_2$. The set $\mathbf{X} \cup \{U_1, U_2\}$ is not a JH-solid set of maximal subgroups: the minimal normal subgroup T_3 is supplemented by U_1 and U_2 but no maximal subgroup of $\mathbf{X} \cup \{U_1, U_2\}$ has core $\big((U_1)_G \cap (U_2)_G\big)T_3 = T_3$.

On the other hand, it is easy to see that no chief factor of G is $\mathbf{X}$-Frattini, and that any two G-isomorphic chief factors are supplemented by exactly one element of $\mathbf{X}$, so the conditions of Theorem 1.2.36 (2) hold. In other words, $\mathbf{X}$ is a JH-solid set of maximal subgroups of G.

1.3 Crowns

The concept of crown of a soluble group was introduced in [Gas62]. In this seminal paper, W. Gaschütz analyses the structure of the chief factors of a soluble group G as G-modules. Associated with a G-module $\mathfrak{a}$, there exists a section of the group, called $\mathfrak{a}$-*Kopf*, or crown in English, such that, viewed as a G-module, is completely reducible and homogeneous with a composition series of length the number of complemented chief factors G-isomorphic to $\mathfrak{a}$ in any chief series of G. These crowns are complemented sections of G.

The study of non-soluble chief factors made by J. Lafuente in [Laf84a], and, in particular, the introduction of the concept of G-connected chief factors, allowed him to discover that some sections associated with non-abelian chief factors can be constructed enjoying similar properties to Gaschütz's crowns. This originated the concept of crown of a non-abelian chief factor.

Given a group G, fixing a JH-solid set of maximal subgroups $\mathbf{X}$ of G and restricting ourselves to $\mathbf{X}$-supplemented chief factors, we can presume, after the results of Section 1.2, that most of the known results on crowns hold for the so-called $\mathbf{X}$-crowns. The aim of this section is to present results in this direction.

Let us start with the following observations. Let G be a group and H/K a non-abelian chief factor of G. If there exists a maximal subgroup M of G of type 3 complementing H/K, then the primitive group G/M_G has two minimal normal subgroups, namely HM_G/M_G and C/M_G, where $C = \mathrm{C}_G(H/K)$ and $HM_G \cap C = M_G$. In this case $\mathrm{C}_G^*(H/K) = HC$. By Remark 1.2.8 (2b), there exists a monolithic maximal supplement S of HM_G/M_G such that $S_G = C$. Analogously, since $\mathrm{C}_G(C/M_G) = HM_G$, there exists a monolithic maximal supplement T of C/M_G such that $T_G = HM_G$. This means that, although the sets

$$\mathcal{E}_1 = \{N : C^*/N \text{ is a precrown associated with a chief factor}$$
$$G\text{-connected to } H/K\}$$

$$\mathcal{E}_2 = \{M_G : M \text{ is a maximal subgroup of } G \text{ supplementing}$$
$$\text{a chief factor } G\text{-connected to } H/K\}, \text{and}$$

$$\mathcal{E}_3 = \{M_G : M \text{ is a maximal subgroup of } G \text{ supplementing}$$
$$\text{a chief factor } G\text{-isomorphic to } H/K\}$$

in general are different, in fact

$$\bigcap\{N : N \in \mathcal{E}_1\} = \bigcap\{N : N \in \mathcal{E}_2\} = \bigcap\{N : N \in \mathcal{E}_3\}.$$

If we replace the set of all maximal subgroups for a proper JH-solid subset, the above equalities are not longer true.

Let G be a primitive group of type 3 with minimal normal subgroups A and B. If M and S are monolithic maximal subgroups with $M_G = A$ and $S_G = B$, and $\mathbf{X} = \{M, S\}$, then $\mathbf{X}$ is JH-solid and

$$\mathcal{E}_4 = \{M_G : M \text{ is a maximal subgroup in } \mathbf{X}$$
$$\text{supplementing a chief factor } G\text{-connected to } A\}$$
$$= \{A, B\}$$

and

$$\mathcal{E}_5 = \{M_G : M \text{ is a maximal subgroup in } \mathbf{X}$$
$$\text{supplementing a chief factor } G\text{-isomorphic to } A\}$$
$$= \{B\}.$$

Then

$$\bigcap\{N : N \in \mathcal{E}_4\} = 1 < B = \bigcap\{N : N \in \mathcal{E}_5\}.$$

These observations motivate the following definitions.

Definitions 1.3.1. 1. *Let H/K be a supplemented chief factor of a group G and consider the set $\mathcal{E}$ composed of all cores of the monolithic maximal subgroups of G which supplement chief factors G-connected to H/K. Write $R = \bigcap\{N : N \in \mathcal{E}\}$ and $C^* = \mathrm{C}_G^*(H/K)$. Then we say that the factor C^*/R is the* crown *of G associated with H/K.*

2. *Let* $\mathbf{X}$ *be a JH-solid set of monolithic maximal subgroups of a group* G *and* H/K *an* $\mathbf{X}$-*supplemented chief factor of* G. *Write* $C^* = C_G^*(H/K)$ *and consider the normal subgroup*

$$R_{\mathbf{X}} = \bigcap\{M_G : M \in \mathbf{X} \text{ and } M \text{ supplements a chief factor}$$

$$G\text{-connected to } H/K\}.$$

Then $C^*/R_{\mathbf{X}}$ *is the* $\mathbf{X}$-*crown of* G *associated with* H/K.

Obviously a crown of G associated with a supplemented chief factor of G is just an $\mathbf{X}$-crown of G for the set $\mathbf{X} = \text{Max}^*(G)$ of all monolithic maximal subgroups of G.

Theorem 1.3.2. *Let* $\mathbf{X}$ *be a JH-solid set of monolithic maximal subgroups of a group* G *and* H/K *an* $\mathbf{X}$-*supplemented chief factor of* G. *Write* $C^*/R_{\mathbf{X}}$ *for the* $\mathbf{X}$-*crown of* G *associated with* H/K. *Then*

$$C^*/R_{\mathbf{X}} = \text{Soc}(G/R_{\mathbf{X}}).$$

Furthermore

1. *every minimal normal subgroup of* $G/R_{\mathbf{X}}$ *is an* $\mathbf{X}$-*supplemented chief factor of* G *which is* G-*connected to* H/K, *and*
2. *no* $\mathbf{X}$-*supplemented chief factor of* G *over* C^* *or below* $R_{\mathbf{X}}$ *is* G-*connected to* H/K.

In other words, there exist m *normal subgroups* $A_1, \ldots, A_m$ *of* G *such that*

$$C^*/R_{\mathbf{X}} = A_1/R_{\mathbf{X}} \times \cdots \times A_m/R_{\mathbf{X}}$$

where $A_i/R_{\mathbf{X}}$ *is an* $\mathbf{X}$-*supplemented chief factor* G-*connected to* H/K, *for* $i = 1, \ldots, m$, *and* m *is the number of* $\mathbf{X}$-*supplemented chief factors* G-*connected to* H/K *in each chief series of* G. *Moreover*, $\Phi(G/R_{\mathbf{X}}) = O_{q'}(G/R_{\mathbf{X}}) = 1$, *for each prime* q *dividing the order of* $|H/K|$.

Proof. We can write $R_{\mathbf{X}} = R = N_1 \cap \cdots \cap N_r$, such that C^*/N_i are $\mathbf{X}$-precrowns associated with chief factors G-connected to H/K and r is minimal with this property. Consider the group monomorphism

$$\psi \colon C^*/R = C^*/(N_1 \cap \cdots \cap N_r) \longrightarrow C^*/N_1 \times \cdots \times C^*/N_r$$
$$c(N_1 \cap \cdots \cap N_r) \longmapsto (cN_1, \ldots, cN_r)$$

for any $c \in C^*$. Observe that ψ is compatible with the action of G:

$$\big(c(N_1 \cap \cdots \cap N_r)^{\psi}\big)^g = (cN_1, \ldots, cN_r)^g = (c^g N_1, \ldots, c^g N_r)$$
$$= \big(c^g(N_1 \cap \cdots \cap N_r)\big)^{\psi}.$$

From minimality of r, we have that $C^* = N_i(N_1 \cap \cdots \cap N_{i-1})$, for $i \le r$, and then

$$(N_1 \cap \cdots \cap N_{i-1})/(N_1 \cap \cdots \cap N_i) \cong_G C^*/N_i.$$

Therefore the chain

$$R = (N_1 \cap \cdots \cap N_r) \le (N_1 \cap \cdots \cap N_{r-1}) \le \cdots \le N_1 \le C^*$$

is a piece of chief series of G and each chief factor is G-connected to H/K.

Hence the order $|C^*/R| = |H/K|^r$ and ψ is an isomorphism. By Corollary 1.2.37, every chief factor of G between R and C^* is $\mathbf{X}$-supplemented in G. Therefore, there exist r normal subgroups $A_1, \ldots, A_r$ of G such that

$$C^*/R = A_1/R \times \cdots \times A_r/R,$$

where A_i/R is a $\mathbf{X}$-supplemented chief factor G-connected to H/K, $i = 1, \ldots, r$.

Suppose that H_0/K_0 is a $\mathbf{X}$-supplemented chief factor of G which is G-connected to H/K and let $M \in \mathbf{X}$ be a supplement of H_0/K_0 in G. Then $H_0 \le C^*$. Observe that since $R \le M_G$, then $H_0 \not\le R$. Therefore no $\mathbf{X}$-supplemented chief factor of G over C^* or below R is G-connected to H/K.

By Theorem 1.2.36, the number of $\mathbf{X}$-supplemented chief factors G-connected to H/K in each chief series of G is an invariant of the group and coincides with the length of any piece of chief series of G between R and C^*.

If B/R is a minimal normal subgroup of G/R and $B \cap C^* = R$, then $B \le C_G(A_1/R)$ which is contained in C^* by Proposition 1.2.15. This contradiction implies that $C^*/R = \operatorname{Soc}(G/R)$. Since every minimal normal subgroup of G/R is supplemented in G/R, we have that $\Phi(G/R) = 1 = O_{q'}(G/R)$, for each prime q dividing the order of $|H/K|$. $\square$

Corollary 1.3.3 ([Laf84a]). *Two supplemented chief factors of a group G define the same crown of G if and only if they are G-connected.*

Let C^*/R be the $\mathbf{X}$-crown of G associated with an $\mathbf{X}$-supplemented chief factor H/K. Applying Theorem 1.3.2, we have that $C^*/R = (R_{\mathbf{X}}/R) \times (C_0/R)$, and the $\mathbf{X}$-crown of G associated to H/K is isomorphic to C_0/R which is a direct product of $\mathbf{X}$-supplemented components of C^*/R.

Corollary 1.3.4. *Let $\mathbf{X}$ be a JH-solid set of monolithic maximal subgroups of a group G. Let H/K be an $\mathbf{X}$-supplemented chief factor of a group G and write C^*/R for the $\mathbf{X}$-crown of G associated with H/K. Then*

1. if H/K is abelian and p is the prime dividing $|H/K|$, then $C^* = C_G(H/K) = C$ and

$$C/R = \operatorname{Soc}(G/R) = \operatorname{F}(G/R) = O_p(G/R)$$

is a completely reducible and homogeneous G-module over $\operatorname{GF}(p)$ whose composition factors are G-isomorphic to H/K and the length of a composition series of C/R, as G-module, is the number of $\mathbf{X}$-complemented G-chief factors G-isomorphic to H/K in each chief series of G;

2. *if H/K is non-abelian, then $\{A_j/R : j = 1, \ldots, m\}$ is the set of all minimal normal subgroups of G/R; in particular, if C^*/R is a chief factor of G, then $R = \mathrm{C}_G(H/K)$ and $G/R \cong [H/K] * G$ is a primitive group of type 2.*

Proof. Applying Theorem 1.3.2, $C^*/R = Soc(G/R)$.

1. If H/K is abelian, then H/K is a p-group for some prime p and $C^* = C = \mathrm{C}_G(H/K)$ is the common centraliser of the chief factors of G between C and R. Then $\mathrm{C}_{G/R}(C/R) = C/R = F(G/R)$ and Statement 1 follows from Theorem 1.3.2.

2. Suppose now that H/K is non-abelian. Then $\{A_j/R : j = 1, \ldots, m\}$, as in Theorem 1.3.2, are the minimal normal subgroups of G/R. Finally observe that if C^*/R is a chief factor, then C^*/R is the **X**-precrown of G associated with H/K and $R = \mathrm{C}_G(H/K)$.   □

Our main goal is now to prove that in every group G, we can order in some sense the **X**-crowns of G to obtain a chief series of G in which some G-isomorphic images of the **X**-crowns are placed one after the other, possibly separated by **X**-Frattini chief factors, and all **X**-supplemented chief factors which are G-connected are consecutive.

We need a technical proposition to explore how the crowns of the quotient group are related to the crowns of the original group. We will use it in inductive arguments.

Proposition 1.3.5. *Let* **X** *be a JH-solid set of monolithic maximal subgroups of a group G and let N be a normal subgroup of G contained in some maximal subgroup of G in* **X**.

1. *For any **X**-crown C^*/R of G, either*
    a) $C^* \leq RN$ or
    *b) $RN < C^*$ and $(C^*/N)/(RN/N)$ is an **X**$/N$-crown of G/N.*
2. *For any **X**$/N$-crown $(C_0^*/N)/(R_0/N)$ of G/N, there is an **X**-crown C^*/R of G such that $C_0^* = C^*$ and $R_0 = RN$.*

Proof.   1. Assume that C^* is not contained in RN. Then, applying Corollary 1.3.4, there exists a minimal normal subgroup A/R of G/R such that A is not contained in RN. Therefore AN/RN is a chief factor of G which is G-isomorphic to A/R. Hence $RN < AN \leq \mathrm{C}_G^*(AN/RN) = C^*$. Applying Theorem 1.3.2, AN/RN is **X**-supplemented and clearly $(C^*/N)/(NR/N)$ is the **X**$/N$-crown of G/N associated with the chief factor $AN/N/RN/N$ of G/N.

2. Let $(C_0^*/N)/(R_0/N)$ be the **X**$/N$-crown of G/N associated with an **X**$/N$-supplemented chief factor $(H/N)/(K/N)$ of G/N. Then $(H/N)/(K/N)$ is G-isomorphic to the chief factor H/K of G and H/K is **X**-supplemented in G. Consider the **X**-crown C^*/R of G associated with H/K. It follows that $C_0^*/N = \mathrm{C}_{G/N}^*\big((H/N)/(K/N)\big) = \mathrm{C}_G^*(H/K)/N$ and then $C_0 = C^*$.

On the other hand, it is clear that $RN \leq R_0$. In addition, every chief factor of a given chief series of G between RN and R_0 is **X**-supplemented in G and G-connected to H/K. Since, by Theorem 1.3.2, the number of **X**/N-supplemented chief factors of each chief series of G/N which are G/N-connected to $(H/N)/(K/N)$ is exactly the number of chief factors of G/N between R_0/N and C^*/N, we have that $RN = R_0$. $\qquad\square$

Lemma 1.3.6. *Let G be a group with $\Phi(G) = 1$. There exists a crown C^*/R and a non-trivial normal subgroup D of G such that $C^* = R \times D$.*

Proof. We argue by induction on the order of G. Let M be a minimal normal subgroup of G. Since $\Phi(G) = 1$, it follows that M is supplemented in G and we can consider the crown C_0^*/R_0 and a precrown C_0^*/N_0 associated with M in G. We know that $C_0^* = N_0 \times M$.

If $N_0 = R_0$, then the normal subgroup $D = M$ and the crown $C^*/R = C_0^*/R_0$ fulfils our requirements.

Assume that $R_0 < N_0$. This means that $R_0 \times M < C_0^*$. Write $F/M = \Phi(G/M)$. By Proposition 1.3.5, $(C_0^*/M)/(R_0M/M)$ is a crown of G/M associated with the chief factors of G/M, i.e. the chief factors of G over M, which are G-connected to M. Since, by Theorem 1.3.2, $\Phi\bigl((G/M)/(R_0M/M)\bigr) = 1$, we have that $F \leq R_0M$ and then $F = M \times (F \cap R_0)$. Put $N = F \cap R_0$. Suppose that $N \neq 1$, and let A be a minimal normal subgroup of G contained in N. Recall that all monolithic maximal subgroups of G form a JH-solid set and their intersection is $\Phi(G)$. Since obviously $MA \cap \Phi(G) = 1$, we can apply Corollary 1.2.38 and deduce that the chief factor MA/M is supplemented in G. But this contradicts the fact that $MA/M \leq F/M = \Phi(G/M)$. Therefore $F = M$ and $\Phi(G/M) = 1$. By induction, there exists a crown $(C_1^*/M)/(R_1/M)$ and a non-trivial normal subgroup D_1/M of G/M, such that $C_1^*/M = (R_1/M) \times (D_1/M)$.

Suppose first that $(C_1^*/M)/(R_1/M)$ is the crown associated with the chief factors G-connected to M. Then $C_1^* = C_0^*$ and $R_1 = R_0 \times M$. In this case, we take $D = D_1$ and $C^*/R = C_0^*/R_0$. Note that $M = D_1 \cap R_1 = D_1 \cap (R_0 \times M) = (D_1 \cap R_0) \times M$. Hence $D_1 \cap R_0 = 1$.

Suppose now that the chief factors of G between M and D_1 are not G-connected to M. If $C_0^*/M \leq (R_0M/M)(D_1/M)$, then $C_0^* = R_0(C_0^* \cap D_1)$. Then $C_0^*/R_0 \cong_G (C_0^* \cap D_1)/(R_0 \cap D_1)$ and $M \leq C_0^* \cap D_1$. Hence all chief factors of G between $(R_0 \cap D_1) \times M$ and $C_0^* \cap D_1$ are G-connected to M by Theorem 1.3.2. Since no chief factor of G between M and D_1 is G-connected to M, we deduce that $C_0^* \cap D_1 = (R_0 \cap D_1) \times M$. Then $C_0^* = R_0M$, against our assumption. Hence, by Proposition 1.3.5, we have that $(R_0M/M)(D_1/M) < C_0^*/M$ and then $R_0 \leq R_0M \leq R_0D_1 \leq C_0^*$. Applying Theorem 1.3.2, every chief factor of G between R_0M and R_0D_1 is G-connected to M. Since $D_1R_0/MR_0 \cong_G D_1/M(D_1 \cap R_0)$ and we are assuming that all chief factors of G between M and D_1 are not G-connected to M, we have that $D_1 = M(D_1 \cap R_0)$. In this case, take $D = D_1 \cap R_0 \neq 1$ and $C^* = C_1^*$. This completes the proof. $\qquad\square$

We prove now the corresponding result for a JH-solid set $\mathbf{X}$ of monolithic maximal subgroups of a group G.

Proposition 1.3.7. *Let G be a group and $\mathbf{X}$ a JH-solid set of monolithic maximal subgroups of G such that $\Phi_{\mathbf{X}}(G) = 1$. There exists an $\mathbf{X}$-crown $C^*/R_{\mathbf{X}}$ of G and a non-trivial normal subgroup D of G such that $C^* = R_{\mathbf{X}} \times D$.*

Proof. Observe first that $\Phi(G) \leq \Phi_{\mathbf{X}}(G) = 1$. By Lemma 1.3.6, there exists a crown C^*/R and a non-trivial normal subgroup D of G such that $C^* = R \times D$. Consider the G-isomorphism $\varphi \colon C^*/R \longrightarrow D$. If $C^*/R = (A_1/R) \times \cdots \times (A_r/R)$, then all the images $(A_i/R)^{\varphi} = N_i$ are minimal normal subgroups of G below D, the N_i are G-connected, C^*/R is the crown of G associated with them and $D = N_1 \times \cdots \times N_r$. Moreover, by Theorem 1.2.38, every chief factor of G below D is $\mathbf{X}$-supplemented in G. Hence $R = R_{\mathbf{X}}$ and $C^*/R = C^*/R_{\mathbf{X}}$ is the $\mathbf{X}$-crown of G associated with the N_i. $\qquad\square$

Theorem 1.3.8 (see [För88]). *Let $\mathbf{X}$ be a non-empty JH-solid set of monolithic maximal subgroups of a group G and.*

1. *Let $C_1^*/R_1, \ldots, C_n^*/R_n$ denote the $\mathbf{X}$-crowns of G. Then there exists a permutation $\sigma \in \mathrm{Sym}(n)$ and a chain of normal subgroups of G*

$$1 = C_{(0)} \leq R_{(1)} < C_{(1)} \leq R_{(2)} < C_{(2)} \leq \cdots < C_{(n-1)} \leq R_{(n)} < C_{(n)} \leq G$$

such that $G/C_{(n)} = \Phi_{\mathbf{X}/C_{(n)}}(G/C_{(n)})$ (including the case $G = C_{(n)}$) and for $i = 1, \ldots, n$, we have

$$R_{(i)}/C_{(i-1)} = \Phi_{\mathbf{X}/C_{(i-1)}}(G/C_{(i-1)}), \quad C_{i^\sigma}^* = R_{i^\sigma}C_{(i)}, \quad R_{i^\sigma} \cap C_{(i)} = R_{(i)}.$$

2. *Moreover, if N is a normal subgroup of G and $C_{(k-1)} \leq N \leq R_{(k)}$, for some $k \in \{1, \ldots, n\}$,*

$$1 = N/N = C_{(k-1)}N/N \leq R_{(k)}/N < C_{(k)}/N$$
$$\leq R_{(k+1)}/N < \cdots < C_{(n)}/N \leq G/N$$

is a chain of G/N enjoying the corresponding property.

Proof. 1. We use induction on $|G|$. Clearly $\Phi_{\mathbf{X}}(G)$ is contained in each R_i. Moreover, every $\mathbf{X}$-supplemented chief factor of G is G-isomorphic to an $\mathbf{X}/\Phi_{\mathbf{X}}(G)$-supplemented chief factor of $G/\Phi_{\mathbf{X}}(G)$. Hence, by Proposition 1.3.5 (2), we can assume without loss of generality that $\Phi_{\mathbf{X}}(G) = R_{(1)} = 1$.

By Proposition 1.3.7, there exists an $\mathbf{X}$-crown C_k^*/R_k of G and a normal subgroup $C_{(1)}$ of G such that $C_k^* = R_k \times C_{(1)}$. If $G = C_{(1)}$, the result is trivial. If $C_{(1)}$ is a proper subgroup of G and $\mathbf{X}/C_{(1)} = \emptyset$, then $\Phi_{\mathbf{X}/C_{(1)}}(G/C_{(1)}) = G/C_{(1)}$ or, in other words, no maximal subgroup of G in $\mathbf{X}$ contains $C_{(1)}$. Hence no chief factor of G over $C_{(1)}$ is $\mathbf{X}$-supplemented.

In this case there exists exactly one $\mathbf{X}$-crown of G and the theorem holds trivially. Assume that $\bar{\mathbf{X}} = \mathbf{X}/C_{(1)}$ is non-empty, i.e. $C_{(1)}$ is contained in some maximal subgroup of G in $\mathbf{X}$. Then we can apply the inductive hypothesis to the quotient group $\bar{G} = G/C_{(1)}$. Observe that if $C_j^* \le R_j C_{(1)}$, for some $j \ne k$, then $C_j^*/R_j \cong_G (C_j^* \cap C_{(1)})/(R_j \cap C_{(1)})$ and the chief factors of G between C_j^* and R_j are G-connected to some chief factors of G below $C_{(1)}$ and therefore to the chief factors in C_k^*/R_k, which is not possible by Theorem 1.3.2. Hence, by Proposition 1.3.5, $R_j C_{(1)} < C_j^*$, for all $j \in \{1, \dots, n\} \setminus \{k\}$. Therefore $\{\bar{C}_j^*/\bar{R}_j : j \ne k\}$ are the $\mathbf{X}$-crowns of $\bar{G}$ and, by induction, there exists a bijection $\tau \colon \{2, \dots, n\} \longrightarrow \{1, \dots, n\} \setminus \{k\}$, and a chain of normal subgroups of $\bar{G}$

$$1 = \bar{C}_{(1)} \le \bar{R}_{(2)} < \bar{C}_{(2)} \le \bar{R}_{(3)} < \bar{C}_{(3)} \le \cdots < \bar{C}_{(n-1)} \le \bar{R}_{(n)} < \bar{C}_{(n)} \le \bar{G}$$

such that $\bar{G}/\bar{C}_{(n)} = \Phi_{\bar{\mathbf{X}}/\bar{C}_{(n)}}(\bar{G}/\bar{C}_{(n)})$, and for $i = 2, \dots, n+1$, we have

$$\bar{R}_{(i)}/\bar{C}_{(i-1)} = \Phi_{\mathbf{X}/\bar{C}_{(i-1)}}(\bar{G}/\bar{C}_{(i-1)}), \qquad \bar{C}_{i^\tau}^* = \bar{R}_{i^\tau}\bar{C}_{(i)}, \qquad \bar{R}_{i^\tau} \cap \bar{C}_{(i)} = \bar{R}_{(i)}.$$

Now, just take the inverse images $R_{(j)}/C_{(1)} = \bar{R}_{(j)}$ and $C_{(j)}/C_{(1)} = \bar{C}_{(j)}$, for $j = 2, \dots, n$. The required permutation is σ such that $1^\sigma = k$ and $i^\sigma = i^\tau$, for $i = 2, \dots, n$.

2. Assume that N is a normal subgroup of G such that $C_{(k-1)} \le N \le R_{(k)}$. Every $\mathbf{X}$-supplemented chief factor H/K of G such that $N \le K$ is G-isomorphic to an $\mathbf{X}/N$-supplemented chief factor of G/N and therefore is G-connected to some chief factor between $R_{(j)}/N$ and $C_{(j)}/N$, for some $j \ge k$. The $\mathbf{X}/N$-crown of G/N associated with $(H/N)/(K/N)$ is $(C_{j^\sigma}^*/N)/(R_{j^\sigma}/N)$ and clearly we have that $C_{j^\sigma}^*/N = (R_{j^\sigma}/N)(C_{(j)}/N)$ and $(R_{j^\sigma}/N) \cap (C_{(j)}/N) = R_{(j)}/N$. In addition, $R_{(i)}/C_{(i-1)}$ is equal to $\Phi_{\mathbf{X}/C_{(i-1)}}(G/C_{(i-1)})$. Hence

$$(R_{(i)}/N)/(C_{(i-1)}/N) = \Phi_{(\mathbf{X}/N)/(C_{(i-1)}/N)}\big((G/N)/(C_{(i-1)}/N)\big)$$

for all $i = k+1, \dots, n$. Now $R_{(k)}/C_{(k-1)} = \Phi_{\mathbf{X}/C_{(k-1)}}(G/C_{(k-1)})$ implies that $\Phi_{\mathbf{X}/N}(G/N) = R_{(k)}/N$. $\qquad\qquad\square$

Now, the result we were looking for becomes clear.

Corollary 1.3.9 (see [Gas62] and [För88]). *Let $\mathbf{X}$ be a JH-solid set of monolithic maximal subgroups of a group G. If $C_1^*/R_1, \dots, C_n^*/R_n$ are the $\mathbf{X}$-crowns of G, there exists a permutation $\sigma \in \mathrm{Sym}(n)$ and a chief series of G*

$$1 = F_{1,0} < F_{1,1} < \cdots < F_{1,m_1} = N_{1,0} < N_{1,1} < \cdots < N_{1,k_1}$$
$$= F_{2,0} < F_{2,1} < \cdots < F_{2,m_2} = N_{2,0} < N_{2,1} < \cdots < N_{2,k_2}$$
$$\cdots$$
$$= F_{n,0} < F_{n,1} < \cdots < F_{n,m_n} = N_{n,0} < N_{n,1} < \cdots < N_{n,k_n}$$
$$= F_{n+1,0} < F_{n+1,1} < \cdots < F_{n+1,m_{n+1}} = G$$

such that

1. *the $F_{i,j}/F_{i,j-1}$ are $\mathbf{X}$-Frattini chief factors of G,*
2. *the $N_{i,j}/N_{i,j-1}$ are $\mathbf{X}$-supplemented chief factors of G satisfying that $N_{i,j}/N_{i,j-1}$ is G-connected to $N_{i',j'}/N_{i',j'-1}$ if and only if $i = i'$; moreover $C_{i^\sigma}/R_{i^\sigma}$ is the $\mathbf{X}$-crown associated with $N_{i,j}/N_{i,j-1}$;*
3. *$F_{i,m_i} / F_{i,j} = \Phi_{\mathbf{X}/F_{i,j}} (G/F_{i,j})$, for each $i = 1, \ldots, n+1$ and $j = 1, \ldots, m_i - 1$.*

Let $\mathbf{X}$ be a JH-solid set of monolithic maximal subgroups of a group G. Then if $C^*/R_{\mathbf{X}}$ is the $\mathbf{X}$-crown of G associated with a chief factor H/K, and $R_{\mathbf{X}} = G_0 < G_1 < \cdots < G_n = C^*$ is a piece of chief series of G, then the subgroup

$$V = \bigcap_{i=1}^{n} \{M_i : M_i \text{ is an } \mathbf{X}\text{-supplement of } G_i/G_{i-1}\}.$$

is a supplement (if H/K is abelian, then V is a complement) of $C^*/R_{\mathbf{X}}$ in G, by repeated applications of Lemma 1.2.16 (2). However, this supplement depends on the choice of the chief series and on the choice of the maximal subgroups and it is not preserved by epimorphic images. The following example is illustrative of these problems.

Example 1.3.10. Denote by N the elementary abelian group of order 3^2. The cyclic group Z of order 2 acts on N by inversion. Form the semidirect product $G = [N]Z$ and write $A = \langle a \rangle$, $N = \langle a, b \rangle$, and $Z = \langle z \rangle$. Consider the JH-solid set of maximal subgroups $\mathbf{X} = \{M_1 = \langle a, z \rangle, M_2 = \langle b, az \rangle, M_3 = \langle ab, z \rangle, M_4 = \langle a^2 b, z \rangle\}$. The $\mathbf{X}$-crown of G associated with any of the chief factors below N is $N = C_G(A)$. All subgroups of the form $V_{ij} = M_i \cap M_j$, $i \neq j$, are complements of N in G. Note that $\bigcap_{i=1}^{4} M_i = 1$.

Consider now the group G/A. Observe that $\mathbf{X}/A = \{M_1/A\}$ and the $\mathbf{X}/A$-crown of G/A associated with N/A is N/A itself. Notice that the subgroup $V_{23}A/A = \langle a, bz \rangle/A$ is a complementchief factor!complemented of N/A in G/A which does not belong to $\mathbf{X}/A$.

Proposition 1.3.11. *Let G be a group and $\mathbf{X}$ a JH-solid set of monolithic maximal subgroups of G. Assume that if U and S are two distinct elements of $\mathbf{X}$, then $U_G \neq S_G$. Let $C^*/R_{\mathbf{X}}$ be the $\mathbf{X}$-crown of G associated with the $\mathbf{X}$-supplemented chief factor F. Consider the set*

$$\mathbf{X}_F = \{M \in \mathbf{X} : M \text{ supplements a chief factor } G\text{-connected to } F\}.$$

We define the subgroup $T = \mathrm{T}(G, \mathbf{X}, F) = \bigcap\{M : M \in \mathbf{X}_F\}$. Clearly $T_G = R_{\mathbf{X}}$.

1. *Assume that if U and S are two distinct elements of $\mathbf{X}$ and both supplement a chief factor H/K of G, then $M = (U \cap S)H \in \mathbf{X}$. Then the subgroup T satisfies the following properties.*

a) *For any piece of chief series of G, $R_{\mathbf{X}} = G_0 < G_1 < \cdots < G_n = C^*$ and any family $\{M_i \in \mathbf{X} : i = 1, \ldots, n\}$ such that M_i is a supplement of G_i/G_{i-1}, for each $i = 1, \ldots, n$, we have*

$$T(G, \mathbf{X}, F) = \bigcap_{i=1}^{n} M_i$$

and $T(G, \mathbf{X}, F)$ is a supplement (a complement, if F is abelian) of $C^*/R_{\mathbf{X}}$ in G.

b) *For any normal subgroup N of G such that F is G-connected with an $\mathbf{X}/N$-supplemented chief factor F_1 of G/N, then $T(G/N, \mathbf{X}/N, F_1) = TN/N$.*

2. *Conversely, assume that the subgroup T satisfies the above Conditions 1a and 1b. Then, if U and S are elements of $\mathbf{X}_F$ such that $U_G \neq S_G$, and both supplement a chief factor H/K of G, then $M = (U \cap S)H \in \mathbf{X}_F$.*

Proof. 1. a) Fix a piece of chief series of G, $R_{\mathbf{X}} = G_0 < G_1 < \cdots < G_n = C^*$ and a family $\{M_i \in \mathbf{X} : i = 1, \ldots, n\}$ such that M_i is a supplement of G_i/G_{i-1}, for each $i = 1, \ldots, n$ and write $D = \bigcap_{i=1}^{n} M_i$. If $\mathbf{X}_F = \{M_i \in \mathbf{X} : i = 1, \ldots, n\}$, then there is nothing to prove.

Assume that there exists $U \in \mathbf{X}_F \setminus \{M_i \in \mathbf{X} : i = 1, \ldots, n\}$. Then U supplements G_j/G_{j-1}, for some $j = 1, \ldots, n$. Since U and M_j are distinct monolithic $\mathbf{X}_F$-supplements of the same chief factor G_j/G_{j-1} and $U_G \neq M_{jG}$, we have that G_j/G_{j-1} is abelian by Lemma 1.2.17 (2), and so is $C^*/R_{\mathbf{X}}$. Therefore U and M_j complement G_j/G_{j-1}. By hypothesis, $M = (U \cap M_j)G_j \in \mathbf{X}_F$. Now we have that $M_j \cap M = (M_j \cap U)(M_j \cap G_j) = (M_j \cap U)G_{j-1} = M_j \cap U$ and analogously $U \cap M = M_j \cap U$. Then $D \cap U = D \cap M$. Observe that M complements a chief factor G_k/G_{k-1}, for some $k > j$. If $M = M_k$, then $D \cap U = D$. If $M \neq M_k$, repeat the previous argument replacing U by M and M_j by M_k. Observe also that G_n/G_{n-1} is self-centralising in G/G_{n-1} and so the latter group is primitive. Hence $(M_n)_G = G_{n-1}$. Therefore M_n is the unique maximal subgroup of G in $\in \mathbf{X}_F$ complementing the last chief factor. Since the other possible maximal subgroups in $\mathbf{X}_F$ do not change the intersection, it follows that $T(G, \mathbf{X}, F) = \bigcap_{i=1}^{n} M_i$.

Moreover, if we apply repeatedly Lemma 1.2.16 (2), we deduce that the subgroup $T = \bigcap_{i=1}^{n} M_i$ is a supplement (complement if the crown is abelian) of $C^*/R_{\mathbf{X}}$ in G.

b) Let N be a minimal normal subgroup of G such that G/N has an $\mathbf{X}/N$-supplemented chief factor F_1 which is G-connected to F. By Proposition 1.3.5, we have that $R_{\mathbf{X}}N < C^*$ and $(C^*/N)/(R_{\mathbf{X}}N/N)$ is the $\mathbf{X}/N$-crown of G/N associated with F_1. If $N \leq R_{\mathbf{X}}$, it is clear that $T/N = T(G/N, \mathbf{X}/N, F_1)$. Assume that N is G-connected to F, i.e. $R_{\mathbf{X}} < R_{\mathbf{X}}N$. We consider a piece of chief series of G

$$R_{\mathbf{X}} = G_0 < G_1 = R_{\mathbf{X}}N < G_2 \cdots < G_n = C^*.$$

By Statement 1a we have that

$$T = \mathrm{T}(G, \mathbf{X}, F) = \bigcap_{i=1}^{n} \{M_i : M_i \text{ is an } \mathbf{X}\text{-supplement of } G_i/G_{i-1}\}.$$

Since $N \leq \bigcap_{i=2}^{n} M_i$ and $G = M_1 N$, we have that

$$TN = \left(\bigcap_{i=1}^{n} M_i\right) N = M_1 N \cap \left(\bigcap_{i=2}^{n} M_i\right) = \bigcap_{i=2}^{n} M_i,$$

and

$$TN/N = \bigcap_{i=2}^{n} (M_i/N) = \mathrm{T}(G/N, \mathbf{X}/N, F).$$

An inductive argument proves the validity of the Statement 1b for any normal subgroup N of G such that F is G-connected with a chief factor of G/N.

2. Assume that the subgroup T satisfies Statement 1a and Statement 1b and suppose that U and S are elements of $\mathbf{X}_F$ such that $U_G \neq S_G$, and both supplement the same chief factor H/K of G. Since U and S are monolithic and $U_G \neq S_G$, H/K is abelian by Lemma 1.2.17 (2).

Observe that $C^* = C = \mathrm{C}_G(H/K) = HU_G = HS_G$ and $K = U_G \cap H = S_G \cap H$. Suppose that $R_\mathbf{X} < U_G \cap S_G$. Let $N/R_\mathbf{X}$ be a chief factor of G such that $N \leq U_G \cap S_G$. It is clear that $F_2 = (HN/N)/(KN/N)$ is a chief factor of G/N and the $\mathbf{X}/N$-crown of G/N associated with F_2 is C/N. We see that in the group G/N all hypotheses hold for $\mathbf{X}/N$ and $\mathrm{T}(G/N, \mathbf{X}/N, F_2) = TN/N$. To see that TN/N satisfies Statement 1a, let $1 = N/N = G_1/N < \cdots < G_n/N = C/N$ be a piece of chief series of G/N and M_i/N an $\mathbf{X}/N$-complement of $(G_i/N)/(G_{i-1}/N)$ for $i = 2, \ldots, n$. Let M_1 be an $\mathbf{X}$-complement of $N/R_\mathbf{X}$. Then $R_\mathbf{X} < N = G_1 < \cdots < G_n = C$ is a piece of chief series of G and M_i is an $\mathbf{X}$-complement of G_i/G_{i-1}, for $i = 1, \ldots, n$. Since G satisfies Statement 1a, we have that $T = \bigcap_{i=1}^{n} M_i$ and then $TN = M_1 N \cap (\bigcap_{i=2}^{n} M_i) = \bigcap_{i=2}^{n} M_i$. Since T satisfies Statement 1b, we have $TN/N = \bigcap_{i=2}^{n} (M_i/N)$ and TN/N satisfies Statement 1a. Clearly, TN/N satisfies Statement 1b.

Arguing by induction, we have that the maximal subgroup

$$M/N = \big((U/N) \cap (S/N)\big)(HN/N) = \big((U \cap S)H\big)/N \in \mathbf{X}_F/N$$

and then $M = (U \cap S)H \in \mathbf{X}_F$. Hence, we can assume that $U_G \cap S_G = R_\mathbf{X} = 1$. This implies that $K = 1$ and H is a minimal normal subgroup of G. Observe that also U_G and S_G are minimal normal subgroups of G. We can consider these three different pieces of chief series of G below C:

$$1 < H < C \qquad 1 < U_G < C \qquad 1 < S_G < C.$$

By Statement 1a, applied to the second or the third piece of chief series, we have that $T = U \cap S = \bigcap\{M : M \in \mathbf{X}_F\}$. In other words, for all $M \in \mathbf{X}_F$, we have that $U \cap S \leq M$. Since $\mathbf{X}$ is JH-solid, the number of $\mathbf{X}$-complemented chief factors of G which are G-isomorphic to H is the same in any chief series by Theorem 1.2.36. Hence, there exists an $\mathbf{X}$-complement of C/H in G. If M is such a complement, then $H \leq M$. Therefore $(U \cap S)H \leq M$. But $(U \cap S)H$ is a maximal subgroup of G, by Lemma 1.2.17. Therefore $M = (U \cap S)H \in \mathbf{X}_F$.

$\square$

1.4 Systems of maximal subgroups

JH-solid sets of monolithic maximal subgroups are characterised by their excellent adequacy to the Jordan-Hölder correspondence, as we saw in Theorem 1.2.36, but are not strong enough to fulfil some expected properties when working with supplements of $\mathbf{X}$-crowns. A supplement of a particular $\mathbf{X}$-crown C^*/R of a group G is obtained by the intersection of an $\mathbf{X}$-supplement of each chief factor in a piece of chief series of G passing through R and C^*, applying repeatedly Lemma 1.2.16. If we want these supplements of $\mathbf{X}$-crowns to be preserved by epimorphic images and to be independent of the choice of the chief series and of the choice of maximal subgroups, the JH-solid set of monolithic maximal subgroups $\mathbf{X}$ have to satisfy some rather stronger conditions characterised in Proposition 1.3.11. A *subsystem of maximal subgroups* of a group G is in fact a JH-solid set of monolithic maximal subgroups of G, with different cores, and satisfying the properties stated in Proposition 1.3.11.

Why are we interested in supplements of $\mathbf{X}$-crowns? The answer will be clear in Section 4.3 where the *subgroups of prefrattini type* are introduced. W. Gaschütz constructed his celebrated prefrattini subgroups, in [Gas62], by intersecting complements of (abelian) crowns. Several generalisations of prefrattini subgroups are constructed by intersecting some cleverly chosen maximal subgroups. The key is these "clever" choice of supplements. Within the limits of the soluble groups, maximal subgroups into which a fixed Hall system reduces are used. But the extension of these ideas to a general non necessarily soluble group required of a new arithmetical-free method of choice of maximal subgroups. Subsystems of maximal subgroups are the answer and, supporting this idea, we will show that in a soluble group G, given a system of maximal subgroups $\mathbf{X}$ of G, there exists a Hall system Σ of G such that $\mathbf{X}$ is the set of all maximal subgroups of G into which Σ reduces. Thus, the original method for soluble groups due to Gaschütz is included in our theory.

In this way from soluble to finite, we lose the arithmetical properties. This is no surprising since they characterise solubility. But we find deep relations between maximal subgroups hidden behind the luxuriant Hall theory.

Definition 1.4.1. *Let G be a group. We say that two maximal subgroups U, S of G are* core-related *in G if $U_G = S_G$.*

It is clear that the core-relation is an equivalence relation in the set $\mathrm{Max}(G)$ of all maximal subgroups of a group G.

By Theorem 1.1.10, the core-relation coincides with conjugacy in soluble groups. Moreover, by Lemma 1.2.17 (2), two monolithic maximal subgroups supplementing the same non-abelian chief factor are core-related.

Definitions 1.4.2. *1. Let* $\mathbf{X}$ *be a, possibly empty, set of monolithic maximal subgroups of* G. *We will say that* $\mathbf{X}$ *is a* subsystem of maximal subgroups of G *provided the following two properties are satisfied:*
   a) if $U,\ S \in \mathbf{X}$ *and* $U \neq S$, *then* $U_G \neq S_G$, *and*
   b) if $U,\ S \in \mathbf{X}$, $U \neq S$ *and both complement the same abelian chief factor* H/K *of* G, *then* $M = (U \cap S)H \in \mathbf{X}$.
 2. If a subsystem of maximal subgroups $\mathbf{X}$ *is a complete set of representatives of the core-relation in the set* $\mathrm{Max}^*(G)$ *of all monolithic maximal subgroups of* G, *then we will say that* $\mathbf{X}$ *is a* system of maximal subgroups of G.

Since Condition 1b of the above definition only has an effect on maximal subgroups of type 1, we have that every subset of representatives of the core-relation in the set of maximal subgroups of type 2 is a subsystem of maximal subgroups.

If $\mathbf{X}$ is a subsystem of maximal subgroups of a group G, then $\mathbf{X}$ can be written as the disjoint union set $\mathbf{X} = \mathbf{X}_1 \cup \mathbf{X}_2$, where $\mathbf{X}_k = \{U \in \mathbf{X} : U$ is a maximal subgroup of type $k\}$ for $k = 1,\ 2$. On the other hand, if $F_1, \ldots,\ F_n$ are representatives of the G-isomorphism classes of abelian chief factors of G, then $\mathbf{X}_1$ is a disjoint union set $\mathbf{X}_1 = \bigcup_{i=1}^{n} \mathbf{X}_{F_i}$, for $\mathbf{X}_{F_i} = \{U \in \mathbf{X} : U$ complements a chief factor G-isomorphic to $F_i\}$.

Clearly a subsystem of maximal subgroups is, in particular, a JH-solid set of monolithic maximal subgroups by Lemma 1.2.17.

Let $\mathbf{X}$ be a subsystem of maximal subgroups of a group G. If $g \in G$, denote $\mathbf{X}^g = \{S^g : S \in \mathbf{X}\}$. It is clear that $\mathbf{X}^g$ is again a subsystem of maximal subgroups of G.

We say that two subsystems of maximal subgroups $\mathbf{X}_1$ and $\mathbf{X}_2$ of a group G are *conjugate* in G, if there exists an element $g \in G$ such that $\mathbf{X}_1{}^g = \mathbf{X}_2$.

Proposition 1.4.3. *Let* G *be a group and* φ *an epimorphism of* G. *If* $\mathbf{X}$ *is a subsystem of maximal subgroups of* G, *then the set* $\mathbf{X}^\varphi = \{M^\varphi : \mathrm{Ker}(\varphi) \leq M \in \mathbf{X}\}$ *is a subsystem of maximal subgroups of* G^φ.

   Conversely, if $\mathbf{Y}$ *is a subsystem of maximal subgroups of* G^φ, *then the set* $\mathbf{Y}^{\varphi^{-1}} = \{M \leq G : \mathrm{Ker}(\varphi) \leq M, M/\mathrm{Ker}(\varphi) \in \mathbf{Y}\}$ *is a subsystem of maximal subgroups of* G.

Proof. Let M^φ, S^φ be two distinct maximal subgroups of G^φ in $\mathbf{X}^\varphi$. Then M, S are two distinct maximal subgroups of G in $\mathbf{X}$ and then $M_G \neq S_G$. Moreover $\mathrm{Ker}(\varphi) \leq M \cap S$. It is clear that this implies that $(M^\varphi)_{G^\varphi} \neq (S^\varphi)_{G^\varphi}$.

If M^φ and S^φ are two maximal subgroups complementing an abelian chief factor H^φ/K^φ of G^φ, then H/K is an abelian chief factor of G which is complemented by M and S. Therefore $(M \cap S)H \in \mathbf{X}$. Hence $(M^\varphi \cap S^\varphi)H^\varphi \in \mathbf{X}^\varphi$.

For the converse, just notice that for any subgroup $H \leq G$ such that $\operatorname{Ker}(\varphi) \leq H$, we have $\big(H/\operatorname{Ker}(\varphi)\big)_{G^\varphi} = H_G/\operatorname{Ker}(\varphi)$. $\qquad\qquad\square$

Notation 1.4.4. Bearing in mind Notation 1.2.19, if G is a group, N is a normal subgroup of G, and $\varphi\colon G \longrightarrow G/N$ is the canonical epimorphism, we write

$$\mathbf{X}^\varphi = \mathbf{X}/N = \{M/N : M \in \mathbf{X} \text{ and } N \leq M\}$$

for a subsystem of maximal subgroups $\mathbf{X}$ of G.

Corollary 1.4.5. *Let G be a group factorised as $G = MN$, where M is a subgroup of G and N is a normal subgroup of G. If $\mathbf{X}$ is subsystem of maximal subgroups of G and $\mathbf{Y}$ is a subsystem of maximal subgroups of M, then*

$$(\mathbf{X} \cap M)/(N \cap M) = \{(S \cap M)/(N \cap M) : S \in \mathbf{X}, N \leq S\}$$

is a subsystem of maximal subgroups of $M/(N \cap M)$ and

$$\mathbf{Y}N/N = \{SN/N : S \in \mathbf{Y}, N \cap M \leq S\}$$

is a subsystem of maximal subgroups of G/N.

Lemma 1.4.6. *Let C/R be the crown of a complemented abelian chief factor F of a group G.*

1. *Suppose that N is a normal subgroup of G such that $R \leq N < C$. If T is a complement of C/N in G, then the set*

$$\mathbf{Y}(F, N, T) = \{TM : N \leq M < C \text{ and } C/M \text{ is a chief factor of } G\}$$

   is a subsystem of maximal subgroups of G.
   Moreover any chief factor of G between C and N is complemented by some maximal subgroup of $\mathbf{Y}(F, N, T)$ and $T = \bigcap\{U : U \in \mathbf{Y}(F, N, T)\}$.
2. *Let H/K be a chief factor of G such that $R \leq K < H < C$, T a complement of C/H in G, and U a complement of H/K in G. Then $S = T \cap U$ is a complement of C/K in G such that $T = SH$ and $\mathbf{Y}(F, H, T) \cup \{U\} \subseteq \mathbf{Y}(F, K, S)$.*
3. *If $\mathbf{X}$ is a subsystem of maximal subgroups of G such that F is $\mathbf{X}$-supplemented in G, and $T = \mathrm{T}(G, \mathbf{X}, F)$ is the complement of $C/R_{\mathbf{X}}$ defined in Proposition 1.3.11, then*

$$\begin{aligned}
\mathbf{Y}(F, R_{\mathbf{X}}, T) &= \mathbf{X}_F \\
&= \{U \in \mathbf{X} : U \text{ complements a chief factor } G\text{-isomorphic to } F\}.
\end{aligned}$$

Proof.   1. Since F is abelian, $C = \mathrm{C}_G(F)$. Write $\mathbf{Y} = \mathbf{Y}(F, N, T)$ and consider $U = TM \in \mathbf{Y}$, for some normal subgroup M such that $N \leq M$ and C/M is a chief factor of G. It is clear that U complements C/M in G. Hence U is a maximal subgroup of G. Since $U_G < C$, it follows that $U_G = M$.

Let $U_1 = TM_1$ and $U_2 = TM_2$ be two elements of $\mathbf{Y}$, with M_1 and M_2 as in the definition of the elements of $\mathbf{Y}$. We have seen in the preceding paragraph that $(U_i)_G = M_i$, $i = 1, 2$. Clearly $U_1 \neq U_2$ implies that $(U_1)_G = M_1 \neq M_2 = (U_2)_G$. Suppose that U_1 and U_2 complement the same chief factor H/K of G. Observe that $C = HM_1 = HM_2 = M_1 M_2$ and $M_1 \cap H = K = M_2 \cap H$. The subgroup $M_3 = (M_1 \cap M_2)H$ is a normal subgroup of G and $N \leq M_3 \leq C$. Moreover

$$
\begin{aligned}
C/M_3 &= HM_1/(M_1 \cap M_2)H \\
&\cong_G M_1/(M_1 \cap M_2)(M_1 \cap H) \\
&= M_1/(M_1 \cap M_2) \\
&\cong_G F
\end{aligned}
$$

and G/M_3 is a chief factor of G. By Lemma 1.2.17, the subgroup $(U_1 \cap U_2)H$ is maximal in G. Since $M_1 \cap TM_2 \leq C \cap TM_2 = M_2(C \cap T) = M_2$, it follows that $M_1 \cap TM_2 = M_1 \cap M_2$. Hence

$$(U_1 \cap U_2)H = (TM_1 \cap TM_2)H = T(M_1 \cap TM_2)H = T(M_1 \cap M_2)H = TM_3 \in \mathbf{Y}.$$

Consequently, $\mathbf{Y}$ is a subsystem of maximal subgroups of G.

Let H/K be a chief factor of G such that $N \leq K < H \leq C$. Let U be a complement of H/K in G and write $M = U_G$. Then $C = HM$ and $K = M \cap H$. Then $TM \in \mathbf{Y}(F, N, T)$. Now $(TM)H = TC = G$ and $TM \cap H \leq TM \cap C = M$. Hence $TM \cap H = M \cap H = K$ and TM complements H/K in G.

Clearly, $T \leq \bigcap \{U : U \in \mathbf{Y}(F, N, T)\}$. If $N = G_k \leq G_{k-1} \leq \cdots \leq G_0 = G$ is a piece of chief factor of G and, for $i = 1, \ldots, k$, U_i is a maximal subgroup in $\mathbf{Y}(F, N, T)$ complementing G_{i-1}/G_i, then $T = \bigcap_{i=1}^{k} U_i$, by Proposition 1.3.11. Hence, $T = \bigcap \{U : U \in \mathbf{Y}(F, N, T)\}$.

2. Applying Corollary 1.3.4, C/K is completely reducible G-module. Hence, by [DH92, A, 4.6], $C = HA$ for some normal subgroup A of G containing K such that $H \cap A = K$. By Lemma 1.2.16 (2), the subgroup $S = T \cap U$ is a complement of C/K in G and $SH = (T \cap U)H = T \cap UH = T$. If C/M is a chief factor of G such that $H \leq M$, then $SM = TM \in \mathbf{Y}(F, H, T)$. Hence $\mathbf{Y}(F, H, T) \subseteq \mathbf{Y}(F, K, S)$. Moreover C/U_G is a chief factor of G such that $K \leq U_G$. Since SU_G complements C/U_G in G, it follows that $U = SU_G$ is a maximal subgroup of G in $\mathbf{Y}(F, K, S)$.

3. Let TM be a maximal subgroup of G in $\mathbf{Y}(F, R_{\mathbf{X}}, T)$. The chief factor C/M is complemented by some maximal subgroup, U say, in $\mathbf{X}$. Since $U_G = M$, because C/M is self-centralising in G/M, it follows that $TM \leq U$. Hence $U = TM \in \mathbf{X}$. Therefore $\mathbf{Y}(F, R_{\mathbf{X}}, T) \subseteq \mathbf{X}_F$. If $U \in \mathbf{X}_F$, then $T \leq U$ and U complements $C/U_G \cong_G F$. Clearly $R_{\mathbf{X}} \leq U_G$ and $U = TU_G$. Hence $U \in \mathbf{Y}(F, R_{\mathbf{X}}, T)$. Therefore $\mathbf{X}_F \subseteq \mathbf{Y}(F, R_{\mathbf{X}}, T)$.   □

Theorem 1.4.7. *Let G be a group. Every subsystem of maximal subgroups of G is contained in a system of maximal subgroups of G. In particular, every group possesses a system of maximal subgroups.*

Proof. Let $\mathbf{X}$ be a subsystem of maximal subgroups of G. Then $\mathbf{X} = \mathbf{X}_1 \cup \mathbf{X}_2$, where

$$\mathbf{X}_k = \{U \in \mathbf{X} : U \text{ is a maximal subgroup of type } k\}, \qquad \text{for } k = 1, 2.$$

Also, if $F_1, \ldots, F_n$ are representatives of the G-isomorphism classes of complemented abelian chief factors of G, we have that $\mathbf{X}_1 = \bigcup_{i=1}^n \mathbf{X}_{F_i}$, where $\mathbf{X}_{F_i} = \{U \in \mathbf{X} : U \text{ complements a chief factor } G\text{-isomorphic to } F_i\}$.

Fix a complemented abelian chief factor F which is $\mathbf{X}$-complemented in G. Consider its $\mathbf{X}$-crown $C/R_{\mathbf{X}}$ and the subgroup $T^0 = T(G, \mathbf{X}, F)$ as in the previous lemma. Then $\mathbf{Y}(F, R_{\mathbf{X}}, T^0) = \mathbf{X}_F$. If C/R is the crown of F and $R = G_r \leq G_{r-1} \leq \ldots \leq G_0 = R_{\mathbf{X}} \leq \ldots \leq C$ is a piece of chief series of G, applying Lemma 1.4.6 (2), we construct a series of subsystems of maximal subgroups

$$\mathbf{X}_F = \mathbf{Y}(F, G_0, T^0) \subseteq \mathbf{Y}(F, G_1, T^1) \subseteq \ldots \subseteq \mathbf{Y}(F, G_r, T^r) = \mathbf{Y}(F, R, T),$$

and T is a complement of the crown C/R such that $T^0 = TR_{\mathbf{X}}$.

Note that every complemented chief factor G-isomorphic to F lies between R and C and hence it is complemented by a maximal subgroup in $\mathbf{Y}(F, R, T)$ by Lemma 1.4.6 (1). Hence, $\mathbf{Y}(F, R, T)$ is a complete set of representatives of the core-relation in the set of all maximal subgroups of G which complement a chief factor G-isomorphic to F.

Now, it is rather clear that

$$\mathbf{Y}_1 = \bigcup_{i=1}^n \mathbf{Y}(F_i, R_i, T_i)$$

is a subsystem of maximal subgroups of G which is a complete set of representatives of the core-relation in the set of all maximal subgroups of G of type 1. Moreover $\mathbf{X}_1 \subseteq \mathbf{Y}_1$.

For the maximal subgroups of type 2, just note that we only have to complete $\mathbf{X}_2$ to a complete set of representatives $\mathbf{Y}_2$ of the core-relation in the set of all maximal subgroups of type 2 of G.

Consequently $\mathbf{Y} = \mathbf{Y}_1 \cup \mathbf{Y}_2$ is a system of maximal subgroups of G and $\mathbf{X} \subseteq \mathbf{Y}$. $\qquad \square$

Corollary 1.4.8. *Let G be a group factorised as $G = MN$, where M is a subgroup of G and N is a normal subgroup of G. If $\mathbf{Y}$ is a subsystem of maximal subgroups of M, then there exists a system of maximal subgroups $\mathbf{X}$ of G such that*

$$\mathbf{Y}/(M \cap N) = (\mathbf{X} \cap M)/(N \cap M)$$

Proof. By Corollary 1.4.5, the set

$$\mathbf{Y}N/N = \{SN/N : S \in \mathbf{Y}, N \cap M \le S\}$$

is a subsystem (a system, in fact) of maximal subgroups of G/N. By Proposition 1.4.3. the set

$$\mathbf{X}_0 = \{S \le G : N \le S, S/N \in \mathbf{Y}N/N\}$$

is a subsystem of maximal subgroups of G. By Theorem 1.4.7 there exists a system of maximal subgroups $\mathbf{X}$ of G such that $\mathbf{X}_0 \subseteq \mathbf{X}$.

Observe that if $S \in \mathbf{X}_0$, then $S = UN$ for some $U \in \mathbf{Y}$ such that $N \cap M \le U$. Moreover $S \cap M = UN \cap M = U(N \cap M) = U$. Hence

$$\mathbf{Y}/(M \cap N) = (\mathbf{X}_0 \cap M)/(M \cap N) \subseteq (\mathbf{X} \cap M)/(M \cap N).$$

Observe that $(\mathbf{X} \cap M)/(M \cap N)$ is a system of maximal subgroups of $M/(N \cap M)$ and so is $\mathbf{Y}/(M \cap N)$. Hence equality holds.   □

The following results analyse the behaviour of systems of maximal subgroups in some particular maximal subgroups called *critical subgroups*. These subgroups turn out to be crucial in the introduction of normalisers associated with some classes of groups in Chapter 4.

Definition 1.4.9. *Let G be a group. A monolithic maximal subgroup M of G is said to be a* critical subgroup *of G if M supplements the subgroup $\mathrm{F}'(G) = \mathrm{Soc}\big(G \bmod \varPhi(G)\big)$.*

Since

$$\mathrm{F}'(G)/\varPhi(G) = \mathrm{Soc}\big(G/\varPhi(G)\big) = N_1/\varPhi(G) \times \cdots \times N_n/\varPhi(G)$$

for normal subgroups N_i of G such that each $N_i/\varPhi(G)$ is a chief factor of G, we can say that a maximal subgroup M of G is critical if there exists a chief factor of G of the form $N/\varPhi(G)$ supplemented by M.

If the group G is soluble, then $\mathrm{F}'(G) = \mathrm{F}(G)$, the Fitting subgroup of G. In this case, this definition coincides with that of [DH92, III, 6.4 (a)].

Proposition 1.4.10. *Let G be a group and N a normal subgroup of G. If M is a subgroup of G, then $\mathrm{F}'(M)N/N$ is contained in $\mathrm{F}'(MN/N)$. Consequently, if U is critical in M and $M \cap N$ is contained in U, then UN/N is critical in MN/N.*

Proof. Write $F/N = \varPhi(MN/N)$ and recall that $\varPhi(M) \le F$. Let $K/\varPhi(M)$ be a minimal normal subgroup of $M/\varPhi(M)$. We have that $\varPhi(M) \le K \cap F \le K$ and $K \cap F$ is normal in M. Hence either $\varPhi(M) = K \cap F$ or $K \le F$ by minimality of $K/\varPhi(M)$. If $K \le F$, then $KN/N \le \mathrm{F}'(MN/N)$. Assume that $\varPhi(M) = K \cap F$. It follows that KF/F is a minimal normal subgroup of MN/F. Hence $KN/N \le \mathrm{F}'(MN/N)$ and $\mathrm{F}'(M)N/N \le \mathrm{F}'(MN/N)$.

Assume that U is critical in M. Then $M = U\mathrm{F}'(M)$ and $MN/N = (UN/N)\big(\mathrm{F}'(M)N/N\big) = (UN/N)\,\mathrm{F}'(MN/N)$. If $M \cap N \le U$, UN/N is maximal in MN/N. Hence, in this case, UN/N is critical in MN/N.

Proposition 1.4.11. *Let M be a critical subgroup of a group G. Suppose that H/K is a chief factor of G covered by M and avoided by $\Phi(G)$. Then we have the following.*

1. *The section $(H \cap M)/(K \cap M)$ is a chief factor of M such that $M \cap C_G(H/K) = C_M\big((H \cap M)/(K \cap M)\big)$.*
2. $\operatorname{Aut}_G(H/K) \cong \operatorname{Aut}_M\big((H \cap M)/(K \cap M)\big)$.
3. $[H/K] * G \cong [(H \cap M)/(K \cap M)] * M$.
4. *If U is a monolithic maximal subgroup of G which supplements H/K in G, then $U \cap M$ is a maximal subgroup of M which supplements $(H \cap M)/(K \cap M)$ in M.*

Proof. First of all, since $H = K(M \cap H)$, it follows that H/K is M-isomorphic to $(H \cap M)/(K \cap M)$. Therefore $M \cap C_G(H/K) = C_M\big((H \cap M)/(K \cap M)\big)$. We shall prove now that $G = M\, C_G(H/K)$. Since M is critical in G, M is a supplement in G of a chief factor of G of the form $N/\Phi(G)$. Note that $H\Phi(G)/K\Phi(G)$ is G-isomorphic to H/K. Hence, by considering $H\Phi(G)/K\Phi(G)$ instead of H/K if necessary, we can assume that $\Phi(G) \leq K$.

If $G = MK$, then $G = M\, C_G(H/K)$. Assume that $K \leq M$. Then $H \leq M$, since M covers H/K. Therefore $[H, N] \leq \Phi(G)$ and thus $N \leq C_G(H/K)$. Consequently, in both cases, $G = M\, C_G(H/K)$.

Now Statements 1 and 2 follow from [DH92, A, 13.9].

3. If H/K is non-abelian, then clearly $[H/K]*G \cong [(H \cap M)/(K \cap M)]*M$. If H/K is abelian, then the correspondence

$$\alpha \colon [H/K] * G \longrightarrow [(H \cap M)/(K \cap M)] * M,$$

given by

$$\big(xK, y\, C_G(H/K)\big)^\alpha = \Big(x(K \cap M), y\, C_M\big((H \cap M)/(K \cap M)\big)\Big)$$

for any $x \in H$, $y \in M$, is an isomorphism. Hence $[H/K]*G \cong [(H \cap M)/(K \cap M)]*M$.

4. Note that $H = K(M \cap H)$ because M covers H/K.

Let us prove first that if X is a monolithic maximal subgroup of G such that $X \cap M = U \cap M$ and $N \leq X$, then $X \cap M$ is a maximal subgroup of M which supplements $(H \cap M)/(K \cap M)$ in M.

Note that $X = X \cap MN = (X \cap M)N$. Let T be a subgroup such that $X \cap M \leq T \leq M$. Then $N \cap M \leq X \cap M \leq T$ and $X = (X \cap M)N \leq TN \leq MN = G$. By maximality of X in G, we have that either $X = TN$ or $TN = G$. If $X = TN$, then $X \cap M = TN \cap M = T(N \cap M) = T$. If $G = TN$, then $M = M \cap TN = T(M \cap N) = T$. Hence $X \cap M$ is a maximal subgroup of M.

Now consider the subgroup $(X \cap M)(H \cap M)$. Suppose that $(X \cap M)(H \cap M) = X \cap M$. This is to say that $M \cap H \leq M \cap X = U \cap M$ and then $H = K(M \cap H) \leq U$, which is a contradiction. Hence, by maximality of

$X \cap M$ in M, we have that $M = (X \cap M)(H \cap M) = (U \cap M)(H \cap M)$. Moreover $K \cap M$ is contained in $U \cap H \cap M$. Therefore $U \cap M$ supplements $(H \cap M)/(K \cap M)$ in M.

Clearly if $N \leq U$, we can apply the above arguments to $X = U$. Suppose that $G = UN$. If $U_G = M_G$, then $K = (U_G) \cap H = M_G \cap H \leq M \cap H$ and then $H = K$. This contradiction yields $U_G \neq M_G$. Applying Lemma 1.2.17, the subgroup $X = (U \cap M)N$ is a maximal subgroup of G. Also by Lemma 1.2.17 (2), we have that $N/\Phi(G)$ is abelian. In particular $M \cap N = \Phi(G) \leq U \cap M$. Therefore $X \cap M = (U \cap M)(N \cap M) = U \cap M$ and $U \cap M$ supplements $(H \cap M)/(K \cap M)$ in M by the above arguments. $\square$

Corollary 1.4.12. *Let M be a critical subgroup of a group G. Assume that U is a monolithic maximal subgroup of G such that $U_G \neq M_G$. Then $M \cap U$ is a monolithic maximal subgroup of M.*

Proof. Assume that U supplements a chief factor H/K of G. Suppose that H/K is supplemented by M. By Lemma 1.2.17 (2), the chief factor H/K is abelian. In this case U complements the chief factor C/U_G, for $C = \mathrm{C}_G(H/K)$, and this chief factor is covered by M, since $M_G \neq U_G$. Hence, we can assume that U supplements a chief factor covered by M. Since this chief factor is avoided by $\Phi(G)$, we have that $M \cap U$ is a maximal subgroup of M, by Proposition 1.4.11 (4). $\square$

Theorem 1.4.13. *Let $\mathbf{X}$ be a subsystem of maximal subgroups of a group G and M a critical subgroup of G in $\mathbf{X}$. Consider the set*

$$\mathbf{X}_M = \{S \cap M : S \in \mathbf{X}, S \neq M\},$$

with no repetitions. Then

1. *if $G = MN$, for some chief factor $N/\Phi(G)$ of G, then $\mathbf{X}_M = \{S \cap M : N \leq S \in \mathbf{X}\}$;*
2. *$\mathbf{X}_M$ is a subsystem of maximal subgroups of M.*

Proof.   1. Let S be an element of $\mathbf{X}$ such that $S \cap M \in \mathbf{X}_M$ and $G = SN$. Then $N/\Phi(G)$ is abelian by Lemma 1.2.17 (2), $S^* = (S \cap M)N \in \mathbf{X}$ and $S^* \cap M = S \cap M$.

2. Assume that $G = MN$, for some chief factor $N/\Phi(G)$ of G. Applying Corollary 1.4.12, all elements of $\mathbf{X}_M$ are monolithic maximal subgroups of M. Consider two distinct maximal subgroups $S \cap M$ and $U \cap M$ in $\mathbf{X}_M$. By Statement 1, we can assume that $N \leq S \cap U$. By Theorem 1.2.22, we have that $S_G = N(S \cap M)_M \neq U_G = N(U \cap M)_M$. Hence $(S \cap M)_M \neq (U \cap M)_M$.

Suppose that $S \cap M$ and $U \cap M$ are distinct elements of $\mathbf{X}_M$, for S, $U \in \mathbf{X}$, and both complement the same abelian chief factor H/K of M. We can assume that $N \leq S \cap U$. Then $S = N(S \cap M)$ and $U = N(U \cap M)$. Since $M \cap N \leq S \cap M$, it follows that $H \cap M \cap N \leq H \cap M \cap S = K$. Therefore $H(M \cap N)/K(M \cap N) \cong_M H/K$. Clearly, $S \cap M$ and $U \cap M$ complement the chief factor $H(M \cap N)/K(M \cap N)$ of M. By Theorem 1.2.22, U and S complement the chief factor HN/KN of G. Thus, $(S \cap U)HN = (S \cap U)H$ is a maximal subgroup in $\mathbf{X}$, inasmuch as $\mathbf{X}$ is a subsystem of maximal subgroups of G. Therefore $(S \cap U)H \cap M = (S \cap U \cap M)H \in \mathbf{X}_M$.

Consequently, $\mathbf{X}_M$ is a subsystem of maximal subgroups of M.   □

Theorem 1.4.14. *Let M be a critical subgroup of a group G. Assume that $\mathbf{Y}$ is a system of maximal subgroups of M. Then there exists a system of maximal subgroups $\mathbf{X}$ of G such that $M \in \mathbf{X}$ and $\mathbf{X}_M \subseteq \mathbf{Y}$.*

Proof. Without loss of generality we can assume that $\Phi(G) = 1$. Since M is critical in G, it follows that $G = NM$, for some minimal normal subgroup N of G.

Suppose that N is non-abelian and consider the following set of monolithic maximal subgroups of G

$$\mathbf{X} = \{SN : M \cap N \leq S \in \mathbf{Y}\} \cup \{M\}.$$

If U is a maximal subgroup of G and $N \cap U_G = 1$, then $G = UN$ and $U_G = C_G(N) = M_G$, since N is non-abelian. If $N \leq U_G$, then $U \cap M$ is a maximal subgroup of M and there exists $S \in \mathbf{Y}$, such that $N \cap M \leq S_M = (U \cap M)_M$. Now observe that $SN_G = S_M N = U_G$. Therefore $\mathbf{X}$ is a complete set of representatives of the core-relation in G.

Suppose now that S_1 and S_2 are maximal subgroups of M in $\mathbf{Y}$ such that $M \cap N \leq S_1 \cap S_2$ and the maximal subgroups $U_1 = S_1 N$ and $U_2 = S_2 N$ of G complement the same abelian chief factor H/K of G. We see that $(U_1 \cap U_2)H \in \mathbf{X}$. Changing if necessary H/K by HN/KN, we can assume that $N \leq K$. Now S_1 and S_2 complement the abelian chief factor $(H \cap M)/(K \cap M)$ of M. Since $\mathbf{Y}$ is a system of maximal subgroups of M, the subgroup $(S_1 \cap S_2)(H \cap M)$ is in $\mathbf{Y}$. Since $N \cap M \leq H \cap M \leq (S_1 \cap S_2)(H \cap M)$, we have that $(S_1 \cap S_2)(H \cap M)N = (S_1 \cap S_2)H$ is a maximal subgroup of G in $\mathbf{X}$. Clearly $(S_1 \cap S_2)H = (U_1 \cap U_2)H$. This shows that $\mathbf{X}$ is a system of maximal subgroups of G and $M \in \mathbf{X}$.

Finally, if $S \in \mathbf{Y}$ and $M \cap N \leq S$, then $M \cap SN = S$. Hence $\mathbf{X}_M \subseteq \mathbf{Y}$.

Assume now that N is abelian. Hence $M \cap N = 1$. Write $\mathbf{Y} = \mathbf{Y}_1 \cup \mathbf{Y}_2$, where $\mathbf{Y}_i$ is the set of maximal subgroups of type i in $\mathbf{Y}$, for $i = 1, 2$. Let $\{F_1, \ldots, F_n\}$ be a complete set of representatives of the M-isomorphism classes of abelian chief factors of M. Then $\mathbf{Y}_1 = \bigcup_{i=1}^{n} \mathbf{Y}_{F_i}$, where $\mathbf{Y}_{F_i} = \{S \in \mathbf{Y} : S$ complements a chief factor M-isomorphic to $F_i\}$.

Applying Theorem 1.2.22, $\mathbf{X}_2 = \{SN : S \in \mathbf{Y}_2\}$ is a complete set of representatives of the core-relation in the set of all maximal subgroups of type 2 of G. Note that $(\mathbf{X}_2)_M = \{SN \cap M : S \in \mathbf{Y}_2\} = \mathbf{Y}_2$.

Since $M \cong G/N$, we can find a complete set $\{L_1, \ldots, L_n\}$ of representatives of the G/N-isomorphism (G-isomorphism) classes of abelian chief factors of G/N such that $L_i \cong F_i$, $1 \leq i \leq n$.

If N is not isomorphic to L_i for all $i = 1, \ldots, n$, then all complements of N in G are core-related. In this case $\mathbf{X}_1 = \{SN : S \in \mathbf{Y}_1\} \cup \{M\}$ is a subsystem of maximal subgroups of G containing a representative of each equivalence class of the core-relation in the set of all maximal subgroups of G of type 1. Therefore $\mathbf{X} = \mathbf{X}_1 \cup \mathbf{X}_2$ is a system of maximal subgroups of G such that $\mathbf{X}_M = \mathbf{Y}$.

Suppose that N is G-isomorphic to some of L_i, $1 \leq i \leq n$. Let us assume that $N \cong_G L_n$. For each $i \in \{1, \ldots, n-1\}$, denote $\mathbf{X}_{L_i} = \{SN : S \in \mathbf{Y}_{F_i}\}$. Then $(\mathbf{X}_{L_i})_M = \mathbf{Y}_{F_i}$ and $\mathbf{X}_{L_i}$ is a subsystem of maximal subgroups of G containing a representative of each equivalence class of the core-relation in the set of all complements of chief factors of G which are G-isomorphic to L_i.

If L_i is a Frattini chief factor of G/N, then all complements of N in G are core-related and $\mathbf{X}_1 = \{SN : S \in \mathbf{Y}_1\} \cup \{M\}$ is a system of maximal subgroups of G satisfying the condition of the theorem. Therefore we may assume that L_n is complemented, and so there exists a $\mathbf{Y}$-complemented chief factor A/B of M such that A/B is M-isomorphic to F_n.

Let C/R be the crown of G associated with N and AN/BN in G. By Proposition 1.3.5, RN is a proper subgroup of C and $(C/N)/(RN/N)$ is the crown of $(AN/N)/(BN/N)$ in G/N. Applying Proposition 1.3.11, $(C/N)/(RN/N)$ in G/N is complemented in G/N. Let T be a subgroup of M such TN/N is a complement of $(C/N)/(RN/N)$ in G/N. Since TN is a complement of C/RN in G and M is a complement of RN/R in G, it follows that $T = TN \cap M$ is a complement of C/R in G by Lemma 1.4.6 (2). In addition, applying Lemma 1.4.6 (1), the set $\mathbf{Y}(AN/BN, RN, TN)$, composed of all subgroups TK where K is a normal subgroup of G such that $RN \leq K$ and C/K is a chief factor of G, is a subsystem of maximal subgroups of G and

$$\mathbf{Y}(L_n, RN, TN) \cup \{M\} \subseteq \mathbf{Y}(L_n, R, T)$$
$$= \{TK : R \leq K \text{ and } C/K \text{ is chief factor of } G\}.$$

Write $\mathbf{X}_{L_n} = \mathbf{Y}(L_n, R, T)$. Then $\mathbf{X}_{L_n}$ is a subsystem of maximal subgroups of G by Lemma 1.4.6 (1).

Consider a subgroup $U \in \mathbf{X}_{L_n}$, $U \neq M$. We see that $U \cap M \in \mathbf{Y}_{F_n}$. Suppose that $U = TK$ for some normal subgroup K of G such that $R \leq K$ and C/K is a chief factor of G. If K is contained in M_G, then $U = TK = M$ against our assumption. Hence we have that $G = MK$ and $C = M_G K$. In particular, $U_G \neq M_G$. Moreover, $(C \cap M)/(K \cap M)$ is a chief factor of M which is M-isomorphic to F_n and is complemented in M by the maximal subgroup $U \cap M$ of M by Proposition 1.4.11 (1) and (4). Note that $(C \cap M)/(K \cap M)$ is $\mathbf{Y}$-complemented in M because $\mathbf{Y}$ is a system of maximal subgroups of M. Consider a maximal subgroup $Y \in \mathbf{Y}_{F_n}$ which complements the chief factor $(C \cap M)/(K \cap M)$ in M. Applying Proposition 1.3.11, we have that T is contained in Y. Hence $U \cap M = T(K \cap M) \leq Y$. Maximality of $U \cap M$ in M forces $U \cap M = Y$. Therefore, $(\mathbf{X}_{L_n})_M \subseteq \mathbf{Y}_{F_n}$.

If U is a maximal subgroup of G which complements a chief factor isomorphic to L_n, then U complements the chief factor $C/U_G \cong_G L_n$. The maximal subgroup TU_G is in $\mathbf{X}_{L_n}$ and $(TU_G)_G = U_G$. Thus, $\mathbf{X}_{L_n}$ is a complete set of representatives for the core-relation in the set of all complements of chief factors G-isomorphic to L_n.

Consider the union set $\mathbf{X}_1 = \bigcup_{i=1}^{n} \mathbf{X}_{F_i}$ and $\mathbf{X} = \mathbf{X}_1 \cup \mathbf{X}_2$ is a system of maximal subgroups of G such that $\mathbf{X}_M \subseteq \mathbf{Y}$ and $M \in \mathbf{X}$.   □

Theorem 1.4.15. *If N is a normal subgroup of a group G and $\mathbf{X}^*$ is system of maximal subgroups of G/N, then there exists a system of maximal subgroups $\mathbf{X}$ of G such that $\mathbf{X}/N = \mathbf{X}^*$.*

Proof. We argue by induction of the order of G. It is clear that $N \neq 1$. Assume that N is a minimal normal subgroup of G. It is clear that we can suppose that $N \cap \Phi(G) = 1$. Let M be a critical subgroup of G such that $G = MN$. If α is the isomorphism $G/N \cong M/(M \cap N)$, then $(\mathbf{X}^*)^\alpha = \{(U \cap M)/(N \cap M) : U/N \in \mathbf{X}^*\}$ is a system of maximal subgroups of $M/(N \cap M)$. By induction, there exists a system of maximal subgroups $\mathbf{X}(M)$ of M such that $\mathbf{X}(M)/(N \cap M) = (\mathbf{X}^*)^\alpha$. By Theorem 1.4.14 there exists a system of maximal subgroups $\mathbf{X}$ of G such that $\mathbf{X}_M \subseteq \mathbf{X}(M)$.

The set $(\mathbf{X}/N)^\alpha = \{(S \cap M)/(N \cap M) : S \in \mathbf{X}, N \leq S\}$ is a system of maximal subgroups of $M/(M \cap N)$ by Corollary 1.4.5. Notice that $(\mathbf{X}/N)^\alpha \subseteq \mathbf{X}(M)/(M \cap N) = (\mathbf{X}^*)^\alpha$ and then $(\mathbf{X}/N)^\alpha = (\mathbf{X}^*)^\alpha$. Consequently $\mathbf{X}/N = \mathbf{X}^*$ and the theorem is true.

Now assume that L is a minimal normal subgroup of G and L is a proper subgroup of N. By inductive hypothesis the theorem is true for the group G/L. Since $\mathbf{X}^{**} = \{(S/L)/(N/L) : S/N \in \mathbf{X}^*\}$ is a system of maximal subgroups of $(G/L)/(N/L)$, there exists a system of maximal subgroups $\mathbf{X}_0$ of G/L such that $\mathbf{X}_0/(N/L) = \mathbf{X}^{**}$. On the other hand, since for L the theorem is true, there exists a system of maximal subgroups $\mathbf{X}$ of G such that $\mathbf{X}/L = \mathbf{X}_0$. If $H \in \mathbf{X}$ and $N \leq H$, then $L \leq H$ and $H/L \in \mathbf{X}_0$, $(H/L)/(N/L) \in \mathbf{X}^{**}$, and then $H/N \in \mathbf{X}^*$. Consequently, $\mathbf{X}^* = \mathbf{X}/N$.   □

Corollary 1.4.16. *Given a system of maximal subgroups $\mathbf{X}$ of a group G and a critical subgroup M of G such that $M \in \mathbf{X}$, there exists a system of maximal subgroups $\mathbf{Y}$ of M, such that $\mathbf{X}_M \subseteq \mathbf{Y}$.*

Proof. Assume that M supplements a chief factor $N/\Phi(G)$ of G. Denote by α the isomorphism $\alpha : G/N \longrightarrow M/(N \cap M)$. Then $(\mathbf{X} \cap N)/(M \cap N)$ is a system of maximal subgroups of $M/(N \cap M)$. By Theorem 1.4.15, there exists a system of maximal subgroups $\mathbf{Y}$ of M such that $\mathbf{Y}/(N \cap M) = (\mathbf{X} \cap M)/(M \cap N)$. Let $U \in \mathbf{X}$ with $U \neq M$. If $G = UN$, then $N/\Phi(G)$ is abelian by Lemma 1.2.17 (2), and $V = (U \cap M)N \in \mathbf{X}$. In this case, we have that $U \cap M = V \cap M$. Hence we can assume that $N \leq U$. Then $(U \cap M)/(N \cap M) \in (\mathbf{X} \cap N)/(M \cap N)$ and $U \cap M \in \mathbf{Y}$. Therefore $\mathbf{X}_M \subseteq \mathbf{Y}$.   □

The soluble case is particularly interesting in this context. Given a Hall system Σ of a soluble group G, we consider the set

$$\mathbf{S}(\Sigma) = \{S \in \mathrm{Max}(G) : \Sigma \text{ reduces into } S\}.$$

Maximal subgroups are always pronormal (see [DH92, Section I, 6]) and therefore if M is a maximal subgroup of G, then Σ reduces into exactly one conjugate of M by a theorem due to Mann (see [DH92, I, 6.6]). Then $\mathbf{S}(\Sigma)$ is a
complete set of representatives of the core-relation. By [DH92, I, 4.22], $\mathbf{S}(\Sigma)$
is indeed a system of maximal subgroups of G. the following result shows that
all systems of maximal subgroups of the soluble group G arise in this manner.

Theorem 1.4.17. *Let $\mathbf{X}$ be a subsystem of maximal subgroups of a soluble
group G. Then there exists a Hall system Σ of G such that Σ reduces into
each maximal subgroup of G in $\mathbf{X}$.*

Proof. We argue by induction on the order of G. Let N be a minimal normal
subgroup of G. Then $\mathbf{X}/N$ is a subsystem of maximal subgroups of G/N
by Proposition 1.4.3. By induction there exists a Hall system Σ of G such
that the Hall system $\Sigma N/N$ reduces into each maximal subgroup of G/N
in $\mathbf{X}/N$. Hence Σ reduces into each maximal subgroup of G containing N
and belonging to $\mathbf{X}$ by [DH92, I, 4.17 b]. In particular, we can assume that
$\Phi(G) = 1$.

If no complement of N in G is in $\mathbf{X}$, then Σ reduces into each maximal subgroup of G in $\mathbf{X}$. Thus, we can assume that the set of complements
$\{T_1, \ldots, T_r\}$ of N in $\mathbf{X}$ is non-empty, i.e. $r \geq 1$. We can also assume that T_1
is not normal in G. By [DH92, I, 4.16] there exists an element $n \in N$ such
that $\Sigma_0 = \Sigma^n$ reduces into T_1. Then $\Sigma_0 N/N = \Sigma N/N$. This means that we
can assume without loss of generality that $\Sigma_0 = \Sigma$. If $r = 1$, then it is clear
that Σ reduces into each maximal subgroup of G in $\mathbf{X}$. Suppose that $r > 1$.
For $j \neq 1$, the subgroup $M = (T_1 \cap T_j)N$ is a maximal subgroup of G in $\mathbf{X}$.
Since $\Sigma N/N$ reduces into M/N, it is clear that Σ reduces into M. Let p be
the prime dividing the order of N and consider the Hall p'-subgroup Q of G
in Σ. We know, by Lemma 1.2.17 (1), that M complements a p-chief factor
of G. Hence $T_1 \cap T_j$ has p-index in G and so $Q \leq (T_1 \cap T_j)^a$ for some $a \in N$.
This implies that Σ reduces into T_1^a and into T_j^a by [DH92, I, 4.20]. Since T_1
is pronormal in G, we have that $a \in T_1 \cap N = 1$ and then Σ reduces into T_j.
Thus, in any case Σ reduces into T_i, for $i = 1, \ldots, r$ and then Σ reduces into
each maximal subgroup of G in $\mathbf{X}$. $\square$

Corollary 1.4.18. *If G is a soluble group then:*

1. the map

$$\{\text{Hall systems of } G\} \longrightarrow \{\text{Systems of maximal subgroups of } G\}$$

such that the image of a Hall system Σ of G is the set $\mathbf{X}(\Sigma)$ given by

$$\mathbf{X}(\Sigma) = \{S \in \mathrm{Max}(G) : \Sigma \text{ reduces into } S\},$$

is surjective.
2. All systems of maximal subgroups of G are conjugate.

  3. *The number of systems of maximal subgroups of G is the index of the stabiliser* $\mathrm{N}_G\big(\mathbf{X}(\varSigma)\big) = \bigcap\{\mathrm{N}_G(S) : S \in \mathbf{X}(\varSigma)\}$.

Corollary 1.4.19. *A group G is soluble if and only if all systems of maximal subgroups of G are conjugate.*

Proof. Only the sufficiency of the condition is in doubt. Suppose that all systems of maximal subgroups are conjugate in G. If G is non-soluble, there exists a non-abelian chief factor H/K of G; then $G\big/\mathrm{C}_G(H/K)$ is a primitive group of type 2 by Proposition 1.1.14. Take S and U two maximal subgroups of G such that $S_G = U_G = \mathrm{C}_G(H/K)$, i.e. $S\big/\mathrm{C}_G(H/K)$ and $U\big/\mathrm{C}_G(H/K)$ are two core-free maximal subgroups of $G\big/\mathrm{C}_G(H/K)$. There exist two systems of maximal subgroups of G, $\mathbf{X}$ and $\mathbf{Y}$, such that $S \in \mathbf{X}$ and $U \in \mathbf{Y}$ by Theorem 1.4.14. Since $\mathbf{Y} = \mathbf{X}^g$ for some $g \in G$, then $U = S^g$ and all core-free maximal subgroups of $G\big/\mathrm{C}_G(H/K)$ are conjugate. But this contradicts the fact of being a primitive group of type 2 (see Remark 1.1.11 (4)). Therefore G is soluble. $\qquad\square$

2

Classes of groups and their properties

2.1 Classes of groups and closure operators

A *group theoretical class* or *class of groups* $\mathfrak{X}$ is a collection of groups with the property that if $G \in \mathfrak{X}$, then every group isomorphic to G belongs to $\mathfrak{X}$. The groups which belong to a class $\mathfrak{X}$ are referred to as $\mathfrak{X}$-*groups*.

Following K. Doerk and T. O. Hawkes [DH92], we denote the empty class of groups by $\emptyset$ whereas the Fraktur (Gothic) font is used when a single capital letter denotes a class of groups. If $\mathcal{S}$ is a set of groups, we use $(\mathcal{S})$ to denote the smallest class of groups containing $\mathcal{S}$, and when $\mathcal{S} = \{G_1, \ldots, G_n\}$, a finite set, $(G_1, \ldots, G_n)$ rather than $(\{G_1, \ldots, G_n\})$.

Since certain natural classes of groups recur frequently, it is convenient to have a short fixed alphabet of classes:

- $\emptyset$ denotes the empty class of groups;
- $\mathfrak{A}$ denotes the class of all abelian groups;
- $\mathfrak{N}$ denotes the class of all nilpotent groups;
- $\mathfrak{U}$ denotes the class of all supersoluble groups;
- $\mathfrak{S}$ denotes the class of all soluble groups;
- $\mathfrak{J}$ denotes the class of all simple groups;
- $\mathbb{P}$ denotes either the class $\mathfrak{A} \cap \mathfrak{J}$ of all cyclic groups of prime order or the set of all primes;
- $\mathfrak{P}$ denote the class of all primitive groups;
- $\mathfrak{P}_i$ denotes the class of all primitive groups of type i, $1 \leq i \leq 3$;
- $\mathfrak{E}$ denotes the class of all finite groups.

The group classes are, of course, partially ordered by inclusion and the notation

$$\mathfrak{X} \subseteq \mathfrak{Y}$$

will be used to denote the fact that $\mathfrak{X}$ is a subclass of the class $\mathfrak{Y}$.

Sometimes it is preferable to deal with group theoretical properties or properties of groups: A *group theoretical property* $\mathcal{P}$ is a property pertaining to groups such that if a group G has $\mathcal{P}$, then every isomorphic image of G has $\mathcal{P}$. The groups which have a given group theoretical property form a class of groups and to belong to a given group theoretical class is a group theoretical property. Consequently, there is a one-to-one correspondence between the group classes and the group theoretical properties; for this reason we will often not distinguish between a group theoretical property and the class of groups that possess it.

Note that we do not require that a class of groups contains groups of order 1.

Definition 2.1.1. *Let G be a group and let $\mathfrak{X}$ be a class of groups.*

1. We define

$$\pi(G) = \{p : p \in \mathbb{P} \text{ and } p \mid |G|\}, \quad and$$
$$\pi(\mathfrak{X}) = \bigcup\{\pi(G) : G \in \mathfrak{X}\}.$$

2. We also define

$$\kappa \mathfrak{X} = \{S \in \mathfrak{J} : S \text{ is a composition factor of an } \mathfrak{X}\text{-group}\}$$

and

$$\operatorname{char} \mathfrak{X} = \{p : p \in \mathbb{P} \text{ and } C_p \in \mathfrak{X}\};$$

we say that $\operatorname{char}(\mathfrak{X})$ *is the* characteristic *of* $\mathfrak{X}$.

Obviously $\operatorname{char} \mathfrak{X}$ is contained in $\pi(\mathfrak{X})$, but the equality does not hold in general. If $\mathfrak{X} = \big(G : G = O^{p'}(G)\big)$ is the class of all p'-perfect groups for some prime p, then $\operatorname{char} \mathfrak{X} = \{p\} \neq \pi(\mathfrak{X}) = \mathbb{P}$. Note that $\operatorname{char} \mathfrak{X}$, regarded as a subclass of $\mathfrak{J}$, is contained in $\kappa \mathfrak{X}$. The class of all p'-perfect groups shows that the inclusion is proper.

Definition 2.1.2. *If $\mathfrak{X}$ and $\mathfrak{Y}$ are two classes of groups, the* product class $\mathfrak{X}\mathfrak{Y}$ *is defined as follows: a group G belongs to $\mathfrak{X}\mathfrak{Y}$ if and only if there is a normal subgroup N of G such that $N \in \mathfrak{X}$ and $G/N \in \mathfrak{Y}$. Groups in the class $\mathfrak{X}\mathfrak{Y}$ are called $\mathfrak{X}$-by-$\mathfrak{Y}$-groups.*

If $\mathfrak{X} = \emptyset$ or $\mathfrak{Y} = \emptyset$, we have the obvious interpretation $\mathfrak{X}\mathfrak{Y} = \emptyset$.

It should be observed that this binary algebraic operation on the class of all classes of groups is neither associative nor commutative. For instance, let G be the alternating group of degree 4. Then $G \in (\mathfrak{CC})\mathfrak{C}$, where $\mathfrak{C}$ is the class of all cyclic groups. However G has no non-trivial normal cyclic subgroups, so $G \notin \mathfrak{C}(\mathfrak{CC})$.

On the other hand, the inclusion $\mathfrak{X}(\mathfrak{Y}\mathfrak{Z}) \subseteq (\mathfrak{X}\mathfrak{Y})\mathfrak{Z}$ is universally valid and, indeed, follows at once from our definition.

For the powers of a class $\mathfrak{X}$, we set $\mathfrak{X}^0 = (1)$, and for $n \in \mathbb{N}$ make the inductive definition $\mathfrak{X}^n = (\mathfrak{X}^{n-1})\mathfrak{X}$. A group in $\mathfrak{X}^2$ is sometimes denoted meta-$\mathfrak{X}$.

The past decades have seen the introduction of a very large number of classes of groups and it would be quite impossible to use a systematic alphabet for them. However, one soon observes that many of these classes are obtainable from simpler classes by certain uniform procedures. From this observation stems the importance for our purposes of the concept of closure operation. The first systematic use of closure operations in group theory occurs in papers of P. Hall [Hal59, Hal63] although the ideas are implicit in earlier papers of R. Baer and also in B. I. Plotkin [Plo58].

By an *operation* we mean a function c assigning to each class of groups $\mathfrak{X}$ a class of groups $c\,\mathfrak{X}$ subject to the following conditions:

1. $c\,\emptyset = \emptyset$, and
2. $\mathfrak{X} \subseteq c\,\mathfrak{X} \subseteq c\,\mathfrak{Y}$ whenever $\mathfrak{X} \subseteq \mathfrak{Y}$.

Should it happen that $\mathfrak{X} = c\,\mathfrak{X}$, the class $\mathfrak{X}$ is said to be c-*closed*. By 1 and 2, the classes $\emptyset$ and $\mathfrak{E}$ are c-closed when c is any operation.

A partial ordering of operations is defined as follows: $c_1 \leq c_2$ means that $c_1\,\mathfrak{X} \subseteq c_2\,\mathfrak{X}$ for every class of groups $\mathfrak{X}$. Products of operations are formed according to the rule

$$(c_1\,c_2)\mathfrak{X} = c_1(c_2\,\mathfrak{X}).$$

An operation c is called a *closure operation* if it is idempotent, that is, if

3. $c = c^2$.

If c is a closure operation, then by Condition 2 and Condition 3, the class $c\,\mathfrak{X}$ is the uniquely determined, smallest c-closed class that contains $\mathfrak{X}$. Thus if a and b are closure operations, $a \leq b$ if and only if b-closure invariably implies a-closure.

A closure operation can be determined by specifying the classes of groups that are closed. Let $\mathcal{S}$ be a class of classes of groups and suppose that every intersection of members of $\mathcal{S}$ belongs to $\mathcal{S}$: for example, $\mathcal{S}$ might consist of the closed classes of a closure operation. $\mathcal{S}$ determines a closure operation c defined as follows: for any class of groups $\mathfrak{X}$, let $c\,\mathfrak{X}$ be the intersection of all those members of $\mathcal{S}$ that contain $\mathfrak{X}$. The c-closed classes are precisely the members of $\mathcal{S}$.

Now we list some of the most commonly used closure operations. For a class $\mathfrak{X}$ of groups, we define:

$s\,\mathfrak{X} = (G : G \leq H$ for some $H \in \mathfrak{X})$;

$Q\,\mathfrak{X} = (G :$ there exist $H \in \mathfrak{X}$ and an epimorphism from H onto $G)$;

$s_n\,\mathfrak{X} = (G : G$ is subnormal in H for some $H \in \mathfrak{X})$;

$R_0\,\mathfrak{X} = (G :$ there exist $N_i \trianglelefteq G \ (i = 1, \ldots, r)$

$$\text{with } G/N_i \in \mathfrak{X} \text{ and } \bigcap_{i=1}^{r} N_i = 1).$$

Note that a group $G \in \mathrm{R}_0 \, \mathfrak{X}$ if and only if G is isomorphic with a subdirect product of a direct product of a finite set of $\mathfrak{X}$-groups ([DH92, II, 1.18]).

$$\mathrm{N}_0 \, \mathfrak{X} = \big(G : \text{there exist } K_i \text{ subnormal in } G \ (i = 1, \ldots, r)$$
$$\text{with } K_i \in \mathfrak{X} \text{ and } G = \langle K_1, \ldots, K_r \rangle\big);$$
$$\mathrm{D}_0 \, \mathfrak{X} = (G : G = H_1 \times \cdots \times H_r \text{ with each } H_i \in \mathfrak{X});$$
$$\mathrm{E}_{\varPhi} \, \mathfrak{X} = (G : \text{there exists } N \trianglelefteq G \text{ with } N \leq \varPhi(G) \text{ and } G/N \in \mathfrak{X}).$$

The operations S_n and Q, and N_0 and R_0 are dual in the well-known duality between normal subgroup and factor group: this will become more apparent in the context of Fitting classes and formations in next sections.

Lemma 2.1.3 ([DH92, II, 1.6]). *The operations defined in the above list are all closure operations.*

We shall say that a class $\mathfrak{X}$ is *subgroup-closed* if $\mathfrak{X} = \mathrm{S}\,\mathfrak{X}$, that is, if every subgroup of an $\mathfrak{X}$-group is again an $\mathfrak{X}$-group; if $\mathfrak{X} = \mathrm{Q}\,\mathfrak{X}$, we shall say that $\mathfrak{X}$ is an *homomorph*, that is, every epimorphic image of an $\mathfrak{X}$ is an $\mathfrak{X}$-group. If $\mathfrak{X} = \mathrm{S}_n \, \mathfrak{X}$, we might say that $\mathfrak{X}$ is *subnormal subgroup-closed* and if $\mathfrak{X} = \mathrm{R}_0 \, \mathfrak{X}$, we could say that $\mathfrak{X}$ is *residually closed*. An $\mathrm{E}_{\varPhi}$-closed class is called *saturated*.

The product of two closure operations need not be a closure operation since it may easily fail to be idempotent. This leads us to make the following definition. Let $\{\mathrm{A}_\lambda : \lambda \in \varLambda\}$ be a set of operations (not necessarily closure operations). We define $\mathrm{c} = \langle \mathrm{A}_\lambda : \lambda \in \varLambda \rangle$, the *closure operation generated by the* A_λ, as that closure operation whose closed classes are the classes of groups that are A_λ-closed for every $\lambda \in \varLambda$. That is, $\mathrm{c}\,\mathfrak{X} = \bigcap \{\mathfrak{Y} : \mathfrak{X} \subseteq \mathfrak{Y} = \mathrm{A}_\lambda \, \mathfrak{Y} \text{ for all } \lambda \in \varLambda\}$ for any class $\mathfrak{X}$ of groups.

It is easily verified that c is the uniquely determined least closure operation such that $\mathrm{A}_\lambda \leq \mathrm{c}$ for every $\lambda \in \varLambda$.

Of particular interest are $\langle \mathrm{A} \rangle$, the closure operation generated by the operation A, and also $\langle \mathrm{A}, \mathrm{B} \rangle$. In the latter case $\mathrm{A}\,\mathrm{B}$ and $\mathrm{B}\,\mathrm{A}$ may differ from $\langle \mathrm{A}, \mathrm{B} \rangle$, even although A and B are closure operations.

Now follows a simple but useful criterion for the product of two closure operations to be a closure operation.

Proposition 2.1.4 ([DH92, II, 1.16]). *If A and B are closure operations, any two of the following statements are equivalent:*

1. $\mathrm{A}\,\mathrm{B}$ *is a closure operation;*
2. $\mathrm{B}\,\mathrm{A} \leq \mathrm{A}\,\mathrm{B}$;
3. $\mathrm{A}\,\mathrm{B} = \langle \mathrm{A}, \mathrm{B} \rangle$.

Next we give a list of some situations in which the criterion may be applied.

Lemma 2.1.5 ([DH92, II, 1.17 and 1.18]).

1. $\mathrm{Q}\,\mathrm{E}_{\varPhi} \leq \mathrm{E}_{\varPhi}\,\mathrm{Q}$. *Thus $\mathrm{E}_{\varPhi}\,\mathrm{Q}$ is a closure operation.*
2. $\mathrm{D}_0\,\mathrm{S} \leq \mathrm{S}\,\mathrm{D}_0$. *Hence $\mathrm{S}\,\mathrm{D}_0$ is a closure operation.*

3. $D_0 E_\Phi \leq E_\Phi D_0$. *Hence* $E_\Phi D_0$ *is a closure operation.*

4. $R_0 Q \leq Q R_0$, *whence* $Q R_0$ *is a closure operation. Moreover,* $R_0 \leq S D_0$, *whence every* $S D_0$-*closed class is* R_0-*closed.*

We shall adhere to the conventions about the empty class exposed in [DH92, II, p. 271].

2.2 Formations: Basic properties and results

Some of the most important classes of groups are formations. They are considered in some detail in the present section. We gather together facts of a general nature about formations and we give some important examples. Some classical results are also included.

Definition 2.2.1. *A* formation *is a class of groups which is both* Q-*closed and* R_0-*closed, that is, a class of groups* $\mathfrak{F}$ *is a formation if* $\mathfrak{F}$ *has the following two properties:*

1. *If* $G \in \mathfrak{F}$ *and* $N \trianglelefteq G$, *then* $G/N \in \mathfrak{F}$;
2. *If* N_1, $N_2 \trianglelefteq G$ *with* $N_1 \cap N_2 = 1$ *and* $G/N_i \in \mathfrak{F}$ *for* $i = 1, 2$, *then* $G \in \mathfrak{F}$.

By Lemma 2.1.5, $Q R_0 = \langle Q, R_0 \rangle$. Hence a class $\mathfrak{F}$ is a formation if and only if $\mathfrak{F} = Q R_0 \mathfrak{F}$. If $\mathfrak{X}$ is a class of groups, we shall sometimes write form $\mathfrak{X}$ instead of $Q R_0 \mathfrak{X}$ for the *formation generated by* $\mathfrak{X}$.

Note that a class of groups which is simultaneously closed under S, Q, and D_0 is a formation by Lemma 2.1.5. Therefore the class $\mathfrak{N}_c$ of nilpotent groups of class at most c, the class $\mathfrak{S}^{(d)}$ of soluble groups of derived length at most d, the class $\mathfrak{E}(n)$ of groups of exponent at most n, the class $\mathfrak{U}$ of supersoluble groups, and the class $\mathfrak{A}$ of abelian groups are the most classical examples of formations. They are $\langle S, Q, D_0 \rangle$-closed classes of groups.

The following elementary fact is useful in establishing the structure of minimal counterexamples in proofs involving Q- and R_0-closed classes.

Proposition 2.2.2 ([DH92, II, 2.5]). *Let* $\mathfrak{X}$ *and* $\mathfrak{Y}$ *be classes of groups.*

1. *Let* $\mathfrak{X} = Q \mathfrak{X}$, $\mathfrak{Y} = R_0 \mathfrak{Y}$, *and let* G *be a group of minimal order in* $\mathfrak{X} \backslash \mathfrak{Y}$. *Then* G *is monolithic (i.e.* G *has a unique minimal normal subgroup). If, in addition,* $\mathfrak{Y}$ *is saturated, then* G *is primitive.*

2. *Let* G *be a group of minimal order in* $R_0 \mathfrak{X} \backslash \mathfrak{X}$. *Then* G *has a normal subgroups* N_1 *and* N_2 *such that* $G/N_i \in \mathfrak{X}$ *for* $i = 1, 2$ *and* $N_1 \cap N_2 = 1$. *If* $\mathfrak{X} = Q \mathfrak{X}$, *then* N_1 *and* N_2 *can be chosen to be minimal normal subgroups of* G.

The next lemma provides some more examples of formations.

Lemma 2.2.3. *1. If* S *is a non-abelian simple group, then* $D_0\big((S) \cup (1)\big) = D_0(S, 1)$ *is a* $\langle S_n, N_0 \rangle$-*closed formation. Hence* form$(S) = D_0(S, 1)$.

2. *If $\mathfrak{F}$ and $\mathfrak{G}$ are formations and $\mathfrak{F} \cap \mathfrak{G} = (1)$, then $\mathrm{D_0}(\mathfrak{F} \cup \mathfrak{G}) = \mathrm{R_0}(\mathfrak{F} \cup \mathfrak{G})$.*

3. *Let $\emptyset \neq \mathfrak{F}$ be a formation and let S be a non-abelian simple group. Then*
$$\mathrm{Q\,R_0}(\mathfrak{F}, S) = \mathrm{D_0}(\mathfrak{F}, S) = \mathrm{D_0}\big(\mathfrak{F} \cup (S)\big).$$

Proof. 1. Write $\mathfrak{D} = \mathrm{D_0}(S, 1)$. Applying [DH92, A, 4.13], every normal subgroup of a $\mathfrak{D}$-group is a direct product of a subset of direct components isomorphic with S. Hence $\mathfrak{D}$ is S_n-closed. In addition, every normal subgroup N of a group $G \in \mathfrak{D}$ satisfies $G = N \times C_G(N)$. Hence $G/N \in \mathfrak{D}$ and $\mathfrak{D}$ is Q-closed.

Assume that $\mathrm{R_0}\,\mathfrak{D} \neq \mathfrak{D}$ and derive a contradiction. Let G be a group of minimal order in $\mathrm{R_0}\,\mathfrak{D} \setminus \mathfrak{D}$. Then, by Proposition 2.2.2, G has minimal normal subgroups N_1 and N_2 such that $G/N_i \in \mathfrak{D}$, $i = 1, 2$, and $N_1 \cap N_2 = 1$. Consider the normal subgroup $N_2 N_1/N_1$ of G/N_1. Since $G/N_1 \in \mathfrak{D}$, it follows that $G/N_1 = N_2 N_1/N_1 \times R/N_1$ and $N_2 N_1/N_1$ and R/N_1 are direct products of copies of S. In particular, $G = (N_1 N_2)R$ and $R \cap N_1 N_2 = N_1$. It implies that $R \cap N_2 = 1$ and $G = RN_2$. But $G/N_2 \in \mathfrak{D}$ and so $R \in \mathfrak{D}$. Hence $G \in \mathfrak{D}$, contrary to our initial supposition. Consequently $\mathfrak{D}$ is $\mathrm{R_0}$-closed and hence $\mathfrak{D}$ is a formation. It is clear then that $\mathfrak{D} = \mathrm{form}(S)$.

Finally we show that $\mathfrak{D}$ is $\mathrm{N_0}$-closed. Let N_1 and N_2 be normal subgroups of a group $G = N_1 N_2$ such that $N_i \in \mathfrak{D}$, $i = 1, 2$. Then $M = N_1 \cap N_2 \in \mathfrak{D}$ and $G/M \in \mathrm{D_0}\,\mathfrak{D} = \mathfrak{D}$. Moreover if $C_i = C_{M_i}(M)$, it is clear that $C_1 \cap C_2 \leq C_M(M) = 1$ and $|C_i| = |N_i : M|$, $i = 1, 2$. Hence $C_1 C_2 = C_G(M)$ is isomorphic to G/M. Consequently $G = M \times C_G(M) \in \mathfrak{D}$. We can conclude that $\mathfrak{D}$ is $\mathrm{N_0}$-closed.

2. Clearly $\mathrm{D_0}(\mathfrak{F} \cup \mathfrak{G}) \subseteq \mathrm{R_0}(\mathfrak{F} \cup \mathfrak{G})$. Let $G \in \mathrm{R_0}(\mathfrak{F} \cup \mathfrak{G})$. Then G has normal subgroups N_i, $i = 1, \ldots, n$, such that $G/N_i \in \mathfrak{F}$ and G has normal subgroups M_i, $i = 1, \ldots, m$, such that $G/M_i \in \mathfrak{G}$. Moreover $\big(\bigcap_{i=1}^{n} N_i\big) \cap \big(\bigcap_{j=1}^{m} M_j\big) = 1$. Put $N = \bigcap_{i=1}^{n} N_i$ and $M = \bigcap_{j=1}^{m} M_j$. Then $G/N \in \mathrm{R_0}\,\mathfrak{F} = \mathfrak{F}$ and $G/M \in \mathrm{R_0}\,\mathfrak{G} = \mathfrak{G}$. Hence $G/MN \in \mathrm{Q}\mathfrak{F} \cap \mathrm{Q}\mathfrak{G} = \mathfrak{F} \cap \mathfrak{G} = (1)$. It follows that $G = MN \cong M \times N$ and $G \in \mathrm{D_0}(\mathfrak{F} \cup \mathfrak{G})$. Hence $\mathrm{D_0}(\mathfrak{F} \cup \mathfrak{G}) = \mathrm{R_0}(\mathfrak{F} \cup \mathfrak{G})$.

3. Denote $\mathfrak{D} = \mathrm{D_0}(\mathfrak{F}, S) = \mathrm{D_0}\big(\mathfrak{F} \cup (S)\big)$. Clearly we may assume $S \notin \mathfrak{F}$. In this case, $\mathrm{D_0}(S, 1) \cap \mathfrak{F} = (1)$ and $\mathfrak{D} = \mathrm{D_0}\big(\mathfrak{F}, \mathrm{D_0}(S, 1)\big) = \mathrm{R_0}\big(\mathfrak{F}, \mathrm{D_0}(S, 1)\big)$ by Statement 2. In particular, $\mathfrak{D}$ is $\mathrm{R_0}$-closed.

Let $G \in \mathfrak{D}$ and N a normal subgroup of G. Since $G \in \mathfrak{D}$, we have that $G = M_1 \times M_2$, $M_1 \in \mathfrak{F}$ and $M_2 \in \mathrm{D_0}(S, 1)$. If N is contained in either M_1 or M_2, then $G/N \in \mathfrak{D}$ and if $M_1 \cap N = M_2 \cap N = 1$, then $N \leq Z(G) = Z(M_1) \times Z(M_2)$. Since groups in $\mathrm{D_0}(S, 1)$ have trivial centre, we have that $N \leq M_1$, with contradicts $N \cap M_1 = 1$. Hence either $N \leq M_1$ or $N \leq M_2$. In both cases, $G/N \in \mathfrak{D}$. This implies that $\mathfrak{D}$ is Q-closed and so $\mathfrak{D}$ is indeed a formation. $\qquad\square$

An important result in the theory of formations is the theorem of D. W. Barnes and O. H. Kegel that shows that a if a group with a prescribed action appears as a Frattini chief factor of a group in a given formation, then it will also appear as a complemented chief factor of a group in the same formation. The proof of this result depends on the following lemma.

Lemma 2.2.4 ([BBPR96a]). *Let the group $G = NB$ be the product of two subgroups N and B. Assume that N is normal in G. Since B acts by conjugation on N, we can construct the semidirect product, $X = [N]B$, with respect to this action. Then the natural map $\alpha\colon X \longrightarrow G$ given by $(nb)^\alpha = nb$, for every $n \in N$ and $b \in B$, is an epimorphism, $\mathrm{Ker}(\alpha) \cap N = 1$ and $\mathrm{Ker}(\alpha) \leq \mathrm{C}_X(N)$.*

Corollary 2.2.5 ([BK66]). *Let $\mathfrak{F}$ be a formation. Let M and N be normal subgroups of a group $G \in \mathfrak{F}$. Assume that $M \leq \mathrm{C}_G(N)$ and form the semidirect product $H = [N](G/M)$ with respect to the action of G/M on N by conjugation. Then $H \in \mathfrak{F}$.*

Proof. Consider G acting on N by conjugation and construct $X = [N]G$, the corresponding semidirect product. By Lemma 2.2.4, there exists an epimorphism $\alpha\colon X \longrightarrow G = NG$ such that $\mathrm{Ker}(\alpha) \cap N = 1$. Since $X/\mathrm{Ker}(\alpha) \cong G \in \mathfrak{F}$ and $X/N \cong G \in \mathfrak{F}$, it follows that $X \in \mathrm{R}_0\,\mathfrak{F} = \mathfrak{F}$. Now M is a normal subgroup of X contained in G and $X/M \cong [N](G/M)$. Hence $X/M \in \mathrm{Q}\,\mathfrak{F} = \mathfrak{F}$. □

Let G be a group in a formation $\mathfrak{F}$ and let N be an abelian normal subgroup of G. Suppose that U is a subgroup of G such that $G = UN$. Then, by Lemma 2.2.4, G is an epimorphic image of $X = [N]U$, where U acts on N by conjugation. If $Z = N \cap U$, we have that $Z \leq \mathrm{C}_G(N)$ and it is a normal subgroup of X. Moreover, $X/Z \cong [N](U/Z) \cong [N](G/N) \in \mathfrak{F}$ by Corollary 2.2.5. Since X has a normal subgroup, X_1 say, such that $X/X_1 \cong G \in \mathfrak{F}$ and $X_1 \cap U = 1$, it follows that $X \in \mathfrak{F}$. In particular, $U \in \mathfrak{F}$.

This result is a particular case of the following theorem of R. M. Bryant, R. A. Bryce, and B. Hartley.

Theorem 2.2.6 ([BBH70]). *Let U be a subgroup of a group G such that $G = UN$ for some nilpotent normal subgroup N of G. If G belongs to a formation $\mathfrak{F}$, then U is an $\mathfrak{F}$-group.*

The proof of this result also involves an application of Lemma 2.2.4. We need to prove a preliminary lemma.

Assume that G is a group and N a normal subgroup of G. Let N^* be a copy of the subgroup N and consider G acting by conjugation on N^*. Denote $X = [N^*]G$ the semidirect product of N^* with G with respect to this action.

If G is a group and n is a positive integer, denote $\mathrm{K}_1(G) = G$ and $\mathrm{K}_n(G) = [G, \mathrm{K}_{n-1}(G)]$ ([Hup67, III, 1.9]).

Lemma 2.2.7. *With the above notation*

$$\mathrm{K}_n([N, N^*]N) \leq \mathrm{K}_{n+1}(N^*)\,\mathrm{K}_n(N) \qquad \text{for all } n \in \mathbb{N}.$$

Proof. We use induction on n. We write a star $(^*)$ to denote the image by the G-isomorphism between N and N^*. Let x, $y \in N$. Then $[x, y^*] = x^{-1}(y^*)^{-1}xy^* = x^{-1}(y^{-1})^*xy^* = \big((y^{-1})^*\big)^x y^* = \big((y^{-1})^x y\big)^* = [x, y]^* = [x^*, y^*]$. This argument shows that if A and B are subgroups of N, then

$[A, B^*] = [A^*, B^*]$. In particular, $[N, N^*] = (N^*)'$ and so $\mathrm{K}_1([N, N^*]N) = [N, N^*]N = (N^*)'N = \mathrm{K}_2(N^*)\,\mathrm{K}_1(N)$. Now assume that the lemma holds for a given value of $n \geq 1$. Then

$$
\begin{aligned}
\mathrm{K}_{n+1}([N, N^*]N) &= \big[\mathrm{K}_n([N, N^*]N), [N, N^*]N\big] && \text{by definition} \\
&\leq \big[\mathrm{K}_{n+1}(N^*)\,\mathrm{K}_n(N), [N, N^*]N\big] && \text{by inductive hypothesis} \\
&= \big[\mathrm{K}_{n+1}(N^*), [N, N^*]N\big] \\
&\quad \cdot \big[\mathrm{K}_n(N), [N, N^*]N\big] && \text{by [DH92, A, 7.4 (f)]} \\
&= \big[\mathrm{K}_{n+1}(N^*), [N, N^*]\big]\big[\mathrm{K}_{n+1}(N^*), N\big] \\
&\quad \cdot \big[\mathrm{K}_n(N), [N, N^*]\big]\big[\mathrm{K}_n(N), N\big] && \text{by [DH92, A, 7.4 (f)]} \\
&\leq \mathrm{K}_{n+2}(N^*)\,\mathrm{K}_{n+1}(N) && \text{because } \big[\mathrm{K}_n(N), [N, N^*]\big] \\
& && \quad = [\mathrm{K}_n(N^*), \mathrm{K}_2(N^*)]
\end{aligned}
$$

because of the preceeding argument and applying [Hup67, III, 2.11].This completes the induction step and with it the proof of the lemma. $\qquad\square$

Proof (of Theorem 2.2.6). Assume that the result is not true and let G be a counterexample of minimal order. Then there exists a nilpotent normal subgroup N of G and a proper subgroup U of G such that $G = NU$, $G \in \mathfrak{F}$, and $U \notin \mathfrak{F}$. Among the pairs (N, U) of subgroups of G satisfying the above condition, we choose a pair such that $|G : U| + \mathrm{cl}(N)$ is minimal (here $\mathrm{cl}(N)$ denotes the nilpotency class of N). Let V be a maximal subgroup of G containing U. Then $V = U(V \cap N)$ and $G = VN$. If $U \neq V$, then $|G : V| + \mathrm{cl}(N) < |G : U| + \mathrm{cl}(N)$ and so $V \in \mathfrak{F}$ by the choice of the pair (N, U). Therefore $U \in \mathfrak{F}$ by minimality of G, contrary to the choice of G. Therefore $U = V$ is a maximal subgroup of G. If $Z = \mathrm{Z}(N)$ were not contained in U, then $G = U\,\mathrm{Z}(N)$ and U would be in $\mathfrak{F}$ by the above argument. This would contradict the choice of G. Consequently $\mathrm{Z}(N)$ is contained in U. Denote $X = [N^*]U$ the semidirect product of a copy of N with U as usual. By Lemma 2.2.4, there exists an epimorphism $\alpha\colon X \longrightarrow UN = G$ and $\mathrm{Ker}(\alpha) \cap N^* = \mathrm{Ker}(\alpha) \cap U = 1$. It is clear that Z is a normal subgroup of G and $X/Z \cong [N^*](U/Z)$. Now we consider the group $T = [N^*](G/Z)$. Note that $T \in \mathfrak{F}$ by Corollary 2.2.5 and $[N^*](U/Z)$ is a supplement of $\langle (N/Z)^T \rangle$ in T. Moreover $\langle (N/Z)^T \rangle = [N/Z, T](N/Z) = [N/Z, N^*][N/Z, G/Z](N/Z) = [N/Z, N^*](N/Z)$. If $c = \mathrm{cl}(N)$, we have that $\mathrm{K}_c\big(\langle (N/Z)^T \rangle\big) = \mathrm{K}_c([N, N^*]N)Z/Z$ is contained in $\mathrm{K}_{c+1}(N^*)\,\mathrm{K}_c(N)Z/Z$ by Lemma 2.2.7. Since $\mathrm{K}_{c+1}(N^*) = 1$ and $\mathrm{K}_c(N) \leq Z$, it follows that $\mathrm{K}_c\big(\langle (N/Z)^T \rangle\big) = 1$ and $\langle (N/Z)^T \rangle$ is a normal nilpotent subgroup of T whose nilpotency class is less than c. Consequently, since $T \in \mathfrak{F}$, we have that $[N^*](U/Z) \in \mathfrak{F}$ by the minimal choice of G. Hence $X \in \mathrm{R}_0\mathfrak{F} = \mathfrak{F}$. This contradicts the choice of G and shows that U is, like G, and $\mathfrak{F}$-group. $\qquad\square$

Let $\mathfrak{F}$ be a non-empty formation. Each group G has a smallest normal subgroup whose quotient belongs to $\mathfrak{F}$; this is called the $\mathfrak{F}$-*residual* of G and it is denoted by $G^{\mathfrak{F}}$. Clearly $G^{\mathfrak{F}}$ is a characteristic subgroup of G and $G^{\mathfrak{F}} = \bigcap\{N \trianglelefteq G : G/N \in \mathfrak{F}\}$. Consequently $G^{\mathfrak{F}} = 1$ if and only if $G \in \mathfrak{F}$.

The following proposition will be useful for later applications.

Proposition 2.2.8. *Let $\mathfrak{F}$ be a non-empty formation and let G be a group. If N is normal subgroup of G, we have:*

1. $(G/N)^{\mathfrak{F}} = G^{\mathfrak{F}}N/N$.
2. *If U is a subgroup of $G = UN$, then $U^{\mathfrak{F}}N = G^{\mathfrak{F}}N$.*
3. *If N is nilpotent and $G = UN$, then $U^{\mathfrak{F}}$ is contained in $G^{\mathfrak{F}}$.*

Proof.   1.   Denote $R/N = (G/N)^{\mathfrak{F}}$. It is clear that $G/R \in \mathfrak{F}$. Hence $G^{\mathfrak{F}}N$ is contained in R. Moreover $G/G^{\mathfrak{F}}N \in \mathfrak{F}$. It implies that $(G/N)/(G^{\mathfrak{F}}N/N) \in \mathfrak{F}$ and so $R/N \leq G^{\mathfrak{F}}N/N$. Therefore $R = G^{\mathfrak{F}}N$.

2.   Let θ denote the canonical isomorphism from $G/N = UN/N$ to $U/(U \cap N)$. Then $\left((G/N)^{\mathfrak{F}}\right)^{\theta} = \left(U/(U \cap N)\right)^{\mathfrak{F}}$, which is equal to $U^{\mathfrak{F}}(U \cap N)/(U \cap N)$ by Statement 1. Hence $U^{\mathfrak{F}}N/N = (G/N)^{\mathfrak{F}} = G^{\mathfrak{F}}N/N$ and $U^{\mathfrak{F}}N = G^{\mathfrak{F}}N$.

3.   We have $G/G^{\mathfrak{F}} = (UG^{\mathfrak{F}}/G^{\mathfrak{F}})(NG^{\mathfrak{F}}/G^{\mathfrak{F}}) \in \mathfrak{F}$. Applying Theorem 2.2.6, it follows that $UG^{\mathfrak{F}}/G^{\mathfrak{F}} \in \mathfrak{F}$. Therefore $U^{\mathfrak{F}}$ is contained in $U \cap G^{\mathfrak{F}}$.   □

Remark 2.2.9. We shall use henceforth the property of the $\mathfrak{F}$-residual stated in Statement 1 without further comment.

In general, the product class of two formations is not a formation in general ([DH92, IV, 1.6]). Fortunately we know a way of modifying the definition of a product to ensure that the corresponding product of two formations is again a formation. It was due to W. Gaschütz ([Gas69]).

Definition 2.2.10. *Let $\mathfrak{F}$ and $\mathfrak{G}$ be formations. We define $\mathfrak{F} \circ \mathfrak{G} := (G : G^{\mathfrak{G}} \in \mathfrak{F})$, and call $\mathfrak{F} \circ \mathfrak{G}$ the formation product of $\mathfrak{F}$ with $\mathfrak{G}$.*

This product enjoys the following properties ([DH92, IV, pages 337–338]).

Proposition 2.2.11. *Let $\mathfrak{F}$, $\mathfrak{G}$, and $\mathfrak{H}$ be formations. Then:*

1. $\mathfrak{F} \circ \mathfrak{G} \subseteq \mathfrak{F}\mathfrak{G}$*, and $\mathfrak{G} \subseteq \mathfrak{F} \circ \mathfrak{G}$ if $\mathfrak{F}$ is non-empty,*
2. *if $\mathfrak{F}$ is s_n-closed, then $\mathfrak{F} \circ \mathfrak{G} = \mathfrak{F}\mathfrak{G}$,*
3. $\mathfrak{F} \circ \mathfrak{G}$ *is a formation,*
4. $G^{\mathfrak{F} \circ \mathfrak{G}} = (G^{\mathfrak{G}})^{\mathfrak{F}}$ *for all $G \in \mathfrak{E}$, and*
5. $(\mathfrak{F} \circ \mathfrak{G}) \circ \mathfrak{H} = \mathfrak{F} \circ (\mathfrak{G} \circ \mathfrak{H})$.

Example 2.2.12. Let $\mathfrak{F}$ and $\mathfrak{G}$ be formations such that $\pi(\mathfrak{F}) \cap \pi(\mathfrak{G}) = \emptyset$. Denote $\pi_1 = \pi(\mathfrak{F})$ and $\pi_2 = \pi(\mathfrak{G})$. Then $\mathfrak{F} \times \mathfrak{G} = \left(G : G = \mathrm{O}_{\pi_1}(G) \times \mathrm{O}_{\pi_2}(G), \mathrm{O}_{\pi_1}(G) \in \mathfrak{F}, \mathrm{O}_{\pi_2}(G) \in \mathfrak{G}\right)$ is a formation. Moreover, if $\mathfrak{F}$ and $\mathfrak{G}$ are saturated, then $\mathfrak{F} \times \mathfrak{G}$ is saturated and, if $\mathfrak{F}$ and $\mathfrak{G}$ are subgroup-closed, then $\mathfrak{F} \times \mathfrak{G}$ is also subgroup-closed.

Proof. Note that $\mathfrak{F} \times \mathfrak{G} = (\mathfrak{F} \circ \mathfrak{G}) \cap (\mathfrak{G} \circ \mathfrak{F})$. Hence $\mathfrak{F} \times \mathfrak{G}$ is a formation by Proposition 2.2.11 (3).

Assume that $\mathfrak{F}$ and $\mathfrak{G}$ are saturated, then $\mathfrak{F} \circ \mathfrak{G}$ and $\mathfrak{G} \circ \mathfrak{F}$ are saturated by [DH92, IV, 3.13]. Hence $\mathfrak{F} \times \mathfrak{G}$ is saturated. $\qquad\square$

Remark 2.2.13. Example 2.2.12 could be generalised along the following lines: Let $\mathcal{I}$ be a non-empty set. For each $i \in \mathcal{I}$, let $\mathfrak{F}_i$ be a subgroup-closed saturated formation. Assume that $\pi(\mathfrak{F}_i) \cap \pi(\mathfrak{F}_j) = \emptyset$ for all $i, j \in \mathcal{I}$, $i \neq j$. Denote $\pi_i = \pi(\mathfrak{F}_i)$, $i \in \mathcal{I}$. Then

$$\bigtimes_{i \in \mathcal{I}} \mathfrak{F}_i := \big(G = O_{\pi_{i_1}}(G) \times \cdots \times$$
$$O_{\pi_{i_n}}(G) : O_{\pi_{i_j}}(G) \in \mathfrak{F}_{i_j}, 1 \leq j \leq n, \{i_1, \ldots, i_n\} \subseteq \mathcal{I}\big)$$

is a subgroup-closed saturated formation.

One of the most important results in the theory of classes of groups is the one stating the equivalence between saturated and local formations. W. Gaschütz introduced the local method to generate saturated formations in the soluble universe. Later, his student U. Lubeseder [Lub63] proved that every saturated formation in the soluble universe can be described in that way. Lubeseder's proof requires elementary ideas from the theory of modular representations, which are dispensed with in the account of the theorem in Huppert's book [Hup67]. In 1978 P. Schmid [Sch78] showed that solubility is not necessary for Lubeseder's result, although his proof reinstates the facts about blocks used by Lubeseder and also makes essential use of a theorem of W. Gaschütz, about the existence of certain non-split extensions. In an unpublished manuscript, R. Baer has investigated a different definition of local formation. It is more flexible than the one studied by P. Schmid in that the simple components, rather than the primes dividing its order, are used to label chief factors and its automorphism group. Hence the actions allowed on the insoluble chief factors can be independent of those on the abelian chief factors. R. Baer's approach leads to a family of formations called Baer-local formations. Local formations are a special case of Baer-local formations. Moreover, in the universe of soluble groups the two definitions coincide. The price to be paid for the greater generality of Baer's approach is that the Baer-local formations are no longer saturated. However, there is a suitable substitute for saturation. We say that a formation is *solubly saturated* if it is closed under taking extensions by the Frattini subgroup of the soluble radical. Of course solubly saturation is weaker than saturation. But it evidently coincides with saturation for classes of finite soluble groups, and it plays a precisely analogous role in Baer's generalisation: the Baer-local formations are precisely the solubly saturated ones.

Another approach to the Gaschütz-Lubeseder theorem in the finite universe is due to L. A. Shemetkov (see [She78, She97, She00]). He uses functions assigning a certain formation to each group (he recently calls them *satellites*) and introduces the notion of composition formation. It turns out that the composition formations are exactly the Baer-local formations ([She97]).

Any function $f\colon \mathbb{P} \longrightarrow \{\text{formations}\}$ is called a *formation function*. Given a formation function f, we define the class $\mathrm{LF}(f)$ as the class of all groups satisfying the following condition:

$G \in \mathrm{LF}(f)$ if, for all chief factors H/K of G and for all primes p dividing $|H/K|$, we have that $\mathrm{Aut}_G(H/K) = G/\mathrm{C}_G(H/K) \in f(p)$.
$$\tag{2.1}$$

The class $\mathrm{LF}(f)$ is a formation ([DH92, IV, 3.3]).

Definition 2.2.14. *A class of groups $\mathfrak{F}$ is called a* local formation *if there exists a formation function f such that $\mathfrak{F} = \mathrm{LF}(f)$.*

Theorem 2.2.15 (Gaschütz-Lubeseder-Schmid, [DH92, IV, 4.6]). *A formation $\mathfrak{F}$ is saturated if and only if $\mathfrak{F}$ is local.*

A map $f\colon \mathfrak{J} \longrightarrow \{\text{classes of groups}\}$ is called a *Baer function* provided that $f(J)$ is a formation for all simple groups J.

If f is a Baer function, then the class of all groups G satisfying that $\mathrm{Aut}_G(H/K)$ belongs to $f(J)$ if H/K is a chief factor of G whose composition factor is isomorphic to J is a formation. We call this formation the *Baer-local formation defined by f*, and we denote it by $\mathrm{BLF}(f)$. A class $\mathfrak{B}$ is called a *Baer-local formation* if $\mathfrak{B} = \mathrm{BLF}(f)$ for some Baer function f.

Theorem 2.2.16 ([DH92, IV, 4.12]). *The solubly saturated formations are precisely the Baer-local formations.*

Example 2.2.17. Let $\mathfrak{Q}$ be the solubly saturated formation locally defined by the Baer function f given by

$$f(S) = \begin{cases} (1) & \text{when } S \cong C_p, \text{ and} \\ \mathrm{D}_0(1, S) & \text{when } S \in \mathfrak{J} \setminus \mathbb{P}. \end{cases}$$

The formation in Example 2.2.17 is characterised as the class $\mathfrak{Q}$ of all groups G such that $G = \mathrm{C}_G^*(H/K)$ for every chief factor H/K of G, i.e. the class of all groups which only induce inner automorphisms on each chief factor (see [Ben70]). Groups in $\mathfrak{Q}$ are called *quasinilpotent*. It is clear that a nilpotent group is just a soluble quasinilpotent group. $\mathfrak{Q}$ is also s_n-closed and N_0-closed, that is, $\mathfrak{Q}$ is a Fitting class (see Section 2.3). Each group G has a largest normal $\mathfrak{Q}$-subgroup. This subgroup is called the *generalised Fitting subgroup* of G, and it is denoted by $\mathrm{F}^*(G)$. Applying [HB82b, X, 13.9, 13.10], $\mathrm{F}^*(G)$ is the intersection of the innerisers of the chief factors of G.

The main properties of the generalised Fitting subgroup are analysed in many books, for example in Section 13 of Chapter X of the book of B. Huppert and N. Blackburn [HB82b] or, more recently, in Section 6.5 of the book of H. Kurzweil and B. Stellmacher [KS04]. Let us summarise here the most relevant.

Definitions 2.2.18. *1. A group G is said to be* quasisimple *if G is perfect, i.e. $G' = G$, and $G/\operatorname{Z}(G)$ is a non-abelian simple group.*

2. A subgroup H of a group G is said to be a component *of G if H is a quasisimple subnormal subgroup of G.*

3. The cosocle *of a group G, $\operatorname{Cosoc}(G)$, is the intersection of all maximal normal subgroups of G.*

4. A group G is said to be comonolithic *if G has a unique maximal normal subgroup.*

5. If G is a comonolithic group and $M = \operatorname{Cosoc}(G)$ is the unique maximal normal subgroup of G, then the quotient G/M is said to be the head *of G.*

It is clear that if G is a quasisimple group, then G is comonolithic and $\operatorname{Cosoc}(G) = \operatorname{Z}(G)$. Also it is easy to see that if K is a normal subgroup of a quasisimple group G, then G/K is also a quasisimple group.

The next result, due to H. Wielandt, will be extremely useful.

Theorem 2.2.19 ([Wie39]). *If H and K are subnormal subgroups of a group G, H is perfect and comonolithic and H is not contained in K, then K normalises H.*

Proposition 2.2.20 (see [KS04, 6.5.3]). *If H and K are components of a group G, then either $H = K$ or $[H, K] = 1$.*

Definition 2.2.21. *The* layer *of a group G is the subgroup $\operatorname{E}(G)$ generated by all components of G, i.e. the product of all components of G.*

Proposition 2.2.22. *Let G be a group.*

1. We have that $\operatorname{F}^*(G) = \operatorname{F}(G)\operatorname{E}(G)$ and $[\operatorname{F}(G), \operatorname{E}(G)] = 1$; in fact

$$\operatorname{C}_{\operatorname{F}^*(G)}\big(\operatorname{E}(G)\big) = \operatorname{F}(G)$$

(see [HB82b, X, 13.15]).

2. $\operatorname{E}(G)$ is the central product of all components of G, but not the product of any proper subset of them (see [HB82b, X, 13.18] or [KS04, 6.5.6]).

3. $\operatorname{F}^*(G)/\operatorname{F}(G) = \operatorname{Soc}\big(\operatorname{C}_G(\operatorname{F}(G))\operatorname{F}(G)/\operatorname{F}(G)\big)$ (see [HB82b, X, 13.13]).

4. $\operatorname{C}_G\big(\operatorname{F}^*(G)\big) \leq \operatorname{F}^*(G)$ (see [HB82b, X, 13.12] or [KS04, 6.5.8]).

2.3 Schunck classes and projectors

The starting point of the theory of classes of groups is the attempt to develop a generalised Sylow theory, which leads to an investigation into the problem of the existence of certain conjugacy classes of subgroups in finite groups.

Perhaps the most well-known existence and conjugacy theorem is Sylow's theorem which says, in its simplest form, that if p is a prime and G is a group, then the maximal p-subgroups of G are conjugate in G.

The beginnings of this particular area of finite group theory came with P. Hall's generalisation of Sylow's theorem for soluble groups.

Theorem 2.3.1 ([Hal28]). *Let G be a soluble group and π any set of primes. Then the maximal π-subgroups of G are conjugate in G.*

In a soluble group G, the π-subgroups of G with π'-index in G are exactly the maximal π-subgroups of G and they are referred as the *Hall π-subgroups of G*. Of course, this is the terminology we shall use here and we also use $\operatorname{Hall}_\pi(G)$ to denote the set of all Hall π-subgroups of G.

By considering the order and index of Hall π-subgroups, it is easy to see that they satisfy the following three conditions.

Let N be a normal subgroup of a soluble group G. Then:

1. $\operatorname{Hall}_\pi(G/N) = \{SN/N : S \in \operatorname{Hall}_\pi(G)\}$.
2. $\operatorname{Hall}_\pi(N) = \{S \cap N : S \in \operatorname{Hall}_\pi(G)\}$.
3. If $T/N \in \operatorname{Hall}_\pi(G/N)$ and $S \in \operatorname{Hall}_\pi(T)$, then $S \in \operatorname{Hall}_\pi(G)$.

In particular, Hall π-subgroups behave well as we pass from G to a factor group G/N or to a normal subgroup N. It is these three properties that have led to wide generalisations, the first and third properties leading to the theory of saturated formations and Schunck classes and the associated projectors and the second property to the theory of Fitting classes and injectors.

Both generalisations lead to conjugacy classes of subgroups in soluble groups which share another important property of Hall subgroups:

If $S \in \operatorname{Hall}_\pi(G)$ and $S \leq H \leq G$, then $S \in \operatorname{Hall}_\pi(H)$.

The results of P. L. M. Sylow and P. Hall seemed to be suggestive of certain arithmetic properties of groups. In 1937, P. Hall [Hal37] discovered the so-called Hall systems of a soluble group G by choosing a set of Hall p'-subgroups of G, one for each prime p, and taking their intersections. He proved that if Σ and Σ^* are two Hall systems of G, there exists an element $g \in G$ such that $\Sigma^* = \Sigma^g$. That is, G acts transitively by conjugation on the set of its Hall systems. Therefore the number of Hall systems of a soluble group is the index in G of the stabiliser of a Hall system with respect to the action of G. This stabiliser is what P. Hall called the *system normaliser*. P. Hall observed that all system normalisers are nilpotent, they are preserved under epimorphisms, and form a conjugacy class of subgroups. It is important to remark that system normalisers, defined in terms of the genuine Sylow structure of a soluble group, cover the central chief factors of the group and avoid the eccentric ones. Hence they are the natural connection between the two characterisations of soluble groups, the arithmetic and the normal (or commutator) structure, and afford a "measure of the nilpotence" of the group.

Despite of system normalisers, there was a little evidence to suggest the huge proliferation of results in the area. However, in 1961, R. W. Carter [Car61] introduced another conjugacy class of subgroups in each soluble group.

A *Carter subgroup* of a group G is a nilpotent subgroup C of G such that $N_G(C) = C$. He proves:

Theorem 2.3.2 (R. W. Carter). *A soluble group G has a Carter subgroup and any two Carter subgroups of G are conjugate in G.*

It is clear that a Carter subgroup of a group G is a maximal nilpotent subgroup of G. However, if G is a non-nilpotent soluble group, then G has a maximal nilpotent subgroup which is not a Carter subgroup. Consequently, regarding maximality, the Carter subgroups are not to the class $\mathfrak{N}$ of all nilpotent groups as the Hall subgroups are to the class $\mathfrak{S}_\pi$ of all soluble π-groups.

However, there is a close relation between the abovementioned conjugacy classes: in a group G of nilpotent length 2, the Carter subgroups of G are exactly the system normalisers of G. Carter's theorem would then follow from this observation using induction on the nilpotent length.

W. Gaschütz viewed the Carter subgroups as analogues of the Sylow and Hall subgroups of a soluble groups and in 1963 published a seminal paper [Gas63] where a broad extension of the Hall and Carter subgroups was presented. The theory of formations was born. The new "covering subgroups" had many of the properties of the Sylow and Hall subgroups, but the theory was not arithmetic one, based on the orders of subgroups. Instead, the important idea was concerned with group classes having the same properties. He introduces the concepts of formation and $\mathfrak{F}$-covering subgroup, for a class $\mathfrak{F}$ of groups. He then proved that if $\mathfrak{F}$ is a formation of soluble groups, then every soluble group has an $\mathfrak{F}$-covering subgroup if and only if $\mathfrak{F}$ is saturated and, in this case, the $\mathfrak{F}$-covering subgroups form a unique conjugacy class of subgroups. These $\mathfrak{F}$-covering subgroups coincided with the Sylow p-subgroups, the Hall π-subgroups, and the Carter subgroups in the respective classes$\mathfrak{S}_p$, $\mathfrak{S}_\pi$, and $\mathfrak{N}$. Subsequently, H. Schunck in his Kiel Dissertation [Sch66], written under the direction of W. Gaschütz and H. Schubert, discovered precisely which classes $\mathfrak{Z}$, of soluble groups, always gave rise to $\mathfrak{Z}$-covering subgroups; he showed that these classes can be characterised in terms of their primitive groups and that they form a considerably larger family of classes than the saturated formations [Sch67]. They are known as Schunck classes and are the main concern of this section.

Two years later, W. Gaschütz [Gas69] defined the notion of $\mathfrak{Z}$-projector of some class of soluble groups $\mathfrak{Z}$ and showed that for Schunck classes $\mathfrak{Z}$ the notions of $\mathfrak{Z}$-projector and $\mathfrak{Z}$-covering subgroup coincided. Since then the term "projector" has been widely adopted in this context in preference to "covering subgroup."

The first serious attempt to broaden the study of Schunck classes and their projectors and take it outside the soluble universe was made by P. Förster [För84b], [För85b], and [För85c]. However, it should be remarked that the study of projective classes outside the soluble universe had been observed and treated previously by R. P. Erickson [Eri82] and P. Schmid [Sch74].

In the first part of the section we gather some of the basic facts about Schunck classes and projectors. The book of K. Doerk and T. O. Hawkes

[DH92] presents, in its Chapter III, an excellent treatment of this theme. Hence we refer to it for the proof of some of the results we include here. In the second part, we study the relationship between Schunck classes and formations and some Schunck classes which are close to saturated formations.

Definitions 2.3.3. *Let $\mathfrak{H}$ be a class of groups.*

1. *A subgroup X of a group G is said to be $\mathfrak{H}$-maximal subgroup of G if $X \in \mathfrak{H}$ and if $X \leq K \in \mathfrak{H}$, then $X = K$.*
   Denote by $\mathrm{Max}_{\mathfrak{H}}(G)$ the set of all $\mathfrak{H}$-maximal subgroups of G.
2. *A subgroup U of a group G is called an $\mathfrak{H}$-projector of G if UN/N is $\mathfrak{H}$-maximal in G/N for all $N \trianglelefteq G$.*
   We shall use $\mathrm{Proj}_{\mathfrak{H}}(G)$ to denote the (possibly empty) set of $\mathfrak{H}$-projectors of a group G.
3. *An $\mathfrak{H}$-covering subgroup of a group G is a subgroup E of G satisfying the following two conditions:*
   a) $E \in \mathrm{Max}_{\mathfrak{H}}(G)$, and
   b) if $T \leq G$, $E \leq T$, $N \trianglelefteq T$, and $T/N \in \mathfrak{H}$, then $T = NE$.
   The set of all $\mathfrak{H}$-covering subgroups of a group G will be denoted by $\mathrm{Cov}_{\mathfrak{H}}(G)$.

Consider the case where $\mathfrak{H} = \mathfrak{E}_\pi$, the class of all π-groups. Then, for each soluble group G,

$$\mathrm{Max}_{\mathfrak{H}}(G) = \mathrm{Proj}_{\mathfrak{H}}(G) = \mathrm{Cov}_{\mathfrak{H}}(G) = \mathrm{Hall}_\pi(G) \neq \emptyset.$$

However, the set $\mathrm{Hall}_\pi(G)$ can be empty for a non-soluble group G. In fact, P. Förster [För85b] showed that if π a non-empty set of primes such that, for each group G, $\mathrm{Hall}_\pi(G) \neq \emptyset$ then, either $\pi = \{p\}$, p a prime, or $\pi = \mathbb{P}$.

Definitions 2.3.4. *1. A class $\mathfrak{H}$ is called* projective *if $\mathrm{Proj}_{\mathfrak{H}}(G) \neq \emptyset$ for each group G.*
2. *A class $\mathfrak{H}$ will be called a* Gaschütz class *if $\mathrm{Cov}_{\mathfrak{H}}(G) \neq \emptyset$ for each group G.*
3. *A class $\mathfrak{H}$ is said to be a* Schunck class *if $\mathfrak{H}$ is a homomorph that comprises precisely those groups whose primitive epimorphic images are in $\mathfrak{H}$.*

Remark 2.3.5. If $\mathfrak{H}$ is a Schunck class, then $\mathfrak{H}$ is a saturated homomorph, that is, $\mathrm{E}_\Phi \, \mathfrak{H} = \mathfrak{H} = \mathrm{Q} \, \mathfrak{H}$.

It is clear that a saturated formation is a Schunck class. However, the family of all Schunck classes is considerably larger than the one of all saturated formations. Moreover, the fundamental role of the local definition of saturated formations, and therefore the arithmetic properties, are substituted in the case of Schunck classes by the primitive quotients of the group, and therefore by the role of maximal subgroups. In 1974, K. Doerk [Doe71, Doe74] introduced the concept of the *boundary* of a Schunck class, which plays a fundamental role in the study of Schunck classes.

Definitions 2.3.6.   *1. For a class $\mathfrak{H}$ of groups, define*

$$\mathrm{b}(\mathfrak{H}) := (G \in \mathfrak{E} \setminus \mathfrak{H} : G/N \in \mathfrak{H} \text{ for all } 1 \neq N \trianglelefteq G).$$

Obviously, $\mathrm{b}(\emptyset) = \mathrm{b}(\mathfrak{E}) = \emptyset$.
$\mathrm{b}(\mathfrak{H})$ *is said to be the* boundary *of $\mathfrak{H}$.*
We say that a class of groups $\mathfrak{B}$ is a boundary *if $\mathfrak{B} = \mathrm{b}(\mathfrak{H})$ for some class of groups $\mathfrak{H}$.*
2. If $\mathfrak{Y}$ is a class of groups, define

$$\mathrm{h}(\mathfrak{Y}) := \big(G \in \mathfrak{E} : \mathrm{Q}(G) \cap \mathfrak{Y} = \emptyset\big),$$

that is, the class of $\mathfrak{Y}$-perfect groups.
Clearly $\mathrm{h}(\emptyset) = \mathfrak{E}$ and $\mathrm{h}(\mathfrak{E}) = \emptyset$ if $1 \in \mathfrak{Y}$. Moreover $\mathfrak{Y} \cap \mathrm{h}(\mathfrak{Y}) = \emptyset$ and $\mathrm{h}(\mathfrak{Y})$ is a homomorph.

Theorem 2.3.7.   *1. Let $\mathfrak{H}$ be a homomorph. Then $\mathrm{h}\big(\mathrm{b}(\mathfrak{H})\big) = \mathfrak{H}$.*
2. Let $\mathfrak{B}$ be a boundary. Then $\mathrm{b}\big(\mathrm{h}(\mathfrak{B})\big) = \mathfrak{B}$.

Proof.   1. Clearly $\mathfrak{H} \subseteq \mathrm{h}\big(\mathrm{b}(\mathfrak{H})\big)$. Suppose that $\mathrm{h}\big(\mathrm{b}(\mathfrak{H})\big)$ is not contained in $\mathfrak{H}$ and let G be a group in $\mathrm{h}\big(\mathrm{b}(\mathfrak{H})\big) \setminus \mathfrak{H}$ of minimal order. Since $\mathrm{h}\big(\mathrm{b}(\mathfrak{H})\big)$ is a homomorph, it follows that $G \in \mathrm{b}(\mathfrak{H})$. This is a contradiction. Therefore $\mathfrak{H} = \mathrm{h}\big(\mathrm{b}(\mathfrak{H})\big)$.

2. If $\mathfrak{B} = \mathrm{b}(\mathfrak{X})$ for some class of groups $\mathfrak{X}$, it follows that every proper epimorphic image of a group in $\mathfrak{B}$ does not belong to $\mathfrak{B}$. Hence $\mathfrak{B} \subseteq \mathrm{b}\big(\mathrm{h}(\mathfrak{B})\big)$.

Assume that $G \in \mathrm{b}\big(\mathrm{h}(\mathfrak{B})\big)$. Then $G \notin \mathrm{h}(\mathfrak{B})$ and so there exists a normal subgroup N of G such that $G/N \in \mathfrak{B}$. Suppose that $N \neq 1$. In this case $G/N \in \mathrm{h}(\mathfrak{B})$ by definition of boundary. This contradicts our choice of G. Consequently $N = 1$ and $G \in \mathfrak{B}$. This means that $\mathfrak{B} = \mathrm{b}\big(\mathrm{h}(\mathfrak{B})\big)$. $\square$

Theorem 2.3.8. *Let $\emptyset \neq \mathfrak{H}$ be a class of groups. $\mathfrak{H}$ is a Schunck class if and only if $\mathfrak{H}$ is a homomorph and $\mathrm{b}(\mathfrak{H}) \subseteq \mathfrak{P}$.*

Proof. If $\mathfrak{H}$ is a Schunck class, then $\mathfrak{H}$ is a homomorph. Suppose that $G \in \mathrm{b}(\mathfrak{H})$ but G is not primitive. Then every epimorphic image of G belongs to $\mathfrak{H}$. Hence $G \in \mathfrak{H}$, contrary to the choice of G. Consequently G is primitive and $\mathrm{b}(\mathfrak{H}) \subseteq \mathfrak{P}$.

Conversely suppose that $\mathfrak{H}$ is a homomorph and $\mathrm{b}(\mathfrak{H}) \subseteq \mathfrak{P}$. Let G be a group whose epimorphic primitive images lie in $\mathfrak{H}$. Suppose that G does not belong to $\mathfrak{H}$. Then $G \in \mathrm{b}(\mathfrak{H})$ by [DH92, III, 2.2 (c)]. In this case G is primitive. This implies $G \in \mathfrak{H}$, which contradicts the fact that $G \in \mathrm{b}(\mathfrak{H})$. Therefore $G \in \mathfrak{H}$. $\square$

Corollary 2.3.9. *For each class $\mathfrak{X}$, the class*

$$\mathrm{P}\,\mathrm{Q}\,\mathfrak{X} = \big(G : \mathrm{Q}(G) \cap \mathfrak{P} \subseteq \mathrm{Q}\,\mathfrak{X}\big)$$

is the smallest Schunck class containing $\mathfrak{X}$. Therefore $\mathfrak{X}$ is a Schunck class if and only if $\mathfrak{X} = \mathrm{P}\,\mathrm{Q}\,\mathfrak{X}$.

Proof. Clearly $\mathfrak{X} \subseteq \mathrm{P\,Q}\,\mathfrak{X}$ and $\mathrm{P\,Q}\,\mathfrak{X}$ is a homomorph. Moreover if $G \in \mathrm{b}(\mathrm{P\,Q}\,\mathfrak{X})$, then $G \notin \mathrm{P\,Q}\,\mathfrak{X}$. Hence $\mathrm{Q}(G) \cap \mathfrak{P}$ is not contained in $\mathrm{Q}\,\mathfrak{X}$. Since $G/N \in \mathrm{P\,Q}\,\mathfrak{X}$ for all $1 \neq N \lhd G$, it follows that $\mathrm{Q}(G/N) \cap \mathfrak{P} \subseteq \mathrm{Q}\,\mathfrak{X}$. Therefore G should be primitive. Applying Theorem 2.3.8, $\mathrm{P\,Q}\,\mathfrak{X}$ is a Schunck class. Now if $\mathfrak{H}$ is a Schunck class and $\mathfrak{X} \subseteq \mathfrak{H}$, then $\mathrm{Q}\,\mathfrak{X} \subseteq \mathrm{Q}\,\mathfrak{H} = \mathfrak{H}$. Hence $\mathrm{P\,Q}\,\mathfrak{X} \subseteq \mathrm{P\,Q}\,\mathfrak{H} = \mathfrak{H}$. $\qquad\square$

Remark 2.3.10. The above corollary shows, in particular, that $\mathrm{P\,Q}$ is a closure operation.

For another closure operation for Schunck classes related to crowns, the reader is referred to [Haw73] and [Laf84a].

Combining Theorem 2.3.7 and Theorem 2.3.8, we have:

Corollary 2.3.11. *If $\mathfrak{Z}$ is a boundary composed of primitive groups, then $\mathrm{h}(\mathfrak{Z})$ is a Schunck class.*

In general, Schunck classes are not R_0-closed, as the following example shows:

Example 2.3.12. Let E be a non-abelian simple group. Then $\mathfrak{Z} = (E \times E)$ is a boundary composed of a primitive group. Hence $\mathrm{h}(\mathfrak{Z})$ is a Schunck class by Corollary 2.3.11. Clearly $E \in \mathrm{h}(\mathfrak{Z})$ and $E \times E \in \mathrm{R}_0\,\mathrm{h}(\mathfrak{Z}) \setminus \mathrm{h}(\mathfrak{Z})$.

This example also shows that $\mathrm{h}(\mathfrak{Z})$ is not D_0-closed.

Suppose that $\mathfrak{H}$ is a projective class. If G is a group in $\mathfrak{H}$, then $\mathrm{Proj}_{\mathfrak{H}}(G) = \{G\}$. Hence, for each normal subgroup N of G, we have that $\mathrm{Proj}_{\mathfrak{H}}(G) = \{G/N\}$ by definition of $\mathfrak{H}$-projector. Therefore $G/N \in \mathfrak{H}$. Moreover, if G is a group such that every primitive epimorphic images of G is in $\mathfrak{H}$, then G must be an $\mathfrak{H}$-group because otherwise an $\mathfrak{H}$-projector E of G would be contained in a maximal subgroup M of G. Since $G/\mathrm{Core}_G(M)$ is primitive, it would follow that $G/\mathrm{Core}_G(M) \in \mathfrak{H}$, and so $G = E\,\mathrm{Core}_G(M) = M$. This contradiction yields that $G \in \mathfrak{H}$ and $\mathfrak{H}$ is a Schunck class. It is proved in [DH92, III, 3.10] that the converse is also true.

Theorem 2.3.13 ([DH92, III, 3.10]). *A class $\mathfrak{H} \neq \emptyset$ is projective if and only if it is a Schunck class.*

Förster's proof of the above theorem depends on the following property of the projectors and covering subgroups. This property, usually called $\mathfrak{H}$-inductivity, allows him to translate the question of the universal existence of $\mathfrak{H}$-projectors and $\mathfrak{H}$-covering subgroups to the groups in the boundary of $\mathfrak{H}$ (see [DH92, III, 3.8]).

Proposition 2.3.14 ([DH92, III, 3.7]). *Let $\mathfrak{H}$ be a homomorph. Let f denote a function which assigns to each group G a possibly empty set $\mathrm{f}(G)$ of subgroups of G. If f is either of the functions $\mathrm{Proj}_{\mathfrak{H}}(\cdot)$ or $\mathrm{Cov}_{\mathfrak{H}}(\cdot)$, then it satisfies the following two conditions:*

1. $G \in f(G)$ if and only if $G \in \mathfrak{H}$;
2. whenever $N \trianglelefteq G$, $N \leq V \leq G$, $U \in f(V)$, and $V/N \in f(G/N)$, then $U \in f(G)$.

W. Gaschütz [Gas69] actually proved that in the soluble universe the Schunck classes are exactly the Gaschütz classes. However, in the general finite universe, they are no longer coincidental. For instance, the alternating group of degree 5 has no $\mathfrak{N}$-covering subgroups. However, we have:

Theorem 2.3.15 ([DH92, III, 3.11]). *A Schunck class whose boundary contains no primitive groups of type 2 is a Gaschütz class.*

The conjugacy question can be also resolved partially by examining the groups in the boundary. This approach works well for covering subgroups (see [DH92, III, 3.13]), but in the case of projectors, we must work with Schunck classes of monolithic boundary (see [DH92, III, 3.19]). In this context, the following result turns out to be crucial. It will also be used in other chapters.

Proposition 2.3.16 ([För84b]). *Let $\mathfrak{H}$ be a Schunck class. Then $b(\mathfrak{H}) \cap \mathfrak{P}_3 = \emptyset$ if and only if $\mathfrak{H}$ satisfies the following property:*

Let H be an $\mathfrak{H}$-maximal subgroup of G such that $G = H \, F^*(G)$. Then H is an $\mathfrak{H}$-projector of G. $\qquad\qquad (2.2)$

Proof. Assume that $b(\mathfrak{H}) \cap \mathfrak{P}_3 = \emptyset$. Let G be a group with an $\mathfrak{H}$-maximal subgroup H such that $G = H \, F^*(G)$. We prove that H is an $\mathfrak{H}$-projector of G by induction on $|G|$. First we claim:

For all $N \trianglelefteq G$, the hypotheses are inherited from H, G to H, HN. $\quad(2.3)$

Since $G = H \, F^*(G)$, we have that $HN = H\bigl(HN \cap F^*(G)\bigr) = H \, F^*(HN)$ as $HN \cap F^*(G)$ is a normal quasinilpotent subgroup of HN.

For all minimal normal subgroups M of G such that $G/M \notin \mathfrak{H}$, the hypotheses are inherited from H, G to HM/M, G/M. $\quad(2.4)$

It follows that $G/M = (HM/M)\bigl(F^*(G)M/M\bigr)$ and $F^*(G)M/M$ is a normal quasinilpotent subgroup of G/M. Hence $G/M = (HM/M) \, F^*(G/M)$.

Assume that K/M is an $\mathfrak{H}$-maximal subgroup of G/M such that $HM/M \leq K/M$. Since $G/M \notin \mathfrak{H}$, we have that K is a proper subgroup of G. Moreover, if $K \in \mathfrak{H}$, we have $H = K$ by the $\mathfrak{H}$-maximality of K. Therefore we may assume that $K \notin \mathfrak{H}$. Since $F^*(G)$ is contained in the inneriser of M, it follows that $G = H \, F^*(G) = HM \, C_G(M)$ and so $K = HM \, C_K(M)$.

Assume that M is abelian. Then $K = H \, C_K(M)$ and M is a minimal normal subgroup of K. Since $K \notin \mathfrak{H}$, we have that there exists a normal subgroup C of K such that $K/C \in b(\mathfrak{H}) \subseteq \mathfrak{P}_1 \cup \mathfrak{P}_2$ by [DH92, III, 2.2c]. It is clear that M is not contained in C. Hence $K/C \in \mathfrak{P}_1$ and $\operatorname{Soc}(K/C) = MC/C$. Consequently $MC = C_K(MC/C) = C_K(M)$. Moreover $HC \neq K$ because $K/C \notin \mathfrak{H}$. This implies that HC is a maximal subgroup of K, $K = (HC)(MC)$ and $HC \cap MC = C$. On the other hand, $HC \cap M = 1$ and

$HC \cong HCM/M = K/M \in \mathfrak{H}$. The $\mathfrak{H}$-maximality of H in G implies that C is contained in H and so $K = HM$.

Suppose that M is not abelian. Put $C = \mathrm{C}_G(M)$. Then G/C is a primitive group of type 2 and $\mathrm{Soc}(G/C) = MC/C$ by Proposition 1.1.14. Suppose, by way of contradiction, that $G/C \in \mathfrak{H}$. Then $K/\mathrm{C}_K(M) = HM\mathrm{C}_K(M)/\mathrm{C}_K(M) \cong G/C \in \mathfrak{H}$. Let N be a normal subgroup of K such that $K/N \in \mathrm{b}(\mathfrak{H})$ ([DH92, III, 2.2 (c)]). Since $N \cap M = 1$, we have that $\mathrm{Soc}(K/N) = NM/N$. Hence $N = \mathrm{C}_G(NM/M) = C$. This contradicts $K/C \in \mathfrak{H}$. Therefore $G/C \notin \mathfrak{H}$ and HC is a proper subgroup of G. By induction, H is an $\mathfrak{H}$-projector of HC. Hence $H(HC \cap M)/(HC \cap M)$ is an $\mathfrak{H}$-projector of $HC/(HC \cap M) \cong G/M$. Therefore HM/M is an $\mathfrak{H}$-projector of G/M. In particular $HM/M = K/M$. This completes the proof of (2.4).

Assume that $G \in \mathfrak{H}$. Then $H = G$ is an $\mathfrak{H}$-projector of G. Hence we may assume that $G \notin \mathfrak{H}$. If $G/M \in \mathfrak{H}$ for all minimal normal subgroups of G, it follows that $G \in \mathrm{b}(\mathfrak{H})$. Hence G is a monolithic primitive group, $\mathrm{Soc}(G)$ is a minimal normal subgroup of G and $\mathrm{C}_G\big(\mathrm{Soc}(G)\big) \leq \mathrm{Soc}(G)$ by Proposition 1.1.12 and Proposition 1.1.14. Then $G = H\,\mathrm{Soc}(G)$, and from $H \in \mathrm{Max}_{\mathfrak{H}}(G)$ and the fact that $\mathrm{Soc}(G)$ is the unique minimal normal subgroup of G, we derive the claim of the proposition: $H \in \mathrm{Proj}_{\mathfrak{H}}(G)$.

Therefore we may suppose that $G/M \notin \mathfrak{H}$ for some minimal normal subgroup M of G. Then, in view of (2.3) and (2.4), the inductive hypothesis can be applied to yield that $H \in \mathrm{Proj}_{\mathfrak{H}}(HM)$ and $HM/M \in \mathrm{Proj}_{\mathfrak{H}}(G/M)$. By $\mathfrak{H}$-inductivity, $H \in \mathrm{Proj}_{\mathfrak{H}}(G)$.

Conversely assume that $\mathfrak{H}$ satisfies Property 2.2. Suppose that $\mathrm{b}(\mathfrak{H}) \cap \mathfrak{P}_3 \neq \emptyset$ and derive a contradiction. Consider $G \in \mathrm{b}(\mathfrak{H}) \cap \mathfrak{P}_3$. Then, by Theorem 1, $S = \mathrm{Soc}(G) = A \times B$, where A and B are the two unique minimal normal subgroups of G and both are complemented by a core-free maximal subgroup U of G. Consider the subgroup $T = U \cap S$. Then T is isomorphic to A and B. Since U is primitive by Corollary 1.1.13, it follows that T is not contained in $\Phi(U)$. Let Y be a proper subgroup of U such that $U = TY$. Write $X = YB$. Then $XA = YS = YTS = US = G$. Hence $X/(X \cap A) \cong G/A \in \mathfrak{H}$. Let L be a minimal supplement of $X \cap A$ in X. Clearly $X \cap A \cap L$ is contained in $\Phi(L)$ and so $L \in \mathrm{E}_\Phi \mathfrak{H} = \mathfrak{H}$. Let H be an $\mathfrak{H}$-maximal subgroup of G containing L. Since $G = XA = LA$, it follows that $G = HA$. Applying (2.2), H is an $\mathfrak{H}$-projector of G. Since $G/B \in \mathfrak{H}$, we have that $G = HB$. Therefore H is a core-free maximal subgroup of G such that $H \cap B = H \cap A = 1$. In particular, $L = H$. This implies that $X = G$ and so $Y = U$, contrary to the choice of Y. Consequently $\mathrm{b}(\mathfrak{H}) \cap \mathfrak{P}_3 = \emptyset$. $\qquad\square$

The same arguments to those used in the proof of Proposition 2.3.16 lead to the following result.

Proposition 2.3.17. *Let $\mathfrak{H}$ be a Schunck class. If H is an $\mathfrak{H}$-maximal subgroup of a group G such that $G = H\,\mathrm{F}(G)$, then H is an $\mathfrak{H}$-projector of G.*

We now direct our attention towards certain formations that may be naturally associated with a Schunck class. In fact, our next objective is to prove that a Schunck class $\mathfrak{H}$ contains a unique largest formation. This result was proved independently by U. Kattwinkel [Kat77] and K.-U. Schaller [Sch77] in the soluble universe and by J. Lafuente [Laf84a] in the general case. We begin with a definition.

Definition 2.3.18. *Let $\mathfrak{H}$ be a class of groups.*

A chief factor H/K of a group G is said to be $\mathfrak{H}$-central in G if $[H/K]*G$ is in $\mathfrak{H}$. Otherwise, the chief factor H/K is said to be $\mathfrak{H}$-eccentric in G.

Note that if $\mathfrak{H}$ is a saturated formation and H is the canonical local definition of $\mathfrak{H}$ (see [DH92, IV, 3.9]), then a chief factor H/K of a group G is $\mathfrak{H}$-central in G if and only if H/K is H-central in G in the sense of [DH92, IV, 3.1].

Let $\mathfrak{H}$ be a class of groups. Denote by $f(\mathfrak{H})$ the class of all groups G in which every chief factor is $\mathfrak{H}$-central. The class $f_1(\mathfrak{H})$ is defined to be the class of all groups such that $[H/K]\big(G/\operatorname{C}_G(H/K)\big) \in \mathfrak{H}$ for every chief factor H/K of G.

It follows that $f_1(\mathfrak{H})$ is contained in $f(\mathfrak{H})$ but the equality does not hold in general.

Example 2.3.19. Let S be a non-abelian simple group. Consider the class $\mathfrak{H}$ of all groups with no quotient isomorphic to the direct product $S \times S$ of two copies of S, i.e. the Schunck class of all $(S \times S)$-perfect groups, is a Schunck class whose boundary is $b(\mathfrak{H}) = (S \times S)$. The group $S \times S \in f(\mathfrak{H}) \setminus f_1(\mathfrak{H})$. Note that $f_1(\mathfrak{H})$ is contained in $\mathfrak{H}$.

Theorem 2.3.20. *Let $\mathfrak{H}$ be a class of groups. Then:*

1. *$f(\mathfrak{H})$ and $f_1(\mathfrak{H})$ are formations.*
2. *If $\mathfrak{F}$ is a formation contained in $\mathfrak{H}$, then $\mathfrak{F}$ is contained in $f(\mathfrak{H})$.*
3. *Let $\mathfrak{H}$ be a Schunck class. Then $b(\mathfrak{H}) \cap \mathfrak{P}_3 = \emptyset$ if and only if $f(\mathfrak{H})$ is the largest formation contained in $\mathfrak{H}$.*
4. *Let $\mathfrak{H}$ be a Schunck class. Then $f_1(\mathfrak{H})$ is the largest formation contained in $\mathfrak{H}$.*

Proof.   1. Clearly $f(\mathfrak{H})$ and $f_1(\mathfrak{H})$ are formations.

2. Suppose, arguing for contradiction, that $\mathfrak{F}$ is a formation contained in $\mathfrak{H}$ such that $\mathfrak{F}$ is not contained in $f(\mathfrak{H})$. Let G be a group in $\mathfrak{F} \setminus f(\mathfrak{H})$ of minimal order. Then G has a unique minimal normal subgroup N and $G/N \in f(\mathfrak{H})$ by [DH92, II, 2.5]. Assume that N is non-abelian. Then $\operatorname{C}_G(N) = 1$ and G is isomorphic to $[N]*G$. Hence $[N]*G \in \mathfrak{H}$. Now if N is an abelian, we have that $[N]\big(G/\operatorname{C}_G(N)\big) \in \mathfrak{F} \subseteq \mathfrak{H}$ by Corollary 2.2.5. Therefore $G \in f(\mathfrak{H})$, contrary to our supposition. Hence $\mathfrak{F}$ is contained in $f(\mathfrak{H})$ and Statement 2 holds.

3. Let $\mathfrak{H}$ be a Schunck class such that $b(\mathfrak{H}) \cap \mathfrak{P}_3 = \emptyset$. Assume that $f(\mathfrak{H})$ is not contained in $\mathfrak{H}$. Then a group of minimal order in the non-empty class $f(\mathfrak{H}) \setminus \mathfrak{H}$ is in the boundary of $\mathfrak{H}$. Hence G is a monolithic primitive group. If G is a primitive group of type 1, then G is isomorphic to

$[\mathrm{Soc}(G)]\big(G/\,\mathrm{C}_G\big(\mathrm{Soc}(G)\big)\big) \in \mathfrak{H}$ by Proposition 1.1.12 and if G is a primitive group of type 2, then $G = [\mathrm{Soc}(G)] * G \in \mathfrak{H}$. In both cases, we have that $G \in \mathfrak{H}$. This contradiction yields $f(\mathfrak{H}) \subseteq \mathfrak{H}$.

Conversely assume that $f(\mathfrak{H})$ is the largest formation contained in $\mathfrak{H}$. If G is in the boundary of $\mathfrak{H}$ and G is a primitive group of type 3, then G/A and G/B are $\mathfrak{H}$-groups, where A and B are the minimal normal subgroups of G. This implies that G/A and G/B belong to $f(\mathfrak{H})$ because all their factors are $\mathfrak{H}$-central. Hence $G \in \mathfrak{H}$, contrary to assumption. Consequently, $\mathrm{b}(\mathfrak{H}) \cap \mathfrak{P}_3 = \emptyset$.

4. Consider a group $G \in f_1(\mathfrak{H})$ and let N be a normal subgroup of G such that G/N is primitive. If G/N is a primitive group of type 1 or 3, then G/N belongs to $\mathfrak{H}$ by Proposition 1.1.12 (3). If $X = G/N$ is a primitive group of type 2 and $Z = \mathrm{Soc}(G/N)$, then $[Z]X$ is a primitive group of type 3 by Proposition 1.1.12 (3). Hence $[Z]X \in \mathfrak{H}$ and so $X \in \mathfrak{H}$ since $\mathfrak{H}$ is Q-closed. Consequently every primitive epimorphic image of G belongs to $\mathfrak{H}$ and then G is an $\mathfrak{H}$-group.

Let $\mathfrak{F}$ be a formation contained in $\mathfrak{H}$. Then $\mathfrak{F}$ is contained in $f(\mathfrak{H})$ by Statement 2. Let G be an $\mathfrak{F}$-group. Then every abelian chief factor of G is $\mathfrak{H}$-central in G. Suppose that H/K is a non-abelian chief factor of G. Denote $X = [H/K]\big(G/\,\mathrm{C}_G(H/K)\big)$. Then X is a primitive group of type 3 by Proposition 1.1.12 (3) with two minimal normal subgroups X_1 and X_2 such that $X_1 \cap X_2 = 1$ and $X/X_i \in \mathfrak{F}$, $1 \le i \le 2$. Hence $X \in \mathrm{R_0}\mathfrak{F} = \mathfrak{F}$. Therefore $G \in f_1(\mathfrak{H})$ and $\mathfrak{F}$ is contained in $f_1(\mathfrak{H})$. □

Example 3.1.37 shows that a class of groups $\mathfrak{H}$ does not contain a unique largest formation in general.

Example 2.3.21. Every Schunck class whose boundary consists of primitive groups of type 2 is a saturated formation.

Proof. By Theorem 2.3.20 (3), $f(\mathfrak{H})$ is contained in $\mathfrak{H}$. Now if G is a group in $\mathfrak{H}$ and H/K is an abelian chief factor of G, then $[H/K] * G$ is an $\mathfrak{H}$-group because it is not in the boundary of $\mathfrak{H}$. Since every non-abelian chief factor of G is $\mathfrak{H}$-central in G, it follows that $G \in f(\mathfrak{H})$ and $\mathfrak{H}$ is a saturated formation.

We bring the section to a close by studying a concrete family of Schunck classes with an eye to a subsequent application in Chapter 4.

Consider a formation $\mathfrak{F}$. Then, by Lemma 2.1.5 (1), $\mathfrak{H} = \mathrm{E}_\Phi\mathfrak{F}$ is a Schunck class and it is the smallest Schunck class containing $\mathfrak{F}$. Note that a primitive group is in $\mathfrak{H}$ if and only if it is in $\mathfrak{F}$. Hence $\mathfrak{H}$ has monolithic boundary and so $f(\mathfrak{H})$ is the largest formation contained in $\mathfrak{H}$ by Theorem 2.3.20 (3). It follows that $\mathfrak{H} = \mathrm{E}_\Phi f(\mathfrak{H})$, but $\mathfrak{F}$ is not equal to $f(\mathfrak{H})$ in general: if $\mathfrak{F} = \mathfrak{A}$ is the formation of all abelian groups, then $f(\mathrm{E}_\Phi\,\mathfrak{F})$ is the class of all nilpotent groups.

Schunck classes $\mathfrak{H}$ of the form $\mathrm{E}_\Phi\mathfrak{F}$ for some formation $\mathfrak{F}$ can be characterised by the property that each group not in $\mathfrak{H}$ always has a special critical subgroup.

Definition 2.3.22. *Let $\mathfrak{H}$ be a class of groups.*

1. *A maximal subgroup U of a group G is said to be $\mathfrak{H}$-normal in G if the primitive group $G/\operatorname{Core}_G(U)$ is in $\mathfrak{H}$. Otherwise, U is said to be $\mathfrak{H}$-abnormal in G.*
2. *A maximal subgroup U of a group G is said to be $\mathfrak{H}$-critical in G if U is an $\mathfrak{H}$-abnormal critical subgroup of G.*

Note that an $\mathfrak{H}$-critical subgroup is a monolithic maximal subgroup supplementing an $\mathfrak{H}$-eccentric chief factor.

Lemma 2.3.23. *Let $\mathfrak{H}$ be a Schunck class and let G be a group, N a normal subgroup of G, and $M \leq G$. If U is $\mathfrak{H}$-critical in M and $M \cap N \leq U$, then UN/N is $\mathfrak{H}$-critical in MN/N.*

Proof. By Proposition 1.4.10, UN/N is critical in MN/N. Since $N \cap M \leq U$, we have that $\operatorname{Core}_M(U)N/N = \operatorname{Core}_{MN/N}(UN/N)$ and so $M/\operatorname{Core}_M(U) \cong (MN/N)/\operatorname{Core}_{MN/N}(UN/N)$ is not in $\mathfrak{H}$. In other words, UN/N is an $\mathfrak{H}$-critical subgroup of MN/N.   □

Let S be a non-abelian simple group and the Schunck class $\mathfrak{H}$ of all groups with no quotient isomorphic to the direct product $S \times S$ of two copies of S. All monolithic maximal subgroups of the group $G = S \times S$ are $\mathfrak{H}$-normal in G. Hence $G \notin \mathfrak{H}$ and G has no $\mathfrak{H}$-critical subgroups.

The following theorem characterises the Schunck classes of the form $\mathfrak{H} = \mathrm{E}_\Phi \mathfrak{F}$, for some formation $\mathfrak{F}$, among the Schunck classes for which every group which is not in $\mathfrak{H}$ possesses $\mathfrak{H}$-critical subgroups.

Theorem 2.3.24. *For a Schunck class $\mathfrak{H}$, the following statements are pairwise equivalent:*

1. *every group which is not in $\mathfrak{H}$ possesses $\mathfrak{H}$-critical subgroups;*
2. *$\mathfrak{H} = \mathrm{E}_\Phi \mathrm{Q} \, \mathrm{R}_0 \mathcal{P}(\mathfrak{H})$, where $\mathcal{P}(\mathfrak{H})$ is the class of all primitive groups in $\mathfrak{H}$;*
3. *$\mathfrak{H} = \mathrm{E}_\Phi \mathfrak{F}$, for some formation $\mathfrak{F}$;*
4. *a group G belongs to $\mathfrak{H}$ if and only if every minimal normal subgroup of $G/\Phi(G)$ is $\mathfrak{H}$-central in G.*

Proof. 1 implies 2. Since, for every group G, $\Phi(G)$ is the intersection of all normal subgroups N of G such that G/N is primitive and $\mathfrak{H}$ is a homomorph, it follows that $\mathfrak{H} \subseteq \mathrm{E}_\Phi \mathrm{Q} \, \mathrm{R}_0 \mathcal{P}(\mathfrak{H})$.

Let $G \in \mathrm{R}_0 \mathcal{P}(\mathfrak{H})$. Then there exist normal subgroups $N_1, \ldots, N_t$ of G such that $\bigcap_{i=1}^{t} N_i = 1$ and G/N_i is a primitive group in $\mathfrak{H}$, $1 \leq i \leq n$. Consequently, $\Phi(G) = 1$. Suppose that $G \notin \mathfrak{H}$ and let U be an $\mathfrak{H}$-critical subgroup of G. Then U is monolithic and $G = UN$ for some minimal normal subgroup of G. Moreover $N \cap N_i = 1$ for some i. Therefore NN_i/N_i is a chief factor of G which is G-isomorphic to N. If G/N_i is a primitive group of type 1, then $G/N_i \cong [N]\big(G/\operatorname{C}_G(N)\big) \cong G/\operatorname{Core}_G(U) \in \mathfrak{H}$ by Proposition 1.1.12. If G/N_i

is a primitive group of type 2, then $N_i = \mathrm{C}_G(N) = \mathrm{Core}_G(U)$ by Proposition 1.1.14 and $G/N_i = G/\mathrm{Core}_G(U) \in \mathfrak{H}$. Suppose that G/N_i is a primitive group of type 3. Then, by Proposition 1.1.13, NN_i/N_i and $C = \mathrm{C}_G(N)/N_i$ are the minimal normal subgroups of G/N_i and G/C and G/NN_i are primitive groups of type 2. Moreover $N_i = C \cap NN_i$. Assume that $G = UN_i$. Then $N\,\mathrm{Core}_G(U)$ is contained in $N_i\,\mathrm{Core}_G(U)$ because $N\,\mathrm{Core}_G(U)/\mathrm{Core}_G(U) = \mathrm{Soc}\big(N\,\mathrm{Core}_G(U)/\mathrm{Core}_G(U)\big)$. Hence N is abelian. This contradiction shows that N_i is contained in U. Hence $G/\mathrm{Core}_G(U) \in \mathrm{Q}(G/N_i) \subseteq \mathfrak{H}$. In any case, we have that $G/\mathrm{Core}_G(U)$ is an $\mathfrak{H}$-group, contrary to the choice of U. Therefore $G \in \mathfrak{H}$ and the equality $\mathfrak{H} = \mathrm{E}_{\varPhi}\mathrm{Q}\,\mathrm{R}_0\mathcal{P}(\mathfrak{H})$ holds.

Since for every class $\mathfrak{X}$ of groups, the class $\mathrm{Q}\,\mathrm{R}_0\mathfrak{X}$ is a formation, it is clear that 2 implies 3.

3 implies 4. Let G be a group in $\mathfrak{H}$. If $N/\varPhi(G)$ is a minimal normal subgroup of $G/\varPhi(G)$, then $N/\varPhi(G)$ is a supplemented chief factor of G and the primitive group associated with $N/\varPhi(G)$ is isomorphic to a quotient group of G. Hence $[N/\varPhi(G)] * G \in \mathfrak{H}$ and the chief factor $N/\varPhi(G)$ is $\mathfrak{H}$-central in G.

Conversely, assume that every minimal normal subgroup of $G/\varPhi(G)$ is $\mathfrak{H}$-central in G. Without loss of generality we may suppose that $\varPhi(G) = 1$. Let N be a normal subgroup of G such that G/N is a monolithic primitive group. Then $\mathrm{Soc}(G/N) = AN/N$ for some minimal normal subgroup A of G. Since A is $\mathfrak{H}$-central in G, it follows that $G/N \in \mathfrak{H}$ and so $G/N \in \mathfrak{F}$. Therefore $G/\varPhi(G) = G \in \mathrm{Q}\,\mathrm{R}_0\mathfrak{F} = \mathfrak{F}$.

4 implies 1. Let G be a group which is not in $\mathfrak{H}$. Assume first that $\varPhi(G) = 1$. Suppose that all critical subgroups of G are $\mathfrak{H}$-normal in G. This means that each minimal normal subgroup is $\mathfrak{H}$-central in G. By hypothesis, G is in $\mathfrak{H}$. This is a contradiction. Therefore G has an $\mathfrak{H}$-critical subgroup.

For the case $\varPhi(G) \neq 1$, consider the group $G^* = G/\varPhi(G)$ which is not in $\mathfrak{H}$ either. Since $\varPhi(G^*) = 1$, the G^* possesses an $\mathfrak{H}$-critical subgroup $U^* = U/\varPhi(G)$. Clearly U is $\mathfrak{H}$-critical in G. □

The "soluble" version of Theorem 2.3.24 was proved by P. Förster in [För78].

2.4 Fitting classes, Fitting sets, and injectors

The theory of Fitting classes began when B. Fischer in his Habilitationsschrift [Fis66] wanted to see how far it is possible to dualise the theory of saturated formations and projectors by interchanging the roles of normal subgroups and quotients groups. From this point of view the closure operations s_n and N_0 are the natural duals of Q and R_0, and so a Fitting class, i.e. a $\langle\mathrm{s}_n, \mathrm{N}_0\rangle$-closed class, should be regarded as the dual of a formation. However, in the soluble universe, it turns out that Fitting classes parallel Schunck classes more closely in the dual theory because they are precisely the classes for which a theory of injectors, dual of projectors, is possible. At the time of Fischer's initial investigation the projectors were still known by covering subgroups and by close analogy the dual concept chosen by Fischer was the so-called Fischer subgroup: if $\mathfrak{F}$ is a class of groups, a Fischer $\mathfrak{F}$-subgroup belongs to $\mathfrak{F}$ and contains each $\mathfrak{F}$-subgroup that it normalises. For an arbitrary Fitting class $\mathfrak{F}$, Fischer was able to prove that the existence of Fischer $\mathfrak{F}$-subgroups in every soluble group. However, he was not able to prove that the Fischer subgroups of a soluble group are all conjugate. Some years later, R. S. Dark [Dar72] gave an example of a Fitting class $\mathfrak{F}$ and a soluble group which has two conjugacy classes of Fischer $\mathfrak{F}$-subgroups.

As it turned out, the definition of projector, rather than covering subgroup, is the right thing to dualise in order to guarantee conjugacy. In 1967 the concept of injector appears in the celebrated paper "Injektoren endlicher auflösbarer Gruppen" by B. Fischer, W. Gaschütz, and B. Hartley [FGH67]. They prove that a class of soluble groups $\mathfrak{F}$ is a Fitting class if and only if every soluble group has an $\mathfrak{F}$-injector. Moreover, the $\mathfrak{F}$-injectors then form a single conjugacy class.

When $\mathfrak{F}$ is the Fitting class of all soluble π-groups, π a set of primes, the $\mathfrak{F}$-injectors of a soluble group, like its $\mathfrak{F}$-projectors, turn out to be the Hall π-subgroups. This is the only situation in which the injectors and projectors coincide, and so the two theories are quite independent generalisations of the classical Sylow and Hall subgroups.

In fact, as we see in Section 2.2, in the general, non-necessarily soluble, universe, projective classes and Schunck classes remain equivalent concepts. However, in Chapter 7, we shall show that there exist non-injective Fitting classes.

Definition 2.4.1. *A* Fitting class *is a class of groups which is both* s_n-*closed and* N_0-*closed, that is, a class of groups* $\mathfrak{F}$ *is a* Fitting class *if* $\mathfrak{F}$ *has the following two properties:*

1. *if* $G \in \mathfrak{F}$ *and* N *is a subnormal subgroup of* G, *then* $N \in \mathfrak{F}$, *and*
2. *if* N_1 *and* N_2 *are subnormal subgroups of a group* G *and* $G = \langle N_1, N_2 \rangle$, *then* $G \in \mathfrak{F}$.

Hence a class $\mathfrak{F}$ is a Fitting class if and only if $\mathfrak{F} = \langle \mathrm{s}_n, \mathrm{N}_0 \rangle \mathfrak{F}$.

As usual for classes defined by closure operations, the intersection of a family of Fitting classes is again a Fitting class, and the union of a family of Fitting classes which is a directed set with respect to the partial order of inclusion is also a Fitting class.

In particular, if $\mathfrak{Z}$ is a class of groups, the intersection Fit $\mathfrak{Z}$ of all Fitting classes containing $\mathfrak{Z}$ is the smallest Fitting class containing $\mathfrak{Z}$; Fit $\mathfrak{Z} = \langle \mathrm{s}_n, \mathrm{N}_0 \rangle \mathfrak{Z}$ is the *Fitting class generated by* $\mathfrak{Z}$. Note that if S is a non-abelian simple group, then $\mathrm{Fit}(S) = \mathrm{form}(S) = \mathrm{D}_0(1, S)$ by Example 2.2.3 (1).

Historically, the first example of a Fitting class is the class $\mathfrak{N}$ of all nilpotent groups. This was proved by H. Fitting in 1938. The formations $\mathfrak{N}_c$ and $\mathfrak{S}^{(d)}$ are also Fitting classes and, in general, since $\mathrm{D}_0 \leq \mathrm{N}_0$ and $\mathrm{R}_0 \leq \mathrm{SD}_0$, a subgroup closed Fitting class is R_0-closed. However a formation does not need to be a Fitting class. The formations $\mathfrak{A}$ of all abelian groups and $\mathfrak{U}$ of all supersoluble groups are not N_0-closed. Nevertheless, the following result can be used in some contexts as a substitute of the R_0-closure. It is known as the "quasi-R_0-lemma."

Lemma 2.4.2 ([DH92, IX, 1.13]). *Let N_1 and N_2 be normal subgroups of a group G such that $N_1 \cap N_2 = 1$ and G/N_1N_2 is nilpotent. Suppose that $\mathfrak{F}$ is a Fitting class such that $G/N_1 \in \mathfrak{F}$. Then $G \in \mathfrak{F}$ if and only if $G/N_2 \in \mathfrak{F}$.*

Definition 2.4.3. *If $\mathfrak{F}$ is a Fitting class and G is a group, then the subgroup*

$$G_{\mathfrak{F}} = \langle S : S \text{ is a subnormal } \mathfrak{F}\text{-subgroup of } G\}$$

is a normal $\mathfrak{F}$-subgroup of G, and it is called the $\mathfrak{F}$-radical of G.

Remark 2.4.4. If N is a normal subgroup of G and $\mathfrak{F}$ is a Fitting class, then $N_{\mathfrak{F}} = N \cap G_{\mathfrak{F}}$.

As might be expected, the class product of Fitting classes need not be a Fitting class in general (see Step 7 in [DH92, IX, 2.14 (b)]). A special product can be defined, which is dual to the formation product of Definition 2.2.10, which preserves the Fitting class property.

Definitions and notation 2.4.5. *Let $\mathfrak{X}$ and $\mathfrak{Y}$ be Fitting classes.*

1. *$\mathfrak{X} \diamond \mathfrak{Y}$ is the class of all groups G such that $G/G_{\mathfrak{X}} \in \mathfrak{Y}$. (This product, called* Fitting product, *was introduced by Gaschütz, see [DH92, IX, 1.10])*
2. *$\mathfrak{X} \cdot \mathfrak{Y}$ is the class of all groups G such that $G = G_{\mathfrak{X}}G_{\mathfrak{Y}}$.*

Proposition 2.4.6 (see [DH92, IX, 1.12]). *Let $\mathfrak{F}$, $\mathfrak{G}$, and $\mathfrak{H}$ be Fitting classes. Then:*

1. *$\mathfrak{F} \diamond \mathfrak{G} \subseteq \mathfrak{F}\mathfrak{G}$, and $\mathfrak{F} \subseteq \mathfrak{F} \diamond \mathfrak{G}$ if $\mathfrak{G}$ is non-empty,*
2. *if the class $\mathfrak{G}$ is a homomorph, then $\mathfrak{F} \diamond \mathfrak{G} = \mathfrak{F}\mathfrak{G}$,*
3. *$\mathfrak{F} \diamond \mathfrak{G}$ is a Fitting class,*
4. *for all $G \in \mathfrak{E}$, the $\mathfrak{G}$-radical of $G/G_{\mathfrak{F}}$ is $G_{\mathfrak{F} \diamond \mathfrak{G}}/G_{\mathfrak{F}}$, and*
5. *$(\mathfrak{F} \diamond \mathfrak{G}) \diamond \mathfrak{H} = \mathfrak{F} \diamond (\mathfrak{G} \diamond \mathfrak{H})$.*

On the other hand, if $\mathfrak{X}$ and $\mathfrak{Y}$ are Fitting classes, then the class $\mathfrak{X} \cdot \mathfrak{Y}$ is not necessarily a Fitting class (see [DH92, page 575]).

If $\mathfrak{X}$ and $\mathfrak{Y}$ are Fitting classes such that $\mathfrak{X} \subseteq \mathfrak{Y}$ and $\mathfrak{F}$ is a Fitting class, we write that $\mathfrak{F} \in \mathrm{Sec}(\mathfrak{X}, \mathfrak{Y})$ if $\mathfrak{X} \subseteq \mathfrak{F} \subseteq \mathfrak{Y}$; in this case we say that $\mathfrak{F}$ is in the *section* of $\mathfrak{X}$ and $\mathfrak{Y}$. The most known section of Fitting classes is the Lockett section.

In [Loc71], Lockett exploited the aberrant behaviour of radicals in direct products and show how to associate with each Fitting class $\mathfrak{X}$ another containing it, called $\mathfrak{X}^*$, such that $(G \times H)_{\mathfrak{X}^*} = G_{\mathfrak{X}^*} \times H_{\mathfrak{X}^*}$. Lockett's universe was the soluble one, but the definition of $\mathfrak{X}^*$, its Fitting character and its behaviour with respect to direct products still hold in the general finite universe (see [DH92, X, Section 1]). Thus $\mathfrak{X}^*$ is the class of all groups G such that $(G \times G)_{\mathfrak{X}}$ is subdirect in $G \times G$. We now define $\mathfrak{X}_*$ as the intersection of all Fitting classes $\mathfrak{F}$ such that $\mathfrak{F}^* = \mathfrak{X}^*$. Obviously $\mathfrak{X}_*$ is a Fitting class and it has the remarkable property that $(\mathfrak{X}_*)^* = \mathfrak{X}^*$ by [DH92, X, 1.13].

Definition 2.4.7. *Let $\mathfrak{X}$ be a Fitting class.*

1. *$\mathfrak{X}$ is a* Lockett class *if $\mathfrak{X} = \mathfrak{X}^*$.*
2. *The* Lockett section *of $\mathfrak{X}$ is $\operatorname{Locksec}(\mathfrak{X}) = \operatorname{Sec}(\mathfrak{X}_*, \mathfrak{X}^*)$.*

Observe that if $\mathfrak{X}$ is a Fitting class, each group $G \notin \mathfrak{X}$ such that every proper subnormal subgroup of G is in $\mathfrak{X}$ has to be comonolithic, by the N_0-closure of G, and $G_{\mathfrak{X}} = \operatorname{Cosoc}(G)$. Hence the following definition makes sense.

As we have seen in Section 2.3 boundaries play an important role in the study of Schunck classes. In fact, they provide a method to construct Schunck classes by exploiting the one-to-one correspondence between homomorphs and boundaries given by the maps b and h (Theorem 2.3.7). It is clear how the analogous maps b and h for Fitting classes must be defined.

Definitions and notation 2.4.8. *Let $\mathfrak{X}$ be a Fitting class.*

1. *The* boundary *of $\mathfrak{X}$, $b(\mathfrak{X})$, is the class of all groups $X \notin \mathfrak{X}$ such that every proper subnormal subgroup of X is an $\mathfrak{X}$-group.*
2. *$\bar{b}(\mathfrak{X}) = \big(G \in b(\mathfrak{X}) : G = G'\big)$.*
3. *For a prime p, we denote $b_p\mathfrak{X} = (G \in b(\mathfrak{X}) : G/\operatorname{Cosoc}(G) \in \mathfrak{S}_p)$.*
4. *$\mathfrak{X}^b$ denotes the Fitting class generated by the cosocles of all groups $G \in b(\mathfrak{X})$:*
$$\mathfrak{X}^b = \operatorname{Fit}\big(\operatorname{Cosoc}(G) : G \in b(\mathfrak{X})\big).$$

Definition 2.4.9. *If $\mathfrak{Y}$ is a class of groups, denote*
$$h(\mathfrak{Y}) = \big(H : s_n(H) \cap \mathfrak{F} = \emptyset\big).$$

Remark 2.4.10. It reasonable to think that to use the same notation for distinct concepts of boundary introduces considerable ambiguity. However, we shall rely on the context to make the meaning clear. The same applies to the map h.

Definition 2.4.11. *A* preboundary *is a class $\mathfrak{m}$ of groups satisfying the following properties:*

1. *$\mathfrak{m}$ is* subnormally independent, *that is, if M is a proper subnormal subgroup of a group $X \in \mathfrak{m}$, then $M \notin \mathfrak{m}$;*
2. *if $X \in \mathfrak{m}$, then X is comonolithic.*

The maps b and h bear the same relation to the closure operation s_n as the maps b and h of Section 2.3 bear to the closure operation Q. The following theorem is the Fitting class version of Theorem 2.3.7.

Theorem 2.4.12 ([DH92, XI, 4.4]).

1. If $\mathfrak{F}$ is a Fitting class, then $\mathrm{h}\big(\mathrm{b}(\mathfrak{F})\big) = \mathfrak{F}$.
2. If $\mathfrak{B}$ is a boundary of a Fitting class, then $\mathrm{b}\big(\mathrm{h}(\mathfrak{B})\big) = \mathfrak{B}$.
3. If $\mathfrak{F}$ is a Fitting class such that $\mathfrak{F} = \mathfrak{F}\mathfrak{S}$, then $\mathrm{b}(\mathfrak{F})$ is a preboundary of perfect groups and if $\mathfrak{B}$ is a preboundary of perfect groups, then $\mathrm{h}(\mathfrak{B})$ is a Fitting class such that $\mathrm{h}(\mathfrak{B}) = \mathrm{h}(\mathfrak{B})\mathfrak{S}$

Therefore if $\mathfrak{T}$ is Fitting class, then $\mathfrak{T}\mathfrak{S} = \mathfrak{T}$ if and only if $\mathrm{b}(\mathfrak{F})$ is a preboundary of perfect groups.

Lemma 2.4.13. *Let E be a comonolithic perfect group. Then*

$$\mathrm{N}(E) = [E, \mathrm{Cosoc}(E)]$$

is the smallest normal subgroup of E contained in $\mathrm{Cosoc}(E)$ such that

$$\mathrm{Cosoc}(E)/\,\mathrm{N}(E) = \mathrm{Z}\big(E/\,\mathrm{N}(E)\big).$$

Proof. Put $M = \mathrm{Cosoc}(E)$. Observe first that $\mathrm{N}(E)$ is a normal subgroup of E such that $N = \mathrm{N}(E) \leq M$ and $M/N \leq \mathrm{Z}(E/N)$. Since E is perfect, we have that $M/N = \mathrm{Z}(E/N)$ by the maximality of M. Let N_1 be a normal subgroup of G such that $N_1 \leq M$ and $M/N_1 = \mathrm{Z}(E/N_1)$. Then $[E, M] = \mathrm{N}(E)$ is contained in N_1. $\qquad\qquad\square$

Hence, if E is a comonolithic perfect group, then $E/\,\mathrm{N}(E)$ is quasisimple.

Definition 2.4.14. *Let $\mathfrak{F}$ be a Fitting class. A comonolithic perfect subnormal subgroup E of a group G is said to be an $\mathfrak{F}$-component of G if $E \notin \mathfrak{F}$ and $\mathrm{N}(E) = [E, \mathrm{Cosoc}(E)] \in \mathfrak{F}$.*
The subgroup generated by of all $\mathfrak{F}$-components of G is denoted by $\mathrm{E}_{\mathfrak{F}}(G)$.

Note that for the trivial Fitting class $\mathfrak{F} = (1)$, we have that the (1)-components of any group G are exactly the usual components and $\mathrm{E}_{(1)}(G) = \mathrm{E}(G)$ (see Definition 2.2.18 (2) and Definition 2.2.21).

Definitions and notation 2.4.15. *Let G be a group and $\mathfrak{m}$ a preboundary. We denote*

1. $\mathrm{b}_{\mathfrak{m}}(G)$ *for the set of all subnormal subgroups X of G such that $X \in \mathfrak{m}$.*
2. $\mathrm{E}_{\mathfrak{m}}(G)$ *for the subgroup generated by all subnormal subgroups X of G such that $X \in \mathrm{b}_{\mathfrak{m}}(G)$.*

If $\mathfrak{F}$ is a Fitting class such that $\mathfrak{F}\mathfrak{S} = \mathfrak{F}$ and X is an $\mathfrak{F}$-component of a group G, then X is a comonolithic perfect subnormal subgroup such that $\mathrm{N}(X) \leq X_{\mathfrak{F}} \leq \mathrm{Cosoc}(X)$. However, $X_{\mathfrak{F}} = \mathrm{Cosoc}(X)$, since $\mathrm{Cosoc}(X)/\,\mathrm{N}(X)$ is abelian. In other words, $X \in \mathrm{b}(\mathfrak{F})$. Therefore if $\mathfrak{m} = \mathrm{b}(\mathfrak{F})$, then $\mathrm{E}_{\mathfrak{m}}(G) = \mathrm{E}_{\mathfrak{F}}(G)$, for every group G, and $\mathrm{b}_{\mathfrak{m}}(G)$ is the set of all $\mathfrak{F}$-components of G.

W. Anderson introduced the concept of Fitting sets in a successful attempt to localise the theory of Fitting classes to individual groups. He could adapt the general method of B. Fischer, W. Gaschütz, and B. Hartley to prove the existence of injectors, for Fitting sets, in each soluble group (see [DH92, VIII, 2.9]). In the proofs of both theorems, a lemma due to B. Hartley involving Carter subgroups turns out to be crucial (see [DH92, VIII, 2.8]). I. Hawthorn published in [Haw98] a completely original proof which only depends on some easy results on strongly closed p-subgroups. We present here this proof of the fundamental result of B. Fischer, W. Gaschütz, B. Hartley, and W. Anderson and we even go a bit further.

Definition 2.4.16. *Let G be a group. A* Fitting set *$\mathcal{F}$ of G is a non-empty set of subgroups of G such that*

1. *if $H \in \mathcal{F}$ and $g \in G$, then $H^g \in \mathcal{F}$,*
2. *if $H \in \mathcal{F}$ and S is a subnormal subgroup of H, then $S \in \mathcal{F}$, and*
3. *if N_1 and N_2 are normal $\mathcal{F}$-subgroups of the product $N_1 N_2$, then $N_1 N_2 \in \mathcal{F}$.*

If $\mathfrak{F}$ is a Fitting class and G is a group, then the set $\mathrm{Tr}_{\mathfrak{F}}(G) = \{H \leq G : H \in \mathfrak{F}\}$ (which is called the *trace* of $\mathfrak{F}$ in G) of all $\mathfrak{F}$-subgroups of G is a Fitting set of G. But not every Fitting set arise in this manner (see [DH92, VIII, 2.2]).

Definition 2.4.17. *If $\mathcal{F}$ is a Fitting set of G, then the subgroup*

$$G_{\mathcal{F}} = \langle S : S \text{ is a subnormal } \mathcal{F}\text{-subgroup of } G \rangle$$

is a normal $\mathcal{F}$-subgroup of G and it is called the $\mathcal{F}$-radical of G (see [DH92, VIII, 2.4]).

Remark 2.4.18. Let $\mathcal{F}$ be a Fitting set of a group G. If $H \leq G$, then the set

$$\mathcal{F}_H = \{S \leq H : S \in \mathcal{F}\}$$

is a Fitting set of H. When there is no danger of confusion we shall usually denote $\mathcal{F}_H$ simply by $\mathcal{F}$.

Definitions 2.4.19. *Let $\mathcal{F}$ be a non-empty set of subgroups of a group G.*

1. *The subgroups in $\mathcal{F}$ are called $\mathcal{F}$-subgroups of G. An $\mathcal{F}$-subgroup is said to be $\mathcal{F}$-maximal in G if for any $\mathcal{F}$-subgroup T such that $S \leq T$, we have that $S = T$.*
2. *An $\mathcal{F}$-subgroup S is said to be an $\mathcal{F}$-injector of G if $S \cap N$ is $\mathcal{F}$-maximal in N for any subnormal subgroup N of G.*

The, possibly empty, set of $\mathcal{F}$-injectors of a group G will be denoted by $\mathrm{Inj}_{\mathcal{F}}(G)$.

If $\mathfrak{F}$ is a Fitting class, we talk about $\mathfrak{F}$-maximal subgroups and of $\mathfrak{F}$-injectors. The, possibly empty, set of $\mathfrak{F}$-injectors of a group G will be denoted by $\mathrm{Inj}_{\mathfrak{F}}(G)$.

Definitions 2.4.20. *A set of subgroups $\mathcal{F}$ of a group G is said to be an injective set if G possesses $\mathcal{F}$-injectors.*

A class $\mathfrak{F}$ of groups is said to be an injective class in a universe $\mathfrak{X}$ if every group $G \in \mathfrak{X}$ possesses $\mathfrak{F}$-injectors.

Definition 2.4.21. *Let G be a group and p a prime. Consider a p-subgroup P_0 of G and suppose that $P_0 \leq P$, for $P \in \mathrm{Syl}_p(G)$. We say that P_0 is strongly closed in P with respect to G, if $P_0^g \cap P \leq P_0$, for all $g \in G$.*

Remark 2.4.22. Let G be a group and p a prime. Let P_0 be a p-subgroup of G such that $P_0 \leq P \in \mathrm{Syl}_p(G)$. Suppose that P_0 is strongly closed in P with respect to G. Then:

1. P_0 is a normal subgroup of P.
2. $P_0 \cap O_p(G)$ is a normal subgroup of G.

Lemma 2.4.23. *Let G be a group and p a prime. Let P_0 be a p-subgroup of G such that $P_0 \leq P \in \mathrm{Syl}_p(G)$. Suppose that P_0 is strongly closed in P with respect to G.*

1. *If $P_0 \leq P^x$, for some $x \in G$, then P_0 is strongly closed in P^x with respect to G.*
2. *If N is a normal subgroup of G, then P_0N/N is strongly closed in PN/N with respect to G/N.*

Proof.   1. Observe that $P_0^{x^{-1}} = P_0^{x^{-1}} \cap P \leq P_0$. Hence $x \in N_G(P_0)$. If $g \in G$, we have

$$P_0^g \cap P^x = (P_0^{gx^{-1}} \cap P)^x \leq P_0^x = P_0.$$

This means that P_0 is strongly closed in P^x with respect to G.

2. Observe that, for each $g \in G$, there exists an element $x \in N$ such that

$$P_0^g \cap PN = P_0^g \cap P^x = (P_0^{gx^{-1}} \cap P)^x \leq P_0^x \leq P_0N.$$

The assertion easily follows.   □

Lemma 2.4.24 (M. E. Harris, [Har72]). *Let G be a soluble group and p a prime. Let P_0 be a p-subgroup of G such that $P_0 \leq P \in \mathrm{Syl}_p(G)$. If P_0 is strongly closed in P with respect to G, then P_0 is a normally embedded subgroup of G (see [DH92, Section I, 7]).*

Proof. We use induction on the order of G. If M is a non-trivial normal subgroup of G, for any $H \leq G$ we write $\bar{H}$ to denote the subgroup HM/M of the quotient group $\bar{G} = G/M$.

By Lemma 2.4.23 (2), we have that $\bar{P}_0$ is strongly closed in $\bar{P}$ with respect to $\bar{G}$. By induction, the subgroup $\bar{P}_0$ is normally embedded in $\bar{G}$, that is, there exists a normal subgroup $\bar{N}$ of $\bar{G}$, such that $\bar{P} \cap \bar{N} = \bar{P}_0$. This means that there exists a normal subgroup N of G such that $P_0 M = (P \cap N)M$. Then $P \cap P_0 M = P \cap (P \cap N)M = (P \cap N)(P \cap M)$.

If either $\mathrm{Core}_G(P_0) \neq 1$ or $O_{p'}(G) \neq 1$, then put either $M = \mathrm{Core}_G(P_0)$ or $M = O_{p'}(G)$. In this case, $P_0 = P \cap N$ and the assertion follows. Hence we may assume that $\mathrm{Core}_G(P_0) = O_{p'}(G) = 1$. Then, by [KS04, 6.4.4], we have that $C_G(O_p(G)) \leq O_p(G)$ inasmuch as G is soluble. If $M = P_0 \cap O_p(G) \neq 1$, then M is a non-trivial normal subgroup of G by Remark 2.4.22 (2). This contradicts $\mathrm{Core}_G(P_0) = 1$. Hence we can assume that P_0 and $O_p(G)$ have trivial intersection. Since P_0 is normal in P by Remark 2.4.22 (1), it follows that $P_0 \leq C_G(O_p(G)) \leq O_p(G)$. Hence $P_0 = 1$ and the lemma follows. □

Applying a result of P. Lockett (see [DH92, I, 7.8]) we have the following lemma.

Lemma 2.4.25. *Let G be a soluble group and p and q two primes. Let P_0 be a p-subgroup of G such that $P_0 \leq P \in \mathrm{Syl}_p(G)$ and Q_0 a q-subgroup of G such that $Q_0 \leq Q \in \mathrm{Syl}_q(G)$. If P_0 is strongly closed in P with respect to G and Q_0 is strongly closed in Q with respect to G, then there exists an element $g \in G$ such that $P_0^g Q_0 = Q_0 P_0^g$.*

Theorem 2.4.26 (B. Fischer, W. Gaschütz, B. Hartley, and W. Anderson). *If G is a soluble group and $\mathcal{F}$ is a Fitting set of G, then G has a unique conjugacy class of $\mathcal{F}$-injectors.*

Proof (I. Hawthorn). We apply induction on the order of G and assume the result is true for all groups of smaller order.

Since G is soluble, there exists a prime p such that $O^p(G)$ is a proper subgroup of G. By induction, $O^p(G)$ possesses a unique conjugacy class of $\mathcal{F}$-injectors. Let S be one of them. Note that if $g \in G$, the subgroup S^g is also an $\mathfrak{F}$-injector of $O^p(G)$ and then there exists an element $h \in O^p(G)$ such that $S^g = S^h$. Consequently the Frattini argument holds and $G = N_G(S) O^p(G)$. In fact, if P is a Sylow p-subgroup of $N_G(S)$, then $G = P O^p(G)$.

Let R be the subgroup generated by the $\mathcal{F}$-subgroups of PS containing S. Since any such subgroup is subnormal in PS, we have that $R \in \mathcal{F}$.

Let T be an $\mathcal{F}$-subgroup of G such that S is contained in T. Observe that $T \cap O^p(G)$ is an $\mathcal{F}$-subgroup. The $\mathcal{F}$-maximality of S in $O^p(G)$ implies that $S = T \cap O^p(G)$. Hence T is contained in $N_G(S)$. Therefore any Sylow p-subgroup of T is conjugate in $N_G(S)$ to a subgroup of P. Since $T/S \cong T O^p(G)/O^p(G)$ is a p-group, it follows that T is conjugate in $N_G(S)$ to a group of the form $P_0 S$, for some subgroup P_0 of P. Hence, all extensions of S which are elements of $\mathcal{F}$ are conjugate in $N_G(S)$ to subgroups of R. In particular if G has in $\mathcal{F}$-injector, then it is conjugate to R.

It remains to show that R is an $\mathcal{F}$-injector of G. Since R is $\mathcal{F}$-maximal in G, it is enough to prove that R contains an $\mathcal{F}$-injector of M for every maximal normal subgroup M of G.

Suppose that $|G : M| = q$, q a prime, and let T be an $\mathcal{F}$-injector of M. The subgroups $T \cap M \cap O^p(G) = T \cap O^p(G)$ and $S \cap M \cap O^p(G) = M \cap S$ are

$\mathcal{F}$-injectors of the normal subgroup $M \cap O^p(G)$. Therefore they are conjugate in $M \cap O^p(G)$. Choose T in such a way that $T \cap O^p(G) = M \cap S = U$. Let $P_1 \in \mathrm{Syl}_p(T)$ and $Q_1 \in \mathrm{Syl}_q(S)$ so that $T = P_1 U$ and $S = Q_1 U$. Since S and T are subgroups of $N_G(U)$, there exist a Sylow p-subgroup P of $N_G(U)$ such that $P_1 \leq P$ and a Sylow q-subgroup Q of $N_G(U)$ such that $Q_1 \leq Q$. If $g \in N_G(U)$, then $(P_1^g \cap P)U \leq T^g \in \mathcal{F}$. Since $(P_1^g \cap P)U$ and T are subnormal subgroups of PU, we have that $\langle P_1^g \cap P, P_1 \rangle U$ is an $\mathcal{F}$-subgroup of PU. Moreover $T \leq \langle P_1^g \cap P, P_1 \rangle U \leq \langle T^g, T \rangle \leq M$. The $\mathcal{F}$-maximality of T in M yields $P_1^g \cap P \leq P_1$. This is to say that P_1 is strongly closed in P with respect to $N_G(U)$. Analogously it can be shown that Q_1 is strongly closed in Q with respect to $N_G(U)$. By Lemma 2.4.25, there exists an element $g \in N_G(U)$ such that the product $P_1^g Q_1$ is a subgroup of $N_G(U)$.

Consider the subgroup $K = P_1^g Q_1 U = (P_1 U)^g (Q_1 U) = T^g S$. Observe that $K \cap O^p(G) = T^g S \cap O^p(G) = (T^g \cap O^p(G))S = (T \cap O^p(G))^g S = U^g S = US = S$ and similarly $K \cap M = T^g$. Hence S and T^g are normal $\mathfrak{F}$-subgroups of K and therefore K is an $\mathcal{F}$-group. Since S is contained in K, we have that R contains a conjugate of K. This concludes the proof. $\qquad \square$

Theorem 2.4.27. *Let $\mathcal{F}$ be a Fitting set of a group G such that $G/G_\mathcal{F}$ is soluble. Then G has a unique conjugacy class of $\mathcal{F}$-injectors.*

Proof. Denote $N = G_\mathcal{F}$. The set $\mathcal{F}^* = \{H/N : H \in \mathcal{F}, N \leq H\}$ is a Fitting set of the soluble group G/N. Moreover, using the arguments of [DH92, VIII, 2.17 (a)], we have that

$$\mathcal{F}_0 = \{S \leq G : SN/N \in \mathcal{F}^* \text{ and } S \text{ is subnormal in } SN\}$$

is a Fitting set of G. Observe that $\mathcal{F}_0 \subseteq \mathcal{F}$ and for any subnormal subgroup S of G, we have that $S_{\mathcal{F}_0} = S_\mathcal{F}$.

We apply now the arguments of [DH92, VIII, 2.17 (b)], which hold in the non-soluble case, to conclude that if V/N is an $\mathcal{F}^*$-injector of G/N, then V is an $\mathcal{F}_0$-injector of G. We claim that, indeed, V is an $\mathcal{F}$-injector of G. To see that, we prove that for any subnormal subgroup S of G, the subgroup $V \cap S$ is $\mathcal{F}$-maximal in S. Suppose that there exists $W \in \mathcal{F}$ such that $V \cap S \leq W \leq S$. Then $(V \cap S)N/N = (V/N) \cap (SN/N) \leq WN/N \leq SN/N$. Since $S_\mathcal{F} = S_{\mathcal{F}_0} \leq V \cap S \in \mathrm{Inj}_{\mathcal{F}_0}(S)$, then $S_\mathcal{F} \leq W$. Recall that $N \cap S = S_\mathcal{F}$, by [DH92, VIII, 2.4 (d)]. Therefore $WN \cap S = W(N \cap S) = WS_\mathcal{F} = W$. Hence W is subnormal in WN and then $WN \in \mathcal{F}$. Consequently, $WN/N \in \mathcal{F}^*$. Since $(V/N) \cap (SN/N)$ is $\mathcal{F}^*$-maximal in SN/N, we have that $(V \cap S)N = WN$. This implies that

$$V \cap S = (V \cap S)(N \cap S) = (V \cap S)N \cap S = WN \cap S = W,$$

and then $V \cap S$ is $\mathcal{F}$-maximal in S. Thus, we deduce that $V \in \mathrm{Inj}_\mathcal{F}(G)$ as claimed.

On the other hand, applying [DH92, VIII, 2.15], if $V \in \mathrm{Inj}_\mathcal{F}(G)$, then V/N is an $\mathcal{F}^*$-injector of the soluble group G/N. By Theorem 2.4.26, the $\mathcal{F}^*$-injectors of G/N are conjugate in G/N. Consequently the $\mathcal{F}$-injectors of G form a conjugacy class of subgroups of G. $\qquad \square$

Corollary 2.4.28 ([BCMV84]). *If $\mathfrak{F}$ is a Fitting class, every group in $\mathfrak{FS}$ has a unique conjugacy class of $\mathfrak{F}$-injectors.*

One line taken in the study of Fitting classes in the soluble universe has been their classification according to the embedding properties of their injectors, and in this direction the pursuit of those with normal injectors has been especially fruitful. In this context the following definition makes sense.

Definition 2.4.29. *Let $\mathfrak{X}$ be a class of groups which is closed under taking subnormal subgroups, and let $1 \neq \mathfrak{F}$ be a Fitting class contained in $\mathfrak{X}$.*

1. *We say that $\mathfrak{F}$ is normal in $\mathfrak{X}$ or $\mathfrak{X}$-normal if $G_{\mathfrak{F}}$ is $\mathfrak{F}$-maximal in G for all $G \in \mathfrak{X}$.*
2. *$\mathfrak{F}$ is said to be dominant in $\mathfrak{X}$ or $\mathfrak{X}$-dominant if for all $H \in \mathfrak{X}$ any two $\mathfrak{F}$-maximal subgroups of H containing $H_{\mathfrak{F}}$ are conjugate in H.*

If $\mathfrak{X} = \mathfrak{E}$, we simply say that $\mathfrak{F}$ is a normal (respectively dominant) Fitting class.

It is clear that if $\mathfrak{F}$ is $\mathfrak{X}$-normal, then every group G has a unique $\mathfrak{F}$-injector, namely the $\mathfrak{F}$-radical. Moreover, applying [DH92, IX, 4.2], if $\mathfrak{F}$ is $\mathfrak{X}$-dominant, then every $\mathfrak{X}$-group has a unique conjugacy class of $\mathfrak{F}$-injectors, namely the $\mathfrak{F}$-maximal subgroups of H containing $H_{\mathfrak{F}}$.

The first investigation in normal Fitting classes was carried out by D. Blessenohl and W. Gaschütz in [BG70]. They quickly settle the question of which Schunck classes of soluble groups have normal projectors — these turn out to be the classes of all π-perfect groups (the projector in G being $O^{\pi}(G)$) — and then go to lay the foundations for the much more complex dual theory (see [DH92, X, Section 3]).

2.5 Fitting formations

We have seen that many of the examples of Fitting classes are formations too. Naturally such classes are called *Fitting formations*. The class $\mathfrak{N}$, of all nilpotent groups, the classes $\mathfrak{E}_{\pi}$, of all π-groups, the class $\mathfrak{E}_{\pi}\mathfrak{E}_{\pi'}$, of all groups with a normal Hall π-subgroup, for any set π of prime numbers, are examples of Fitting formations.

We will be interested in the following example:

Example 2.5.1. Let $\mathcal{I}$ be a non-empty set. For each $i \in \mathcal{I}$, let $\mathfrak{F}_i$ be a subgroup-closed Fitting formation. Assume that $\pi(\mathfrak{F}_i) \cap \pi(\mathfrak{F}_j) = \emptyset$ for all $i, j \in \mathcal{I}, i \neq j$. Then $\times_{i \in \mathcal{I}} \mathfrak{F}_i$ is a saturated Fitting formation (see Remark 2.2.13).

The most remarkable milestone in the theory of Fitting formations was settled by R. A. Bryce and J. Cossey in 1982.

Theorem 2.5.2 (R. A. Bryce and J. Cossey, [BC82]). *Every subgroup-closed Fitting class of finite soluble groups is a saturated formation.*

The way towards the proof of this impressive result started ten years before. In [BC72] the same authors proved the following.

Theorem 2.5.3 (R. A. Bryce and J. Cossey, [BC72]). *Every subgroup-closed Fitting formation of finite soluble groups is saturated.*

An outline of the proof of these two results appears in Chapter XI of [DH92].

Unfortunately the above theorem is not true in the general universe of all finite groups as it is pointed out in [DH92, IX, 1.6]. In [BBE98], the authors found necessary and sufficient conditions for a subgroup-closed Fitting formation to be saturated.

Theorem 2.5.4 ([BBE98]). *For a subgroup-closed Fitting formation $\mathfrak{F}$ the following are equivalent:*

1. *If $G \in \mathfrak{F}$ is a primitive group of type 2 and E_p is the maximal Frattini extension of G with p-elementary abelian kernel, then $E_p \in \mathfrak{F}$, for every prime p dividing $|\mathrm{Soc}(G)|$,*
2. *$\mathfrak{F}$ is saturated.*

Up to now, no classification of the Fitting formations is known. However many of the known Fitting formations are gathered in two types: solubly saturated Fitting formations and Fitting formations defined by Fitting families of modules.

The search for a soluble non-saturated Fitting formation led to T. O. Hawkes to the introduction of what he called (see [Haw70]) the class of p-soluble groups, p a prime, whose absolute arithmetic p-rank is a p'-number. After that, and extending Hawkes's methods, many examples of soluble non-saturated Fitting formations have been introduced by different authors. The method presented by J. Cossey and C. Kanes in [CK87] and modified by Cossey in [Cos89] includes all previous constructions. Motivated by the local (or Baer) functions, the criterion to decide whether a particular p-soluble group belongs to one of these Cossey-Kanes classes is defined by imposing some conditions of a certain class of modules associated with the p-chief factors. That the classes so defined are Fitting formations is a consequence of the closure properties of the family of modules.

Definition 2.5.5. *Let K be a field. We associate with each group G of a suitable universe $\mathfrak{V}$ a class $\mathfrak{M}(G)$ of irreducible KG-modules. The class $\mathfrak{M} = \bigcup_G \mathfrak{M}(G)$ is said to be a Fitting family of modules over K if it satisfies the following properties:*

1. *If $V \in \mathfrak{M}(G)$ and N is a normal subgroup of G such that $N \leq \mathrm{C}_G(V)$, then V, regarded in the natural way as a $K(G/N)$-module, is in $\mathfrak{M}(G/N)$.*

2. If $V \in \mathfrak{M}(H)$ and H is an epimorphic image of G, then V, regarded in the natural way as a KG-module, is in $\mathfrak{M}(G)$.

3. If $V \in \mathfrak{M}(G)$, N is a subnormal subgroup of G and U is an irreducible constituent of V_N, then $U \in \mathfrak{M}(N)$.

4. If N_1 and N_2 are normal subgroups of G such that $G = N_1 N_2$ and V is an irreducible KG-module such that all composition factors of V_{N_i} are in $\mathfrak{M}(N_i)$, for $i = 1$, 2, then $V \in \mathfrak{M}(G)$.

Clearly if $\mathfrak{M}(G)$ is non-empty, then the trivial KG-module K_G is in $\mathfrak{M}(G)$.

With this definition we can construct Fitting formations with the following procedure.

Theorem 2.5.6. Fix a prime r. Let K be an extension field of $k = \mathrm{GF}(r)$. For any r-soluble group G, we denote $\mathfrak{T}_K(G)$ the class of all irreducible KG-modules V such that V is a composition factor of the module $W^K = W \otimes K$, where W is an r-chief factor of G. If, for every r-soluble group G, a class of irreducible KG-modules $\mathfrak{M}(G)$ is defined, and $\mathfrak{M} = \bigcup_G \mathfrak{M}(G)$, the class

$$\mathfrak{T}(1, \mathfrak{M}) = \big(G : G \text{ is } r\text{-soluble and } \mathfrak{T}_K(G) \subseteq \mathfrak{M}(G)\big)$$

is a Fitting formation provided $\mathfrak{M}$ is a Fitting family in the r-soluble universe.

A proof of this theorem is presented in [DH92, IX, 2.18].

Thus, given a Fitting family $\mathfrak{M}$ in the r-soluble universe, r a prime, we have a way to distinguish between the abelian chief factors of a soluble group G: an r-chief factor M of G can be such that all composition factors of M^K are in $\mathfrak{M}(G)$ or not.

The family of modules proposed by J. Cossey and C. Kanes [CK87] is motivated by the class of characters, called π-factorable characters, introduced by I. M. Isaacs in [Isa84]. D. Gajendragadkar introduced in [Gaj79] the idea of π-special characters and established their basic properties. This idea was considerably developed and refined by Isaacs. The definition of π-special modules is derived from the definition of π-special characters and the properties are similar to those of Isaacs and Gajendragadkar.

We therefore specify that *for the rest of this section all groups considered are soluble.*

Definition 2.5.7. Let K be an algebraically closed field of characteristic $r > 0$, π a set of primes, and G a group.

1. An irreducible KG-module V is called π-special if the dimension of V is a π-number and whenever S is a subnormal subgroup of G and U is a composition factor of V_S, then $\det(x \text{ on } U) = 1$ for all π'-elements x of S.

2. Suppose that $\mathcal{P} = \{\pi_i : i \in \mathcal{I}\}$ is a partition of $\mathbb{P}$, the set of all primes. An irreducible KG-module V is called $\mathcal{P}$-factorable if $V = U_{j_1} \otimes \cdots \otimes U_{j_n}$ for some π_{j_i}-special modules U_{j_i}, $\pi_{j_i} \in \mathcal{P}$, $i = 1, \ldots, n$, and $j_i \neq j_k$ when $i \neq k$.

3. *An irreducible KG-module V is called π-factorable if V is $\mathcal{P}$-factorable for $\mathcal{P} = \{\pi, \pi'\}$.*

It turns out that if U and W are respectively π-special and π'-special irreducible KG-modules, then $U \otimes W$ is irreducible. Moreover, if U' and W' are respectively π-special and π'-special irreducible KG-modules, and $U \otimes W \cong U' \otimes W'$, then $U \cong U'$ and $W \cong W'$ (see [CK87, 2.4] for more details and notation). It is also true the following:

Lemma 2.5.8 ([CK87, 2.2]). *Let G be a group, K a field, π a set of primes and V be π-special KG-module. If S is a subnormal subgroup of G, then every irreducible constituent of V_S is π-special.*

The next lemma equips us with the basic arguments to prove closure properties of "Fitting type." Its proof is rather technical and can be seen in [CK87].

Lemma 2.5.9. *Let G be a group, K an algebraically closed field and $\mathcal{P} = \{\pi_i : i \in \mathcal{I}\}$ a partition of the set $\mathbb{P}$ of all primes.*

1. *If V is a $\mathcal{P}$-factorable KG-module and N is a normal subgroup of G, then any irreducible KN-submodule of V_N is a $\mathcal{P}$-factorable KN-module.*
2. *Suppose that M and N are normal subgroups of G such that $G = MN$. Let V be an irreducible KG-module such that all irreducible KM-submodules of V_M and all irreducible KN-submodules of V_N are $\mathcal{P}$-factorable. Then V is a $\mathcal{P}$-factorable KG-module.*

And now we prove the main result.

Theorem 2.5.10. *Let K be an algebraically closed field of characteristic a prime p. Let $\mathcal{P} = \{\pi_i : i \in \mathcal{I}\}$ be a partition of the set $\mathbb{P}$ of all primes. For each $i \in \mathcal{I}$, let $\mathfrak{X}_i$ be a Fitting formation. Denote $\mathcal{X} = \{\mathfrak{X}_i : i \in \mathcal{I}\}$. For every soluble group G, denote by $\mathfrak{M}(G)$ the class of all irreducible $\mathcal{P}$-factorable KG-modules V such that $V = V_1 \otimes \cdots \otimes V_{n(V)}$. Suppose further that*

1. *V_j is a $\pi_{i(j)}$-special KG-module, and*
2. *$G/\,\mathrm{C}_G(V_j) \in \mathfrak{X}_{i(j)}$, for $j = 1, \ldots, n(V)$.*

Then $\mathfrak{M} = \mathfrak{M}(K, \mathcal{P}, \mathcal{X}) = \bigcup_G \mathfrak{M}(G)$ is a Fitting family in the universe $\mathfrak{S}$.

Proof.   1. Let G be a group, let N be a normal subgroup of G, and let V be a KG-module in $\mathfrak{M}(G)$ such that $N \leq \mathrm{C}_G(V)$. Suppose that $V = V_1 \otimes \cdots \otimes V_n$ is a $\mathcal{P}$-factorisation of V where V_j is a $\pi_{i(j)}$-special KG-module and $G/\,\mathrm{C}_G(V_j) \in \mathfrak{X}_{i(j)}$, for each $j = 1, \ldots, n$. Then $\mathrm{C}_G(V) = \bigcap_{j=1}^n \mathrm{C}_G(V_j)$, and so $N \leq \mathrm{C}_G(V_j)$ for each j. Consider V_j as a $K(G/N)$-module. It is clear that V_j is also a $\pi_{i(j)}$-special $K(G/N)$-module. Therefore the $K(G/N)$-module V is $\mathcal{P}$-factorable. Since $(G/N)/\,\mathrm{C}_{G/N}(V_j) \cong G/\,\mathrm{C}_G(V_j)$, we have that $V \in \mathfrak{M}(G/N)$.

2. Let G and H be two groups such that $\varphi \colon G \longrightarrow H$ is an epimorphism, and consider an KH-module $V \in \mathfrak{M}(H)$. Suppose that $V = V_1 \otimes \cdots \otimes V_n$ is a $\mathcal{P}$-factorisation of V where V_j is a $\pi_{i(j)}$-special KH-module and $H/\mathrm{C}_H(V_j) \in \mathfrak{X}_{i(j)}$, for each $j = 1, \ldots, n$. Each V_j is considered as a KG-module via φ. Let S be a subnormal subgroup of G. Since φ is an epimorphism, the image S^φ is a subnormal subgroup of H. Moreover, U is a composition factor of $(V_j)_S$ if and only if U is a composition factor of $(V_j)_{S^\varphi}$. For any $\pi(j)'$-element x of S, we have that x^φ is a $\pi(j)'$-element of S^φ and $\det(x \text{ on } U) = \det(x^\varphi \text{ on } U) = 1$. Therefore the KG-module V_j is $\pi_{i(j)}$-special and V is a $\mathcal{P}$-factorable KG-module. Finally, observe that $\mathrm{Ker}(\varphi) \leq \mathrm{C}_G(V) = \bigcap_{j=1}^n \mathrm{C}_G(V_j)$. Then $G/\mathrm{C}_G(V_j) \cong (G/\mathrm{Ker}(\varphi))/(\mathrm{C}_G(V_j)/\mathrm{Ker}(\varphi)) \cong H/\mathrm{C}_H(V_j) \in \mathfrak{X}_{i(j)}$. Therefore $V \in \mathfrak{M}(G)$.

3. Let G be a group, N be a normal subgroup of G and V a KG-module in $\mathfrak{M}(G)$. Let U be an irreducible KN-submodule of V_N. Then U is a $\mathcal{P}$-factorable KN-module by Lemma 2.5.9 (1). In fact, if $V = V_1 \otimes \cdots \otimes V_n$ is a $\mathcal{P}$-factorisation of V where V_j is a $\pi_{i(j)}$-special KG-module then there exists a KN-submodule U_j of V_j such that $U = U_1 \otimes \cdots \otimes U_n$ is a $\mathcal{P}$-factorisation of U where each U_j is a $\pi_{i(j)}$-special KN-module, $1 \leq j \leq n$. Since each $\mathfrak{X}_{i(j)}$ is a Fitting class and $G/\mathrm{C}_G(V_j) \in \mathfrak{X}_{i(j)}$, then the normal subgroup $N \mathrm{C}_G(V_j)/\mathrm{C}_G(V_j)$ is in $\mathfrak{X}_{i(j)}$. This is to say that $N/\mathrm{C}_N(V_j) \in \mathfrak{X}_{i(j)}$. Since U_j is a KN-submodule of V_j, we have that $\mathrm{C}_N(V_j)$ is a normal subgroup of $\mathrm{C}_N(U_j)$ and then $N/\mathrm{C}_N(U_j)$ is an epimorphic image of $N/\mathrm{C}_N(V_j)$. Since each $\mathfrak{X}_{i(j)}$ is also a formation, we have that $N/\mathrm{C}_N(U_j) \in \mathfrak{X}_{i(j)}$. Therefore we deduce that $U \in \mathfrak{M}(N)$.

4. Let G be a group and suppose that M and N are normal subgroups of G such that $G = MN$. Let V be an irreducible KG-module such that all irreducible KM-submodules of V_M are in $\mathfrak{M}(M)$ and all irreducible KN-submodules of V_N are in $\mathfrak{M}(N)$. By Lemma 2.5.9 (2), V is $\mathcal{P}$-factorable KG-module. Suppose that $V = V_1 \otimes \cdots \otimes V_n$ is a $\mathcal{P}$-factorisation of V where V_j is a $\pi_{i(j)}$-special KG-module. By Clifford's theorem [DH92, B, 7.3], $(V_j)_M$ and $(V_j)_N$ are completely reducible. Suppose that $(V_j)_M = Z_{j,1} \oplus \cdots \oplus Z_{j,r(j)}$ is a decomposition of $(V_j)_M$ in irreducible KM-submodules. By Lemma 2.5.8 every $Z_{j,t}$ is a $\pi_{i(j)}$-special KM-module. Therefore

$$V_M = \bigoplus_{k(1)=1}^{r(1)} \cdots \bigoplus_{k(n)=1}^{r(n)} \left[Z_{1,k(1)} \otimes \cdots \otimes Z_{n,k(n)} \right]$$

Any $Z_{1,k(1)} \otimes \cdots \otimes Z_{n,k(n)}$ is a $\mathcal{P}$-factorisation of an irreducible constituent of V_M. Then $Z_{1,k(1)} \otimes \cdots \otimes Z_{n,k(n)} \in \mathfrak{M}(M)$. Therefore $M/\mathrm{C}_M(Z_{j,l}) \in \mathfrak{X}_{i(j)}$ for any par (j, l). It is clear that $\mathrm{C}_M(V_j) = \bigcap_{i=1}^{r(j)} \mathrm{C}_M(Z_{j,i})$. Since $\mathfrak{X}_{i(j)}$ is a formation, the group $M/\mathrm{C}_M(V_j)$ is in $\mathfrak{X}_{i(j)}$. We can argue analogously with V_N and deduce that $N/\mathrm{C}_N(V_j) \in \mathfrak{X}_{i(j)}$. Moreover since

$$M \mathrm{C}_N(V_j)/\mathrm{C}_M(V_j) \mathrm{C}_N(V_j) \cong M/\mathrm{C}_M(V_j) \in \mathfrak{X}_{i(j)}$$

and

$$N \operatorname{C}_M(V_j)/\operatorname{C}_M(V_j)\operatorname{C}_N(V_j) \cong N/\operatorname{C}_N(V_j) \in \mathfrak{X}_{i(j)},$$

and $\mathfrak{X}_{i(j)}$ is a Fitting class, we have that

$$G/\operatorname{C}_M(V_j)\operatorname{C}_N(V_j) = \\ [M \operatorname{C}_N(V_j)/\operatorname{C}_M(V_j)\operatorname{C}_N(V_j)][N \operatorname{C}_M(V_j)/\operatorname{C}_M(V_j)\operatorname{C}_N(V_j)]$$

is in $\mathfrak{X}_{i(j)}$. Finally, notice that $\operatorname{C}_M(V_j)\operatorname{C}_N(V_j)$ is a normal subgroup of $\operatorname{C}_G(V_j)$ and then $G/\operatorname{C}_G(V_j)$ is isomorphic to a quotient group of $G/\operatorname{C}_M(V_j)\operatorname{C}_N(V_j)$. Since $\mathfrak{X}_{i(j)}$ is a formation, we have that $G/\operatorname{C}_G(V_j) \in \mathfrak{X}_{i(j)}$. This implies that $V \in \mathfrak{M}(G)$. □

Examples and remarks 2.5.11. The Cossey-Kanes construction covers many of the known constructions of Fitting formations. For instance:

1. Let p be a prime and K an algebraically closed field of characteristic p. If $\mathcal{P} = \{\pi_1 = \{p\}, \pi_2 = p'\}$ and $\mathcal{X} = \{\mathfrak{X}_1 = (1), \mathfrak{X}_2 = \mathfrak{S}\}$, then $\mathfrak{M}^p = \mathfrak{M}(K, \mathcal{P}, \mathcal{X})$ is a Fitting family of modules in the universe $\mathfrak{S}$. The Fitting formation $\mathfrak{T} = \mathfrak{T}(\mathfrak{M}^p)$ is the one introduced by Hawkes in [Haw70].

2. The Fitting formations studied by K. L. Haberl and H. Heineken ([HH84] or [DH92, IX, 2.26]) are constructed using a not necessarily algebraically closed field K. Nevertheless they can also be included in the Cossey-Kanes construction thanks to a modification made by Cossey in [Cos89]. According to [HH84, 4.1], every Haberl-Heineken Fitting formation can be seen as a Fitting formation constructed by the Cossey-Kanes method with $\mathfrak{X}_1 = \mathfrak{S}, \mathfrak{X}_2 = (1)$.

3. The non-saturated Fitting formations introduced by T. R. Berger and J. Cossey in [BC78] are defined in terms of the Cossey-Kanes procedure. The Fitting formations of Berger-Cossey are the first examples of non-saturated Fitting formations composed of soluble groups whose p-length is less or equal to 1 for all primes p.

4. A result due to L. G. Kovács, which appears in [CK87, 4.2], characterises the saturation of the Fitting formations $\mathfrak{T}(\mathfrak{M}(K, \mathcal{P}, \mathcal{X}))$. This means that some of the Fitting formations constructed by the Cossey-Kanes procedure can be saturated.

5. Let $\mathfrak{M} = \bigcup_G \mathfrak{M}(G)$ be a Fitting family. In Theorem 2.5.6, assume that, instead of the class $\Gamma_K(G)$, we work with the class $\Delta_K(G)$ of all irreducible KG-modules V such that V is a composition factor of the module $W^K = W \otimes K$, where W is a complemented r-chief factor of G (r is a prime, K is a field with char $K = r$ and G is r-soluble). Then the class

$$\mathfrak{C}(\mathfrak{M}) = (G : G \text{ is } r\text{-soluble and } \Delta_K(G) \subseteq \mathfrak{M}(G))$$

is a Fitting class and a Schunck class (see [CO87]). Moreover, in this paper a criterion to decide which of these classes is a formation is presented.

3

$\mathfrak{X}$-local formations

In 1985 P. Förster [För85b] presented a common extension of the Gaschütz-Lubeseder-Schmid and Baer theorems (see Section 2.2). He introduced the concept of $\mathfrak{X}$-local formation, where $\mathfrak{X}$ is a class of simple groups with a completeness property. If $\mathfrak{X} = \mathfrak{J}$, the class of all simple groups, $\mathfrak{X}$-local formations are exactly the local formations and when $\mathfrak{X} = \mathbb{P}$, the class of all abelian simple groups, the notion of $\mathfrak{X}$-local formation coincides with the concept of Baer-local formation. P. Förster also defined a Frattini-like subgroup $\Phi_{\mathfrak{X}}^*(G)$ in each group G, which enables him to introduce the concept of $\mathfrak{X}$-saturation. Förster's definition of $\mathfrak{X}$-saturation is not the natural one if our aim is to generalise the concepts of saturation and soluble saturation. Since $O_{\mathfrak{J}}(G) = G$ and $O_{\mathbb{P}}(G) = G_{\mathfrak{S}}$, we would expect the $\mathfrak{X}$-Frattini subgroup of a group G to be defined as $\Phi\big(O_{\mathfrak{X}}(G)\big)$, where $O_{\mathfrak{X}}(G)$ is the largest normal subgroup of G whose composition factors belong to $\mathfrak{X}$. We have that $\Phi\big(O_{\mathfrak{X}}(G)\big)$ is contained in $\Phi_{\mathfrak{X}}^*(G)$, but the equality does not hold in many cases. Nevertheless, Förster proved that $\mathfrak{X}$-saturated formations are exactly the $\mathfrak{X}$-local ones. If $\mathfrak{X} = \mathfrak{J}$, then we obtain as a special case the Gaschütz-Lubeseder-Schmid theorem. When $\mathfrak{X} = \mathbb{P}$, Baer's theorem appears as a corollary of Förster's result. Since $\Phi\big(O_{\mathfrak{X}}(G)\big)$ is contained in $\Phi_{\mathfrak{X}}^*(G)$ for every group G, we can deduce from Förster's theorem that every $\mathfrak{X}$-local formation fulfils the following property:

A group G belongs to $\mathfrak{F}$ if and only if $G/\Phi\big(O_{\mathfrak{X}}(G)\big)$ belongs to $\mathfrak{F}$. (3.1)

Therefore from the very beginning the following question naturally arises:

Open question 3.0.1. *Let $\mathfrak{F}$ be a formation with the property* (3.1). *Is $\mathfrak{F}$ $\mathfrak{X}$-local?*

After studying general properties of $\mathfrak{X}$-local formations in Section 3.1, we draw near the solution of Question 3.0.1 in Section 3.2. Products of $\mathfrak{X}$-local formations are the theme of Section 3.3, whereas some partially saturated formations are studied in Section 3.4.

Throughout this chapter, $\mathfrak{X}$ denotes a fixed class of simple groups satisfying $\pi(\mathfrak{X}) = \mathrm{char}\, \mathfrak{X}$.

3.1 X-local formations

This section is devoted to study some basic facts on X-local formations. We investigate the behaviour of X-local formations as classes of groups, focussing our attention on some distinguished X-local formation functions defining them.

We begin with the concept of X-local formation due to Förster [För85b].

Denote by $\mathfrak{J}$ the class of all simple groups. For any subclass $\mathfrak{Y}$ of $\mathfrak{J}$, we write $\mathfrak{Y}' = \mathfrak{J} \setminus \mathfrak{Y}$. Let $E\mathfrak{Y}$ be the class of groups whose composition factors belong to $\mathfrak{Y}$. It is clear that $E\mathfrak{Y}$ is a Fitting class, and so each group G has a largest normal $E\mathfrak{Y}$-subgroup, the $E\mathfrak{Y}$-radical $O_{\mathfrak{Y}}(G)$. A chief factor of G which belongs to $E\mathfrak{Y}$ is called a $\mathfrak{Y}$-*chief factor*, and if, moreover, p divides the order of a $\mathfrak{Y}$-chief factor H/K of G, we shall say that H/K is a $\mathfrak{Y}_p$-*chief factor* of G.

Sometimes it will be convenient to identify the prime p with the cyclic group C_p of order p.

Definition 3.1.1 (P. Förster). *An X-formation function f associates with each $X \in (\mathrm{char}\, \mathfrak{X}) \cup \mathfrak{X}'$ a formation $f(X)$ (possibly empty). If f is an X-formation function, then the X-local formation $\mathrm{LF}_{\mathfrak{X}}(f)$ defined by f is the class of all groups G satisfying the following two conditions:*

1. if H/K is an $\mathfrak{X}_p$-chief factor of G, then $G/\, C_G(H/K) \in f(p)$, and
2. $G/K \in f(E)$, whenever G/K is a monolithic quotient of G such that the composition factor of its socle $\mathrm{Soc}(G/K)$ is isomorphic to E, if $E \in \mathfrak{X}'$.

Remarks 3.1.2.   1.   It is obvious from the definition that $\mathrm{LF}_{\mathfrak{X}}(f)$ is Q-closed.

2. Applying Theorem 1.2.34, it is only necessary to consider the $\mathfrak{X}_p$-chief factors of a given chief series of a group G in order to check whether or not G satisfies Condition 1.

3. If, for some prime $p \in \mathrm{char}\, \mathfrak{X}$, $f(p) = \emptyset$, then every X-chief factor of a group $G \in \mathrm{LF}_{\mathfrak{X}}(f)$ is a p'-group.

4. If, for some $S \in \mathfrak{X}'$, $f(S) = \emptyset$, then a group $G \in \mathrm{LF}_{\mathfrak{X}}(f)$ cannot have a monolithic quotient whose socle is in $E(S)$. Consequently $\mathrm{LF}_{\mathfrak{X}}(f) \subseteq E\big((S)'\big)$.

5. If $f(S) \neq \emptyset$ for some $S \in \mathfrak{X}'$, then $\mathrm{LF}_{\mathfrak{X}}(f) \subseteq E\big((S)'\big) \circ f(S)$.

Remark 3.1.2 (5) is a consequence of the following lemma:

Lemma 3.1.3. *Let G be a group and let $\{M_i : 1 \leq i \leq s\}$ be the set of all minimal normal subgroups of G. Then, for each $1 \leq i \leq s$, G has a normal subgroup N_i such that G/N_i is monolithic and $\mathrm{Soc}(G/N_i)$ is G-isomorphic to M_i. Moreover $G \in R_0(\{G/N_i : 1 \leq i \leq s\})$.*

Proof. For each $1 \leq i \leq s$, we consider an element N_i of maximal order of the family $\{T_i \trianglelefteq G : T_i \cap M_i = 1\}$. Then G/N_i is monolithic, $\mathrm{Soc}(G/N_i)$ is G-isomorphic to M_i and $G \in R_0(\{G/N_i : 1 \leq i \leq s\})$. $\square$

Proof (of Remark 3.1.2 (5)). Assume that $G \in \mathrm{LF}_{\mathfrak{X}}(f)$ and $f(S) \neq \emptyset$ for some $S \in \mathfrak{X}'$. Then every minimal normal subgroup of G/N, for $N = \mathrm{O}_{(S)'}(G)$, is in $\mathrm{E}(S)$. Therefore $G/N \in f(S)$ by the above lemma. In particular, $G \in \mathrm{E}((S)') \circ f(S)$. Remark 3.1.2 (5) is proved.   □

We can now deduce the following result.

Theorem 3.1.4. *Let f be an $\mathfrak{X}$-formation function. Then $\mathrm{LF}_{\mathfrak{X}}(f)$ is a formation.*

Proof. We prove that $\mathrm{LF}_{\mathfrak{X}}(f)$ is R_0-closed. Let N_1 and N_2 be two different minimal normal subgroups of a group G such that $G/N_i \in \mathrm{LF}_{\mathfrak{X}}(f)$ $(i = 1, 2)$. We see that G satisfies Condition 1 of Definition 3.1.1. If $N_1 \in \mathrm{E}(\mathfrak{X}')$, then clearly $G \in \mathrm{LF}_{\mathfrak{X}}(f)$. Hence we may assume that $N_1 \in \mathrm{E}\,\mathfrak{X}$. Let p be a prime dividing $|N_1|$. Then $N_1 N_2/N_1$ is an $\mathfrak{X}_p$-chief factor of G/N_2 and $\mathrm{Aut}_G(N_1) \cong \mathrm{Aut}_{G/N_2}(N_1 N_2/N_2)$ and $G/N_2 \in \mathrm{LF}_{\mathfrak{X}}(f)$. Therefore $G/\mathrm{C}_G(N_1) \in f(p)$. Since $G/N_1 \in \mathrm{LF}_{\mathfrak{X}}(f)$, by appealing to the generalised Jordan-Hölder theorem (1.2.34), we infer that G satisfies Condition 1.

Consider now a monolithic quotient G/K of G such that $\mathrm{Soc}(G/K) \in \mathrm{E}(S)$ for some simple group $S \in \mathfrak{X}'$. If $f(S) = \emptyset$, then $\mathrm{LF}_{\mathfrak{X}}(f) \subseteq \mathrm{E}((S)')$ by Remark 3.1.2 (4). Therefore $G/N_i \in \mathrm{E}((S)')$ for $i \in \{1, 2\}$. This implies $G \in \mathrm{E}((S)')$, contrary to supposition. Hence $f(S) \neq \emptyset$ and so $G/N_i \in \mathrm{E}((S)') \circ f(S)$ by Remark 3.1.2 (5). In particular, $G/K \in \mathrm{E}((S)') \circ f(S)$ because the latter class is a formation. Since $\mathrm{O}_{(S)'}(G/K) = 1$, it follows that $G/K \in f(S)$. Hence G satisfies Condition 2 of Definition 3.1.1.

Consequently $G \in \mathrm{LF}_{\mathfrak{X}}(f)$. Applying Remark 3.1.2 (1) and [DH92, II, 2.6], $\mathrm{LF}_{\mathfrak{X}}(f)$ is a formation.   □

Definition 3.1.5. *A formation $\mathfrak{F}$ is said to be $\mathfrak{X}$-local if $\mathfrak{F} = \mathrm{LF}_{\mathfrak{X}}(f)$ for some $\mathfrak{X}$-formation function f. In this case we say that f is an $\mathfrak{X}$-local definition of $\mathfrak{F}$ or f defines $\mathfrak{F}$.*

Examples 3.1.6. 1. Each formation $\mathfrak{F}$ is $\mathfrak{X}$-local for $\mathfrak{X} = \emptyset$ because $\mathfrak{F} = \mathrm{LF}_{\mathfrak{X}}(f)$, where $f(S) = \mathfrak{F}$ for all $S \in \mathfrak{J}$.

2. If $\mathfrak{X} = \mathfrak{J}$, then an $\mathfrak{X}$-formation function is simply a formation function and the $\mathfrak{X}$-local formations are exactly the local formations.

3. If $\mathfrak{X} = \mathbb{P}$, then an $\mathfrak{X}$-formation function is a Baer function and the $\mathfrak{X}$-local formations are exactly the Baer-local ones.

Remarks 3.1.7. Let f and f_i be $\mathfrak{X}$-formation functions for all $i \in \mathcal{I}$.

1. $\bigcap_{i \in \mathcal{I}} \mathrm{LF}_{\mathfrak{X}}(f_i) = \mathrm{LF}_{\mathfrak{X}}(g)$, where $g(S) = \bigcap_{i \in \mathcal{I}} f_i(S)$ for all $S \in (\mathrm{char}\,\mathfrak{X}) \cup \mathfrak{X}'$.

2. Let $N \trianglelefteq G$ and $G/N \in \mathrm{LF}_{\mathfrak{X}}(f)$. If $N \in \mathrm{E}\,\mathfrak{X}$ and $G/\mathrm{C}_G(N) \in f(p)$ for all $p \mid |N|$, then $G \in \mathrm{LF}_{\mathfrak{X}}(f)$.

Proof.   1. This follows immediately from the definition of $\mathfrak{X}$-local formation.

2. If H/K is an $\mathfrak{X}_p$-chief factor of G above N, then $G/\operatorname{C}_G(H/K) \in f(p)$ because $G/N \in \operatorname{LF}_{\mathfrak{X}}(f)$. Let H/K be an $\mathfrak{X}_p$-chief factor of G below N. Then $\operatorname{C}_G(N) \le \operatorname{C}_G(H/K)$ and so $G/\operatorname{C}_G(H/K) \in \operatorname{Q} f(p) = f(p)$. By the generalised Jordan-Hölder theorem (1.2.34), we have that G satisfies Condition 1 of Definition 3.1.1.

Let K be a normal subgroup of G such that G/K is a monolithic group with $\operatorname{Soc}(G/K) \in \operatorname{E}(S)$, $S \in \mathfrak{X}'$. Then, since $N \in \operatorname{E}\mathfrak{X}$, we have that $N \le K$. Therefore $G/K \in f(S)$ because $G/N \in \operatorname{LF}_{\mathfrak{X}}(f)$.

Consequently $G \in \operatorname{LF}_{\mathfrak{X}}(f)$. $\square$

Definition 3.1.8. *Let $p \in \operatorname{char} \mathfrak{X}$. Then the subgroup $\operatorname{C}^{\mathfrak{X}_p}(G)$ is defined to be the intersection of the centralisers of all $\mathfrak{X}_p$-chief factors of G, with $\operatorname{C}^{\mathfrak{X}_p}(G) = G$ if G has no $\mathfrak{X}_p$-chief factors.*

Remark 3.1.9. Let $\operatorname{LF}_{\mathfrak{X}}(f)$ be an $\mathfrak{X}$-local formation. Then G satisfies Condition 1 of Definition 3.1.1 if and only if $G/\operatorname{C}^{\mathfrak{X}_p}(G) \in f(p)$ for all $p \in \operatorname{char} \mathfrak{X}$ such that $f(p) \ne \emptyset$.

Note that $\left(\operatorname{C}^{\mathfrak{X}_p}(G) \right)^{\theta}$ is contained in $\operatorname{C}^{\mathfrak{X}_p}(G^{\theta})$ for every epimorphism θ of G. Therefore, by [DH92, IV, 1.10], the class $\operatorname{Q}\big(G/\operatorname{C}^{\mathfrak{X}_p}(G) : G \in \mathfrak{F}\big)$ is a formation, for each formation $\mathfrak{F}$.

Let N be a normal subgroup of G and let H/K be a chief factor of G below N. Then, by [DH92, A, 4.13 (c)], H/K is a direct product of chief factors of N. Therefore we have

Proposition 3.1.10. $\operatorname{C}^{\mathfrak{X}_p}(G) \cap N = \operatorname{C}^{\mathfrak{X}_p}(N)$ *for all normal subgroups N of G.*

Let f_1 and f_2 be two $\mathfrak{X}$-formation functions. We write $f_1 \le f_2$ if $f_1(S) \subseteq f_2(S)$ for all $S \in (\operatorname{char} \mathfrak{X}) \cup \mathfrak{X}'$. Note that in this case $\operatorname{LF}_{\mathfrak{X}}(f_1) \subseteq \operatorname{LF}_{\mathfrak{X}}(f_2)$. By Remark 3.1.7 (1), each $\mathfrak{X}$-local formation $\mathfrak{F}$ has a unique $\mathfrak{X}$-formation function $\underline{f}$ defining $\mathfrak{F}$ such that $\underline{f} \le f$ for each $\mathfrak{X}$-formation function f such that $\mathfrak{F} = \operatorname{LF}_{\mathfrak{X}}(f)$. We say that $\underline{f}$ is the *minimal $\mathfrak{X}$-local definition* of $\mathfrak{F}$. This $\mathfrak{X}$-local formation function will always be denoted by the use of a "lower bar."

Moreover if $\mathfrak{Y}$ is a class of groups, the intersection of all $\mathfrak{X}$-local formations containing $\mathfrak{Y}$ is the smallest $\mathfrak{X}$-local formation containing $\mathfrak{Y}$. Such $\mathfrak{X}$-local formation is denoted by $\operatorname{form}_{\mathfrak{X}}(\mathfrak{Y})$. If $\mathfrak{X} = \mathfrak{J}$, we also write $\operatorname{lform}(\mathfrak{Y}) = \operatorname{form}_{\mathfrak{J}}(\mathfrak{Y})$, and if $\mathfrak{X} = \mathbb{P}$, $\operatorname{form}_{\mathbb{P}}(\mathfrak{Y})$ is usually denoted by $\operatorname{bform}(\mathfrak{Y})$.

Theorem 3.1.11. *Let $\mathfrak{Y}$ be a class of groups. Then $\mathfrak{F} = \operatorname{form}_{\mathfrak{X}}(\mathfrak{Y}) = \operatorname{LF}_{\mathfrak{X}}(\underline{f})$, where*

$$\underline{f}(p) = \operatorname{Q}\operatorname{R}_0\big(G/\operatorname{C}_G(H/K) : G \in \mathfrak{Y} \text{ and } H/K \text{ is an } \mathfrak{X}_p\text{-chief factor of } G\big),$$

if $p \in \operatorname{char} \mathfrak{X}$, and

$$\underline{f}(S) = \operatorname{Q}\operatorname{R}_0\big(G/L : G \in \mathfrak{Y}, G/L \text{ is monolithic, and } \operatorname{Soc}(G/L) \in \operatorname{E}(S)\big),$$

if $S \in \mathfrak{X}'$. Moreover $\underline{f}(p) = \operatorname{Q}\big(G/\operatorname{C}^{\mathfrak{X}_p}(G) : G \in \mathfrak{F}\big)$ for all $p \in \operatorname{char} \mathfrak{X}$ such that $\underline{f}(p) \ne \emptyset$.

Proof. Let g be an $\mathfrak{X}$-formation function such that $\mathfrak{F} = \mathrm{LF}_{\mathfrak{X}}(g)$. Since $\mathrm{LF}_{\mathfrak{X}}(\underline{f})$ is an $\mathfrak{X}$-local formation containing $\mathfrak{Y}$, we have $\mathfrak{F} \subseteq \mathrm{LF}_{\mathfrak{X}}(\underline{f})$. Assume that $\mathrm{LF}_{\mathfrak{X}}(\underline{f}) \neq \mathfrak{F}$. Then $\mathrm{LF}_{\mathfrak{X}}(\underline{f}) \setminus \mathfrak{F}$ contains a group G of minimal order. Such a G has a unique minimal normal subgroup N by [DH92, II, 2.5] and $G/N \in \mathfrak{F}$. If N is an $\mathfrak{X}'$-chief factor of G, then $G \in \underline{f}(S)$ for some $S \in \mathfrak{X}'$. This implies that $G \in \mathrm{Q} \mathrm{R}_0 \mathfrak{Y} \subseteq \mathfrak{F}$, a contradiction. Therefore $N \in \mathrm{E}\,\mathfrak{X}$. Let p be a prime divisor of $|N|$. Then $G/\mathrm{C}_G(N) \in \underline{f}(p)$. Now if X is a group in $\mathfrak{Y}$ and H/K is an $\mathfrak{X}_p$-chief factor of X, then $X/\mathrm{C}_X(H/K) \in g(p)$ because $\mathfrak{Y} \subseteq \mathfrak{F}$. Therefore $\underline{f}(p) \subseteq g(p)$, and so $G/\mathrm{C}_G(N) \in g(p)$. Applying Remark 3.1.7 (2), $G \in \mathfrak{F}$, contrary to hypothesis. Consequently $\mathfrak{F} = \mathrm{LF}_{\mathfrak{X}}(\underline{f})$.

Let $p \in \mathrm{char}\,\mathfrak{X}$ and $t(p) = \mathrm{Q}\big(G/\mathrm{C}^{\mathfrak{X}_p}(G) : G \in \mathfrak{F}\big)$. We know that $t(p)$ is a formation. Moreover, if $G \in \mathfrak{F}$ and $\underline{f}(p) \neq \emptyset$, then $G/\mathrm{C}^{\mathfrak{X}_p}(G) \in \underline{f}(p)$. Therefore $t(p) \subseteq \underline{f}(p)$. On the other hand, if $X \in \mathfrak{Y}$, then $X/\mathrm{C}^{\mathfrak{X}_p}(X) \in t(p)$. Hence $X/\mathrm{C}_X(H/K) \in t(p)$ for every $\mathfrak{X}_p$-chief factor H/K of X. This means that $\underline{f}(p) \subseteq t(p)$ and the equality holds. This completes the proof of the theorem. $\qquad\square$

Remark 3.1.12. If $\mathfrak{F}$ is a local formation and $\underline{f}$ is the smallest local definition of $\mathfrak{F}$, then $\underline{f}(p) = \mathrm{Q}\big(G/\mathrm{O}_{p',p}(G) : G \in \mathfrak{F}\big)$ for each $p \in \mathrm{char}\,\mathfrak{F}$ (cf. [DH92, IV, 3.10]). The equality $\underline{f}(p) = \mathrm{Q}\big(G/\mathrm{O}_{p',p}(G) : G \in \mathfrak{F}\big)$ does not hold for $\mathfrak{X}$-local formations in general: Let $\mathfrak{X} = (C_2)$ and consider $\mathfrak{F} = \mathrm{LF}_{\mathfrak{X}}(\underline{f})$, where $\underline{f}(2) = \mathfrak{S}$ and $\underline{f}(S) = \mathfrak{E}$ for all $S \in \mathfrak{X}'$. Then $\mathrm{Alt}(5) \in \mathfrak{F}$ and so $\mathrm{Alt}(5) \in \mathrm{Q}\big(G/\mathrm{O}_{2',2}(G) : G \in \mathfrak{F}\big)$. Since $\underline{f}(2) \subseteq \mathfrak{S}$, it follows that $\mathrm{Alt}(5) \notin \underline{f}(2)$. Consequently $\underline{f}(2) \neq \mathrm{Q}\big(G/\mathrm{O}_{2',2}(G) : G \in \mathfrak{F}\big)$.

Corollary 3.1.13. *Let $\mathfrak{X}$ and $\bar{\mathfrak{X}}$ be classes of simple groups such that $\bar{\mathfrak{X}} \subseteq \mathfrak{X}$. Then every $\mathfrak{X}$-local formation is $\bar{\mathfrak{X}}$-local.*

Proof. Let $\mathfrak{F} = \mathrm{LF}_{\mathfrak{X}}(\underline{f})$ be an $\mathfrak{X}$-local formation. Since $\mathrm{char}\,\bar{\mathfrak{X}} \subseteq \mathrm{char}\,\mathfrak{X}$, we can consider the $\bar{\mathfrak{X}}$-formation function g defined by

$$g(p) = \underline{f}(p) \qquad\qquad \text{if } p \in \mathrm{char}\,\bar{\mathfrak{X}},$$
$$g(E) = \mathfrak{F} \qquad\qquad \text{if } E \in \bar{\mathfrak{X}}'.$$

It is clear that $\mathfrak{F} \subseteq \mathrm{LF}_{\bar{\mathfrak{X}}}(g)$. Suppose that $\mathfrak{F} \neq \mathrm{LF}_{\bar{\mathfrak{X}}}(g)$, and choose a group Y of minimal order in $\mathrm{LF}_{\bar{\mathfrak{X}}}(g) \setminus \mathfrak{F}$. Then Y has a unique minimal normal subgroup N, and $G/N \in \mathfrak{F}$. If $N \in \mathrm{E}(\bar{\mathfrak{X}}')$, then $G \in \mathfrak{F}$, which contradicts the choice of G. Therefore $N \in \mathrm{E}\,\bar{\mathfrak{X}}$ and $G/\mathrm{C}_G(N) \in \underline{f}(p)$ for each prime p dividing $|N|$. Applying Remark 3.1.7 (2), we conclude that $G \in \mathfrak{F}$, contrary to supposition. Hence $\mathfrak{F} = \mathrm{LF}_{\bar{\mathfrak{X}}}(g)$ and $\mathfrak{F}$ is $\bar{\mathfrak{X}}$-local. $\qquad\square$

By [DH92, IV, 3.8], each local formation $\mathfrak{F} = \mathrm{LF}(f)$ can be defined by a formation function g given by $g(p) = \mathfrak{F} \cap \mathfrak{S}_p f(p)$ for all primes p. The corresponding result for $\mathfrak{X}$-local formations is the following:

Theorem 3.1.14. *Let* $\mathfrak{F} = \mathrm{LF}_{\mathfrak{X}}(f)$ *be an* $\mathfrak{X}$*-local formation defined by the* $\mathfrak{X}$*-formation function* f. *Set*

$$f^*(p) = \mathfrak{F} \cap \mathfrak{S}_p f(p) \qquad\qquad \text{for all } p \in \operatorname{char}\mathfrak{X},$$
$$f^*(S) = \mathfrak{F} \cap f(S) \qquad\qquad \text{for all } S \in \mathfrak{X}'.$$

Then:

1. f^* *is an* $\mathfrak{X}$*-formation function such that* $\mathfrak{F} = \mathrm{LF}_{\mathfrak{X}}(f^*)$.
2. $\mathfrak{S}_p f^*(p) = f^*(p)$ *for all* $p \in \operatorname{char}\mathfrak{X}$.

Proof.   1. It is clear that f^* is an $\mathfrak{X}$-formation function. Let $\mathfrak{F}^* = \mathrm{LF}_{\mathfrak{X}}(f^*)$ and let $G \in \mathfrak{F}^*$. If H/K is an $\mathfrak{X}_p$-chief factor of G, then $G/\operatorname{C}_G(H/K) \in \mathfrak{F} \cap \mathfrak{S}_p f(p)$. Since, by [DH92, A, 13.6], $\operatorname{O}_p\big(G/\operatorname{C}_G(H/K)\big) = 1$, it follows that $G/\operatorname{C}_G(H/K) \in f(p)$. Now if G/L is a monolithic quotient of G with $\operatorname{Soc}(G/L) \in \mathrm{E}(S)$ for some $S \in \mathfrak{X}'$, it follows that $G/L \in f(S)$. Therefore $G \in \mathfrak{F}$.

Now if H/K is an $\mathfrak{X}_p$-chief factor of a group $A \in \mathfrak{F}$, then $A/\operatorname{C}_A(H/K) \in \mathrm{Q}\,\mathfrak{F} \cap f(p) \subseteq f^*(p)$. If A/L is a monolithic quotient of A with $\operatorname{Soc}(A/L) \in \mathrm{E}(S)$, $S \in \mathfrak{X}'$, then $A/L \in \mathrm{Q}\,\mathfrak{F} \cap f(S) \subseteq f^*(S)$. This implies that $A \in \mathfrak{F}^*$ and therefore $\mathfrak{F} = \mathfrak{F}^*$.

2. Let $G \in \mathfrak{S}_p f^*(p)$, $p \in \operatorname{char}\mathfrak{X}$. Then $G/\operatorname{O}_p(G) \in f^*(p)$ and so $G \in \mathfrak{S}_p f(p)$ because $\operatorname{O}_p\big(G/\operatorname{O}_p(G)\big) = 1$. Moreover $G/\operatorname{O}_p(G) \in \mathfrak{F}$. If H/K is an $\mathfrak{X}_p$-chief factor of G below $\operatorname{O}_p(G)$, then $\operatorname{O}_p(G) \leq \operatorname{C}_G(H/K)$ by [DH92, B, 3.12 (b)] and so $G/\operatorname{C}_G(H/K) \in \mathrm{Q}\,f(p) = f(p)$. If G/L is a monolithic quotient of G such that $\operatorname{Soc}(G/L) \in \mathrm{E}(S)$, $S \in \mathfrak{X}'$, it follows that $\operatorname{O}_p(G) \leq L$. Therefore $G/L \in \mathrm{Q}\,f^*(p) = f^*(p) \subseteq \mathfrak{F}$ and so $G/L \in f(S)$. This proves that $G \in \mathfrak{F}$. Consequently $G \in f^*(p)$ and $\mathfrak{S}_p f^*(p) = f^*(p)$. $\square$

Definition 3.1.15. *Let* f *be an* $\mathfrak{X}$*-formation function defining an* $\mathfrak{X}$*-local formation* $\mathfrak{F}$. *Then* f *is called:*

1. **integrated** *if* $f(S) \subseteq \mathfrak{F}$ *for all* $S \in (\operatorname{char}\mathfrak{X}) \cup \mathfrak{X}'$,
2. **full** *if* $\mathfrak{S}_p f(p) = f(p)$ *for all* $p \in \operatorname{char}\mathfrak{X}$.

Let $\mathfrak{F} = \mathrm{LF}_{\mathfrak{X}}(f)$ be an $\mathfrak{X}$-local formation. Then the $\mathfrak{X}$-formation function g given by $g(S) = \mathfrak{F} \cap f(S)$ for all $S \in (\operatorname{char}\mathfrak{X}) \cup \mathfrak{X}'$ is an integrated $\mathfrak{X}$-local definition of $\mathfrak{F}$. Moreover f^* is, according to the above theorem, an integrated and full $\mathfrak{X}$-local definition of $\mathfrak{F}$.

It is known (cf. [DH92, IV, 3.7]) that if $\mathfrak{X} = \mathfrak{J}$, then every $\mathfrak{X}$-local formation has a unique integrated and full $\mathfrak{X}$-local definition, the canonical one. This is not true in general. In fact, if $\emptyset \neq \mathfrak{X} \neq \mathfrak{J}$, we can find an $\mathfrak{X}$-local formation with several integrated and full $\mathfrak{X}$-local definitions.

Example 3.1.16. Let $\emptyset \neq \mathfrak{X} \neq \mathfrak{J}$. Then we can consider $X \in \mathfrak{J} \setminus \mathfrak{X}$ and a prime $p \in \operatorname{char}\mathfrak{X}$. The formation $\mathfrak{F} = \mathfrak{S}_p$ is an $\mathfrak{X}$-local formation which can be $\mathfrak{X}$-locally defined by the following integrated and full $\mathfrak{X}$-formation functions:

$$f_1(S) = \begin{cases} \mathfrak{S}_p & \text{if } S \cong C_p, \\ \emptyset & \text{if } S \not\cong C_p, \end{cases}$$

and

$$f_2(S) = \begin{cases} \mathfrak{S}_p & \text{if } S \cong C_p, \\ \mathfrak{S}_p & \text{if } S \cong X, \\ \emptyset & \text{otherwise} \end{cases}$$

for all $S \in (\text{char } \mathfrak{X}) \cup \mathfrak{X}'$.

We say that an $\mathfrak{X}$-formation function f defining an $\mathfrak{X}$-local formation $\mathfrak{F}$ is a *maximal integrated $\mathfrak{X}$-formation function* if $g \leq f$ for each integrated $\mathfrak{X}$-formation function g such that $\mathfrak{F} = \mathrm{LF}_{\mathfrak{X}}(g)$.

The next result shows that every $\mathfrak{X}$-local formation can be $\mathfrak{X}$-locally defined by a maximal integrated $\mathfrak{X}$-formation function F. Moreover F is full.

Theorem 3.1.17. *Let $\mathfrak{F} = \mathrm{LF}_{\mathfrak{X}}(f)$ be an $\mathfrak{X}$-local formation. Then:*

1. *$\mathfrak{F}$ is $\mathfrak{X}$-locally defined by the integrated and full $\mathfrak{X}$-formation function F given by $F(p) = \mathfrak{S}_p \underline{f}(p)$ for all $p \in \text{char } \mathfrak{X}$ and $F(S) = \mathfrak{F}$ for all $S \in \mathfrak{X}'$.*
2. *For each $p \in \text{char } \mathfrak{X}$, $F(p) = (G : C_p \wr G \in \mathfrak{F})$.*
3. *If $\mathfrak{F} = \mathrm{LF}_{\mathfrak{X}}(g)$, then $F(p) = \mathfrak{F} \cap \mathfrak{S}_p g(p)$ for all $p \in \text{char } \mathfrak{X}$.*

Proof.   1.   Since $\underline{f} \leq F$, it follows that $\mathfrak{F} \subseteq \mathrm{LF}_{\mathfrak{X}}(F)$. Suppose, by way of contradiction, that the equality does not hold and let G be a group of minimal order in $\mathrm{LF}_{\mathfrak{X}}(F) \backslash \mathfrak{F}$. Then the group G has a unique minimal normal subgroup, N say, and $G/N \in \mathfrak{F}$. Furthermore $N \in \mathrm{E}\,\mathfrak{X}$ because otherwise $G \in F(S)$ for some $S \in \mathfrak{X}'$ and then $G \in \mathfrak{F}$, contrary to supposition. Let p be a prime dividing $|N|$. Then $G/\mathrm{C}_G(N) \in \mathfrak{S}_p \underline{f}(p)$ and so $G/\mathrm{C}_G(N) \in \underline{f}(p)$ because $\mathrm{O}_p(G/\mathrm{C}_G(N)) = 1$ by [DH92, A, 13.6 (b)]. Then Remark 3.1.7 (2) implies that $G \in \mathfrak{F}$. This contradiction yields $\mathrm{LF}_{\mathfrak{X}}(F) \subseteq \mathfrak{F}$ and then $\mathfrak{F} = \mathrm{LF}_{\mathfrak{X}}(F)$. It is clear that F is full. Let $p \in \text{char } \mathfrak{X}$. If possible, choose a group G of minimal order in $F(p) \backslash \mathfrak{F}$. We know that G has a unique minimal normal subgroup N and, since $\underline{f}(p) \subseteq \mathfrak{F}$, $\mathrm{O}_p(G) \neq 1$. Hence N is a p-group. Moreover $G/N \in \mathfrak{F}$ and $G/\mathrm{C}_G(N) \in \underline{f}(p)$ because $\mathrm{O}_p(G)$ centralises N. But then $G \in \mathfrak{F}$. This contradicts the choice of G, and so we conclude that $F(p) \subseteq \mathfrak{F}$.

2.   Let $p \in \text{char } \mathfrak{X}$ and let $\bar{F}(p)$ denote the class $(G : C_p \wr G \in \mathfrak{F})$. If $G \in F(p)$, then $C_p \wr G \in \mathfrak{S}_p F(p) = F(p) \subseteq \mathfrak{F}$ by Statement 1. Hence $G \in \bar{F}(p)$ and so $F(p) \subseteq \bar{F}(p)$. Now consider a group $G \in \bar{F}(p)$ and set $W = C_p \wr G$. Denote $B = C_p^\natural$ the base group of W and $A = \bigcap\{\mathrm{C}_W(H/K) : H \leq B \text{ and } H/K \text{ is a chief factor of } W\}$. Since $W \in \mathfrak{F}$, it follows that $W/A \in F(p)$. On the other hand, A acts as a group of operators for B by conjugation and A stabilises a chain of subgroups of B. Applying [DH92, A, 12.4], we have that $A/\mathrm{C}_A(B)$ is a p-group. Then A is itself a p-group because $\mathrm{C}_A(B) = B$ by [DH92, A, 18.8]. Consequently $W \in F(p)$ and $G \in \mathrm{Q}\,F(p) = F(p)$. This proves that $\bar{F}(p) = F(p)$.

3. Let g be an $\mathfrak{X}$-formation function such that $\mathfrak{F} = \mathrm{LF}_{\mathfrak{X}}(g)$. Since $\underline{f} \leq g$, it follows that $F(p) = \mathfrak{S}_p \underline{f}(p) \subseteq \mathfrak{F} \cap \mathfrak{S}_p g(p) = g^*(p)$ for all $p \in \mathrm{char}\,\overline{\mathfrak{X}}$. Let X be a group in $g^*(p)$ and set $W = C_p \wr X$. As above, denote by $B = C_p^{\natural}$ the base group of W. Then $W/B \in g^*(p)$. Moreover $W/B \in \mathfrak{F} = \mathrm{LF}_{\mathfrak{X}}(g^*)$ by Theorem 3.1.14. Applying Remark 3.1.7 (2), we conclude that $W \in \mathfrak{F}$. Hence $X \in F(p)$ and $F(p) = g^*(p)$.   □

Let g be an integrated $\mathfrak{X}$-formation function defining an $\mathfrak{X}$-local formation $\mathfrak{F}$. Then $g(p) \subseteq \mathfrak{F} \cap \mathfrak{S}_p g(p) = F(p)$ for all $p \in \mathrm{char}\,\mathfrak{X}$ by Theorem 3.1.17 (3). Therefore $g \leq F$. We shall say that F is the *canonical* $\mathfrak{X}$-local definition of $\mathfrak{F} = \mathrm{LF}_{\mathfrak{X}}(F)$. As in the case of local formations, the canonical $\mathfrak{X}$-local definition will be identified by the use of an uppercase Roman letter. Hence if we write $\mathfrak{F} = \mathrm{LF}_{\mathfrak{X}}(F)$, we are assuming that F is the canonical $\mathfrak{X}$-local definition of $\mathfrak{F}$.

Corollary 3.1.18. *Let $\mathfrak{F}$ be an $\mathfrak{X}$-local formation and $\mathfrak{Y} \subseteq \mathfrak{X}$. Let F_1 and F_2 be the canonical $\mathfrak{Y}$-local and $\mathfrak{X}$-local definitions of $\mathfrak{F}$, respectively. Then $F_1(p) = F_2(p)$ for all $p \in \mathrm{char}\,\mathfrak{Y}$.*

Proof. Applying Corollary 3.1.13, we know that $\mathfrak{F}$ is $\mathfrak{Y}$-local. Let p be a prime in $\mathrm{char}\,\mathfrak{Y}$. Then $p \in \mathrm{char}\,\mathfrak{X}$ and by Theorem 3.1.17 (2) we have that $F_1(p) = (G : C_p \wr G \in \mathfrak{F}) = F_2(p)$.   □

Taking $\mathfrak{Y} = (C_p)$, $p \in \mathrm{char}\,\mathfrak{X}$ in Corollary 3.1.18 and, applying Theorem 3.1.11 and Theorem 3.1.17, we have:

Corollary 3.1.19. *Let $\mathfrak{F}$ be an $\mathfrak{X}$-local formation. If $p \in \mathrm{char}\,\mathfrak{X}$, then*

$$F(p) = \mathfrak{S}_p \,\mathrm{Q}\,\mathrm{R}_0\big(G/\,\mathrm{C}_G(H/K) : G \in \mathfrak{F},\, H/K \text{ is an abelian}$$
$$p\text{-chief factor of } G\big).$$

Corollary 3.1.20. *Let $\mathfrak{F} = \mathrm{LF}_{\mathfrak{X}}(\underline{f}) = \mathrm{LF}_{\mathfrak{X}}(F)$ and $\mathfrak{G} = \mathrm{LF}_{\mathfrak{X}}(g) = \mathrm{LF}_{\mathfrak{X}}(G)$ be $\mathfrak{X}$-local formations. Then any two of the following statements are equivalent:*

1. $\mathfrak{F} \subseteq \mathfrak{G}$
2. $F \leq G$
3. $\underline{f} \leq g$

Corollary 3.1.21 ([BBCER05, Lemma 4.5]). *Let $\mathfrak{F}$ be a formation and let $\{\mathfrak{X}_i : i \in \mathcal{I}\}$ be a family of classes of simple groups such that $\pi(\mathfrak{X}_i) = \mathrm{char}\,\mathfrak{X}_i$ for all $i \in \mathcal{I}$. Put $\mathfrak{X} = \bigcup_{i \in \mathcal{I}} \mathfrak{X}_i$. If $\mathfrak{F}$ is $\mathfrak{X}_i$-local for all $i \in \mathcal{I}$, then $\mathfrak{F}$ is $\mathfrak{X}$-local.*

Proof. First of all, note that $\pi(\mathfrak{X}) = \mathrm{char}\,\mathfrak{X}$.
Applying Theorem 3.1.17, $\mathfrak{F} = \mathrm{LF}_{\mathfrak{X}_i}(F_i)$, where

$$F_i(S) = \begin{cases} (G : C_p \wr G \in \mathfrak{F}) & \text{if } S \cong C_p,\, p \in \mathrm{char}\,\mathfrak{X}_i, \\ \mathfrak{F} & \text{if } S \in \mathfrak{X}_i', \end{cases}$$

for all $i \in \mathcal{I}$.

Let f be the $\mathfrak{X}$-formation function defined by

$$f(S) = \begin{cases} (G : C_p \wr G \in \mathfrak{F}) & \text{if } S \cong C_p, \, p \in \text{char } \mathfrak{X}, \\ \mathfrak{F} & \text{if } S \in \mathfrak{X}'. \end{cases}$$

It is clear that $\mathfrak{F} \subseteq \mathrm{LF}_{\mathfrak{X}}(f)$. Assume that the inclusion is proper and derive a contradiction. Let $G \in \mathrm{LF}_{\mathfrak{X}}(f) \setminus \mathfrak{F}$ of minimal order. Then G has a unique minimal normal subgroup N such that $G/N \in \mathfrak{F}$. It is clear that $N \in \mathrm{E}\,\mathfrak{X}$ because otherwise $G \in \mathfrak{F}$. Hence $N \in \mathrm{E}\,\mathfrak{X}_i$ for some $i \in \mathcal{I}$ and $G/\,\mathrm{C}_G(N) \in f(p) = F_i(p)$ for all $p \in \pi(N)$. Therefore $G \in \mathrm{LF}_{\mathfrak{X}_i}(F_i) = \mathfrak{F}$. This is a contradiction. Consequently $\mathfrak{F} = \mathrm{LF}_{\mathfrak{X}}(f)$ and $\mathfrak{F}$ is an $\mathfrak{X}$-local formation.   □

When $\mathfrak{X}$ is the class of all abelian simple groups, we have $\mathfrak{X} = \bigcup_{p \in \mathbb{P}}(C_p)$. Therefore

Corollary 3.1.22 ([BBCER05, Corollary 4.6]). *A formation $\mathfrak{F}$ is Baer-local if and only if $\mathfrak{F}$ is (C_p)-local for every prime p.*

A natural question arising from the above discussion is whether an $\mathfrak{X}$-local formation has a unique upper bound for all its $\mathfrak{X}$-local definitions, that is, if $\mathfrak{F}$ can be $\mathfrak{X}$-locally defined by an $\mathfrak{X}$-formation function F^0 such that $f \leq F^0$ for each $\mathfrak{X}$-local definition f of $\mathfrak{F}$. If such F^0 exists, we will refer to it as the *maximal $\mathfrak{X}$-local definition* of $\mathfrak{F}$.

In [Doe73], K. Doerk presented a beautiful result showing that in the soluble universe each local formation has a maximal local definition (see also [DH92, V, 3.18]). The same author, P. Schmid [Sch74], and L. A. Shemetkov [She78] posed the problem of whether every local formation of finite groups has a maximal local definition. The answer is "no" as the following example shows:

Example 3.1.23 ([Sal85]). Let $\mathfrak{F} = \mathfrak{S}$ be the local formation of all soluble groups. Then $\mathfrak{F} = \mathrm{LF}(f_1) = \mathrm{LF}(f_2)$, where f_1 and f_2 are the formation functions defined by

$$f_1(2) = \mathrm{D}_0\big(\mathfrak{S}, \mathrm{Alt}(5)\big),$$
$$f_1(p) = \mathfrak{S} \qquad\qquad \text{for each prime } p \neq 2,$$
$$f_2(3) = f_2(5) = \mathrm{D}_0\big(\mathfrak{S}, \mathrm{Alt}(5)\big),$$
$$f_2(p) = \mathfrak{S} \qquad\qquad \text{for each prime } p \neq 3, 5.$$

Assume that $\mathfrak{F}$ has a maximal local definition, F^0 say. Then $f_i \leq F^0$ for $i = 1, 2$. This implies that $\mathrm{Alt}(5) \in \mathrm{LF}(F^0) = \mathfrak{F}$, a contradiction. Therefore $\mathfrak{F}$ does not have a maximal local definition.

Perhaps the most simple example of a local formation with a maximal local ($\mathfrak{J}$-local) definition is given by the class $\mathfrak{E}_\pi$ of all π-groups for a set of primes π. It is rather clear that

$$\widehat{F}(p) = \begin{cases} \mathfrak{E} & \text{if } p \in \pi, \\ \emptyset & \text{if } p \notin \pi, \end{cases}$$

defines the maximal local definition of $\mathfrak{E}_\pi$.

In the following we shall give a description of $\mathfrak{X}$-local formations with a maximal $\mathfrak{X}$-local definition. The main source for this description is P. Förster and E. Salomon [FS85].

The following concept, introduced for local formations in [Sal85], turns out to be crucial.

Definition 3.1.24 ([FS85]). *Let $\mathfrak{F} = \mathrm{LF}_{\mathfrak{X}}(F)$ be an $\mathfrak{X}$-local formation. Denote by $\mathrm{b}_{\mathfrak{X}}(\mathfrak{F})$ the class of all groups $G \in \mathrm{b}(\mathfrak{F})$ such that $\mathrm{Soc}(G) \in \mathrm{E}\,\mathfrak{X}$. A group $G \in \mathrm{b}_{\mathfrak{X}}(\mathfrak{F})$ is called $\mathfrak{X}$-dense with respect to $\mathfrak{F}$ if $G \in \mathrm{b}(F(p))$ for each prime p dividing $|\mathrm{Soc}(G)|$. Further, $\mathrm{b}(\mathfrak{F})$ is said to be $\mathfrak{X}$-wide if there does not exist an $\mathfrak{X}$-dense group $G \in \mathrm{b}_{\mathfrak{X}}(\mathfrak{F})$.*

Note that a group $G \in \mathrm{b}_{\mathfrak{X}}(\mathfrak{F})$ with abelian socle cannot be $\mathfrak{X}$-dense because F is full.

Remark 3.1.25. Let $\mathfrak{F} = \mathrm{LF}_{\mathfrak{X}}(F)$ and $G \in \mathrm{b}_{\mathfrak{X}}(\mathfrak{F})$. G is $\mathfrak{X}$-dense with respect to $\mathfrak{F}$ if and only if there exists an $\mathfrak{X}$-formation function f such that $\mathfrak{F} = \mathrm{LF}_{\mathfrak{X}}(f)$ and $G \in \mathrm{b}(f(p))$ for all primes p dividing $|\mathrm{Soc}(G)|$.

Proof. If G is $\mathfrak{X}$-dense with respect to $\mathfrak{F}$, then we take $f = F$. Conversely, assume that $G \in \mathrm{b}(f(p))$ for all $p \in \pi(\mathrm{Soc}(G))$ for some $\mathfrak{X}$-formation function f such that $\mathfrak{F} = \mathrm{LF}_{\mathfrak{X}}(f)$. Then $G/\mathrm{Soc}(G) \in \mathfrak{F} \cap \mathfrak{S}_p f(p) = F(p)$ for all $p \in \pi(\mathrm{Soc}(G))$ by Theorem 3.1.17 (3). Since $G \notin \mathfrak{F}$, it follows that $G \in \mathrm{b}(F(p))$ for every prime p dividing $|\mathrm{Soc}(G)|$. This is to say that G is $\mathfrak{X}$-dense with respect to $\mathfrak{F}$. $\qquad\square$

Examples 3.1.26.   1. Suppose that $\mathfrak{X}$ contains a non-abelian group S. Then S is $\mathfrak{X}$-dense with respect to any $\mathfrak{X}$-local formation $\mathfrak{F}$ satisfying $S \notin \mathfrak{F}$ and $C_p \in \mathfrak{F}$ for all $p \in \pi(S)$. For example, $\mathfrak{F} = \mathfrak{N}$ or $\mathfrak{S}$.

2. Let $\mathfrak{F} = \mathfrak{N}\mathfrak{F}_0$ for some formation $\mathfrak{F}_0$. Let $\mathfrak{R}_{\mathfrak{X}}$ denote the class of all $\mathfrak{X}$-groups without abelian chief factors; it is clear that $\mathfrak{R}_{\mathfrak{X}} = \mathfrak{R}_{\mathfrak{X}}^2$ is a Fitting formation. It follows that $\mathfrak{F} = \mathrm{LF}_{\mathfrak{X}}(F)$ where $F(p) = \mathfrak{S}_p \mathfrak{F}_0$ for all $p \in \mathrm{char}\,\mathfrak{X}$, and $F(S) = \mathfrak{F}$ for all $S \in \mathfrak{X}'$. Then $\mathrm{b}(\mathfrak{F})$ is $\mathfrak{X}$-wide if and only if $\mathfrak{R}_{\mathfrak{X}} \mathfrak{F}_0 = \mathfrak{F}_0$.

Proof.   1. It is obvious.

2. It is rather clear that $\mathfrak{F} = \mathrm{LF}_{\mathfrak{X}}(F)$. Suppose that $\mathrm{b}(\mathfrak{F})$ is $\mathfrak{X}$-wide and $\mathfrak{R}_{\mathfrak{X}} \mathfrak{F}_0 \neq \mathfrak{F}_0$. Let $G \in \mathfrak{R}_{\mathfrak{X}} \mathfrak{F}_0 \setminus \mathfrak{F}_0$ be a group of minimal order. Then G has a unique minimal normal subgroup N such that $G/N \in \mathfrak{F}_0$. Since $G \notin \mathfrak{F}_0$, then N is a non-abelian $\mathfrak{X}$-group. If $G \in \mathfrak{F}$, then $G \in \mathfrak{F}_0$ because $F(G) = 1$, contrary to supposition. Hence $G \in \mathrm{b}(\mathfrak{F})$. Moreover $G \notin \mathfrak{S}_p \mathfrak{F}_0$ for all $p \in \pi(N)$. This means that $G \in \mathrm{b}(F(p))$ for all $p \in \pi(N)$ and so G is $\mathfrak{X}$-dense with respect to $\mathfrak{F}$. This is a contradiction. Hence $\mathfrak{R}_{\mathfrak{X}} \mathfrak{F}_0 \subseteq \mathfrak{F}_0$ and the equality holds.

Conversely, assume that $\mathfrak{R}_{\mathfrak{X}}\mathfrak{F}_0 = \mathfrak{F}_0$ and suppose that there exists $G \in \mathrm{b}_{\mathfrak{X}}(\mathfrak{F})$ such that $G \in \mathrm{b}\big(F(p)\big) = \mathrm{b}(\mathfrak{S}_p\mathfrak{F}_0)$ for each $p \in \pi\big(\mathrm{Soc}(G)\big)$. Let p and q be two different primes dividing $|\mathrm{Soc}(G)|$. Then $G/N \in \mathfrak{S}_p\mathfrak{F}_0 \cap \mathfrak{S}_q\mathfrak{F}_0$. Therefore $G \in \mathfrak{R}_{\mathfrak{X}}\mathfrak{F}_0 = \mathfrak{F}_0$. This contradicts the fact that $G \in \mathrm{b}(\mathfrak{F})$. Consequently $\mathrm{b}(\mathfrak{F})$ is $\mathfrak{X}$-wide. $\square$

For each prime p, denote $\mathfrak{E}(p)$ the class of all groups such that p divides the order of every chief factor of G. Then it is clear that $\mathfrak{E}(p) = \big(\mathfrak{E}(p)\big)^2$ is a Fitting formation and $\mathfrak{E}(p) \cap \mathfrak{S} = \mathfrak{S}_p$.

Note that if $p \in \mathrm{char}\,\mathfrak{X}$, then $\mathfrak{E}(p) \cap \mathrm{E}\,\mathfrak{X} = \mathrm{E}(\mathfrak{X}_p)$.

Remark 3.1.27. Let $\mathfrak{F} = \mathrm{LF}_{\mathfrak{X}}(f) = \mathrm{LF}_{\mathfrak{X}}(F)$ be an $\mathfrak{X}$-local formation. Then $F(p) = \mathfrak{F} \cap \mathfrak{E}(p)f(p)$ for each $p \in \mathrm{char}\,\mathfrak{X}$.

Proof. Let $p \in \mathrm{char}\,\mathfrak{X}$. By Theorem 3.1.17 (3), $F(p) = \mathfrak{F} \cap \mathfrak{S}_p f(p)$. Therefore $F(p) \subseteq \mathfrak{F} \cap \mathfrak{E}(p)f(p)$. Assume that the equality does not hold and let $G \in \big(\mathfrak{F} \cap \mathfrak{E}(p)f(p)\big) \setminus F(p)$ of minimal order. Then G has a unique minimal normal subgroup N such that $N \in \mathfrak{E}(p)$ and $G/N \in F(p)$. Since F is full, we have that N is not a p-group. Hence $\mathrm{C}_G(N) = 1$ and so $G \in F(p)$ because $G \in \mathfrak{F}$. This contradiction yields $F(p) = \mathfrak{F} \cap \mathfrak{E}(p)f(p)$. $\square$

Let $\mathfrak{F} = \mathrm{LF}_{\mathfrak{X}}(f)$ be an $\mathfrak{X}$-local formation. Denote $\bar{f}$ the following $\mathfrak{X}$-formation function:

$$\bar{f}(p) = \begin{cases} \mathfrak{E}(p)f(p) & \text{if } p \in \mathrm{char}\,\mathfrak{X}, \\ f(S) & \text{if } S \in \mathfrak{X}'. \end{cases}$$

In general, $\mathfrak{F} \neq \mathrm{LF}_{\mathfrak{X}}(\bar{f})$; take $\mathfrak{F} = \mathfrak{N}$, $\mathfrak{X} = \mathfrak{J}$, and $f(p) = (1)$ for all $p \in \mathbb{P}$. However:

Theorem 3.1.28. *Let $\mathfrak{F} = \mathrm{LF}_{\mathfrak{X}}(f) = \mathrm{LF}_{\mathfrak{X}}(F)$ be an $\mathfrak{X}$-local formation with f integrated. The following statements are pairwise equivalent:*

1. $\mathfrak{F} = \mathrm{LF}_{\mathfrak{X}}(\bar{f})$;
2. $\mathfrak{F} = \mathrm{LF}_{\mathfrak{X}}(\bar{F})$;
3. $\mathrm{b}(\mathfrak{F})$ *is $\mathfrak{X}$-wide.*

Proof. 1 implies 2. Let $p \in \mathrm{char}\,\mathfrak{X}$. Then, by Theorem 3.1.17 (3) $F(p) = \mathfrak{F} \cap \mathfrak{S}_p f(p) \subseteq \mathfrak{E}(p)f(p)$. Consequently $\mathfrak{E}(p)F(p) \subseteq \mathfrak{E}(p)f(p)$. It is then clear that $\mathfrak{F} = \mathrm{LF}_{\mathfrak{X}}(\bar{F})$.

2 implies 3. Let $G \in \mathrm{b}_{\mathfrak{X}}(\mathfrak{F})$ be an $\mathfrak{X}$-dense group with respect to $\mathfrak{F}$. Then $\mathrm{Soc}(G) \in \mathrm{E}\,\mathfrak{X}$ and so $\mathrm{Soc}(G) \in \mathfrak{E}(p)$ for all primes p dividing $|\mathrm{Soc}(G)|$. Therefore $G \in \mathfrak{E}(p)F(p)$. Applying Remark 3.1.7 (2), we have that $G \in \mathrm{LF}_{\mathfrak{X}}(\bar{F}) = \mathfrak{F}$, contrary to the choice of G. Hence $\mathrm{b}_{\mathfrak{X}}(\mathfrak{F})$ is wide.

3 implies 1. Suppose that $\mathrm{b}_{\mathfrak{X}}(\mathfrak{F})$ is $\mathfrak{X}$-wide. Since $f \leq \bar{f}$, it follows that $\mathfrak{F} \subseteq \mathrm{LF}_{\mathfrak{X}}(\bar{f})$. Hence the burden of the proof is to show that $\mathrm{LF}_{\mathfrak{X}}(\bar{f}) \subseteq \mathfrak{F}$. Assume that this is not true, and let G be a group of minimal order in $\mathrm{LF}_{\mathfrak{X}}(\bar{f}) \setminus \mathfrak{F}$.

It follows easily that G has a unique minimal normal subgroup, N say, and $G/N \in \mathfrak{F}$. If $N \in \mathrm{E}(\mathfrak{X}')$, then $G \in \bar{f}(S) = \mathfrak{F}$ for some simple group $S \in \mathfrak{X}'$, contrary to supposition. Hence $N \in \mathrm{E}\,\mathfrak{X}$ and so $G/\operatorname{C}_G(N) \in \mathfrak{E}(p)f(p)$ for each prime p dividing $|N|$. If N is abelian, then $G/\operatorname{C}_G(N) \in \mathfrak{F} \cap \mathfrak{E}(p)f(p) = F(p)$ by Remark 3.1.27. Now applying Remark 3.1.7 (2), $G \in \mathfrak{F}$, which is not the case. Hence N is non-abelian and then $\operatorname{C}_G(N) = 1$. Then $G/N \in \mathfrak{F} \cap \mathfrak{E}(p)f(p) = F(p)$ for all primes p dividing $|N|$. Since $G \notin F(p)$, we have that G is $\mathfrak{X}$-dense with respect to $\mathfrak{F}$, and we have reached a final contradiction. Therefore $\mathrm{LF}_{\mathfrak{X}}(\bar{f}) \subseteq \mathfrak{F}$ and the equality holds. $\qquad\square$

The next result shows that the $\mathfrak{X}$-local formations of $\mathfrak{X}$-wide boundary are precisely those for which a partial converse of Theorem 3.1.17 (3) holds.

Theorem 3.1.29. *Let $\mathfrak{F} = \mathrm{LF}_{\mathfrak{X}}(F)$ be an $\mathfrak{X}$-local formation. Then the following statements are equivalent:*

1. $\mathrm{b}(\mathfrak{F})$ *is $\mathfrak{X}$-wide.*
2. *If f is an $\mathfrak{X}$-formation function such that $\mathfrak{F} \cap \mathfrak{S}_p f(p) = F(p)$ for all $p \in \operatorname{char}\mathfrak{X}$, and $f(S) = \mathfrak{F}$ for all $S \in \mathfrak{X}'$, then $\mathfrak{F} = \mathrm{LF}_{\mathfrak{X}}(f)$.*

Proof. 1 implies 2. Let f be an $\mathfrak{X}$-formation function such that $\mathfrak{F} \cap \mathfrak{S}_p f(p) = F(p)$ for all $p \in \operatorname{char}\mathfrak{X}$ and $f(S) = \mathfrak{F}$ for all $S \in \mathfrak{X}'$. Denote $\mathfrak{F}_1 = \mathrm{LF}_{\mathfrak{X}}(f)$. It is clear that $\mathfrak{F} \subseteq \mathfrak{F}_1$ because $F(p) \subseteq \mathfrak{S}_p f(p)$ for all $p \in \operatorname{char}\mathfrak{X}$. Suppose that $\mathfrak{F}_1$ is not contained in $\mathfrak{F}$ and let $G \in \mathfrak{F}_1 \setminus \mathfrak{F}$ of minimal order. As usual, G has a unique minimal normal subgroup N such that $G/N \in \mathfrak{F}$. Moreover $N \in \mathrm{E}\,\mathfrak{X}$ and $G/\operatorname{C}_G(N) \in f(p)$ for all $p \in \pi(N)$. If N were abelian, then $G/\operatorname{C}_G(N) \in \mathfrak{F} \cap f(p) \subseteq F(p)$ and we would have $G \in \mathfrak{F}$ by Remark 3.1.7 (2). This would contradict the choice of G. Hence N should be non-abelian and so $G \in f(p)$ for all $p \in \pi(N)$. This implies that $G/N \in \mathfrak{F} \cap f(p) \subseteq F(p)$. Since $G \in \mathrm{b}(\mathfrak{F})$, we have that $G \notin F(p)$. Hence G is $\mathfrak{X}$-dense with respect to $\mathfrak{F}$ and $\mathrm{b}(\mathfrak{F})$ is not $\mathfrak{X}$-wide. This is a contradiction. Consequently $\mathfrak{F}_1 \subseteq \mathfrak{F}$ and the equality holds.

2 implies 1. Let f be the $\mathfrak{X}$-formation function given by $f(p) = \mathfrak{E}(p)F(p)$ for all $p \in \operatorname{char}\mathfrak{X}$ and $f(S) = F(S) = \mathfrak{F}$ for all $S \in \mathfrak{X}'$. Then, by Remark 3.1.27, we have $\mathfrak{F} \cap \mathfrak{S}_p f(p) = \mathfrak{F} \cap \mathfrak{E}(p)F(p) = F(p)$ for all $p \in \operatorname{char}\mathfrak{X}$. Consequently $\mathfrak{F} = \mathrm{LF}_{\mathfrak{X}}(f)$ by Statement 2. Applying Theorem 3.1.28, we conclude that $\mathrm{b}(\mathfrak{F})$ is $\mathfrak{X}$-wide. $\qquad\square$

Theorem 3.1.30. *Let $\mathfrak{F} = \mathrm{LF}_{\mathfrak{X}}(F)$ be an $\mathfrak{X}$-local formation with a maximal $\mathfrak{X}$-local definition. Then $\mathrm{b}(\mathfrak{F})$ is $\mathfrak{X}$-wide.*

Proof. Let $p \in \operatorname{char}\mathfrak{X}$ and define the following $\mathfrak{X}$-formation function: $F_p(p) = \mathfrak{E}(p)F(p)$ and $F_p(S) = F(S)$ for every $S \in (\operatorname{char}\mathfrak{X}) \cup \mathfrak{X}'$ such that $S \not\cong C_p$. Then $F \leq F_p$. Hence $\mathfrak{F} \subseteq \mathrm{LF}_{\mathfrak{X}}(F_p)$. We suppose that $\mathfrak{F} \neq \mathrm{LF}_{\mathfrak{X}}(F_p)$ and derive a contradiction. Let $G \in \mathrm{LF}_{\mathfrak{X}}(F_p) \setminus \mathfrak{F}$ be a group of minimal order. Then G has a unique minimal normal subgroup N and $G/N \in \mathfrak{F}$. If $N \in \mathrm{E}(\mathfrak{X}')$, then $G \in F(S)$ for some $S \in \mathfrak{X}'$ and so $G \in \mathfrak{F}$, which is a contradiction. Hence

$N \in E\mathfrak{X}$. Suppose that N is abelian. Since $G \notin \mathfrak{F}$, we conclude that N is a p-group. But in this case $G/\operatorname{C}_G(N) \in \mathfrak{E}(p)F(p) \cap \mathfrak{F} = F(p)$ by Remark 3.1.27. Hence $G \in \mathfrak{F}$ by Remark 3.1.7 (2). Consequently N should be non-abelian. Let q be a prime different from p such that q divides the order of N. Then $G \in F_p(q) = F(q) \subseteq \mathfrak{F}$. This is the desired contradiction.

Therefore $\mathfrak{F}_p = \operatorname{LF}_{\mathfrak{X}}(F_p) = \mathfrak{F}$ for all $p \in \operatorname{char} \mathfrak{X}$. Let g be the maximal $\mathfrak{X}$-local definition of $\mathfrak{F}$. Then $\mathfrak{E}(p)F(p) \subseteq g(p)$ for all $p \in \operatorname{char} \mathfrak{X}$. Consequently $\mathfrak{F} = \operatorname{LF}(\bar{F})$. Applying Theorem 3.1.28, $\operatorname{b}(\mathfrak{F})$ is $\mathfrak{X}$-wide. $\square$

Let $\mathfrak{F} = \operatorname{LF}_{\mathfrak{X}}(F)$ be an $\mathfrak{X}$-local formation. Define

$$\underline{F}(S) = \begin{cases} \operatorname{h}\big(\operatorname{b}(F(p)) \cap \mathfrak{F}\big) & \text{if } S = p \in \operatorname{char} \mathfrak{X}, \\ \operatorname{h}\big(\operatorname{b}_S(\mathfrak{F})\big) & \text{if } S \in \mathfrak{X}' \end{cases}$$

$$\widehat{F}(S) = \begin{cases} \big(G : \operatorname{Q}\operatorname{R}_0(F(p) \cup \{G\}) \subseteq \underline{F}(p)\big) & \text{if } S = p \in \operatorname{char} \mathfrak{X}, \\ \operatorname{h}\big(\operatorname{b}_S(\mathfrak{F})\big) & \text{if } S \in \mathfrak{X}' \setminus \mathbb{P}, \\ \big(G : \operatorname{Q}\operatorname{R}_0(F(q) \cup \{G\}) \subseteq \underline{F}(q)\big) & \text{if } S \in \mathfrak{X}' \cap \mathbb{P}. \end{cases}$$

Note that $\operatorname{h}\big(\operatorname{b}_S(\mathfrak{F})\big)$ is a saturated formation for all $S \in \mathfrak{X}' \setminus \mathbb{P}$ by Example 2.3.21. Moreover $\operatorname{Q}\widehat{F}(p) = \widehat{F}(p)$ for each prime p.

Lemma 3.1.31. $\underline{F}(p) \cap \mathfrak{F} = \widehat{F}(p) \cap \mathfrak{F} = F(p)$ for each prime p.

Proof. $F(p) \subseteq \widehat{F}(p) \cap \mathfrak{F} \subseteq \underline{F}(p) \cap \mathfrak{F}$. Now if $p \in \operatorname{char} \mathfrak{X}$, then $\underline{F}(p) \cap \mathfrak{F} \subseteq F(p)$ by using familiar arguments. If $p \in \mathfrak{X}'$, then $F(p) = \mathfrak{F}$. Therefore in both cases $\underline{F}(p) \cap \mathfrak{F} \subseteq F(p)$ and $F(p) = \underline{F}(p) \cap \mathfrak{F}$. $\square$

Lemma 3.1.32. *Let p be a prime. If $\mathfrak{L}$ is a formation contained in $\underline{F}(p)$, then $\operatorname{Q}\operatorname{R}_0\big(F(p) \cup \mathfrak{L}\big)$ is contained in $\underline{F}(p)$.*

Proof. It is enough to prove $\operatorname{R}_0\big(F(p) \cup \mathfrak{L}\big) \subseteq \underline{F}(p)$ since $\underline{F}(p)$ is a homomorph. Suppose that $\operatorname{R}_0\big(F(p) \cup \mathfrak{L}\big)$ is not contained in $\underline{F}(p)$ and take $G \in \operatorname{R}_0\big(F(p) \cup \mathfrak{L}\big) \setminus \underline{F}(p)$ of minimal order. Then $G^{F(p)} \neq 1 \neq G^{\mathfrak{L}}$ and $G \notin \underline{F}(p)$. Furthermore, there exists a normal subgroup K of G such that $G/K \in \operatorname{b}\big(F(p)\big) \cap \mathfrak{F}$ or $G/K \in \operatorname{b}_p(\mathfrak{F})$ according whether $p \in \operatorname{char} \mathfrak{X}$ or $p \in \mathfrak{X}'$. Suppose that $K \cap G^{F(p)} \neq 1$ and let N be a minimal normal subgroup of G such that N is contained in $K \cap G^{F(p)}$. By the choice of G, we have $G/N \in \underline{F}(p)$. Hence $G/K \in \underline{F}(p)$. This is impossible. Consequently $K \cap G^{F(p)} = 1$ and, analogously, $K \cap G^{\mathfrak{L}} = 1$. Assume that $p \in \operatorname{char} \mathfrak{X}$. Then $G/K \in \mathfrak{F}$. Thus $G \in \operatorname{R}_0 \mathfrak{F} = \mathfrak{F}$. This implies that $G/G^{\mathfrak{L}} \in \mathfrak{L} \cap \mathfrak{F} \subseteq \underline{F}(p) \cap \mathfrak{F} = F(p)$ by Lemma 3.1.31 and so $G^{F(p)} \leq G^{\mathfrak{L}}$. Since $G^{F(p)} \cap G^{\mathfrak{L}} = 1$, it follows that $G \in F(p)$. This contradicts the choice of G. Now suppose that $p \in \mathfrak{X}'$. In this case $G/K \in \operatorname{b}_p(\mathfrak{F})$. Let $L/K = \operatorname{Soc}(G/K)$. Then $L = G^{\mathfrak{F}}K = G^{\mathfrak{F}} \times K$ and so $G^{\mathfrak{F}}$ is a minimal normal subgroup of G. Let B be a minimal normal subgroup contained in $G^{\mathfrak{L}}$. Then $G/B \in \operatorname{h}\big(\operatorname{b}_p(\mathfrak{F})\big)$ by the choice of G. Suppose that $G/B \notin \mathfrak{F}$. Then G has a factor group, G/T say, such that $B \leq T$ and $G/T \in \mathrm{b}(\mathfrak{F})$. Set $M/T = \mathrm{Soc}(G/T)$. Then $M = G^{\mathfrak{F}}T$ because $G^{\mathfrak{F}}$ is a minimal normal subgroup of G. Therefore M/T is a p-group and $G/T \in \mathrm{b}_p(\mathfrak{F})$. This is a contradiction. Consequently $\mathrm{Q}\,\mathrm{R}_0\big(F(p) \cup \mathfrak{L}\big)$ is contained in $\underline{F}(p)$. $\square$

Theorem 3.1.33 ([FS85]). *Let $\mathfrak{F} = \mathrm{LF}_{\mathfrak{X}}(F)$ be an $\mathfrak{X}$-local formation. Then $\mathfrak{F}$ possesses a maximal $\mathfrak{X}$-local definition if and only if $\mathrm{b}(\mathfrak{F})$ is $\mathfrak{X}$-wide and, for each prime p, there exists a unique maximal formation contained in $\underline{F}(p)$. In this case, $\widehat{F}$ is an $\mathfrak{X}$-formation function and $\widehat{F}$ is the maximal $\mathfrak{X}$-local definition of $\mathfrak{F}$.*

Proof. First, suppose that $\mathfrak{F}$ possesses a maximal $\mathfrak{X}$-local definition, g say. Then $\mathrm{b}(\mathfrak{F})$ is $\mathfrak{X}$-wide by Theorem 3.1.30. Let p be a prime in char $\mathfrak{X}$. Then $g(p) \cap \mathfrak{F}$ is contained in $F(p)$ by Theorem 3.1.17 (3). Hence $g(p)$ is contained in $\mathrm{h}\big(\mathrm{b}(F(p)) \cap \mathfrak{F}\big) = \underline{F}(p)$.

Assume now that $p \in \mathfrak{X}' \cap \mathbb{P}$ and $g(p)$ is not contained in $\underline{F}(p)$. Let G be a group of least order in $g(p) \setminus \underline{F}(p)$. Then $G \in \mathrm{b}_p(\mathfrak{F})$, and from $\mathfrak{F} = \mathrm{LF}_{\mathfrak{X}}(g)$ we readily get that $G \in \mathfrak{F}$, the desired contradiction. Consequently $g(p) \subseteq \underline{F}(p)$. Let $\mathfrak{L}$ be a formation contained in $\underline{F}(p)$. By Lemma 3.1.32, $\mathrm{Q}\,\mathrm{R}_0\big(F(p) \cup \mathfrak{L}\big) \subseteq \underline{F}(p)$. Consider the $\mathfrak{X}$-formation function defined by setting

$$g_1(q) = \begin{cases} \mathrm{Q}\,\mathrm{R}_0\big(F(p) \cup \mathfrak{L}\big) & \text{if } p = q, \\ F(q) & \text{if } p \neq q \end{cases}$$

and $g_1(S) = g(S)$ for every $S \in \mathfrak{X}' \setminus \mathbb{P}$. Since $g_1(p) \cap \mathfrak{F} \subseteq F(p)$ by Lemma 3.1.31 and Lemma 3.1.32, we immediately have that $\mathfrak{F} = \mathrm{LF}_{\mathfrak{X}}(g_1)$. The maximal character of g implies that $g_1(p) \subseteq g(p)$. Thus $\mathfrak{L} \subseteq g(p)$. Consequently, $g(p)$ is the unique maximal formation contained in $\underline{F}(p)$.

Conversely, suppose that $\mathrm{b}(\mathfrak{F})$ is $\mathfrak{X}$-wide and for each prime p, there exists a unique maximal formation, $g(p)$, contained in $\underline{F}(p)$. Consider the $\mathfrak{X}$-formation function g_1 defined by $g_1(p) = g(p)$ for every prime p and $g_1(S) = \mathrm{h}(\mathrm{b}_S(\mathfrak{F}))$ for every $S \in \mathfrak{X}' \setminus \mathbb{P}$. Clearly $\mathfrak{F} \subseteq \mathrm{LF}_{\mathfrak{X}}(g_1)$ because $F(S) \subseteq g(p)$ for all p and $\mathfrak{F} \subseteq g_1(S)$ for all $S \in \mathfrak{X}' \setminus \mathbb{P}$. If $\mathfrak{F} \neq \mathrm{LF}_{\mathfrak{X}}(g_1)$, then a group $G \in \mathrm{LF}_{\mathfrak{X}}(g_1) \setminus \mathfrak{F}$ of minimal order would be an $\mathfrak{X}$-dense group. Since $\mathrm{b}(\mathfrak{F})$ is $\mathfrak{X}$-wide, we conclude that $\mathfrak{F} = \mathrm{LF}_{\mathfrak{X}}(g_1)$. On the other hand, let j be an $\mathfrak{X}$-formation function such that $\mathfrak{F} = \mathrm{LF}_{\mathfrak{X}}(j)$. Then, for all p, we have $j(p) \cap \mathfrak{F} \subseteq F(p)$. Consequently, $j(p) \subseteq \underline{F}(p)$ and then $j(p) \subseteq g(p)$. Furthermore, it is clear that $j(S) \subseteq g_1(S)$ for every $S \in \mathfrak{X}' \setminus \mathbb{P}$. Consequently, g_1 is the maximal $\mathfrak{X}$-local definition of $\mathfrak{F}$.

Note that in this case $g(p) = \widehat{F}(p)$ and $g(S) = \widehat{F}(S)$ for all $S \in \mathfrak{X}' \setminus \mathbb{P}$. Therefore $\widehat{F}$ is an $\mathfrak{X}$-formation function and it is actually the maximal $\mathfrak{X}$-local definition of $\mathfrak{F}$. $\square$

Proposition 3.1.34. *Let $\mathfrak{Y} \subseteq \mathfrak{X}$ be classes of simple groups. If $\mathfrak{F} = \mathrm{LF}_{\mathfrak{X}}(F)$ has a maximal $\mathfrak{X}$-local definition, then $\mathfrak{F}$ has a unique maximal $\mathfrak{Y}$-local definition. If, in addition, char $\mathfrak{X} = $ char $\mathfrak{Y}$, then the converse is valid if, and only if, $\mathrm{b}(\mathfrak{F})$ is $\mathfrak{X}$-wide.*

Proof. Note that $\mathfrak{F} = \mathrm{LF}_{\mathfrak{Y}}(F_1)$, where $F_1(p) = F(p)$ for all $p \in \mathrm{char}\,\mathfrak{Y}$ and $F_1(S) = \mathfrak{F}$ for all $S \in \mathfrak{Y}'$ (see Corollary 3.1.13). Therefore if $\mathfrak{F}$ has a maximal $\mathfrak{X}$-local definition, then $\mathrm{b}(\mathfrak{F})$ is $\mathfrak{X}$-wide (and so $\mathrm{b}(\mathfrak{F})$ is $\mathfrak{Y}$-wide) and $\widehat{F}(p) = \widehat{F}_1(p)$ for all $p \in \mathrm{char}\,\mathfrak{Y}$ is a formation. We are left to show that $\widehat{F}_1(p)$ is a formation for all $p \in (\mathrm{char}\,\mathfrak{X}) \cap \mathfrak{Y}'$. To see this, we prove that $\underline{F}_1(p) = \mathfrak{G} = \mathrm{h}(\mathrm{b}_q(\mathfrak{F}))$ contains a unique maximal formation. Set $\mathfrak{H} = f(\mathfrak{G}) = (G : H/K$ is $\mathfrak{G}$-central in G for every chief factor of $G)$. Applying Theorem 2.3.20, $\mathfrak{H}$ is a formation. Suppose that $\mathfrak{H}$ is not contained in $\mathfrak{G}$ and let $G \in \mathfrak{H} \setminus \mathfrak{G}$ be a group of minimal order. Then $G \in \mathrm{b}(\mathfrak{G}) = \mathrm{b}_q(\mathfrak{F})$ and so G is a monolithic group. Moreover $X = [N](G/\mathrm{C}_G(N)) \in \mathfrak{G}$. If $X \notin \mathfrak{F}$, then $X \in \mathrm{b}_q(\mathfrak{F})$, because $G/\mathrm{C}_G(N) \in \mathfrak{F}$. Hence $X \in \mathfrak{G} \cap \mathrm{b}_q(\mathfrak{F}) = \emptyset$. This is a contradiction. Therefore $X \in \mathfrak{F}$ and $G/\mathrm{C}_G(N) \in F(p)$. Applying Remark 3.1.7 (2), we conclude that $G \in \mathrm{LF}_{\mathfrak{X}}(F) = \mathfrak{F}$. We have obtained a contradiction. Consequently $\mathfrak{H} \subseteq \mathfrak{G}$. Let now $\mathfrak{L}$ be a formation contained in $\mathfrak{G}$. Then by Theorem 2.3.20 (2), $\mathfrak{L} \subseteq \mathfrak{H}$. This means that $\widehat{F}_1(p)$ is a formation. By Theorem 3.1.33, it follows that $\mathfrak{F}$ has a maximal $\mathfrak{Y}$-local definition.

Now if $\mathrm{char}\,\mathfrak{X} = \mathrm{char}\,\mathfrak{Y}$, then $F(p) = F_1(p)$ for all $p \in \mathrm{char}\,\mathfrak{X}$. Consequently if $\mathfrak{F}$ has a maximal $\mathfrak{Y}$-local definition, then $\widehat{F}(p)$ is a formation for all $p \in \mathrm{char}\,\mathfrak{X}$. By Theorem 3.1.30, $\mathfrak{F}$ has a maximal $\mathfrak{X}$-local definition if, and only if, $\mathrm{b}(\mathfrak{F})$ is $\mathfrak{X}$-wide. $\square$

Examples 3.1.35.   1.   Let $\mathfrak{F} = \mathfrak{S}$ be the $\mathfrak{J}$-local (local) formation of all soluble groups. Then $\mathfrak{F} = \mathrm{LF}_{\mathfrak{J}}(F)$ where $F(p) = \mathfrak{F}$ for all $p \in \mathbb{P}$. Hence $\widehat{F}(p) = \mathfrak{E}$ and so $\widehat{F}$ is a $\mathfrak{J}$-formation function. However, $\mathfrak{F}$ does not have a maximal $\mathfrak{J}$-local definition (see Example 3.1.23).

This example shows that the requirement that $\mathrm{b}(\mathfrak{F})$ be $\mathfrak{X}$-wide cannot be removed from Theorem 3.1.33.

2.   Let $\mathfrak{F}_0$ be the class of all groups whose Frattini chief factors have odd order. Then $\mathfrak{F}_0$ is a formation and $\mathfrak{R}_3\mathfrak{F}_0 = \mathfrak{F}_0$. Let $\mathfrak{F} = \mathfrak{N}\mathfrak{F}_0$. Applying Example 3.1.26 (2), we have that $\mathfrak{F}$ is a $\mathfrak{J}$-local formation with $\mathfrak{J}$-wide boundary. Assume that $\mathfrak{F} \neq \mathfrak{S}_2\mathfrak{F}_0$ and let $G \in \mathfrak{F}\setminus\mathfrak{S}_2\mathfrak{F}_0$ be a group of minimal order. Then G has a unique minimal normal subgroup N. Moreover $G/N \in \mathfrak{S}_2\mathfrak{F}_0$. Since $G \notin \mathfrak{F}_0$, we conclude that N is a p-group for some odd prime p. Hence $\mathrm{F}(G)$ is a p-group. This implies that $G \in \mathfrak{F}_0$ because $G/\mathrm{F}(G)$ has no Frattini 2-chief factors. This is a contradiction. Consequently $\mathfrak{F} = \mathfrak{S}_2\mathfrak{F}_0$ and $\mathfrak{F} = \mathrm{LF}_{\mathfrak{J}}(F)$, where

$$F(p) = \begin{cases} \mathfrak{F} & \text{if } p = 2, \\ \mathfrak{F}_0 & \text{if } p \neq 2. \end{cases}$$

Then $\underline{F}(q) = \mathrm{h}\Big(\mathrm{b}\big(F(q)\big) \cap \mathfrak{F}\Big) = \mathrm{h}(\mathrm{b}(\mathfrak{F}_0) \cap \mathfrak{F}) = \mathrm{h}(\mathrm{b}(\mathfrak{F}_0)) = \mathfrak{F}_0$ for each odd prime q (note $\mathrm{b}(\mathfrak{F}_0) \subseteq \mathfrak{F}$). Consequently

$$\underline{F}(p) = \widehat{F}(p) = \begin{cases} \mathfrak{E} & \text{if } p = 2, \\ \mathfrak{F}_0 & \text{if } p \neq 2 \end{cases}$$

and $\underline{F} = \widehat{F}$ is a $\mathfrak{J}$-formation function. Applying Theorem 3.1.33, we have that $\underline{F}$ is the maximal $\mathfrak{J}$-local definition of $\mathfrak{F}$.

Let $\mathfrak{F} = \mathrm{LF}_{\mathfrak{X}}(F)$ be an $\mathfrak{X}$-local formation. In contrast to the condition that $\mathrm{b}(\mathfrak{F})$ is $\mathfrak{X}$-wide, the other condition from Theorem 3.1.33 — namely, that $\widehat{F}(X)$ be a formation for all $X \in \mathbb{P}$ — is not always easy to check when a concrete formation $\mathfrak{F}$ is given. We give an example of a local formation for which $\widehat{F}$ is not a formation function.

Example 3.1.36 ([FS85]). Let $\mathfrak{R} = \mathfrak{R}_{\mathfrak{J}}$ be the formation composed of all groups whose chief factors are non-abelian. Consider the local formation $\mathfrak{F} = \mathfrak{N}\mathfrak{R}\mathfrak{N}$. Then the canonical definition F of $\mathfrak{F}$ is given by $F(p) = \mathfrak{S}_p\mathfrak{R}\mathfrak{N}$ for all p. Applying Examples 3.1.26 (2), we have that $\mathrm{b}(\mathfrak{F})$ is $\mathfrak{J}$-wide because $\mathfrak{R}_{\mathfrak{J}}(\mathfrak{R}\mathfrak{N}) = \mathfrak{R}\mathfrak{N}$.

Let $S = \mathrm{SL}(2,5)$. By [DH92, B, 10.9], S has an irreducible module V over $\mathrm{GF}(p)$ such that $\mathrm{Ker}(S \text{ on } V) = \mathrm{C}_S(V) = \mathrm{Z}(S)$. Let $X = [V]S$ be the corresponding semidirect product, and let $Y = S \wr_{\mathrm{Z}(X)} X$ be the wreath product of S with respect to the permutation representation of S with X with respect to the permutation representation of X on the set of cosets of $\mathrm{Z}(X) = \mathrm{Z}(S)$ in X. As usual, for any subgroup U of S, $U^{\natural} = U \times \cdots \times U$ ($|X/\mathrm{Z}(X)|$ copies of U) shall denote the canonical subgroup of $S^{\natural}$, the base group of Y, isomorphic to a direct product of $|X/\mathrm{Z}(X)|$ copies of S. Note that $\mathrm{Z}(X) \le \mathrm{Z}(Y)$ and $\mathrm{Z}(S)^{\natural}X/\mathrm{Z}(X)$ is X-isomorphic to the regular wreath product $C_2 \wr_{\mathrm{reg}} X/\mathrm{Z}(X)$ and this is isomorphic to the semidirect product of the regular $X/\mathrm{Z}(X)$-module over $\mathrm{GF}(2)$ with $X/\mathrm{Z}(X)$. Therefore there exists a normal subgroup Z of Y such that $Z \le \mathrm{Z}(S)^{\natural}$ and $\mathrm{Z}(S)^{\natural}/Z$ is a cyclic group of order 2.

We consider now $G = Y/Z$. It is clear that S is isomorphic to a quotient of G. Let $A = \mathrm{Z}(X)Z/Z$ and $B = \mathrm{Z}(S)^{\natural}/Z$. It is clear that A and B are subgroups of order 2 contained in $\mathrm{Z}(G)$ such that $A \cap B = 1$. Hence there exists $D \le \mathrm{Z}(G)$ of order 2 such that $D \cap A = D \cap B = 1$. In particular $G \in \mathrm{R}_0(G/A, G/D)$.

Assume that p is a prime and $p > 5$. Then Förster and Salomon [FS85, Example 4.1] proved that G/A, $G/D \in \widehat{F}(p)$.

However since $\widehat{F}(p)$ is Q-closed, S is isomorphic to a quotient of G and $S \in \mathrm{b}(F(p)) \cap \mathfrak{F}$, it follows that $G \notin \widehat{F}(p)$. This shows that $\widehat{F}(p)$ is not a formation and hence $\mathfrak{F} = \mathfrak{N}\mathfrak{R}\mathfrak{N}$ does not have a maximal $\mathfrak{J}$-local definition as $\mathfrak{J}$-local formation.

The above example can be modified to show that $\mathrm{h}(\mathrm{b}_q(\mathfrak{F}))$, $\mathfrak{F}$ an $\mathfrak{X}$-local formation and $q \in \mathfrak{X}' \cap \mathbb{P}$, does not always contain a unique largest formation.

Example 3.1.37. Let $\mathfrak{F} = \mathfrak{S}_p\mathfrak{R}\mathfrak{N}$ as in the above example. Suppose that $\mathfrak{X} = \emptyset$. Put $q = 2$ and take G, A, D as in Example 3.1.36. Then $\mathrm{Q}\,\mathrm{R}_0(G/A) \cup \mathrm{Q}\,\mathrm{R}_0(G/D) \subseteq \mathrm{h}(\mathrm{b}_2(\mathfrak{F}))$, but $G \in \mathrm{R}_0(G/A, G/D)$ does not belong to $\mathrm{h}(\mathrm{b}_2(\mathfrak{F}))$. Consequently $\mathfrak{F}$ is an $\emptyset$-local formation without a maximal $\emptyset$-local definition.

Proof. First of all, we know that $S = \mathrm{SL}(2,5)$ is a quotient of G and $S \in \mathrm{b}_2(\mathfrak{F})$. Therefore $G \notin \mathrm{h}\big(\mathrm{b}_2(\mathfrak{F})\big)$. Moreover, $G \in \mathrm{R}_0(G/A, G/D)$. Now let $\mathfrak{B}_1 = \mathrm{b}_2(\mathfrak{F}) \cap \mathfrak{NRN}$ and $\mathfrak{B}_2 = \mathrm{b}_2(\mathfrak{F}) \setminus \mathfrak{NRN}$. Thus $\mathrm{b}_2(\mathfrak{F}) = \mathfrak{B}_1 \cup \mathfrak{B}_2$ and $\mathrm{h}\big(\mathrm{b}_2(\mathfrak{F})\big) = \mathrm{h}(\mathfrak{B}_1) \cap \mathrm{h}(\mathfrak{B}_2)$. Förster and Salomon [FS85, Example 4.1] proved that $\mathrm{Q}\,\mathrm{R}_0(G/A) \cup \mathrm{Q}\,\mathrm{R}_0(G/D) \subseteq \mathrm{h}(\mathfrak{B}_1)$. Moreover $\mathfrak{B}_2$ is a class composed by primitive groups. Hence $\mathrm{h}(\mathfrak{B}_2)$ is a Schunck class by Corollary 2.3.11. Note that $[H/K] * (G/A) \in \mathrm{h}(\mathfrak{B}_2)$ for each chief factor H/K of G/A (and the same applies to G/D). This implies that G/A and G/B belong to $f\big(\mathrm{h}(\mathfrak{B}_2)\big)$, which is the largest formation contained in $\mathrm{h}(\mathfrak{B}_2)$ by Theorem 2.3.20 (3). Hence $\mathrm{Q}\,\mathrm{R}_0(G/A) \cup \mathrm{Q}\,\mathrm{R}_0(G/D) \subseteq \mathrm{h}(\mathfrak{B}_2)$. $\qquad\square$

In [DH92, pages 364 and 365], the authors study the effect of some closure operations on a local formation. More precisely, they prove:

> Let $\mathfrak{F} = \mathrm{LF}(f)$ be a local formation and let c be one of the closure operations s, s_n, or N_0.
> 1. If $f(p) = \mathrm{c}\,f(p)$ for all $p \in \mathbb{P}$, then $\mathfrak{F} = \mathrm{c}\,\mathfrak{F}$, and
> 2. if $\mathfrak{F} = \mathrm{c}\,\mathfrak{F}$, and F is the canonical local definition of $\mathfrak{F}$, then $F(p) = \mathrm{c}\,F(p)$ for all $p \in \mathbb{P}$.

The natural question is: can the above results be extended to $\mathfrak{X}$-local formations? If $\mathrm{c} = \mathrm{s}$, 1 is not always true (compare with [För85b, Lemma 3.13]).

Example 3.1.38. Let $\mathfrak{X} = (C_2)$ and $\mathfrak{F} = \mathrm{LF}_{\mathfrak{X}}(f)$, where $f(2) = (1)$ and $f(S) = \mathfrak{E}$ if $S \ncong C_2$. It is clear that $\mathrm{s}\,f(S) = f(S)$ for all $S \in (\mathrm{char}\,\mathfrak{X}) \cup \mathfrak{X}'$, but $\mathfrak{F}$ is not s-closed because $\mathrm{Alt}(5) \in \mathfrak{F}$ but $\mathrm{Alt}(4) \notin \mathfrak{F}$.

Our next result shows that 1 is true for $\mathrm{c} = \mathrm{s}_n$ or N_0.

Proposition 3.1.39. *Let $\mathfrak{F} = \mathrm{LF}_{\mathfrak{X}}(f)$ be an $\mathfrak{X}$-local formation and let c be one of the closure operations s_n or N_0. If $f(S) = \mathrm{c}\,f(S)$ for all $S \in (\mathrm{char}\,\mathfrak{X}) \cup \mathfrak{X}'$, then $\mathfrak{F} = \mathrm{c}\,\mathfrak{F}$.*

Proof. Let $\mathrm{c} = \mathrm{s}_n$. Let $G \in \mathfrak{F}$, and let N be a normal subgroup of G. We prove that $N \in \mathfrak{F}$ by induction on $|G|$. Let A be a minimal normal subgroup of G. Then $NA/A \in \mathfrak{F}$. If B were another minimal normal subgroup of G, then $NB/B \in \mathfrak{F}$. This would imply $N \in \mathfrak{F}$. Consequently we may assume that $A = \mathrm{Soc}(G)$ is the unique minimal normal subgroup of G. Let $p \in \mathrm{char}\,\mathfrak{X}$. Then $N/\big(N \cap \mathrm{C}^{\mathfrak{X}_p}(G)\big) \cong N\,\mathrm{C}^{\mathfrak{X}_p}(G)/\mathrm{C}^{\mathfrak{X}_p}(G)$ and $N\,\mathrm{C}^{\mathfrak{X}_p}(G)/\mathrm{C}^{\mathfrak{X}_p}(G)$ is a normal subgroup of $G/\mathrm{C}^{\mathfrak{X}_p}(G) \in f(p)$. Since $N \cap \mathrm{C}^{\mathfrak{X}_p}(G) = \mathrm{C}^{\mathfrak{X}_p}(N)$ by Proposition 3.1.10, it follows that $N/\mathrm{C}^{\mathfrak{X}_p}(N) \in f(p)$.

Assume now that N/L is a monolithic quotient of N such that $T/L = \mathrm{Soc}(N/L) \in \mathrm{E}(S)$ for some simple group $S \in \mathfrak{X}'$. If A is not contained in L, then T/L is contained in $AL/L \neq 1$ and so $A \in \mathrm{E}(S)$. Since G is a monolithic $\mathfrak{F}$-group, it follows that $G \in f(S)$. Hence $N \in \mathrm{s}_n\,f(S) = f(S)$ and $N/L \in \mathrm{Q}\,f(S) = f(S)$. Suppose that A is contained in L. We have that $N/A \in \mathfrak{F}$ by induction. Therefore $N/L \in f(S)$ because N/L is isomorphic to a monolithic quotient of N/A whose socle belongs to $\mathrm{E}(S)$. Therefore $N \in \mathfrak{F}$ and $\mathfrak{F} = \mathrm{s}_n\,\mathfrak{F}$.

Now suppose $\text{c} = \text{N}_0$. Applying [DH92, II, 2.11], it is enough to show that $G \in \mathfrak{F}$ provided that $G = N_1 N_2$, where N_i is a normal subgroup of G and $N_i \in \mathfrak{F}_i$, $i \in \{1, 2\}$. We argue by induction on $|G|$. It is rather clear that we may assume that G has a unique minimal normal subgroup, A say, and $G/A \in \mathfrak{F}$. Let $p \in \text{char } \mathfrak{X}$. Then $G/\operatorname{C}^{\mathfrak{X}_p}(G) = \left(N_1 \operatorname{C}^{\mathfrak{X}_p}(G)/\operatorname{C}^{\mathfrak{X}_p}(G)\right)\left(N_2 \operatorname{C}^{\mathfrak{X}_p}(G)/\operatorname{C}^{\mathfrak{X}_p}(G)\right)$. Moreover $N_i \operatorname{C}^{\mathfrak{X}_p}(G)/\operatorname{C}^{\mathfrak{X}_p}(G) \cong N_i/\left(N_i \cap \operatorname{C}^{\mathfrak{X}_p}(G)\right) = N_i/\operatorname{C}^{\mathfrak{X}_p}(N_i) \in f(p)$. Hence $G/\operatorname{C}^{\mathfrak{X}_p}(G) \in \text{N}_0\, f(p) = f(p)$.

Suppose that G/L is a monolithic quotient of G such that $\operatorname{Soc}(G/L) \in \text{E}(S)$ for some simple group $S \in \mathfrak{X}'$. If $L \neq 1$, then $G/L \in \mathfrak{F}$ by induction. This implies $G/L \in f(S)$. Thus we may assume that $L = 1$. In this case $A \in \text{E}(S)$. It is clear that $\operatorname{Soc}(N_i) \in \text{E}(S)$ for $i \in \{1, 2\}$. Therefore, applying Remark 3.1.2 (5), $N_i \in f(S)$ because $N_i \in \mathfrak{F}$, $i \in \{1, 2\}$. Consequently $G \in \text{N}_0\, f(S) = f(S)$ and $G \in \mathfrak{F}$. We conclude that $\mathfrak{F}$ is N_0-closed. $\qquad\square$

The next proposition shows that Statement 2 holds for $\mathfrak{X}$-local formations.

Proposition 3.1.40. *Let $\mathfrak{F} = \operatorname{LF}_{\mathfrak{X}}(F)$ be an $\mathfrak{X}$-local formation. If c is one of the closure operations s, s_n, or N_0 and $\mathfrak{F} = \text{c}\mathfrak{F}$, then $F(S) = \text{c}\, F(S)$ for all $S \in (\text{char } \mathfrak{X}) \cup \mathfrak{X}'$.*

Proof. If $S \in \mathfrak{X}'$, then $F(S) = \mathfrak{F}$. Hence we have to prove that $F(p) = \text{c}\, F(p)$ for all $p \in \text{char } \mathfrak{X}$.

Assume $\text{c} = \text{s}$ and $p \in \text{char } \mathfrak{X}$. Let $G \in F(p)$, and let H be a subgroup of G. Then if $W = C_p \wr G$, we know that $W \in \mathfrak{F}$. Hence $BH \in \mathfrak{F}$, where B is the base group of W. Therefore $BH/\operatorname{C}^{\mathfrak{X}_p}(BH) \in F(p)$. Now $\operatorname{C}^{\mathfrak{X}_p}(BH)$ centralises every chief factor of BH below B. Since $B \leq \operatorname{C}^{\mathfrak{X}_p}(BH)$ and $\operatorname{C}_W(B) = B$, we have that $\operatorname{C}^{\mathfrak{X}_p}(BH)/B$ is a p-group by [DH92, A, 12.4]. Thus $H \in F(p)$ and $F(p)$ is subgroup-closed.

The case $\text{c} = \text{s}_n$ is analogous.

Now assume that $\text{c} = \text{N}_0$. By [DH92, II, 2.11], it will suffice to show that if $G = N_1 N_2$ with N_i a normal subgroup of G and $N_i \in F(p)$, $i = 1, 2$, then $G \in F(p)$. Let $W = C_p \wr G$ with B as the base group of W. Note that $W = (BN_1)(BN_2)$, $BN_i \trianglelefteq W$, and $BN_i \in \mathfrak{S}_p F(p) = F(p) \subseteq \mathfrak{F}$ for $i = 1, 2$. Therefore $W \in \text{N}_0\, \mathfrak{F} = \mathfrak{F}$. By Theorem 3.1.17 (3), $G \in F(p)$. $\qquad\square$

Given a group G, denote by $\text{s}_{\mathfrak{X}}(G)$ the set of all subgroups H of G such that $H \in \text{E}\mathfrak{X}$. If $\mathfrak{L}$ is a class of groups, write $\mathfrak{L}(\mathfrak{X}) = (G : \text{s}_{\mathfrak{X}}(G) \subseteq \mathfrak{L})$. It is clear that $\mathfrak{L}(\mathfrak{X})$ is the unique largest subgroup-closed class such that $\mathfrak{L}(\mathfrak{X}) \cap \text{E}\mathfrak{X} \subseteq \mathfrak{L}$.

If $\mathfrak{F}$ is a formation, then $\mathfrak{F}(\mathfrak{X})$ is clearly a formation, but if $\mathfrak{F}$ is an $\mathfrak{X}$-local formation, then $\mathfrak{F}(\mathfrak{X})$ is not an $\mathfrak{X}$-local formation in general as the next example shows.

Example 3.1.41. Consider $\mathfrak{X} = \mathfrak{J}$, the class of all simple groups, let $G = \operatorname{Alt}(5)$, and let $\mathfrak{F} = \mathfrak{N}^2 \operatorname{D}_0(1, G)$. In this case, $\mathfrak{F}(\mathfrak{X})$ is the class of all groups U such that every subgroup of U belongs to $\mathfrak{F}$. Hence G belongs to $\mathfrak{F}(\mathfrak{X})$. If $\mathfrak{F}(\mathfrak{X})$ were an $\mathfrak{X}$-local formation, then $[V]G$ would be an $\mathfrak{F}(\mathfrak{X})$-group for every irreducible and faithful $GF(2)G$-module V. In particular, if D is the dihedral group of order 10, then $VD \in \mathfrak{F}$. This would be a contradiction. Hence $\mathfrak{F}(\mathfrak{X})$ is not an $\mathfrak{X}$-local formation.

The next result provides precise conditions to ensure that $\mathfrak{F}(\mathfrak{X})$ is again an $\mathfrak{X}$-local formation.

Theorem 3.1.42 ([BB91]). *Let $\mathfrak{F}$ be an $\mathfrak{X}$-local formation. The following statements are pairwise equivalent:*

1. *For each primitive group G of type 2 in $\mathfrak{F}(\mathfrak{X})$ such that $\mathrm{Soc}(G) \in \mathrm{E}\,\mathfrak{X}$, and for every irreducible and faithful G-module V over $GF(p)$, $p \in \pi\big(\mathrm{Soc}(G)\big)$, the corresponding semidirect product $[V]G$ is an $\mathfrak{F}(\mathfrak{X})$-group.*
2. *For each primitive group G of type 2 in $\mathfrak{F}(\mathfrak{X})$ such that $\mathrm{Soc}(G) \in \mathrm{E}\,\mathfrak{X}$ and for every irreducible and faithful G-module V over $GF(p)$, $p \in \pi\big(\mathrm{Soc}(G)\big)$, and for every $X \in \mathrm{s}_{\mathfrak{X}}(G)$ such that $G = X\,\mathrm{Soc}(G)$, the semidirect product $[V]X$ is an $\mathfrak{F}$-group.*
3. *$\mathfrak{F}(\mathfrak{X})$ is an $\mathfrak{X}$-local formation.*

Proof. 2 implies 3. Suppose $\mathfrak{F} = \mathrm{LF}_{\mathfrak{X}}(F)$. Define $F^*(p) = F(p)(\mathfrak{X})$, for each prime $p \in \mathrm{char}\,\mathfrak{X}$ and $F^*(E) = F(E)(\mathfrak{X})$, for every $E \in \mathfrak{X}'$. Then F^* is an $\mathfrak{X}$-formation function. We see that $\mathfrak{F}(\mathfrak{X}) = \mathrm{LF}_{\mathfrak{X}}(F^*)$. Assume that $\mathfrak{F}(\mathfrak{X})$ is not contained in $\mathrm{LF}_{\mathfrak{X}}(F^*)$ and derive a contradiction. We choose a group $G \in \mathfrak{F}(\mathfrak{X}) \setminus \mathrm{LF}_{\mathfrak{X}}(F^*)$ of minimal order. Using familiar arguments, we have that G is a monolithic group. Denote $N = \mathrm{Soc}(G)$. If $N \in \mathrm{E}(\mathfrak{X}')$, then $G \in F(E)(\mathfrak{X})$ for some $E \in \mathfrak{X}'$ and so $G \in \mathrm{LF}_{\mathfrak{X}}(F^*)$, which is a contradiction. Hence $N \in \mathrm{E}\,\mathfrak{X}$. Suppose that N is abelian. Then N is a p-group for some prime $p \in \mathrm{char}\,\mathfrak{X}$. Let X be a subgroup of G such that $X \in \mathrm{E}\,\mathfrak{X}$. Without loss of generality, we may assume that N is contained in X. Certainly $X \in \mathfrak{F}(\mathfrak{X})$ as $\mathfrak{F}(\mathfrak{X})$ is subgroup-closed. If X is a proper subgroup of G, then $X \in \mathrm{LF}_{\mathfrak{X}}(F^*)$ by the choice of G. This implies that $X/\mathrm{C}_X^h(N) \in F^*(p)$, where $\mathrm{C}_X^h(N)$ is the intersection of the centralisers in X of all chief factors of X below N. Applying [DH92, A, 2.11], $\mathrm{C}_X^h(N)/\mathrm{C}_X(N)$ is a p-group. Hence $X/\mathrm{C}_X(N) \in F^*(p)$ and so $X/\mathrm{C}_X(N) \in F(p)$. If $X = G$, then $G/\mathrm{C}_G(N) \in F(p)$ because $\mathfrak{F}$ is $\mathfrak{X}$-local. Consequently $G/\mathrm{C}_G(N) \in F^*(p)$. Applying Remark 3.1.7 (2), we have that $G \in \mathrm{LF}_{\mathfrak{X}}(F^*)$ and we have the desired contradiction. Therefore N is a non-abelian group. Let p be a prime dividing the order of N and let $X \in \mathrm{E}\,\mathfrak{X}$. Assume that $T = XN$ is a proper subgroup of G. Arguing as above, $T = XN \in \mathrm{LF}_{\mathfrak{X}}(F^*)$ and $\mathrm{C}_T^h(N) \cong \mathrm{C}_T^h(N)/\mathrm{C}_T(N)$ is a p-group (note that $\mathrm{C}_T(N) = 1$). Hence $T/\mathrm{C}_T^h(N) \in F^*(p)$. Since $X\,\mathrm{C}_T^h(N)/\mathrm{C}_T^h(N) \in \mathrm{s}_{\mathfrak{X}}\big(T/\mathrm{C}_T^h(N)\big)$, it follows that $X\,\mathrm{C}_T^h(N)/\mathrm{C}_T^h(N)$ is in $F(p)$ and so $X \in F(p)$. Suppose that $T = G$ and consider an irreducible and faithful G-module V over $GF(p)$ (such V exists by [DH92, B, 10.9]). By Statement 2, the semidirect product $[V]X$ is an $\mathfrak{F}$-group. It implies that $X \in F(p)$. Therefore $G \in F^*(p)$ and $G \in \mathrm{LF}_{\mathfrak{X}}(F^*)$ and we have the desired contradiction.

On the other hand, taking into account that $\mathrm{LF}_{\mathfrak{X}}(F^*)$ is subgroup-closed, it is easy to see that $\mathrm{LF}_{\mathfrak{X}}(F^*)$ is contained in $\mathfrak{F}(\mathfrak{X})$. Consequently $\mathfrak{F}(\mathfrak{X})$ is an $\mathfrak{X}$-local formation.

3 implies 1. Taking into account that the $\mathfrak{F}(\mathfrak{X})$ can be locally defined by an $\mathfrak{X}$-formation function, it is clear that if G is a primitive group of type 2 in $\mathfrak{F}(\mathfrak{X})$ and $\mathrm{Soc}(G) \in \mathrm{E}\,\mathfrak{X}$, then the semidirect product $[V]G$ is an $\mathfrak{F}(\mathfrak{X})$-group for every irreducible and faithful G-module V over $\mathrm{GF}(p)$, $p \in \pi\big(\mathrm{Soc}(G)\big)$. Hence Statement 1 holds.

Finally, it is clear that 1 implies 2. The circle of implications is now complete.   □

Example 3.1.43. Assume that $\mathfrak{X}$ is the class of all simple groups and consider the class $\mathfrak{F} = \big(G : \mathrm{Alt}(5) \notin \mathrm{Q}(G)\big)$. Then $\mathrm{b}(\mathfrak{F}) = \big(\mathrm{Alt}(5)\big)$. Hence $\mathfrak{F}$ is a saturated formation by Example 2.3.21. If G is a primitive group of type 2 in $\mathfrak{F}(\mathfrak{X})$, then every subgroup of $[V]X$ is an $\mathfrak{F}$-group, for every subgroup X of G such that $G = X\,\mathrm{Soc}(G)$ and for every irreducible and faithful G-module V over $\mathrm{GF}(p)$, $p \in \pi\big(\mathrm{Soc}(G)\big)$. Consequently $\mathfrak{F}(\mathfrak{X})$ is a saturated formation. It is clear that $\mathfrak{F}(\mathfrak{X})$ is the largest subgroup-closed formation contained in $\mathfrak{F}$.

3.2 A generalisation of Gaschütz-Lubeseder-Schmid-Baer theorem

In this section we study two different Frattini-like subgroups associated with a class of simple groups which lead to the corresponding notion of saturation. We then present an extension of Gaschütz-Lubeseder-Schmid and Baer theorems.

We begin with the following definition due to P. Förster.

Definition 3.2.1 ([För85b]). *Let G be a group. For a prime p, we define $\Phi_{\mathfrak{X}}^p(G)$ as follows:*

- *If $\mathrm{O}_{p'}(G) = 1$,*

$$\Phi_{\mathfrak{X}}^p(G) = \begin{cases} \Phi(G) & \text{if } \mathrm{Soc}\big(G/\Phi(G)\big) \text{ and } \Phi(G) \text{ belong to } \mathrm{E}\,\mathfrak{X}, \\ \Phi\big(\mathrm{O}_{\mathfrak{X}}(G)\big) & \text{otherwise.} \end{cases}$$

- *In general, $\Phi_{\mathfrak{X}}^p(G)$ is the subgroup of G such that $\Phi_{\mathfrak{X}}^p(G)/\mathrm{O}_{p'}(G) = \Phi_{\mathfrak{X}}^p\big(G/\mathrm{O}_{p'}(G)\big)$.*
- *Finally put $\Phi_{\mathfrak{X}}^*(G) = \mathrm{O}_{\mathfrak{X}}(G) \cap \bigcap_{p \in \mathrm{char}\,\mathfrak{X}} \Phi_{\mathfrak{X}}^p(G)$.*

If q is a prime such that $q \notin \mathrm{char}\,\mathfrak{X}$, then $\Phi_{\mathfrak{X}}^*(G)$ is a q'-group because $\pi(\mathfrak{X}) = \mathrm{char}\,\mathfrak{X}$. Hence $\Phi_{\mathfrak{X}}^*(G) \leq \mathrm{O}_{q'}(G) \leq \Phi_{\mathfrak{X}}^q(G)$. Consequently $\Phi_{\mathfrak{X}}^*(G) = \mathrm{O}_{\mathfrak{X}}(G) \cap \bigcap_{p \in \mathbb{P}} \Phi_{\mathfrak{X}}^p(G)$.

The basic properties of $\Phi_{\mathfrak{X}}^*(G)$ are displayed in the next result.

Proposition 3.2.2. *Let G be a group.*

1. *$\Phi_{\mathfrak{X}}^*(G)$ and $\Phi_{\mathfrak{X}}^p(G)$, p a prime, are characteristic subgroups of G.*
2. *$\Phi\big(O_{\mathfrak{X}}(G)\big) \leq \Phi_{\mathfrak{X}}^*(G) \leq O_{\mathfrak{X}}(G) \cap \Phi(G)$.*
3. *Let p be a prime. If $O_{p'}(G) = 1$, then $\Phi_{\mathfrak{X}}^*(G) = \Phi_{\mathfrak{X}}^p(G)$.*
4. *Let p be a prime. If N is a normal subgroup of G contained in $\Phi_{\mathfrak{X}}^p(G)$, then $O_{p'}(G/N) = O_{p'}(G)N/N$.*
5. *If N is a normal subgroup of G contained in $\Phi_{\mathfrak{X}}^*(G)$, then $\Phi_{\mathfrak{X}}^*(G/N) = \Phi_{\mathfrak{X}}^*(G)/N$.*

Proof.   1. It is clear.

2. Let p be a prime. Then $\Phi_{\mathfrak{X}}^*(G)\,O_{p'}(G)/\,O_{p'}(G)$ is isomorphic to a subgroup of $\Phi\big(G/\,O_{p'}(G)\big)$, which is a p-group. Hence $\Phi_{\mathfrak{X}}^*(G) \cap O_{p'}(G)$ is a normal Hall p'-subgroup of $\Phi_{\mathfrak{X}}^*(G)$ and so $\Phi_{\mathfrak{X}}^*(G)$ is p-nilpotent. Therefore $\Phi_{\mathfrak{X}}^*(G)$ is nilpotent.

Assume, arguing by contradiction, that $\Phi_{\mathfrak{X}}^*(G)$ is not contained in $\Phi(G)$. Then there exists a maximal subgroup M of G such that $G = M\Phi_{\mathfrak{X}}^*(G)$. Since $\Phi_{\mathfrak{X}}^*(G)$ is nilpotent, we can find a prime p and a Sylow p-subgroup P of $\Phi_{\mathfrak{X}}^*(G)$ such that $G = MP$. In particular, $O_{p'}(G)$ is contained in M. Hence $\Phi_{\mathfrak{X}}^p(G)/\,O_{p'}(G)$ is a subgroup of $M/\,O_{p'}(G)$ and so $\Phi_{\mathfrak{X}}^*(G) \leq M$. This contradiction leads to $\Phi_{\mathfrak{X}}^*(G) \leq \Phi(G)$. Now

$$\Phi\big(O_{\mathfrak{X}}(G)\big)\,O_{p'}(G)/\,O_{p'}(G) \leq \Phi\big(O_{\mathfrak{X}}(G)\,O_{p'}(G)\big)\,O_{p'}(G)/\,O_{p'}(G)$$

$$\leq \Phi\big(O_{\mathfrak{X}}(G)\,O_{p'}(G)/\,O_{p'}(G)\big) \leq \Phi\left(O_{\mathfrak{X}}\big(G/\,O_{p'}(G)\big)\right) \leq \Phi_{\mathfrak{X}}^p(G)/\,O_{p'}(G)$$

for each prime p. Consequently $\Phi\big(O_{\mathfrak{X}}(G)\big) \leq \Phi_{\mathfrak{X}}^*(G)$.

3. Suppose that $O_{p'}(G) = 1$ for some prime p. Since $\Phi_{\mathfrak{X}}^p(G)$ is contained in $\Phi(G)$, it follows that $\Phi_{\mathfrak{X}}^p(G)$ is a p-group. Hence if q is a prime, $q \neq p$, we have that $\Phi_{\mathfrak{X}}^p(G) \leq O_{q'}(G) \leq \Phi_{\mathfrak{X}}^q(G)$. Therefore $\Phi_{\mathfrak{X}}^p(G) \leq \Phi_{\mathfrak{X}}^*(G)$ and so $\Phi_{\mathfrak{X}}^*(G) = \Phi_{\mathfrak{X}}^p(G)$.

4. Let N be a normal subgroup of G such that $N \leq \Phi_{\mathfrak{X}}^p(G)$ for some prime p. Put $Q/N = O_{p'}(G/N)$ and $M = N \cap O_{p'}(G)$. Then $N\,O_{p'}(G)/\,O_{p'}(G) \leq \Phi_{\mathfrak{X}}^p(G)/\,O_{p'}(G) \leq \Phi\big(G/\,O_{p'}(G)\big)$, which is a p-group. Therefore N/M is a p-group. Since $(Q/M)/(N/M)$ is a p'-group, it follows that $Q/M = (N/M)(H/M)$ for some Hall p'-subgroup H/M of Q/M. It is clear that H is a Hall p'-subgroup of $Q \trianglelefteq G$. Moreover the Hall p'-subgroups of Q are conjugate. Therefore $G = N_G(H)N$ by the Frattini argument. Since $N\,O_{p'}(G)/\,O_{p'}(G)$ is contained in $\Phi\big(G/\,O_{p'}(G)\big)$, it follows that $G = N_G(H)$ and $H \leq O_{p'}(G)$. Consequently $Q/N = O_{p'}(G)N/N$.

5. Let N be a normal subgroup of G contained in $\Phi_{\mathfrak{X}}^*(G)$. Let p be a prime. Suppose that $O_{p'}(G/N) = 1$. Then $O_{p'}(G)$ is contained in N by Statement 4. Moreover $\Phi_{\mathfrak{X}}^p(G/N)$ is $\Phi(G/N) = \Phi(G)/N$ or $\Phi\big(O_{\mathfrak{X}}(G/N)\big) = \Phi\big(O_{\mathfrak{X}}(G)/N\big)$. Suppose that $\Phi_{\mathfrak{X}}^p(G/N) \neq \Phi\big(O_{\mathfrak{X}}(G)/N\big)$. Then $\mathrm{Soc}\left((G/N)/\big(\Phi(G)/N\big)\right)$ and $\Phi(G)/N$ belongs to $\mathrm{E}\,\mathfrak{X}$ and for $\mathrm{Soc}\left((G/\,O_{p'}(G))/\big(\Phi(G)/\,O_{p'}(G)\big)\right)$ and $\Phi\big(G/\,O_{p'}(G)\big)$ the same is true. Hence we have that $\Phi_{\mathfrak{X}}^p\big(G/\,O_{p'}(G)\big) = \Phi(G)/\,O_{p'}(G)$ and $\Phi_{\mathfrak{X}}^p(G/N) = \Phi_{\mathfrak{X}}^p(G)/N$.

Assume now that $\Phi_{\mathfrak{X}}^p(G/N) = \Phi\big(O_{\mathfrak{X}}(G)/N\big)$. Then $\Phi_{\mathfrak{X}}^p(G)/O_{p'}(G) = \Phi\big(O_{\mathfrak{X}}(G)/O_{p'}(G)\big)$. By [DH92, A, 9.3 (c)], it follows that $\Phi_{\mathfrak{X}}^p(G)$ is nilpotent. Hence $\Phi_{\mathfrak{X}}^p(G)$ is contained in $\Phi\big(O_{\mathfrak{X}}(G)\big)O_{p'}(G)$ by [DH92, A, 9.11]. Therefore $\Phi_{\mathfrak{X}}^p(G)/N$ is contained in $\Phi\big(O_{\mathfrak{X}}(G)\big)N/N \le \Phi\big(O_{\mathfrak{X}}(G/N)\big)$. Since $N/O_{p'}(G)$ is contained in $\Phi\big(O_{\mathfrak{X}}(G)/O_{p'}(G)\big)$, it follows that $\Phi_{\mathfrak{X}}^p(G/N)$ is isomorphic to $\Phi_{\mathfrak{X}}^p(G)/N$. It leads to $\Phi_{\mathfrak{X}}^p(G)/N = \Phi_{\mathfrak{X}}^p(G/N)$.

Assume now that $O_{p'}(G/N) = O_{p'}(G)N/N \ne 1$. Denote with bars the images in $\bar{G} = G/O_{p'}(G)$. Since $O_{p'}(\bar{G}/\bar{N}) = 1$ and $\bar{N} \le \Phi_{\mathfrak{X}}^p(\bar{G})$, it follows that $\Phi_{\mathfrak{X}}^p(\bar{G}/\bar{N}) = \Phi_{\mathfrak{X}}^p(\bar{G})/\bar{N}$. By definition of $\Phi_{\mathfrak{X}}^p(G)$, we have that $\Phi_{\mathfrak{X}}^p(\bar{G}) = \overline{\Phi_{\mathfrak{X}}^p(G)}$. Therefore the image of $\Phi_{\mathfrak{X}}^p(\bar{G}/\bar{N})$ under the natural isomorphism between $\bar{G}/\bar{N}$ and $G/N\,O_{p'}(G)$ is $\Phi_{\mathfrak{X}}^p(G)/N\,O_{p'}(G)$. This implies that $\Phi_{\mathfrak{X}}^p\big(G/N\,O_{p'}(G)\big) = \Phi_{\mathfrak{X}}^p(G)/N\,O_{p'}(G)$. On the other hand, by definition we have $\Phi_{\mathfrak{X}}^p(G/N)/O_{p'}(G/N) = \Phi_{\mathfrak{X}}^p(G/N)/N\,O_{p'}(G)/N = \Phi_{\mathfrak{X}}^p\big((G/N)/(N\,O_{p'}(G)/N)\big)$. Now the image of $\Phi_{\mathfrak{X}}^p\big((G/N)/(N\,O_{p'}(G)/N)\big)$ under the natural isomorphism between the groups $(G/N)/(N\,O_{p'}(G)/N)$ and $G/N\,O_{p'}(G)$ is the subgroup $\Phi_{\mathfrak{X}}^p\big(G/N\,O_{p'}(G)\big)$. Therefore we have that $\Phi_{\mathfrak{X}}^p(G)/N = \Phi_{\mathfrak{X}}^p(G/N)$.

Consequently $\Phi_{\mathfrak{X}}^p(G)/N = \Phi_{\mathfrak{X}}^p(G/N)$ for all primes p and so $\Phi_{\mathfrak{X}}^*(G)/N = \Phi_{\mathfrak{X}}^*(G/N)$. $\qquad\square$

Remark 3.2.3. If N is a normal subgroup of a group G, then $\Phi(G)N/N \le \Phi(G/N)$ and $\Phi(N) \le \Phi(G)$ ([DH92, A, 9.2]). This is not true for $\Phi_{\mathfrak{X}}^*(G)$ in general, as the next examples show.

Examples 3.2.4.   1.   Let $H = \mathrm{SL}(2,5)$. Then H has an irreducible module V over $\mathrm{GF}(2)$ such that $\mathrm{Ker}(H \text{ on } V) = Z(H)$ (cf. [DH92, B, 10.9]). Let $G = [V]H$ be the corresponding semidirect product. Put $\mathfrak{X} = (C_2)$. Then $\Phi_{\mathfrak{X}}^*(G) = \Phi(G) = \Phi(H)$ and $\Phi_{\mathfrak{X}}^*(G/V) = 1$.

2.   If $G_1 = G \times \mathrm{Alt}(5)$, where G and $\mathfrak{X}$ are as in 1, it follows that $\Phi(H) = \Phi_{\mathfrak{X}}^*(G) \nleq \Phi_{\mathfrak{X}}^*(G_1) = 1$.

If $\mathfrak{X} = \mathfrak{J}$, then $\Phi_{\mathfrak{X}}^*(G) = \Phi(G)$ for every group G by Proposition 3.2.2 (2). However, if $\emptyset \ne \mathfrak{X} \ne \mathfrak{J}$, then we can find a group G such that $\Phi\big(O_{\mathfrak{X}}(G)\big)$ is a proper subgroup of $\Phi_{\mathfrak{X}}^*(G)$ as the next example shows.

Example 3.2.5 ([BBCER05]). Assume that $\emptyset \ne \mathfrak{X} \ne \mathfrak{J}$. Then there exist a non-abelian simple group $S \in \mathfrak{X}'$ and a prime $p \in \pi(S)$ such that $p \in \mathrm{char}\,\mathfrak{X}$. It is certainly true that $\mathrm{char}\,\mathfrak{X}$ is the set of all prime numbers. Suppose that $\mathrm{char}\,\mathfrak{X} \ne \mathbb{P}$ and take $p \in \mathrm{char}\,\mathfrak{X}$ and $q \notin \mathrm{char}\,\mathfrak{X}$. If S is the alternating group of degree $p + q$, then $S \in \mathfrak{X}'$ and $p \in \mathrm{char}\,\mathfrak{X} \cap \pi(S)$. Let T be the group algebra $\mathrm{GF}(p)S$ and consider $G = [T]S$, the corresponding semidirect product. It is rather clear that $\Phi(G) = \mathrm{Rad}\,T$. Since $O_{p'}(G) = 1$ and $\Phi(G)$ and $\mathrm{Soc}\big(G/\Phi(G)\big)$ belong to $\mathrm{E}\,\mathfrak{X}$, we have that $\Phi_{\mathfrak{X}}^*(G) = \Phi_{\mathfrak{X}}^p(G) = \Phi(G)$ by Proposition 3.2.2 (3). It is certainly true that $\Phi(G) \ne 1$ because $\mathrm{Rad}\,T \ne 1$. However, $O_{\mathfrak{X}}(G) = T$ and $\Phi(T) = 1$.

This example shows, in particular, that $\Phi_{\mathfrak{X}}^*(G)$ is not always the Frattini subgroup of the soluble radical when $\mathfrak{X}$ is the class of all abelian simple groups.

In [BBCER05] another Frattini-like subgroup associated with a class of simple groups is introduced and analysed. It is smaller than Förster's one and coincides with the Frattini subgroup of the $\mathrm{E}\,\mathfrak{X}$-radical except in a very few number of cases. We present here a slight variation of this subgroup as it appears in [BBCER05].

Definition 3.2.6. *Let p be a prime. A group G belongs to $\mathrm{A}_{\mathfrak{X}_p}(\mathfrak{P}_2)$ provided that G is monolithic and there exists an elementary abelian normal p-subgroup N of G such that*

1. *$N \leq \Phi(G)$ and G/N is a primitive group of type 2,*
2. *$\mathrm{Soc}(G/N) \in \mathrm{E}\,\mathfrak{X} \setminus \mathfrak{E}_{p'}$, and*
3. *$\mathrm{C}_G^h(N) \leq N$, where*

$$\mathrm{C}_G^h(N) := \bigcap \{\mathrm{C}_G(H/K) : H/K \text{ is a chief factor of } G \text{ below } N\}.$$

The next result shows that $\mathrm{A}_{\mathfrak{X}_p}(\mathfrak{P}_2) \neq \emptyset$ if $\mathfrak{X}$ contains non-abelian simple groups.

Proposition 3.2.7. *Let G be a primitive group of type 2 such that $\mathrm{Soc}(G) \in \mathrm{E}\,\mathfrak{X}$. Then, for each prime $p \in \pi(\mathrm{Soc}(G))$, there exists a group $E_p \in \mathrm{A}_{\mathfrak{X}_p}(\mathfrak{P}_2)$ and a minimal normal p-subgroup T_p of E_p contained in $\Phi(E_p)$ such that $E_p/\mathrm{C}_{E_p}(T_p)$ is isomorphic to G.*

Proof. Note that $p \in \mathrm{char}\,\mathfrak{X}$ because $\pi(\mathfrak{X}) = \mathrm{char}\,\mathfrak{X}$. Let E_p be the maximal Frattini extension of G with p-elementary abelian kernel $\mathrm{A}_p(G)$. Then $E_p/\mathrm{A}_p(G) \cong G$ and $\mathrm{A}_p(G) = \Phi(E_p)$ (see [GS78]). Moreover, by [GS78, Theorem 1], we have that $\mathrm{Ker}(G \text{ on } \mathrm{Soc}(\mathrm{A}_p(G))) = \mathrm{O}_{p',p}(G) = 1$. Hence there exists a minimal normal subgroup T_p of E_p such that $T_p \leq \mathrm{A}_p(G)$ and $\mathrm{C}_{E_p}(T_p) = \mathrm{A}_p(G)$. If E_p is monolithic, then clearly $E_p \in \mathrm{A}_{\mathfrak{X}_p}(\mathfrak{P}_2)$ and the proposition is proved. Suppose that E_p is not monolithic. By Lemma 3.1.3, there exists a normal subgroup N of E_p such that $N \cap T_p = 1$, E_p/N is monolithic, and $\mathrm{Soc}(E_p/N) = T_pN/N$. Now $N \leq \mathrm{C}_{E_p}(T_p) = \mathrm{A}_p(G) = \Phi(E_p)$ and $\mathrm{C}_{E_p/N}(T_pN/N) = \mathrm{C}_{E_p}(T_p)/N = \Phi(E_p)/N = \Phi(E_p/N)$. Therefore $E_p/N \in \mathrm{A}_{\mathfrak{X}_p}(\mathfrak{P}_2)$ and T_pN/N is a minimal normal subgroup of E_p/N such that $(E_p/N)/\mathrm{C}_{E_p/N}(T_pN/N) \cong E_p/\mathrm{C}_{E_p}(T) \cong G$. □

Definition 3.2.8. *The $\mathfrak{X}$-Frattini subgroup of a group G is the subgroup $\Phi_{\mathfrak{X}}(G)$ defined as follows:*

$$\Phi_{\mathfrak{X}}(G) := \begin{cases} \Phi(\mathrm{O}_{\mathfrak{X}}(G)) & \text{if } G \notin \mathrm{A}_{\mathfrak{X}_p}(\mathfrak{P}_2) \text{ for all } p \in \mathrm{char}\,\mathfrak{X}, \\ \Phi(G) & \text{otherwise.} \end{cases}$$

It is clear that $\Phi_{\mathfrak{X}}(G)$ is a characteristic subgroup of G. Moreover if $\mathfrak{X} = \mathfrak{J}$, then obviously $\Phi_{\mathfrak{X}}(G) = \Phi(G)$ and if $\mathfrak{X} = \mathbb{P}$, then $A_{\mathfrak{X}_p}(\mathfrak{P}_2) = \emptyset$ for all $p \in \operatorname{char} \mathfrak{X}$. Hence $\Phi_{\mathfrak{X}}(G) = \Phi(G_{\mathfrak{S}})$ for every group G. Moreover,

Proposition 3.2.9. *Let G be a group. Then $\Phi_{\mathfrak{X}}(G)$ is contained in $\Phi_{\mathfrak{X}}^*(G)$.*

Proof. We know, by Proposition 3.2.2 (2), that $\Phi\big(O_{\mathfrak{X}}(G)\big)$ is contained in $\Phi_{\mathfrak{X}}^*(G)$. Suppose now that $G \in A_{\mathfrak{X}_p}(\mathfrak{P}_2)$ for some prime $p \in \operatorname{char} \mathfrak{X}$. Then $O_{p'}(G) = 1$ and $\Phi(G)$ is a p-group. Since $\Phi(G)$ and $\operatorname{Soc}(G/\Phi(G))$ belong to $\operatorname{E} \mathfrak{X}$, $\Phi_{\mathfrak{X}}^p(G) = \Phi(G)$. In addition, $\Phi_{\mathfrak{X}}^p(G) = \Phi_{\mathfrak{X}}^*(G)$ by Proposition 3.2.2 (3). Therefore $\Phi(G) = \Phi_{\mathfrak{X}}(G) = \Phi_{\mathfrak{X}}^*(G)$. □

Remarks 3.2.10. 1. Example 3.2.5 shows that the equality $\Phi_{\mathfrak{X}}(G) = \Phi_{\mathfrak{X}}^*(G)$ does not hold in general.

2. If $\mathfrak{X}_1 \subseteq \mathfrak{X}_2$, then $\Phi_{\mathfrak{X}_1}(G) \leq \Phi_{\mathfrak{X}_2}(G)$ for all groups G.

By definition, if $G \notin A_{\mathfrak{X}_p}(\mathfrak{P}_2)$ for $p \in \operatorname{char} \mathfrak{X}$, then $\Phi_{\mathfrak{X}}(G) = \Phi\big(O_{\mathfrak{X}}(G)\big)$. We do not know whether in groups belonging to $A_{\mathfrak{X}_p}(\mathfrak{P}_2)$ for some $p \in \operatorname{char} \mathfrak{X}$ the above equality holds. This raises the following question:

Open question 3.2.11. *Let $\mathfrak{X}$ be a class of simple groups such that $\operatorname{char} \mathfrak{X} = \pi(\mathfrak{X})$ and let $p \in \operatorname{char} \mathfrak{X}$. If $G \in A_{\mathfrak{X}_p}(\mathfrak{P}_2)$, is it true that $\Phi(G) = \Phi\big(O_{\mathfrak{X}}(G)\big)$?*

Moreover, the compatibility of $\Phi_{\mathfrak{X}}(G)$ with quotients of G is not visible and doubtful. In fact, we do not know whether $\Phi_{\mathfrak{X}}(G/N) = \Phi_{\mathfrak{X}}(G)/N$ for $N \trianglelefteq G$ such that $N \leq \Phi_{\mathfrak{X}}(G)$.

In the sequel, using the ideas contained in the paper [BBCER05], we shall prove that the $\mathfrak{X}$-local formations are exactly those formations which are closed under extensions by the Frattini-like subgroups studied above. It leads to extensions of the Gaschütz-Lubeseder-Schmid and Baer theorems.

We begin with the following definitions.

Definitions 3.2.12. *Let $\mathfrak{F}$ be a formation. We say that:*

1. *$\mathfrak{F}$ is $\mathfrak{X}$-saturated (N) if $\mathfrak{F}$ contains a group G whenever it contains $G/\Phi\big(O_{\mathfrak{X}}(G)\big)$.*
2. *$\mathfrak{F}$ is $\mathfrak{X}$-saturated (F) if $G \in \mathfrak{F}$ provided that $G/\Phi_{\mathfrak{X}}^*(G) \in \mathfrak{F}$.*
3. *$\mathfrak{F}$ is $\mathfrak{X}$-saturated if $G \in \mathfrak{F}$ provided that $G/\Phi_{\mathfrak{X}}(G) \in \mathfrak{F}$.*
4. *G has property $\mathfrak{X}_*$ if $\mathfrak{F}$ contains every group $G \in A_{\mathfrak{X}_p}(\mathfrak{P}_2)$, $p \in \operatorname{char} \mathfrak{X}$, whenever it contains $G/\Phi(G)$.*

Remarks 3.2.13. Let $\mathfrak{F}$ be a formation.

1. $\mathfrak{F}$ is $\mathfrak{X}$-saturated if and only if $\mathfrak{F}$ is $\mathfrak{X}$-saturated (N) and $\mathfrak{F}$ has property $\mathfrak{X}_*$.

2. If $\mathfrak{F}$ is $\mathfrak{X}$-saturated (F), then $\mathfrak{F}$ is $\mathfrak{X}$-saturated.

3. If $\mathfrak{X} = \mathfrak{J}$, then $\mathfrak{F}$ is $\mathfrak{X}$-saturated if and only if $\mathfrak{F}$ is saturated.

4. If $\mathfrak{X} \subseteq \mathbb{P}$, then $\mathfrak{F}$ is $\mathfrak{X}$-saturated if and only if $\mathfrak{F}$ is $\mathfrak{X}$-saturated (N).

The main result in this section is the following.

Theorem 3.2.14. *Let $\mathfrak{F}$ be a formation. The following statements are pairwise equivalent:*

1. *$\mathfrak{F}$ is $\mathfrak{X}$-local.*
2. *$\mathfrak{F}$ is $\mathfrak{X}$-saturated (F).*
3. *$\mathfrak{F}$ is $\mathfrak{X}$-saturated.*
4. *$\mathfrak{F}$ is $\mathfrak{X}$-saturated (N) and $\mathfrak{F}$ has property $\mathfrak{X}_*$.*

We begin with some preliminary results.

Lemma 3.2.15. *Let p be a prime in $\operatorname{char}\mathfrak{X}$, let G be a group, and let N be a normal subgroup of G such that $N \leq \mathrm{O}_\mathfrak{X}(G)$. Then $\mathrm{C}^{\mathfrak{X}_p}\big(G/\Phi(N)\big) = \mathrm{C}^{\mathfrak{X}_p}(G)/\Phi(N)$.*

Proof. Put $A/\Phi(N) = \mathrm{C}^{\mathfrak{X}_p}\big(G/\Phi(N)\big)$. It is clear that A is a normal subgroup of G such that $\Phi(N) \leq \mathrm{O}_{p',p}(G) \leq \mathrm{C}^{\mathfrak{X}_p}(G) \leq A$. We prove that $A \leq \mathrm{C}^{\mathfrak{X}_p}(G)$; we consider A acting on G and N by conjugation, and define the following formation function:

$$f(q) = \begin{cases} (1) & \text{for } q = p, \\ \mathfrak{E} & \text{for } q \neq p. \end{cases}$$

Next we see that A acts f-hypercentrally on N (cf. [DH92, IV, 6.2]). Let H/K be an A-composition factor of G between $A \cap N$ and N. Since $[A, N] \leq A \cap N$, it is true that $\mathrm{C}_A(H/K) = A$. Let H/K be a chief factor of G between $\Phi(N)$ and $A \cap N$ such that p divides $|H/K|$. Then H/K is an $\mathfrak{X}_p$-chief factor of G because $N \leq \mathrm{O}_\mathfrak{X}(G)$. Hence $\mathrm{C}_A(H/K) = A$ and so A centralises every A-composition factor of N between K and H. It yields that A acts f-hypercentrally on $N/\Phi(N)$. By [DH92, IV, 6.7], A acts f-hypercentrally on N.

Let H/K be an $\mathfrak{X}_p$-chief factor of G below $\Phi(N)$. Since H/K is a minimal normal subgroup of G/K and $H/K \leq A/K$, we can apply [DH92, A, 4.13] to conclude that $H/K = L_1/K \times \cdots \times L_r/K$, where L_i/K is a minimal normal subgroup of A/K for all $1 \leq i \leq r$. Since L_i/K is an A-composition factor of N and p divides $|L_i/K|$, it follows that $A \leq \mathrm{C}_G(L_i/K)$. Hence $\mathrm{C}_A(H/K) = A$. Consequently A centralises all $\mathfrak{X}_p$-chief factors of G below $\Phi(N)$ and so $A \leq \mathrm{C}^{\mathfrak{X}_p}(G)$. $\square$

Theorem 3.2.16. *If $\mathfrak{F}$ is an $\mathfrak{X}$-local formation, then $\mathfrak{F}$ is $\mathfrak{X}$-saturated (F).*

Proof. Let G be a group such that $G/\Phi_\mathfrak{X}^*(G) \in \mathfrak{F}$. We prove that $G \in \mathfrak{F}$ by induction on $|G|$. Let p be a prime in $\operatorname{char}\mathfrak{X}$. Then $G/\Phi_\mathfrak{X}^p(G) \in \mathfrak{F}$ and $\Phi_\mathfrak{X}^p(G)/\mathrm{O}_{p'}(G) = \Phi_\mathfrak{X}^p\big(G/\mathrm{O}_{p'}(G)\big)$ by Proposition 3.2.2 (3). Consequently, if $\mathrm{O}_{p'}(G) \neq 1$, we have $G/\mathrm{O}_{p'}(G) \in \mathfrak{F}$. This implies that every $\mathfrak{X}_p$-chief factor H/K of G is G-isomorphic to an $\mathfrak{X}_p$-chief factor of $G/\mathrm{O}_{p'}(G)$. Hence $G/\mathrm{C}_G(H/K) \in F(p)$, where F is the canonical $\mathfrak{X}$-local definition of $\mathfrak{F}$.

We may assume that $\mathrm{O}_{p'}(G) = 1$ for some prime $p \in \operatorname{char}\mathfrak{X}$. In this case $\Phi_\mathfrak{X}^p(G) = \Phi_\mathfrak{X}^*(G)$ is a p-group. Suppose that $\Phi_\mathfrak{X}^*(G) = \Phi\big(\mathrm{O}_\mathfrak{X}(G)\big)$. Then p divides $\big|\mathrm{O}_\mathfrak{X}(G)/\Phi\big(\mathrm{O}_\mathfrak{X}(G)\big)\big|$ and so G has an $\mathfrak{X}_p$-chief factor above

$\Phi\big(O_{\mathfrak{X}}(G)\big)$. In particular, $F(p) \neq \emptyset$. Since $G/\Phi(O_{\mathfrak{X}}(G)) \in \mathfrak{F}$, we have that $\big(G/\Phi(O_{\mathfrak{X}}(G))\big)\big/\mathrm{C}^{\mathfrak{X}_p}\big(G/\Phi(O_{\mathfrak{X}}(G))\big) \in F(p)$. By Lemma 3.2.15, it follows $G/\mathrm{C}^{\mathfrak{X}_p}(G) \in F(p)$. We conclude then that G satisfies Condition 1 in Definition 3.1.1.

Assume now that $\Phi_{\mathfrak{X}}^*(G) \neq \Phi\big(O_{\mathfrak{X}}(G)\big)$, then $\Phi(G)$ and $\mathrm{Soc}\big(G/\Phi(G)\big) = \mathrm{F}'(G)/\Phi(G)$ belong to $\mathrm{E}\,\mathfrak{X}$ and $\Phi_{\mathfrak{X}}^*(G) = \Phi(G)$. Note that in this case p divides the order of every $\mathfrak{X}$-chief factor of G below $\mathrm{F}'(G)$. Let T be the intersection of the centralisers in G of the $\mathfrak{X}_p$-chief factors of G between $\Phi(G)$ and $\mathrm{F}'(G)$. Then $G/T \in F(p)$ because $G/\Phi(G) \in \mathfrak{F}$. Moreover, $T/\Phi(G)$ centralises $\mathrm{F}'(G)/\Phi(G)$ because $\mathrm{F}'(G)/\Phi(G)$ is a direct product of $\mathfrak{X}_p$-chief factors of G. By [För85b, Satz 1.2], $T/\Phi(G)$ is a p-group. This yields T is a p-group and so $G \in F(p)$.

Consequently, in both cases, G satisfies Condition 1 in Definition 3.1.1.

Let L be a normal subgroup of G such that G/L is monolithic and $\mathrm{Soc}(G/L)$ belongs to $\mathrm{E}(S)$ for some $S \in \mathfrak{X}'$. Then $\Phi_{\mathfrak{X}}^*(G) \leq O_{\mathfrak{X}}(G) \leq L$ and so $G/L \in \mathfrak{F} = F(S)$. Hence G satisfies Condition 2 in Definition 3.1.1 and therefore $G \in \mathfrak{F}$. This is to say that $\mathfrak{F}$ is $\mathfrak{X}$-saturated (F). $\square$

Lemma 3.2.17. *Let p be a prime and let $\mathfrak{F}$ be a (C_p)-saturated formation.*

1. *Let X be a group, and let M, N be $\mathrm{GF}(p)X$-modules with N irreducible and X acting faithfully on M. If $[M]X \in \mathfrak{F}$, then $[N]X \in \mathfrak{F}$.*
2. *Let N be an elementary abelian normal p-subgroup of a group G. Assume that $[N](G/N) \in \mathfrak{F}$ and that $C_p \in \mathfrak{F}$. Then $G \in \mathfrak{F}$.*

Proof. 1 and 2 follow from the proofs of [DH92, IV, 4.1] and [DH92, IV, 4.15], respectively, taking into account that the Hartley group used there plays the role of the normal p-subgroup. $\square$

Lemma 3.2.18. *Let $\mathfrak{F}$ be a (C_p)-saturated formation, p a prime. If $X \in \mathrm{R}_0\big(G/\mathrm{C}_G(H/K) : G \in \mathfrak{F}$ and H/K is an abelian p-chief factor of $G\big)$, then $[N]X \in \mathfrak{F}$ for every irreducible $\mathrm{GF}(p)X$-module.*

Proof. The group X has a set $\{N_1, \ldots, N_n\}$ of normal subgroups satisfying:

1. X/N_i is isomorphic to $G_i/\mathrm{C}_{G_i}(H_i/K_i)$, where $G_i \in \mathfrak{F}$ and H_i/K_i is an abelian p-chief factor of G_i,
2. $\bigcap_{i=1}^n N_i = 1$.

By Corollary 2.2.5, $[H_i/K_i](X/N_i) \in \mathfrak{F}$, $1 \leq i \leq n$. Note that H_i/K_i can be regarded as X-modules over $\mathrm{GF}(p)$ and $\mathrm{Ker}(X$ on $H_i/K_i) = N_i$, $1 \leq i \leq n$. Moreover, the semidirect product $[H_i/K_i]X$ has normal subgroups H_i/K_i and N_i satisfying $[H_i/K_i]X/(H_i/K_i)$, $[H_i/K_i]X/N_i \in \mathfrak{F}$. Therefore $[H_i/K_i]X \in \mathrm{R}_0\,\mathfrak{F} = \mathfrak{F}$, $1 \leq i \leq n$. Put $M = H_1/K_1 \times \cdots \times H_n/K_n$. Then M is an X-module and $\mathrm{Ker}(X$ on $M) = \bigcap_{i=1}^n N_i = 1$. Hence X acts faithfully on M. Consider the set $\{M_1, \ldots, M_n\}$ of normal subgroups of $[M]X$: $M_1 = H_2/K_2 \times \cdots \times H_n/K_n, \ldots, M_n = H_1/K_1 \times \cdots \times H_{n-1}/K_{n-1}$ and $M_i = $

$H_1/K_1 \times \cdots \times H_{i-1}/K_{i-1} \times H_{i+1}/K_{i+1} \times \cdots \times H_n/K_n$, $2 \leq i \leq n-1$. Then $\bigcap_{j=1}^n M_j = 1$ and $[M]X/M_j \cong [H_j/K_j]X \in \mathfrak{F}$. Therefore $[M]X \in \mathrm{R_0}\,\mathfrak{F} = \mathfrak{F}$. By Lemma 3.2.17, $[N]X \in \mathfrak{F}$ for every irreducible $\mathrm{GF}(p)X$-module.   □

Theorem 3.2.19. *If $\mathfrak{F}$ is an $\mathfrak{X}$-saturated formation, then $\mathfrak{F}$ is $\mathfrak{X}$-local.*

Proof. By Remark 3.2.10 (3.2), $\mathfrak{F}$ is a (C_p)-saturated formation for all $p \in$ char $\mathfrak{X}$.

Bearing in mind Theorem 3.1.17, the natural candidate f for an $\mathfrak{X}$-local definition of $\mathfrak{F}$ is given by

$$f(p) = \mathfrak{S}_p \,\mathrm{Q}\,\mathrm{R_0}\big(G/\,\mathrm{C}_G(H/K) : G \in \mathfrak{F} \text{ and}$$

$$\qquad H/K \text{ is an abelian } p\text{-chief factor of } G\big) \qquad \text{for } p \in \text{char } \mathfrak{X},$$

$$f(S) = \mathfrak{F} \qquad\qquad\qquad\qquad\qquad\qquad\qquad \text{for } S \in \mathfrak{X}'.$$

It is clear that f is an $\mathfrak{X}$-formation function.

Put $\mathfrak{H} = \mathrm{LF}_{\mathfrak{X}}(f)$. Suppose that $\mathfrak{F}$ is not contained in $\mathfrak{H}$ and let $G \in \mathfrak{F} \setminus \mathfrak{H}$ of minimal order. We shall show that this supposition leads to a contradiction. Since $\mathfrak{H}$ is a formation, it follows that G has a unique minimal normal subgroup, N say, and that $G/N \in \mathfrak{H}$. If N has composition type $S \in \mathfrak{X}'$, then $G \in f(S) = \mathfrak{F}$. This is impossible. Therefore N is an $\mathfrak{X}$-chief factor of G. If N is non-abelian, then G is a primitive group of type 2. Let p be a prime divisor of $|N|$. Then $p \in$ char $\mathfrak{X}$ and, by Proposition 3.2.7, there exists $E \in \mathrm{A}_{\mathfrak{X}_p}(\mathfrak{P}_2)$ such that $E/\,\mathrm{C}_E(T) \cong G$ for some minimal normal subgroup T of G. Moreover T is a p-group. Since $\Phi(E) = \mathrm{C}_E(T) = \Phi_{\mathfrak{X}}(E)$ and $\mathfrak{F}$ is $\mathfrak{X}$-saturated, it follows that $E \in \mathfrak{F}$. This means that $G \in f(p)$. Then we conclude that $G \in \mathfrak{F}$ because $\mathrm{O}_p(G) = 1$. But $G \notin \mathfrak{F}$ by supposition, and so we must have that N is a p-group for some prime $p \in$ char $\mathfrak{X}$. In this case, $G/\,\mathrm{C}_G(N) \in f(p)$ and so $G \in \mathfrak{H}$ by Remark 3.1.7 (2), and we reach a contradiction. Therefore $\mathfrak{F} \subseteq \mathfrak{H}$.

Suppose that $\mathfrak{H}$ is not contained in $\mathfrak{F}$, and let G be a group of minimal order in $\mathfrak{H} \setminus \mathfrak{F}$. Then, as usual, G has a unique minimal normal subgroup N and $G/N \in \mathfrak{F}$. Moreover neither $N \in \mathrm{E}(\mathfrak{X}')$ nor N is a non-abelian $\mathrm{E}\,\mathfrak{X}$-group because $G \notin \mathfrak{F}$. Consequently, N is an abelian p-group for some prime $p \in$ char $\mathfrak{X}$. In particular, $f(p) \neq \emptyset$ and therefore $\mathfrak{H}$ contains the cyclic group of order p. By Corollary 2.2.5, $A = [N](G/N) \in \mathfrak{H}$. Assume that $N < \mathrm{C}_G(N)$. Then $M = (G/N) \cap \mathrm{C}_A(N)$ is a non-trivial normal subgroup of A. Since $|A/M| < |G|$, we have that $A/M \in \mathfrak{F}$ by minimality of G. Hence $A \cong A/(N \cap M) \in \mathrm{R_0}\,\mathfrak{F} = \mathfrak{F}$. We can apply Lemma 3.2.17 (2) and deduce that $G \in \mathfrak{F}$. This is a contradiction. Hence we must have $\mathrm{C}_G(N) = N$ and so $G/N \in f(p)$. Since $\mathrm{O}_p(G/N) = 1$ by [DH92, B, 3.12 (b)], it follows that $G/N \in \mathrm{Q}\,\mathrm{R_0}\big(B/\,\mathrm{C}_B(H/K) : B \in \mathfrak{F}$ and H/K is an abelian p-chief factor of $B\big)$. This yields that $G/N \cong X/T$ for some normal subgroup T of

$$X \in \mathrm{R_0}\big(B/\,\mathrm{C}_B(H/K) : B \in \mathfrak{F} \text{ and } H/K \text{ is an abelian } p\text{-chief factor of } B\big).$$

Now N can be regarded as an irreducible X-module over $\mathrm{GF}(p)$ such that $T = \mathrm{Ker}(X \text{ on } N)$. By Lemma 3.2.18, we have $[N]X \in \mathfrak{F}$. Consequently $G \cong$

$[N](G/N) \cong [N](X/T)$ belongs to $\mathfrak{F}$. We have reached a contradiction. Hence $\mathfrak{H} \subseteq \mathfrak{F}$ and the equality holds. $\square$

Return for the moment to Theorem 3.2.14. It can be deduced at once from Theorem 3.2.16, Theorem 3.2.19, and Remarks 3.2.13.

Note that the Gaschütz-Lubeseder-Schmid theorem is a special case of Theorem 3.2.14 when $\mathfrak{X} = \mathfrak{J}$, the class of all simple groups.

Another generalisation of Gaschütz's concept of local formation in the general finite universe is due to L. A. Shemetkov, who introduced in 1973 the notion of composition formation. The most general version of these kind of formations was presented in [She01]. Let us describe Shemetkov's approach. Let $\mathfrak{Y} \neq \emptyset$ be a class of simple groups. A function which associates with every group $A \in \mathfrak{Y}$ a formation $f(A)$ and with every group $B \in \mathfrak{Y}'$ a formation $\emptyset \neq f(\mathfrak{Y}')$ is called a $C_{\mathfrak{Y}}$-*satellite*. If f is a $C_{\mathfrak{Y}}$-satellite, then the class $\mathrm{CF}_{\mathfrak{Y}}(f)$ of all groups G satisfying:

1. if H/K is a $\mathfrak{Y}$-chief factor of G and S is the composition factor of H/K, then $G/C_G(H/K) \in f(S)$, and
2. $G/O_{\mathfrak{Y}}(G) \in f(\mathfrak{Y}')$

is a formation.

We say that a formation $\mathfrak{F}$ is a $\mathfrak{Y}$-*composition formation* if $\mathfrak{F} = \mathrm{CF}_{\mathfrak{Y}}(f)$ for some $C_{\mathfrak{Y}}$-satellite f.

Remark 3.2.20. Let $\emptyset \neq \mathfrak{Y}$ be a class of simple groups. Denote $\mathfrak{X} = \mathrm{char}\,\mathfrak{Y} = \{C_p : p \in \mathrm{char}\,\mathfrak{Y}\}$. Then the $\mathfrak{Y}$-composition formations are exactly the $\mathfrak{X}$-saturated ones.

Proof. Let $\mathfrak{F} = \mathrm{CF}_{\mathfrak{Y}}(f)$ be a $\mathfrak{Y}$-composition formation. Then it is clear that $\mathfrak{F} = \mathrm{LF}_{\mathfrak{X}}(f_0)$, where f_0 is the $\mathfrak{X}$-formation function defined by

$$f_0(S) = \begin{cases} f(p) & \text{if } S \cong C_p \in \mathfrak{X}, \\ \mathfrak{F} & \text{if } S \in \mathfrak{X}'. \end{cases}$$

By Theorem 3.2.14, $\mathfrak{F}$ is $\mathfrak{X}$-saturated.

Conversely, suppose that $\mathfrak{F}$ is an $\mathfrak{X}$-saturated formation. Then, by Theorem 3.2.14, $\mathfrak{F} = \mathrm{LF}_{\mathfrak{X}}(F)$, where F is the canonical $\mathfrak{X}$-local definition of $\mathfrak{F}$. We define a $C_{\mathfrak{Y}}$-satellite f by the following formula:

$$f(S) = \begin{cases} F(p) & \text{if } S \cong C_p \in \mathfrak{X}, \\ \mathfrak{F} & \text{if } S \in \mathfrak{X}'. \end{cases}$$

Then $\mathfrak{F} = \mathrm{CF}_{\mathfrak{Y}}(f)$. $\square$

Assume that $\mathfrak{X} \subseteq \mathbb{P}$, then $\mathfrak{F}$ is $\mathfrak{X}$-saturated if and only if $\mathfrak{F}$ is (C_p)-saturated for all $p \in \mathrm{char}\,\mathfrak{X}$ by Theorem 3.2.14, Corollary 3.1.13 and Corollary 3.1.21. Therefore we have:

Corollary 3.2.21 ([She97, Theorem 3.2], [She01, Lemma 7]). *Let $\mathfrak{F}$ be a formation, $\emptyset \neq \mathfrak{Y}$ a non-empty class of simple groups and $\pi = \mathrm{char}\,\mathfrak{Y}$. The following statements are pairwise equivalent:*

1. *$\mathfrak{F}$ is closed under extensions by the Frattini subgroup of a normal soluble π-subgroup.*
2. *$\mathfrak{F}$ contains each group G provided that $\mathfrak{F}$ contains $G/\Phi(\mathrm{F}(G)_\pi)$, where $\mathrm{F}(G)_\pi$ is the Hall π-subgroup of the Fitting subgroup of G.*
3. *A group G belongs to $\mathfrak{F}$ if and only if $G/\Phi(\mathrm{O}_p(G))$ belongs to $\mathfrak{F}$ for all $p \in \pi$.*
4. *$\mathfrak{F}$ is a $\mathfrak{Y}$-composition formation.*

When $\mathfrak{Y} = \mathbb{P}$, the class of all abelian simple groups, we have:

Corollary 3.2.22 ([För84a, Korollar 3.11]). *Let $\mathfrak{F}$ be a formation. The following statements are pairwise equivalent:*

1. *$\mathfrak{F}$ is solubly saturated.*
2. *A group G belongs to $\mathfrak{F}$ if and only if $G/\Phi(\mathrm{F}(G)) \in \mathfrak{F}$.*
3. *$\mathfrak{F}$ contains a group G provided that $\mathfrak{F}$ contains $G/\Phi(\mathrm{O}_p(G))$ for every prime p.*

Final remark 3.2.23. In the sequel we make use of the fact that the concepts of "$\mathfrak{X}$-saturated formation" and "$\mathfrak{X}$-local formation" are equivalent without appealing to Theorem 3.2.14.

3.3 Products of $\mathfrak{X}$-local formations

As a point of departure, consider the following observations: if $\mathfrak{F}$ and $\mathfrak{G}$ are saturated formations, then the formation product $\mathfrak{F} \circ \mathfrak{G}$ is again saturated ([DH92, IV, 3.13 and 4.8]). However, the formation product of two solubly saturated formations is not solubly saturated in general as the following example shows.

Example 3.3.1 ([Sal85]). Let $\mathfrak{F} = \mathrm{D}_0(1, \mathrm{Alt}(5))$ and $\mathfrak{G} = \mathfrak{S}_2$. Then it is clear that $\mathfrak{F}$ and $\mathfrak{G}$ are solubly saturated. Assume that $\mathfrak{H} = \mathfrak{F} \circ \mathfrak{G}$ is solubly saturated. Then $\mathfrak{H} = \mathrm{LF}_{\mathbb{P}}(H)$, where H is the canonical $\mathbb{P}$-local definition of $\mathfrak{H}$. Since $\mathfrak{G} \subseteq \mathfrak{H}$, it follows that $H(2) \neq \emptyset$. Consider $G = \mathrm{SL}(2,5)$. Then $G/\mathrm{Z}(G) \in \mathfrak{H}$ and $G/\mathrm{C}_G(\mathrm{Z}(G)) \in H(2)$. Applying Remark 3.1.7 (2), we have that $G \in \mathfrak{H}$. This is not true. Hence $\mathfrak{H}$ is not solubly saturated.

Taking the above example into account, the following question arises:

Which are the precise conditions on two $\mathfrak{X}$-local formations $\mathfrak{F}$ and $\mathfrak{G}$ to ensure that $\mathfrak{F} \circ \mathfrak{G}$ is an $\mathfrak{X}$-local formation?

The problem of the existence of solubly saturated factorisations of solubly saturated formations was taken up by Salomon [Sal85]. A complete answer to the general question was obtained in [BBCER06].

In the first part of the section we are concerned with the above question. We stay close to the treatment presented in [BBCER06].

In the following $\mathfrak{F}$ and $\mathfrak{G}$ are formations and $\mathfrak{H} = \mathfrak{F} \circ \mathfrak{G}$

If $p \in \mathrm{char}\,\mathfrak{X}$, denote

$$G_{\mathfrak{X}}(p) = \mathfrak{S}_p \, \mathrm{Q}\, \mathrm{R}_0 \big(G / \mathrm{C}_G(H/K) : G \in \mathfrak{G} \text{ and }$$

$$H/K \text{ is an } \mathfrak{X}_p\text{-chief factor of } G \big).$$

By Theorem 3.1.11, the smallest $\mathfrak{X}$-local formation $\mathrm{form}_{\mathfrak{X}}(\mathfrak{G})$ containing $\mathfrak{G}$ is $\mathfrak{X}$-locally defined by the $\mathfrak{X}$-formation function G given by $G(p) = G_{\mathfrak{X}}(p)$, $p \in \mathrm{char}\,\mathfrak{X}$, and $G(S) = \mathfrak{F}$ for every $S \in \mathfrak{X}'$.

The next theorem provides an $\mathfrak{X}$-local definition of $\mathrm{form}_{\mathfrak{X}}(\mathfrak{H})$.

Theorem 3.3.2. *Assume that $\mathfrak{F}$ is an $\mathfrak{X}$-local formation defined by an integrated $\mathfrak{X}$-formation function f. Then the smallest $\mathfrak{X}$-local formation $\mathrm{form}_{\mathfrak{X}}(\mathfrak{H})$ containing $\mathfrak{H}$ is $\mathfrak{X}$-locally defined by the $\mathfrak{X}$-formation function h given by*

$$h(p) = \begin{cases} f(p) \circ \mathfrak{G} & \text{if } \mathfrak{S}_p \subseteq \mathfrak{F} \\ G_{\mathfrak{X}}(p) & \text{if } \mathfrak{S}_p \not\subseteq \mathfrak{F} \end{cases} \qquad p \in \mathrm{char}\,\mathfrak{X}$$

$$h(S) = \mathfrak{H} \qquad\qquad\qquad\qquad\qquad S \in \mathfrak{X}'$$

Proof. It is clear that h is an $\mathfrak{X}$-formation function. We set $\bar{\mathfrak{H}} = \mathrm{LF}_{\mathfrak{X}}(h)$ and first prove that $\mathfrak{H} \subseteq \bar{\mathfrak{H}}$. Assume that $\mathfrak{H} \setminus \bar{\mathfrak{H}}$ contains a group G of minimal order. Then G has a unique minimal normal subgroup N and $G/N \in \bar{\mathfrak{H}}$. Let $A = G^{\mathfrak{G}} \trianglelefteq G$. If $A = 1$, then $G \in \mathfrak{G} \subseteq \bar{\mathfrak{H}}$, contrary to supposition. Therefore N is contained in A. If N were an $\mathfrak{X}'$-chief factor of G, since $G/N \in \bar{\mathfrak{H}}$, G would satisfy the first condition to belong to $\bar{\mathfrak{H}}$. Since $G \in \mathfrak{H}$, the second condition would also be satisfied, bearing in mind that $h(S) = \mathfrak{H}$ for every simple group $S \in \mathfrak{X}'$. This would imply that $G \in \bar{\mathfrak{H}}$. Hence $N \in \mathrm{E}\,\mathfrak{X}$. Applying [DH92, A, 4.13], $N = N_1 \times \cdots \times N_n$, where N_i is a minimal normal subgroup of A, $1 \le i \le n$. Since $A \in \mathfrak{F}$, it follows that $f(p) \ne \emptyset$ for each prime p dividing $|N|$. Moreover $A / \mathrm{C}_N(N_i) \in f(p)$, for all $i \in \{1, \ldots, n\}$, and $p \mid |N|$. Consequently $\big(G / \mathrm{C}_G(N) \big)^{\mathfrak{G}} \cong A / \mathrm{C}_A(N) \in \mathrm{R}_0 \, f(p) = f(p)$ and so $G / \mathrm{C}_G(N) \in f(p) \circ \mathfrak{G} = h(p)$ for all $p \mid |N|$. Hence, applying Remark 3.1.7 (2), we have that $G \in \bar{\mathfrak{H}}$. This contradiction proves that $\mathfrak{H} \subseteq \bar{\mathfrak{H}}$. Since $\bar{\mathfrak{H}}$ is $\mathfrak{X}$-local, it follows that $\mathrm{form}_{\mathfrak{X}}(\mathfrak{H}) \subseteq \bar{\mathfrak{H}}$.

On the other hand, we know by Theorem 3.1.17 that $\mathrm{form}_{\mathfrak{X}}(\mathfrak{H}) = \mathrm{LF}_{\mathfrak{X}}(H)$, where H is the $\mathfrak{X}$-formation function defined by

$$\begin{cases} H(p) = H_{\mathfrak{X}}(p) & \text{if } p \in \mathrm{char}(\mathfrak{X}) \\ H(E) = \mathfrak{H} & \text{if } E \in \mathfrak{X}' \end{cases}$$

Suppose that $\bar{\mathfrak{H}}$ is not contained in $\mathrm{form}_{\mathfrak{X}}(\mathfrak{H})$ and choose a group $Z \in \bar{\mathfrak{H}} \setminus \mathrm{form}_{\mathfrak{X}}(\mathfrak{H})$ of minimal order. Then Z has a unique minimal normal subgroup N and $Z/N \in \mathrm{form}_{\mathfrak{X}}(\mathfrak{H})$. Moreover it is clear that $N \in \mathrm{E}\,\mathfrak{X}$. Let p be a prime dividing $|N|$. If $\mathfrak{S}_p \not\subseteq \mathfrak{F}$, then $h(p) = G_{\mathfrak{X}}(p)$. Since $Z \in \bar{\mathfrak{H}}$, we have that $Z/\mathrm{C}_Z(N) \in G_{\mathfrak{X}}(p) \subseteq H(p)$. Assume we are in the case $\mathfrak{S}_p \subseteq \mathfrak{F}$. Then $Z/\mathrm{C}_Z(N) \in h(p) = f(p) \circ \mathfrak{G}$ and $C_{p'}\big(Z/\mathrm{C}_Z(N)\big) \in \mathfrak{S}_p\big(f(p) \circ \mathfrak{G}\big) \subseteq \mathfrak{S}_p f(p) \circ \mathfrak{G}$. By Theorem 3.1.17, we know that $\mathfrak{S}_p f(p) \subseteq \mathfrak{F}$ and, hence, $C_{p'}\big(Z/\mathrm{C}_Z(N)\big) \in \mathfrak{F} \circ \mathfrak{G} \subseteq \mathrm{form}_{\mathfrak{X}}(\mathfrak{H})$. This implies that $Z/\mathrm{C}_Z(N) \in H_{\mathfrak{X}}(p) = H(p)$ by Theorem 3.1.17. Applying Remark 3.1.7 (2), we can conclude that $Z \in \mathrm{form}_{\mathfrak{X}}(\mathfrak{H})$. This contradiction shows that $\bar{\mathfrak{H}} \subseteq \mathrm{form}_{\mathfrak{X}}(\mathfrak{H})$ and, hence, $\bar{\mathfrak{H}} = \mathrm{form}_{\mathfrak{X}}(\mathfrak{H})$.   □

The following definition was introduced in [Sal85] for Baer-local formations.

Definition 3.3.3. *We say that the boundary* $\mathrm{b}(\mathfrak{H})$ *is* $\mathfrak{X}\mathfrak{G}$-*free if every group* $G \in \mathrm{b}(\mathfrak{H})$ *such that* $\mathrm{Soc}(G)$ *is a* p-*group for some prime* $p \in \mathrm{char}\,\mathfrak{X}$ *satisfies that* $G/\mathrm{C}_G\big(\mathrm{Soc}(G)\big) \notin G_{\mathfrak{X}}(p)$.

Remark 3.3.4. Note that in Example 3.3.1, $\mathrm{b}(\mathfrak{H})$ is not $\mathbb{P}\mathfrak{G}$-free.

The next result provides a test for $\mathfrak{X}$-locality of $\mathfrak{H}$ in terms of its boundary.

Theorem 3.3.5. *Assume that* $\mathfrak{F}$ *is* $\mathfrak{X}$-*local. Then* $\mathfrak{H}$ *is an* $\mathfrak{X}$-*local formation if and only if* $\mathrm{b}(\mathfrak{H})$ *is* $\mathfrak{X}\mathfrak{G}$-*free.*

Proof. Suppose that $\mathfrak{H}$ is $\mathfrak{X}$-local. Then $\mathfrak{H} = \mathrm{LF}_{\mathfrak{X}}(H)$, where H is the canonical $\mathfrak{X}$-local definition of $\mathfrak{H}$. Let G be a group in $\mathrm{b}(\mathfrak{H})$ such that $\mathrm{Soc}(G)$ is a p-group for some $p \in \mathrm{char}\,\mathfrak{X}$. If $G/\mathrm{C}_G\big(\mathrm{Soc}(G)\big)$ were in $G_{\mathfrak{X}}(p)$, then we would have that $G/\mathrm{C}_G\big(\mathrm{Soc}(G)\big) \in H_{\mathfrak{X}}(p) = H(p)$, since $\mathfrak{G} \subseteq \mathfrak{H}$. By Remark 3.1.7 (2), it would imply that $G \in \mathfrak{H}$. This would be a contradiction. Therefore $G/\mathrm{C}_G\big(\mathrm{Soc}(G)\big) \notin G_{\mathfrak{X}}(p)$ and $\mathrm{b}(\mathfrak{H})$ is $\mathfrak{X}\mathfrak{G}$-free.

Conversely, suppose that $\mathrm{b}(\mathfrak{H})$ is $\mathfrak{X}\mathfrak{G}$-free. Consider an integrated $\mathfrak{X}$-local definition f of $\mathfrak{F}$. By Theorem 3.3.2, $\mathrm{form}_{\mathfrak{X}}(\mathfrak{H}) = \mathrm{LF}_{\mathfrak{X}}(h)$, where

$$h(p) = \begin{cases} f(p) \circ \mathfrak{G} & \text{if } \mathfrak{S}_p \subseteq \mathfrak{F} \\ G_{\mathfrak{X}}(p) & \text{if } \mathfrak{S}_p \not\subseteq \mathfrak{F} \end{cases} \qquad p \in \mathrm{char}\,\mathfrak{X}$$

$$h(S) = \mathfrak{H} \qquad\qquad\qquad\qquad S \in \mathfrak{X}'$$

We shall prove that $\mathfrak{H} = \mathrm{form}_{\mathfrak{X}}(\mathfrak{H})$. Assume that this is not the case and choose a group G of minimal order in $\mathrm{form}_{\mathfrak{X}}(\mathfrak{H}) \setminus \mathfrak{H}$. Then $G \in \mathrm{b}(\mathfrak{H})$ and so G has a unique minimal normal subgroup, N say, and $G/N \in \mathfrak{H}$. If N were an $\mathfrak{X}'$-group, we would have that $G \in h(S)$ for some $S \in \mathfrak{X}'$. This would imply that $G \in \mathfrak{H}$, contrary to supposition. Hence N is an $\mathfrak{X}$-chief factor of G. Let p be a prime dividing $|N|$. Since $p \in \mathrm{char}\,\mathfrak{X}$, it follows that $G/\mathrm{C}_G(N) \in h(p)$. Since $h(p) \subseteq \mathfrak{S}_p\mathfrak{H}$ and $\mathrm{O}_p\big(G/\mathrm{C}_G(N)\big) = 1$, we have that $G/\mathrm{C}_G(N) \in \mathfrak{H}$. Therefore $\mathrm{C}_G(N) \neq 1$ and so N is an abelian p-group.

Assume that $\mathfrak{S}_p$ is not contained in $\mathfrak{F}$. Then $h(p) = G_{\mathfrak{X}}(p)$. We conclude that $\mathrm{b}(\mathfrak{H})$ is not $\mathfrak{X}\mathfrak{G}$-free. This contradiction shows that $\mathfrak{S}_p$ is contained in $\mathfrak{F}$. Then $G/\mathrm{C}_G(N) \in f(p) \circ \mathfrak{G}$. It follows that $G^{\mathfrak{G}}/\mathrm{C}_{G^{\mathfrak{G}}}(N) \in f(p)$. Since $G^{\mathfrak{G}}/N \in \mathfrak{F}$, we can apply Remark 3.1.7 (2) to conclude that $G^{\mathfrak{G}} \in \mathfrak{F}$, that is, $G \in \mathfrak{H}$. This contradiction shows that $\mathrm{form}_{\mathfrak{X}}(\mathfrak{H})$ is contained in $\mathfrak{H}$ and, therefore, $\mathfrak{H}$ is $\mathfrak{X}$-local. $\square$

Example 3.3.6. Let S be a non-abelian simple group with trivial Schur multiplier. Consider $\mathfrak{F} = \mathrm{D}_0(1, S)$, the formation of all groups which are a direct product of copies of S together with the trivial group. Let $\mathfrak{X}$ be a class of simple groups such that $S \notin \mathfrak{X}$. Notice that $\mathfrak{F}$ is $\mathfrak{X}$-local. Let $\mathfrak{G}$ be any formation. Suppose that $G \in \mathrm{b}(\mathfrak{H})$, $N = \mathrm{Soc}(G)$ is the minimal normal subgroup of G, and N is a p-group for some $p \in \mathrm{char}\,\mathfrak{X}$. If $G/\mathrm{C}_G(N) \in G_{\mathfrak{X}}(p)$, then $N \leq \mathrm{Z}(G^{\mathfrak{G}})$ because $1 \neq G^{\mathfrak{G}} \leq \mathrm{C}_G(N)$. Since $G/N \in \mathfrak{H}$, it follows that $G^{\mathfrak{G}}/N \in \mathfrak{F}$. Assume that $G^{\mathfrak{G}}/N \neq 1$. This implies that $G^{\mathfrak{G}}/N$, a direct product of copies of S, has non-trivial Schur multiplier, contrary to [Suz82, Exercise 4 (c), page 265]. Thus $G^{\mathfrak{G}} = N$ and then $G \in \mathrm{form}_{\mathfrak{X}}(\mathfrak{H})$ by Remark 3.1.7 (2). Therefore if $\mathrm{form}_{\mathfrak{X}}(\mathfrak{G}) \subseteq \mathfrak{N}_{p'}\mathfrak{G}$ for all primes $p \in \mathrm{char}(\mathfrak{X})$, it follows that $G \in \mathfrak{G}$, and this contradicts our choice if G. Hence $\mathrm{b}(\mathfrak{H})$ is $\mathfrak{X}\mathfrak{G}$-free and $\mathfrak{H}$ is $\mathfrak{X}$-local by Theorem 3.3.5. Consequently, $\mathfrak{H}$ is $\mathfrak{X}$-local for all formations $\mathfrak{G}$ satisfying $\mathrm{form}_{\mathfrak{X}}(\mathfrak{G}) \subseteq \mathfrak{N}_{p'}\mathfrak{G}$ for all primes $p \in \mathrm{char}(\mathfrak{X})$.

As an application of Theorem 3.3.5 we have:

Theorem 3.3.7. *Assume that $\mathfrak{F}$ is $\mathfrak{X}$-local and $\mathfrak{G}$ is a formation satisfying one of the following conditions:*

1. $\mathfrak{G}$ is $\mathfrak{X}$-local, or
2. $\mathfrak{S}_p\mathfrak{G} = \mathfrak{G}$ for all $p \in \mathrm{char}\,\mathfrak{X} \setminus \mathrm{char}\,\mathfrak{F}$.

Then $\mathfrak{H}$ is $\mathfrak{X}$-local if $\mathfrak{F}$ and $\mathfrak{G}$ satisfy the following condition:

$$\text{If } p \in \mathrm{char}\,\mathfrak{X} \cap \pi(\mathfrak{F}) \text{ and } \mathfrak{S}_p \subseteq \mathfrak{G}, \text{ then } \mathfrak{S}_p \subseteq \mathfrak{F}. \tag{3.2}$$

Proof. Consider the canonical $\mathfrak{X}$-local definition F of $\mathfrak{F}$. We will obtain a contradiction by assuming that $\mathfrak{H}$ is not $\mathfrak{X}$-local. Then, by Theorem 3.3.5, there exists a group $G \in \mathrm{b}(\mathfrak{H})$ such that $N = \mathrm{Soc}(G)$ is the unique minimal normal subgroup of G, N is a p-group for some prime $p \in \mathrm{char}\,\mathfrak{X}$ and $G/\mathrm{C}_G(N) \in G_{\mathfrak{X}}(p)$. Since $G_{\mathfrak{X}}(p) \subseteq \mathfrak{S}_p\mathfrak{G}$ and $\mathrm{O}_p(G/\mathrm{C}_G(N)) = 1$, it follows that $G/\mathrm{C}_G(N) \in \mathfrak{G}$. Then $G^{\mathfrak{G}} \leq \mathrm{C}_G(N)$. Since $G^{\mathfrak{G}} \neq 1$, it follows that $N \leq G^{\mathfrak{G}}$. Hence $N \leq \mathrm{Z}(G^{\mathfrak{G}})$. Moreover $G^{\mathfrak{G}}/N \in \mathfrak{F}$ because $G/N \in \mathfrak{H}$. Suppose that N is not contained in $\Phi(G^{\mathfrak{G}})$. Then there exists a maximal subgroup M of $G^{\mathfrak{G}}$ such that $G^{\mathfrak{G}} = MN$. Notice that M is normal in $G^{\mathfrak{G}}$. Then $\mathrm{O}^p(G^{\mathfrak{G}})$ is contained in M and is a normal subgroup of G. If $\mathrm{O}^p(G^{\mathfrak{G}}) \neq 1$, it follows that $N \leq \mathrm{O}^p(G^{\mathfrak{G}}) \leq M$. This contradiction proves that $G^{\mathfrak{G}}$ is a p-group. Assume that $p \notin \mathrm{char}\,\mathfrak{F}$. In this case, since $G^{\mathfrak{G}}/N \in \mathfrak{F}$, it follows that $N = G^{\mathfrak{G}}$. This means that $G/N \in \mathfrak{G}$. If $\mathfrak{G}$ is $\mathfrak{X}$-local, we conclude that $G \in \mathfrak{G}$ by Remark 3.1.7 (2). If $\mathfrak{G}$ is not $\mathfrak{X}$-local, we have $G \in \mathfrak{S}_p\mathfrak{G} = \mathfrak{G}$ because $p \notin \mathrm{char}\,\mathfrak{F}$.

In both cases, we reach a contradiction. Hence we have that $p \in$ char $\mathfrak{F}$. In this case $F(p) \neq \emptyset$. In particular, $\mathfrak{S}_p \subseteq \mathfrak{F}$ as $\mathfrak{F}$ is $\mathfrak{X}$-local. Therefore $G^{\mathfrak{G}} \in \mathfrak{F}$. This contradiction proves that N is contained in $\Phi(G^{\mathfrak{G}})$. This implies that p divides $|G^{\mathfrak{G}}/N|$ and so $p \in \pi(\mathfrak{F})$. If $p \in$ char $\mathfrak{F}$, then $F(p) \neq \emptyset$ and $G^{\mathfrak{G}} \in \mathfrak{F}$ as $\mathfrak{F}$ is $\mathfrak{X}$-local and Remark 3.1.7 (2) can be applied. Suppose that $p \notin$ char $\mathfrak{F}$. If $\mathfrak{G}$ is $\mathfrak{X}$-local, we have that $\mathfrak{S}_p \subseteq \mathfrak{G}$ because $G_{\mathfrak{X}}(p) \neq \emptyset$. The same holds if $\mathfrak{S}_p \mathfrak{G} = \mathfrak{G}$. Hence if $p \notin$ char $\mathfrak{F}$, we have that $\mathfrak{S}_p$ is contained in $\mathfrak{G}$. By Condition (3.2), $\mathfrak{S}_p \subseteq \mathfrak{F}$. This contradiction completes the proof. $\qquad \square$

Since local formations are $\mathfrak{X}$-local for every class of simple groups $\mathfrak{X}$ (see Corollary 3.1.13), we obtain as a special case of Theorem 3.3.7 the following results:

Corollary 3.3.8. *Suppose that either of the following conditions is fulfilled:*

1. *$\mathfrak{F}$ is local and $\mathfrak{G}$ is $\mathfrak{X}$-local.*
2. *$\mathfrak{F}$ is local and $\mathfrak{S}_p \mathfrak{G} = \mathfrak{G}$ for all $p \in$ char $\mathfrak{X} \setminus$ char $\mathfrak{F}$.*

Then $\mathfrak{H}$ is an $\mathfrak{X}$-local formation.

Proof. If $\mathfrak{F}$ is local, then condition (3.2) in Theorem 3.3.7 is satisfied, since $\mathfrak{S}_p \subseteq \mathfrak{F}$ for every $p \in \pi(\mathfrak{F})$. $\qquad \square$

Corollary 3.3.9 ([DH92, IV, 3.13 and 4.8]). *$\mathfrak{H}$ is a local formation if either of the following conditions is satisfied:*

1. *$\mathfrak{F}$ and $\mathfrak{G}$ are both local.*
2. *$\mathfrak{F}$ is local and $\mathfrak{S}_p \mathfrak{G} = \mathfrak{G}$ for all $p \notin$ char $\mathfrak{F}$.*

Example 3.3.6 shows that there are cases in which a product of an $\mathfrak{X}$-local formation and a non $\mathfrak{X}$-local formation is $\mathfrak{X}$-local. This observation leads to the following question:

Are there $\mathfrak{X}$-local products of non $\mathfrak{X}$-local formations?

The local version of the above question is the one appearing in *The Kourovka Notebook* ([MK90]) as Question 9.58. It was posed by L. A. Shemetkov and A. N. Skiba and answered affirmatively in several papers (see [BBPR98, Ved88, Vor93]).

The next example gives a positive answer to the above question when $|$char $\mathfrak{X}| \geq 2$.

Example 3.3.10 ([BBPR98]). Assume that p and q are different primes in char $\mathfrak{X}$. Consider the formations $\mathfrak{F} = \mathfrak{S}_p \mathfrak{A}_q \cap \mathfrak{A}_q \mathfrak{S}_p$ and $\mathfrak{G} = \mathfrak{S}_q \mathfrak{A}_p$, where $\mathfrak{A}_r$ denotes the formation of all abelian r-groups for a prime r. $\mathfrak{F}$ is not (C_q)-local and $\mathfrak{G}$ is not (C_p)-local. Therefore, by Corollary 3.1.13, $\mathfrak{F}$ and $\mathfrak{G}$ are not $\mathfrak{X}$-local. However $\mathfrak{H} = \mathfrak{F} \circ \mathfrak{G}$ is local and so it is $\mathfrak{X}$-local.

Note that if the formation of all p-groups, p a prime, were a product of two proper subformations, Question 9.58 in [MK90] would be solved automatically. Perhaps it was the reason to put forward the following question in *The Kourovka Notebook* [MK90]:

> **Question 10.72 (Shemetkov).** *To prove indecomposability of $\mathfrak{S}_p$, p a prime, into a product of two non-trivial subformations.*

This question was solved positively by L. A. Shemetkov and A. N. Skiba in [SS89].

We present a general version of this conjecture as a corollary of a more general result at the end of the section.

On the other hand, bearing in mind Example 3.3.10, the following question naturally arises:

> Which are the precise conditions on two formations $\mathfrak{F}$ and $\mathfrak{G}$ to ensure that $\mathfrak{H} = \mathfrak{F} \circ \mathfrak{G}$ is X-local?

Our next results answer this question.

Notation 3.3.11. If $\mathfrak{Y}$ is a class of groups, denote $\mathfrak{Y}^{\mathfrak{G}} = (Y^{\mathfrak{G}} : Y \in \mathfrak{Y})$.

Lemma 3.3.12. *If T is a group such that $T \notin \mathfrak{G}$ and $\mathfrak{S}_p(T) \subseteq \mathfrak{H}$ for some prime p, then $\mathfrak{S}_p(T^{\mathfrak{G}}) \subseteq \mathfrak{F}$.*

Proof. Let Z be a group in $\mathfrak{S}_p(T^{\mathfrak{G}})$. Then Z has a normal subgroup P such that P is a p-group and Z/P is isomorphic to $T^{\mathfrak{G}} \neq 1$. Assume that p^s is the exponent of the abelian p-group P/P'. Consider $Q = P \wr_{\mathrm{nat}} H$, where $H = \langle (1, 2, \ldots, p^s) \rangle$ is a cyclic group of order p^s regarded as a subgroup of the symmetric group of degree p^s. Here the wreath product is taken with respect to the natural permutation representation of H of degree p^s. Set $D = \{(a, \ldots, a) : a \in P\}$ the diagonal subgroup of $P^{\natural}$, the base group of Q. Since $a^{p^s} \in P'$, we have that D is contained in $[P^{\natural}, H]$ by [DH92, A, 18.4]. In particular D is contained in Q'. Let $Y = Q \wr T$ be the regular wreath product of Q with T. Since $Q \in \mathfrak{S}_p(T) \subseteq \mathfrak{H}$, it follows that $Q \in \mathfrak{H}$. Therefore $Y^{\mathfrak{G}} \in \mathfrak{F}$. Now, by Proposition 2.2.8, we know that $Y^{\mathfrak{G}} = (B \cap Y^{\mathfrak{G}})T^{\mathfrak{G}}$, where $B = Q^{\natural}$ is the base group of Y. Now, by [DH92, A, 18.8], $BT^{\mathfrak{G}}$ is isomorphic to $(Q^n) \wr T^{\mathfrak{G}}$, where $n = |T : T^{\mathfrak{G}}|$ and $C' \leq [C, T^{\mathfrak{G}}]$, for $C = (Q^n)^{\natural}$, by virtue of [DH92, A, 18.4]. This implies that $B' = [B, T^{\mathfrak{G}}] \leq [B, Y^{\mathfrak{G}}] \leq B \cap Y^{\mathfrak{G}}$. Hence $B'T^{\mathfrak{G}}$ is contained in $Y^{\mathfrak{G}}$. Applying Theorem 2.2.6, $B'T^{\mathfrak{G}} \in \mathfrak{F}$. Therefore $(Q')^n) \wr T^{\mathfrak{G}} \in \mathfrak{F}$. Since P is isomorphic to a subgroup of Q', it follows that $P^n \wr T^{\mathfrak{G}} \in \mathfrak{F}$ by Theorem 2.2.6. Since P can be regarded as a subgroup of P^n, we have that $P \wr T^{\mathfrak{G}}$ is a subgroup of $P^n \wr T^{\mathfrak{G}}$ supplementing the Fitting subgroup of $P^n \wr T^{\mathfrak{G}}$. Applying again Theorem 2.2.6, we have that $P \wr T^{\mathfrak{G}} \in \mathfrak{F}$. By [DH92, A, 18.9], Z is isomorphic to a subgroup of $P \wr T^{\mathfrak{G}}$ supplementing the Fitting subgroup of $P \wr T^{\mathfrak{G}}$. Therefore $Z \in \mathfrak{F}$ by virtue of Theorem 2.2.6. This completes the proof of the lemma. $\qquad\square$

Theorem 3.3.13. $\mathfrak{H}$ *is an $\mathfrak{X}$-local formation if and only if the following two conditions hold:*

1. *If $p \in (\mathrm{char}\,\mathfrak{X}) \cap \big(\mathrm{char}\,\mathrm{form}_{\mathfrak{X}}(\mathfrak{H})\big)$ and $H_{\mathfrak{X}}(p)$ is not contained in $\mathfrak{G}$, then $\mathfrak{S}_p H_{\mathfrak{X}}(p)^{\mathfrak{G}} \subseteq \mathfrak{F}$.*
2. *If $p \in (\mathrm{char}\,\mathfrak{X}) \cap \big(\mathrm{char}\,\mathrm{form}_{\mathfrak{X}}(\mathfrak{H})\big)$, $G \in \mathrm{b}(\mathfrak{H})$, and $N = \mathrm{Soc}(G)$ is a p-group, then $[N](G/N) \notin \mathfrak{H}$.*

Proof. Assume that $\mathfrak{H}$ is an $\mathfrak{X}$-local formation, that is, $\mathfrak{H} = \mathrm{form}_{\mathfrak{X}}(\mathfrak{H})$. We know that $\mathfrak{H} = \mathrm{LF}_{\mathfrak{X}}(H)$, where H is the $\mathfrak{X}$-formation function defined in Theorem 3.1.17. Consider a prime $p \in \mathrm{char}(\mathfrak{X})$ and assume there exists a group $T \in H_{\mathfrak{X}}(p) \setminus \mathfrak{G}$. We have that $\mathfrak{S}_p(T) \subseteq \mathfrak{S}_p H_{\mathfrak{X}}(p) = H_{\mathfrak{X}}(p) \subseteq \mathfrak{H}$. Hence, by Lemma 3.3.12, we have that $\mathfrak{S}_p(T^{\mathfrak{G}}) \subseteq \mathfrak{F}$. Now consider a group G in $\mathfrak{S}_p H_{\mathfrak{X}}(p)^{\mathfrak{G}}$. Then G has a normal p-subgroup N such that $G/N \cong \bar{T}^{\mathfrak{G}}$, where $\bar{T} \in H_{\mathfrak{X}}(p)$. If $\bar{T}^{\mathfrak{G}} \neq 1$, we have just proved that $\mathfrak{S}_p(\bar{T}^{\mathfrak{G}}) \subseteq \mathfrak{F}$ and, therefore, $G \in \mathfrak{F}$. If $\bar{T}^{\mathfrak{G}} = 1$, then $G \in \mathfrak{S}_p$. Consider the group $A := G \times T^{\mathfrak{G}}$. We have that $A \in \mathfrak{S}_p(T^{\mathfrak{G}}) \subseteq \mathfrak{F}$ and, therefore, $G \in \mathrm{Q}(\mathfrak{F}) = \mathfrak{F}$. We conclude that $\mathfrak{S}_p H_{\mathfrak{X}}(p)^{\mathfrak{G}} \subseteq \mathfrak{F}$ and Statement 1 holds.

Let G be a group in $\mathrm{b}(\mathfrak{H})$ such that $N = \mathrm{Soc}(G)$ is a p-group for a prime $p \in (\mathrm{char}\,\mathfrak{X}) \cap \big(\mathrm{char}\,\mathrm{form}_{\mathfrak{X}}(\mathfrak{H})\big)$. Note that N is a minimal normal subgroup of G. If $H := [N](G/N) \in \mathfrak{H}$, we would have that $H/\mathrm{C}_H(N) \in H_{\mathfrak{X}}(p)$ and, therefore, $G/\mathrm{C}_G(N) \in H_{\mathfrak{X}}(p)$. Since $G/N \in \mathfrak{H}$, this would imply by Remark 3.1.7 (2) that $G \in \mathrm{LF}_{\mathfrak{X}}(H) = \mathfrak{H}$. This contradiction proves Condition 2.

To prove the sufficiency, assume that $\mathfrak{H}$ is the product of $\mathfrak{F}$ and $\mathfrak{G}$ and $\mathfrak{H}$ satisfies Conditions 1 and 2. We will obtain a contradiction by supposing that $\mathrm{form}_{\mathfrak{X}}(\mathfrak{H}) \setminus \mathfrak{H}$ contains a group G of minimal order. Such a G has a unique minimal normal subgroup, N, and $G/N \in \mathfrak{H}$. This is to say that $G \in \mathrm{b}(\mathfrak{H})$. If $N \in \mathrm{E}(\mathfrak{X}')$, then there exists $S \in \mathfrak{X}'$ such that $G \in H(S) = \mathfrak{H}$, contrary to supposition. Therefore $N \in \mathrm{E}\mathfrak{X}$. Let p be a prime dividing $|N|$. Then $G/\mathrm{C}_G(N) \in H_{\mathfrak{X}}(p)$. In particular $p \in (\mathrm{char}\,\mathfrak{X}) \cap (\mathrm{char}\,\mathrm{form}_{\mathfrak{X}}(\mathfrak{H}))$. If N were non-abelian, then $\mathrm{C}_G(N) = 1$ and $G \in H_{\mathfrak{X}}(p)$. This would imply that $G \in \mathfrak{H}$ because $\mathrm{O}_p(G) = 1$. It would contradict the choice of G. Therefore N is an abelian p-group. Applying Corollary 2.2.5, $A = [N](G/N) \in \mathrm{form}_{\mathfrak{X}}(\mathfrak{H})$. Suppose that the intersection B of $\mathrm{C}_A(N)$ with G/N is non-trivial. Then $B \trianglelefteq A$ and $A/B \in \mathfrak{H}$ by the choice of G. Since $G/N \in \mathfrak{H}$, we have that $A \in \mathrm{R}_0 \mathfrak{H} = \mathfrak{H}$. This contradicts Statement 2. Hence $B = 1$ and $N = \mathrm{C}_G(N)$. In particular $G \in H_{\mathfrak{X}}(p) \setminus \mathfrak{G}$. Applying Statement 1, we have that $\mathfrak{S}_p H_{\mathfrak{X}}(p)^{\mathfrak{G}} \subseteq \mathfrak{F}$. We deduce then that $G^{\mathfrak{G}} \in \mathfrak{F}$ and so $G \in \mathfrak{H}$. We have reached a final contradiction. Therefore $\mathrm{form}_{\mathfrak{X}}(\mathfrak{H}) \subseteq \mathfrak{H}$ and $\mathfrak{H}$ is $\mathfrak{X}$-local. $\square$

Remark 3.3.14. If $\mathfrak{X} = \mathfrak{J}$, then Condition 1 implies Condition 2 in the above theorem.

Proof. Assume that $\mathfrak{H}$ satisfies Condition 1. Let $G \in \mathrm{b}(\mathfrak{H})$ such that $N = \mathrm{Soc}(G)$ is the unique minimal normal subgroup of G. Suppose that N is a p-group for some $p \in (\mathrm{char}\,\mathfrak{X}) \cap \big(\mathrm{char}\,\mathrm{form}_{\mathfrak{X}}(\mathfrak{H})\big)$.

Suppose that $\Phi(G) = 1$. Then G is a primitive group, $C_G(N) = N$ and G is isomorphic to $[N](G/N)$. Therefore, $[N](G/N) \notin \mathfrak{H}$ and the remark follows. Now assume that $\Phi(G) \neq 1$. Consider $T/N := O_{p'}(G/N)$. Since T/N is p-nilpotent and $N \leq \Phi(G)$, we have by [Hup67, VI, 6.3] that T is p-nilpotent. This implies that $T = N$ because otherwise we would find a non-trivial normal p'-subgroup of G. Hence, $O_{p'}(G/N) = 1$. Consequently, $G \in H_{\mathfrak{X}}(p)$ by [DH92, IV, 3.7]. By Condition 1, $\mathfrak{S}_p(G^{\mathfrak{G}}) \subseteq \mathfrak{F}$. In particular, $G^{\mathfrak{G}} \in \mathfrak{F}$. We conclude that $G \in \mathfrak{H}$, which contradicts our supposition. $\square$

Corollary 3.3.15 ([BBPR98, Theorem A]). *A formation product $\mathfrak{H}$ of two formations $\mathfrak{F}$ and $\mathfrak{G}$ is local if and only if $\mathfrak{H}$ satisfies the following condition:*

If $p \in \operatorname{char} \operatorname{lform}(\mathfrak{H})$ and $H_{\mathfrak{J}}(p)$ is not contained in $\mathfrak{G}$, then

$$\mathfrak{S}_p H_{\mathfrak{J}}(p)^{\mathfrak{G}} \subseteq \mathfrak{F}.$$

When a product is $\mathfrak{X}$-local, the formation $\mathfrak{G}$ has a very nice property.

Corollary 3.3.16. *If $\mathfrak{H} = \mathfrak{F} \circ \mathfrak{G}$ is $\mathfrak{X}$-local, then $\operatorname{form}_{\mathfrak{X}}(\mathfrak{G}) \subseteq \mathfrak{N}_{p'}\mathfrak{G}$ for all primes $p \in \operatorname{char}(\mathfrak{X}) \setminus \pi(\mathfrak{F})$.*

Proof. Let $p \in \operatorname{char}(\mathfrak{X}) \setminus \pi(\mathfrak{F})$. By Theorem 3.3.13, we have that $H_{\mathfrak{X}}(p) \subseteq \mathfrak{G}$. Consider the canonical $\mathfrak{X}$-formation function G defining $\operatorname{form}_{\mathfrak{X}}(\mathfrak{G})$. Suppose that $\operatorname{form}_{\mathfrak{X}}(\mathfrak{G})$ is not contained in $\mathfrak{N}_{p'}\mathfrak{G}$, and let $G \in \operatorname{form}_{\mathfrak{X}}(\mathfrak{G}) \setminus \mathfrak{N}_{p'}\mathfrak{G}$ be a group of minimal order. Then $G \in \mathfrak{H}$ and G has a unique minimal normal subgroup, N say. In addition, $N \leq G^{\mathfrak{G}}$ and $G/N \in \mathfrak{N}_{p'}\mathfrak{G}$. If $N \in \mathrm{E}\,\mathfrak{X}'$, it follows that $G \in G(S)$ for some $S \in \mathfrak{X}'$. But, in this case, $G \in \mathfrak{G}$. This is a contradiction. Hence N is an $\mathrm{E}\,\mathfrak{X}$-group. Consider $q \in \pi(N)$. If N were non-abelian, then G would belong to $G(q) \subseteq \mathfrak{S}_q\mathfrak{G}$. Hence $G \in \mathfrak{G}$ because $O_q(G) = 1$. This would contradict our assumption. Therefore N is an elementary abelian q-group for some prime $q \in \operatorname{char} \mathfrak{X}$. Assume that $\Phi(G) = 1$. Then G is a primitive group and $N = C_G(N)$. Therefore $G \in G(q)$. If $p \neq q$, then $G \in \mathfrak{N}_{p'}\mathfrak{G}$ because $G(q) \subseteq \mathfrak{S}_q\mathfrak{G}$ and if $p = q$, then $G \in \mathfrak{S}_p H_{\mathfrak{X}}(p) = H_{\mathfrak{X}}(p) \subseteq \mathfrak{G}$. In both cases, we reach a contradiction. Hence N is contained in $\Phi(G)$. If $p \neq q$, then $F(G)$ is a p'-group and $G/F(G) \cong (G/N)/F(G/N) \in \mathfrak{G}$. Hence, $G \in \mathfrak{N}_{p'}\mathfrak{G}$, contrary to supposition. Assume that $p = q$. Then, since $G/N \in \mathfrak{N}_{p'}\mathfrak{G}$, it follows that $(G/N)^{\mathfrak{G}} = G^{\mathfrak{G}}/N$ is a p'-group. Thus $G^{\mathfrak{G}}/N$ is contained in $O_{p'}(G/N)$ which is trivial by [Hup67, VI, 6.3]. Hence $N = G^{\mathfrak{G}}$. Since $G \in \mathfrak{H}$, we have that $G^{\mathfrak{G}} = N \in \mathfrak{F}$ and $p \in \pi(\mathfrak{F})$. This final contradiction proves that $\operatorname{form}_{\mathfrak{X}}(\mathfrak{G}) \subseteq \mathfrak{N}_{p'}\mathfrak{G}$. $\square$

If $\mathfrak{X} = \mathfrak{J}$, we have:

Corollary 3.3.17 ([She84]). *If $\mathfrak{H} = \mathfrak{F} \circ \mathfrak{G}$ is local, then $\operatorname{lform}(\mathfrak{G})$ is contained in $\mathfrak{N}_{p'}\mathfrak{G}$ for all primes $p \notin \pi(\mathfrak{F})$.*

This result motivates the following definition.

Definition 3.3.18. *Let ω be a non-empty set of primes, and let $\mathfrak{F}$ be a formation.*

1. *(see [She84]) $\mathfrak{F}$ is said to be ω-local if* lform$(\mathfrak{F})$ *is contained in* $\mathfrak{N}_{\omega'}\mathfrak{F}$.
2. *(see [SS00a]) $\mathfrak{F}$ is called ω-saturated if the condition* $G/\big(\varPhi(G)\cap \mathrm{O}_\omega(G)\big) \in \mathfrak{F}$ *always implies* $G \in \mathfrak{F}$.

When $\omega = \{p\}$, we shall say p-local *(respectively, p-saturated) instead of* $\{p\}$-*local (respectively, $\{p\}$-saturated).*

Remarks 3.3.19. Let $\emptyset \neq \omega$ be a set of primes and let $\mathfrak{F}$ be a formation.

1. $\mathfrak{F}$ is ω-local if and only if $\mathfrak{F}$ is p-local for all $p \in \omega$. Hence $\mathfrak{F}$ is local if and only if $\mathfrak{F}$ is p-local for all primes p.

2. If $\mathfrak{F}$ is an ω-local formation, then $\mathfrak{F}$ is ω-saturated.

3. If $\mathfrak{F}$ is ω-saturated, then $\mathfrak{N}_{\omega'}\mathfrak{F}$ is local. Therefore every ω-saturated formation is ω-local (see [SS95]).

4. Every formation composed of ω'-groups is ω-saturated.

5. Every ω-saturated formation is $\mathfrak{X}_\omega$-saturated, where $\mathfrak{X}_\omega$ is the class of all simple ω-groups.

Proof. 1, 2, and 4 are clear. To prove Statement 3, suppose that $\mathfrak{F}$ is ω-saturated. If q is a prime such that $q \in \omega'$, then $\mathfrak{H} = \mathfrak{N}_{\omega'}\mathfrak{F}$ is q-saturated. Assume that p is a prime in ω such that $\mathfrak{H}$ is not p-saturated. Then there exists a group G such that $G/\big(\varPhi(G) \cap \mathrm{O}_p(G)\big) \in \mathfrak{H}$ but $G \notin \mathfrak{H}$. Let us choose G of least order. Then G has a unique minimal normal subgroup N, N is contained in $\varPhi(G) \cap \mathrm{O}_p(G)$, and $G/N \in \mathfrak{H}$. Since $\mathfrak{F}$ is contained in $\mathfrak{H}$, it follows that $G^\mathfrak{F} \neq 1$ and so N is also contained in $G^\mathfrak{F}$. Now $\mathrm{O}_{p'}(G/N) = 1$ and $G^\mathfrak{F}/N$ is a p'-group because $G/N \in \mathfrak{H}$. This implies that $G^\mathfrak{F} = N$. But then $G/N \in \mathfrak{F}$ and so $G \in \mathfrak{F}$ because $\mathfrak{F}$ is p-saturated. This contradiction shows that $\mathfrak{H}$ is p-saturated for all $p \in \omega$. Therefore $\mathfrak{H}$ is saturated. In particular, lform$(\mathfrak{F}) \subseteq \mathfrak{N}_{\omega'}\mathfrak{F}$ and $\mathfrak{F}$ is ω-local.

5 follows directly from the fact that $\varPhi_{\mathfrak{X}_\omega}(G) \subseteq \varPhi(G) \cap \mathrm{O}_\omega(G)$ for every group G. $\qquad\square$

The family of $\mathfrak{X}_\omega$-saturated formations does not coincide with the one of ω-saturated formations in general. This follows from the fact that there exist Baer-local formations which are not ω-saturated for any non-empty set of primes ω.

Example 3.3.20 ([BBCER03]). Let $\mathfrak{Y} = \{\mathrm{Alt}(n) : n \geq 5\}$ and $\mathfrak{F} = \mathrm{E}\,\mathfrak{Y}$. It is clear that $\mathfrak{F}$ is a Baer-local formation. In particular, $\mathfrak{F}$ is $\mathfrak{X}$-saturated for every $\mathfrak{X} \subseteq \mathbb{P}$ by Corollary 3.1.13.

Assume that $\mathfrak{F}$ is p-saturated for a prime p. If $p \geq 5$, set $k := p$; otherwise, set $k := 5$. As p divides $|\mathrm{Alt}(k)|$, there exists a group E with a normal elementary abelian p-subgroup $A \neq 1$ such that $A \leq \varPhi(E)$ and $E/A \cong \mathrm{Alt}(k)$ ([DH92, B, 11.8]). Then $E/\big(\varPhi(E) \cap \mathrm{O}_p(E)\big) \cong E/A \in \mathfrak{F}$. Therefore $E \in \mathfrak{F}$, and we have a contradiction.

This implies that $\mathfrak{F}$ is not ω-saturated for any non-empty set of primes ω. In particular, $\mathfrak{F}$ is (C_2)-saturated but not 2-saturated.

Suppose that $\mathfrak{G}$ is a p-saturated formation, p a prime. Then $\mathrm{lform}(G) \subseteq \mathfrak{N}_{p'}\mathfrak{G}$. Therefore $G(p) \subseteq \mathfrak{N}_{p'}\mathfrak{G}$ and so $G(p) = G_{\mathfrak{J}}(p) \subseteq \mathfrak{G}$. The converse is also true as the following lemma shows.

Lemma 3.3.21. $\mathfrak{G}$ *is p-saturated if and only if $G(p) \subseteq \mathfrak{G}$.*

Proof. Only the sufficiency is in doubt. Suppose that $\mathfrak{G}$ is not p-saturated and $G_{\mathfrak{J}}(p) \subseteq \mathfrak{G}$. Let G be a group of minimal order satisfying $G/\big(\Phi(G) \cap O_p(G)\big) \in \mathfrak{G}$ and $G \notin \mathfrak{G}$. G is a monolithic group and $N := \mathrm{Soc}(G) \leq \Phi(G) \cap O_p(G)$. We have that $O_{p',p}(G/N) = O_{p',p}(G)/N$, since $N \leq \Phi(G)$. Moreover, $G/N \in \mathfrak{G}$ and, therefore, $G/O_{p',p}(G) \in G_{\mathfrak{J}}(p)$, bearing in mind that $p \in \pi(G/N)$. Since $O_{p',p}(G) = O_p(G)$, $G \in G_{\mathfrak{J}}(p) \subseteq \mathfrak{G}$. This is not possible. $\quad\square$

Theorem 3.3.22. *Let $\mathfrak{F}$ and $\mathfrak{G}$ be formations. Let $\mathfrak{M}$ be a p-saturated formation contained in $\mathfrak{F} \circ \mathfrak{G}$, where p is a prime. If $M_{\mathfrak{J}}(p)$ is not contained in $\mathfrak{G}$, then $\mathfrak{S}_p M_{\mathfrak{J}}(p)^{\mathfrak{G}} \subseteq \mathfrak{F}$.*

Proof. Assume that $\mathfrak{M}$ is p-saturated. Then $M_{\mathfrak{J}}(p)$ is contained in $\mathfrak{M}$ by Lemma 3.3.21. There exists a group $T \in M_{\mathfrak{J}}(p) \setminus \mathfrak{G}$. We have that $\mathfrak{S}_p(T) \subseteq M_{\mathfrak{J}}(p) \subseteq \mathfrak{M} \subseteq \mathfrak{F} \circ \mathfrak{G}$. Hence $\mathfrak{S}_p(T^{\mathfrak{G}}) \subseteq \mathfrak{F}$ by Lemma 3.3.12.

Now consider a group G in $\mathfrak{S}_p M_{\mathfrak{J}}(p)^{\mathfrak{G}}$. Then G has a normal p-subgroup N such that $G/N \cong \bar{T}^{\mathfrak{G}}$, where $\bar{T} \in M_{\mathfrak{J}}(p)$. If $\bar{T}^{\mathfrak{G}} \neq 1$, we have just proved that $\mathfrak{S}_p(\bar{T}^{\mathfrak{G}}) \subseteq \mathfrak{F}$ and, therefore, $G \in \mathfrak{F}$. If $\bar{T}^{\mathfrak{G}} = 1$, then $G \in \mathfrak{S}_p$. Consider the group $A := G \times T^{\mathfrak{G}}$. We have that $A \in \mathfrak{S}_p(T^{\mathfrak{G}}) \subseteq \mathfrak{F}$ and, therefore, $G \in \mathrm{Q}(\mathfrak{F}) = \mathfrak{F}$. We conclude that $\mathfrak{S}_p M_{\mathfrak{J}}(p)^{\mathfrak{G}} \subseteq \mathfrak{F}$. $\quad\square$

Corollary 3.3.23. *Let $\mathfrak{F}$ and $\mathfrak{G}$ be formations and let p be a prime. Then the following statements are equivalent:*

1. *$\mathfrak{H} = \mathfrak{F} \circ \mathfrak{G}$ is a p-saturated formation.*
2. *If $H_{\mathfrak{J}}(p)$ is not contained in $\mathfrak{G}$, then $\mathfrak{S}_p H_{\mathfrak{J}}(p)^{\mathfrak{G}} \subseteq \mathfrak{F}$.*

Proof. 1 implies 2 by virtue of Theorem 3.3.22. Let us prove that 2 implies 1. We shall derive a contradiction by supposing that $H_{\mathfrak{J}}(p) \setminus \mathfrak{H}$ contains a group G of minimal order. Then G has a unique minimal normal subgroup N, and $G/N \in \mathfrak{H}$. Since $H_{\mathfrak{J}}(p)$ is contained in $\mathfrak{S}_p \mathfrak{H}$, it follows that N is a p-group. It is clear that $H_{\mathfrak{J}}(p)$ is not contained in $\mathfrak{G}$. Hence $\mathfrak{S}_p H_{\mathfrak{J}}(p)^{\mathfrak{G}} \subseteq \mathfrak{F}$. Note that $N \leq G^{\mathfrak{G}}$ and $G^{\mathfrak{G}}/N \in H_{\mathfrak{J}}(p)^{\mathfrak{G}}$. Therefore $G^{\mathfrak{G}} \in \mathfrak{S}_p H_{\mathfrak{J}}(p)^{\mathfrak{G}} \subseteq \mathfrak{F}$. This contradiction shows that $H_{\mathfrak{J}}(p) \subseteq \mathfrak{H}$ and that $\mathfrak{H}$ is p-saturated by Lemma 3.3.21. $\quad\square$

Theorem 3.3.22 also confirms a more general version of the abovementioned conjecture of L. A. Shemetkov concerning the non-decomposability of the formation of all p-groups (p a prime) as formation product of two non-trivial subformations.

Corollary 3.3.24. *Let $\mathfrak{F}$, $\mathfrak{G}$, and $\mathfrak{M}$ be formations such that $\mathfrak{M}$ is contained in $\mathfrak{F} \circ \mathfrak{G}$ and $\mathfrak{M}$ is p-saturated. If $\mathfrak{F} \subseteq \mathfrak{S}_p$ and $\mathfrak{F} \neq \mathfrak{S}_p$, then $\mathfrak{M} \subseteq \mathfrak{G}$.*

Proof. If $M_{\mathfrak{F}}(p) = \emptyset$, it follows that $\mathfrak{M} \subseteq \mathfrak{E}_{p'}$. In this case, we have that $\mathfrak{M} \subseteq \mathfrak{E}_{p'} \cap (\mathfrak{F} \circ \mathfrak{G}) \subseteq \mathfrak{E}_{p'} \cap (\mathfrak{S}_p \circ \mathfrak{G})$. Therefore, $\mathfrak{M} \subseteq \mathfrak{G}$. If $M_{\mathfrak{F}}(p) \neq \emptyset$, we have that $\mathfrak{M} \subseteq \mathfrak{E}_{p'} M_{\mathfrak{F}}(p)$. If $M_{\mathfrak{F}}(p)$ is contained in $\mathfrak{G}$, then $\mathfrak{M} \subseteq (\mathfrak{E}_{p'} M_{\mathfrak{F}}(p)) \cap (\mathfrak{S}_p \mathfrak{G}) \subseteq (\mathfrak{E}_{p'} \mathfrak{G}) \cap (\mathfrak{S}_p \mathfrak{G}) = \mathfrak{G}$ and the result holds. Suppose that $M_{\mathfrak{F}}(p)$ is not contained in $\mathfrak{G}$. Then $\mathfrak{S}_p M_{\mathfrak{F}}(p)^{\mathfrak{G}}$ is contained in $\mathfrak{F}$ by Theorem 3.3.22. In particular, $\mathfrak{S}_p \subseteq \mathfrak{F}$, and we have a contradiction. $\qquad\square$

3.4 ω-local formations

The family of ω-local formations, ω a set of primes, emerges naturally in the study of local formations that are products of two formations as it was observed in Section 3.3. There it is also proved that the ω-local formations are exactly the ones which are closed under extensions by the Hall ω-subgroup of the Frattini subgroup. In this section ω-saturated formations are studied by using a functional approach. This method was initially proposed by L. A. Shemetkov in [She84] for p-local formations, and further developed in [SS00a, SS00b, BBS97].

The second part of the section is devoted to study the relation between ω-saturated formations and $\mathfrak{X}$-local formations, where $\mathfrak{X}$ is a class of simple groups which is naturally associated with ω.

Definition 3.4.1 ([SS00a]). *Let ω be a non-empty set of primes. Every function of the form*

$$f \colon \omega \cup \{\omega'\} \longrightarrow \{formations\}$$

is called an ω-local satellite.

If f is an ω-local satellite, define the class

$$\mathrm{LF}_\omega(f) = \big(G : G/G_{\omega d} \in f(\omega') \text{ and } G/\operatorname{O}_{p',p}(G) \in f(p) \text{ for all } p \in \omega \cap \pi(G)\big),$$

where $G_{\omega d}$ is the product of all normal subgroups N of G such that every composition factor of N is divisible by at least one prime in ω ($G_{\omega d} = 1$ if $\pi(\operatorname{Soc}(G)) \cap \omega = \emptyset$).

If f is an ω-local satellite, we write $\operatorname{Supp}(f) = \{p \in \omega \cup \{\omega'\} : f(p) \neq \emptyset\}$. Denote $\pi_1 = \operatorname{Supp}(f) \cap \omega$, $\pi_2 = \omega \setminus \pi_1$. Then $\mathrm{LF}_\omega(f) = \big(\bigcap_{p \in \pi_2} \mathfrak{E}_{p'}\big) \cap \big(\bigcap_{p \in \pi_1} \mathfrak{E}_{p'} \mathfrak{S}_p \circ f(p)\big) \cap \mathfrak{E}_{\omega d} \circ f(w')$. Here $\mathfrak{E}_{\omega d}$ is the class of all groups G such that every composition factor of G is divisible by at least one prime in ω. Since the intersection and the formation product of two formations are again formations, the above formula implies that $\mathrm{LF}_\omega(f)$ is a formation.

Theorem 3.4.2 ([SS00a]). *Let ω be a non-empty set of primes and let $\mathfrak{F}$ be a formation. The following statements are equivalent:*

1. 𝔉 *is* ω-*saturated*.
2. $\mathfrak{F} = \mathrm{LF}_\omega(f)$, *where* $f(p) = \mathfrak{F}_3(p)$, $p \in \omega$, *and* $f(\omega') = \mathfrak{F}$.

Proof. 1 implies 2. It is clear that $\mathfrak{F} \subseteq \mathrm{LF}_\omega(f)$. Suppose that the equality does not hold and derive a contradiction. Choose a group $G \in \mathrm{LF}_\omega(f) \setminus \mathfrak{F}$ of minimal order. Then, as usual, G has a unique minimal normal subgroup N and $G/N \in \mathfrak{F}$. Since $G/G_{\omega d} \in f(\omega') = \mathfrak{F}$, it follows that $G_{\omega d} \neq 1$. This implies that $\pi(N) \cap \omega \neq \emptyset$. Let $p \in \omega$ be a prime dividing $|N|$. If N were non-abelian, then $G \in \mathfrak{F}_3(p)$. Since, by Lemma 3.3.21, $\mathfrak{F}_3(p) \subseteq \mathfrak{F}$, we would have $G \in \mathfrak{F}$. This would be a contradiction. Therefore N is an abelian p-group. Moreover $N \cap \Phi(G) = 1$ because $\mathfrak{F}$ is ω-saturated. Hence $N = C_G(N)$ and $G/N \in \mathfrak{F}_3(p)$. This implies that $G \in \mathfrak{S}_p\mathfrak{F}_3(p) = \mathfrak{F}_3(p)$, and we have a contradiction. Consequently $\mathfrak{F} = \mathrm{LF}_\omega(f)$.

2 implies 1. Suppose that $\mathfrak{F}$ is not ω-saturated. Then there exists a prime $p \in \omega$ and a group G such that $G/(\Phi(G) \cap O_p(G)) \in \mathfrak{F}$ but $G \notin \mathfrak{F}$. Denote $L = \Phi(G) \cap O_p(G)$. Then $(G/L)_{\omega d} = G_{\omega d}/L$ and $O_{q',q}(G/L) = O_{q',q}(G)/L$ for all primes q. Hence $G/G_{\omega d} \in f(\omega')$ and $G/O_{q',q}(G) \in f(q)$ for all $q \in \omega \cap \pi(G)$ because $G/L \in \mathfrak{F}$. Consequently $G \in \mathfrak{F}$. This contradiction completes the proof of the theorem. □

Remark 3.4.3. An ω-saturated formation can be ω-locally defined by two distinguished ω-local satellites: the minimal ω-local satellite and the canonical one. Moreover, if $\mathfrak{Y}$ is a class of groups, the intersection of all ω-local formations containing $\mathfrak{Y}$ is the smallest ω-local formation containing $\mathfrak{Y}$. Such ω-local formation is denoted by $\mathrm{lform}_\omega(\mathfrak{Y})$. It is clear that $\mathrm{lform}_\omega(\mathfrak{Y}) = \mathrm{LF}_\omega(f)$, where f is given by:

$$\begin{aligned}
f(p) &= \mathrm{Q\,R_0}\big(G/O_{p',p}(G) : G \in \mathfrak{Y}\big) && \text{if } p \in \pi(\mathfrak{Y}) \cap \omega, \\
f(p) &= \emptyset, && p \in \omega \setminus \pi(\mathfrak{Y}), \\
f(\omega') &= \mathrm{Q\,R_0}(G/G_{\omega d} : G \in \mathfrak{Y})
\end{aligned}$$

(see [SS00a] for details).

Let ω be a non-empty set of primes. One can ask the following question:

Is it possible to ensure the existence of a class $\mathfrak{X}(\omega)$ of simple groups such that $\mathrm{char}\,\mathfrak{X}(\omega) = \pi\big(\mathfrak{X}(\omega)\big)$ *satisfying that a formation is ω-saturated if and only if it is $\mathfrak{X}(\omega)$-saturated?*

The following example shows that the answer is negative.

Example 3.4.4 ([BBCER03]). Consider the formation

$$\mathfrak{F} := (G : \text{all abelian composition factors of } G \text{ are isomorphic to } C_2).$$

Suppose that $\mathfrak{F}$ is $\mathfrak{X}$-saturated for a class $\mathfrak{X}$ containing a non-abelian simple group E and $\pi(\mathfrak{X}) = \mathrm{char}\,\mathfrak{X}$. Let $p \neq 2$ be a prime dividing $|E|$. Then $p \in \mathrm{char}\,\mathfrak{X}$. Since $E \in \mathfrak{F}$, it follows that if $\mathfrak{F} = \mathrm{LF}_\mathfrak{X}(f)$, then $f(p) \neq \emptyset$. This means that $C_p \in \mathfrak{F}$. This is a contradiction. Hence $\mathfrak{X}$ should be composed of abelian simple groups. Since $\mathfrak{F}$ is solubly saturated, we have that $\mathfrak{F}$ is $\mathfrak{X}$-saturated exactly for the classes of simple groups $\mathfrak{X}$ contained in $\mathbb{P}$ by Corollary 3.1.13. Since $\mathfrak{F}$ is clearly 2-saturated, if we assume the existence of a class $\mathfrak{X}(2)$ fulfilling the property, it follows that $\mathfrak{X}(2) \subseteq \mathbb{P}$. This is not possible because the formation in Example 3.3.20 is $\mathfrak{X}(2)$-saturated but not 2-saturated.

The following theorem shows that an $\mathfrak{X}$-local formation always contains a largest ω-local formation for $\omega = \operatorname{char} \mathfrak{X}$.

Theorem 3.4.5 ([BBCS05]). *Let $\mathfrak{X}$ be a class of simple groups such that $\omega = \operatorname{char} \mathfrak{X} = \pi(\mathfrak{X})$. Let $\mathfrak{F} = \operatorname{LF}_{\mathfrak{X}}(F)$ be an $\mathfrak{X}$-local formation. Then the ω-local formation $\mathfrak{F}_\omega = \operatorname{LF}_\omega(f)$, where $f(p) = F(p)$ for every $p \in \omega$ and $f(\omega') = \mathfrak{F}$, is the largest ω-local formation contained in $\mathfrak{F}$.*

Proof. Suppose, for a contradiction, that $\mathfrak{F}_\omega$ is not contained in $\mathfrak{F}$. Let G be a group of minimal order in $\mathfrak{F}_\omega \setminus \mathfrak{F}$. Then, as usual, G has a unique minimal normal subgroup N, and $G/N \in \mathfrak{F}$. If $G_{\omega d} = 1$, we would have that $G \in f(\omega') = \mathfrak{F}$, contradicting the choice of G. Assume that $G_{\omega d} \neq 1$. Then N is contained in $G_{\omega d}$. This means that there exists a prime $p \in \omega$ dividing $|N|$. Hence $G/\operatorname{C}_G(N) \in f(p) = F(p)$. If N is a p-group, it follows that N is an $\mathfrak{X}$-chief factor of G. By Remark 3.1.7 (2), we conclude that $G \in \operatorname{LF}_{\mathfrak{X}}(F) = \mathfrak{F}$, against the choice of G. Hence N is non-abelian and so $\operatorname{C}_G(N) = 1$ and $G \in F(p)$. Since $F(p) = \mathfrak{S}_p \underline{f}(p)$ and $\operatorname{O}_p(G) = 1$, it follows that $G \in \underline{f}(p) \subseteq \mathfrak{F}$. This contradiction proves that $\mathfrak{F}_\omega \subseteq \mathfrak{F}$.

Now let $\mathfrak{G} = \operatorname{LF}_\omega(g)$ be an ω-local formation contained in $\mathfrak{F}$. Suppose, if possible, that $\mathfrak{G}$ is not contained in $\mathfrak{F}_\omega$ and let A be a group of minimal order in the supposed non-empty class $\mathfrak{G} \setminus \mathfrak{F}_\omega$. Then A has a unique minimal normal subgroup B, and $A/B \in \mathfrak{F}_\omega$. Since $A \in \mathfrak{G} \subseteq \mathfrak{F}$, we have that $A/A_{\omega d} \in \mathfrak{F} = f(\omega')$. Suppose that $p \in \omega \cap \pi(B)$. If B is an $\mathfrak{X}$-chief factor of A, it follows that $A/\operatorname{C}_A(B) \in F(p) = f(p)$. If B is an $\mathfrak{X}'$-chief factor of A, then B is non-abelian and $A \cong A/\operatorname{C}_A(B) \in g(p)$. Then $\operatorname{O}_p(A) = 1$ and so, by [DH92, B, 10.9], A has a faithful irreducible representation over $\operatorname{GF}(p)$. Let M be the corresponding module and $G = [M]A$ the corresponding semidirect product. Let us see that $G \in \mathfrak{G}$. Since M is contained in $G_{\omega d}$, it follows that $G/G_{\omega d} \in g(\omega')$ because $A/A_{\omega d} \in g(\omega')$. Moreover, we have that $G/\operatorname{C}_G(M) \cong A \in g(p)$. We can conclude that $G \in \mathfrak{G}$ and, consequently, $G = [M]A \in \mathfrak{F}$. This implies that $A \cong G/\operatorname{C}_G(M) \in f(p)$. Now we can state that $A \in \mathfrak{F}_\omega$, contradicting the choice of A. Therefore $\mathfrak{G}$ is contained in $\mathfrak{F}_\omega$. $\qquad\square$

An immediate application of Theorem 3.4.5 is the following corollary:

Corollary 3.4.6 ([BBCER03]). *Let ω be a set of primes and let $\mathfrak{X}_\omega$ be the class of all simple ω-groups. If $\mathfrak{F}$ is an $\mathfrak{X}_\omega$-local formation composed of ω-separable groups, then $\mathfrak{F}$ is ω-local.*

Proof. Suppose that $\mathfrak{F}$ is an $\mathfrak{X}_\omega$-local formation. According to Theorem 3.4.5, $\mathfrak{F} = \mathrm{LF}_{\mathfrak{X}_\omega}(F)$ contains a largest ω-local formation $\mathfrak{F}_\omega$, where $f(p) = F(p)$ for every $p \in \omega$ and $f(\omega') = \mathfrak{F}$. Suppose that the inclusion is proper, and let G be a group of minimal order in $\mathfrak{F} \setminus \mathfrak{F}_\omega$. Then G has a unique minimal normal subgroup N, and $G/N \in \mathfrak{F}_\omega$. It is clear that $G/G_{\omega d} \in f(\omega') = \mathfrak{F}$. If $p \in \pi(N) \cap \omega$, it follows that N is an ω-group, since G is ω-separable. Hence, N is an $\mathfrak{X}_\omega$-chief factor of G and, therefore, $G/\mathrm{C}_G(N) \in F(p) = f(p)$. Taking into account that $G/N \in \mathfrak{F}_\omega$, we conclude that $G \in \mathfrak{F}_\omega$. This contradiction proves that $\mathfrak{F} = \mathfrak{F}_\omega$ is ω-local. $\qquad\square$

Corollary 3.4.7 ([BBCER03]). *Let $\mathfrak{F}$ be a formation composed of ω-separable groups. Then $\mathfrak{F}$ is ω-saturated if and only if $\mathfrak{F}$ is $\mathfrak{X}_\omega$-saturated, where $\mathfrak{X}_\omega$ is the class of all simple ω-groups.*

The following consequence of Theorem 3.4.5 is of interest.

Corollary 3.4.8 ([Sal85]). *Every solubly saturated formation contains a maximal saturated formation with respect to inclusion.*

Remarks 3.4.9. 1. The converse of Corollary 3.4.8 does not hold. It is enough to consider $\mathfrak{F} = \mathrm{D}_0\big(\mathfrak{S}_2, \mathrm{Alt}(5)\big)$. By Lemma 2.2.3, $\mathfrak{F}$ is a formation. The group $\mathrm{SL}(2,5)$ shows that $\mathfrak{F}$ is not solubly saturated. However $\mathfrak{S}_2$ is the maximal saturated formation contained in $\mathfrak{F}$.

2. There exist formations not containing a maximal saturated formation as the Example 5.5 in [Sal85] shows: Let $\mathfrak{F}$ be the class of all soluble groups G such that Sylow subgroups corresponding to different primes permute. By [Hup67, VI, 3.2], $\mathfrak{F}$ is a formation. Let q be a prime and consider the formation function f_q given by: $f_q(p) = \mathfrak{S}_{\{p,q\}}$ for all $p \in \mathbb{P}$. Then the saturated formation $\mathfrak{F}_q = \mathrm{LF}(f_q)$ is contained in $\mathfrak{F}$ by [Hup67, VI, 3.1]. Let q_1 and q_2 be two different primes and let $\mathfrak{F}_{q_1,q_2}$ be the smallest saturated formation containing $\mathfrak{F}_{q_1}$ and $\mathfrak{F}_{q_2}$. Then $C_{q_1} \times C_{q_2} \in F(p)$ for all $p \in \mathbb{P}$, where F is the canonical local definition of $\mathfrak{F}_{q_1,q_2}$. This is due to the fact that $C_{q_1} \in F_{q_1}(p)$ and $C_{q_2} \in F_{q_2}(p)$, where F_{q_1} and F_{q_2} are the canonical local definitions of $\mathfrak{F}_{q_1}$ and $\mathfrak{F}_{q_2}$, respectively. Let q_3 be a prime, $q_3 \neq q_1, q_2$. By [DH92, B, 10.9], $C_{q_1} \times C_{q_2}$ has an irreducible and faithful module M over $\mathrm{GF}(q_3)$. Let $G = [M](C_{q_1} \times C_{q_2})$ be the corresponding semidirect product. Then $G \in \mathfrak{F}_{q_1,q_2}$, but $G \notin \mathfrak{F}$. This shows that $\mathfrak{F}$ does not contain a maximal saturated formation with respect to the inclusion.

A natural question arising from the above results is the following:

What are the precise conditions to ensure that an $\mathfrak{X}$-local formation is ω-local for $\omega = \mathrm{char}\,\mathfrak{X}$?

The next result gives the answer.

Theorem 3.4.10. *Let $\mathfrak{F} = \mathrm{LF}_{\mathfrak{X}}(f) = \mathrm{LF}(F)$ be an $\mathfrak{X}$-local formation and $\omega = \mathrm{char}\,\mathfrak{X}$. The following conditions are pairwise equivalent:*

1. $\mathfrak{F}$ *is* ω-*local.*
2. $\underline{f}(S) \subseteq \underline{f}(p)$ *for every* $S \in \mathfrak{X}'$ *and* $p \in \pi(S) \cap \omega$.
3. $\mathfrak{S}_p\underline{f}(S) \subseteq \mathfrak{F}$ *for every* $S \in \mathfrak{X}'$ *and* $p \in \pi(S) \cap \omega$.

Proof. 1 implies 2. Assume that $\mathfrak{F}$ is ω-local. Then, by Theorem 3.4.5, $\mathfrak{F} = \mathrm{LF}_\omega(f)$, where

$$f(p) = F(p) = \mathfrak{S}_p\underline{f}(p) \qquad\qquad \text{if } p \in \omega,$$
$$f(\omega') = \mathfrak{F}.$$

Let $S \in \mathfrak{X}'$ and $p \in \pi(S) \cap \omega$. Then S is non-abelian. By Theorem 3.1.11, $\underline{f}(S) = \mathrm{Q}\,\mathrm{R}_0\big(G/L : G \in \mathfrak{F}, G/L \text{ is monolithic, and } \mathrm{Soc}(G/L) \in \mathrm{E}(S)\big)$.

Let G be a group in $\mathfrak{F}$ and let L be a normal subgroup of G such that G/L is monolithic and $\mathrm{Soc}(G/L) \in \mathrm{E}(S)$. Since G/L is a primitive group of type 2, $L = \mathrm{C}_G\big(\mathrm{Soc}(G/L)\big)$. Moreover $G/L \in \mathfrak{F}$. This implies that $G/L \in F(p) = \mathfrak{S}_p\underline{f}(p)$. Hence $G/L \in \underline{f}(p)$ because $\mathrm{O}_p(G/L) = 1$. Consequently $\underline{f}(S) \subseteq \underline{f}(p)$ for all $p \in \pi(S) \cap \omega$.

2 implies 3. Let $S \in \mathfrak{X}'$ and $p \in \pi(S) \cap \omega$. Then $\mathfrak{S}_p\underline{f}(S) \subseteq \mathfrak{S}_p\underline{f}(p) = F(p) \subseteq \mathfrak{F}$.

3 implies 2. Applying Theorem 3.4.5, it is known that $\mathfrak{F}_\omega = \mathrm{LF}_\omega(f)$, where

$$f(p) = F(p) \qquad\qquad \text{if } p \in \omega, \text{ and}$$
$$f(\omega') = \mathfrak{F},$$

is the largest ω-local formation contained in $\mathfrak{F}$. Suppose, by way of contradiction, that $\mathfrak{F}$ is not ω-local. Then $\mathfrak{F}_\omega \neq \mathfrak{F}$. Let G be a group of minimal order in $\mathfrak{F} \setminus \mathfrak{F}_\omega$. By a familiar argument, G has a unique minimal normal subgroup N, and $G/N \in \mathfrak{F}_\omega$. If $\pi(N) \cap \omega = \emptyset$, then $G_{\omega d} = 1$ and so $G \in \mathfrak{F}_\omega$, which contradicts the fact that $G \notin \mathfrak{F}_\omega$. Therefore $\pi(N) \cap \omega \neq \emptyset$. Let p be a prime in $\pi(N) \cap \omega$. If N is an $\mathfrak{X}_p$-chief factor of G, $G/\mathrm{C}_G(N) \in F(p) = f(p)$. Assume that N is an $\mathfrak{X}'$-chief factor of G and $N \in \mathrm{E}(S)$. Then S is non-abelian and so $\mathrm{O}_p(G) = 1$. By [DH92, B, 10.9], G has an irreducible and faithful module M over $\mathrm{GF}(p)$. Let $Z = [M]G$ be the corresponding semidirect product. Since $G \in \underline{f}(S)$, it follows that $Z \in \mathfrak{S}_p\underline{f}(S) \subseteq \mathfrak{F}$. This implies that $G \cong Z/\mathrm{C}_Z(M) \in F(p) = f(p)$. Consequently $G/\mathrm{C}_G(N) \in f(p)$ for all $p \in \pi(N) \cap \omega$ and $G \in \mathfrak{F}_\omega$. This contradicts our initial supposition. Therefore $\mathfrak{F} = \mathfrak{F}_\omega$ and $\mathfrak{F}$ is ω-local. $\qquad\square$

4

Normalisers and prefrattini subgroups

The aim of this chapter is to obtain information about the structure of a finite group through the study of $\mathfrak{H}$-normalisers and subgroups of prefrattini type.

In the soluble universe, after the introduction of saturated formations and covering subgroups by W. Gaschütz, R. W. Carter, and T. O. Hawkes introduced in [CH67] a conjugacy class of subgroups associated to saturated formations $\mathfrak{F}$ of full characteristic, the $\mathfrak{F}$-normalisers, defined in terms of a local definition of $\mathfrak{F}$, which generalised Hall's system normalisers. The Carter-Hawkes's $\mathfrak{F}$-normalisers keep all essential properties of system normalisers and, in the case of the saturated formation $\mathfrak{N}$ of the nilpotent groups, the $\mathfrak{N}$-normalisers of a group are exactly Hall's system normalisers.

In this context, and having in mind the known characterisation of $\mathfrak{F}$-normalisers by means of $\mathfrak{F}$-critical subgroups, it is natural to think about $\mathfrak{H}$-normalisers associated with Schunck classes $\mathfrak{H}$ for which the existence of $\mathfrak{H}$-critical subgroups is assured in each soluble group not in $\mathfrak{H}$. A. Mann [Man70] chose this characterisaton as his starting point and was able to extend introduced the normaliser concept to certain Schunck classes following this arithmetic-free way.

Concerning the prefrattini subgroups, we said in Sections 1.3 and 1.4 that the classical prefrattini subgroups of soluble groups were introduced by W. Gaschütz ([Gas62]). A prefrattini subgroup is defined by W. Gaschütz as an intersection of complements of the crowns of the group. They form a characteristic conjugacy class of subgroups which cover the Frattini chief factors and avoid the complemented ones. Gaschütz's original prefrattini subgroups have been widely investigated and variously generalised. The first extension is due to T. O. Hawkes ([Haw67]). He introduced the idea of obtaining analogues to Gaschütz's prefrattini subgroups, associated with a saturated formation $\mathfrak{F}$, by taking intersections of certain maximal subgroups defined in terms of $\mathfrak{F}$ into which a Hall system of the group reduces. Note that Hawkes restricts the set of maximal subgroups considered to the set of $\mathfrak{F}$-abnormal maximal subgroups. He observed that all of the relevant properties of the original idea were kept and, furthermore, he presented an original new theorem of factorisation of the $\mathfrak{F}$-normaliser and the new prefrattini subgroup associated to the same Hall system.

The extension of this theory to Schunck classes, still in the soluble realm, was done by P. Förster in [För83].

Another generalisation of the Gaschütz work in the soluble universe is due to H. Kurzweil [Kur89]. He introduced the H-prefrattini subgroups of a soluble group G, where H is a subgroup of G. The H-prefrattini subgroups are conjugate in G and they have the cover-avoidance property; if $H = 1$ they coincide with the classical prefrattini subgroups of Gaschütz and if $\mathfrak{F}$ is a saturated formation and H is an $\mathfrak{F}$-normaliser of G the H-prefrattini subgroups are those described by Hawkes.

The first attempts to develop a theory of prefrattini subgroups outside the soluble universe appeared in the papers of A. A. Klimowicz in [Kli77] and A. Brandis in [Bra88]. Both defined some types of prefrattini subgroups in π-soluble groups. They manage to adapt the arithmetical methods of soluble groups to the complements of crowns of p-chief factors, for $p \in \pi$, of π-soluble groups. Also the extension of prefrattini subgroups to a class of non finite groups with a suitable Sylow structure, made by M. J. Tomkinson in [Tom75], has to be mentioned.

All these types of prefrattini subgroups keep the original properties of Gaschütz: they form a conjugacy class of subgroups, they are preserved by epimorphic images and they avoid some chief factors, exactly those associated to the crowns whose complements are used in their definition, and cover the rest. Moreover, some other papers (see [Cha72, Mak70, Mak73]) analysed their excellent permutability properties, following the example of the theorem of factorisation of Hawkes.

At the beginning of the decade of the eighties of the past twentieth century, when the classification of simple groups was almost accomplished, H. Wielandt proposed, as a main aim after the classification, to progress in the universe of non-necessarily soluble groups trying to extend the magnificent results obtained in the soluble realm. As we have mentioned in Section 2.3, R. P. Erickson, P. Förster and P. Schmid answered this Wielandt's challenge analysing the projective classes in the non-soluble universe. It seems natural to progress in that direction and think about normalisers and prefrattini subgroups in the general finite universe. This was the starting point A. Ballester-Bolinches' Ph. Doctoral Thesis at the Universitat de València in 1989 [BB89b].

This chapter has two main themes which are organised in three sections. The first two sections are devoted to study the theory of normalisers of finite, non-necessarily soluble, groups. The second subject under investigation is the theory of prefrattini subgroups outside the soluble universe. This is presented in Section 4.3.

4.1 $\mathfrak{H}$-normalisers

Obviously the definition of $\mathfrak{H}$-normalisers in the general universe has to be motivated by the characterisation of $\mathfrak{H}$-normalisers of soluble groups by chains of $\mathfrak{H}$-critical subgroups.

In this section, $\mathfrak{H}$ will be a Schunck class of the form $\mathfrak{H} = \mathrm{E}_\Phi\mathfrak{F}$, for some formation $\mathfrak{F}$. Thus, by Theorem 2.3.24, the existence of $\mathfrak{H}$-critical subgroups is assured in every group which does not belong to $\mathfrak{H}$.

Here we present the extension of the theory of $\mathfrak{H}$-normalisers to general non-necessarily soluble groups done by A. Ballester-Bolinches in his Ph. Doctoral Thesis [BB89b] and published in [BB89a]. Previous ways of extending the soluble theory had been looked at. J. Beidleman and B. Brewster [BB74] studied normalisers associated to saturated formations in the π-soluble universe, π a set of primes, and L. A. Shemetkov [She76] introduced normalisers associated to saturated formations in the general universe of all finite groups by means of critical supplements of the residual.

The definition of $\mathfrak{H}$-normaliser presented here is obviously motivated by the most abstract characterisation of the classical $\mathfrak{H}$-normalisers.

Definition 4.1.1. *Let G be a group. A subgroup D of G is said to be an $\mathfrak{H}$-normaliser of G if either $D = G$ or there exists a chain of subgroups*

$$D = H_n \leq H_{n-1} \leq \cdots \leq H_1 \leq H_0 = G \tag{4.1}$$

such that H_i is $\mathfrak{H}$-critical subgroup of H_{i-1}, for each $i \in \{1, \ldots, n\}$, and H_n contains no $\mathfrak{H}$-critical subgroup.

The condition on H_n is equivalent to say that $D \in \mathfrak{H}$. Moreover $D = G$ if and only if $G \in \mathfrak{H}$.

The non-empty set of all $\mathfrak{H}$-normalisers of G will be denoted by $\mathrm{Nor}_{\mathfrak{H}}(G)$.

If we restrict ourselves to the universe of soluble groups, this definition is equivalent to the classical ones of R. W. Carter and T. O. Hawkes in [CH67] and A. Mann in [Man70] (see [DH92, V, 3.8]).

In this section, we analyse the main properties of $\mathfrak{H}$-normalisers, primarily motivated by their behaviour in the soluble universe. In particular, we study their relationship with systems of maximal subgroups and projectors.

Each $\mathfrak{H}$-normaliser of a soluble group is associated with a particular Hall system of the group ([Man70]). Obviously this is no longer true in the general case. But bearing in mind the relationship between systems of maximal subgroups and Hall systems (see Theorem 1.4.17 and Corollary 1.4.18), it seems natural to wonder about the relationship between $\mathfrak{H}$-normalisers and systems of maximal subgroups.

Assume that D is an $\mathfrak{H}$-normaliser of a group G constructed by the chain

$$D = H_n \leq H_{n-1} \leq \cdots \leq H_1 \leq H_0 = G \tag{4.2}$$

such that H_i is $\mathfrak{H}$-critical subgroup of H_{i-1}, for each $i \in \{1, \ldots, n\}$, and H_n contains no $\mathfrak{H}$-critical subgroup. Let $\mathbf{X}(D)$ be a system of maximal subgroups of D. Applying several times Theorem 1.4.14, we can obtain a system of maximal subgroups $\mathbf{X}$ of G such that there exist systems of maximal subgroups $\mathbf{X}_i$ of H_i, for $i = 0, 1, \ldots, n$, with $\mathbf{X}_0 = \mathbf{X}$, $\mathbf{X}_n = \mathbf{X}(D)$ and for each i, $H_i \in \mathbf{X}_{i-1}$ and $(\mathbf{X}_{i-1})_{H_i} = \{H_i \cap S : S \in \mathbf{X}_{i-1}, S \neq H_i\} \subseteq \mathbf{X}_i$. This motivates the following definition.

Definition 4.1.2. *Let D be an $\mathfrak{H}$-normaliser of a group G constructed by a chain (4.2) and let $\mathbf{X}$ be a system of maximal subgroups of G such that there exist systems of maximal subgroups $\mathbf{X}_i$ of H_i, $i = 0, 1, \ldots, n$, with $\mathbf{X}_0 = \mathbf{X}$, $\mathbf{X}_n = \mathbf{X}(D)$ and for each i, $H_i \in \mathbf{X}_{i-1}$ and $(\mathbf{X}_{i-1})_{H_i} = \{H_i \cap S : S \in \mathbf{X}_{i-1}, S \neq H_i\} \subseteq \mathbf{X}_i$. We will say that D is an $\mathfrak{H}$-normaliser of G associated with $\mathbf{X}$.*

By the previous paragraph, every $\mathfrak{H}$-normaliser is associated with some system of maximal subgroups. Next we see that every system of maximal subgroups has an associated $\mathfrak{H}$-normaliser.

Proposition 4.1.3. *Given a system of maximal subgroups $\mathbf{X}$ of a group G, there exists an $\mathfrak{H}$-normaliser of G associated with $\mathbf{X}$.*

Proof. We argue by induction on the order of G. We can assume that $G \notin \mathfrak{H}$. Let M be an $\mathfrak{H}$-critical maximal subgroup of G such that $M \in \mathbf{X}$. By Corollary 1.4.16, there exists a system of maximal subgroups $\mathbf{Y}$ of M, such that $\mathbf{X}_M \subseteq \mathbf{Y}$. By induction, there exists an $\mathfrak{H}$-normaliser D of M associated with $\mathbf{Y}$. Then D is an $\mathfrak{H}$-normaliser of G associated with $\mathbf{X}$. $\square$

Remarks 4.1.4.   1.   An $\mathfrak{H}$-normaliser can be associated with some different systems of maximal subgroups. Consider the symmetric group of order 5, $G = \operatorname{Sym}(5)$, and $\mathfrak{H} = \mathfrak{N}$ the class of nilpotent groups. Write $D = \langle (12), (45) \rangle$. The subgroups $M_1 = D\langle (123) \rangle$ and $M_2 = D\langle (345) \rangle$ are $\mathfrak{N}$-critical maximal subgroups of G and $\mathbf{X}_1 = \{M_1, \operatorname{Alt}(5)\}$ and $\mathbf{X}_2 = \{M_2, \operatorname{Alt}(5)\}$ are systems of maximal subgroups of G. Observe that D is an $\mathfrak{N}$-normaliser of G associated with $\mathbf{X}_1$ and $\mathbf{X}_2$.

2.   Given a system of maximal subgroups $\mathbf{X}$ of a group G, there is not a unique $\mathfrak{H}$-normaliser of G associated with $\mathbf{X}$. In the soluble group

$$G = \langle a, b : a^9 = b^2 = 1, a^b = a^{-1} \rangle,$$

the Hall system $\Sigma = \{G, \langle a \rangle, \langle b \rangle\}$ reduces into the $\mathfrak{N}$-critical subgroup $M = \langle a^3, b \rangle$ and then the $\mathfrak{N}$-normalisers $D_1 = \langle b \rangle$ and $D_2 = \langle a^3 b \rangle$ are associated with the system of maximal subgroups defined by Σ: $\mathbf{X}(\Sigma) = \{\langle a \rangle, \langle a^3, b \rangle\}$.

For a non-soluble example, consider the Example of 1 and observe that $D_1 = \langle (12), (45) \rangle$, $D_2 = \langle (13), (45) \rangle$ and $D_3 = \langle (23), (45) \rangle$ are $\mathfrak{N}$-normalisers associated with $\mathbf{X}_1$.

One of the basic properties of $\mathfrak{H}$-normalisers of soluble groups is that they are preserved by epimorphic images (see [DH92, V, 3.2]). This is also true in the general case.

Proposition 4.1.5. *Let G be a group. Let N be a normal subgroup of G. If D is an $\mathfrak{H}$-normaliser of G associated with a system of maximal subgroups $\mathbf{X}$, then DN/N is an $\mathfrak{H}$-normaliser of G/N associated with $\mathbf{X}/N$.*

In particular, the $\mathfrak{H}$-normalisers of a group are preserved under epimorphic images.

Proof. We argue by induction on the order of G. Suppose first that N is a minimal normal subgroup of G. If $G \in \mathfrak{H}$, $D = G$ and there is nothing to prove. If $G \notin \mathfrak{H}$, then G has an $\mathfrak{H}$-critical subgroup $M \in \mathbf{X}$ such that D is an $\mathfrak{H}$-normaliser of M associated with a system of maximal subgroups $\mathbf{Y}$ of M and $\mathbf{X}_M \subseteq \mathbf{Y}$. If N is contained in M, then DN/N is, applying induction, an $\mathfrak{H}$-normaliser of M/N associated with the system of maximal subgroups $\mathbf{Y}/N$ of M/N. Since $\mathbf{X}/N_{M/N} = \mathbf{X}_M/N$ is contained in $\mathbf{Y}/N$ and M/N is $\mathfrak{H}$-critical in G/N by Lemma 2.3.23, it follows that DN/N is an $\mathfrak{H}$-normaliser of G/N associated with $\mathbf{X}/N$. Suppose that $G = MN$. By induction, $D(M \cap N)/(M \cap N)$ is an $\mathfrak{H}$-normaliser of $M/(M \cap N)$ associated with $\mathbf{Y}/(M \cap N)$. Therefore, by virtue of the canonical isomorphism between G/N and $M/(M \cap N)$, it follows that DN/N is an $\mathfrak{H}$-normaliser of G/N associated with $\mathbf{X}/N$ (note that the image of $\mathbf{X}/N = \{YN/N : Y \in \mathbf{X}_M\}$ under the above isomorphism is just $\mathbf{Y}/(M \cap N)$).

Assume now that N is not a minimal normal subgroup of G and let A be a minimal normal subgroup of G contained in N. Then, by induction, DA/A is an $\mathfrak{H}$-normaliser of G/A associated with $\mathbf{X}/A$ and $(DN/A)/(N/A)$ is $\mathfrak{H}$-normaliser of $(G/A)/(N/A)$ associated with $(\mathbf{X}/A)/(N/A)$. Consequently, DN/N is an $\mathfrak{H}$-normaliser of G/N associated with $\mathbf{X}/N$.

The proof of the proposition is now complete.

It is well-known that $\mathfrak{H}$-normalisers of soluble groups cover the $\mathfrak{H}$-central chief factors and avoid the $\mathfrak{H}$-eccentric ones (see [DH92, V, 3.3]). The cover-avoidance property is a typical property of the soluble universe that we cannot expect to be satisfied in the general one.

We present here some results to show partial aspects of the cover-avoidance property of $\mathfrak{H}$-normalisers in the general universe.

Lemma 4.1.6. *Let M be an $\mathfrak{H}$-critical subgroup of a group G. If H/K is an $\mathfrak{H}$-central chief factor of G, then M covers H/K and $[H/K] * G \cong [(H \cap M)/(K \cap M)] * M$. In particular $(H \cap M)/(K \cap M)$ is an $\mathfrak{H}$-central chief factor of M.*

Proof. If M does not cover H/K, then $K = H \cap \operatorname{Core}_G(M)$ and M supplements H/K. Moreover $H \operatorname{Core}_G(M)/\operatorname{Core}_G(M)$ is the socle of the monolithic primitive group $G/\operatorname{Core}_G(M)$. Since $H \operatorname{Core}_G(M)/\operatorname{Core}_G(M) \cong_G H/K$, then

$G/\operatorname{Core}_G(M) \cong [H/K] * G \in \mathfrak{H}$, contrary to the $\mathfrak{H}$-abnormality of M in G. Hence M covers H/K. Since H/K is $\mathfrak{H}$-central in G, then $\operatorname{C}_G(H/K)$ is not contained in $\operatorname{Core}_G(M)$ and therefore $G = M \operatorname{C}_G(H/K)$. Now the result follows from [DH92, A, 13.9]. □

Corollary 4.1.7. *Let D be an $\mathfrak{H}$-normaliser of a group G. If H/K is an $\mathfrak{H}$-central chief factor of G, then D covers H/K and $(H \cap D)/(K \cap D)$ is an $\mathfrak{H}$-central chief factor of D. Moreover, $\operatorname{Aut}_G(H/K) \cong \operatorname{Aut}_D\big((H \cap D)/(K \cap D)\big)$.*

Proposition 4.1.8. *Let D be an $\mathfrak{H}$-normaliser of a group G. If H/K is a supplemented chief factor of G covered by D, then $[H/K] * G \cong [(H \cap D)/(K \cap D)] * D \in \mathfrak{H}$.*

Proof. If $D = G$ the result is clear. Suppose that D is an $\mathfrak{H}$-critical subgroup of G. Since H/K is avoided by $\Phi(G)$ and covered by D, then $(H \cap D)/(K \cap D)$ is a chief factor of D, $\operatorname{Aut}_G(H/K) \cong \operatorname{Aut}_D\big((H \cap D)/(K \cap D)\big)$ and $[H/K] * G \cong [(H \cap D)/(K \cap D)] * D$, by Statements (1), (2), and (3) of Proposition 1.4.11. Thus, if H/K is non-abelian, then $[H/K] * G$ is isomorphic to a quotient group of D and therefore $[H/K] * G \in \mathfrak{H}$. If H/K is abelian, then H/K it is complemented by a maximal subgroup M of G. By Proposition 1.4.11 (4), we have that $M \cap D$ is a maximal subgroup of D, and $(H \cap D)/(K \cap D)$ is a chief factor of D complemented by $M \cap D$. Since $D \in \mathfrak{H}$, the primitive group associated with $(H \cap D)/(K \cap D)$ is isomorphic to a quotient group of D and therefore $[(H \cap D)/(K \cap D)] * D \in \mathfrak{H}$.

In the general case, we consider the chain (4.2) of subgroups of G. If H/K is a supplemented chief factor of G covered by D, then H/K is covered by H_1 and avoided by $\Phi(G)$. By Proposition 1.4.11, $(H \cap H_1)/(K \cap H_1)$ is a supplemented chief factor of H_1. Now, since D is an $\mathfrak{H}$-normaliser of H_1, then $[(H \cap H_1)/(K \cap H_1)] * H_1 \cong [(H \cap D)/(K \cap D)] * D$ by induction. Since clearly $[(H \cap H_1)/(K \cap H_1)] * H_1 \cong [H/K] * G$, we deduce that $[H/K] * G \cong [(H \cap D/(K \cap D)] * D \in \mathfrak{H}$. □

Corollary 4.1.9. *Let D be an $\mathfrak{H}$-normaliser of a group G. Then, among all supplemented chief factors of G, D covers exactly the $\mathfrak{H}$-central ones.*

We show next that nothing can be said about the $\mathfrak{H}$-eccentric chief factors of G.

Example 4.1.10. Let S be the alternating group of degree 5. Consider the class $\mathfrak{F} = (G : S \notin \operatorname{Q}(G))$. Then $\operatorname{b}(\mathfrak{F}) = (S)$. Hence $\mathfrak{F}$ is a saturated formation by Example 2.3.21. Let E be the maximal Frattini extension of S with 3-elementary abelian kernel (see [DH92, Appendix β] for details). The group E possesses a 3-elementary abelian normal subgroup N such that $N \leq \Phi(E)$, and $E/N \cong S$. Let M be a maximal subgroup of E, such that $M/N \cong \operatorname{Alt}(4)$. Then M is $\mathfrak{F}$-critical in E and, since M is soluble, and then $M \in \mathfrak{F}$, we have that M is an $\mathfrak{F}$-normaliser of E. Observe also that if a minimal normal subgroup K of E in N is $\mathfrak{F}$-central in E, then $K \leq \operatorname{Z}(E)$. Recall that $N \cong \operatorname{A}_3(S)$, the

3-Frattini module, and we can think of N as an $\mathrm{GF}(3)[S]$-module. If we denote $S(N)$ the socle of such module, we have that $\mathrm{Ker}\big(S \text{ on } S(N)\big) = O_{3',3}(S) = 1$, by a theorem of R. Griess and P. Schmid [GS78]. Therefore there exists an $\mathfrak{F}$-eccentric minimal normal subgroup K of E, such that $K \leq N$. It is clear that M covers K.

Note that the group E has at least three conjugacy classes of $\mathfrak{F}$-normalisers. Moreover, none of these $\mathfrak{F}$-normalisers has the cover-avoidance property in E.

Lemma 4.1.11. *Let G be a group. Consider a system of maximal subgroups* **X** *of G and an $\mathfrak{H}$-normaliser D of G associated with* **X**. *Then, for any monolithic $\mathfrak{H}$-abnormal maximal subgroup $H \in$* **X**, *we have that D is contained in H.*

Proof. We prove the assertion by induction on $|G|$. Let H be a monolithic $\mathfrak{H}$-abnormal maximal subgroup in **X**. Assume that G has an $\mathfrak{H}$-central minimal normal subgroup, N say. By Corollary 4.1.7, N is contained in $D \cap H$. Moreover, applying Proposition 4.1.5, D/N is an $\mathfrak{H}$-normaliser of G associated with **X**$/N$. By induction, $D/N \leq H/N$ and then $D \leq H$. Thus, we can assume that every minimal normal subgroup of G is $\mathfrak{H}$-eccentric in G. If N is contained in H, then, again by Proposition 4.1.5 and induction, we have that $D \leq DN \leq H$. Therefore we assume that $\mathrm{Core}_G(H) = 1$ and G is a monolithic primitive group. There exists a unique minimal normal subgroup N of G. Observe that $\mathrm{F}'(G) = N$ and so H is $\mathfrak{H}$-critical in G. Since $H \in$ **X**, we have that D is contained in H by construction of D. $\qquad\square$

Lemma 4.1.12. *If a maximal subgroup M of a group G contains an $\mathfrak{H}$-normaliser of G, then M is $\mathfrak{H}$-abnormal in G.*

Proof. Suppose that D is an $\mathfrak{H}$-normaliser of the group G and D is contained in the maximal subgroup M of G. If H/K is a chief factor supplemented by M and H/K is $\mathfrak{H}$-central in G, then D covers H/K, by Corollary 4.1.9, and so does M, a contradiction. Hence H/K is $\mathfrak{H}$-eccentric in G and M is $\mathfrak{H}$-abnormal in G. $\qquad\square$

The previous lemmas allow us to discover the relationship between $\mathfrak{H}$-normalisers and monolithic maximal subgroups. The corresponding result in the soluble universe is in [DH92, V, 3.4].

Corollary 4.1.13. *Let M be a monolithic maximal subgroup of a group G. Then M is $\mathfrak{H}$-abnormal in G if and only if M contains an $\mathfrak{H}$-normaliser of G.*

It is not true in general that an $\mathfrak{H}$-abnormal maximal subgroup M of a group G contains an $\mathfrak{H}$-normaliser of G.

Example 4.1.14. Consider the saturated formation $\mathfrak{F}$ composed of all S-perfect groups, for $S \cong \mathrm{Alt}(5)$, the alternating group of degree 5 as in Example 4.1.10. Let G be the direct product $G = S_1 \times S_2$ of two copies S_1, S_2 of S. Clearly each core-free maximal subgroup is $\mathfrak{F}$-abnormal in G. Suppose, arguing by contradiction, that U is a core-free maximal subgroup of G and there exists $E \in \mathrm{Nor}_{\mathfrak{F}}(G)$ such that E is contained in U. Let M be an $\mathfrak{F}$-critical maximal subgroup of G such that E is contained in M and E is an $\mathfrak{F}$-normaliser of M. Since M is monolithic, we can assume that $S_1 = \mathrm{Core}_G(M)$. Therefore $M = S_1 \times (M \cap S_2)$. It is clear that $M \cap S_2 \neq 1$. Let N be a minimal normal subgroup of M contained in $M \cap S_2$. Since N is a supplemented $\mathfrak{F}$-central chief factor of M, then N is covered by E by virtue of Corollary 4.1.9. Consequently, $N \leq M \cap S_2 \cap U = 1$. This contradiction yields that no core-free maximal subgroup of G contains an $\mathfrak{F}$-normaliser of G.

The fundamental connection between $\mathfrak{H}$-normalisers and $\mathfrak{H}$-projectors of a soluble group is that every $\mathfrak{H}$-projector contains an $\mathfrak{H}$-normaliser (see [Man70, Theorem 9] and [DH92, V, 4.1]). This is no longer true in the general case: any Sylow 5-subgroup of $G = \mathrm{Alt}(5)$, the alternating group of degree 5, is an $\mathfrak{N}$-projector of G and contains no $\mathfrak{N}$-normaliser of G.

However we can prove some interesting results that confirm the close relation between $\mathfrak{H}$-normalisers and $\mathfrak{H}$-projectors, especially when saturated formations $\mathfrak{H}$ are considered.

Definitions 4.1.15. *Let G be a group.*

1. *A maximal subgroup M of G is said to be $\mathfrak{H}$-crucial in G if M is $\mathfrak{H}$-abnormal and $M / \mathrm{Core}_G(M) \in \mathfrak{H}$.*
2. *If $G \notin \mathfrak{H}$, an $\mathfrak{H}$-normaliser D of G is said to be $\mathfrak{H}$-crucial in G if there exists a chain of subgroups*

$$D = H_n \leq H_{n-1} \leq \cdots \leq H_1 \leq H_0 = G \tag{4.3}$$

such that H_i is $\mathfrak{H}$-crucial $\mathfrak{H}$-critical subgroup of H_{i-1}, for each $i \in \{1, \ldots, n\}$, and H_n contains no $\mathfrak{H}$-critical subgroup.

Proposition 4.1.16. *If D is an $\mathfrak{H}$-crucial $\mathfrak{H}$-normaliser of a group G, then D is an $\mathfrak{H}$-projector of G.*

Proof. Clearly $G \notin \mathfrak{H}$. Suppose first that D is maximal in G. Then we have that $D / \mathrm{Core}_G(D)$ is an $\mathfrak{H}$-maximal subgroup of the group $G / \mathrm{Core}_G(D)$ and $G / \mathrm{Core}_G(D)$ is a primitive group in the boundary of $\mathfrak{H}$. Since $D / \mathrm{Core}_G(D)$ is an $\mathfrak{H}$-projector of $G / \mathrm{Core}_G(D)$, then D is an $\mathfrak{H}$-projector of G by Proposition 2.3.14.

Suppose that D is not maximal in G, and let M be an $\mathfrak{H}$-crucial $\mathfrak{H}$-critical subgroup of G such that D is an $\mathfrak{H}$-crucial $\mathfrak{H}$-normaliser of M. By induction, D is an $\mathfrak{H}$-projector of M. By Proposition 2.3.14, D is an $\mathfrak{H}$-projector of G. $\square$

Lemma 4.1.17. *Let G be a group and E an $\mathfrak{H}$-maximal subgroup of G such that $G = E \, \mathrm{F}(G)$, then E is an $\mathfrak{H}$-normaliser of G.*

Proof. We proceed by induction on $|G|$. If $E = G$, there is nothing to prove. We can assume that $G \notin \mathfrak{H}$ and E is then a proper subgroup of G. Let M be a maximal subgroup of G containing E. Since $M = E\,\mathrm{F}(M)$ and E is $\mathfrak{H}$-maximal in M, then E is an $\mathfrak{H}$-normaliser of M, by induction. Applying Proposition 2.3.17, E is an $\mathfrak{H}$-projector of G and then M is $\mathfrak{H}$-critical in G. Therefore E is an $\mathfrak{H}$-normaliser of G.   □

Let $\mathfrak{F}$ be a saturated formation. It is known that in a soluble group in $\mathfrak{N}\mathfrak{F}$, the $\mathfrak{F}$-projectors and the $\mathfrak{F}$-normalisers coincide (see [DH92, V, 4.2]). The above lemma allows us to extend this result to Schunck classes in the general universe.

Theorem 4.1.18. *If G is a group in $\mathfrak{N}\mathfrak{H}$, then the $\mathfrak{H}$-projectors and the $\mathfrak{H}$-normalisers of G coincide.*

Proof. We prove by induction on the order of G that the $\mathfrak{H}$-normalisers of G are $\mathfrak{H}$-crucial in G. If $G \in \mathfrak{H}$, the result is trivial. Thus, we can assume that $G \notin \mathfrak{H}$. Let M be an $\mathfrak{H}$-critical subgroup of G. Then $G = M\,\mathrm{F}(G)$ and $M \cap \mathrm{F}(G)$ is contained in $\mathrm{Core}_G(M)$ because $\mathrm{F}(G)/\Phi(G)$ is abelian. Hence $M/\mathrm{Core}_G(M)$ is a quotient group of $M/\bigl(M \cap \mathrm{F}(G)\bigr) \cong G/\mathrm{F}(G)$, and then $M/\mathrm{Core}_G(M) \in \mathfrak{H}$. Therefore M is $\mathfrak{H}$-crucial in G. If $D \in \mathrm{Nor}_{\mathfrak{H}}(G)$, then there exists an $\mathfrak{H}$-critical subgroup M of G such that $D \in \mathrm{Nor}_{\mathfrak{H}}(M)$. Since $M \in \mathfrak{N}\mathfrak{H}$, we have that D is an $\mathfrak{H}$-crucial $\mathfrak{H}$-normaliser of M by induction. Therefore D is an $\mathfrak{H}$-crucial $\mathfrak{H}$-normaliser of G.

Therefore we can apply Proposition 4.1.22 to conclude that each $\mathfrak{H}$-normaliser of G is an $\mathfrak{H}$-projector of G.

Now, let E be an $\mathfrak{H}$-projector of G. Since $G \in \mathfrak{N}\mathfrak{H}$, it follows that $G = E\,\mathrm{F}(G)$. By Lemma 4.1.17, E is an $\mathfrak{H}$-normaliser of G.   □

The previous result can be used to show that, for saturated formations $\mathfrak{F}$, the $\mathfrak{F}$-normalisers of groups with soluble $\mathfrak{F}$-residual can be described in terms of projectors. The corresponding result for soluble groups appears in [DH92, V, 4.3].

Theorem 4.1.19.   *1. Let $\mathfrak{F}$ be a formation and $\mathfrak{H} = \mathrm{E}_\Phi\mathfrak{F}$. Then, for any group G, if D is an $\mathfrak{N}\mathfrak{F}$-normaliser of G, the $\mathfrak{H}$-projectors of D are $\mathfrak{H}$-normalisers of G.*

   2. Let $\mathfrak{F}$ be a saturated formation and let G be a group such that the $\mathfrak{F}$-residual $G^{\mathfrak{F}}$ is a soluble group of nilpotent length r. We construct the chain of subgroups

$$D_r \leq D_{r-1} \leq D_{r-2} \leq \cdots \leq D_1 \leq D_0 = G$$

   where D_i is an $\mathfrak{N}^{r-i}\mathfrak{F}$-projector of D_{i-1}, for $i \in \{1, \dots, r\}$. Then D_r is an $\mathfrak{F}$-normaliser of G.

Proof.   1. By Corollary 3.3.9, $\mathfrak{N}\mathfrak{F}$ is a saturated formation. Moreover, $\mathfrak{H}$ is contained in $\mathfrak{N}\mathfrak{F}$.

If $G \in \mathfrak{N}\mathfrak{F}$, then $G \in \mathfrak{N}\mathfrak{H}$ and so $\operatorname{Proj}_{\mathfrak{H}}(G) = \operatorname{Nor}_{\mathfrak{H}}(G)$ by Theorem 4.1.18. Thus we can assume that $G \notin \mathfrak{N}\mathfrak{F}$. Let D be an $\mathfrak{N}\mathfrak{F}$-normaliser of G. Then there exists a chain of subgroups (4.2), such that H_{i-1} is an $\mathfrak{N}\mathfrak{F}$-critical subgroup of H_i, for each index i. Since $\mathfrak{H} \subseteq \mathfrak{N}\mathfrak{F}$, every $\mathfrak{H}$-normaliser of D is an $\mathfrak{H}$-normaliser of G. Since $D \in \mathfrak{N}\mathfrak{F} \subseteq \mathfrak{N}\mathfrak{H}$, we have that $\operatorname{Proj}_{\mathfrak{H}}(D) = \operatorname{Nor}_{\mathfrak{H}}(D)$ by Theorem 4.1.18. Hence each $\mathfrak{H}$-projector of D is an $\mathfrak{H}$-normaliser of G.

2. Let $\mathfrak{F}$ be a saturated formation and let G be a group whose $\mathfrak{F}$-residual, $G^{\mathfrak{F}}$, is a soluble group of nilpotent length r. This is to say that $G \in \mathfrak{N}^r\mathfrak{F}$. We construct the chain of subgroups

$$D_{r-1} \leq D_{r-2} \leq \cdots \leq D_1 \leq D_0 = G$$

where D_i is an $\mathfrak{N}^{r-i}\mathfrak{F}$-projector of D_{i-1}, for $i \in \{1, \ldots, r-1\}$. Since $G \in \mathfrak{N}(\mathfrak{N}^{r-1}\mathfrak{F})$, then the $\mathfrak{N}^{r-1}\mathfrak{F}$-projectors and the $\mathfrak{N}^{r-1}\mathfrak{F}$-normalisers of G coincide by Theorem 4.1.18. Therefore D_1 is an $\mathfrak{N}^{r-1}\mathfrak{F}$-normaliser of G. By Statement 1, the $\mathfrak{N}^{r-2}\mathfrak{F}$-projectors of D_1 are $\mathfrak{N}^{r-2}\mathfrak{F}$-normalisers of G. Thus, D_2 is an $\mathfrak{N}^{r-2}\mathfrak{F}$-normaliser of G. Repeating this argument, we obtain that D_{r-1} is an $\mathfrak{N}\mathfrak{F}$-normaliser of G. Hence, every $\mathfrak{F}$-projector of D_{r-1} is an $\mathfrak{F}$-normaliser of G by Statement 1. Consequently D_r is an $\mathfrak{F}$-normaliser of G. □

The next result yields a sufficient condition for a subgroup of a group in $\mathfrak{N}\mathfrak{H}$ to contain an $\mathfrak{H}$-normaliser.

Theorem 4.1.20. *Let G be a group in $\mathfrak{N}\mathfrak{H}$ and E a subgroup of G that covers all $\mathfrak{H}$-central chief factors of a given chief series of G. Then E contains an $\mathfrak{H}$-normaliser of G.*

Proof. We argue by induction on the order of G. Clearly we can assume that $G \notin \mathfrak{H}$ and that E is a proper subgroup of G. If M is a maximal subgroup of G such that $E \leq M$, then M is an $\mathfrak{H}$-abnormal subgroup of G and $G = M\operatorname{F}(G)$ because E covers the section $G/\operatorname{F}(G)$. This is to say that M is $\mathfrak{H}$-critical in G. Moreover M is has the cover-avoidance property and the intersections of M with all normal subgroups of a chief series of G give a chief series of M. If H/K is a chief factor of G in that series covered by M, then $(M \cap H)/(M \cap K)$ is a chief factor of M such that $[H/K] * G \cong [(M \cap H)/(M \cap K)] * M$ by Proposition 1.4.11. Consequently, E covers all $\mathfrak{H}$-central chief factors of a chief series of M. By induction, E contains an $\mathfrak{H}$-normaliser of M which is an $\mathfrak{H}$-normaliser of G. □

We end this section with the analysis of the relation between the $\mathfrak{F}$-normalisers and the $\mathfrak{F}$-hypercentre, $\mathfrak{F}$ a saturated formation.

Recall that a normal subgroup N of a group G is said to be *$\mathfrak{F}$-hypercentral in G* if every chief factor of G below N is $\mathfrak{F}$-central in G. The product of $\mathfrak{F}$-hypercentral normal subgroups of a group is again an $\mathfrak{F}$-hypercentral normal subgroup of the group (see [DH92, IV, 6.4]). Thus every group G possesses a unique maximal normal $\mathfrak{F}$-hypercentral subgroup called the $\mathfrak{F}$-*hypercentre* of G and denoted by $Z_{\mathfrak{F}}(G)$.

Let G be a group. By Corollary 4.1.7, the $\mathfrak{F}$-hypercentre of G is contained in every $\mathfrak{F}$-normaliser of G. Therefore $Z_{\mathfrak{F}}(G)$ is contained in $\mathrm{Core}_G(D)$, for every $\mathfrak{H}$-normaliser D of G. However, the equality does not hold in general.

Example 4.1.21. Consider E and $\mathfrak{F}$ as in Example 4.1.10. By [GS78, Example 1 (b)], $Z_{\mathfrak{F}}(E) = 1$. If M is a maximal subgroup of E such that $M/N \cong \mathrm{Alt}(4)$, then M is an $\mathfrak{F}$-normaliser of E and $\mathrm{Core}_E(M) = N \neq 1$.

In the next section, we shall see that the equality holds in groups with soluble $\mathfrak{F}$-residual.

Next we describe the $\mathfrak{F}$-hypercentre of a group in terms of the $\mathfrak{F}$-residual of the group and an $\mathfrak{F}$-normaliser. A similar description appears in [DH92, IV, 6.14] for $\mathfrak{F}$-maximal subgroups supplementing the $\mathfrak{F}$-residual. Note that, in general, the $\mathfrak{F}$-normalisers are not $\mathfrak{F}$-maximal subgroups.

Proposition 4.1.22. *Let $\mathfrak{F}$ be a saturated formation. If D is an $\mathfrak{F}$-normaliser of a group G, then $Z_{\mathfrak{F}}(G) = \mathrm{C}_D(G^{\mathfrak{F}})$.*

Proof. Applying [DH92, IV, 6.10]), we have that $[G^{\mathfrak{F}}, Z_{\mathfrak{F}}(G)] = 1$. Therefore $Z_{\mathfrak{F}}(G)$ is contained in $\mathrm{C}_D(G^{\mathfrak{F}})$. Next we prove that $\mathrm{C}_D(G^{\mathfrak{F}})$ is an $\mathfrak{F}$-hypercentral normal subgroup of G. Since $G = DG^{\mathfrak{F}}$, the $\mathrm{C}_D(G^{\mathfrak{F}})$ is normal in G. Let H/K be a chief factor of G below $\mathrm{C}_D(G^{\mathfrak{F}})$. Then $G^{\mathfrak{F}} \leq \mathrm{C}_G(H/K)$. This implies that $G = D\,\mathrm{C}_G(H/K)$. Consequently H/K is a chief factor of D by [DH92, A, 13.9]). Since $D \in \mathfrak{F}$, the chief factor H/K is $\mathfrak{F}$-central in D and then in G by [DH92, A, 13.9]). Consequently $\mathrm{C}_D(G^{\mathfrak{F}})$ is an $\mathfrak{F}$-hypercentral normal subgroup of G and hence it is contained in $Z_{\mathfrak{F}}(G)$.   □

4.2 Normalisers of groups with soluble residual

In this section we assume that $\mathfrak{F}$ is a saturated formation. Most of the properties of $\mathfrak{F}$-normalisers of soluble groups, such as conjugacy, cover-avoidance property, relation with $\mathfrak{F}$-projectors, do not hold in the general case (see examples of the previous section). However $\mathfrak{F}$-normalisers of groups G in which the $\mathfrak{F}$-residual $G^{\mathfrak{F}}$ is soluble (i.e. groups in the class $\mathfrak{S}\mathfrak{F}$) do really satisfy these classical properties. The purpose of the section is to give a full account of these results. We remark that no use of the corresponding results for soluble groups occurs in our arguments.

The following elementary result will be used frequently in the section. Let M be an $\mathfrak{F}$-abnormal maximal subgroup of a group G. Then $G = MG^{\mathfrak{F}}$. Assume, in addition, that $G^{\mathfrak{F}}$ is soluble. Then every chief factor of G supplemented by M is abelian. In particular, M is a maximal subgroup of G of type 1.

Our starting point is a result of P. Schmid which proves that the $\mathfrak{F}$-projectors of a group with soluble $\mathfrak{F}$-residual form a conjugacy class of subgroups.

Theorem 4.2.1 ([Sch74]). *Let $\mathfrak{F}$ be a saturated formation. Let G be group whose $\mathfrak{F}$-residual $G^{\mathfrak{F}}$ is soluble. Then $\mathrm{Proj}_{\mathfrak{F}}(G)$ is a conjugacy class of subgroups of G.*

Proof. We argue by induction on $|G|$. Obviously we can assume that $G^{\mathfrak{F}} \neq 1$. Let N be a minimal normal subgroup of G such that $N \leq G^{\mathfrak{F}}$ and suppose that E and D are $\mathfrak{F}$-projectors of G. By induction, $X = EN = D^g N$ for some $g \in G$. Since N is abelian, we have that E and D^g are $\mathfrak{F}$-projectors of X, by Lemma 4.1.17 and Theorem 4.1.18. If X is a proper subgroup of G, then E and D^g are conjugate in X by induction. Thus we can assume that $G = EN$, for every minimal normal subgroup N which is contained in $G^{\mathfrak{F}}$. Since $G/N \cong E/(E \cap N) \in \mathrm{Q}\mathfrak{F} = \mathfrak{F}$, we have that $N = G^{\mathfrak{F}}$. This is to say that $G^{\mathfrak{F}}$ is an abelian minimal normal subgroup of G and every $\mathfrak{F}$-projector of G is a maximal subgroup of G. Let p be the prime dividing $|G|$. Let F be the canonical local definition of $\mathfrak{F} = \mathrm{LF}(F)$, and consider the $F(p)$-residual $T = G^{F(p)}$ of G. Clearly T contains N. Since $G/N \in \mathfrak{C}_{p'}F(p)$ (see [DH92, IV, 3.2]), it follows that T/N is a p'-group. Moreover, since F is full, we have that $\mathrm{O}^p(T) = T$. Hence, for any $E \in \mathrm{Proj}_{\mathfrak{F}}(G)$, we have that $T = N(T \cap E)$ and $T \cap E$ is a Hall p'-subgroup of T. By the Schur-Zassenhaus theorem [Hup67, I, 18.1 and 18.2], the Hall p'-subgroups of T are a conjugacy class of subgroups of T. If $T \cap E$ is normal in G, then $T \cap E = \mathrm{O}^p(T) = T$. This is a contradiction. Hence $E = \mathrm{N}_G(T \cap E)$ and then $\mathrm{Proj}_{\mathfrak{F}}(G)$ is a conjugacy class of subgroups of G. $\qquad\square$

Assume that G is a group with soluble $\mathfrak{F}$-residual, $\mathfrak{F}$ a saturated formation. Then $\mathrm{Proj}_{\mathfrak{F}}(G) = \mathrm{Cov}_{\mathfrak{F}}(G)$. This can be proved by reducing the problem to the case $G \in \mathrm{b}(\mathfrak{F})$ (note that if E is an $\mathfrak{F}$-projector of G, then E is an $\mathfrak{F}$-projector of EN for every minimal normal subgroup N of G by Proposition 2.3.16). In such case, the equality is obviously true because G is a primitive group of type 1 (see [DH92, III, 3.9]).

We show next that in groups with soluble $\mathfrak{F}$-residual, the $\mathfrak{F}$-normalisers can be joined to the group by means of some special chains.

Lemma 4.2.2. *Let G be a group whose $\mathfrak{F}$-residual $G^{\mathfrak{F}}$ is soluble. If D is an $\mathfrak{F}$-normaliser of G, there exists a chain of subgroups*

$$D = H_n \leq H_{n-1} \leq \cdots \leq H_1 \leq H_0 = G \qquad (4.4)$$

such that H_i is $\mathfrak{H}$-critical maximal subgroup of H_{i-1} of type 1, for each $i \in \{1, \ldots, n\}$, and H_n contains no $\mathfrak{F}$-critical subgroup.

Proof. We prove the assertion by induction on $|G|$. We can assume that $G \notin \mathfrak{F}$. If M is an $\mathfrak{F}$-critical subgroup of G containing D as $\mathfrak{F}$-normaliser, then M is a maximal subgroup of type 1. Moreover $M^{\mathfrak{S}} \leq G^{\mathfrak{S}}$ by Proposition 2.2.8 (3). Hence $M^{\mathfrak{S}}$ is soluble. By induction, D can be joined to M by means of a chain of $\mathfrak{F}$-critical maximal subgroups of type 1. This completes the proof the lemma. $\qquad \square$

Lemma 4.2.3 (see [Ezq86]). *If M is a maximal subgroup of a group G which supplements the Fitting subgroup $\mathrm{F}(G)$, then every subgroup with the cover-avoidance property in M is a subgroup with the cover-avoidance property in G.*

Proof. Let D be a subgroup with the cover-avoidance property in M. Let H/K be a chief factor of G covered by M. Observe that $G = M \mathrm{F}(G) = M \mathrm{C}_G(H/K)$. Then $(H \cap M)/(K \cap M)$ is a chief factor of M. If D covers $(H \cap M)/(K \cap M)$, then $H \cap M = (K \cap M)(H \cap D)$. Since $H = K(H \cap M)$, we have that $H = K(H \cap D)$ and D covers H/K. If D avoids $(H \cap M)/(K \cap M)$, then $D \cap H \leq K$ and D avoids H/K. Finally D avoids all chief factors avoided by M. $\qquad \square$

Theorem 4.2.4. *Let G be a group whose $\mathfrak{F}$-residual $G^{\mathfrak{S}}$ is soluble. If D is an $\mathfrak{F}$-normaliser of G, then D covers all the $\mathfrak{F}$-central chief factors of G and avoids all the $\mathfrak{F}$-eccentric ones.*

Proof. We use induction on the order of G to prove that $\mathfrak{F}$-normalisers are subgroups with the cover-avoidance property in G. Let $D \neq G$ be an $\mathfrak{F}$-normaliser of G and suppose that D is maximal in G. If H/K is a non-abelian chief factor of G, then D covers H/K since D is of type 1. If H/K is abelian and D does not cover H/K, then $G = DH$ and $K \leq D$. In the group G/K, the minimal normal subgroup H/K is abelian and complemented by the maximal subgroup D/K. Then D avoids H/K.

If D is not maximal in G, there exists an $\mathfrak{F}$-critical maximal subgroup M of G such that $D \in \mathrm{Nor}_{\mathfrak{F}}(M)$. By induction, D has the cover-avoidance property in M. Since M supplements $\mathrm{F}(G)$, D has the cover-avoidance property in G by Lemma 4.2.3.

If H/K is an $\mathfrak{F}$-central chief factor of G, then, by Corollary 4.1.7, D covers H/K. Suppose that H/K is an $\mathfrak{F}$-eccentric chief factor of G which is covered by D. Suppose that D is defined by a chain (4.4) as in Lemma 4.2.2. Observe that $G = H_1 \mathrm{F}(G) = H_1 \mathrm{C}_G(H/K)$ and H_1 covers H/K. Hence, $(H \cap H_1)/(K \cap H_1)$ is a chief factor of H_1 such that $\mathrm{Aut}_G(H/K) \cong \mathrm{Aut}_{H_1}((H \cap H_1)/(K \cap H_1))$. By repeating the argument we obtain that $(H \cap D)/(K \cap D)$ is an $\mathfrak{F}$-eccentric chief factor of D. Since $D \in \mathfrak{F}$, all chief factors of D are $\mathfrak{F}$-central. This contradiction yields that H/K is avoided by D. $\qquad \square$

Combining Corollary 4.1.7 and Theorem 4.2.4, a chief series of an $\mathfrak{F}$-normaliser D of a group G with soluble $\mathfrak{F}$-residual can be obtained by intersecting D with the members of a given chief series of G.

Our next result partially extends a result of J. D. Gillam (see [DH92, V, 3.3]) on the cover-avoidance property of $\mathfrak{F}$-normalisers. We wonder whether the cover-avoidance property characterises the $\mathfrak{F}$-normalisers of groups whose $\mathfrak{F}$-residual is soluble. The answer in general is negative even in soluble groups (see an example in [DH92, page 401]). Gillam's result characterises the $\mathfrak{F}$-normaliser of a soluble group associated with a particular Hall system by the cover-avoidance property together with the permutability with the Hall system. Obviously this is not possible in our context. However, Theorem 4.1.20 allows us to show that the characterisation of the $\mathfrak{F}$-normalisers by the cover-avoidance property, holds in groups whose $\mathfrak{F}$-residual is nilpotent.

Corollary 4.2.5. *If $\mathfrak{F}$ is a saturated formation and G is a group in $\mathfrak{N}\mathfrak{F}$, then, for a subgroup D of G, the following sentences are equivalent:*

1. *D is an $\mathfrak{F}$-normaliser of G,*
2. *D covers the $\mathfrak{F}$-central chief factors of G and avoids the $\mathfrak{F}$-eccentric ones.*

We have seen in Example 4.1.21 that, in general, the $\mathfrak{F}$-hypercentre of a group G is not the core in G of an $\mathfrak{F}$-normaliser of G. The equality in groups with soluble $\mathfrak{F}$-residual follows from the cover-avoidance property of the $\mathfrak{F}$-normalisers.

Proposition 4.2.6. *Let G be a group such that the $\mathfrak{F}$-residual $G^{\mathfrak{F}}$ is a soluble group. If D is an $\mathfrak{F}$-normaliser of G, then $\mathrm{Z}_{\mathfrak{F}}(G) = \mathrm{Core}_G(D)$.*

Proof. If $\mathrm{Z}_{\mathfrak{F}}(G) = 1$, the core of any $\mathfrak{F}$-normaliser is trivial by Theorem 4.2.4. If $\mathrm{Z}_{\mathfrak{F}}(G)$ is non-trivial, the group $G/\mathrm{Z}_{\mathfrak{F}}(G)$ has trivial $\mathfrak{F}$-hypercentre and the quotient $D\,\mathrm{Z}_{\mathfrak{F}}(G)/\mathrm{Z}_{\mathfrak{F}}(G)$ is an $\mathfrak{F}$-normaliser of $G/\mathrm{Z}_{\mathfrak{F}}(G)$ by Proposition 4.1.5. Consequently $\mathrm{Core}_G(D) \leq \mathrm{Z}_{\mathfrak{F}}(G)$. $\qquad\square$

Our next major objective is to show that the connections between $\mathfrak{F}$-normalisers and $\mathfrak{F}$-projectors of groups with soluble $\mathfrak{F}$-residual are similar to the ones of the soluble case. In particular every $\mathfrak{F}$-normaliser is contained in an $\mathfrak{F}$-projector. Since, by Theorem 4.2.1, the $\mathfrak{F}$-projectors of groups in $\mathfrak{S}\mathfrak{F}$ form a conjugacy class of subgroups, every $\mathfrak{F}$-projector contains an $\mathfrak{F}$-normaliser.

Theorem 4.2.7. *Let $\mathfrak{F}$ be a saturated formation. If $G \in \mathfrak{N}\mathfrak{F}$ and H is a subgroup of G such that $G = H\,\mathrm{F}(G)$, then each $\mathfrak{F}$-projector of H is of the form $H \cap E$, for some $\mathfrak{F}$-projector E of G.*

Proof. Clearly we can assume that $\mathrm{F}(G) \neq 1$, $G \neq H$, and $G \notin \mathfrak{F}$. Moreover, arguing by induction on the order of G, we can assume that H is a maximal subgroup of G. Since $H/\big(H \cap \mathrm{F}(G)\big) \in \mathfrak{F}$, each $\mathfrak{F}$-projector D of H satisfies $H = D\big(H \cap \mathrm{F}(G)\big)$. Then $G = D\,\mathrm{F}(G)$. If E is an $\mathfrak{F}$-maximal subgroup of G such that $D \leq E$, then $E \in \mathrm{Proj}_{\mathfrak{F}}(G)$ by Proposition 2.3.17. It is rather easy to show that D and $E \cap H$ cover and avoid the same chief factors of a given chief series of G. Consequently $D = E \cap H$. $\qquad\square$

Theorem 4.2.8. *Let $\mathfrak{F}$ be a saturated formation. If G is a group whose $\mathfrak{F}$-residual $G^{\mathfrak{F}}$ is soluble, and H is a subgroup of G such that $G = H\,\mathrm{F}(G)$, then there exist an $\mathfrak{F}$-projector A of H and an $\mathfrak{F}$-projector E of G such that $A = H \cap E$.*

Proof. By Theorem 4.2.7, we can assume that $G \notin \mathfrak{NF}$. The quotient group $\bar{G} = G/\operatorname{F}(G)$ has soluble non-trivial $\mathfrak{F}$-residual $\bar{G}^{\mathfrak{F}} = G^{\mathfrak{F}}\operatorname{F}(G)/\operatorname{F}(G)$, Since $\bar{G}^{\mathfrak{F}} \neq 1$, we can consider a chief factor of G of the form $\bar{G}^{\mathfrak{F}}/\bar{K}$. Since $\mathfrak{F}$ is saturated, then $\bar{G}^{\mathfrak{F}}/\bar{K}$ is a complemented abelian chief factor of $\bar{G}$. Let $M/\operatorname{F}(G)$ be a complement of $\bar{G}^{\mathfrak{F}}/\bar{K}$ in $\bar{G}$. Then M is an $\mathfrak{F}$-crucial maximal subgroup of G. If $N/\operatorname{Core}_G(M) = \operatorname{Soc}(G/\operatorname{Core}_G(M))$, then H covers $N/\operatorname{Core}_G(M)$ and $(N \cap H)/(\operatorname{Core}_G(M) \cap H)$ is an $\mathfrak{F}$-eccentric chief factor of H. Moreover, $H = (N \cap H)(M \cap H)$ and $(N \cap H)/(\operatorname{Core}_G(M) \cap H)$ is an abelian chief factor of H. Consequently $M \cap H$ is an $\mathfrak{F}$-crucial maximal subgroup of H. On the other hand, $M = (M \cap H)\operatorname{F}(M)$ and so $M^{\mathfrak{F}}\operatorname{F}(M) = (M \cap H)^{\mathfrak{F}}\operatorname{F}(M)$ by Proposition 2.2.8 (2). Analogously $G^{\mathfrak{F}}\operatorname{F}(G) = H^{\mathfrak{F}}\operatorname{F}(G)$. This implies that $M^{\mathfrak{F}}$ is soluble. By induction, there exist $A \in \operatorname{Proj}_{\mathfrak{F}}(M \cap H)$ and $E \in \operatorname{Proj}_{\mathfrak{F}}(M)$ such that $A = H \cap E \cap M = H \cap E$. By Proposition 2.3.16, the $\mathfrak{F}$-projectors of any $\mathfrak{F}$-crucial monolithic maximal subgroup of a group are $\mathfrak{F}$-projectors of the group. Since $M \cap H$ is $\mathfrak{F}$-crucial in H, we have that A is an $\mathfrak{F}$-projector of H, and since M is $\mathfrak{F}$-crucial in G, then E is an $\mathfrak{F}$-projector of G.   □

Theorem 4.2.9. *Let $\mathfrak{F}$ be a saturated formation. Let G be a group whose $\mathfrak{F}$-residual $G^{\mathfrak{F}}$ is soluble. Then each $\mathfrak{F}$-normaliser of G is contained in an $\mathfrak{F}$-projector of G and each $\mathfrak{F}$-projector contains an $\mathfrak{F}$-normaliser.*

Proof. We argue by induction of the order of G. We can assume that $G \notin \mathfrak{F}$. Let D be an $\mathfrak{F}$-normaliser of G. There exists an $\mathfrak{F}$-critical subgroup M of G such that $D \in \operatorname{Nor}_{\mathfrak{F}}(M)$. Since $M^{\mathfrak{F}}$ is soluble, there exists an $\mathfrak{F}$-projector A of M such that D is contained in A. Since M is critical in G, we can apply Theorem 4.2.8 to conclude that there exist $B \in \operatorname{Proj}_{\mathfrak{F}}(M)$ and $E \in \operatorname{Proj}_{\mathfrak{F}}(G)$ such that $B = M \cap E$. By Theorem 4.2.1, the subgroups A and B are conjugate in M. Hence there exists an element $x \in M$ such that $A = B^x$. Thus, $A = M \cap E^x$ and D is contained in E^x which is an $\mathfrak{F}$-projector of G.

By Theorem 4.2.1, the $\mathfrak{F}$-projectors of G form a conjugacy class of subgroups. Hence, every $\mathfrak{F}$-projector contains an $\mathfrak{F}$-normaliser.   □

Assume that $\mathfrak{F}$ is a saturated formation. Let G be a group whose $\mathfrak{F}$-residual $G^{\mathfrak{F}}$ is soluble. If Σ is a Hall system of $G^{\mathfrak{F}}$, then we denote $\operatorname{N}_G(\Sigma) = \bigcap\{\operatorname{N}_G(H) : H \in \Sigma\}$. Sometimes $\operatorname{N}_G(\Sigma)$ is said to be the *absolute system normaliser in G of Σ*.

In [Yen70], it is proved that if G is a soluble group, then the $\mathfrak{F}$-projectors of T are $\mathfrak{F}$-normalisers of G. Our next objective is to show that this result holds not only in soluble groups but also in groups whose $\mathfrak{F}$-residual is soluble. As a consequence we will obtain the conjugacy of $\mathfrak{F}$-normalisers in such groups.

In general, if N is a soluble normal subgroup of a group G and Σ is a Hall system of N, then Σ^g is also a Hall system of N, for all $g \in G$. Since Hall systems of a soluble group are conjugate, there exists an element $x \in N$ such that $\Sigma^g = \Sigma^x$. Hence, by the Frattini argument, we have that $G = \operatorname{N}_G(\Sigma)N$. Then $\operatorname{N}_G(\Sigma) \cap N$ is a system normaliser of N. Hence $\operatorname{N}_G(\Sigma) \cap N$ is nilpotent by [DH92, I, 5.4] and $\operatorname{N}_G(\Sigma)/\operatorname{N}_N(\Sigma)$ is isomorphic to G/N. If, in addition,

N contains $G^{\mathfrak{F}}$, it follows that $G/N \in \mathfrak{F}$ and so $N_G(\Sigma)$ belongs to $\mathfrak{N}\mathfrak{F}$. In that case, $N_G(\Sigma)^{\mathfrak{F}}$ is contained in $N_N(\Sigma)$ and so Σ reduces into $N_G(\Sigma)^{\mathfrak{F}}$.

The next lemma will be used in subsequent proofs.

Lemma 4.2.10. *Let G be a group whose $\mathfrak{F}$-residual $G^{\mathfrak{F}}$ is soluble. Consider a Hall system Σ of $G^{\mathfrak{F}}$ and write $T = N_G(\Sigma)$. If N is a normal subgroup of G, then $TN/N = N_{G/N}(\Sigma N/N)$.*

Therefore, if E is an $\mathfrak{F}$-projector of T, then EN/N is an $\mathfrak{F}$-projector of $N_{G/N}(\Sigma N/N)$.

Proof. We argue by induction on the order of G. Clearly TN/N is contained in $N_{G/N}(\Sigma N/N)$. Assume that N is a minimal normal subgroup of G.

Suppose that $N \cap G^{\mathfrak{F}} = 1$. Note that G acts transitively by conjugation on the set of Hall systems of $G^{\mathfrak{F}} N/N$. Hence $|G/N : N_{G/N}(\Sigma N/N)|$ is the number of Hall systems of $G^{\mathfrak{F}} N/N$. Moreover, by the same argument, the number of Hall systems of $G^{\mathfrak{F}}$ is $|G : T|$. Hence $|G/N : N_{G/N}(\Sigma N/N)| = |G : T|$. Now $|G : TN| \leq |G : T| = |G/N : N_{G/N}(\Sigma N/N)| \leq |G/N : TN/N|$. This implies that $TN/N = N_{G/N}(\Sigma N/N)$.

Assume now that $N \leq G^{\mathfrak{F}} = R$. Since system normalisers are preserved under epimorphisms by [DH92, I, 5.8], we have that $N_{R/N}(\Sigma N/N) = N_R(\Sigma)N/N$. Hence, since $G = RT$, we have that $|G/N : N_{G/N}(\Sigma N/N)| = |R/N : N_{R/N}(\Sigma N/N)| = |R : N_R(\Sigma)N| = |R : (T \cap R)N| = |R : R \cap TN| = |G : TN| = |G/N : TN/N|$ and then $TN/N = N_{G/N}(\Sigma N/N)$.

If N is not a minimal normal subgroup of G and A is a minimal normal subgroup of G contained in N, it follows that $TA/A = N_{G/A}(\Sigma A/A)$. By induction, $(TN/A)/(N/A) = N_{(G/A)/(N/A)}\big((\Sigma N/A)/(N/A)\big)$. Then $TN/N = N_{G/N}(\Sigma N/N)$. $\qquad\square$

The following result is also useful.

Proposition 4.2.11 ([Hal37]). *Let G be a soluble group and N a normal subgroup of G. Let Σ^* be a Hall system of N such that $\Sigma^* = \Sigma \cap N$ for some Hall system Σ of G. Put $M = N_G(\Sigma^*)$. We have*

1. *$N_G(\Sigma)$ is contained in M,*
2. *$\Sigma_1 = \Sigma \cap M$ is a Hall system of M, and*
3. *$N_M(\Sigma_1) = N_G(\Sigma)$.*

Proof. 1. For any Hall subgroup H^* of N in Σ^*, there exists a Hall subgroup H of G in Σ such that $H^* = H \cap N$. If $x \in N_G(\Sigma)$, then $H^{*x} = (H \cap N)^x = H^x \cap N = H \cap N = H^*$, since N is normal in G. Then $x \in N_G(\Sigma^*)$. Hence $N_G(\Sigma) \leq N_G(\Sigma^*) = M$.

2. Let p be any prime dividing the order of G, H the Hall p'-subgroup of N in Σ^* and P the Sylow p-subgroup of G in Σ. There exists a Hall p'-subgroup S of G in Σ such that $S \cap N = H$. Since S normalises H and $G = PS$, it follows that $T = N_G(H) \cap P$ is a Sylow p-subgroup of $N_G(H)$. Moreover, for any prime $q \neq p$, P is contained in the Hall q'-subgroup S_q of G in Σ.

Hence, $T \leq S_q$. The subgroup $S_q \cap N$ is the Hall q'-subgroup of N in Σ^* and $S_q \cap N$ is normal in S_q. Therefore T normalises $S_q \cap N$. This means that $T \leq N_G(\Sigma^*) = M$ and $T = M \cap P$.

For two different primes p_i, $i = 1, 2$, dividing the order of G consider the corresponding Sylow subgroups $P_i \in \mathrm{Syl}_{p_i}(G)$ of G in Σ and $T_i = P_i \cap M$, $i = 1, 2$. Note that $P_1 P_2$ is a subgroup of G and $\langle T_1, T_2 \rangle$ is contained in $P_1 P_2 \cap M$. Hence, $\langle T_1, T_2 \rangle$ is a $\{p_1, p_2\}$-subgroup and so $\langle T_1, T_2 \rangle = T_1 T_2$. Therefore $\Sigma \cap M = \Sigma_1$ is a Hall system of M.

3. Clearly, Σ^{*g} is a Hall system of N, for all $g \in G$. Therefore, there exists $x \in N$, such that $\Sigma^{*g} = \Sigma^{*x}$. The Frattini argument implies that $G = MN$. Therefore, if $P \in \mathrm{Syl}_p(G) \cap \Sigma$, then $(P \cap M)N/N = PN/N \in \mathrm{Syl}_p(G/N)$. Hence $(P \cap M)(P \cap N) = P \cap (P \cap M)N = P \cap PN = P$.

If $x \in N_G(\Sigma)$, then $x \in M$ and, for any Sylow subgroup $P \in \Sigma$, we have that $(P \cap M)^x = (P \cap M)$. Hence $N_G(\Sigma) \leq N_M(\Sigma_1)$. Conversely, if $x \in N_M(\Sigma_1)$, for any Sylow subgroup $P \in \Sigma$, we have that $x \in N_G(P \cap M)$ and $x \in M \leq N_G(P \cap N)$. Hence $x \in N_G(P)$. Consequently $N_M(\Sigma_1) \leq N_G(\Sigma)$ and the equality holds. $\qquad \square$

Lemma 4.2.12. *Let G be a group with a soluble normal subgroup H such that $G^{\mathfrak{F}} \leq H$. Let Σ be a Hall system of H. Denote $R = N_G(\Sigma)$. Then each $\mathfrak{F}$-projector of R is contained in an $\mathfrak{F}$-projector of $N_G(\Sigma \cap G^{\mathfrak{F}})$.*

Proof. Assume that the result is not true and let G be a minimal counterexample. Let H be a normal subgroup of G of minimal index $|H : G^{\mathfrak{F}}|$ among all normal subgroups for which the assertion does not hold. Let H/K a chief factor of G such that $G^{\mathfrak{F}} \leq K$. Note that $\Sigma \cap K$ is a Hall system of K and denote $B = N_G(\Sigma \cap K)$. Since the lemma is true for G, K, and $\Sigma \cap K$, we have that each $\mathfrak{F}$-projector of B is contained in an $\mathfrak{F}$-projector of $N_G(\Sigma \cap G^{\mathfrak{F}})$. By Proposition 4.2.11 (2), we have that $\Sigma^* = \Sigma \cap (H \cap B)$ is a Hall system of $H \cap B = N_H(\Sigma \cap K)$. On the other hand, since $G = N_G(\Sigma \cap K)K = BH$, and then $B/(B \cap H) \cong G/H \in \mathfrak{F}$, the subgroups B, $H \cap B$, and the Hall system Σ^* satisfy the hypotheses of the lemma. If B is a proper subgroup of G, each $\mathfrak{F}$-projector of $Q = N_B(\Sigma^*)$ is contained in an $\mathfrak{F}$-projector of $N_B(\Sigma^* \cap B^{\mathfrak{F}})$. Note that $N_{H \cap B}(\Sigma^*) = N_H(\Sigma)$ by Proposition 4.2.11 (3). Moreover $N_G(\Sigma) \leq Q$. Since $G = H N_G(\Sigma)$, we have that $B = (H \cap B) N_G(\Sigma)$. Consequently $Q = N_G(\Sigma)(Q \cap H \cap B) = N_G(\Sigma) N_{H \cap B}(\Sigma^*) = N_G(\Sigma) = R$. This contradiction yields $B = G$. In other words, every Sylow subgroup of K is normal in G. In particular, $G \in \mathfrak{N}\mathfrak{F}$. Suppose that p is the prime divisor of the order of H/K. If P is the Sylow p-subgroup of H in Σ, we have that $H = PK$ and $R = N_G(\Sigma) = N_G(P)$. Let E be an $\mathfrak{F}$-projector of R, then $G = HR = KR = K(ER^{\mathfrak{F}}) = EK = E\,\mathrm{F}(G)$ because $R^{\mathfrak{F}}$ is contained in $G^{\mathfrak{F}}$. By Theorem 4.2.7, E is contained in an $\mathfrak{F}$-projector of $G = N_G(\Sigma \cap G^{\mathfrak{F}})$. This is the final contradiction. $\qquad \square$

Theorem 4.2.13. *Let G be a group whose $\mathfrak{F}$-residual $G^{\mathfrak{F}}$ is soluble. Consider a Hall system Σ of $G^{\mathfrak{F}}$ and denote $T = N_G(\Sigma)$. Suppose that M is an*

$\mathfrak{F}$-abnormal maximal subgroup of G. If Σ reduces into $M \cap G^{\mathfrak{F}}$, then there exists an $\mathfrak{F}$-projector of T contained in an $\mathfrak{F}$-projector of $\mathrm{N}_M(\Sigma \cap M^{\mathfrak{F}})$.

Proof. We split the proof in two steps.

1. *There exists an $\mathfrak{F}$-projector of T contained in M.*

We use induction on the order of G. Note that, by [DH92, I, 4.17a] and Lemma 4.2.10, the hypotheses of the lemma hold in $G/\operatorname{Core}_G(M)$ and $M/\operatorname{Core}_G(M)$. If $\operatorname{Core}_G(M)$ is non-trivial, then, by induction, there exists an $\mathfrak{F}$-projector of $T\operatorname{Core}_G(M)/\operatorname{Core}_G(M)$, $D/\operatorname{Core}_G(M)$ say, contained in $M/\operatorname{Core}_G(M)$. We know that the $\mathfrak{F}$-residual of $T\operatorname{Core}_G(M)/\operatorname{Core}_G(M)$ is nilpotent and therefore the $\mathfrak{F}$-projectors of $T\operatorname{Core}_G(M)/\operatorname{Core}_G(M)$ are conjugate by Theorem 4.2.1. If E is an $\mathfrak{F}$-projector of T, then there exists $g \in T$ such that $D = E^g \operatorname{Core}_G(M)$. Hence E^g is an $\mathfrak{F}$-projector of T contained in M. Assume now that $\operatorname{Core}_G(M) = 1$. Since M is $\mathfrak{F}$-abnormal in G, the group G is a primitive group of type 1 and $G = MN$, where N is the minimal normal subgroup of G. Clearly we can assume that G is not an $\mathfrak{F}$-group. Then $N \leq G^{\mathfrak{F}}$ and, by Proposition 2.2.8 (3), $M \cap G^{\mathfrak{F}} = M^{\mathfrak{F}}$. If $M^{\mathfrak{F}} = 1$, then M is an $\mathfrak{F}$-group and $\mathrm{N}_M(\Sigma \cap M^{\mathfrak{F}}) = M$. Then M is an $\mathfrak{F}$-projector of G and $G \in \mathfrak{N}\mathfrak{F}$. In this case $G = T$ and our claim is true. Suppose that $M^{\mathfrak{F}} \neq 1$. We see that in this case T is contained in M. Consider an element $am \in T$, with $a \neq 1, a \in N$, and $m \in M$. If p is the prime divisor of $|N| = |G^{\mathfrak{F}} : M^{\mathfrak{F}}|$ and S^p is the Hall p'-subgroup of $G^{\mathfrak{F}}$ in Σ, then $(S^p)^{am} = S^p$. Moreover, $S^p \leq M^{\mathfrak{F}} \leq M$ and then $(S^p)^a \leq M$. If $x \in S^p$, then $[x, a] \in M \cap N = 1$. Consequently a centralises S^p and $N \leq \mathrm{Z}(G^{\mathfrak{F}})$ by [DH92, I, 5.5]. Thus $G^{\mathfrak{F}}$ is contained in $\mathrm{C}_G(N)$ which is equal to N by Theorem 1. Hence $M^{\mathfrak{F}} \leq N \cap M = 1$. This contradiction shows that $a = 1$ and T is contained in M.

2. *Conclusion.*

Let D be an $\mathfrak{F}$-projector of T contained in M. Since $T^{\mathfrak{F}} \leq G^{\mathfrak{F}}$, we have that $G = TG^{\mathfrak{F}} = DG^{\mathfrak{F}}$. Put $R = M \cap G^{\mathfrak{F}}$; by hypothesis $\Sigma \cap R$ is a Hall system of R. Then we have that $D \leq \mathrm{N}_M(\Sigma) \leq \mathrm{N}_M(\Sigma \cap R) = D\bigl(G^{\mathfrak{F}} \cap \mathrm{N}_M(\Sigma \cap R)\bigr) = D\,\mathrm{N}_R(\Sigma \cap R)$. Since system normalisers of soluble groups are nilpotent, it follows that $\mathrm{N}_R(\Sigma \cap R)$ is a nilpotent normal subgroup of $\mathrm{N}_M(\Sigma \cap R)$. Hence $\mathrm{N}_M(\Sigma \cap R) \in \mathfrak{N}\mathfrak{F}$ and D supplements the Fitting subgroup of $\mathrm{N}_M(\Sigma \cap R)$. By Theorem 4.2.8, D is contained in an $\mathfrak{F}$-projector E of $\mathrm{N}_M(\Sigma \cap R)$. Since $M^{\mathfrak{F}} \leq R$ and R is soluble, we can apply Lemma 4.2.12 to M, R, and $\Sigma \cap R$ and deduce that each $\mathfrak{F}$-projector of $\mathrm{N}_M(\Sigma \cap R)$ is contained in an $\mathfrak{F}$-projector of $\mathrm{N}_M(\Sigma \cap M^{\mathfrak{F}})$. Therefore E, and then D, is contained in an $\mathfrak{F}$-projector of $\mathrm{N}_M(\Sigma \cap M^{\mathfrak{F}})$. $\square$

Theorem 4.2.14. *Let G be a group whose $\mathfrak{F}$-residual $G^{\mathfrak{F}}$ is soluble. Consider a Hall system Σ of $G^{\mathfrak{F}}$ and denote $T = \mathrm{N}_G(\Sigma)$. If D is an $\mathfrak{F}$-projector of T, then D covers all $\mathfrak{F}$-central chief factors of G and avoids the $\mathfrak{F}$-eccentric ones.*

Proof. By Lemma 4.2.10, it is enough to prove that D covers the $\mathfrak{F}$-central minimal normal subgroups of G and avoids the $\mathfrak{F}$-eccentric ones. Let N be a $\mathfrak{F}$-central minimal normal subgroup of G. Then $N \leq \mathrm{C}_G(G^{\mathfrak{F}})$. It implies that N is contained in T and $G = DG^{\mathfrak{F}} = D\,\mathrm{C}_G(N)$. Hence N is a minimal normal subgroup of ND and $[N]*(ND) \cong [N]*G \in \mathfrak{F}$. Since $\mathfrak{F}$ is a saturated formation and $ND/N \in \mathfrak{F}$, we have that $ND \in \mathfrak{F}$. Since D is $\mathfrak{F}$-maximal in T, we have that $N \leq D$. Suppose now that N is $\mathfrak{F}$-eccentric in G. Then $N \leq G^{\mathfrak{F}}$ and N is abelian. If D does not avoid N, then $N \cap D \neq 1$. By [DH92, I, 5.5], we deduce that $N \leq \mathrm{Z}(G^{\mathfrak{F}})$, and then N is $\mathfrak{F}$-central in G, contrary to supposition. Therefore N is avoided by D. $\square$

Now we can give a characterisation of the $\mathfrak{F}$-normalisers of a group G whose $\mathfrak{F}$-residual is soluble in terms of the $\mathfrak{F}$-projectors of the absolute system normalisers of the Hall systems of $G^{\mathfrak{F}}$.

Theorem 4.2.15. *Let G be a group whose $\mathfrak{F}$-residual $G^{\mathfrak{F}}$ is soluble. For every Hall system Σ of $G^{\mathfrak{F}}$, every $\mathfrak{F}$-projector of $\mathrm{N}_G(\Sigma)$ is an $\mathfrak{F}$-normaliser of G.*
Thus

$$\mathrm{Nor}_{\mathfrak{F}}(G) = \bigcup \{ E \in \mathrm{Proj}_{\mathfrak{F}}\big(\mathrm{N}_G(\Sigma)\big) : \Sigma \text{ is a Hall system of } G^{\mathfrak{F}} \},$$

and $\mathrm{Nor}_{\mathfrak{F}}(G)$ is a conjugacy class of subgroups of G.

Proof. We can assume that G is not an $\mathfrak{F}$-group. Let Σ be a Hall system of $G^{\mathfrak{F}}$ and let M be an $\mathfrak{F}$-critical subgroup of G such that Σ reduces into $M \cap G^{\mathfrak{F}}$. By Theorem 4.2.13 there exists an $\mathfrak{F}$-projector D of $\mathrm{N}_G(\Sigma)$ contained in an $\mathfrak{F}$-projector D^* of $\mathrm{N}_M(\Sigma \cap M^{\mathfrak{F}})$. Arguing by induction, D^* is an $\mathfrak{F}$-normaliser of M, and then of G. Applying Theorem 4.2.4, for D^*, and Theorem 4.2.14, for D, we have that both cover simultaneously all $\mathfrak{F}$-central chief factors of G and avoid the $\mathfrak{F}$-eccentric ones. Therefore D and D^* have the same order and $D = D^*$. Since $\mathrm{N}_G(\Sigma) \in \mathfrak{N}\mathfrak{F}$, the $\mathfrak{F}$-projectors of $\mathrm{N}_G(\Sigma)$ are a conjugacy class of subgroups by Theorem 4.2.1. Therefore, every $\mathfrak{F}$-projector of $\mathrm{N}_G(\Sigma)$ is an $\mathfrak{F}$-normaliser of G.

Conversely, if D is an $\mathfrak{F}$-normaliser of G and $D \neq G$, then D is an $\mathfrak{F}$-normaliser of an $\mathfrak{F}$-critical subgroup M of G. By induction, there exists a Hall system Σ^* of $M^{\mathfrak{F}}$ such that $D \in \mathrm{Proj}_{\mathfrak{F}}\big(\mathrm{N}_M(\Sigma^*)\big)$. Since, by Proposition 2.2.8 (3), $M^{\mathfrak{F}}$ is contained in $G^{\mathfrak{F}}$, we can find a Hall system Σ of $G^{\mathfrak{F}}$ which reduces into $M \cap G^{\mathfrak{F}}$ and $\Sigma \cap M^{\mathfrak{F}} = \Sigma^*$ by [DH92, I, 4.16]. Applying Theorem 4.2.13, $\mathrm{N}_M(\Sigma^*)$ contains an $\mathfrak{F}$-projector of $\mathrm{N}_G(\Sigma)$. Since $\mathrm{Proj}_{\mathfrak{F}}\big(\mathrm{N}_M(\Sigma^*)\big)$ is a conjugacy class of subgroups of $\mathrm{N}_M(\Sigma^*)$, it follows that there exists an $\mathfrak{F}$-projector E of $\mathrm{N}_G(\Sigma^g)$, for some $g \in G$, contained in D. Thus, D is an $\mathfrak{F}$-projector of $\mathrm{N}_G(\Sigma^g)$ by Theorem 4.2.4 and Theorem 4.2.14. Consequently,

$$\bigcup \{ E \in \mathrm{Proj}_{\mathfrak{F}}\big(\mathrm{N}_G(\Sigma)\big) : \Sigma \text{ is a Hall system of } G^{\mathfrak{F}} \} = \mathrm{Nor}_{\mathfrak{F}}(G).   \quad \square$$

Corollary 4.2.16. *Let G be a group whose $\mathfrak{F}$-residual $G^{\mathfrak{F}}$ is soluble. If H is an $\mathfrak{F}$-projector of G complementing $G^{\mathfrak{F}}$ in G, then H normalises some Sylow p-subgroup of $G^{\mathfrak{F}}$, for each prime p dividing the order of $G^{\mathfrak{F}}$.*

Proof. By Theorem 4.2.9, H contains an $\mathfrak{F}$-normaliser of G. Since in this case both complement $G^{\mathfrak{F}}$, then H is an $\mathfrak{F}$-normaliser of G. By Theorem 4.2.15, there exists a Hall system Σ of $G^{\mathfrak{F}}$ such that $H \leq \mathrm{N}_G(\Sigma)$. This means that H normalises every Sylow subgroup of $G^{\mathfrak{F}}$ in Σ. $\square$

The following useful splitting theorem is a generalisation of a theorem due to G. Higman on complementation of abelian normal subgroups. The corresponding result for finite soluble groups was obtained by R. W. Carter and T. O. Hawkes (see [CH67] and [DH92, IV, 5.18]).

Theorem 4.2.17. *Let $\mathfrak{F}$ be a saturated formation and let G be group whose $\mathfrak{F}$-residual $G^{\mathfrak{F}}$ is abelian. Then $G^{\mathfrak{F}}$ is complemented in G and two any complements are conjugate in G. The complements are the $\mathfrak{F}$-normalisers of G.*

Proof. First we prove that an $\mathfrak{F}$-normaliser of G is a complement of $G^{\mathfrak{F}}$. Suppose that this is not true and let G be a minimal counterexample. Put $R = G^{\mathfrak{F}}$. Then there exists $D \in \mathrm{Nor}_{\mathfrak{F}}(G)$ such that $D \cap R \neq 1$. Observe that, since R is abelian and $G = RD$, the subgroup $R \cap D$ is normal in G.

Assume that there exists an $\mathfrak{F}$-eccentric minimal normal subgroup N of G such that $N \leq R$. The quotient DN/N is an $\mathfrak{F}$-normaliser of G/N and $R/N = (G/N)^{\mathfrak{F}}$. By minimality of G, we have that $R \cap D = N$. But then D covers N and N has to be $\mathfrak{F}$-central in G by Theorem 4.2.4. This is a contradiction. Hence every minimal normal subgroup of G below R is $\mathfrak{F}$-central in G. Then, if N is any minimal normal subgroup of G below R, we have that $N \leq D$ and, by minimality of G, $R \cap D = N$. Consequently, N is the unique minimal normal subgroup of G below R.

Let M be an $\mathfrak{F}$-critical subgroup of G such that $D \in \mathrm{Nor}_{\mathfrak{F}}(M)$. Since $M^{\mathfrak{F}}$ is contained in R, we have that $M^{\mathfrak{F}}$ is an abelian normal subgroup of G. If $M^{\mathfrak{F}} \neq 1$, then N is contained in $M^{\mathfrak{F}}$ and, by minimality of G, we have that $M^{\mathfrak{F}} \cap D = 1$. This is a contradiction. Hence $M \in \mathfrak{F}$ and then $M = D$. This implies that R/N is chief factor of G complemented by D. Let p be the prime dividing the order of N. Then R is an abelian p-group. Suppose that F is the integrated and full local definition of $\mathfrak{F}$. Then $F(p) \neq \emptyset$ and $R \leq G^{F(p)}$. Observe that $\mathfrak{F}$ is contained in $\mathfrak{E}_{p'}F(p)$ and that $G^{F(p)}/R$ is therefore a p'-group. Thus $R \in \mathrm{Syl}_p(G^{F(p)})$. By the Schur-Zassenhaus Theorem [Hup67, I, 18.1 and 18.2], there exists a complement Q of R in $G^{F(p)}$. Observe that R/N is a chief factor avoided by D. Therefore R/N is $\mathfrak{F}$-eccentric in G. Consequently $G/\mathrm{C}_G(R/N) \notin F(p)$, and $G^{F(p)}$ is not contained in $\mathrm{C}_G(R/N)$. Consider the p'-group Q acting on the normal p-subgroup R by conjugation. Then $R = [R, Q] \times \mathrm{C}_R(Q)$ by [DH92, A, 12.5]. Observe that both $\mathrm{C}_R(Q) = \mathrm{C}_R(QR) = \mathrm{C}_R(G^{F(p)})$ and $[R, Q] = [R, QR] = [R, G^{F(p)}]$ are normal subgroups of G. Since N is the unique minimal normal subgroup of G below R, then either $\mathrm{C}_R(Q) = 1$ or $[R, Q] = 1$. Since N is $\mathfrak{F}$-central in G, we have that $N \leq \mathrm{C}_R(G^{F(p)}) = \mathrm{C}_R(Q)$. Consequently, $G^{F(p)} = QR \leq \mathrm{C}_G(R) \leq \mathrm{C}_G(R/N)$, contrary to supposition. Therefore each $\mathfrak{F}$-normaliser complements $G^{\mathfrak{F}}$ in G.

Consider now a subgroup H of G such that $G = HG^{\mathfrak{F}}$ and $H \cap G^{\mathfrak{F}} = 1$. Since every chief factor of G below $G^{\mathfrak{F}}$ is $\mathfrak{F}$-eccentric, the subgroup H covers all $\mathfrak{F}$-central chief factors of a chief series of G through $G^{\mathfrak{F}}$. By Theorem 4.1.20, there exists $D \in \mathrm{Nor}_{\mathfrak{F}}(G)$ such that $D \leq H$. Therefore $D = H \in \mathrm{Nor}_{\mathfrak{F}}(G)$.

Finally, by Theorem 4.2.15, $\mathrm{Nor}_{\mathfrak{F}}(G)$ is a conjugacy class of subgroups of G. Hence the complements of $G^{\mathfrak{F}}$ are the $\mathfrak{F}$-normalisers of G and they are conjugate. $\qquad\square$

A consequence of Theorem 4.2.17 is the following result due to P. Schmid.

Corollary 4.2.18 ([Sch74]). *For every group G, we have that*

$$G^{\mathfrak{F}} \cap Z_{\mathfrak{F}}(G) \leq (G^{\mathfrak{F}})' \cap Z(G^{\mathfrak{F}}).$$

Proof. Theorem 4.2.17, applied to the group $G/(G^{\mathfrak{F}})'$, leads to $Z_{\mathfrak{F}}(G) \cap G^{\mathfrak{F}} \leq (G^{\mathfrak{F}})'$. By [DH92, IV, 6.10]), we have that $[G^{\mathfrak{F}}, Z_{\mathfrak{F}}(G)] = 1$. Therefore $G^{\mathfrak{F}} \cap Z_{\mathfrak{F}}(G) \leq (G^{\mathfrak{F}})' \cap Z(G^{\mathfrak{F}})$. $\qquad\square$

Next, we use Corollary 4.2.18 to give a short proof of a well-known result of L. A. Shemetkov ([She72]).

Theorem 4.2.19. *Let G be a group such that for some prime p, the Sylow p-subgroups of $G^{\mathfrak{F}}$ are abelian. Then every chief factor of G below $G^{\mathfrak{F}}$ whose order is divisible by p is an $\mathfrak{F}$-eccentric chief factor of G.*

Proof. Suppose that the theorem is false and let G be a minimal counterexample. Then $G^{\mathfrak{F}} \neq 1$. Let N be a minimal normal subgroup of G such that $N \leq G^{\mathfrak{F}}$. From minimality of G, every chief factor of G between N and $G^{\mathfrak{F}}$ whose order is divisible by p is $\mathfrak{F}$-eccentric, the prime p divides $|N|$ and N is an $\mathfrak{F}$-central chief factor of G. Then $N \leq G^{\mathfrak{F}} \cap Z_{\mathfrak{F}}(G) \leq (G^{\mathfrak{F}})' \cap Z(G^{\mathfrak{F}})$ by Corollary 4.2.18. Let P be a Sylow p-subgroup of $G^{\mathfrak{F}}$. Since P is abelian, we have that $N \leq (G^{\mathfrak{F}})' \cap Z(G^{\mathfrak{F}}) \cap P = 1$ by Taunt's Theorem (see [Hup67, VI, 14.3]). This contradiction concludes the proof. $\qquad\square$

We round the section off with another interesting splitting theorem.

Theorem 4.2.20. *Let G be a group such that every chief factor of G below $G^{\mathfrak{F}}$ is $\mathfrak{F}$-eccentric. Assume that $G^{\mathfrak{F}}$ is p-nilpotent for every prime p in $\pi = \pi(|G : G^{\mathfrak{F}}|)$, Then*

1. *(P. Schmid, [Sch74]) $G^{\mathfrak{F}}$ is complemented in G and any two complements are conjugate;*
2. *(A. Ballester-Bolinches, [BB89a]) the complements of $G^{\mathfrak{F}}$ in G are the $(\mathfrak{F} \cap \mathfrak{S}_{\pi})$-normalisers of G.*

Proof. First we note that the class $\mathfrak{L} = \mathfrak{F} \cap \mathfrak{S}_{\pi}$ is a saturated formation and $G^{\mathfrak{L}} = G^{\mathfrak{F}}$.

We argue by induction on the order of G. Consider $N = O^{\pi}(G^{\mathfrak{F}})$ and suppose that $N \neq 1$. The quotient group $G^{\mathfrak{F}}/N = (G/N)^{\mathfrak{F}}$ is a nilpotent π-group.

By induction, $G^{\mathfrak{F}}/N$ is complemented in G/N and any two complements are conjugate. If L/N is a complement of $G^{\mathfrak{F}}/N$ in G/N, then N is a normal Hall π'-subgroup of L. By the Schur-Zassenhaus Theorem [Hup67, I, 18.1 and 18.2], there exists a Hall π-subgroup H of L and two Hall π-subgroups of L are conjugate in L. Observe that $H \cap G^{\mathfrak{F}} = 1$ and then $G^{\mathfrak{F}}$ is complemented in G. Moreover if A and B are two complements of $G^{\mathfrak{F}}$ in G, then AN/N and BN/N are conjugate in G/N. Without loss of generality we can assume that $AN = BN$. Since A and B are Hall π-subgroups of AN and N is a normal Hall π'-subgroup of AN, it follows that A and B are conjugate by the Schur-Zassenhaus Theorem. If E is an $\mathfrak{L}$-normaliser of G, then EN/N is an $\mathfrak{L}$-normaliser of G/N by Proposition 4.1.5. By induction, $E \cap G^{\mathfrak{F}} \leq N$. Since E is a π-group and N is a π'-group, we have that $E \cap G^{\mathfrak{F}} = 1$ and E complements $G^{\mathfrak{F}}$ in G.

Therefore we can assume that $N = 1$, i.e. $G^{\mathfrak{F}}$ is a nilpotent π-group, and G is a π-group in $\mathfrak{N}\mathfrak{F}$. Here the $\mathfrak{L}$-normalisers and the $\mathfrak{F}$-normalisers of G coincide. Since every chief factor of G below $G^{\mathfrak{F}}$ is $\mathfrak{F}$-eccentric in G, if D is an $\mathfrak{F}$-normaliser of G, then $D \cap G^{\mathfrak{F}} = 1$, by Corollary 4.2.5, and D is a complement of $G^{\mathfrak{F}}$ in G. Any complement E of $G^{\mathfrak{F}}$ is an $\mathfrak{F}$-group. By Lemma 4.1.17, E is contained in an $\mathfrak{F}$-normaliser. Hence E is an $\mathfrak{F}$-normaliser of G. Thus, the complements of $G^{\mathfrak{F}}$ in G are the $\mathfrak{F}$-normalisers of G, and they are conjugate, by Theorem 4.2.15. $\qquad\square$

Postscript

K. Doerk (see [DH92, V, 3.18]) used the $\mathfrak{F}$-normalisers to show that a saturated formation $\mathfrak{F}$ has a unique upper bound for all local definitions, that is, a maximal local definition, in the soluble universe. In fact, he proved that the formation function g given by

$$g(p) = \big(G : \text{the } \mathfrak{F}\text{- normalisers of } G \text{ are in } F(p)\big),$$

for all primes p, is the maximal local definition of $\mathfrak{F}$.

As we have seen in Chapter 3, the situation in the general finite universe is not so clear cut. However, it is possible to use the $\mathfrak{F}$-normalisers of finite, non-necessarily soluble, groups to give necessary and sufficient conditions for a saturated formation $\mathfrak{F}$ to have a maximal local definition ([BB89a], [BB91]).

4.3 Subgroups of prefrattini type

The introduction of systems of maximal subgroups in [BBE91] made possible the extension of prefrattini subgroups to finite, non-necessarily soluble, groups. Later, in [BBE95], we introduced the concept of a *weakly solid* (or simply *w-solid) set of maximal subgroups* following some ideas due to M. J. Tomkinson [Tom75]. Equipped with these new notions, we were able to present a common generalisation of all prefrattini subgroups of the literature. These new subgroups enjoy most of the properties of the soluble case, for instance they are preserved by epimorphic images and enjoy excellent factorisation properties. Unfortunately, we cannot expect to keep cover-avoidance property and conjugacy. In fact, conjugacy characterises solubility, and conjugacy and cover-avoidance property are equivalent in some sense (see Corollary 4.3.14). In fact we can repeat here the comment said in the introduction of Section 1.4: we lose the arithmetical properties, but we find deep relations between maximal subgroups which are general to all finite groups.

We present here a distillation of the preceding concepts. Observe, for instance, that the definition of system of maximal subgroups given in [BBE91] is different, but equivalent, to the one in Section 1.4. In fact this presentation allows us to speak of a particular subgroup of prefrattini type, which is defined by the intersection of all maximal subgroups in a subsystem of maximal subgroups. This point of view is new since all precedents of prefrattini subgroups in the past were families of subgroups of the group. To recover this classical idea of a set of prefrattini subgroups, we include the concept of w-solid set as a union-set of subsystems of maximal subgroups.

Definitions 4.3.1. *Let* $\mathbf{X}$ *be a (possibly empty) set of monolithic maximal subgroups of a group* G.

1. *We will say that* $\mathbf{X}$ *is a* weakly solid (w-solid) *set of maximal subgroups of* G *if*

   for any $U, S \in \mathbf{X}$ *such that* $\mathrm{Core}_G(U) \neq \mathrm{Core}_G(S)$ *and both complement the same abelian chief factor* H/K *of* G, *then* $M = (U \cap S)H \in \mathbf{X}$.   (4.5)

2. $\mathbf{X}$ *is said to be* solid *if it satisfies* (4.5) *and whenever a chief factor is* $\mathbf{X}$*-supplemented in* G, *then all its monolithic supplements are in* $\mathbf{X}$.

Next we give a varied selection of examples of w-solid and solid sets.

Examples 4.3.2.   1.   The set $\mathrm{Max}^*(G)$, of all monolithic maximal subgroups of a group G, is solid.

2. Consider a subgroup L of a group G; the set $\mathbf{X}_L$ of all monolithic maximal subgroups of G containing L is w-solid.

3. Given a w-solid (respectively solid) set $\mathbf{X}$ of maximal subgroups of a group G and a class $\mathfrak{H}$ of groups, then the set $\mathbf{X}_{\mathfrak{H}}^a$ of all $\mathfrak{H}$-abnormal subgroups in $\mathbf{X}$ and the set $\mathbf{X}_{\mathfrak{H}}^n$ of all $\mathfrak{H}$-normal subgroups in $\mathbf{X}$ are w-solid (respectively solid) as well.

If $\mathbf{X}$ is a system of maximal subgroups, then $\mathbf{X}_{\mathfrak{H}}^a$ and $\mathbf{X}_{\mathfrak{H}}^n$ are subsystems of maximal subgroups.

Let M be a monolithic maximal subgroup of G. Recall that the normal index of M in G, defined by W. E. Deskins in [Des59] and denoted by $\eta(G, M)$, is indeed $\eta(G, M) = \left| \mathrm{Soc}\big(G/\mathrm{Core}_G(M)\big) \right|$.

3. The following families of monolithic maximal subgroups of a group G are w-solid:

a) Fixed a prime p, the set $\mathbf{X}_p$ of all monolithic maximal subgroups M of G such that $|G : M|$ is a p-power. In fact, if G is p-soluble, then $\mathbf{X}_p$ is indeed solid. However this is not true in the non-soluble case; in $G = \mathrm{Alt}(5)$ the set $\mathbf{X}_5$ is composed of all maximal subgroups isomorphic to $\mathrm{Alt}(4)$ and clearly it is not solid.

b) Fixed a set of primes π, the set $\mathbf{X}^\pi$ of all monolithic maximal subgroups M of G such that $|G : M|$ is a π'-number.

c) the set of all monolithic maximal subgroups of G of composite index in G.

d) the set of all monolithic maximal subgroups M of the group G such that $\eta(G, M) \neq |G : M|$.

If G is a group, the set $\mathcal{S}(G)$ composed of all systems of maximal subgroups of G is non-empty by Theorem 1.4.7. If $\mathbf{X}$ is a w-solid set of maximal subgroups of G and $\mathbf{Y} \in \mathcal{S}(G)$, then $\mathbf{X} \cap \mathbf{Y}$ is a subsystem of maximal subgroups of G. Applying Theorem 1.4.7, we have that $\mathbf{X} = \bigcup\{\mathbf{X} \cap \mathbf{Y} : \mathbf{Y} \in \mathcal{S}(G)\}$.

Definitions 4.3.3. *1. Let G be a group. Let $\mathbf{X}$ be a non-empty subsystem of maximal subgroups of G. Define*

$$W(G, \mathbf{X}) = \bigcap\{M : M \in \mathbf{X}\}.$$

For convenience, we define $W(G, \emptyset) = G$.

We will say that W is a subgroup of prefrattini type of G if $W = W(G, \mathbf{X})$ for some subsystem $\mathbf{X}$ of maximal subgroups of G.

2. If $\mathbf{X}$ be a w-solid set of maximal subgroups of G, we say that

$$\mathrm{Pref}_\mathbf{X}(G) = \{W(G, \mathbf{X} \cap \mathbf{Y}) : \mathbf{Y} \in \mathcal{S}(G),\ \mathbf{X} \cap \mathbf{Y} \neq \emptyset\}$$

is the set of all $\mathbf{X}$-prefrattini subgroups of G.

We show in the following that the known prefrattini subgroups are associated with w-solid sets of maximal subgroups.

Examples 4.3.4. 1. The $\mathrm{Max}^*(G)$-prefrattini subgroups are simply called *prefrattini subgroups* of G. We write

$$\mathrm{Pref}(G) = \{W(G, \mathbf{X}) : \mathbf{X} \in \mathcal{S}(G)\}.$$

In other words, a prefrattini subgroup of a group G is a subgroup of the form $W(G, \mathbf{X})$, where $\mathbf{X}$ is a system of maximal subgroups of G. If G is a soluble group, we can apply Corollary 1.4.18 and conclude that the prefrattini subgroups of G are those introduced by W. Gaschütz in [Gas62] which originated this theory.

2. Let $\mathfrak{H}$ be a Schunck class. The $\mathrm{Max}^*(G)_\mathfrak{H}^a$-prefrattini subgroups of a group G are the $\mathfrak{H}$-prefrattini subgroups defined in [BBE91]. If G is soluble, they are the $\mathfrak{H}$-prefrattini subgroups studied by P. Förster in [För83] and, if $\mathfrak{H}$ is a saturated formation, the $\mathrm{Max}^*(G)_\mathfrak{H}^a$-prefrattini subgroups of G are the ones introduced by T. O. Hawkes in [Haw67].

3. If G is a soluble group, then $\mathrm{Pref}_{\mathbf{X}_L}(G)$ is the set of all L-prefrattini subgroups introduced by H. Kurzweil in [Kur89].

4. The $\mathbf{X}_p$-prefrattini subgroups of a p-soluble group are the p-prefrattini subgroups studied by A. Brandis in [Bra88].

Notation 4.3.5. If $\mathfrak{H}$ is a Schunck class, G is a group, and $\mathbf{X}$ is a system of maximal subgroups of G, we denote

$$W(G, \mathfrak{H}, \mathbf{X}) = W(G, \mathbf{X}_{\mathfrak{H}}^a),$$

and say that $W(G, \mathfrak{H}, \mathbf{X})$ is the $\mathfrak{H}$-prefrattini subgroup of G associated with $\mathbf{X}$. We write

$$\mathrm{Pref}_{\mathfrak{H}}(G) = \{W(G, \mathfrak{H}, \mathbf{X}) : \mathbf{X} \in \mathcal{S}(G)\}$$

for the set of all $\mathfrak{H}$-prefrattini subgroups of G.

Theorem 4.3.6. *Consider a group G, $\mathbf{X}$ a subsystem of maximal subgroups of G and $W = W(G, \mathbf{X})$. Then*

$$W = \bigcap\{\mathrm{T}(G, \mathbf{X}, F) : F \text{ is an } \mathbf{X}\text{-supplemented chief factor of } G\}.$$

Moreover W has the following properties.

1. *Let $1 = G_0 < G_1 < \cdots < G_n = G$ be a chief series of G; write $\mathcal{I} = \{i : 1 \leq i \leq n \text{ such that } G_i/G_{i-1} \text{ is } \mathbf{X}\text{-supplemented}\}$; then, if $\mathcal{I}$ is non-empty,*

$$W = \bigcap_{i \in \mathcal{I}}\{S_i : S_i \text{ is an } \mathbf{X}\text{-supplement of } G_i/G_{i-1}\}.$$

2. *If N is a normal subgroup of G, then $WN/N = W(G/N, \mathbf{X}/N)$.*

Proof. Applying Proposition 1.3.11, we can deduce that

$$W = \bigcap\{\mathrm{T}(G, \mathbf{X}, F) : F \text{ is an } \mathbf{X}\text{-supplemented chief factor of } G\}.$$

Now Assertion 1 follows from Theorem 1.2.36 and Theorem 1.3.8.

In proving Assertion 2, suppose first that N is a minimal normal subgroup of G and let $1 = G_0 < G_1 = N < \cdots < G_n = G$ be a chief series of G. Clearly we can assume that $\mathbf{X}$ is non-empty. Then $\mathcal{I} = \{i : 1 \leq i \leq n \text{ such that } G_i/G_{i-1} \text{ is } \mathbf{X}\text{-supplemented}\}$ is non-empty and $W = \bigcap_{i \in \mathcal{I}}\{S_i : S_i \text{ is an } \mathbf{X}\text{-supplement of } G/G_{i-1}\}$ by Statement 1. If N is an $\mathbf{X}$-Frattini, then N is contained in S_i for all $i \in \mathcal{I}$ and then $W/N = W(G/N, \mathbf{X}/N)$. Otherwise, N is contained in S_i for all $i \in \mathcal{I} \setminus \{1\}$ and $G = NS_1$. The case $\mathcal{I} = \{1\}$ leads to $W = S_1$ and $\mathbf{X}/N = \emptyset$. Then $G = WN$ and $WN/N = W(G/N, \mathbf{X}/N)$. Suppose that $\mathcal{I} \setminus \{1\} \neq \emptyset$. Then $WN = \bigcap_{i \in \mathcal{I} \setminus \{1\}} S_i$ and then $WN/N = W(G/N, \mathbf{X}/N)$. Therefore Assertion 2 holds when N is a minimal normal subgroup of G.

A familiar inductive argument proves the validity of Statement 2 for any normal subgroup N of G. $\square$

Remark 4.3.7. Theorem 4.3.6 does not hold when $\mathbf{X}$ is simply a JH-solid set (see Example 1.3.10). This is the reason why we introduce the prefrattini subgroups associated with subsystems of maximal subgroups and not with JH-solid sets of maximal subgroups.

All classical examples of prefrattini subgroups in the soluble universe, including Kurzweil's, enjoy the conjugacy and the cover-avoidance property. Now we prove that, roughly speaking, it can be said that conjugacy and cover-avoidance property of soluble chief factors are equivalent properties for subgroups of prefrattini type. In fact, conjugacy of prefrattini subgroups characterises solubility. The consideration of primitive non-soluble groups, whose core-free maximal subgroups are neither conjugate nor CAP-subgroups, causes that in the general non-soluble universe these properties fail.

Proposition 4.3.8. *Let G be a group and $\mathbf{X}$ a subsystem of maximal subgroups of G. Put $W = W(G, \mathbf{X})$. Let H/K be a chief factor of G.*

1. *If H/K is $\mathbf{X}$-Frattini, then $W(G, \mathbf{X})$ covers H/K.*
2. *If H/K possesses $\mathbf{X}$-complement in G, then $W(G, \mathbf{X})$ avoids H/K.*

Proof. Assume that H/K is an $\mathbf{X}$-Frattini chief factor of G. Then $H/K \leq MK/K$ for all $M \in \mathbf{X}$. Hence,

$$H/K \leq \bigcap \{MK/K : M \in \mathbf{X}\} = W(G/K, \mathbf{X}/K) = WK/K,$$

by Proposition 4.3.6, and $W(G, \mathbf{X})$ covers H/K.

If a maximal subgroup M of G belongs to $\mathbf{X}$, then $W \leq M$. Hence, if M complements H/K, W avoids H/K.   □

Corollary 4.3.9. *Let G be a group, $\mathbf{X}$ a solid set of maximal subgroups of G and H/K an abelian chief factor of G. Then*

1. *H/K is either covered or avoided by all $W \in \mathrm{Pref}_{\mathbf{X}}(G)$;*
2. *H/K is covered by some $W \in \mathrm{Pref}_{\mathbf{X}}(G)$ if and only if H/K is an $\mathbf{X}$-Frattini chief factor of G.*

The above result justifies the following definition.

Definition 4.3.10. *Let G be a group and $\mathbf{X}$ a w-solid set of maximal subgroups of G. We say that $\mathrm{Pref}_{\mathbf{X}}(G)$ satisfies ACAP if whenever F is an abelian chief factor of G,*

1. *then F is either covered or avoided by all $W \in \mathrm{Pref}_{\mathbf{X}}(G)$, and*
2. *F is covered by some $W \in \mathrm{Pref}_{\mathbf{X}}(G)$ if and only if F is an $\mathbf{X}$-Frattini chief factor of G.*

Clearly if $\mathrm{Pref}_{\mathbf{X}}(G)$ satisfies ACAP, any $W \in \mathrm{Pref}_{\mathbf{X}}(G)$ covers all abelian $\mathbf{X}$-Frattini chief factors of G and avoids all abelian $\mathbf{X}$-complemented.

By the above corollary, if $\mathbf{X}$ is a solid set of maximal subgroups of a group G, then $\mathrm{Pref}_{\mathbf{X}}(G)$ satisfies ACAP. We give some more examples.

Examples 4.3.11. 1. By Lemma 1.5 of [Kur89], if L is a subgroup of a soluble group G, the set $\mathrm{Pref}_{\mathbf{X}_L}(G)$ of all L-prefrattini subgroups of G satisfies ACAP (note that $\mathbf{X}_L$ is w-solid, but not solid in general).

2. Let G be the group as in Example 1.3.10. We consider the set $\mathbf{X} = \{\langle a, z \rangle, \langle b, z \rangle, \langle ab, z \rangle, \langle a^2 b, z \rangle\}$. Then $\mathbf{X}$ is a subsystem of maximal subgroups of G and $\mathrm{W}(G, \mathbf{X}) = \langle z \rangle$. We consider the system $\mathbf{Y}$ of maximal subgroups defined by the Hall system $\Sigma = \{N, \langle abz \rangle\}$ (see Theorem 1.4.17). Then $\mathrm{W}(G, \mathbf{X} \cap \mathbf{Y}) = \langle ab, z \rangle$. It is clear that the $\mathbf{X}$-prefrattini subgroups of G do not satisfy ACAP.

Proposition 4.3.12. *Let G be a group, and let $\mathbf{X}$ be a w-solid set of maximal subgroups of G. Assume that $\mathrm{Pref}_{\mathbf{X}}(G)$ satisfies ACAP. Let $\mathbf{X}_1$, $\mathbf{X}_2$ be two systems of maximal subgroups of G and H/K an abelian chief factor of G. Then, there exists an $\mathbf{X}$-complement of H/K in $\mathbf{X}_1$ if and only if there exists an $\mathbf{X}$-complement of H/K in $\mathbf{X}_2$.*

Proof. Put $\{i, j\} = \{1, 2\}$. Suppose that M_i is an $\mathbf{X}$-complement of H/K in $\mathbf{X}_i$ but for all maximal subgroups $S \in \mathbf{X} \cap \mathbf{X}_j$ such that $K \leq S$, we have $H \leq S$. Denote by W_k the $(\mathbf{X} \cap \mathbf{X}_k)$-prefrattini subgroup of G, $k = 1, 2$. Applying Theorem 4.3.6, $W_i \leq M_i$. Then $K = W_i K \cap H$. Since $\mathrm{Pref}_{\mathbf{X}}(G)$ satisfies ACAP, we have $K = W_j K \cap H$. However $W_j K/K$ is the $\mathbf{X}/K$-prefrattini subgroup of G/K associated with $\mathbf{X}_j/K$ by Theorem 4.3.6 (2). Then $W_j K/K = \bigcap \{S/K : S \in \mathbf{X} \cap \mathbf{X}_j, K \leq S\}$. Our assumption implies $H/K \leq W_j K/K$. This contradiction proves that H/K has an $\mathbf{X}$-complement in $\mathbf{X}_j$. $\qquad\square$

Theorem 4.3.13. *Let $\mathbf{X}$ be a w-solid set of maximal subgroups of group G. For $\mathbf{Y} = \mathbf{X}_{\mathfrak{S}}^n$, the set of all $\mathfrak{S}$-normal maximal subgroups in $\mathbf{X}$, the following statements are equivalent:*

1. *$\mathrm{Pref}_{\mathbf{Y}}(G)$ satisfies ACAP;*
2. *$\mathrm{Pref}_{\mathbf{Y}}(G)$ is a set of conjugate subgroups of subgroups of G.*

Proof. 1 implies 2. Assume that Assertion 2 does not hold and choose for G a counterexample of least order. If H is any non-trivial normal subgroup of G, then $\mathbf{X}/H$ is w-solid set of maximal subgroups of G/H and $(\mathbf{X}/H)_{\mathfrak{S}}^n = \mathbf{Y}/H$. It is clear that $\mathrm{Pref}_{\mathbf{Y}/H}(G/H)$ satisfies ACAP. Hence the minimal choice of G implies that $\mathrm{Pref}_{\mathbf{Y}/H}(G/H)$ is a set of conjugate subgroups of G/H.

Let N be a minimal normal subgroup of G. If N is $\mathbf{Y}$-Frattini, then N is covered by every $\mathbf{Y}$-prefrattini subgroup of G by Theorem 4.3.6. In that case, the Since the theorem holds in G/N, $\mathrm{Pref}_{\mathbf{Y}}(G)$ is a conjugacy class of subgroups of G, contrary to supposition. Hence N is $\mathbf{Y}$-supplemented in G. In particular N is $\mathfrak{S}$-central in G and therefore N is abelian. Let $M \in \mathbf{Y}$ such that $G = MN$ and $M \cap N = 1$, and let $\mathbf{S}$ be a system of maximal subgroups of G such that $M \in \mathbf{S}$ (Theorem 1.4.7). Denote by A the $(\mathbf{Y} \cap \mathbf{S})$-prefrattini subgroup of G. Then $A \leq M$ by Theorem 4.3.6. Since by hypothesis $\mathrm{Pref}_{\mathbf{Y}}(G)$

is not a set of conjugate subgroups of G, there exists a system $\mathbf{S}_0$ of maximal subgroups of G such that $A_0 = \mathrm{W}(G, \mathbf{Y} \cap \mathbf{S}_0)$ and A are not conjugate in G.

Let φ be the isomorphism between G/N and M. We have $(\mathbf{X}/N)^n_{\mathfrak{S}} = \mathbf{Y}/N$ and $(\mathbf{Y}/N)^\varphi = (\mathbf{X} \cap M)^n_{\mathfrak{S}} = \mathbf{Y} \cap M$ by Lemma 1.2.23. Denote $C = \mathrm{Core}_G(M) = \mathrm{C}_M(N)$. Suppose that $C \neq 1$. Since the theorem holds in G/C, there exists $x \in G$ such that $A_0^x C = AC \leq M$. Without loss of generality we can assume that $x = 1$. In particular $A_0 \leq M$. Then $AN \cap M = A$ and $A_0 N \cap M = A_0$ are $(\mathbf{X} \cap M)^n_{\mathfrak{S}}$-prefrattini subgroups of M. The minimal choice of G implies that A and A_0 are conjugate in M. This contradiction leads to $C = 1$. Since M is $\mathfrak{S}$-normal in G, we have G is a primitive soluble group. By Corollary 1.4.18, there exists $g \in G$ such that $\mathbf{S}_0^g = \mathbf{S}$. If $U \in \mathbf{Y} \cap \mathbf{S}$, then U complements the chief factor $\mathrm{Soc}\big(G/\mathrm{Core}_G(U)\big)$. By Proposition 4.3.12, there exists $V \in \mathbf{Y} \cap \mathbf{S}_0$ such that V complements $\mathrm{Soc}\big(G/\mathrm{Core}_G(U)\big)$. Since $G/\mathrm{Core}_G(U)$ is a soluble primitive group, $\mathrm{Core}_G(U) = \mathrm{Core}_G(V)$ and U and V are conjugate in G by Theorem 1.1.10. This implies that $\mathbf{Y} \cap \mathbf{S} = \mathbf{Y} \cap \mathbf{S}_0^g = (\mathbf{Y} \cap \mathbf{S}_0)^g$. Applying Theorem 4.3.6, $A = \bigcap\{U : U \in \mathbf{Y} \cap \mathbf{S}\} = \bigcap\{U : U \in (\mathbf{Y} \cap \mathbf{S}_0)^g\} = \bigcap\{V^g : V \in \mathbf{Y} \cap \mathbf{S}_0\} = A_0^g$. This contradiction proves the implication.

2 implies 1. Note that all non-abelian chief factors of G are $\mathbf{Y}$-Frattini. This means that $\mathbf{Y}$-prefrattini subgroups are conjugate CAP-subgroups indeed. $\qquad\square$

Corollary 4.3.14. *Let $\mathbf{X}$ be a w-solid set of maximal subgroups of a soluble group G. The following statements are equivalent:*

1. $\mathrm{Pref}_{\mathbf{X}}(G)$ *is a set of conjugate subgroups of G, and*
2. *every $W \in \mathrm{Pref}_{\mathbf{X}}(G)$ is a CAP-subgroup of G which covers all $\mathbf{X}$-Frattini chief factors of G and avoids the $\mathbf{X}$-complemented ones.*

In general the prefrattini subgroups of a group are not conjugate: in any non-abelian simple group the prefrattini subgroups are the maximal subgroups. We prove next that the solubility of a group is characterised by the conjugacy of its prefrattini subgroups.

Theorem 4.3.15. *A group G is soluble if and only if the set $\mathrm{Pref}(G)$ of all prefrattini subgroups is a conjugacy class of subgroups of G.*

Proof. If G is a soluble group, then the conjugation of the prefrattini subgroups of G follows directly from Theorem 4.3.6 and Corollary 1.4.18.

Conversely, assume that G is a group such that the set $\mathrm{Pref}(G)$ of all prefrattini subgroups of G is a conjugacy class of subgroups of G. We prove that G is soluble by induction on the order of G. By Theorem 4.3.6 (2), we have that, for every normal subgroup N of G, the set $\mathrm{Pref}(G/N)$ of all prefrattini subgroups of G/N is a conjugacy class of subgroups of G/N. Therefore G/N is soluble for each minimal normal subgroup N of G and G is a monolithic primitive group. Suppose that G is not soluble. Then $N = \mathrm{Soc}(G)$ is not abelian. Let W/N be a prefrattini subgroup of G/N associated with an arbitrary system of maximal subgroups $\mathbf{X}^*$ of G/N. Let P_1 be a non-trivial Sylow p_1-subgroup of N, for some prime p_1; there exists a maximal subgroup M of G such that $\mathrm{N}_G(P_1) \leq M$. Clearly $\mathrm{Core}_G(M) = 1$. The set $\mathbf{X}_1 = \{H \leq G : N \leq H, H/N \in \mathbf{X}^*\} \cup \{M\}$ is a system of maximal subgroups of G. Applying Theorem 4.3.6, $W \cap M$ is the prefrattini subgroup of G associated with $\mathbf{X}_1$. Let P_2 be a non-trivial Sylow p_2-subgroup of N, for a prime p_2 such that $p_1 \neq p_2$. This is always possible since N is non-abelian. Consider now a maximal subgroup S of G such that $\mathrm{N}_G(P_2) \leq S$ and the system of maximal subgroups $\mathbf{X}_2 = \{H \leq G : N \leq H, H/N \in \mathbf{X}^*\} \cup \{S\}$ of G. As above, we have that $W \cap S = \mathrm{W}(G, \mathbf{X}_2)$. Consequently $W \cap M$ and $W \cap S$ are conjugate in G. This implies that $W \cap M$ contains a Sylow p_2-subgroup of N. Since p_2 is arbitrary, we have that $W \cap M$ contains a Sylow p-subgroup of N for any prime p dividing the order of N. This implies that $N \leq M$, which is a contradiction. Hence G is soluble. $\qquad\square$

Finally in this section, we touch on the question of the description of the core and the normal closure of subgroups of prefrattini type. For solid sets $\mathbf{X}$ of maximal subgroups, the core of the $\mathbf{X}$-prefrattini subgroups is the $\mathbf{X}$-Frattini subgroup defined in Definition 1.2.18 (1).

Proposition 4.3.16. *If $\mathbf{X}$ is a solid set of maximal subgroups of a group G and W is an $\mathbf{X}$-prefrattini subgroup of a group G, then $\mathrm{Core}_G(W) = \Phi_{\mathbf{X}}(G)$.*

Proof. Let $\mathbf{Y}$ be a system of maximal subgroups of G. Consider $W = \mathrm{W}(G, \mathbf{X} \cap \mathbf{Y})$. Since $\mathbf{X}$ is solid, we have that $\Phi_{\mathbf{X}}(G) = \bigcap\{\mathrm{Core}_G(M) : M \in \mathbf{X} \cap \mathbf{Y}\} = \mathrm{Core}_G(W)$.

The classical Frattini subgroup of a group G, $\Phi(G)$, is clearly the $\mathrm{Max}^*(G)$-Frattini subgroup of G. The $\mathrm{Max}^*(G)_{\mathfrak{N}}^a$-Frattini subgroup is denoted by $\mathrm{L}(G)$ in [Bec64]. H. Bechtell also denotes the $\mathrm{Max}(G)_{\mathfrak{N}}^n$-Frattini subgroup by $\mathrm{R}(G)$. Following his notation, if $\mathfrak{H}$ is a Schunck class and G is a group, we denote

$$\mathrm{L}_{\mathfrak{H}}(G) = \bigcap\{M : M \text{ is } \mathfrak{H}\text{-abnormal monolithic maximal subgroup of } G\}$$

the $\mathrm{Max}^*(G)_{\mathfrak{H}}^a$-Frattini subgroup of G, and similarly

$$\mathrm{R}_{\mathfrak{H}}(G) = \bigcap\{M : M \text{ is } \mathfrak{H}\text{-normal monolithic maximal subgroup of } G\}$$

the $\mathrm{Max}^*(G)_{\mathfrak{H}}^n$-Frattini subgroup of G.

Theorem 4.3.17. *Let $\mathfrak{F}$ be a saturated formation and let $\mathbf{X}$ be a system of maximal subgroups of a group G, then*

$$\mathrm{Core}_G\big(\mathrm{W}(G, \mathbf{X}_{\mathfrak{F}}^a)\big) = \mathrm{Z}_{\mathfrak{F}}\big(G \bmod \Phi(G)\big) = \mathrm{L}_{\mathfrak{F}}(G).$$

Proof. Denote $W = \mathrm{W}(G, \mathbf{X}_{\mathfrak{F}}^a)$. Applying Theorem 4.3.6, $\mathrm{L}_{\mathfrak{F}}(G) \leq W$. Hence $\mathrm{L}_{\mathfrak{F}}(G)$ is contained in $\mathrm{Core}_G(W)$. Conversely, if S is an $\mathfrak{F}$-abnormal monolithic maximal subgroup of G in $\mathbf{X}$, then we have $W \leq S$ and $\mathrm{Core}_G(W) \leq$

$\mathrm{Core}_G(S)$. Then $\mathrm{Core}_G(W)$ is contained in every $\mathfrak{F}$-abnormal monolithic maximal subgroup of G. Hence $\mathrm{Core}_G(W) \le \mathrm{L}_{\mathfrak{F}}(G)$.

To prove that $\mathrm{Z}_{\mathfrak{F}}\big(G/\Phi(G)\big) = \mathrm{L}_{\mathfrak{F}}(G)/\Phi(G)$ suppose first that $\Phi(G) = 1$. Since every chief factor of G below $\mathrm{Z}_{\mathfrak{F}}(G)$ is $\mathfrak{F}$-central in G, it follows that $\mathrm{Z}_{\mathfrak{F}}(G) \le \mathrm{L}_{\mathfrak{F}}(G)$. To prove the converse observe that if $\Phi(G) = 1$, then $\mathrm{L}_{\mathfrak{F}}(G) \cap G^{\mathfrak{F}} = 1$. Assume not and let N be a minimal normal subgroup of G such that $N \le \mathrm{L}_{\mathfrak{F}}(G) \cap G^{\mathfrak{F}}$. Since $\Phi(G) = 1$, it follows that N is supplemented in G by a monolithic $\mathfrak{F}$-normal maximal subgroup M. Hence $G^{\mathfrak{F}} \le M$. This contradiction leads to $\mathrm{L}_{\mathfrak{F}}(G) \cap G^{\mathfrak{F}} = 1$. Consider a chief factor H/K of G such that $H \le \mathrm{L}_{\mathfrak{F}}(G)$. Since $G^{\mathfrak{F}} \cap \mathrm{L}_{\mathfrak{F}}(G) = 1$, then $HG^{\mathfrak{F}}/KG^{\mathfrak{F}}$ is a chief factor of G which is G-isomorphic to H/K. This means that H/K is $\mathfrak{F}$-central in G. Therefore $\mathrm{L}_{\mathfrak{F}}(G) \le \mathrm{Z}_{\mathfrak{F}}(G)$ and equality holds.

If $\Phi(G) \ne 1$, then consider the quotient group $G^* = G/\Phi(G)$. Since $\Phi(G^*) = 1$, we obtain the required equality. $\qquad\square$

Proposition 4.3.18. *Let G be a group. If $\mathfrak{F}$ is a saturated formation and $\mathbf{X}$ is a system of maximal subgroups of G, then*

$$\mathrm{Core}_G\big(\mathrm{W}(G, \mathbf{X}_{\mathfrak{F}}^n)\big) = \mathrm{R}_{\mathfrak{F}}(G) = \Phi(G \bmod G^{\mathfrak{F}}).$$

Proof. First notice that $G^{\mathfrak{F}}$ is contained in every $\mathfrak{F}$-normal maximal subgroup of G and if $G \in \mathfrak{F}$, then every maximal subgroup of G is $\mathfrak{F}$-normal. Therefore, $\mathrm{R}_{\mathfrak{F}}(G)/G^{\mathfrak{F}} = \mathrm{R}_{\mathfrak{F}}(G/G^{\mathfrak{F}}) = \Phi(G/G^{\mathfrak{F}})$. Since $G^{\mathfrak{F}}$ is contained in $\mathrm{W}(G, \mathbf{X}_{\mathfrak{F}}^n)$, we have $\mathrm{W}(G, \mathbf{X}_{\mathfrak{F}}^n)/G^{\mathfrak{F}} = \mathrm{W}(G/G^{\mathfrak{F}}, \mathbf{X}_{\mathfrak{F}}^n/G^{\mathfrak{F}})$ by Theorem 4.3.6 (2) and so $\mathrm{Core}_G\big(\mathrm{W}(G, \mathbf{X}_{\mathfrak{F}}^n)\big)/G^{\mathfrak{F}} = \Phi(G/G^{\mathfrak{F}})$. $\qquad\square$

Definition 4.3.19. *Let G be a group and suppose that $\mathbf{X}$ is a solid set of maximal subgroups of G. A normal subgroup N of G is said to be*

1. *an $\mathbf{X}$-profrattini normal subgroup of G if either $N = 1$ or every chief factor of G of the form N/K is an $\mathbf{X}$-Frattini chief factor of G, and*
2. *an $\mathbf{X}$-parafrattini normal subgroup of G if either $N = 1$ or every chief factor of G of the form N/K is a non-$\mathbf{X}$-complemented chief factor of G, that is, no maximal subgroup in $\mathbf{X}$ is a complement of N/K in G.*

For $\mathbf{X} = \mathrm{Max}^*(G)$, the solid set of all monolithic maximal subgroups of G, we say simply profrattini and parafrattini.

Examples and remarks 4.3.20. 1. If N is an $\mathbf{X}$-profrattini normal subgroup of G, then N is an $\mathbf{X}$-parafrattini normal subgroup of G. The converse does not hold in general. It is enough to consider a non-abelian simple group S. It is clear that S is $\mathbf{X}$-parafrattini for all solid sets $\mathbf{X}$ of maximal subgroups of S. However S is not $\mathbf{X}$-profrattini.

If N is soluble, N is $\mathbf{X}$-profrattini if and only if N is $\mathbf{X}$-parafrattini.

2. If $\mathfrak{F}$ is a totally nonsaturated formation (see [BBE91]), then $G^{\mathfrak{F}}$ is a profrattini normal subgroup of G for every group G.

3. If $\mathbf{X}$ is a solid set of maximal subgroups of a group G, a quasinilpotent normal subgroup N of G is $\mathbf{X}$-profrattini if and only if $N \le \Phi_{\mathbf{X}}(G)$.

Proof. Assume that N is a quasinilpotent $\mathbf{X}$-profrattini normal subgroup of G but $N \not\leq \Phi_{\mathbf{X}}(G)$. Then there exists a maximal subgroup U of G such that $K \leq U$, $U \in \mathbf{X}$ and $G = UN$. We have that $G/\operatorname{Core}_G(U) = \bigl(N \operatorname{Core}_G(U)/\operatorname{Core}_G(U)\bigr)\bigl(U/\operatorname{Core}_G(U)\bigr)$ and $N \operatorname{Core}_G(U)/\operatorname{Core}_G(U)$ is quasinilpotent. Therefore

$$N \operatorname{Core}_G(U)/\operatorname{Core}_G(U) = \mathrm{F}^*\bigl(G/\operatorname{Core}_G(U)\bigr) = \operatorname{Soc}\bigl(G/\operatorname{Core}_G(U)\bigr).$$

But this contradicts N being $\mathbf{X}$-profrattini. Hence $N \leq \Phi_{\mathbf{X}}(G)$. The converse holds trivially. □

Theorem 4.3.21. *Let G be a group and suppose that $\mathbf{X}$ is a solid set of maximal subgroups of G.*

1. *If N, M are both $\mathbf{X}$-profrattini normal subgroups of G, then NM is an $\mathbf{X}$-profrattini normal subgroup of G.*
2. *If N, M are both $\mathbf{X}$-parafrattini normal subgroups of G, then NM is an $\mathbf{X}$-parafrattini normal subgroup of G.*

Proof. Let $(NM)/K$ be a chief factor of G. The normal subgroups KM and KN lie between K and NM. If $K = KN = KM$, then $NM \leq K$, which is imimpossible. Hence, either $NM = NK$ or $NM = MK$. Suppose that $NM = NK$ (the other case is analogous). By Lemma 1.2.16, if S supplements (respectively, complements) $NM/K = NK/K$, then S also supplements (respectively, complements) the chief factor $N/(N \cap K)$. If N is a $\mathbf{X}$-profrattini (respectively, $\mathbf{X}$-parafrattini) normal subgroup of G, then $S \notin \mathbf{X}$. Hence MN is also $\mathbf{X}$-profrattini (respectively, $\mathbf{X}$-parafrattini) normal subgroup of G. □

Remark 4.3.22. Let G be a group and $\mathbf{X}$ be a solid set of maximal subgroups of G. Suppose that N is a normal subgroup of G satisfying the property that either $N = 1$ or every chief factor N/K of G is $\mathbf{X}$-complemented in G. If M is a normal subgroup of G with the same property, then MN does not have this property in general. For instance, consider $G = A \times B$ where $A = \langle a : a^4 = 1 \rangle$, $B = \langle b : b^2 = 1 \rangle$, and $\mathbf{X} = \operatorname{Max}^*(G)$. Then B and $D = \langle a^2 b \rangle$ are two complemented minimal normal subgroups of G. However BD/B is a Frattini chief factor of G.

Definitions 4.3.23. *Let G be a group and $\mathbf{X}$ be a solid set of maximal subgroups of G.*

1. *The $\mathbf{X}$-profrattini subgroup of G is the normal subgroup*

   $$\operatorname{Pro}_{\mathbf{X}}(G) = \langle N : N \text{ is an } \mathbf{X}\text{-profrattini normal subgroup of } G \rangle.$$

2. *The $\mathbf{X}$-parafrattini subgroup of G is the normal subgroup*

   $$\operatorname{Para}_{\mathbf{X}}(G) = \langle N : N \text{ is an } \mathbf{X}\text{-parafrattini normal subgroup of } G \rangle.$$

For $\mathbf{X} = \mathrm{Max}^*(G)$, *the solid set of all monolithic maximal subgroups of* G, *we write simply* $\mathrm{Pro}(G)$ *and* $\mathrm{Para}(G)$.

It is clear that $\mathrm{Pro}_{\mathbf{X}}(G) \leq \mathrm{Para}_{\mathbf{X}}(G)$. If $\mathbf{X}$ is a solid set of maximal subgroups of G composed of maximal subgroups of type 1, then $\mathrm{Pro}_{\mathbf{X}}(G) = \mathrm{Para}_{\mathbf{X}}(G)$. In particular, the equality holds when G is soluble. There are non-soluble groups such that $\mathrm{Pro}_{\mathbf{X}}(G) = \mathrm{Para}_{\mathbf{X}}(G)$. Consider a prime p and a cyclic group Z of order p^2. Let $G = S \wr Z$ be the regular wreath product of S with Z, where S is a non-abelian simple group. Then $\mathrm{Pro}(G) = \mathrm{Para}(G)$ is the unique maximal normal subgroup of G.

It is clear that for each normal subgroup $\mathrm{Para}_{\mathbf{X}}(G) < N$ (respectively, $\mathrm{Pro}_{\mathbf{X}}(G) < N$) there is at least one G-chief factor N/K which is $\mathbf{X}$-supplemented (respectively, $\mathbf{X}$-complemented) in G. We can say much more than this.

Proposition 4.3.24. *Let* G *be a primitive group of type 2 which splits over* $\mathrm{Soc}(G) = N$ *by a maximal subgroup* S *of* G. *Then* $\mathrm{Soc}(S)$ *is non-abelian.*

Proof. Let A be an abelian minimal normal subgroup of S. Then A is an elementary abelian p-group for some prime p. Since $S \leq \mathrm{N}_G(A)$, then $\mathrm{N}_G(A) = S$ since proper containment leads to a contradiction that A is normal in G, by maximality of S in G. Hence $N \cap \mathrm{C}_G(A) = 1$. If p divides $|N|$, a contradiction arises since A would be contained in a Sylow p-subgroup $P = [T]A$ of NA with $T = P \cap N$. Hence, $T \cap \mathrm{Z}(P) \neq 1$ and there exists an element $x \in \mathrm{C}_N(A)$ such that $x \neq 1$. This is not possible. Consequently p does not divide $|N|$. Let q be a prime dividing $|N|$. By [Gor80, 6.2.2], there exists a unique A-invariant Sylow q-subgroup Q of N. For any element $s \in S$, Q^s is also A-invariant. Consequently, $Q = Q^s$ and $S \leq \mathrm{N}_G(Q)$. Since $N \cap S = 1$, Q is not contained in S and so $G = QS = NS$. This implies $N = Q$, a contradiction. □

Corollary 4.3.25. *Denote by* $\mathfrak{K}$ *the class of all groups* G *such that every chief factor of* G *is complemented in* G *by a maximal subgroup of* G. *Then* $\mathfrak{K}$ *is composed of soluble groups.*

Proof. Suppose that $\mathfrak{K}$ is not contained in $\mathfrak{S}$ and consider a group of minimal order $G \in \mathfrak{K} \setminus \mathfrak{S}$. Then $G \in \mathrm{b}(\mathfrak{S})$ and G is a primitive group of type 2. By hypothesis, $N = \mathrm{Soc}(G)$ is a non-abelian minimal normal subgroup which is complemented in G by a core-free maximal soluble subgroup S of G. But $\mathrm{Soc}(S)$ abelian contradicts Proposition 4.3.24. □

Proposition 4.3.26. *Let* G *be a group and let* $\mathbf{X}$ *be a solid set of maximal subgroups of* G.

1. *Denote by* $\mathcal{N}$ *the set of all normal subgroups* N *of* G *satisfying the property that every chief factor of* G *between* N *and* G *is* $\mathbf{X}$-*supplemented in* G. *If* $N, M \in \mathcal{N}$, *then* $N \cap M \in \mathcal{N}$.

2. *Denote by $\mathcal{K}$ the set of all normal subgroups N of G satisfying the property that every chief factor of G between N and G is $\mathbf{X}$-complemented in G. If N, $M \in \mathcal{K}$, then $N \cap M \in \mathcal{K}$.*

Proof. Consider a chief series of G from M to $M \cap N$.

$$N \cap M \leq \cdots \leq M. \tag{4.6}$$

1. Consider a chief factor H/K of G in (4.6). Then HN/KN is a chief factor of G between N and G. Since $N \in \mathcal{N}$, it follows that HN/KN is $\mathbf{X}$-supplemented in G by $S \in \mathbf{X}$, say. This means that $G = S(HN)$ and $KN \leq S \cap NH$. Hence $G = SH$ and $K \leq S \cap H$. Hence H/K is $\mathbf{X}$-supplemented in G by S. Therefore Assertion 1 follows from Theorem 1.2.36.

2. Note that by Corollary 4.3.25, the groups G/N and G/M are soluble. Then $G/(N \cap M)$ is soluble. Therefore all chief factors in (4.6) are abelian.

The Assertion 2 now follows by applying the same arguments as those used in the proof of Statement 1 replacing "supplemented" by "complemented." $\square$

Corollary 4.3.27. *Let G be a group and $\mathbf{X}$ a solid set of maximal subgroups of G. Then*

1. *$\operatorname{Pro}_{\mathbf{X}}(G) = \bigcap\{N : N \in \mathcal{N}\} \in \mathcal{N}$ and every chief factor of G between $\operatorname{Pro}_{\mathbf{X}}(G)$ and G is $\mathbf{X}$-supplemented in G;*
2. *$\operatorname{Para}_{\mathbf{X}}(G) = \bigcap\{N : N \in \mathcal{K}\} \in \mathcal{K}$ and every chief factor of G between $\operatorname{Para}_{\mathbf{X}}(G)$ and G is $\mathbf{X}$-complemented in G.*

Proof. 1. Denote $K = \bigcap\{N : N \in \mathcal{N}\}$. By Proposition 4.3.26, $K \in \mathcal{N}$. If K/L is an $\mathbf{X}$-supplemented chief factor of G, then $L \in \mathcal{N}$ by Theorem 1.2.34 and this is not possible. Therefore every chief factor of G of the form K/L is $\mathbf{X}$-Frattini. Hence $K \leq \operatorname{Pro}_{\mathbf{X}}(G)$. Assume that $K < \operatorname{Pro}_{\mathbf{X}}(G)$. Let $\operatorname{Pro}_{\mathbf{X}}(G)/N$ be a chief factor of G such that $K \leq N$. Then $\operatorname{Pro}_{\mathbf{X}}(G)/N$ should be $\mathbf{X}$-Frattini. This contradicts Proposition 4.3.26.

The proof for 2 is analogous. $\square$

Corollary 4.3.28. *If $\mathbf{X}$ is a solid set of maximal subgroups of a group G, then $G/\operatorname{Para}_{\mathbf{X}}(G)$ is a soluble group.*

Proof. Note that $G/\operatorname{Para}_{\mathbf{X}}(G) \in \mathfrak{K}$. Apply now Corollary 4.3.25. $\square$

It is clear from the above result that $G^{\mathfrak{S}}$, the soluble residual of G, is contained in $\operatorname{Para}_{\mathbf{X}}(G)$.

Corollary 4.3.29. *Let $\mathbf{X}$ be a solid set of maximal subgroups of a group G. Then $\operatorname{Para}_{\mathbf{X}}(G) = \operatorname{Pro}_{\mathbf{X}}(G)G^{\mathfrak{S}}$.*

Proof. It is clear that $\operatorname{Pro}_{\mathbf{X}}(G)G^{\mathfrak{S}} \leq \operatorname{Para}_{\mathbf{X}}(G)$. Suppose there exists a chief factor $F = \operatorname{Para}_{\mathbf{X}}(G)/N$ of G with $\operatorname{Pro}_{\mathbf{X}}(G)G^{\mathfrak{S}} \leq N$. By definition of $\operatorname{Para}_{\mathbf{X}}(G)$, the chief factor F is non-$\mathbf{X}$-complemented in G. On the other hand, F is abelian and $\mathbf{X}$-supplemented in G because $\operatorname{Pro}_{\mathbf{X}}(G)G^{\mathfrak{S}} \leq N$. Such F cannot exist. Hence $\operatorname{Para}_{\mathbf{X}}(G) = \operatorname{Pro}_{\mathbf{X}}(G)G^{\mathfrak{S}}$. $\square$

Theorem 4.3.30. *Let G be a group and let* $\mathbf{X}$ *be a solid set of maximal subgroups of G. Then N is an* $\mathbf{X}$*-parafrattini normal subgroup of G if and only if $N = \langle N \cap W^g : g \in G \rangle$ for each $W \in \mathrm{Pref}_{\mathbf{X}}(G)$.*

Proof. Suppose that $N = \langle N \cap W^g : g \in G \rangle$ for each $W \in \mathrm{Pref}_{\mathbf{X}}(G)$. Let N/K be a chief factor of G. Assume that N/K is $\mathbf{X}$-complemented in G. Then there exists a maximal subgroup $M \in \mathbf{X}$ of G such that $G = MN$ and $N \cap M = K$. If W is an $\mathbf{X}$-prefrattini subgroup of G such that $W \leq M$, it follows that $W \cap N \leq M \cap N = K$. Hence $N = \langle N \cap W^g : g \in G \rangle \leq K$, contrary to supposition. Therefore N/K is non-$\mathbf{X}$-complemented in G. Hence N is $\mathbf{X}$-parafrattini.

Conversely, assume that N is an $\mathbf{X}$-parafrattini normal subgroup of G. We may suppose that $N \neq 1$. Let $W \in \mathrm{Pref}_{\mathbf{X}}(G)$ and $L = \langle N \cap W^g : g \in G \rangle$. Suppose $L < N$. Let N/H be a chief factor of G such that $L \leq H$. Since N is $\mathbf{X}$-parafrattini, we have that N/H is non-$\mathbf{X}$-complemented in G. Note that $W \cap N \leq L \leq H$. Hence W avoids N/H. This implies that N/H is $\mathbf{X}$-supplemented. Let $\mathbf{S}$ be the system of maximal subgroups of G such that $W = \mathrm{W}(G, \mathbf{X} \cap \mathbf{S})$ and M be an $\mathbf{X}$-supplement of N/H in G such that $M \in \mathbf{S}$. Consider a chief series of G passing through H and N. Let $S_1, \ldots, S_r$ be the $\mathbf{X}$-supplements of the chief factors of G above N such that $S_i \in \mathbf{S}$ ($1 \leq i \leq r$). Then $WN/N = \bigcap_{i=1}^{r} S_i/N$ and $WH/H = \bigcap_{i=1}^{r}(S_i/H) \cap (M/H)$ by Theorem 4.3.6. Therefore $WH = \bigcap_{i=1}^{r}(S_i \cap M) = WN \cap M = W(M \cap N)$. Since $W \cap N \cap M = W \cap N = W \cap H$, it follows that $|H| = |M \cap N|$ and so $H = M \cap N$. Hence M is an $\mathbf{X}$-complement of N/H in G. This contradicts our assumption. Consequently $N = \langle N \cap W^g : g \in G \rangle$ for each $W \in \mathrm{Pref}_{\mathbf{X}}(G)$. $\square$

The following result describes the normal closure of an $\mathbf{X}$-prefrattini subgroup.

Corollary 4.3.31. *Let $\mathbf{X}$ be a solid set of maximal subgroups of group G. If $W \in \mathrm{Pref}_{\mathbf{X}}(G)$, we have that $\langle W^G \rangle = \langle W^g : g \in G \rangle = \mathrm{Para}_{\mathbf{X}}(G)$.*

Proof. Write $P = \mathrm{Para}_{\mathbf{X}}(G)$. Each abelian chief factor of G which is $\mathbf{X}$-complemented in G is avoided by every $\mathbf{X}$-prefrattini subgroup of G by Corollary 4.3.9. Since every chief factor H/K such that $P \leq K < H \leq G$ is abelian and $\mathbf{X}$-complemented in G, it follows that $W \leq \mathrm{Para}_{\mathbf{X}}(G)$ for all $W \in \mathrm{Pref}_{\mathbf{X}}(G)$. From Theorem 4.3.30, $\langle W^G \rangle = P$. $\square$

In [Haw67] an elegant theorem of factorisation of prefrattini subgroups of soluble groups is proved. There T. O. Hawkes makes a strong use of the cover-avoidance property. Here we present a similar factorisation in the general non-soluble universe but, obviously, with no use of the cover-avoidance property.

Theorem 4.3.32. *Let G be a group and let $\mathfrak{H}$ be a Schunck class of the form $\mathfrak{H} = \mathrm{E}_{\Phi}\mathfrak{F}$, for some formation $\mathfrak{F}$. Consider a system of maximal subgroups $\mathbf{X}$ of G. Then, if $\mathbf{Y}$ is a w-solid set of maximal subgroups of G, we have*

$$W\big(G, (\mathbf{X} \cap \mathbf{Y})^a_{\mathfrak{H}}\big) = D\,W(G, \mathbf{X} \cap \mathbf{Y}),$$

where D is an $\mathfrak{H}$-normaliser of G associated with $\mathbf{X}$.

Proof. We argue by induction on the order of G. Obviously we can suppose that $\Phi(G) = 1$. Denote $W^* = W\big(G, (\mathbf{X} \cap \mathbf{Y})^a_{\mathfrak{H}}\big)$ and $W = W\big(G, \mathbf{X} \cap \mathbf{Y}\big)$. By Theorem 4.3.6, W^* is contained in W.

By Lemma 4.1.11, we know that D is contained in every $\mathfrak{H}$-abnormal maximal subgroup of G in $\mathbf{X}$. Hence $\langle D, W \rangle \leq W^*$. If $G \in \mathfrak{H}$, then $G = D$ and $(\mathbf{X} \cap \mathbf{Y})^a_{\mathfrak{H}} = \emptyset$. Thus, $W^* = G = D$. Therefore we may assume that $G \notin \mathfrak{H}$. Consider an $\mathfrak{H}$-critical maximal subgroup M of G in $\mathbf{X}$ such that D is an $\mathfrak{H}$-normaliser of M associated with a system of maximal subgroups $\mathbf{X}(M)$ such that $\mathbf{X}_M \subseteq \mathbf{X}(M)$. Then M supplements a minimal normal subgroup N of G. If G is a simple group, then every maximal subgroup of G is $\mathfrak{H}$-abnormal and then $W^* = W$ and the theorem is true in this case. Hence we can assume that N is a proper subgroup of G and $N \cap M \neq M$.

If N is an $(\mathbf{X} \cap \mathbf{Y})$-Frattini minimal normal subgroup, then $N \leq W \leq W^*$ and the assertion follows by induction. Hence we may suppose that $M \in \mathbf{X} \cap \mathbf{Y}$.

Moreover, arguing as in Lemma 1.2.23, we have that $\mathbf{Y}_M/(M \cap N) = \{(S \cap M)/(M \cap N) : N \leq S \in \mathbf{Y}\}$ is a w-solid set of maximal subgroups of $M/(M \cap N)$. Thus $\mathbf{Y}_M = \{S \cap M : N \leq S \in \mathbf{Y}\}$ is a w-solid set of maximal subgroups of M. By induction,

$$W\big(M, (\mathbf{X}(M) \cap \mathbf{Y}_M)^a_{\mathfrak{H}}\big) = D\,W\big(M, \mathbf{X}(M) \cap \mathbf{Y}_M\big). \tag{4.7}$$

Consider a chief series Γ of G through N. Then, by Theorem 4.3.6 (1), we have that $W = M \cap S_{i_1} \cap \cdots \cap S_{i_r}$, where the S_{i_j} are $(\mathbf{X} \cap \mathbf{Y})$-supplements of chief factors in Γ over N. Observe that $\Gamma \cap M$ gives a piece of chief series of M over $N \cap M$. Moreover, again by Theorem 4.3.6 (1),

$$W\big(M, \mathbf{X}(M) \cap \mathbf{Y}_M\big)(N \cap M)/(N \cap M) = \bigcap_{j=1}^{r} (M \cap S_{i_j})/(N \cap M)$$

and then

$$W = W\big(M, \mathbf{X}(M) \cap \mathbf{Y}_M\big)(N \cap M).$$

Similarly,

$$W^* = W\big(M, (\mathbf{X}(M) \cap \mathbf{Y}_M)^a_{\mathfrak{H}}\big)(N \cap M).$$

Hence, by taking the product with $N \cap M$ in both sides of the equality (4.7) we obtain the required factorisation. $\qquad\square$

Motivated by [Tom75, Theorem 5.3], we present the following factorisation involving $\mathfrak{H}$-normal maximal subgroups.

Theorem 4.3.33. *Let G be a group and let $\mathfrak{F}$ be a saturated formation. Consider a system of maximal subgroups $\mathbf{X}$ of G. Then if $\mathbf{Y}$ is a w-solid set of maximal subgroups of G we have that*

$$W\big(G, (\mathbf{X} \cap \mathbf{Y})^n_{\mathfrak{F}}\big) = W(G, \mathbf{X} \cap \mathbf{Y})G^{\mathfrak{F}}.$$

Proof. Since $G^{\mathfrak{F}}$ is contained in M, for every $\mathfrak{F}$-normal maximal subgroup M of G, it follows that $G^{\mathfrak{F}} \leq \mathrm{W}\big(G, (\mathbf{X} \cap \mathbf{Y})^n_{\mathfrak{F}}\big)$. Since $G/G^{\mathfrak{F}} \in \mathfrak{F}$, it is clear that $(\mathbf{X} \cap \mathbf{Y})^n_{\mathfrak{F}}/G^{\mathfrak{F}} = (\mathbf{X} \cap \mathbf{Y})/G^{\mathfrak{F}}$. Therefore $\mathrm{W}\big(G, (\mathbf{X} \cap \mathbf{Y})^n_{\mathfrak{F}}\big)/G^{\mathfrak{F}} = \mathrm{W}\big(G/G^{\mathfrak{F}}, (\mathbf{X} \cap \mathbf{Y})^n_{\mathfrak{F}}/G^{\mathfrak{F}}\big) = \mathrm{W}(G, \mathbf{X} \cap \mathbf{Y})G^{\mathfrak{F}}/G^{\mathfrak{F}}$ by Theorem 4.3.6 (2). $\qquad\square$

Corollary 4.3.34. *Let G be a group and let $\mathfrak{F}$ be a saturated formation. Consider a system of maximal subgroups $\mathbf{X}$ of G. Then if $\mathbf{Y}$ is a w-solid set of maximal subgroups of G we have that*

$$G = \mathrm{W}\big(G, (\mathbf{X} \cap \mathbf{Y})^n_{\mathfrak{F}}\big)W\big(G, (\mathbf{X} \cap \mathbf{Y})^a_{\mathfrak{F}}\big).$$

Proof. Just notice that if D is an $\mathfrak{F}$-normaliser of G, then $G = DG^{\mathfrak{F}}$. Now apply the factorisations presented in Theorem 4.3.32 and Theorem 4.3.33. $\quad\square$

The theory of prefrattini subgroups was continued by X. Soler-Escrivà in her Ph. Doctoral Thesis at the Universidad Pública de Navarra, [SE02]. Her work is another example of the progress produced by using non-arithmetical properties, even in soluble groups. In its place all relations between maximal subgroups of a group and maximal subgroups of its critical subgroups are used thoroughly (see [ESE05]). This leads to the existence and properties of some distributive lattices, generated by three types of pairwise permutable subgroups, namely hypercentrally embedded subgroups (see [CM98]), $\mathfrak{F}$-normalisers, and subgroups of prefrattini type (see [ESE]).

5

Subgroups of soluble type

Consider a subgroup H of a soluble group G. Since every minimal normal subgroup of G is abelian, the following implication holds:

If $G = MH$, M is a minimal normal subgroup of G,

$$\text{and } H \text{ is a proper subgroup of } G, \text{ then } H \cap M = 1. \quad (5.1)$$

It is precisely this property which makes the theory of $\mathfrak{H}$-projectors and $\mathfrak{H}$-covering subgroups, $\mathfrak{H}$ a Schunck class, so much easier in the soluble universe than in the general finite one.

Salomon, in his Doctoral Dissertation [Sal87], has introduced and studied notions of $\mathfrak{H}$-projectors and $\mathfrak{H}$-covering subgroups (different from the usual ones) which lead to a theory of these subgroups in arbitrary groups resembling the theory of $\mathfrak{H}$-projectors and $\mathfrak{H}$-covering subgroups in finite soluble groups. His first basic idea is to give a definition of $\mathfrak{H}$-projectors along the following lines:

Recall that if $\mathfrak{H}$ is a class of groups, a subgroup H of a group G is called an $\mathfrak{H}$-*projector* of G if $H \in \mathrm{s}(G) \cap \mathfrak{H}$ and if $N \trianglelefteq G$ and $HN/N \le K/N \in \mathrm{s}(G/N) \cap \mathfrak{H}$, then $HN/N = K/N$. Here $\mathrm{s}(X)$ is the set of all subgroups of a group X.

Salomon tries to replace the set $\mathrm{s}(G)$ of all subgroups of a group G by suitable subsets $\mathrm{d}(G)$ of $\mathrm{s}(G)$ which are such that any element H of $\mathrm{d}(G)$ enjoys the above property (5.1); he also tries to develop a theory of the "projectors" so obtained by following the classical approach.

It is clear that in order to carry out this, one cannot take just any $\mathrm{d}(G) \subseteq \mathrm{s}(G)$ satisfying (5.1). First of all, since the definition of $\mathfrak{H}$-projector involves not only the group G itself but also its quotients, such sets $\mathrm{d}(G)$ have to be chosen not only for G but at least for all quotients of G; in fact, the classical theory of projectors suggests that it would be reasonable to demand that such a choice be made for all finite groups simultaneously. Put differently, to begin with, one chooses a "subgroup functor" d which associates with any group G a set of subgroups $\mathrm{d}(G)$ (we refer to elements of $\mathrm{d}(G)$ as *subgroups of soluble type*) subject, of course, to the condition that (5.1) holds for any G and each H in d(G). Moreover, it is also plausible that a subgroup functor d ought to satisfy certain formal properties such as the following: If $H \in$ d(G) and $N \trianglelefteq G$, then $HN/N \in$ d(G/N). Not surprisingly, it turns out that the properties relevant here are closely related to properties of ordinary projectors and covering subgroups.

The theory developed in this Chapter is largely the work of Salomon [Sal87] and Förster [Förb], [Föra].

5.1 Subgroup functors and subgroups of soluble type: elementary properties

The purpose of this section is to establish the necessary formal properties of the various functors for subgroups of soluble type. The functors t and t′ introduced by Salomon in [Sal87] are studied here. Two similar functors r and r′ defined by Förster [Förb] are also studied.

A *subgroup functor* is a function f which assigns to each group G a possibly empty set f(G) of subgroups of G satisfying $\theta\big(\mathrm{f}(G)\big) = f\big(\theta(G)\big)$ for any isomorphism $\theta \colon G \longrightarrow G^*$.

Examples 5.1.1.   1. *Functor* s: it assigns to each group G the set s(G) of all subgroups of G.

2. *Functor* s_n: it associates with each group G the set $s_n(G)$ of all subnormal subgroups of G.

3. Let p be a prime. Let s_p be the function assigning to each group G the set $s_p(G)$ of all subgroups U of G containing a Sylow p-subgroup of G; s_p is a subgroup functor.

4. Let $\mathfrak{H}$ be a Schunck class, then $\mathrm{Proj}_{\mathfrak{H}}()$ and $\mathrm{Cov}_{\mathfrak{H}}()$ are subgroup functors.

Given two subgroup functors e and f, we write $\mathrm{e} \leq \mathrm{f}$ if e(G) $\subseteq$ f(G) for each group G.

Definition 5.1.2. *Let* f *be a subgroup functor. We say that* f *is* inherited *if* f *enjoys the following properties:*

1. If $U \in$ f(T) and $T \in$ f(G), $U \leq T \leq G$, then $U \in$ f(G).
2. If $U \in$ f(G) and $N \trianglelefteq G$, then $UN/N \in$ f(G/N).
3. If $U/N \in$ f(G/N), $N \trianglelefteq G$, then $U \in$ f(G).

If, moreover, f *satisfies*

4. $U \leq G$ and $T \in$ f(G) implies $U \cap T \in$ f(U),

we say that f *is* w-inherited.

Obviously the functors s and s_n are w-inherited. s_p is inherited but not w-inherited.

Lemma 5.1.3. *Let* f *be an inherited functor. Then*

1. *If* $U \in f(G)$ *and* $N \trianglelefteq G$, *then* $UN \in f(G)$.
2. *If* N *is a subnormal subgroup of* G *and* $N = N_1 \trianglelefteq N_2 \trianglelefteq \cdots \trianglelefteq N_k = G$ *is a chain from* N *to* G *such that* $1 \in f(N_i)$ *for all* $i \in \{2, \ldots, k\}$, *then* $N \in f(G)$.

In particular, $s_n(G) \subseteq f(G)$ *for all groups* G *if* $1 \in f(X)$ *for all groups* X.

Proof. 1. follows from Definition 5.1.2 (2 and 3).

2. By 1, $N_i \in f(N_{i+1})$ for all $i \in \{1, \ldots, k-1\}$. Applying Definition 5.1.2 (1), it follows that $N \in f(G)$. □

Lemma 5.1.4. *Let* f *be a* w-*inherited subgroup functor. If* G *is a group,* $U \in f(G)$ *and* N *is a subnormal subgroup of* G *such that* $U \leq N_G(N)$, *then* $UN \in f(G)$.

In particular, if $1 \in f(G)$, *then* $s_n(G) \subseteq f(G)$.

Proof. We argue by induction on $|G|$. Let N_1 be the normal closure of N in G and, for $i > 1$, denote $N_i = \langle N^{N_{i-1}} \rangle$, the normal closure of N in N_{i-1}. Since N is subnormal in G, there exists $n \geq 1$ such that $N = N_n$. Suppose that $G = UN_1$, then $N_2 = N_1$ because U normalises N. Repeating the argument with every N_k, it follows that N is normal in G. By Lemma 5.1.3 (1), we have $UN \in f(G)$. Therefore we may assume that UN_1 is a proper subgroup of G. By induction, $UN \in f(UN_1)$. Now $UN_1 \in f(G)$ yields $UN \in f(G)$ by Definition 5.1.2 (1). □

Lemma 5.1.5. *If* f *is a* w-*inherited subgroup functor and* X_1, $X_2 \in f(G)$, *then* $X_1 \cap X_2 \in f(G)$.

Proof. Since f is w-inherited, $X_1 \cap X_2 \in f(X_1)$. Hence $X_1 \cap X_2 \in f(G)$ because $X_1 \in f(G)$ and f is inherited. □

Definition 5.1.6. *Let* f *be a* w-*inherited subgroup functor and* $\{X_i : i \in \mathcal{I}\} \subseteq f(G)$. *Then the intersection of all subgroups of* G *belonging to* $f(G)$ *and containing* $\bigcup_{i \in \mathcal{I}} X_i$ *is the smallest subgroup of* G *in* $f(G)$ *containing* $\bigcup_{i \in \mathcal{I}} X_i$. *This subgroup is denoted by* $\langle X_i : i \in \mathcal{I} \rangle_f$ *and called the* f-*join of* $\{X_i : i \in \mathcal{I}\}$.

Theorem 5.1.7. *Let* f *be a* w-*inherited subgroup functor. For each group* G, $f(G)$ *is a lattice under the operations "*$\cap$*" and "*$\langle \cdot, \cdot \rangle_f$*."*

Definitions 5.1.8. *For any group G, we define the following subgroup functors:*

$$\mathrm{t}(G) = \{H \leq G : T \trianglelefteq S \leq G,\ S/T \text{ strictly semisimple,}$$
$$\text{then } (H \cap S)T \trianglelefteq S\}$$
$$\mathrm{r}(G) = \{H \leq G : T \trianglelefteq S \leq G,\ S/T \text{ strictly semisimple, } H \leq \mathrm{N}_G(S),$$
$$\text{then } (H \cap S)T \trianglelefteq S\}$$
$$\mathrm{t'}(G) = \{H \leq G : T \trianglelefteq S \operatorname{sn} G,\ S/T \text{ strictly semisimple,}$$
$$\text{then } (H \cap S)T \trianglelefteq S\}$$
$$\mathrm{r'}(G) = \{H \leq G : T \trianglelefteq S \operatorname{sn} G,\ S/T \text{ strictly semisimple, } H \leq \mathrm{N}_G(S),$$
$$\text{then } (H \cap S)T \trianglelefteq S\}$$
$$\mathrm{t''}(G) = \{H \leq G : T \trianglelefteq S \operatorname{sn} G,\ S/T \text{ simple, then } (H \cap S)T \trianglelefteq S\}$$
$$\mathrm{r''}(G) = \{H \leq G : T \trianglelefteq S \operatorname{sn} G,\ S/T \text{ simple, } H \leq \mathrm{N}_G(S),$$
$$\text{then } (H \cap S)T \trianglelefteq S\}$$

here strictly semisimple *groups* are those which can be written as direct products of isomorphic simple groups, while semisimple *groups* are direct products of not necessarily isomorphic simple groups.

If H is a subgroup of G such that $H \in \mathrm{e}(G)$ for some $\mathrm{e} \in \{\mathrm{t,r,t',r',t'',r''}\}$, we shall say that H is a subgroup of soluble type.

The functors t and t' have been introduced by Salomon in his Dissertation [Sal87]. There are certain problems Salomon encounters with these two choices of e: the first of these functors does not really give sets $\mathrm{t}(G)$ "large enough" whenever G is "highly non-soluble," while for t' one of the crucial properties, namely, if $H \in \mathrm{e}(G)$ and $N \trianglelefteq G$, then $H \in \mathrm{e}(HN)$, is missing. Later, Förster [Förb] overcame these problems by introducing the remaining functors. As will be seen in the next section "the r-functors" enjoy the advantage of producing relevant subgroups of primitive groups.

Remarks 5.1.9.   1. If G is soluble, then $\mathrm{t}(G) = \mathrm{r}(G) = \mathrm{t'}(G) = \mathrm{r'}(G) = \mathrm{t''}(G) = \mathrm{r''}(G) = \mathrm{s}(G)$.

2. Condition "$(H \cap S)T \trianglelefteq S$" in the definitions of t'' and r'' implies that H either covers or avoids the simple section S/T, that is, $(H \cap S)T \in \{S, T\}$.

3. In defining t, r, t', r' (t'', r'', respectively) we might have taken the strictly semisimple (simple) groups S/T as direct products of non-abelian simple groups (as non-abelian simple, respectively). Moreover, since every subnormal subgroup of a direct product of non-abelian simple groups is in fact normal by [DH92, A, 4.13] the condition "$(H \cap S)T \trianglelefteq S$" can be replaced by "$(H \cap S)T \operatorname{sn} S$."

4. If H is a subgroup of G such that $H \in \mathrm{e}(G)$ for some $\mathrm{e} \in \{\mathrm{t,r,t',r',t''}, \mathrm{r''}\}$, and N is a normal subgroup of G, then $HN/N \in \mathrm{e}(G/N)$.

5. $\mathrm{s}_n \leq \mathrm{t} \leq \mathrm{t'} \leq \mathrm{t''}$, $\mathrm{r} \leq \mathrm{r'} \leq \mathrm{r''}$, $\mathrm{t} \leq \mathrm{r}$, $\mathrm{t'} \leq \mathrm{r'}$, $\mathrm{t''} \leq \mathrm{r''}$.

6. $\mathrm{r}(G)$ ($\mathrm{r'}(G)$, respectively) is the set of all subgroups H of G with the following property:

If $T \trianglelefteq S \leq G$ ($T \trianglelefteq S \operatorname{sn} G$, respectively), S/T is strictly semisimple, and $H \leq \mathrm{N}_G(S) \cap \mathrm{N}_G(T)$, then $(H \cap S)T \trianglelefteq S$.

Proof. Only a proof for Statement 6 is needed. Let S/T denote a strictly semisimple section of G, with simple component E, say, and let $H \leq \mathrm{N}_G(S)$. Then T^*, the $\mathrm{D}_0(1, E)$-residual of S, is a characteristic subgroup of S contained in T with strictly semisimple quotient S/T^* contained in $\mathrm{D}_0(1, E)$; in particular, $H \leq \mathrm{N}_G(T^*)$. Now Statement 6 follows directly from this observation. □

Proposition 5.1.10. *Let H and N be subgroups of a group G such that N is quasinilpotent and H normalises N. Suppose that the following condition holds:*

$B \trianglelefteq A \operatorname{sn} N$, *$A/B$ strictly semisimple, and $H \leq \mathrm{N}_G(A)$ implies that*
$(H \cap A)B \trianglelefteq A.$ $\qquad(5.2)$

Then $H \cap N$ is subnormal in N.

Proof. Proceeding by induction on $|G| + |N|$, we may clearly assume that $G = HN$, and hence that N is normal in G. In the case $N = 1$ or N is a minimal normal subgroup of G, the claim is immediate from condition (5.2); so without loss of generality, there is a normal subgroup M of G such that $1 \neq M < N$. Taking into account that the class of all quasinilpotent groups is an s_n-closed homomorph and that condition (5.2) is inherited by M and N/M, we may apply the inductive hypothesis twice to get that $H \cap M$ is a subnormal subgroup of M and $HM \cap N$ is a subnormal subgroup of N; in particular, $HM \cap N$ is quasinilpotent. Therefore, if $HM \cap N$ is a proper subgroup of N, then another application of the inductive hypothesis yields that $H \cap N = H \cap (HM \cap N)$ is subnormal in $HM \cap N$, and hence that $H \cap N$ is subnormal in N. Thus, without loss of generality, $N = HM \cap N \leq HM$ and $G = HN = HM$. Therefore we have:

$$H < G = HM \text{ whenever } M \trianglelefteq G \text{ and } 1 \neq M < N. \qquad(5.3)$$

Assume that the Fitting subgroup of N, $\mathrm{F}(N)$, is non-trivial. Then N contains an abelian minimal normal subgroup M of G; since $N = (H \cap N)M$, we can deduce that $H \cap N$ is normal in N because N centralises M. This is due to the fact that M is a direct product of minimal normal subgroups of M which are central in N because N is quasinilpotent. It remains to deal with the case when $\mathrm{F}(N) = 1$. Then N is a direct product of non-abelian simple groups by Proposition 2.2.22 (3). By Condition (5.2), $H \cap M$ is normal in M for every minimal normal subgroup of G contained in N. Therefore we obtain:

N is a direct product of non-abelian simple groups, and $H \cap M = 1$,
whenever M is a minimal normal subgroup of G contained in N. $\quad(5.4)$

Now, N cannot have two non-isomorphic simple direct factors E and F: otherwise for each X in $\{E, F\}$ we could find minimal normal subgroups M_X of G contained in N with X as composition factor, so by Condition (5.3) and Condition (5.4) both M_E and M_F should be normal complements of H in G, leading to the contradiction that $M_E \cong (M_E \times M_F) \cap H \cong M_F$. Consequently, N is strictly semisimple, and our claim holds by assumption. $\square$

Proposition 5.1.11. *The functors* t' *and* t'' *are inherited. The functor* t *is w-inherited.*

Proof. We give a proof for t.

1. Let G be a group and let X and U be subgroups of G such that $X \in t(U)$ and $U \in t(G)$. Consider a strictly semisimple section S/T of G. We prove that $(S \cap X)T$ is normal in G. Clearly we may assume that S/T is a direct product of non-abelian simple groups. Since $U \in t(G)$, it follows that $(S \cap U)T$ is normal in S. Hence $(S\cap U)/(T\cap U)$ is either 1 or a strictly semisimple section of U. If $S\cap U = T\cap U$, then $S\cap X = T\cap X$ and $(S \cap X)T = T$. Thus we may assume that $S\cap U \neq T\cap U$. Since $X \in t(U)$, we have that $(S \cap X)(T \cap U)$ is normal in $S\cap U$. This implies that $(S\cap X)T$ is normal in $(S\cap U)T \trianglelefteq S$. Since S/T is a direct product of non-abelian simple groups, we have $(S\cap X)T$ is normal in S by [DH92, A, 4.13]. Consequently $X \in t(G)$.

2. Let $U \in t(G)$ and N normal in G. Suppose we have $T/N \trianglelefteq S/N \leq G/N$ and $(S/N)/(T/N)$ strictly semisimple. Then $N \leq T \trianglelefteq S \leq G$ and S/T is strictly semisimple, so $U \in t(G)$ gives that $(UN \cap S)T = (U \cap S)NT = (U\cap S)T \trianglelefteq S$. Therefore $UN/N \in t(G/N)$.

3. Consider a normal subgroup N of G. Let $U/N \in t(G/N)$. Suppose that S/T is a non-abelian and strictly semisimple section of G. If $TN = SN$, then $S = (S \cap N)T = (S\cap U)T$ and $(S\cap U)T$ is normal in S. Hence we may assume that $TN \neq SN$. In particular, $SN/TN \cong S/((S \cap N)T)$ and $(SN/N)/(TN/N)$ is a strictly semisimple section of G/N. Since $U/N \in t(G/N)$, it follows that $((SN/N) \cap (U/N))(TN/N)$ is normal in SN/N and so $(SN\cap U)T$ is normal in SN. This implies that $(SN\cap U)T\cap S = (S\cap U)T$ is a normal subgroup of S. Therefore $U \in t(G)$.

The same statements can be obtained for the functors t' and t'' just by adding the assumption S subnormal in G for t', and S subnormal in G and S/T simple for t'' in the above proof.

Finally, it is clear that t is w-inherited because every strictly semisimple section of every subgroup of a group is actually a strictly semisimple section of the group itself. $\square$

Example 5.1.12. Let E be a non-abelian simple group and let $K = E \wr E$ be the regular wreath product. Then K can be written as semidirect product $K = FE$, where F is the base group and E is its canonical complement in K. Consider now $G = E \wr_E K$, the wreath product of E with K with respect to the transitive permutation representation of K on the right cosets of E. Let B be the base group of G. Then $B = E_1 \times \cdots \times E_n$, $n = |K : E|$, where $E_i \cong E$ $(i = 1, \ldots, n)$, E_1 is E-invariant and $E_1E = E_1 \times E$. It is rather easy to see that F (B, respectively) is the unique minimal normal subgroup of K (G, respectively). Consequently, G has a unique maximal normal subgroup, namely BF.

Let H be the diagonal subgroup of $E_1 E$. We claim that $H \in t'(G)$. Let $T \trianglelefteq S$ sn G with S/T non-abelian and strictly semisimple. If $S \neq G$, S is contained in the unique maximal normal subgroup BF of G. Then (by construction of H and BF) $H \cap BF = 1$, whence $H \cap S = 1$ and $(H \cap S)T = T \trianglelefteq S$. If $S = G$, then $T \in \{G, BF\}$ and

$$(S \cap H)T = HT = HBF = G = S.$$

Assume, by way of contradiction, that $H \in r'(HB)$. Then $HN/N \in r'(HB/N)$ whenever $N \trianglelefteq HB$. But $B = E_1 \times C$, where $C = E_2 \times \cdots \times E_n$ and $HB/C = (E \times E_1)C/C \cong E \times E_1$ via the canonical isomorphism. This maps the diagonal subgroup H of $E \times E_1$ onto the non-normal subgroup HC/C of EB/C, which contradicts $HC/C \in r'(EB/C)$ (see Proposition 5.1.10).

Therefore $H \notin r'(HB)$ and, in particular, $H \notin t'(HB)$. This shows that t' is not w-inherited.

We do not know, however, whether r is w-inherited.

Remark 5.1.13. Since t is w-inherited, then, by Theorem 5.1.7, $t(G)$ is closed under the operations "$\cap$" and "$\langle \cdot, \cdot \rangle_t$." In general, $\langle U, V \rangle_t \neq \langle U, V \rangle$: let $n \geq 5$ and let $G = \Sigma_n$ be the symmetric group of degree n. Let $X_1 = \langle (1,2) \rangle$ and $X_2 = \langle (2,3) \rangle$. Then $X_i \in t(G)$, $i = 1, 2$, $\langle X_1, X_2 \rangle \cong \Sigma_3$, and $\langle X_1, X_2 \rangle_t = G$, as we can see by looking at the sections $\mathrm{Alt}(n)/1$ of $\mathrm{Sym}(n)$.

Proposition 5.1.14. *If H is a proper subgroup of a group $G = HM$, with $H \in r'(G)$, and M is a minimal normal subgroup of G, then $H \cap M = 1$.*

Proof. By Proposition 5.1.10, $H \cap M$ is normal in M. Since H is a proper subgroup of G and M is a minimal normal subgroup of G, we have that $H \cap M = 1$. □

Remark 5.1.15. The statement of Proposition 5.1.14 does not hold for elements of $t''(G)$. Consider the regular wreath product $G = E \wr T$ of a non-abelian simple group E and a group T of order 2. Denote by D the T-invariant diagonal subgroup of the base group B, and put $H = TD$. Then $G = HB$, B is a minimal normal subgroup of G, $H \in t''(G)$, but $H \cap B \neq 1$.

Since the functors t'' and r'' do not have the crucial property described in Proposition 5.1.14, they are not so interesting for us as the functors t, r, t', and r'. Nevertheless, we shall see at the end of the section (Theorem 5.1.25) that the composition factors of the subgroups in $r''(G)$ are composition factors of the whole group.

Definition 5.1.16. *A subgroup functor* f *is called* inductive *(respectively, weakly inductive) if it satisfies Conditions 1 and 2 (respectively, 1 and 2'):*

1. $G \in f(G)$
2. *If $H \leq K \leq G$, $N \trianglelefteq G$, $H \in f(K)$, $N \leq K$, and $K/N \in f(G/N)$, then $H \in f(G)$.*

2′. $H \leq G$, $N \trianglelefteq G$, $H \in f(HN)$, *and* $HN/N \in f(G/N)$ *implies that* $H \in f(G)$.

It is clear that w-inherited functors f such that $f(X)$ is non-empty for all groups X are inductive and inductive functors are also weakly inductive.

Proposition 5.1.17. t′ *and* t″ *are inductive functors.*

Proof. We see that t″ is inductive. The same proof applies to t′.

First of all, the defining Condition 1 for inductivity of t″ holds trivially. To verify Condition 2, let $H \leq K \leq G$ and $N \trianglelefteq G$ such that $N \leq K$, $H \in t''(K)$ and $K/N \in t''(G/N)$. We show that $(S \cap H)T$ is normal in G whenever S/T is a non-abelian simple section of G such that S is subnormal in G. Suppose that $SN = TN$. Then $S = (S \cap N)T = (S \cap K)T$ and $(S \cap K)/(T \cap K)$ is a simple section of K such that $S \cap K$ is subnormal in K. Since $H \in t''(K)$, it follows that $(S \cap H)(T \cap K)$ is normal in $S \cap K$. Hence $(S \cap H)T$ is normal in $(S \cap K)T = S$. Suppose that $SN \neq TN$. Then $(SN/N)/(TN/N)$ is a simple section of G/N and SN/N is subnormal in G/N. Since $K/N \in t''(G/N)$, we have that $(K \cap SN)T$ is normal in SN. Hence $(K \cap S)T = (S \cap (K \cap SN))T = S \cap (K \cap SN)T$ is normal in $S \cap SN = S$. Further, if $K \cap T \neq K \cap S$, then $(S \cap K)/(T \cap K)$ is a simple section of K and $S \cap K$ is subnormal in K. As $H \in t''(K)$, this gives that $(H \cap (K \cap S))(K \cap T)$ is normal in $K \cap S$. This is also true if $K \cap T = K \cap S$. Consequently we obtain:

$$(H \cap S)T = (K \cap H \cap S)(K \cap T)T \trianglelefteq (K \cap S)T \trianglelefteq S.$$

This proves the inductivity of t″.   □

Note that, in the special case when $K = HN$, the same argument still works for the functors r, r′ and r″, for $H \leq N_G(S)$ and $N \trianglelefteq G$ imply that $HN \leq N_G(SN)$; hence we get:

Proposition 5.1.18. r, r′, *and* r″ *are weakly inductive. Moreover, if* $H \in r(G)$ *and* L *is a subgroup of* G *containing* H, *then* $H \in r(L)$.

Lemma 5.1.19. *Let* $f \in \{t', r'\}$ *and let* G *be a group.*

1. *If* H *is a subgroup of* G *and* N *is a soluble subgroup of* G *normalised by* H, *then* $H \in f(HN)$.
2. *Let* $H \leq K \leq G = HN$, *where* N *is a direct product of non-abelian simple groups. If* $H \in f(G)$ *and* $K \cap N$ *is normal in* N, *then* $H \in f(K)$.

Proof.   1. For a proof of $H \in f(HN)$, we may use induction on $|G|$ and the properties of f verified earlier in this section (Propositions 5.1.17 and 5.1.18) to see that without loss of generality $1 \neq G = HN$, N is a minimal normal subgroup of G, $H \cap N = 1$, and H is a core-free maximal subgroup of G. Hence G is a primitive group of type 1 and N is the unique minimal normal subgroup of G. Let $T \trianglelefteq S$ be a non-abelian strictly semisimple section of G

such that S is subnormal in G. Assume that S is a proper subgroup of G. Then S is contained in some maximal normal subgroup M of G. Since $M \neq 1$, it follows that N is contained in M and $M = N(H \cap M)$. Since M is a proper subgroup of G, $H \cap M \in f(M)$ by induction. Hence $(S \cap H)T = (S \cap H \cap M)T$ is normal in S. Therefore we may assume that $S = G$. Then $T \neq 1$ because otherwise G would be non-abelian and simple. Consequently N is contained in T and so $S = (S \cap H)T$. This means that $H \in f(G)$.

2. By hypothesis $K \cap N$ is normal in N, so $K \cap N$ is a normal subgroup of $KN = G$. Since N is a direct product of non-abelian simple groups, it follows that $N = (K \cap N) \times M$ for a normal subgroup M of G. Then $G/M = KM/M \cong K$. It is clear therefore that $H \in f(K)$ because $HM/M \in f(G/M)$. $\square$

Theorem 5.1.20. *Let* $f \in \{t', r'\}$ *and let* $G = H F^*(G)$ *be a group which is a product of a subgroup* $H \in f(G)$ *and* $F^*(G)$, *the generalised Fitting subgroup of* G. *If* M *is a normal subgroup of* G *and* M *is quasinilpotent, then* $H \in f(HM)$.

Proof. We argue by induction on $|G|$. If M is soluble, the result follows from Lemma 5.1.19 (1). We may suppose that M is not contained in $F = G_{\mathfrak{S}}$, the soluble radical of G. If $F = 1$, then $F^*(G)$ and M are both direct products of non-abelian simple groups by Proposition 2.2.22 (3). Moreover $HM \cap F^*(G)$ is normal in $F^*(G)$ because $H \cap F^*(G)$ is normal in $F^*(G)$ by Proposition 5.1.10 and [DH92, A, 4.13]. Applying Lemma 5.1.19 (2), we conclude that $H \in f(HM)$. Therefore we can suppose that $F \neq 1$. By induction $HF/F \in f\big((HF/F)(MF/F)\big)$. This yields $H(HM \cap F)/(HM \cap F) \in f\big(HM/(HM \cap F)\big)$. Since $H \in f\big(H(HM \cap F)\big)$ by Lemma 5.1.19 (1), we conclude that $H \in f(HM)$ by weakly inheritness. $\square$

Lemma 5.1.21. *Let* $G = E_1 \times \cdots \times E_n$ *be a direct product of* n *copies of a non-abelian simple group* E. *Let* H *be a subgroup of* G *such that* H *covers or avoids every simple section of* G. *Then* H *is a normal subgroup of* G.

Proof. We argue by induction on $|G|$. It is clear we can assume $\mathrm{Core}_G(H) = 1$. Since either $E_j \cap H = 1$ or $E_j \leq H$ for all $j \in \{1, \ldots, n\}$, it follows that $E_j \cap H = 1$ because H does not contain any normal subgroup of G. Denote $N_j = \mathsf{X}_{i \neq j} E_i$, $1 \leq j \leq n$.

Suppose that $G = HN_j$ for each $j \in \{1, \ldots, n\}$. In particular, H is a subdirect subgroup of G. Hence $H = R_1 \times \cdots \times R_t$, where $R_k \cong R_l \cong E$, $1 \leq k, l \leq t$. Let $g \neq 1$ be an element of R_1. Without loss of generality, we may assume that $\pi_1(g) \neq 1$, where $\pi_1 \colon G \longrightarrow E_1$ is the projection of G over its first component. Then $1 \neq \pi_1(R_1)$ is a normal subgroup of $\pi_1(H) = E_1$, and so $\pi_1(R_1) = E_1$. Let $1 \neq h \in E_1$ such that $g^h = g$. Since $g \in R_1 \cap R_1^h$, it follows that R_1^h is contained in H.

Assume $[h, R_1] = 1$. Then $1 = \pi_1([h, R_1]) = [h, E_1]$ and $h \in \mathrm{Z}(E_1)$. This contradiction shows that $[h, R_1] \neq 1$. Let t be an element of R_1 such that $t^h \neq t$. Then $1 \neq t^h t^{-1} \in H \cap E_1 = 1$ (note that if $t = (e_1, e_2, \ldots, e_n)$, $t^h = (e_1^h, e_2, \ldots, e_n)$).

This contradiction proves that HN_j is a proper subgroup of G for some $j \in \{1, \ldots, n\}$. Then G/N_j is a simple section of G which is not covered by H. Hence H is contained in N_j. By induction H is normal in N_j and so H is subnormal in G. This implies that H is a normal subgroup of G [DH92, A, 4.13] and the result follows. $\qquad\square$

Proposition 5.1.22. *Let H be a subgroup of G. Then $H \in \mathrm{t}(G)$ if, and only if, every simple section of G is covered or avoided by H.*

Proof. The cover-avoidance property dealt with here is obviously a special case of the defining property of $H \in \mathrm{t}(G)$.

Conversely, let H be a subgroup of G enjoying this cover-avoidance property, and let $T \trianglelefteq S \leq G$ and S/T strictly semisimple and non-abelian. Let $X = (H \cap S)T/T$. Then X is a subgroup which covers or avoids every simple section of S/T. By Lemma 5.1.21, X is normal in S/T. Consequently $H \in \mathrm{t}(G)$. $\qquad\square$

Remark 5.1.23. r is not characterised by the corresponding property: consider the direct product of two copies of a non-abelian simple group E and a diagonal subgroup H of G: as H is not normal in G, $H \notin \mathrm{r}(G)$ by Proposition 5.1.10, yet H covers or avoids any (non-abelian) simple section S/T of G with $H \leq \mathrm{N}_G(S)$.

In soluble groups, each composition factor of a subgroup is a composition factor of the whole group. It is not true in general. However, this property holds for subgroups of soluble type as the next result shows. We need first a lemma.

Lemma 5.1.24. *Let G be a group and $H \in \mathrm{r}''(G)$. Assume that M is a subnormal subgroup of G such that $M = \mathsf{X}_{i=1}^{n} E_i$, where the E_i $(i = 1, \ldots, n)$ are pairwise isomorphic non-abelian simple groups; further assume that H normalises M. Let $\pi_i \colon M \longrightarrow E_i$ $(i = 1, \ldots, n)$ denote the projection of M over E_i. Put $\mathcal{I} = \big\{i \in \{1, \ldots, n\} : (H \cap M)\pi_i \neq 1\big\}$. Then $H \cap M$ is subdirect in $\mathsf{X}_{i \in \mathcal{I}} E_i$.*

Proof. Note that $M/\mathrm{Ker}(\pi_i)$ is a simple section of G. Since $H \in \mathrm{r}''(G)$, it follows that H covers or avoids $M/\mathrm{Ker}(\pi_i)$. If $M \cap H = M \cap \mathrm{Ker}(\pi_i)$, we have $\pi_i(M \cap H) = 1$. Therefore $\mathcal{I} = \big\{i \in \{1, \ldots, n\} : \pi_i(M \cap H) \neq 1\big\} = \big\{i \in \{1, \ldots, n\} : M = (M \cap H)\,\mathrm{Ker}(\pi_i)\big\}$. It follows that $H \cap M$ is a subdirect subgroup of $\mathsf{X}_{i \in \mathcal{I}} E_i$. $\qquad\square$

Theorem 5.1.25. *Let G be a group. Every composition factor of a subgroup $H \in \mathrm{r}''(G)$ is isomorphic to a composition factor of G.*

Proof. Proceeding by induction on $|G|$, we consider a minimal normal subgroup of G. Since $HM/M \in \mathrm{r}''(G/M)$, we may assume that our claim holds for $H/(H \cap M)$ $(\cong HM/M)$. If M is abelian, the result follows. Hence we may assume that M is non-abelian. By Lemma 5.1.24, $H \cap M$ is a subdirect subgroup of a suitable normal subgroup of M. Then $H \cap M \in \mathrm{D}_0(1, E)$, where E is the composition factor of M. $\qquad\square$

5.2 Existence criteria

The main goal in this section is to prove some existence results for various subgroups of soluble type. We begin with a result on t-subgroups which may be viewed as a non-existence result: as a consequence of this theorem, in a monolithic primitive group with non-abelian socle the minimal normal subgroup cannot be complemented by a t-subgroup unless the corresponding quotient of the group is soluble. In fact, the results on this type of subgroups show that the non-soluble t-subgroups share some properties with non-soluble subnormal subgroups. This is already apparent from the first theorem of this section, which may be thought as a partial generalisation of the following theorem of Wielandt ([Wie39], see Theorem 2.2.19):

> If H and K are subnormal subgroups of a group G and H is a perfect comonolithic group, then either H is contained in K or K normalises H.

We supplement this result by a theorem which leads to a criterion for the existence of t-complements of the minimal normal subgroup in primitive groups of type 1; it turns out that in order that such a complement exists, the corresponding quotient of the group has to be p-soluble, p the prime dividing the order of the minimal normal subgroup.

The remaining part of this section deals with results on the functors r, t', and r'.

Theorem 5.2.1. *Let* $H = H' \in t(G)$, *and consider a non-abelian, strictly semisimple normal subgroup M of G. Then*

$$[H, M] \leq H \cap M \trianglelefteq HM.$$

Proof. Let (H, M, G) represent a counterexample such that $|G| + |H|$ is minimal. Then, since t is w-inherited, it follows that:

1. $G = HM$ *and H is a proper subgroup of G.*

   Now, since $H \in t(G)$ and M is strictly semisimple, $H \cap M \trianglelefteq HM = G$. Since G is a counterexample, $[H, M]$ is not contained in $H \cap M$.

2. $H \cap M = 1$, *M is a minimal normal subgroup of G and $\mathrm{Core}_G(H) = 1 = \mathrm{C}_G(M)$.*

   Suppose that $H \cap M \neq 1$. Then the triple $\big(H/(H \cap M), M/(H \cap M), G/(H \cap M)\big)$ satisfies the hypotheses of the theorem. By minimality of G, $[H, M]$ is contained in $H \cap M$, contrary to our supposition. Hence $H \cap M = 1$. Note that $M = M_1 \times \cdots \times M_t$, where M_i are minimal normal subgroups of G for all $i \in \{1, \dots, s\}$. Suppose that M is not a minimal normal subgroup of G. Then HM_j is a proper subgroup of G and so $[H, M_j] \leq H \cap M_j = 1$ for all $j \in \{1, \dots, s\}$. Therefore $[H, M] = 1$. This contradiction proves that M is a minimal normal subgroup of G. If $\mathrm{Core}_G(H) \neq 1$, then the minimality of G yields $[H, M] \leq H \cap M = 1$, and we have a contradiction. Thus $\mathrm{Core}_G(H) = 1$ and $\mathrm{C}_H(M) = 1$ because $\mathrm{C}_H(M)$ is a normal subgroup of $HM = G$. Now put

$C = C_G(M)$ and $D = H \cap CM$. Then $CM = CM \cap HM = DM$. Suppose that $D = 1$. Then C is contained in M and so $C = 1$ because M is non-abelian. Assume now that $D \neq 1$. By Lemma 5.1.3, $HC \in t(G)$. Hence $HC \cap M$ is normal in $(HC)M = G$. Since M is a minimal normal subgroup of G, either $HC \cap M = 1$ or $HC \cap M = M$. If $HC \cap M = 1$, then $HC = H$. This implies $C = 1$. Suppose that $HC \cap M = M$. Then $G = HC = HM$, $DC = (H \cap CM)C = HC \cap CM = CM$, and

$$C \cong CM/M = DM/M \cong D \cong DC/C = CM/C \cong M,$$

whence $CM \cong M \times M$ is a strictly semisimple subgroup of G. Since $H \in t(G)$, $D = H \cap CM$ is a normal subgroup of CM. Hence D is normal in $(CM)H = G$. This yields a contradiction against $\mathrm{Core}_G(H) = 1$. Therefore $C_G(M) = 1$.

3. *Let $H_{\mathfrak{S}}$ denote the soluble radical of H. Then $H/H_{\mathfrak{S}}$ is a non-abelian simple group. Moreover there is no proper* t*-subgroup K of G with $H = KH_{\mathfrak{S}}$.*

Let $A/H_{\mathfrak{S}}$ be a minimal normal subgroup of $H/H_{\mathfrak{S}}$. It is clear that A is not soluble. Hence there exists an integer n such that $(A^{(n)})' = A^{(n)} \neq 1$. Moreover $A^{(n)} \in t(A)$ and so $A \in t(G)$. By the inheritness of t, it follows $A^{(n)} \in t(G)$. If A were a proper subgroup of H, then $A^{(n)} < H$. The minimal choice of (G, H) would yield $A^{(n)} \leq C_G(M) = 1$, contrary to our supposition. Therefore $A = H$ and $H/H_{\mathfrak{S}}$ is a non-abelian simple group.

Suppose there exists a subgroup $K \in t(G)$ such that $H = KH_{\mathfrak{S}}$. Then K is not soluble and so there exists an integer b such that $1 \neq K^{(b)}$ is a perfect subgroup of K. Moreover $K^{(b)} \in t(G)$. Since $|G| + |K^{(b)}| < |G| + |H|$, it follows $K^{(b)}$ is contained in $C_G(M) = 1$. This is a contradiction.

Applying [Hup67, VI, 4.7], there exists a Sylow 2-subgroup H_2 of H and a Sylow 2-subgroup M_2 of M such that $G_2 = M_2 H_2$ is a Sylow 2-subgroup of G. By the Odd Order Theorem ([FT63]), $H_2 \neq 1$ and $M_2 \neq 1$. Moreover M_2 is a normal subgroup of G_2 and $H_{\mathfrak{S}}$ is a proper subgroup of $H_2 H_{\mathfrak{S}}$.

4. $[M_2, H] = 1$.

By way of contradiction, assume that $T = C_{M_2}(H)$ is a proper subgroup of M_2. Then T is a proper subgroup of $N = N_{M_2}(T)$. It is clear that N/T is a normal subgroup of $H_2 N/T$ because H_2 normalises N and T. Hence $Z(H_2 N/T) \cap (N/T) \neq 1$. Let $g \in N$ such that $1 \neq gT \in Z(H_2 N/T) \cap (N/T)$. Clearly $H_{\mathfrak{S}} T/T < (H_2 T/T)(H_{\mathfrak{S}} T/T) \leq ((H^g T \cap HT)/T)(H_{\mathfrak{S}} T/T)$. Thus $H_{\mathfrak{S}}$ is a proper subgroup of $H_{\mathfrak{S}}(H \cap H^g T)$. We prove that $H \cap H^g T \in t(G)$. Let $C = N_M(T)$. Then T is normal in C and C is H-invariant. Since t is w-inherited, it follows that $HT/T \in t(HN/T)$ and $H^g T/N \in t(HN/T)$. Hence $(HT/T) \cap (H^g T/T) \in t(HN/T)$ because t has the intersection property. Then $(HT \cap H^g T)/T \in t(HT/T)$ and hence $HT \cap H^g T \in t(HT)$. This implies that $H \cap H^g T \in t(H)$. By the inheritness of t, $H \cap H^g T \in t(G)$. Now $H \cap H^g T$ does not avoid the simple section $H/H_{\mathfrak{S}}$. Hence $H = H_{\mathfrak{S}}(H \cap H^g T)$ by Proposition 5.1.22. Applying Step 3, $H \cap H^g T = H$. Therefore $g \in N_G(HT)$. Moreover $[\langle g \rangle, H, H] = [H, \langle g \rangle, H] \leq [HT, \langle g \rangle, H] \leq [HT \cap C, H] = [T, H] = 1$ by Step 1. By the Three Subgroups Lemma ([Hup67, III, 1.10]), we infer that

$[H, \langle g \rangle] = [H', \langle g \rangle] = [H, H, \langle g \rangle] = 1$. Thus $g \in C_N(H) \leq C_{M_2}(H) = T$. This contradiction against the choice of g establishes Step 4.

5. *Final contradiction.*

Since $C_G(M) = 1$, it follows that G is, up to isomorphism, a subgroup of $\mathrm{Aut}(M)$. Now $O_{2'}(M) = 1$. Hence a theorem of G. Glauberman [Gla66] applies. It says that $C_{\mathrm{Aut}(M)}(M_2)$ is 2-nilpotent. In particular, $C_{\mathrm{Aut}(M)}(M_2)$ is soluble by the Odd Order Theorem [FT63]. By Step 4, H is a subgroup of $C_G(M_2) \leq C_{\mathrm{Aut}(M)}(M_2)$. This is a contradiction because H is perfect and non-trivial. $\square$

Before proving a result similar to Theorem 5.2.1 for abelian normal subgroups M, we have to recall some terminology and results.

Let G be a group. If H^* is a normal subgroup of a proper subgroup H of G and $H^g \cap H$ is a subgroup of H^* for all $g \in G \setminus H$, then G is called a *Frobenius-Wielandt* group with respect to H and H^*. H. Wielandt [Wie58] (cf. [Hup67, V, 7.5]) has shown that the Frobenius-Wielandt kernel, defined by $G^* = G \setminus \bigcup_{g \in G}(H \setminus H^*)^g$, is a normal subgroup of $G = HG^*$ and $H \cap G^* = H^*$. If $H^* = 1$, then G is called simply a *Frobenius group*.

Let $\Omega = \{xH : x \in G\}$ and let H act on Ω by left multiplication. We have that $|\Omega| = |G : H|$ and if $x \in G \setminus H$, then the orbit of xH has size divisible by $|H : H^*|$. Consequently $\gcd(|H/H^*|, |G : H|) = 1$.

Lemma 5.2.2. *Suppose that $H = H' \in \mathrm{t}(G)$, and let M be an abelian normal subgroup of G. Further, assume that H is comonolithic, with maximal normal subgroup $L = \mathrm{Cosoc}(H)$, and that there is no a proper subgroup K of H such that $K \in \mathrm{t}(H)$ and $H = KL$. Then one of the following holds:*

1. $[H, M]$ *is contained in* $H \cap M$.
2. $H \cap M$ *is a subgroup of* $C_H(M/N)$, *with* N *defined by* $N/(H \cap M) = C_{M/(H \cap M)}(H)$. *Moreover* $C_H(M/N)$ *is a proper subgroup of* H *and* HM/N *is a Frobenius-Wielandt group with respect to* HN/N *and* LN/N, *with Frobenius-Wielandt kernel* LM/N.

Proof. Assuming that $[H, M]$ is not contained in $H \cap M$, we will verify Statement 2. As in the proof of Theorem 5.2.1, an induction argument yields that without loss of generality $G = HM$ and $N = H \cap M = 1$. Consider $g \in G$ such that $H^g \cap H$ is not contained in L; put $g = hm$, where $h \in H$ and $m \in M$. Then $H = L(H \cap H^g)$. Moreover $H \cap H^g \in \mathrm{t}(H)$. The hypothesis of the Lemma yields $H \cap H^g = H$. Hence $H = H^m \cap H = C_H(m)$ by [DH92, A, 16.3]; that is, $m \in C_M(H) = N = 1$. Thus $g \in H$. We have shown that $H^g \cap H$ is a subgroup of L for all $g \in G \setminus H$, and G is a Frobenius-Wielandt group with respect to H and L. Note that, if G^* is the Frobenius-Wielandt kernel, then ML is contained in G^*. Therefore $G^* = M(G^* \cap H) = ML$ and Statement 2 follows. $\square$

Lemma 5.2.3. *Let $G = HM$ be a group, where M is an abelian normal p-subgroup of G for some prime p. Assume that H is a subgroup of G such that*

$H \cap M = 1$. If H is p-soluble and $H_{p'} \in t(H_{p'}M)$ for every Hall p'-subgroup $H_{p'}$ of H, then $H \in t(G)$.

Proof. We use induction on $|G|$. Let A be a minimal normal subgroup of G contained in M. Clearly HA/A satisfies the hypothesis of the Lemma in G/A. By induction, $HA/A \in t(G/A)$. If HA is a proper subgroup of G, it follows that $H \in t(HA)$ by induction. Inductivity of t implies that $H \in t(G)$. Hence we may assume that M is a minimal normal subgroup of G and $\mathrm{Core}_G(H) = 1$. That is to say that G is a primitive group of type 1.

Let $T \trianglelefteq S \leq G$ where S/T is a non-abelian simple group. In order to show that $(S \cap H)T \in \{T, S\}$, it will suffice to deal with the case when $G = SM$. This can be seen by means of a reduction argument since $SM = (H \cap SM)M$, $H \cap SM$ is p-soluble and every Hall p'-subgroup B of $H \cap SM$ belongs to $t(BM)$. Suppose that $S = G$. Then either $T = 1$ or M is contained in T. In both cases $(H \cap S)T \in \{S, T\}$. Hence we may assume that S is a proper subgroup of G. Applying Remark 1.1.11 (1), it follows that $S = H^m$ for some $m \in M$. On the other hand, S/T is a p'-group because G is p-soluble. This implies that T contains every Sylow p-subgroup of S. Let K be a Hall p'-subgroup of H such that K^m contains a Hall p'-subgroup of $H \cap S$, $(H \cap S)_{p'}$ say. Since $S = K^m T$, it follows that $K^m/(K^m \cap T)$ is a simple section of $K^m M = KM$. Since $K \in t(KM)$ either $(K \cap K^m)(K^m \cap T) = K^m$ or $K \cap K^m \leq K^m \cap T$. Assume the latter holds. Then $(S \cap H)T = (S \cap H)_{p'}T \leq (K \cap K^m)T = T$. Suppose that $(K \cap K^m)(K^m \cap T) = K^m$. Then $S = S^m T = (K \cap K^m)T \leq (H \cap S)T \leq S$ and so $S = (H \cap S)T$. Applying Proposition 5.1.22, $H \in t(G)$ and the proof of the lemma is complete.   □

Theorem 5.2.4. *Let $G = HM$ be a factorised group, where M is an abelian normal p-subgroup of G for some prime p. Assume that $H \cap M = \mathrm{C}_H(M) = 1$. Then the following three conditions are pairwise equivalent:*

1. *$H \in t(G)$.*
2. *a) H is p-soluble and*
   b) if $L \trianglelefteq K \leq H$, K/L non-abelian simple, and $m \in M$, then $(K \cap K^m)L \in \{L, K\}$.
3. *a) H is p-soluble and*
   b) $H_{p'} \in t(H_{p'}M)$ for every Hall p'-subgroup $H_{p'}$ of H.

Proof. 1 implies 2. First of all, if K is a subgroup of H, then $K = (H \cap M)K = H \cap MK \in t(MK)$ as t is w-inherited. Hence, if $L \trianglelefteq K \leq H$, K/L non-abelian simple, and $m \in M$, then $(K \cap K^m)L \in \{L, K\}$ because $K^m \in t(MK)$.

Next we see that H is p-soluble. Let S/T be a non-abelian simple section of H. Then $S \in t(SM)$ by the above argument. Choose R as a supplement of T in S contained in $t(SM)$ of minimal order. As $s_n \leq t$, $R \cap T$ is the unique maximal normal subgroup of R; the corresponding quotient is a non-abelian simple group: $R/(R \cap T) \cong RT/T = S/T$. In particular, R is perfect. Hence the above lemma applies to R, $R \cap T$, and SM. Since the hypothesis that $C_H(M) = 1$ and $R \neq 1$ rule out the case that $[R, M] \leq R \cap M = 1$, it follows that $RM/C_M(R)$ is a Frobenius-Wielandt group with respect to $R\,C_M(R)/\,C_M(R)$ and $(R \cap T)\,C_M(R)/\,C_M(R)$. Now from our comments on Frobenius-Wielandt groups, we get that $\gcd(|R/(R \cap T)|, |M/\,C_M(R)|) = 1 = \gcd(|S/T|, p)$. Consequently S/T is a p'-group and H is p-soluble.

2 implies 3. Applying the well-known theorem of Hall-Chunikhin ([Hup67, VI, 1.7]), H has Hall p'-subgroups. Let $H_{p'}$ be one of them. We prove that $H_{p'} \in t(H_{p'}M)$. Suppose that $T \trianglelefteq S \leq H_{p'}M$ where S/T is non-abelian and simple. Clearly $S = S_{p'}\,O_p(S)$ for all $S_{p'} \in \mathrm{Hall}_{p'}(S)$, and $O_p(S) = S \cap M$ is contained in T. Put $K = H_{p'} \cap SM \in \mathrm{Hall}_{p'}(SM)$ and note that $SM = KM$ as well as $H_{p'} \cap S = K \cap S$.

Suppose that $(H_{p'} \cap S)T \neq T$ (i.e. $(K \cap S)T \neq T$); we prove that in this case $(H_{p'} \cap S)T = S$ (i.e. $(K \cap S)T = S$). The p'-subgroup $K \cap S$ of S is contained in some Hall p-subgroup of S. Since every Hall p-subgroup of S is a Hall p'-subgroup of SM, the latter can be written as K^m for some $m \in M$ because the Hall p'-subgroups of SM are conjugate. In particular, $K \cap S = K \cap S \cap K^m = K \cap K^m$. From $O_p(S) = S \cap M \leq T$ we infer that $K^m/(K^m \cap TM) \cong K^m(TM)/TM = SM/TM \cong S/T$, so $K^m/(K^m \cap TM)$ is a non-abelian simple group. Since T is proper in $(K \cap S)T$, it follows that MT is a proper subgroup of $(K \cap S)TM$. Assume that $K^m \cap TM = (K \cap K^m)(K^m \cap TM) = (K \cap S)(K^m \cap TM) = K^m \cap ((K \cap S)TM)$. Then $K \cap S$ would be contained in $K^m \cap TM$. This would imply that $(K \cap S)T \leq (K^m \cap TM)T = K^mT \cap TM = S \cap TM = T(S \cap M) = T(T \cap M) = T$, contrary to our supposition. Hence $K^m \cap TM$ is properly contained in $(K \cap K^m)(K^m \cap TM)$. By Statement 2b, $K^m = (K \cap K^m)(K^m \cap TM)$. Consequently $S = K^m\,O_p(S) = K^mT = (K \cap K^m)(K^m \cap TM)T = (K \cap K^m)(K^mT \cap TM) = (K \cap K^m)(S \cap TM) = (K \cap S)T(S \cap M) = (K \cap S)T(M \cap T) = (K \cap S)T$.

Consequently $H_{p'}$ either covers or avoids each simple section of $H_{p'}M$. Applying Proposition 5.1.22, $H_{p'} \in t(H_{p'}M)$ and Statement 3 follows.

3 implies 1. It follows from Lemma 5.2.3.

The circle of implications is now complete. □

Remark 5.2.5. Under the hypothesis of Theorem 5.2.4, H cannot have a non-abelian simple p'-subgroup. For a non-abelian simple p'-subgroup K of H, Lemma 5.2.2 and Theorem 5.2.4 together with the fact that t is w-inherited would give the conclusion that $KM/\,C_M(K)$ is a Frobenius group with respect to $K\,C_M(K)/\,C_M(K)$ ($\cong K$). Then 2 divides $|K|$ by the Odd Order Theorem [FT63]. In addition, applying [Hup67, V, 8.7], the Sylow 2-subgroups of K are cyclic or generalised quaternion. This contradicts [Hup67, V, 22.9].

The next result shows that non-soluble t-subgroups are close to subnormal subgroups.

Corollary 5.2.6. *If $H \in t(G)$ and all composition factors of H are non-abelian, then H is subnormal in G.*

Proof. Consider a counterexample G of least order. Clearly, $1 < H < G$. A reduction argument based on the w-inheritness of t yields $G = HM$ and $H \cap M = 1$ for every minimal normal subgroup M of G. In particular, $\mathrm{Core}_G(U) = 1$ for each maximal subgroup U of G containing H and G is a primitive group. Let K be a minimal subnormal subgroup of H. By hypothesis, K is a non-abelian group. Applying Theorem 5.2.1, it follows that K centralises every minimal normal subgroup of G. If G is a primitive group of type 2 or 3, then K is contained in $\mathrm{C}_G\big(\mathrm{Soc}(G)\big) = 1$ by Theorem 1. This is a contradiction. If G is a primitive group of type 1, then H is p-soluble for the prime p dividing $|\mathrm{Soc}(G)|$ by Theorem 5.2.4 and so K is a p'-subgroup. This contradicts Remark 5.2.5. Therefore no counterexample exists and H is a subnormal subgroup of G. □

A standard argument proves the following:

Corollary 5.2.7. *If $G = HM$ where $H \in \mathrm{t}(G)$ and M is an abelian normal p-subgroup of G such that $H \cap M = \mathrm{C}_H(M) = \mathrm{C}_M\big(O_{p'}(H)\big) = 1$, then:*

Any subgroup K of G with $G = KM$ and $K \cap M = 1$ is a conjugate of H in G.

Examples 5.2.8.   1. Any involution in $\mathrm{Sym}(6) \setminus \mathrm{Alt}(6)$ generates a t-subgroup of $\mathrm{Sym}(6)$ complementing the non-abelian minimal normal subgroup $\mathrm{Alt}(6)$; however, not all of these involutions are conjugate.

2. Any involution in $\mathrm{Sym}(5) \setminus \mathrm{Alt}(5)$ also generates a t-subgroup of $\mathrm{Sym}(5)$ complementing the non-abelian minimal normal subgroup $\mathrm{Alt}(5)$. In this case, these involutions are conjugate in $\mathrm{Sym}(5)$.

3. Let $n \geq 7$, $f = (n, n-1)$ and the section S/T of $\mathrm{Sym}(n)$, where $S = \langle f \rangle \times \mathrm{Alt}(n-2)$ and $T = \langle f \rangle$. Let g be an involution of $\mathrm{Alt}(n-2)$. Then $gf \in \mathrm{Sym}(n) \setminus \mathrm{Alt}(n)$. Put $H = \langle gf \rangle$. Thus $(H \cap S)T = H \times \langle f \rangle$, which is not normal in S. Therefore H is not a t-subgroup (compare with Beispiel 1.14 of [Sal87]).

Corollary 5.2.7 shows that in a primitive group of type 1 all t-subgroups complementing the minimal normal subgroup are conjugate, while in a primitive group of type 2 this need not be true (Example 5.2.8 (1)). However, there is an important restriction on the t-subgroups supplementing the unique minimal normal subgroup in these primitive groups.

Remark 5.2.9. If $H < G = H\,\mathrm{Soc}(G)$ is a primitive group of type 2 and $H \in \mathrm{t}(G)$, then $H \cap \mathrm{Soc}(G) = 1$ and H is soluble.

Proof. Since $H \cap \mathrm{Soc}(G)$ is normal in $H\,\mathrm{Soc}(G) = G$ and $\mathrm{Soc}(G)$ is a minimal normal subgroup of G, it follows that $H \cap \mathrm{Soc}(G) = 1$. Now the soluble residual $H^{\mathfrak{S}}$ is a perfect t-subgroup of G. Applying Theorem 5.2.1, $[H^{\mathfrak{S}}, \mathrm{Soc}(G)] \leq H^{\mathfrak{S}} \cap \mathrm{Soc}(G) = 1$. Hence $H^{\mathfrak{S}} \leq \mathrm{C}_G\big(\mathrm{Soc}(G)\big) = 1$ and H is a soluble group.
□

There are, however, primitive groups of type 1 with non-soluble t-subgroups H such that $H < G = H\,\mathrm{Soc}(G)$.

Lemma 5.2.10. *Assume that G is a Frobenius-Wielandt group with respect to the subgroups H and H^* such that H is non-soluble and complemented by the elementary abelian minimal normal p-subgroup $\mathrm{Soc}(G)$ of G. If H^* is contained in $\Phi(H)$, then G is a primitive group of type 1 and $H \in \mathrm{t}(G)$.*

Proof. Applying Theorem 1, G is a primitive group of type 1. It is known that $\gcd(p, |H/H^*|) = 1$. Hence H is a p'-group because H^* is contained in $\Phi(H)$. This implies that H is a Hall p'-subgroup of G and Condition 2a of Theorem 5.2.4 holds. Let $L \trianglelefteq K \le H$, where K/L is non-abelian and simple. Assume that $(K \cap K^x)L \ne L$ for some $x \in \mathrm{Soc}(G)$. Then $K \cap K^x$ is not contained in H^* (otherwise $K \cap K^x \le K \cap H^* = L \cap H^*$). Hence the definition of Frobenius-Wielandt group yields that $x \in H$; thus $x \in H \cap \mathrm{Soc}(G) = 1$, $K = K^x$, and $(K \cap K^x)L = K$. Consequently Condition 2b of Theorem 5.2.4 holds and $H \in \mathrm{t}(G)$. □

Example 5.2.11. Let $G = MH$ be the semidirect product of $H = \mathrm{SL}(2,5)$ with an elementary abelian 11-subgroup of order 11^2 on which H acts fixed-point-freely (see [Hup67, V, 8.8]). Then G is a Frobenius group. By Lemma 5.2.10, $H \in \mathrm{t}(G)$. Moreover H is perfect and comonolithic. If $1 \ne x \in M$, then neither $H \le H^x$ nor $H^x \le \mathrm{N}_G(H)$.

Therefore the analogue of the result of Wielandt, mentioned at the beginning of the section, does not hold t-subgroups.

To conclude the investigations on t-subgroups, we mention the following:

Proposition 5.2.12. *Let $G = HN$ be a group such that N is a normal subgroup of G, $H \cap N$ is a t-subgroup of N and G/N is soluble. Suppose that $\gcd(|N|, |G/N|) = 1$. Then $H \in \mathrm{t}(G)$.*

Proof. Put $K = H \cap N$ and $\pi = \pi(N)$. Let S/T a non-abelian simple section of G such that T is a proper subgroup of $(H \cap S)T$. We will show that $(H \cap S)T = S$. Since G/N is soluble, N covers S/T; thus S/T is isomorphic to $(S \cap N)/(T \cap N)$. Hence S/T is a π-group. On the other hand, the π-subgroup $(H \cap S)T/T$ of S/T is covered by the normal Hall π-subgroup K of H; thus $(H \cap S)T = (K \cap S)T$. Therefore $T \cap N$ is a proper subgroup of $(K \cap S)(T \cap N)$. Hence $K \in \mathrm{t}(N)$ implies that $S \cap N = (K \cap S)(T \cap N)$, and $(H \cap S)T = (K \cap S)T = (S \cap N)T = S$ follows.

Therefore H either covers or avoids each simple section of G and so $H \in \mathrm{t}(G)$ by Proposition 5.1.22. □

To get more general results than those above appears to be rather difficult:

Open question 5.2.13. *Let $G = HM$ be a group with a soluble subgroup H and a non-abelian minimal normal subgroup M of G satisfying $H \cap M = \mathrm{C}_H(M) = 1$. What are the precise conditions for $H \in \mathrm{t}(G)$?*

In the following, we turn to prove some results for the remaining subgroup functors.

Theorem 5.2.14. *Let G be a monolithic primitive group and let H be a complement of* $\mathrm{Soc}(G)$ *in* G.

1. *If* $\mathrm{Soc}(G)$ *is abelian, H is an r-subgroup of G.*
2. *If* $\mathrm{Soc}(G)$ *is non-abelian, H is an r$'$-subgroup of G.*

Consequently, any complement of the minimal normal subgroup of a monolithic primitive group is an r$'$-subgroup.

Proof. 1. To verify that $H \in \mathrm{r}(G)$, let $T \trianglelefteq S \leq G$ where S/T is non-abelian and strictly semisimple and $H \leq \mathrm{N}_G(S)$; we have to check that $(S \cap H)T$ is normal in S. By Remark 5.1.9 (6), we may also assume that $H \leq \mathrm{N}_G(T)$. Note that H is a maximal subgroup of G because $\mathrm{Soc}(G)$ is abelian. If S is contained in H, then the result follows. Hence we may assume that $\mathrm{N}_G(S) = G$. This implies that $\mathrm{Soc}(G)$ is contained in S and normalises T. Hence $T \cap \mathrm{Soc}(G)$ is a normal subgroup of $H\,\mathrm{Soc}(G) = G$ and so $T \cap \mathrm{Soc}(G) = 1$ or $\mathrm{Soc}(G)$ is contained in T. If $T \cap \mathrm{Soc}(G) = 1$, then T is contained in $\mathrm{C}_G\big(\mathrm{Soc}(G)\big) = \mathrm{Soc}(G)$ by Theorem 1. This would imply $T = 1$ and so $\mathrm{Soc}(G)$ would be non-abelian, contradicting our supposition. Therefore $\mathrm{Soc}(G)$ is contained in T. Then $G = HT$ and $(H \cap S)T = S$.

2. In order to show $H \in \mathrm{r}'(G)$, let $T \trianglelefteq S \,\mathrm{sn}\, G$ with S/T non-abelian and strictly semisimple and $H \leq \mathrm{N}_G(S)$; we have to prove that $(H \cap S)T$ is normal in S. As in Case 1, we may assume that H normalises T. Since T and S are subnormal subgroups of G, they are normalised by $\mathrm{Soc}(G)$ by [DH92, A, 14.3]. Hence T and S are both normal in $H\,\mathrm{Soc}(G) = G$. Since $\mathrm{Soc}(G)$ is a minimal normal subgroup of G, it follows that either $T = 1$ or $\mathrm{Soc}(G) \leq T$. If $T = 1$, then $\mathrm{Soc}(G)$ is contained in S. On the other hand, by [DH92, A, 4.14], S is contained in $\mathrm{Soc}(G)$. Thus $S = \mathrm{Soc}(G)$ and $(H \cap S)T = 1$. Assume that $\mathrm{Soc}(G) \leq T$. Then $G = HT$ and so $(H \cap S)T = S$. □

The next aim is to obtain a similar result for t$'$.

Lemma 5.2.15. *Let G be a monolithic primitive group. Assume that G has a subgroup H such that $G = H\,\mathrm{Soc}(G)$ and $H \cap \mathrm{Soc}(G) = 1$. If $G = S\,\mathrm{Soc}(G)$ with $T \trianglelefteq S \,\mathrm{sn}\, G$ and S/T semisimple, then $(H \cap S)T \in \{S, T\}$.*

Proof. First consider the case when $S = G$. Then T is a normal subgroup of G. If $T \neq 1$, then $\mathrm{Soc}(G)$ is contained in T and then $G = HT$. Hence $(H \cap S)T = S$. If $T = 1$, then $G = \mathrm{Soc}(G)$ and $H = 1$.

Suppose that $1 \neq S$ is a proper subnormal subgroup of G. Then a maximal normal subgroup M of G would contain S as well as the unique minimal normal subgroup $\mathrm{Soc}(G)$ of G; so the contradiction that $G = S\,\mathrm{Soc}(G) = M < G$ would emerge. □

Theorem 5.2.16. *Let G be a group. Assume that $G = HM$ and $H \cap M = 1$ where M is a strictly semisimple normal subgroup of G. Then the following statements are equivalent:*

1. *If E is a simple direct factor of M, and K is a subnormal subgroup of H normalising E, then*

$$\mathrm{I}_K(E) = \{k \in K : K \text{ induces an inner automorphism in } E\} = \mathrm{C}_K(E).$$

2. $H \in \mathrm{t}'(G)$.

Proof. 1 implies 2. We argue by induction on $|G|$. Let A be a minimal normal subgroup of G contained in H. Then it is straightforward to verify that the subgroup H/A satisfies the hypotheses of the theorem in G/A. We can conclude by induction that $H/A \in \mathrm{t}'(G/A)$. In this case we have $H \in \mathrm{t}'(G)$ because t' is inherited. Therefore we can assume that $\mathrm{Core}_G(H) = 1$ and, in particular, $\mathrm{C}_H(M) = 1$. Next we show that we can suppose that M is a minimal normal subgroup of G. Let B be a minimal normal subgroup of G contained in M. If $B < M$, then HB is a proper subgroup of G for which the hypotheses of the theorem clearly hold. Therefore by induction $H \in \mathrm{t}'(HB)$. We also have that $HB/B \in \mathrm{t}'(G/B)$ by induction. Then $H \in \mathrm{t}'(G)$. Thus we can suppose that M is a minimal normal subgroup of G.

Let S/T be a non-abelian and strictly semisimple section of G such that S is subnormal in G. If SM is a proper subgroup of G, then $SM = (H \cap SM)M$ and the hypotheses of the theorem hold in SM because $H \cap SM$ is subnormal in H. Hence by induction $H \cap SM \in \mathrm{t}'(SM)$ and so $(H \cap S)T = (H \cap SM \cap S)T \trianglelefteq S$. Therefore we can assume that $G = SM$. If M is abelian, then $\mathrm{C}_G(M) = M$. This implies that M is the unique minimal normal subgroup of G and G is a monolithic primitive group by Theorem 1. Applying Lemma 5.2.15, $(S \cap H)T \in \{S, T\}$. Assume that M is non-abelian. We prove that $\mathrm{C}_G(M) = 1$. To this end, put $C = \mathrm{C}_G(M)$, $D = H \cap CM$, and suppose that $C \neq 1$. Observe that $D \neq 1$ because $M < CM = C \times M = HM \cap CM = DM$. Let E denote a simple component of M. Since both C and M act upon E by inner automorphisms, from Statement 1 we obtain that $D \leq \mathrm{C}_H(E)$. On the other hand, H (acting by conjugation) permutes the simple components of the minimal normal subgroup M of G transitively, so $D = \bigcap_{h \in H} D^h$ is contained in $\mathrm{C}_H(M) = \mathrm{C}_H\left(\prod_{h \in H} E^h\right) = \bigcap_{h \in H} \mathrm{C}_H(E^h)$. This yields the contradiction that $1 \neq D \leq \mathrm{C}_H(M) = \mathrm{Core}_G(H) = 1$. Therefore $\mathrm{C}_G(M) = 1$ and $M = \mathrm{Soc}(G)$. Hence G is a monolithic primitive group by Theorem 1. Applying Lemma 5.2.15, $(S \cap T)T \in \{S, T\}$.

Consequently $H \in \mathrm{t}'(G)$.

2 implies 1. Assume, arguing by contradiction, that there exists a subnormal subgroup K of $H \in \mathrm{t}'(G)$ such that K normalises a simple direct factor E of M and $I = \mathrm{I}_K(E)$ is not contained in $\mathrm{C}_K(E)$. Then E is non-abelian. Note that I is a normal subgroup of K. Hence I is subnormal in H. Since $\mathrm{s}_n \leq \mathrm{t}'$ and t' is inherited, it follows that $I \in \mathrm{t}'(G)$. Moreover IM is subnormal in G. These facts clearly imply that $I \in \mathrm{t}'(IM)$. On the other hand, $M = E \times A$ for some I-invariant normal complement A of E in M. Since $IA/A \in \mathrm{t}'(IM/A)$ by Remark 5.1.9 (4), we conclude that $I \in \mathrm{t}'(IE)$. Denote $Z = IE$. Then

$Z = E\,\mathrm{C}_Z(E)$ and $E \cap \mathrm{C}_Z(E) = 1$. Hence $Z/\mathrm{C}_Z(E)$ is isomorphic to E. As I is not contained in $\mathrm{C}_Z(E)$, $I\,\mathrm{C}_Z(E)/\mathrm{C}_Z(E) \neq 1$ and so $I\,\mathrm{C}_Z(E) = Z$ because $I\,\mathrm{C}_Z(E)/\mathrm{C}_Z(E) \in \mathrm{t}'\big(Z/\mathrm{C}_Z(E)\big)$. Hence $I \cap \mathrm{C}_Z(E) = \mathrm{C}_K(E)$ is a normal subgroup of Z and $Z/\mathrm{C}_K(E)$ is isomorphic to $E \times E$. This contradicts the fact that $I/\mathrm{C}_K(E) \in \mathrm{t}'\big(Z/\mathrm{C}_K(E)\big)$. $\qquad\square$

Corollary 5.2.17. *If G is a primitive group of type 1, then any complement of $\mathrm{Soc}(G)$ in G is a t'-subgroup.*

The corresponding statement for subgroups of primitive groups of type 2, though, does not always hold.

Example 5.2.18. Let $H = E \wr F$, the regular wreath product of a non-abelian simple group E and a cyclic group F of order 3. Its base group can be written as $B = E_1 \times E_2 \times E_3$ with $E_i \cong E$. A subnormal subgroup K of H is defined by $K = E_1 \times E_2$. Let K act on E such that E_1 induces all inner automorphisms, while E_2 acts trivially, and form the twisted wreath product corresponding to this action: $G = E \wr_K H$. One checks that G is a primitive group of type 2, which, by its very construction, violates Theorem 5.2.16 (1) (taking M as the base group of G).

5.3 Projectors of soluble type

The purpose of this section is to develop a general framework for a theory of Schunck class projectors in finite (not necessarily soluble) groups. A specialisation of that theory will yield a theory resembling the theory of projectors in finite soluble groups.

Hypothesis 5.3.1. *In the sequel, d will denote a subgroup functor subject to the following conditions:*

1. *If $H \in \mathrm{d}(G)$ and $N \trianglelefteq G$, then $HN/N \in \mathrm{d}(G/N)$.*
2. *d is weakly inductive, that is, $G \in \mathrm{d}(G)$ and if $H \in \mathrm{d}(HN)$ and $HN/N \in \mathrm{d}(G/N)$, then $H \in \mathrm{d}(G)$.*
3. *If $H \in \mathrm{d}(G)$ and M is a minimal normal subgroup of G, then $H < G = HM$ implies $H \cap M = 1$.*
4. *If H is a core-free maximal subgroup of a primitive group G of type 3, then $H \notin \mathrm{d}(G)$.*
5. *If $G = H\,\mathrm{F}^*(G)$, $H \in \mathrm{d}(G)$, and M is a quasinilpotent normal subgroup of G, then $H \in \mathrm{d}(HM)$.*
6. *If $H \in \mathrm{d}(G)$ and N is a normal quasinilpotent subgroup of G, then $H \cap N$ is subnormal in N.*

Remark 5.3.2. The results of the previous sections show that any of the functors s_n, t, r, t', and r' satisfies the above conditions.

Definitions 5.3.3. *1. A subgroup U of a group G is said to be* d-maximal *in G if U is a proper subgroup of G, $U \in \mathrm{d}(G)$, and $U < H$, and $H \in \mathrm{d}(G)$ implies that $H = G$.*

2. A group G is said to be a d-primitive *group if G has a* d-maximal *subgroup U such that $\mathrm{Core}_G(U) = 1$.*

Lemma 5.3.4. *Assume that G is a* d-*primitive group. Then G is a monolithic primitive group. If U is a core-free* d-*maximal subgroup of G, then $G = U \mathrm{Soc}(G)$ and $U \cap \mathrm{Soc}(G) = 1$.*

Proof. Let H be a maximal subgroup of G containing U. Put $T = \mathrm{Core}_G(H)$. Since d is weakly inductive, it follows that $UT \in \mathrm{d}(G)$. Hence $U = UT$ because UT is a proper subgroup of G. This implies that $T = 1$ and G is a primitive group. Let M be a minimal normal subgroup of G. Then U is a proper subgroup of UM and $UM \in \mathrm{d}(G)$. This yields $G = UM$ by d-maximality of U. Applying Hypothesis 5.3.1 (3), $U \cap M = 1$. If G were a primitive group of type 3, then $\mathrm{C}_G(M)$ would be non-trivial and so $U\,\mathrm{C}_G(M) = U$ by the d-maximality of U. Then U would be a core-free maximal subgroup of G. This would be a contradiction against Hypothesis 5.3.1 (4). Therefore G is a monolithic primitive group. $\qquad\square$

We say that G is a d-primitive group of type i $(i = 1, 2)$ if G is a primitive group of type i.

Remarks 5.3.5.   1. Applying Remark 5.2.9, a t-primitive group G of type 2 satisfies that $G/\mathrm{Soc}(G)$ is soluble and if G is a t-primitive group of type 1, then $G/\mathrm{Soc}(G)$ is p-soluble, where p is the prime dividing $|\mathrm{Soc}(G)|$, by Theorem 5.2.4.

2. By Theorem 5.2.14, any complement of the minimal normal subgroup of a monolithic primitive group G is an r'-subgroup. In particular, monolithic primitive groups with complemented socle are r'-primitive.

3. Let E be any non-abelian simple group. Applying Theorem 1.1.36, there is a group G with a minimal normal subgroup M such that M is the direct product of copies of $\mathrm{Alt}(6)$, G/M is isomorphic to E, and G does not split over M. It is clear that G is a primitive group of type 2. Applying Lemma 5.3.4, G is not d-primitive.

Definitions 5.3.6. *Let $\mathfrak{H}$ be a class of groups.*

1. *A subgroup X of a group G is said to be* $\mathfrak{H}$-d-maximal *subgroup of G if $X \in \mathfrak{H} \cap \mathrm{d}(G)$ and if $X \leq K \in \mathrm{d}(G) \cap \mathfrak{H}$, then $X = K$.*
   Denote by $\mathrm{Max}_{\mathfrak{H}^{\mathrm{d}}}(G)$ the (possibly empty) set of all $\mathfrak{H}$-d-maximal *subgroups of G.*

2. *A subgroup U of a group G is called an* $\mathfrak{H}$-d-projector *of G if UN/N is* $\mathfrak{H}$-d-maximal *in G/N for all $N \trianglelefteq G$.*
   We shall use $\mathrm{Proj}_{\mathfrak{H}^{\mathrm{d}}}(G)$ to denote the (possibly empty) set of $\mathfrak{H}$-d-*projectors of a group G.*

3. An $\mathfrak{H}$-d-covering subgroup *of a group G is a subgroup E of G satisfying the following two conditions:*
   a) $E \in \mathrm{Max}_{\mathfrak{H}^d}(G)$, *and*
   b) if $T \in \mathrm{d}(G)$, $E \leq T$, $N \trianglelefteq T$, *and* $T/N \in \mathfrak{H}$, *then* $T = NE$.
   The set of all $\mathfrak{H}$-d-covering subgroups of a group G will be denoted by $\mathrm{Cov}_{\mathfrak{H}^d}(G)$.

It is clear that $\mathrm{Max}_{\mathfrak{H}^d}$, $\mathrm{Proj}_{\mathfrak{H}^d}$, and $\mathrm{Cov}_{\mathfrak{H}^d}$ are subgroup functors and $\mathrm{Cov}_{\mathfrak{H}^d} \leq \mathrm{Proj}_{\mathfrak{H}^d} \leq \mathrm{Max}_{\mathfrak{H}^d}$.

Definitions 5.3.7. *1. A class $\mathfrak{H}$ is called* d-projective *if* $\mathrm{Proj}_{\mathfrak{H}^d}(G) \neq \emptyset$ *for each group G.*
2. A class $\mathfrak{H}$ is said to be a d-Schunck class *if $\mathfrak{H}$ is a homomorph that comprises precisely those groups whose* d-*primitive epimorphic images are in $\mathfrak{H}$.*

Clearly a class $\mathfrak{H}$ is a d-Schunck class if and only if $\mathfrak{H}$ is a homomorph whose boundary is composed of d-primitive groups.

Remark 5.3.8. Every d-Schunck class is a Schunck class whose boundary is composed of monolithic primitive groups (Lemma 5.3.4). The converse does not hold: the Schunck class $\mathfrak{H}$ of all groups without epimorphic images isomorphic to the group G of Remark 5.3.5 (3) is not a d-Schunck class.

Our next goal is to prove that the d-projective classes are exactly the d-Schunck classes. We need some previous results.

Lemma 5.3.9. *Let $\mathfrak{H}$ be a homomorph. Let G be a group and N a normal quasinilpotent subgroup of G.*
If $H \in \mathrm{Max}_{\mathfrak{H}^d}(HN)$ and $HN/N \in \mathrm{Max}_{\mathfrak{H}^d}(G/N)$, then $H \in \mathrm{Max}_{\mathfrak{H}^d}(G)$.

Proof. Let T be a subgroup of G such that H is contained in T and $T \in \mathrm{d}(G) \cap \mathfrak{H}$. Since $TN/N \in \mathrm{d}(G/N) \cap \mathfrak{H}$ by Hypothesis 5.3.1 (1) and $HN/N \in \mathrm{Max}_{\mathfrak{H}^d}(G/N)$, it follows that $HN = TN$. By Hypothesis 5.3.1 (6), $T \cap N$ is subnormal in N. Hence $T \cap N$ is a normal quasinilpotent subgroup of HN. By Hypothesis 5.3.1 (5), we know that $H(T \cap N) = T \in \mathrm{d}(HN)$. Therefore $T = H$ by the $\mathfrak{H}$-d-maximality of H in HN. $\qquad\square$

Remark 5.3.10. Lemma 5.3.9 holds for any normal subgroup N of G if the functor d has the following property:

5*. If $H \in \mathrm{d}(G)$ and $N \trianglelefteq G$, then $H \in \mathrm{d}(HN)$.

Note that t and r satisfy Property 5* but t$'$ and r$'$ do not enjoy this property (see Example 5.1.12).

Lemma 5.3.11. *Let $\mathfrak{H}$ be a homomorph and let G be a group. Assume that N is a quasinilpotent normal subgroup of G and K is a subgroup of G such that N is contained in K and $K/N \in \mathrm{Proj}_{\mathfrak{H}^d}(G/N)$. If $H \in \mathrm{Proj}_{\mathfrak{H}^d}(K)$, then $H \in \mathrm{Proj}_{\mathfrak{H}^d}(G)$.*

Proof. We must show that $HM/M \in \mathrm{Max}_{\mathfrak{H}^d}(G/M)$ for all normal subgroups M of G. First of all, $K = HN$ because $HN/N \in \mathrm{Max}_{\mathfrak{H}^d}(K/N)$. Moreover, since $H \in \mathrm{Proj}_{\mathfrak{H}^d}(K)$, it follows that $H(K \cap M)/(K \cap M)$ is $\mathfrak{H}$-d-maximal in $K/(K \cap M)$. This implies that HM/M is $\mathfrak{H}$-d-maximal in KM/M. On the other hand, $(KM/N)/(MN/N)$ is an $\mathfrak{H}$-d-maximal subgroup of $(G/N)/(MN/N)$ because $K/N \in \mathrm{Proj}_{\mathfrak{H}^d}(G/N)$. In particular, $(KM/M)/(MN/M)$ is $\mathfrak{H}$-d-maximal in $(G/M)/(MN/M)$. Since NM/M is quasinilpotent and normal in G/M, we apply Lemma 5.3.9 to conclude that HM/M is $\mathfrak{H}$-d-maximal in G/M.   □

Remark 5.3.12. Lemma 5.3.11 holds for any normal subgroup N of G if d satisfies Property 5*.

Lemma 5.3.13. *Let $G \in \mathrm{b}(\mathfrak{H})$ for some homomorph $\mathfrak{H}$.*

1. $\mathrm{Proj}_{\mathfrak{H}^d}(G) = \mathrm{Cov}_{\mathfrak{H}^d}(G)$, *and this set coincides with*

$$\{U \in \mathrm{d}(G) : U < G = UM \text{ for all minimal normal subgroups } M \text{ of } G\},$$

here $U \cap M = 1$ for all $H \in \mathrm{Proj}_{\mathfrak{H}^d}(G)$ and any minimal normal subgroup M of G.
2. $\mathrm{Proj}_{\mathfrak{H}^d}(G) \neq \emptyset$ *if and only if G is d-primitive.*

Proof.   1. If $\mathrm{Proj}_{\mathfrak{H}^d}(G) = \emptyset$, then $\mathrm{Cov}_{\mathfrak{H}^d}(G) = \emptyset$. Assume $\mathrm{Proj}_{\mathfrak{H}^d}(G) \neq \emptyset$ and let U be an $\mathfrak{H}$-d-projector of G. Since $G \notin \mathfrak{H}$, it follows that U is a proper subgroup of G. Let M be a minimal normal subgroup of G. Then $G/M \in \mathfrak{H}$ by definition of the boundary, and therefore by definition of an $\mathfrak{H}$-d-projector we have $G = UM$. Applying Hypothesis 5.3.1 (3), $U \cap M = 1$. It is clear that in this case U is a d-maximal subgroup of G. Therefore $U \in \mathrm{Cov}_{\mathfrak{H}^d}(G)$. On the other hand, if $G = HM$ and $H \cap M = 1$ for a proper subgroup $H \in \mathrm{d}(G)$ of G and for all minimal normal subgroup of G, then $H \cong G/M \in \mathfrak{H}$ and G and H are the only d-subgroups of G above H. Consequently $H \in \mathrm{Cov}_{\mathfrak{H}^d}(G)$ and Statement 1 holds.

2. Suppose that $\mathrm{Proj}_{\mathfrak{H}^d}(G) \neq \emptyset$ and, as in 1, we take $U \in \mathrm{Proj}_{\mathfrak{H}^d}(G)$. Since $G \notin \mathfrak{H}$, it follows that U is a proper subgroup of G. Let H be a d-maximal subgroup of G containing U, and let $K = \mathrm{Core}_G(H)$. If $K \neq 1$, then $G/K \in \mathfrak{H}$ because G belongs to the boundary of $\mathfrak{H}$. By definition of an $\mathfrak{H}$-d-projector we have $G = UK \leq H < G$. This contradiction shows that $K = 1$ and hence that G is d-primitive.

Conversely, assume that G is d-primitive. Let U be a core-free d-maximal subgroup of G. Then $G = U \mathrm{Soc}(G)$ and $U \cap \mathrm{Soc}(G) = 1$. It is clear then that $U \in \mathfrak{H}$. Moreover U and G are the only d-subgroups of G above U. Consequently $U \in \mathrm{Max}_{\mathfrak{H}^d}(G)$ and so $U \in \mathrm{Proj}_{\mathfrak{H}^d}(G)$. This completes the proof of the lemma.   □

Theorem 5.3.14. *A class $\mathfrak{H} \neq \emptyset$ is* d-*projective if and only if it is a* d-*Schunck class.*

Proof. Let $\mathfrak{H}$ be a d-projective class, and let $G \in \mathfrak{H}$. Since $\text{Proj}_{\mathfrak{H}^d}(G) \neq \emptyset$, it follows from the definition of a projector that $\text{Proj}_{\mathfrak{H}^d}(G) = \{G\}$ and hence that for all $N \trianglelefteq G$ the quotient $G/N = GN/N$ is $\mathfrak{H}^d$-maximal in G/N, in other words, that $G/N \in \mathfrak{H}$. Therefore $\mathfrak{H}$ is a homomorph. By Lemma 5.3.13, $\text{b}(\mathfrak{H})$ is composed of d-primitive groups, and consequently $\mathfrak{H}$ is a d-Schunck class.

Conversely, let $\mathfrak{H}$ be a d-Schunck class. We prove that $\text{Proj}_{\mathfrak{H}^d}(G) \neq \emptyset$ by induction on $|G|$. Clearly we may suppose that $G \notin \mathfrak{H}$ (in particular $G \neq 1$) and that $\text{Proj}_{\mathfrak{H}^d}(A) \neq \emptyset$ for all groups A such that $|A| < |G|$. Let N be a minimal normal subgroup of G. Then $\text{Proj}_{\mathfrak{H}^d}(G/N)$ contains a subgroup, K/N say, and if $|K| < |G|$, there exists a subgroup $H \in \text{Proj}_{\mathfrak{H}^d}(K)$. Then from Lemma 5.3.11 we conclude that $H \in \text{Proj}_{\mathfrak{H}^d}(G)$. There remains the possibility that $G/N \in \text{Proj}_{\mathfrak{H}^d}(G/N)$, which implies that $G/N \in \mathfrak{H}$. But if $G/N \in \mathfrak{H}$ for all minimal normal subgroups N of G, it follows that $G \in \text{b}(\mathfrak{H})$, and then $\text{Proj}_{\mathfrak{H}^d}(G) \neq \emptyset$ by Lemma 5.3.13. The induction argument is therefore complete. $\square$

Corollary 5.3.15. d-*Schunck classes are precisely those Schunck classes $\mathfrak{H}$ such that* $\text{Proj}_{\mathfrak{H}^d}(G) \neq \emptyset$ *for all* $G \in \text{b}(\mathfrak{H})$.

Now we turn our attention to the conjugacy problem for d-projectors and d-covering subgroups. As in the classical case, primitive groups of type 3 have to be excluded from the boundaries. In this context, the information given in the next lemma is useful.

Lemma 5.3.16. *Let $\mathfrak{H}$ be a Schunck class such that $\text{b}(\mathfrak{H}) \cap \mathfrak{P}_3 = \emptyset$. Let $G = HN$, where N is a quasinilpotent normal subgroup G. If H is an $\mathfrak{H}$-d-maximal subgroup of G, then $H \in \text{Proj}_{\mathfrak{H}^d}(G)$.*

Proof. We proceed by induction on $|G|$. Without loss of generality we may assume that $N = \text{F}^*(G)$. Suppose there exists a minimal normal subgroup M of G such that $HM < G$, and $G/M \notin \mathfrak{H}$. Then, by [DH92, III, 2.2 (c)], there exists a normal subgroup T of G such that $M \leq T$ and $G/T \in \text{b}(\mathfrak{H})$. By hypothesis, G/T is a monolithic primitive group; as $G/N = HN/N \in \mathfrak{H}$, we have that N is not contained in T and so NT/T is a non-trivial quasinilpotent normal subgroup of G/T. This implies that $NT/T = \text{Soc}(G/T)$ by Theorem 1 and Proposition 2.2.22 (3). Since $G/T \notin \mathfrak{H}$ but $G/NT \in \mathfrak{H}$, the d-subgroup H of G cannot possibly cover NT/T, and so must avoid this factor. Consequently $HT \cap N = HT \cap NT \cap N = T(H \cap NT) \cap N = T \cap N$, and $HT = HT \cap HN = H(HT \cap N) = H(T \cap N)$ follows. As N is not contained in T, it follows that HT is a proper subgroup of G. Moreover $T \cap N$ is a normal quasinilpotent subgroup of $HT = H(T \cap N)$ and hence $H \in \text{d}(HT)$ by Hypothesis 5.3.1 (5). Let H_1 be an $\mathfrak{H}$-d-maximal subgroup of HT containing H. Then $HT = H_1 T$. Note that HT/T is $\mathfrak{H}$-d-maximal in G/T because $HT/T \in \text{d}(G/T) \cap \mathfrak{H}$ and $(HT/T) \cap (NT/T) = 1$. By induction, $H_1 \in \text{Proj}_{\mathfrak{H}^d}(H_1 T)$ and $H_1 T/T \in \text{Proj}_{\mathfrak{H}^d}(G/T)$. Applying Lemma 5.3.11, $H_1 \in \text{Proj}_{\mathfrak{H}^d}(G)$. Then $H = H_1$ because of $\mathfrak{H}$-d-maximality of H, and the result follows.

Consequently we can suppose that $G/M \in \mathfrak{H}$ for all minimal normal subgroups M of G. Then either $G \in \mathfrak{H}$ — in which case nothing remains to be shown —, or else $G \in \mathrm{b}(\mathfrak{H})$. In the latter case our claim is immediate from the assumption that $G = H \mathrm{F}^*(G)$: then $\mathrm{F}^*(G)$ is the minimal normal subgroup of the monolithic primitive group by Theorem 1 and Proposition 2.2.22 (3). The induction argument is therefore complete.   □

The further results in this section will all require that d has the properties stated in Hypothesis 5.3.1 together with Property 5* introduced in Remark 5.3.10. In this case, $\mathrm{Proj}_{\mathfrak{H}\mathrm{d}}$ is an inductive functor by Lemma 5.3.11 and Remark 5.3.12.

Lemma 5.3.17. *Let $\mathfrak{H}$ be a d-Schunck class such that $\mathrm{b}(\mathfrak{H}) \cap \mathfrak{P}_3 = \emptyset$. If $H \in \mathrm{Proj}_{\mathfrak{H}\mathrm{d}}(G)$ and H is contained in U, where $U \in \mathrm{d}(G)$, then $H \in \mathrm{Proj}_{\mathfrak{H}\mathrm{d}}(U)$.*

Proof. We proceed by induction on $|G|$. Clearly we can assume that $U < G$. Let N be a minimal normal subgroup of G. By induction, HN/N is an $\mathfrak{H}$-d-projector of UN/N because $UN/N \in \mathrm{d}(G/N)$ Hypothesis 5.3.1 (1) and $HN/N \in \mathrm{Proj}_{\mathfrak{H}\mathrm{d}}(G/N)$. The standard isomorphism from UN/N to $U/(U\cap N)$ transforms HN/N into $H(U \cap N)/(U \cap N)$. This yields $H(U \cap N)/(U \cap N) \in \mathrm{Proj}_{\mathfrak{H}\mathrm{d}}(U/(U \cap N))$. Assume that HN is a proper subgroup of G. By Property 5*, $H \in \mathrm{d}(HN)$. Let H_1 be an $\mathfrak{H}$-d-maximal subgroup of HN containing H. By Lemma 5.3.16, $H_1 \in \mathrm{Proj}_{\mathfrak{H}\mathrm{d}}(HN)$. Since $HN/N = H_1N/N$ is an $\mathfrak{H}$-d-projector of G/N, it follows that $H_1 \in \mathrm{Proj}_{\mathfrak{H}\mathrm{d}}(G)$ by Lemma 5.3.11. In particular, $H = H_1$ because $H \in \mathrm{Max}_{\mathfrak{H}\mathrm{d}}(G)$. On the other hand, $N \cap U$ is normal in N as $N \cap U$ is subnormal in N by Hypothesis 5.3.1 (6). Hence $N \cap U$ is a normal subgroup of HN. This implies that $H(N \cap U) \in \mathrm{d}(HN)$ by Hypothesis 5.3.1 (5). Applying induction, $H \in \mathrm{Proj}_{\mathfrak{H}\mathrm{d}}(H(N\cap U))$. Consequently $H \in \mathrm{Proj}_{\mathfrak{H}\mathrm{d}}(U)$ by Hypothesis 5.3.1 (2).

Therefore we can suppose that $G = HN$. Since $U \in \mathrm{d}(G)$ is a proper subgroup of G, it follows that $U \cap N = H \cap N = 1$ by Hypothesis 5.3.1 (3). This implies $U = H$. Therefore $H \in \mathrm{Proj}_{\mathfrak{H}\mathrm{d}}(U)$ and the proof of the lemma is complete.   □

We can state an important consequence of this lemma.

Theorem 5.3.18. *Let $\mathfrak{H}$ be a Schunck class such that $\mathrm{b}(\mathfrak{H}) \cap \mathfrak{P}_3 = \emptyset$. Then $\mathrm{Proj}_{\mathfrak{H}\mathrm{d}}(G) = \mathrm{Cov}_{\mathfrak{H}\mathrm{d}}(G)$ for every group G.*

Proof. Let H be an $\mathfrak{H}$-d-projector of a group G. Let T be a subgroup of G such that $T \in \mathrm{d}(G)$ and $H \leq T$. By Lemma 5.3.17, H is an $\mathfrak{H}$-d-projector of T. Therefore if N is a normal subgroup of T and $T/N \in \mathfrak{H}$, it follows that $T = NH$. Consequently H satisfies Conditions 3a and 3b of Definition 5.3.6 (3) and so $H \in \mathrm{Cov}_{\mathfrak{H}\mathrm{d}}(G)$. The other inclusion is clear.   □

We prove in the following the main conjugacy theorem for projectors.

Theorem 5.3.19. *For any Schunck class $\mathfrak{H}$ such that $\mathrm{b}(\mathfrak{H}) \cap \mathfrak{P}_3 = \emptyset$, the following statements are equivalent:*

1. *For every group $G \in \mathrm{b}(\mathfrak{H})$, $\mathrm{Proj}_{\mathfrak{H}^d}(G)$ is a conjugacy class of G.*
2. *For every group G, $\mathrm{Proj}_{\mathfrak{H}^d}(G)$ is a conjugacy class of G.*

Proof. Only the implication "1 implies 2" is in doubt. We argue by induction on $|G|$ that $\mathrm{Proj}_{\mathfrak{H}^d}(G)$ is a conjugacy class of G. Let N be a minimal normal subgroup of G and H_1 and H_2 $\mathfrak{H}$-d-projectors of G. Since $H_i N / N \in \mathrm{Proj}_{\mathfrak{H}^d}(G/N)$, by induction we have $H_1 N = H_2^g N$ for some $g \in G$. By Theorem 5.3.18, H_1 and H_2^g are $\mathfrak{H}$-d-projectors of $H_1 N$. If $|H_1 N| < |G|$, by induction H_1 is conjugate to H_2^g and hence to H_2. We can therefore suppose that $G = H_1 N$ and hence $G/N \in \mathfrak{H}$ for every minimal normal subgroup of G. Thus, either $G \in \mathfrak{H}$ and then $H_1 = G = H_2$, or $G \in \mathrm{b}(\mathfrak{H})$, and H_1 and H_2 are conjugate by the hypothesis. $\qquad\square$

Two $\mathfrak{H}$-t-projectors of a group need not be conjugate.

Example 5.3.20. Let $\mathfrak{H}$ be the Schunck class defined by $\mathrm{b}(\mathfrak{H}) = \big(\mathrm{Sym}(6)\big)$. Since $\mathrm{Sym}(6)$ is t-primitive, it follows that $\mathfrak{H}$ is a t-Schunck class; $\mathrm{Sym}(6)$ has two non-conjugate t-subgroups complementing $\mathrm{Alt}(6)$, say X and Y, where the involution in X is a 2-cycle, while the involution in Y is a product of three disjoint 2-cycles. Clearly $X, Y \in \mathrm{Proj}_{\mathfrak{H}^t}\big(\mathrm{Sym}(6)\big)$.

As we have mentioned earlier in this section, the subgroup functors t, r enjoy the properties stated in Hypothesis 5.3.1 (1)–(6). They also enjoy Property 5*. Moreover if d is one of these functors, $\mathrm{d}(G) = \mathrm{s}(G)$ for all $G \in \mathfrak{S}$. Therefore $\mathrm{Proj}_{\mathfrak{H}^d}(G) = \mathrm{Proj}_{\mathfrak{H}}(G) = \mathrm{Proj}_{\mathfrak{H} \cap \mathfrak{S}}(G)$ for all $G \in \mathfrak{S}$; thus, we have obtained an alternative approach to projectors and covering subgroups in finite soluble groups.

On the other hand, a conjugacy criterion for $\mathrm{Proj}_{\mathfrak{H}^{r'}}$ (or $\mathrm{Proj}_{\mathfrak{H}^{t'}}$) analogous to Theorem 5.3.19 does not hold; moreover $\mathrm{Cov}_{\mathfrak{H}^{t'}} \lneq \mathrm{Proj}_{\mathfrak{H}^{t'}}$.

Example 5.3.21. Let E be any non-abelian simple group and take an involution t in E. Define $T = \langle \tau \rangle$ with τ being the inner automorphism of E corresponding to t. We consider the regular wreath product $H = E \wr T$ and form the twisted wreath product $G = E \wr_T H$ where T acts naturally on E as a group of inner automorphisms. The base group of G will be denoted by $M = E_1 \times \cdots \times E_n$, $n = |H : T|$. Applying Theorem 5.2.16, $T \in \mathrm{t}'(H)$ and $H \in \mathrm{t}'(G)$. Therefore $T \in \mathrm{t}'(G) \subseteq \mathrm{r}'(G)$ because t' is inherited. Let $D = \langle m\tau \rangle$, where m is the product of the involutions $t_i \in E_i$ such that $E_i^{t_i} = E_i^{\tau}$ and $i\tau = i$. Then D centralises E_i for all i such that $i\tau = i$. This implies that $D \in \mathrm{t}'(MD)$ by Theorem 5.2.16, and hence $D \in \mathrm{t}'(G)$ by inductivity.

Let $\mathfrak{H}$ be the Schunck class of all groups without quotients isomorphic to E or H. Then $\mathrm{b}(\mathfrak{H}) = \{E, H\}$. It is clear that $\mathfrak{H}$ is a t'-Schunck class. Moreover $\{T, D\} \subseteq \mathrm{Proj}_{\mathfrak{H}^{t'}}(G)$. Applying Theorem 5.2.16, $T \notin \mathrm{Proj}_{\mathfrak{H}^{t'}}(TM)$ and $D \in \mathrm{Proj}_{\mathfrak{H}^{t'}}(DM)$. Thus, from $TM = DM$, we get that T and D cannot possibly be conjugate in G. Similarly $\{T, D\} \subseteq \mathrm{Proj}_{\mathfrak{H}^{r'}}(G)$.

We have shown that $\mathfrak{H}$ is a d-Schunck class and G is a group with two non-conjugate $\mathfrak{H}$-d-projectors for $d \in \{t', r'\}$. Note that $\mathrm{Proj}_{\mathfrak{H}^d}(A)$ is a conjugacy class for all $A \in \mathrm{b}(\mathfrak{H})$. Moreover $T \in \mathrm{Proj}_{\mathfrak{H}^{t'}}(G) \setminus \mathrm{Cov}_{\mathfrak{H}^{t'}}(G)$.

We have mentioned in Chapter 2 that Hall π-subgroups, π a set of primes, have been among the motivating examples for the theory of projectors in finite soluble groups: they are, in the soluble universe, the projectors with respect to the class $\mathfrak{S}_\pi$ of all soluble π-groups. When applying the theory of projectors of finite (not necessarily soluble) groups to the class $\mathfrak{E}_\pi$ of all finite π-groups, one obtains that every finite group has $\mathfrak{E}_\pi$-projectors. However, in a non-soluble group, the set of $\mathfrak{E}_\pi$-projectors does not normally form a set of conjugate subgroups, and it does not appear to be true that every π-subgroup is contained in some $\mathfrak{E}_\pi$-projector; certainly, a π-subgroup need not be contained in a conjugate of a given $\mathfrak{E}_\pi$-projector.

In the following example, we see that for each set π of primes there is some class $\mathfrak{E}(\pi)$, which may be thought of as being generated by $\mathfrak{E}_\pi$ in a certain sense, with the following properties: the $\mathfrak{E}(\pi)$-projectors of an arbitrary finite group G form a non-empty set of conjugate subgroups of G and each π-subgroup of G is contained in some $\mathfrak{E}(\pi)$-projector of G.

A worked example

Let π be a set of primes. Define

$$\mathfrak{E}(\pi) := \Big(G : \text{if } N \trianglelefteq G \text{ and } G/N \text{ is t-primitive, then } \pi\big(\mathrm{Soc}(G/N)\big) \cap \pi \neq \emptyset\Big).$$

It is clear that $\mathfrak{E}(\pi) = {}_\mathrm{Q}\,\mathfrak{E}(\pi)$ is contained in $\mathfrak{E}(\pi)$, and the boundary of $\mathfrak{E}(\pi)$ is given by

$$\mathrm{b}\big(\mathfrak{E}(\pi)\big) = \big(G : G \text{ is t-primitive, } G/\mathrm{Soc}(G) \text{ is an } \mathfrak{E}(\pi)\text{-group}$$
$$\text{and } \mathrm{Soc}(G) \text{ is a } \pi'\text{-group}\big).$$

Therefore $\mathfrak{E}(\pi)$ is a t-Schunck class.

Since $\mathrm{t}(G) = \mathrm{s}(G)$ for all soluble groups G, $\mathfrak{E}(\pi) \cap \mathfrak{S} = \mathfrak{S}_\pi$, the class of all soluble π-groups. Assume that G is a t-primitive group of type 2 in $\mathrm{b}\big(\mathfrak{E}(\pi)\big)$. Then $G/\mathrm{Soc}(G)$ is a soluble π-group by Remark 5.2.9. Therefore $\mathrm{Soc}(G)$ is a Hall π'-subgroup of G. By the Schur-Zassenhaus theorem [Hup67, I, 18.1 and 18.2], any two complements of $\mathrm{Soc}(G)$ in G are conjugate. They are the Hall π-subgroups of G and the $\mathfrak{E}(\pi)$-t-projectors of G.

If G is a t-primitive group of type 1 in $\mathrm{b}\big(\mathfrak{E}(\pi)\big)$, then $\mathrm{Soc}(G)$ is a p-group for some prime p and it is complemented in G by a p-soluble subgroup H of G which is an $\mathfrak{E}(\pi)$-projector of G. Applying Corollary 5.2.7, any two complements of $\mathrm{Soc}(G)$ in G are conjugate.

By Theorem 5.3.18 and Theorem 5.3.19, we have:

Theorem 5.3.22. $\mathrm{Proj}_{\mathfrak{E}(\pi)^t}(G) = \mathrm{Cov}_{\mathfrak{E}(\pi)^t}(G)$ *is a class of conjugate subgroups in every group* G.

For each group G, we put

$$\mathrm{h}_\pi(G) = \{H \in \mathrm{t}(G) : H \in \mathfrak{E}(\pi) \text{ and } \pi(|G : H|) \subseteq \pi'\}.$$

Lemma 5.3.23. *If* $G \in \mathfrak{E}(\pi)$ *and* $H \in \mathrm{t}(G)$, $\pi(|G : H|) \subseteq \pi'$, *then* $G = H$.

Proof. Assume the result is false and let G be a counterexample of minimal order. This choice requires that $G = HM$ for any minimal normal subgroup M of G. Hence H is a t-maximal subgroup of G and $\mathrm{Core}_G(H) = 1$. In particular G is t-primitive and $\pi\big(\mathrm{Soc}(G)\big) = \pi(|G : H|) \subseteq \pi'$. We conclude then that $G \in \mathrm{b}\big(\mathfrak{E}(\pi)\big) \cap \mathfrak{E}(\pi) = \emptyset$. This contradiction proves the lemma. $\qquad\square$

Corollary 5.3.24. $\mathrm{h}_\pi(G) = \{G\}$ *if and only if* $G \in \mathfrak{E}(\pi)$.

Theorem 5.3.25. $\mathrm{Proj}_{\mathfrak{E}(\pi)^t}(G) = \mathrm{h}_\pi(G)$ *for every group* G.

Proof. We argue by induction on $|G|$. If $G \in \mathfrak{E}(\pi)$, we have that $\mathrm{h}_\pi(G) = \mathrm{Proj}_{\mathfrak{E}(\pi)^t}(G) = \{G\}$. Therefore suppose that $G \notin \mathfrak{E}(\pi)$. By [DH92, III, 2.2 (c)] there exists a normal subgroup K of G such that $G/K \in \mathrm{b}\big(\mathfrak{E}(\pi)\big)$. Assume that $K \neq 1$. Let H be an $\mathfrak{E}(\pi)$-t-projector of G. Then $HK/K \in \mathrm{Proj}_{\mathfrak{E}(\pi)^t}(G/K)$. By induction, $HK/K \in \mathrm{h}_\pi(G/K)$ and hence $\pi(|G : HK|) \subseteq \pi'$. On the other hand, $HK \in \mathrm{t}(G)$. By Theorem 5.3.22, $H \in \mathrm{Proj}_{\mathfrak{E}(\pi)^t}(HK)$. Since HK is a proper subgroup of G, we conclude that $H \in \mathrm{h}_\pi(HK)$ by induction. Thus $\pi(|G : H|) \subseteq \pi'$ and $H \in \mathrm{h}_\pi(G)$. The same argument, using Remark 5.3.12, shows that $\mathrm{h}_\pi(G) \subseteq \mathrm{Proj}_{\mathfrak{E}(\pi)}(G)$. Therefore the result holds in this case.

Suppose that $K = 1$. Then $G \in \mathrm{b}\big(\mathfrak{E}(\pi)\big)$ and so G is t-primitive, $\pi\big(\mathrm{Soc}(G)\big) \subseteq \pi'$, $G = \mathrm{Soc}(G)H$ and $H \cap \mathrm{Soc}(G) = 1$ for every $\mathfrak{E}(\pi)$-t-projector H of G. It is clear that $\mathrm{Proj}_{\mathfrak{E}(\pi)^t}(G) \subseteq \mathrm{h}_\pi(G)$. Let $A \in \mathrm{h}_\pi(G)$ and let H be an $\mathfrak{E}(\pi)$-t-projector of G. If $G = A\,\mathrm{Soc}(G)$, then A is clearly an $\mathfrak{E}(\pi)$-t-projector of G. Assume that $X = A\,\mathrm{Soc}(G)$ is a proper subgroup of G. Then $A \in \mathrm{Proj}_{\mathfrak{E}(\pi)^t}(X)$ by induction (note that $A \in \mathrm{h}_\pi(X)$ because t is w-inherited and $\mathrm{Soc}(G)$ is a π'-group). On the other hand, $X = \mathrm{Soc}(G)(X \cap H)$. Since $X \cap H \in \mathrm{t}(X)$ and $\pi(|X : X \cap H|) \subseteq \pi'$, it follows that $X \cap H \in \mathrm{h}_\pi(X)$. By induction, $X \cap H \in \mathrm{Proj}_{\mathfrak{E}(\pi)^t}(X)$ and therefore $X \cap H$ and A are conjugate in X by Theorem 5.3.22. Hence, without loss of generality, we may suppose that A is contained in H. Then $A \in \mathrm{h}_\pi(H)$ because $\pi(|G : A|) \subseteq \pi'$ and t is inherited. Applying Corollary 5.3.24, it follows that $A = H$ and $A \in \mathrm{Proj}_{\mathfrak{E}(\pi)^t}(G)$. The proof of the theorem is complete. $\qquad\square$

Theorem 5.3.26. $\mathrm{Proj}_{\mathfrak{E}(\pi)^t}(G) = \mathrm{Proj}_{\mathfrak{E}(\pi)}(G) = \mathrm{Cov}_{\mathfrak{E}(\pi)}(G)$ *for every group* G.

Proof. We only prove that $\mathrm{Proj}_{\mathfrak{E}(\pi)^t}(G) = \mathrm{Proj}_{\mathfrak{E}(\pi)}(G)$. Arguing by induction on $|G|$, we may assume that $G \in \mathrm{b}\big(\mathfrak{E}(\pi)\big)$. Then $\mathrm{Soc}(G)$ is a π'-group which is complemented in G by every $\mathfrak{E}(\pi)$-t-projector of G.

Let $H \in \mathrm{Proj}_{\mathfrak{E}(\pi)^{\mathrm{t}}}(G)$ and let T be an $\mathfrak{E}(\pi)$-maximal subgroup of G containing H. Applying Proposition 2.3.16, $T \in \mathrm{Proj}_{\mathfrak{E}(\pi)}(G)$. Since $H \in \mathrm{h}_{\pi}(T)$ and $T \in \mathfrak{E}(\pi)$, we conclude that $H = T$ by Corollary 5.3.24. Consequently $\emptyset \neq \mathrm{Proj}_{\mathfrak{E}(\pi)^{\mathrm{t}}}(G) \subseteq \mathrm{Proj}_{\mathfrak{E}(\pi)}(G)$.

Assume that G is a t-primitive group of type 1. Then the complements of $\mathrm{Soc}(G)$ in G are precisely the $\mathfrak{E}(\pi)$-t-projectors of G by Corollary 5.2.7. Since every $\mathfrak{E}(\pi)$-projector of G complements $\mathrm{Soc}(G)$ by [DH92, III, 3.9 (i)], it follows that $\mathrm{Proj}_{\mathfrak{E}(\pi)}(G) = \mathrm{Proj}_{\mathfrak{E}(\pi)^{\mathrm{t}}}(G)$.

Suppose that G is a t-primitive group of type 2. Then $\mathrm{Soc}(G)$ is a Hall π'-subgroup of G and $\mathrm{Proj}_{\mathfrak{E}(\pi)^{\mathrm{t}}}(G)$ is the set of all Hall π-subgroups of G. Let T be an $\mathfrak{E}(\pi)$-projector of G. Then $T\,\mathrm{Soc}(G) = G$ and $T = \big(T \cap \mathrm{Soc}(G)\big)T_0$ for some Hall π-subgroup T_0 of T. It is clear that T_0 is a Hall π-subgroup of G. Therefore T_0 is an $\mathfrak{E}(\pi)$-t-projector of G. Then $T_0 \in \mathrm{h}_{\pi}(T)$ by Theorem 5.3.25. Since $T \in \mathfrak{E}(\pi)$, we conclude that $T_0 = T$ by Corollary 5.3.24. Consequently $\mathrm{Proj}_{\mathfrak{E}(\pi)^{\mathrm{t}}}(G) = \mathrm{Proj}_{\mathfrak{E}(\pi)}(G)$. $\qquad\square$

Theorem 5.3.27. *Every π-subgroup of a group G is contained in some $\mathfrak{E}(\pi)$-projector of G.*

Proof. Let A be a π-subgroup of G. We prove by induction on $|G|$ that A is contained in an $\mathfrak{E}(\pi)$-projector of G. It is clear that we may assume that $G \in \mathrm{b}\big(\mathfrak{E}(\pi)\big)$. If G is a t-primitive group of type 2, then the complements of $N = \mathrm{Soc}(G)$ in G are the Hall π-subgroups of G and the $\mathfrak{E}(\pi)$-projectors of G. Let H be one of them. If $AN = G$, then A is an $\mathfrak{E}(\pi)$-projector of G. Assume that $X = AN$ is a proper subgroup of G. Then $X = \mathrm{Soc}(G)(X \cap H)$. By the Schur-Zassenhaus theorem [Hup67, I, 18.1 and 18.2], A and $X \cap H$ are conjugate. Hence A is contained in an $\mathfrak{E}(\pi)$-projector of G. Assume that G is a t-primitive group of type 1 and let p denote the prime dividing $|\mathrm{Soc}(G)|$. By Theorem 5.2.4, G is p-soluble and so, applying the theorem of Hall-Chunikhin ([Hup67, VI, 1.7]), G has Hall p'-subgroups, all of them are conjugate, and every p'-subgroup — in particular, every π-subgroup — is contained in some Hall p'-subgroup. Since the elements of $\mathrm{Proj}_{\mathfrak{E}(\pi)}(G)$ are precisely the complements in G of $\mathrm{Soc}(G)$, they contain Hall p'-subgroups of G and every Hall p'-subgroup of G is contained in some $\mathfrak{E}(\pi)$-projector of G. Therefore A is contained in an $\mathfrak{E}(\pi)$-projector of G. $\qquad\square$

Final remark 5.3.28. Earlier in this example we have made the observation that $\mathfrak{E}(\pi) \cap \mathfrak{S} = \mathfrak{S}_{\pi}$ for which one derives that in soluble groups $\mathrm{Proj}_{\mathfrak{E}(\pi)^{\mathrm{t}}}$ coincides with Hall_{π}, for in such groups t coincides with s. We ought to alert the reader that — in contrast to this observation — in groups from $\mathfrak{T}_{\pi}$, the class of all π-separable groups, $\mathrm{Proj}_{\mathfrak{E}(\pi)^{\mathrm{t}}} \neq \mathrm{Hall}_{\pi}$, that is, $\mathfrak{E}_{\pi}$ is not equal to $\mathfrak{E}(\pi) \cap \mathfrak{T}_{\pi}$ in general: take $\pi = p'$, for some prime p not dividing the order of some non-abelian simple group E. Let $G = [V]E$ be the semidirect product of E with a faithful and irreducible E-module over $\mathrm{GF}(p)$. Then $G \in \big(\mathfrak{E}(\pi) \cap \mathfrak{T}_{\pi}\big) \setminus \mathfrak{E}_{\pi}$.

6

$\mathfrak{F}$-subnormality

How a subgroup can be embedded in a group is always a question of particular interest for clearing up the structure of finite groups.

One of the most important subgroup embedding properties is the subnormality, transitive closure of the relation of normality. This property was extensively studied by H. Wielandt (see [Wie94a]). For an excellent survey of the theory of subnormal subgroups, we refer the reader to J. C. Lennox and S. E. Stonehewer [LS87].

In finite groups the significance of the subnormal subgroups is apparent since they are precisely those subgroups which occur as terms of composition series, the factors of which are of great importance in describing the group structure.

Let $\mathfrak{F}$ be a saturated formation of full characteristic. If G is a soluble group, the $\mathfrak{F}$-normaliser D of G associated with a Hall system Σ of G is contained in the $\mathfrak{F}$-projector E of G in which Σ reduces (see [DH92, V, 4.11] and Theorem 4.2.9). In 1969, T. O. Hawkes [Haw69] analysed how D is embedded in E. It turns out that D can be joined to E by means of a maximal chain of $\mathfrak{F}$-normal subgroups, that is, D is $\mathfrak{F}$-subnormal in E ([DH92, V, 4.12]). The $\mathfrak{F}$-subnormality could be regarded, in the soluble universe, as the natural extension of the subnormality to formation theory. In fact, most of the results concerning subnormal subgroups can be read off by specialising to the case where $\mathfrak{F}$ is the formation of all nilpotent groups.

Our objective in this chapter is to present the main results of the $\mathfrak{F}$-subnormal subgroups. They are primarily connected with the study of subnormal subgroups properties by the methods of formation theory.

6.1 Basic properties

In the sequel, $\mathfrak{F}$ will denote a non-empty formation.

A subgroup U of a group G is called $\mathfrak{F}$-*normal* in G if $G/\operatorname{Core}_G(U) \in \mathfrak{F}$;

otherwise U is said to be $\mathfrak{F}$-*abnormal* in G. This definition was introduced in Definition 2.3.22 (1) for maximal subgroups.

Illustrations 6.1.1. 1. A subgroup U is $\mathfrak{F}$-normal in G if and only if $G^{\mathfrak{F}}$ is contained in U.

   2. A maximal subgroup is normal in G if and only if it is $\mathfrak{N}$-normal in G. In general, a subgroup U is subnormal in G provided that U is $\mathfrak{N}$-subnormal in G.

   3. If $\mathfrak{F} = \mathrm{LF}(F)$ is a saturated formation, a maximal subgroup M of G is $\mathfrak{F}$-normal in G if and only if $G/\operatorname{Core}_G(M) \in F(p)$ for every prime p dividing $\left| \operatorname{Soc}\big(G/\operatorname{Core}_G(M)\big) \right|$.

Definition 6.1.2. *A subgroup H of a group G is said to be $\mathfrak{F}$-subnormal in G if either $H = G$ or there exists a chain of subgroups*

$$H = H_0 < \cdots < H_n = G$$

such that H_{i-1} is an $\mathfrak{F}$-normal maximal subgroup of H_i for $i = 1, \ldots, n$. We shall write $H\,\mathfrak{F}$-sn G; $\operatorname{s}_{n\mathfrak{F}}(G)$ will denote the set of all $\mathfrak{F}$-subnormal subgroups of a group G. It is clear that $\operatorname{s}_{n\mathfrak{F}}$ is a subgroup functor.

Remark 6.1.3. Assume that $\mathfrak{F} = \mathfrak{N}$, the formation of all nilpotent groups. Then $\operatorname{s}_{n\mathfrak{N}}(G) \subseteq \operatorname{s}_n(G)$ for all groups G by Illustration 6.1.1 (2). However the equality does not hold in general because if $G = \mathrm{Alt}(5)$, then $1 \in \operatorname{s}_n(G) \setminus \operatorname{s}_{n\mathfrak{N}}(G)$. Nevertheless, if G is soluble, then $\operatorname{s}_n(G) = \operatorname{s}_{n\mathfrak{N}}(G)$.

To avoid the above situation, O. H. Kegel [Keg78] introduced a little bit different notion of $\mathfrak{F}$-subnormality. It unites the notions of subnormal and $\mathfrak{F}$-subnormal subgroup.

Definition 6.1.4. *A subgroup U of a group G is called K-$\mathfrak{F}$-subnormal subgroup of G if either $U = G$ or there is a chain of subgroups*

$$U = U_0 \leq U_1 \leq \cdots \leq U_n = G$$

such that U_{i-1} is either normal in U_i or U_{i-1} is $\mathfrak{F}$-normal in U_i, for $i = 1, \ldots, n$. We shall write U K-$\mathfrak{F}$-sn G and denote $\operatorname{s}_{n\,K-\mathfrak{F}}(G)$ the set of all K-$\mathfrak{F}$-subnormal subgroups of a group G. Clearly $\operatorname{s}_{n\,K-\mathfrak{F}}$ is a subgroup functor.

Remark 6.1.5. $\operatorname{s}_{n\,K-\mathfrak{N}}(G) = \operatorname{s}_n(G)$ for every group G.

Let e be one of the functors $\operatorname{s}_{n\mathfrak{F}}$ or $\operatorname{s}_{n\,K-\mathfrak{F}}$.

Lemma 6.1.6. *e is inherited, that is, if G is a group, we have*

   1. *If $H \in \mathrm{e}(K)$ and $K \in \mathrm{e}(G)$, then $H \in \mathrm{e}(G)$.*
   2. *If $N \trianglelefteq G$ and $U/N \in \mathrm{e}(G/N)$, then $U \in \mathrm{e}(G)$.*
   3. *If $H \in \mathrm{e}(G)$ and $N \trianglelefteq G$, then $HN/N \in \mathrm{e}(G/N)$.*

Proof. It is obvious from the definitions that Statements 1 and 2 are fulfilled in both cases. We show that Statement 3 is satisfied when $e = s_{n\mathfrak{F}}$. Let H be an $\mathfrak{F}$-subnormal subgroup of G and let N be a normal subgroup of G. Proceeding by induction on $|G|$, we may clearly suppose that $H \neq G$. Let X be an $\mathfrak{F}$-normal maximal subgroup of G such that H is contained in X and H is $\mathfrak{F}$-subnormal in X. If $N \leq X$, then HN/N is $\mathfrak{F}$-subnormal in X/N by induction. Since X/N is $\mathfrak{F}$-normal in G/N, it follows that HN/N is $\mathfrak{F}$-subnormal in G/N by Assertion 1. Therefore we may assume that N is not contained in X and so $G = NX$. By induction, $H(X \cap N)/(X \cap N)$ is $\mathfrak{F}$-subnormal in $X/(X \cap N) \cong G/N$. Hence HN/N is $\mathfrak{F}$-subnormal in G/N. $\square$

Lemma 6.1.7. *Assume that $\mathfrak{F}$ is subgroup-closed.*

1. *If H is a subgroup of a group G and $G^{\mathfrak{F}} \leq H$, then $H \in e(G)$.*
2. *If $H \in e(G)$ and $K \leq G$, then $H \cap K \in e(K)$, that is, e is w-inherited.*
3. *If $\{H_i : 1 \leq i \leq n\} \subseteq e(G)$, then $\bigcap_{i=1}^{n} H_i \in e(G)$.*

Proof. 1. It follows at once from the fact that $X^{\mathfrak{F}} \leq G^{\mathfrak{F}}$ for all subgroups X of G.

2. Let $e = s_{n\mathfrak{F}}$. Proceeding by induction on $|G|$, we may clearly assume that $H \neq G$. Then there exists an $\mathfrak{F}$-normal maximal subgroup M of G such that $H \leq M$ and H is $\mathfrak{F}$-subnormal in M. Since $K^{\mathfrak{F}} \leq G^{\mathfrak{F}} \leq M$, it follows that $M \cap K$ is $\mathfrak{F}$-subnormal in K by Assertion 1. On the other hand, $H \cap K$ is $\mathfrak{F}$-subnormal in $M \cap K$ by induction. Therefore $H \cap K$ is $\mathfrak{F}$-subnormal in K.

3. It follows at once applying Lemma 5.1.5, as e is w-inherited, and using induction on n. $\square$

Example 6.1.8. Lemma 6.1.7 (2) does not remain true if $\mathfrak{F}$ is not subgroup-closed. Let $\mathfrak{F} = \mathrm{LF}(f)$, where $f(2) = \mathfrak{S}_2 \mathrm{Q} \mathrm{R}_0\big(\mathrm{Sym}(3)\big)$, $f(3) = \mathfrak{S}_3\mathfrak{S}_2$ and $f(p) = \emptyset$ for all $p > 3$. If $G = \mathrm{Sym}(4)$ and H is a Sylow 3-subgroup of G, then $H \in s_{n\mathfrak{F}}(G)$ $\big(H \leq \mathrm{Sym}(3) \leq G\big)$. However $H \notin s_{n\mathfrak{F}}\big(\mathrm{Alt}(4)\big)$.

The theory of $\mathfrak{F}$-subnormal subgroups is relevant only in the case of persistence in intermediate subgroups. Therefore

Unless otherwise stated, we stipulate that for the rest of the chapter the formation $\mathfrak{F}$ is closed under the operation of taking subgroups.

Lemma 6.1.9. *Let G be a group.*

1. *If A is a K-$\mathfrak{F}$-subnormal subgroup of G, then $A^{\mathfrak{F}}$ is subnormal in G.*
2. *Let $\mathfrak{H} = \mathrm{E}\,\mathrm{K}(\mathfrak{F})$. Then $A^{\mathfrak{H}} = G^{\mathfrak{H}}$ for every $\mathfrak{F}$-subnormal subgroup of G.*
3. *If $1 \in s_{n\mathfrak{F}}(G)$, then $s_n(G) \subseteq s_{n\mathfrak{F}}(G)$.*
4. *If G is a p-group for some prime p and $1 \in s_{n\mathfrak{F}}(C_p)$, then $s_{n\mathfrak{F}}(G) = s_n(G) = s(G)$.*

Proof. 1. We argue by induction on $|G|$. If $A = G$, then $G^\mathfrak{F}$ is normal in G, and the statement is true. Suppose $A < G$ and let X be an 𝔉-normal maximal subgroup of G containing A such that A is K-𝔉-subnormal in X. Then $A^\mathfrak{F}$ is subnormal in X by induction. Since $A^\mathfrak{F} \leq X^\mathfrak{F}$, it follows that $A^\mathfrak{F}$ is subnormal in $X^\mathfrak{F}$. Moreover $G^\mathfrak{F}$ is contained in X. Hence $X^\mathfrak{F}$ is subnormal in $G^\mathfrak{F}$. This implies that $X^\mathfrak{F}$ is subnormal in G, hence so is $A^\mathfrak{F}$. A similar argument could be applied if X is a normal subgroup of G.

2. Proceeding by induction on $|G|$, we may assume that $A < G$. We argue as in Assertion 1 and use the same notation. It follows that $A^\mathfrak{H} = X^\mathfrak{H}$ for an 𝔉-normal maximal subgroup X of G such that A is 𝔉-subnormal in X. Moreover $G^\mathfrak{H}$ is contained in X as $G/\operatorname{Core}_G(X) \in \mathfrak{F} \subseteq \mathfrak{H}$. Now $X^\mathfrak{F} G^\mathfrak{H}/G^\mathfrak{H}$ belongs to $\mathfrak{H}$ because it is subnormal in $G/G^\mathfrak{H}$, by Statement 1 and Lemma 6.1.6 (3), and $\mathfrak{H}$ is closed under taking subnormal subgroups. Hence $X/X^\mathfrak{F} \cap G^\mathfrak{H}$ belongs to $\mathfrak{H}$. It implies that $X^\mathfrak{H}$ is contained in $G^\mathfrak{H}$. Note that every composition factor of $G^\mathfrak{H}/X^\mathfrak{H}$ belongs to $\kappa(\mathfrak{F})$. Therefore $G^\mathfrak{H} = (G^\mathfrak{H})^\mathfrak{H}$ is contained in $X^\mathfrak{H}$ and so $A^\mathfrak{H} = X^\mathfrak{H} = G^\mathfrak{H}$.

3. Since $\mathrm{s}_{n\mathfrak{F}}$ is a w-inherited functor, the result follows from Lemma 5.1.4.

4. It is enough to show that $1 \in \mathrm{s}_{n\mathfrak{F}}(G)$. Assume that it is not true and let G be a counterexample of minimal order. Let M be a maximal subgroup of G. The minimal choice of G implies that $1 \in \mathrm{s}_{n\mathfrak{F}}(M)$. Since $|G/M| = p$, it follows that $M/M \in \mathrm{s}_{n\mathfrak{F}}(G/M)$. Hence $M \in \mathrm{s}_{n\mathfrak{F}}(G)$ by Lemma 6.1.6 (2). Therefore $1 \in \mathrm{s}_{n\mathfrak{F}}(G)$. This contradiction shows that no counterexample exists.   □

Proposition 6.1.10. *If $G \in \mathrm{E}\,\kappa(\mathfrak{F})$, then $\mathrm{s}_{n\mathfrak{F}}(G) = \mathrm{s}_{n\mathrm{K}\text{-}\mathfrak{F}}(G)$.*

Proof. The inclusion $\mathrm{s}_{n\mathfrak{F}}(G) \subseteq \mathrm{s}_{n\mathrm{K}\text{-}\mathfrak{F}}(G)$ follows from the definitions.

Let $H \in \mathrm{s}_{n\mathrm{K}\text{-}\mathfrak{F}}(G)$. We prove that $H \in \mathrm{s}_{n\mathfrak{F}}(G)$ by induction on $|G|$. We may assume that $H \neq G$. Let N be a minimal normal subgroup of G. Then $G/N \in \mathrm{E}\,\kappa(\mathfrak{F})$ and $HN/N \in \mathrm{s}_{n\mathrm{K}\text{-}\mathfrak{F}}(G/N)$ by Lemma 6.1.6 (3). Consequently $HN/N \in \mathrm{s}_{n\mathfrak{F}}(G/N)$ by induction. This implies that HN is 𝔉-subnormal in G by Lemma 6.1.6 (2). Moreover $HN \in \mathrm{E}\,\kappa(\mathfrak{F})$ by Lemma 6.1.9 (1). Assume that HN is a proper subgroup of G. Since H is K-𝔉-subnormal in HN by Lemma 6.1.7 (2), it follows that H is 𝔉-subnormal in HN by induction. Hence $H \in \mathrm{s}_{n\mathfrak{F}}(G)$, as required. Hence we may suppose that $G = HN$ for every minimal normal subgroup N of G. In particular, $\operatorname{Core}_G(H) = 1$. On the other hand, $H^\mathfrak{F}$ is subnormal in G by Lemma 6.1.9 (1) and so N normalises $H^\mathfrak{F}$ by [DH92, A, 14.3]. Thus $H^\mathfrak{F}$ is normal in G. This implies that $H^\mathfrak{F} \subseteq \operatorname{Core}_G(H) = 1$. Consequently $G/N \in \mathfrak{F}$ for each minimal normal subgroup N of G. If $G \in \mathfrak{F}$, then H is clearly 𝔉-subnormal in G. Hence we may assume that $G \notin \mathfrak{F}$ and therefore $G \in \mathrm{b}(\mathfrak{F})$. This means that G is a monolithic group, and $G^\mathfrak{F} = \operatorname{Soc}(G)$ is the unique minimal normal subgroup of G. Let M be a proper subgroup of G such that $H \in \mathrm{s}_{n\mathrm{K}\text{-}\mathfrak{F}}(M)$ and either $M \trianglelefteq G$ or $G^\mathfrak{F}$ is contained in M. If the second condition holds, then $NH = G$ is contained in M, contrary to supposition. Therefore $M \trianglelefteq G$. Since $G^\mathfrak{F}$ is not contained in M, it follows that $M = 1 = H$ and $G = \operatorname{Soc}(G)$ is a simple group. Therefore

$G \in \mathfrak{F} \cap \mathrm{b}(\mathfrak{F})$. This contradiction leads to $G \in \mathfrak{F}$ and so H is $\mathfrak{F}$-subnormal in G. $\qquad\square$

Proposition 6.1.11. *Let $\mathfrak{F}$ be a saturated formation and let G be a group with an $\mathfrak{F}$-subnormal subgroup H such that $G = H\,\mathrm{F}^*(G)$. If $H \in \mathfrak{F}$, then $G \in \mathfrak{F}$.*

Proof. We argue by induction on $|G|$. Suppose that H is a proper subgroup of G and let M be an $\mathfrak{F}$-normal maximal subgroup of G such that $H \leq M$ and H is $\mathfrak{F}$-subnormal in M. Then $M = H\,\mathrm{F}^*(M)$. By induction, $M \in \mathfrak{F}$. Assume $G \notin \mathfrak{F}$. By Proposition 2.3.16, M is an $\mathfrak{F}$-projector of G. This is impossible because $G = G^{\mathfrak{F}}M$ and $G^{\mathfrak{F}}$ is contained in M. Consequently $G \in \mathfrak{F}$. $\qquad\square$

6.2 $\mathfrak{F}$-subnormal closure

Let $\mathfrak{F}$ be a formation. By Lemma 6.1.7 (3), intersections of $\mathfrak{F}$-subnormal subgroups are $\mathfrak{F}$-subnormal. Therefore for any subset X of a group G, there exists a unique smallest $\mathfrak{F}$-subnormal subgroup of G containing X, the $\mathfrak{F}$-*subnormal closure* of X in G. We write $\mathrm{S}_G(X; \mathfrak{F})$ to denote this subgroup. It is clear that the same argument can be applied to K-$\mathfrak{F}$-subnormal subgroups. Consequently there exists a unique K-$\mathfrak{F}$-subnormal subgroup of G containing X, the K-$\mathfrak{F}$-*closure* of X in G. It is denoted by $\mathrm{S}_G(X; \mathrm{K}\text{-}\mathfrak{F})$.

When $\mathfrak{F} = \mathfrak{N}$, the formation of all nilpotent groups, the subgroup $\mathrm{S}_G(X) = \mathrm{S}_G(X; \mathrm{K}\text{-}\mathfrak{F})$ is the subnormal closure of X in G, that is, the smallest subnormal subgroup of G containing X.

The normal closure of X in G is generated by all of the conjugates of X in G and we might wonder whether or not the subnormal closure is generated by some natural subset of the set of these conjugates. Let us say that two subsets $X, Y \subseteq G$ are *strongly conjugate* if they are conjugate in $\langle X, Y \rangle$. It is rather clear that $\mathrm{S}_G(X)$ must contain all strong conjugates of X. In fact, the following powerful result, due to D. Bartels, is true.

Theorem 6.2.1 ([Bar77]). *Let X be a subset of a group G. Then $\mathrm{S}_G(X) = \langle Y \subseteq G : Y$ is strongly conjugate to X in $G \rangle$.*

The first part of this section is devoted to prove this theorem. First of all, we introduce some notation.

Notation 6.2.2. Let X and Y be subsets of a group G. We write:

- $X \, \sigma \, Y$ if X and Y are strongly conjugate in G.
- $X \, \sigma^{\infty} \, Y$ if there are subsets $X = X_0, X_1, \ldots, X_n = Y$ such that $X_i \, \sigma \, X_{i+1}$ for all i, $0 \leq i < n$ (n natural number).
- $X =_U Y$ if X and Y are conjugate in the subgroup U of G.
- $X =_{"G} Y$ if $\mathrm{S}_G(X) = \mathrm{S}_G(Y) = S$ and $X =_S Y$.
- $\mathrm{K}_G(X) = \langle Y \subseteq G : X \, \sigma \, Y \rangle$.

It is clear that σ^∞ and $=_G$ are equivalence relations on the set of all subsets of G.

Lemma 6.2.3. *Let X and Y be subsets of a group G such that $X \sigma Y$. Then $X =_G Y$.*

Proof. Denote $J := \langle X, Y \rangle$. Since $X \sigma Y$, there exists an element $g \in J$ such that $Y = X^g$. In particular, $\langle X^J \rangle$, the normal closure of X in J, is equal to J. Applying [DH92, A, 14.1], $S_G(X) \cap J$ is subnormal in J and contains X. Since $J = \langle X^J \rangle$, it follows that $S_G(X) \cap J = J$ and so $S_G(J) = S_G(X)$. Analogously $S_G(J) = S_G(Y)$. Therefore $X =_G Y$. □

Lemma 6.2.4. *Let X be a subset of a group G. Then*

$$S_G(X) = \langle Y \subseteq G : X =_G Y \rangle.$$

Proof. Denote $A = \langle Y \subseteq G : X =_G Y \rangle$. Then $A = \langle X^g : g \in S_G(X) \rangle$ by Lemma 6.2.3. It is clear that A is normal in $S_G(X)$. Hence A is subnormal in G. Since A contains X, it follows that $A = S_G(X)$. □

By Lemma 6.2.3, $X \sigma Y$ implies $X =_G Y$. Hence $K_G(X) \subseteq S_G(X)$ for every subgroup X of G.

Lemma 6.2.5. *Let X be a subset of a group G. Then*

1. $K_G(X) = \langle Y \subseteq G : X \sigma^\infty Y \rangle$.
2. $X \sigma^\infty X^g$ for all $g \in K_G(X)$.

Proof. 1. It is clear that $K_G(X) \leq \langle Y \subseteq G : X \sigma^\infty Y \rangle$. Let $Y \subseteq G$ such that $X \sigma^\infty Y$. We have to show that $Y \subseteq K_G(X)$. There is a natural number n and there are subsets $X = X_0, X_1, \ldots, X_n = Y$ such that $X_i \sigma X_{i+1}$ for all i, $0 \leq i < n$. Suppose inductively that we have already shown that $X_0, X_1, \ldots, X_{n-1}$ are contained in $K_G(X)$. Since $K_G(X) = \langle Z : X \sigma Z \rangle$, we may assume that $n > 1$. There exists an element $g \in \langle X_0, X_1, \ldots, X_{n-1} \rangle \subseteq K_G(X)$ such that $X^g = X_0^g = X_{n-1}$. Then $Y \leq K_G(X^g)$, and since σ is G-invariant, it follows that $K_G(X^g) = K_G(X)^g = K_G(X)$, and the induction step is complete.

2. Let Y be a subset of G. Let y be an element of $Y \cup Y^{-1}$ and assume that $X \sigma Y$. Then $X^y \sigma Y^y$ and $Y^y \sigma Y$, whence $X^y \sigma^\infty X$.

If $g \in K_G(X)$, then $g = g_1 \cdots g_t$, where $g_i \in Y_i \cup Y_i^{-1}$, $X \sigma Y_i$, for all i, $1 \leq i \leq t$. If $t = 1$, then $X^{g_1} \sigma^\infty X$ by the above argument. Suppose inductively that $X^{(g_1 \cdots g_{t-1})} \sigma^\infty X$. Then $X^{g_t^{-1}} \sigma^\infty X^{(g_1 \cdots g_{t-1})}$ because $X^{g_t^{-1}} \sigma^\infty X$. Hence $X \sigma^\infty X^g$. □

Proposition 6.2.6. *For any subset X of a group G, the following statements are equivalent:*

1. $\mathrm{K}_G(X) = \mathrm{S}_G(X)$.
2. *The equivalence relations σ^∞ and $=_G$ coincide when restricted to the conjugacy class of X in G.*

Proof. Assume that $\mathrm{K}_G(X) = \mathrm{S}_G(X)$. Then $X =_G Y$ implies that $Y = X^g$ for some $g \in \mathrm{S}_G(X)$. By Lemma 6.2.5 (2), $X \, \sigma^\infty \, Y$. Since $=_G$ is a transitive relation, $X \, \sigma^\infty \, Y$ implies $X =_G Y$ by Lemma 6.2.3. Thus Statement 2 holds.

Conversely, assume Statement 2. Since $\mathrm{K}_G(X) = \langle Y \subseteq G : X \, \sigma^\infty \, Y \rangle$ by Lemma 6.2.5 (1), it follows that $\mathrm{K}_G(X) = \langle Y \subseteq G : X =_G Y \rangle$, which is equal to $\mathrm{S}_G(X)$ by Lemma 6.2.4. □

Lemma 6.2.7. *Let X_0 and X_1 be subsets of a group G such that $X_0 \subseteq \langle X_1 \rangle$. Then $\mathrm{K}_G(X_0) \leq \mathrm{K}_G(X_1)$.*

Proof. Let t be an element of G such that $t \in \langle X_0, X_0^t \rangle$. Then obviously $t \in \langle X_1, X_1^t \rangle$. Hence $X_0 \, \sigma \, Y$ for some $Y \subseteq G$ implies that there is a subset W of G such that $Y \subseteq W$ and $X_1 \, \sigma \, W$. The lemma follows by definition of $\mathrm{K}_G(X_1)$. □

Lemma 6.2.8. *Let G be a group and let N be a normal subgroup of G. Let $X \subseteq G$ and let Y_1/N be a subset of G/N such that $XN/N \, \sigma \, Y_1/N$. Then there exists a subset Y of G such that $X \, \sigma \, Y$ and $Y_1 = YN$.*

Proof. Let

$$\mathcal{A} := \{V \subseteq G : VN/N = Y_1/N \text{ and } X =_G V\}.$$

Since $XN/N \, \sigma \, Y_1/N$, it follows that XN/N and Y_1/N are conjugate in $\mathrm{S}_{G/N}(XN/N) = \mathrm{S}_G(X)N/N$. Hence $Y_1 = X^z N$ for some $z \in \mathrm{S}_G(X)$. It is clear that $X =_G X^z$ and so $X^z = V \in \mathcal{A}$. This shows that $\mathcal{A}$ is non-empty. Let W be an element of $\mathcal{A}$ such that $\langle X, W \rangle$ has minimal order. Since $XN/N \, \sigma \, WN/N$, there exists an element $t \in \langle X, W \rangle$ such that $WN/N = X^t N/N = Y_1/N$. It is clear that $X =_G X^t$. Hence X^t belongs to $\mathcal{A}$. The minimal choice of $\langle X, W \rangle$ implies that $\langle X, X^t \rangle = \langle X, W \rangle$ and so $X \, \sigma \, X^t \, (= Y)$. □

Corollary 6.2.9. *For any subset X of a group G and for any $N \trianglelefteq G$, $\mathrm{K}_G(X)N/N = \mathrm{K}_{G/N}(XN/N)$.*

Proposition 6.2.10. *For any subset X of a group G, the relations σ^∞ and $=_G$ coincide on the conjugacy class of X in G.*

Proof. Assume that the result is false, and let (G, X) be a counterexample with $|G| + |\langle X \rangle|$ as small as possible. Clearly $X \neq \emptyset$ and the conjugacy class of X in G splits into σ^∞-equivalence classes; we denote the set of these equivalence classes by Ω. Since $X \, \sigma^\infty \, Y$ implies $X =_G Y$ for all $Y \subseteq G$ by Lemma 6.2.3, it follows from our choice of (G, X) that Ω contains at least two elements. It is clear that G acts transitively by conjugation on Ω in the obvious way.

Let $K = \mathrm{K}_G(X)$. By Proposition 6.2.6, K is a proper subgroup of G. For any non-trivial normal subgroup N of G, the relations σ^∞ and $=_G$ coincide on the conjugacy class of XN/N in G/N by minimality of G. Hence $\mathrm{K}_{G/N}(XN/N) = \mathrm{S}_{G/N}(XN/N) = \mathrm{S}_G(X)N/N$ by Proposition 6.2.6, and so $KN/N = \mathrm{K}_{G/N}(XN/N)$ is subnormal in G/N. In particular, KN is subnormal in G. Suppose that $Z = KN$ is a proper subgroup of G. Then $K = \mathrm{K}_Z(X) = \mathrm{S}_Z(X)$ by the choice of G. Hence K is subnormal in Z and so is in G. Proposition 6.2.6 implies that the relations σ^∞ and $=_G$ coincide on the conjugacy class of X in G. This is a contradiction against the choice of (G, X). Consequently, $G = KN$ for any non-trivial normal subgroup N of G. From this we conclude that $\mathrm{Core}_G(K) = 1$ and $\langle X^G \rangle$, the normal closure of X in G, is equal to G.

Let p be a prime dividing $|\langle X \rangle|$ and let Q be a Sylow p-subgroup of $\langle X \rangle$. By Lemma 6.2.7, $\mathrm{K}_G(Q)$ is contained in K. Suppose that Q is a proper subgroup of $\langle X \rangle$. The minimal choice of (G, X) implies that $\mathrm{K}_G(Q)$ is subnormal in G. Let N be a minimal normal subgroup of G. By [DH92, A, 14.3], N normalises $\mathrm{K}_G(Q)$. Since $G = KN$, it follows that $\langle \mathrm{K}_G(Q)^G \rangle = \langle \mathrm{K}_G(Q)^K \rangle$ is a subgroup of K. Hence $\langle \mathrm{K}_G(Q)^G \rangle$ is contained in $\mathrm{Core}_G(K) = 1$. This contradiction shows that $Q = \langle X \rangle$ and $\langle X \rangle$ is a p-group.

For any subgroup U of G, let $[U]$ denote the set

$$[U] = \{\omega \in \Omega : \text{there exists } X^g \in \omega \text{ such that } X^g \subseteq U\}.$$

The following statements hold:

1. *For any proper subgroup U of G and for every Sylow p-subgroup P of U, $[U] = [P]$.*

It is clear that $[P] \subseteq [U]$. Conversely, let $\omega \in [U]$ and let $Y \in \omega$ be a subset of U. Let $L = \mathrm{S}_U(Y)$. Since L is subnormal in U, it follows that $L \cap P$ is a Sylow p-subgroup of L. Hence Y^z is contained in P for some $z \in L$. It is clear that $Y =_G Y^z$. Since $=_G$ and σ^∞ coincide on the conjugacy class of Y in U by induction, we have that $Y \sigma^\infty Y^z$. Hence $\omega \in [P]$.

2. *$[U]$ is a proper subset of Ω for any proper subgroup U of G.*

Assume that $[U] = \Omega$. Then $\Omega = [P]$ for some Sylow p-subgroup P of U. Since $\Omega \neq \emptyset$, it follows $P \neq 1$ and so $\mathrm{Z}(P) \neq 1$. Note that if $x \in \mathrm{Z}(P)$ and $\omega \in \Omega$, then $\omega^x = \omega$ because x centralises an element of ω. Hence $\mathrm{Z}(P)$ acts trivially on Ω. Since $\Omega = [P] = [P^g]$ for all $g \in G$, it follows that $\mathrm{Z}(P^g)$ acts trivially on Ω. This implies that $N = \langle \mathrm{Z}(P)^G \rangle$ acts trivially on Ω. Let ω_0 be the element of Ω such that $X \in \omega_0$. If $z \in K$, then $X^z \in \omega_0$ by Lemma 6.2.5 (2). Hence $\omega_0^z = \omega_0$. Let g be an element of G. There exist $z \in K$ and $n \in N$ such that $g = zn$. It follows that $\omega_0^g = \omega_0$ and so $X^g \sigma^\infty X$ for all $g \in G$. Therefore $\mathrm{K}_G(X) = \langle X^G \rangle = G$. This contradiction shows that $[U] \neq \Omega$.

3. *Any maximal subgroup M of G such that $[M] \neq \emptyset$ contains a Sylow p-subgroup of G.*

Let P be a Sylow p-subgroup of M. By Statement 1 and Statement 2, $[M] = [P] \neq \Omega$. Note that if $\omega \in [P]$ and $g \in M \cup \mathrm{N}_G(P)$, then $\omega^g \in [P]$.

Hence $\langle M, \mathrm{N}_G(P)\rangle$ is not transitive on Ω. Therefore $\langle M, \mathrm{N}_G(P)\rangle$ is a proper subgroup of G. In particular, $\mathrm{N}_G(P) \le M$ and so P is a Sylow p-subgroup of G.

4. *Any Sylow p-subgroup P of G is contained in a unique maximal subgroup of G.*

Obviously G is not a p-group. Let P be contained in $L \cap M$, where L and M are maximal subgroups of G. Then $[L] = [M] = [P]$ by Statement 1 and $[P] \ne \Omega$ by Statement 2. This implies that $\langle L, M \rangle$ is not transitive on Ω. Hence $G \ne \langle L, M \rangle$ and $L = M$.

5. *X is contained in a unique maximal subgroup of G.*

Suppose that X is contained in at least two maximal subgroups L and M of G. Choose L and M such that the Sylow p-subgroups of $L \cap M$ have maximal order. There exist Sylow p-subgroups R and S of L and M respectively such that $R \cap S$ is a Sylow p-subgroup of $L \cap M$ containing X. By Statement 4, R and S are Sylow p-subgroups of G. Moreover $R \ne S$ by Statement 4. From this we conclude that $R \cap S$ is a proper subgroup of $R_1 = \mathrm{N}_R(R \cap S)$. Since $N = \mathrm{N}_G(R \cap S)$ is a proper subgroup of G, this implies N is contained in M by our choice of M and L. The same argument with L and S replacing M and R yields $N \le L$. But then $R \cap S < R_1 \le M \cap L$ and $R \cap S$ is a Sylow p-subgroup of $M \cap L$. This contradiction proves Statement 5.

Now from Statement 5 we deduce the final contradiction, thus proving the lemma. We know that K is a proper subgroup of G. Let M be the unique maximal subgroup of G containing X. Since $\langle X^G \rangle = G$, it follows that $M = \mathrm{N}_G(M)$. Let $g \in G \setminus M$. Then $G = \langle X, X^g \rangle$. This implies $X \, \sigma \, X^g$ and therefore we have $G = K$. $\square$

Combining Proposition 6.2.6 and Proposition 6.2.10, we have:

Theorem 6.2.11. $\mathrm{S}_G(X) = \mathrm{K}_G(X)$ *for any subset X of G.*

Let X be a subset of G and $g \in G$ such that $g \in \langle X, X^g \rangle$. Then $g \in \langle \mathrm{S}_G(X), \mathrm{S}_G(X)^g \rangle \le \mathrm{S}_G\big(\mathrm{S}_G(X)\big) = \mathrm{S}_G(X)$. Hence the following result is true.

Corollary 6.2.12. $\mathrm{S}_G(X) = \big\langle g \in G : g \in \langle X, X^g \rangle \big\rangle$.

Let H be a subgroup of a group G. If A is a subgroup of G, containing H, then $H A^{\mathfrak{N}}$ is a subnormal subgroup of A containing H. Now if $g \in G$ and $g \in \langle H, H^g \rangle = J$, then the normal closure of H in J is equal to J. The subnormality of $H J^{\mathfrak{N}}$ in J implies that $J = H J^{\mathfrak{N}}$ and $g \in H \langle H, H^g \rangle^{\mathfrak{N}}$. Moreover there exists $z \in \langle H, H^g \rangle^{\mathfrak{N}}$ such that $J = \langle H, H^z \rangle$. Thus we have shown the following:

Theorem 6.2.13. *Let H be a subgroup of a group G. Then*

$$\mathrm{S}_G(H) = \big\langle H^g : g \in \langle H, H^g \rangle^{\mathfrak{N}} \big\rangle = \big\langle g \in G : g \in H \langle H, H^g \rangle^{\mathfrak{N}} \big\rangle.$$

The descriptions of the subnormal closure provide a proof of the following subnormality criterion due to Wielandt.

Theorem 6.2.14 ([Wie74]). *Let H be a subgroup of a group G. The following statements are pairwise equivalent:*

1. H *is subnormal in* G.
2. H *is subnormal in* $\langle H, g \rangle$ *for all* $g \in G$.
3. H *is subnormal in* $\langle H, H^g \rangle$ *for all* $g \in G$.
4. *If* $g \in G$ *and* $g \in \langle H, H^g \rangle$, *then* $g \in H$.

Moreover, they are equivalent to:

5. *If* $g \in G$ *and* $g \in \langle H, H^g \rangle^{\mathfrak{N}}$, *then* $g \in H$.

Remark 6.2.15. Theorem 6.2.11 does not provide a description of the $\mathfrak{N}$-subnormal closure. Let $G = \mathrm{Alt}(5)$ and $H = \{1\}$. Then $\mathrm{S}_G(H) = H$ and $\mathrm{S}_G(H; \mathfrak{N}) = G$.

If G is a soluble group, then $\mathrm{S}_G(H) = \mathrm{S}_G(H; \mathrm{K}\text{-}\mathfrak{N}) = \mathrm{S}_G(H; \mathfrak{N})$ by Proposition 6.1.10. In this context, the following conjecture arises.

Conjecture 6.2.16 (K. Doerk). Let $\mathfrak{F}$ be a saturated formation and $\pi = \mathrm{char}\,\mathfrak{F}$. Given a subgroup H of a soluble group $G \in \mathfrak{S}_\pi$, the $\mathfrak{F}$-subnormal closure of H in G is the subgroup $\mathrm{S}_G(H; \mathfrak{F}) = \langle g \in G : g \in H\langle H, H^g \rangle^{\mathfrak{F}} \rangle$.

A. Ballester-Bolinches and M. D. Pérez-Ramos [BBPR91] confirmed Conjecture 6.2.16. In fact, they showed that the conjecture is valid for groups with soluble $\mathfrak{F}$-residual, that is, groups in the class $\mathfrak{SF}$.

Henceforth in the rest of the section

$\mathfrak{F} = \mathrm{LF}(F)$ *will denote a subgroup-closed saturated formation of characteristic* π.

The proof of Doerk's conjecture depends heavily on the following extension of Theorem 6.2.14 to subgroup-closed saturated formations.

Theorem 6.2.17 ([BBPR91]). *For a subgroup H of a π-group $G \in \mathfrak{SF}$, the following statements are pairwise equivalent:*

1. H *is* $\mathfrak{F}$-subnormal in G
2. H *is* $\mathfrak{F}$-subnormal in $\langle H, x \rangle$ *for every* $x \in G$.
3. H *is* $\mathfrak{F}$-subnormal in $\langle H, H^x \rangle$ *for every* $x \in G$.
4. *If T is a subgroup of G such that T is contained in $\langle H, T \rangle^{\mathfrak{F}}$, then T is contained in* H.
5. *If $x \in G$ and $x \in \langle H, x \rangle^{\mathfrak{F}}$, it follows that* $x \in H$.
6. *If $x \in G$ and $x \in \langle H, H^x \rangle^{\mathfrak{F}}$, it follows that* $x \in H$.

Proof. 3 implies 1. We argue by induction on $|G|$. We can assume that $G^{\mathfrak{F}} \neq 1$ by Lemma 6.1.7 (1). Let N be a minimal normal subgroup of G such that N is contained in $G^{\mathfrak{F}}$. By induction, HN/N is $\mathfrak{F}$-subnormal in G/N and so HN is $\mathfrak{F}$-subnormal in G by Lemma 6.1.6 (2). If HN were a proper subgroup of G, then H would be $\mathfrak{F}$-subnormal in $HN \in \mathfrak{SF}$ by induction. Applying

Lemma 6.1.6 (1), H is $\mathfrak{F}$-subnormal in G and the implication is true. Hence we can suppose $G = HN$ and $G \neq H$. Since N is soluble, H is a maximal subgroup of G. If H is a normal subgroup of G, then H is $\mathfrak{F}$-subnormal in G because $G \in \text{EK}(\mathfrak{F})$. If H is not normal in G, there exists an element $x \in G$ such that $H \neq H^x$. Then $G = \langle H, H^x \rangle$ and H is $\mathfrak{F}$-subnormal in G by Statement 3.

By Lemma 6.1.7 (2), 1 implies 2 and 2 implies 3. Consequently, 1, 2, and 3 are pairwise equivalent.

It is clear that 4 implies 5 and 5 implies 6 because $X^{\mathfrak{F}} \leq Y^{\mathfrak{F}} \leq G^{\mathfrak{F}}$ if $X \leq Y \leq G$.

1 implies 4. Suppose that H is $\mathfrak{F}$-subnormal in G and T is a subgroup of G such that T is contained in $\langle H, T \rangle^{\mathfrak{F}}$. Then $\langle H, T \rangle = H \langle H, T \rangle^{\mathfrak{F}}$. If H were a proper subgroup of $\langle H, T \rangle$, there would exist an $\mathfrak{F}$-normal maximal subgroup M of $\langle H, T \rangle$ containing H. Since $\langle H, T \rangle^{\mathfrak{F}} \leq M$, we would have $M = \langle H, T \rangle$. This contradiction yields $H = \langle H, T \rangle$ and T is contained in H.

To complete the proof we now show that 6 implies 1. We proceed by induction on $|G|$. Let $x \in G$ and $T = \langle H, H^x \rangle$. If T is a proper subgroup of G, then by induction H is $\mathfrak{F}$-subnormal in T. Since 3 is equivalent to 1, we may assume that $T = G$ for some $x \in G$. By Lemma 6.1.7 (1), $HG^{\mathfrak{F}}$ is $\mathfrak{F}$-subnormal in G. Hence, if $HG^{\mathfrak{F}}$ were a proper subgroup of G, then H would be $\mathfrak{F}$-subnormal in $HG^{\mathfrak{F}}$ by induction. Therefore H would be $\mathfrak{F}$-subnormal in G by Lemma 6.1.6 (1). Therefore we may suppose $G = \langle H, H^x \rangle = HG^{\mathfrak{F}} = H \langle H, H^x \rangle^{\mathfrak{F}}$. In particular, $x = ht$ for some $h \in H$ and $t \in \langle H, H^x \rangle^{\mathfrak{F}} = \langle H, H^t \rangle^{\mathfrak{F}}$. Applying Statement 6, it follows that $t \in H$. Hence $x \in H$ and $H = G$ is $\mathfrak{F}$-subnormal in G. The circle of implications is now complete. $\square$

If H is a subgroup of a group G, denote

$$\text{T}_G(H; \mathfrak{F}) = \langle x \in G : x \in H \langle H, H^x \rangle^{\mathfrak{F}} \rangle.$$

Lemma 6.2.18. *If N is a normal subgroup of a group G and H is a subgroup of G, then*

$$\text{T}_{G/N}(HN/N; \mathfrak{F}) = \text{T}_G(H; \mathfrak{F})N/N.$$

Proof. Denote with bars the images in $\bar{G} = G/N$. It is clear that $\overline{\text{T}_G(H; \mathfrak{F})} = \text{T}_G(H; \mathfrak{F})N/N$ is contained in $\text{T}_{G/N}(HN/N; \mathfrak{F})$. Consider now $\bar{g} \in \langle \bar{H}, \bar{H}^{\bar{g}} \rangle^{\mathfrak{F}}$. Then there exists an element $z \in \langle H, H^g \rangle^{\mathfrak{F}}$ such that $\bar{z} = \bar{g}$. Hence the set $\mathcal{L} = \{ z \in \langle H, H^g \rangle^{\mathfrak{F}} : \bar{z} = \bar{g} \}$ is non-empty. Let $t \in \mathcal{L}$ such that $\langle H, H^t \rangle^{\mathfrak{F}}$ has minimal order. Then $\langle H, H^g \rangle^{\mathfrak{F}} N = \langle H, H^t \rangle^{\mathfrak{F}} N$ and $t = xn$ for some $x \in \langle H, H^t \rangle^{\mathfrak{F}}$ and $n \in N$. It is clear that $x \in \langle H, H^g \rangle^{\mathfrak{F}}$ and $\bar{x} = \bar{t}$. Hence $x \in \mathcal{L}$. The minimal choice of t implies that $\langle H, H^x \rangle^{\mathfrak{F}} = \langle H, H^t \rangle^{\mathfrak{F}}$. Therefore $\bar{g} \in \overline{\text{T}_G(H; \mathfrak{F})}$ and the equality holds. $\square$

Theorem 6.2.19 ([BBPR91]). *Let G be a π-group with soluble $\mathfrak{F}$-residual. Let H be a subgroup of G. Then $\text{S}_G(H; \mathfrak{F}) = \text{T}_G(H; \mathfrak{F}) = \langle T \leq G : T \leq H \langle H, T \rangle^{\mathfrak{F}} \rangle$.*

Proof. Write $S = T_G(H; \mathfrak{F})$. If L is an $\mathfrak{F}$-subnormal subgroup of G containing H, then S is contained in L by Theorem 6.2.17. Thus the first equality holds if we prove that S is $\mathfrak{F}$-subnormal in G. We argue by induction on $|G|$. Since $G^{\mathfrak{F}}$ is soluble and G is a π-group, it follows that $G^{\mathfrak{F}}$ is a proper subgroup of G. Of course, it may be assumed that $G^{\mathfrak{F}} \neq 1$. Let N be a minimal normal subgroup of G contained in $G^{\mathfrak{F}}$. Then, by Lemma 6.2.18, $SN/N = T_{G/N}(HN/N; \mathfrak{F})$. Hence SN/N is $\mathfrak{F}$-subnormal in G/N. By Lemma 6.1.6 (2), SN is $\mathfrak{F}$-subnormal in G. Suppose that $SN = X$ is a proper subgroup of G. Then $S = T_X(H; \mathfrak{F})$ is $\mathfrak{F}$-subnormal in X by induction. By Lemma 6.1.6 (1), S is $\mathfrak{F}$-subnormal in G. Therefore we must have $G = SN$ and thus S is a maximal subgroup of G because N is abelian and S is a proper subgroup of G. This argument also yields $\text{Core}_G(S) = 1$. Therefore G is a primitive group of type 1 and $N = \text{Soc}(G) = C_G(N)$.

Suppose, by way of contradiction, that S is not $\mathfrak{F}$-subnormal in G and let us choose H of minimal order among those subgroups of G such that $T_G(H; \mathfrak{F})$ is not $\mathfrak{F}$-subnormal in G. If M is a maximal subgroup of H satisfying $H = T_H(M; \mathfrak{F})$, then $H \leq T_G(M; \mathfrak{F})$ and $T_G(M; \mathfrak{F})$ is $\mathfrak{F}$-subnormal in G. Consequently S is contained in $T_G(M; \mathfrak{F})$ and S is $\mathfrak{F}$-subnormal in G, contrary to the choice of H. Therefore, each maximal subgroup of H is $\mathfrak{F}$-subnormal in H. This implies that every primitive epimorphic image of H belongs to $\mathfrak{F}$. Hence $H \in \mathfrak{F}$ because $\mathfrak{F}$ is saturated. Let N_0 be a minimal H-invariant subgroup of N. Put $A = HN_0$. If $A = G$, then $H = S$. By Theorem 6.2.17, S is $\mathfrak{F}$-subnormal in G, contrary to supposition. Hence A is a proper subgroup of G. Suppose that A is not an $\mathfrak{F}$-group. Then $N_0 = A^{\mathfrak{F}}$ and $A = S_A(H; \mathfrak{F}) = T_A(H; \mathfrak{F}) \leq S$. This is a contradiction. Therefore $A \in \mathfrak{F}$. Let $\text{Soc}_H(N)$ be the product of all minimal H-invariant subgroups of N. Since $\mathfrak{F}$ is a formation, it follows that $H \, \text{Soc}_H(N) \in \mathfrak{F}$. Suppose that $HN \notin \mathfrak{F}$ and let L be an $\mathfrak{F}$-maximal subgroup of HN containing $H \, \text{Soc}_H(N)$. Then $HN = L(HN)^{\mathfrak{F}}$ and $L \cap (HN)^{\mathfrak{F}} = 1$ by Theorem 4.2.17. But then, since $(HN)^{\mathfrak{F}} \neq 1$, we have that $1 \neq (HN)^{\mathfrak{F}} \cap \text{Soc}_H(N) \leq (HN)^{\mathfrak{F}} \cap L$, which is a contradiction. Therefore $HN \in \mathfrak{F}$. Let $1 = N_0 \trianglelefteq N_1 \trianglelefteq N_2 \trianglelefteq \cdots \trianglelefteq N_r = N$ be an H-composition series of N. Then $H / C_H(N_i/N_{i-1}) \in F(p)$, for $i = 1, \ldots, r$, and p the prime dividing $|N|$. Hence $H^{F(p)} \leq \bigcap \{ C_H(N_i/N_{i-1}) : i = 1, \ldots, r \}$ and so that $H^{F(p)} / C_{H^{F(p)}}(N) = H^{F(p)}$ is a p-group by [DH92, A, 12.4]. Therefore $H \in \mathfrak{S}_p F(p) = F(p)$.

Consider now $g \in \langle H, H^g \rangle^{\mathfrak{F}} \setminus H$. It is clear that H is a proper subgroup of $T = \langle H, H^g \rangle \notin \mathfrak{F}$. Obviously $T = HT^{\mathfrak{F}}$ is contained in S. Denote $T^{\mathfrak{F}} = R$. Let $1 = K_0 \trianglelefteq K_1 \trianglelefteq \cdots \trianglelefteq K_s = N$ be a T-composition series of N. If every T-chief factor K_j/K_{j-1}, $j \in \{1, \ldots, s\}$, is centralised by R, it follows, arguing as above, that R is a p-group and since $H \in F(p)$, it follows that $T \in F(p) \subseteq \mathfrak{F}$, contrary to the choice of T. Consequently, there exists a T-chief factor K_i/K_{i-1}, $i \in \{1, \ldots, s\}$, such that R is not contained in $C_T(K_i/K_{i-1})$. Write $L = K_i T = K_i(RH)$, and denote with bars the images in $\bar{L} = L/K_{i-1}$. We have that $\bar{K}_i \leq \bar{L}^{\mathfrak{F}}$, because otherwise $\bar{K}_i \cap \bar{L}^{\mathfrak{F}} = 1$, and then $\bar{R} \leq C_{\bar{L}}(\bar{K}_i)$, contradicting our choice of $\bar{K}_i$. Therefore $\bar{L}^{\mathfrak{F}} = \bar{K}_i \bar{R}$ by

Proposition 2.2.8. Assume that $|\bar{L}| < |G|$. Then $\mathrm{S}_{\bar{L}}(\bar{H}, \mathfrak{F}) = \mathrm{T}_{\bar{L}}(\bar{H}, \mathfrak{F}) = \bar{L}$ by induction. Applying Lemma 6.2.18, $\mathrm{T}_{\bar{L}}(\bar{H}; \mathfrak{F}) = \mathrm{T}_L(H; \mathfrak{F})K_{i-1}/K_{i-1}$. Hence $\mathrm{T}_L(H; \mathfrak{F})K_{i-1} = K_i RH$. If $\mathrm{T}_L(H; \mathfrak{F}) \cap K_i = 1$, then $\mathrm{T}_L(H; \mathfrak{F}) = RH$ and $K_i = K_{i-1}$. This is a contradiction. Thus $1 \neq K_i \cap \mathrm{T}_L(H; \mathfrak{F}) \leq S \cap N$, which is also impossible. Therefore $|\bar{L}| = |G|$, that is, $G = NT$ and $S = T = \langle H, g \rangle$. Let $n \in N$ such that $[H, n] \neq 1$ and consider $M = \langle H, H^{ng} \rangle$. Since $G = HG^{\mathfrak{F}}$, it follows that $M < G$, because otherwise $ng \in S$, and so $n \in S$, contradicting our supposition. Let L be a maximal subgroup of G containing M. If $L = S$, then $H^n \leq S$. Hence $1 \neq [h, n] = h^{-1}h^n \in S \cap N$, for some $h \in H$. This is impossible. If N were contained in L, then L would contain H^g, and so $S = \langle H, H^g \rangle \leq L$. This would be a contradiction. Hence $\mathrm{Core}_G(L) = 1$ and $L = H(L \cap G^{\mathfrak{F}}) = HL^{\mathfrak{F}}$. Our choice of G implies that $L = \mathrm{T}_L(H; \mathfrak{F}) \leq S$, and we have reached the desired contradiction. Therefore $\mathrm{T}_G(H; \mathfrak{F})$ is $\mathfrak{F}$-subnormal in G and $\mathrm{S}_G(H; \mathfrak{F}) = \mathrm{T}_G(H; \mathfrak{F})$.

On the other hand, it is clear that $\mathrm{S}_G(H; \mathfrak{F})$ is contained in $\mathrm{L}_G(H; \mathfrak{F}) = \langle T \leq G : T \leq H\langle H, T \rangle^{\mathfrak{F}} \rangle$. Now, if K is an $\mathfrak{F}$-subnormal subgroup of G containing H and T is a generator of $\mathrm{L}_G(H; \mathfrak{F})$, it follows that $T \leq K\langle K, T \rangle^{\mathfrak{F}}$. Thus, if $t \in T$, then $t = k_t x_t$ with $k_t \in K$, $x_t \in \langle K, T \rangle^{\mathfrak{F}}$. Denote by $R = \langle x_t : t \in T \rangle$. Then $\langle K, T \rangle = \langle K, R \rangle$ and $R \leq \langle K, R \rangle^{\mathfrak{F}}$. Since K is $\mathfrak{F}$-subnormal in G, it follows that $R \leq K$ by Theorem 6.2.17. Consequently T is contained in K and $\mathrm{L}_G(H; \mathfrak{F}) \leq K$. Since $\mathrm{S}_G(H; \mathfrak{F})$ is $\mathfrak{F}$-subnormal in G, $\mathrm{S}_G(H; \mathfrak{F})$ contains $\mathrm{L}_G(H; \mathfrak{F})$ and the proof of the theorem is complete. $\qquad\square$

Open question 6.2.20. *Let $\mathfrak{F}$ be a saturated formation of characteristic π. Is it possible to find a useful description for the $\mathfrak{F}$-subnormal closure of a subgroup H of a π-group G?.*

6.3 Lattice formations

One of the most striking results in the theory of subnormal subgroups is the celebrated "join" theorem, proved by H. Wielandt in 1939: the subgroup generated by two subnormal subgroups of a finite group is itself subnormal. As a result, the set of all subnormal subgroups of a group is a sublattice of the subgroup lattice.

Let $\mathfrak{F}$ be a formation. One might wonder whether the set of $\mathfrak{F}$-subnormal subgroups of a group forms a sublattice of the subgroup lattice. The answer is in general negative.

Example 6.3.1 ([BBPR91]). Let $\mathfrak{F}$ be the formation of all 2-nilpotent groups and $G = \mathrm{Sym}(4)$. By [DH92, A, 10.9], G has an irreducible and faithful module V over $\mathrm{GF}(3)$. Let $R = [V]G$ be the corresponding semidirect product. If P is a Sylow 2-subgroup of G, then VP is an $\mathfrak{F}$-normal maximal subgroup of R. Since $VP \in \mathfrak{F}$, it follows that P is $\mathfrak{F}$-subnormal in R. However, if $x \in G \setminus \mathrm{N}_R(P)$, then $G = \langle P, P^x \rangle$ is not $\mathfrak{F}$-subnormal in R.

Therefore the following question naturally arises:

Which are the formations $\mathfrak{F}$ for which the set $\mathrm{s}_{n\mathfrak{F}}(G)$ is a sublattice of the subgroup lattice of G for every group G?

This question was first proposed by L. A. Shemetkov in his monograph [She78] in 1978 and it appeared in the *Kourovka Notebook* in 1984 as Problem 9.75 [MK84].

In 1992, A. Ballester-Bolinches, K. Doerk, and M. D. Pérez-Ramos gave in [BBDPR92] the answer to that question in the soluble universe for saturated formations.

On the other hand, O. H. Kegel [Keg78] showed that if $\mathfrak{F}$ is a subgroup-closed formation such that $\mathfrak{F}\mathfrak{F} = \mathfrak{F}$, then the set of all K-$\mathfrak{F}$-subnormal subgroups of a group G is a sublattice of the subgroup lattice of G for every group G. He also asks for other formations enjoying the lattice property for K-$\mathfrak{F}$-subnormal subgroups.

In 1993, A. F. Vasil'ev, S. F. Kamornikov, and V. N. Semenchuk [VKS93] published the extension of the lattice results of [BBDPR92] to the general finite universe. They also proved that the problems of O. H. Kegel and L. A. Shemetkov are equivalent for saturated formations.

Our objective in this section is to give a full account of the above results. *In the sequel, $\mathfrak{F}$ will be a (subgroup-closed) formation.*

Definition 6.3.2. *We say that $\mathfrak{F}$ is a* lattice *(respectively,* K-lattice*) formation if the set of all $\mathfrak{F}$-subnormal (respectively, K-$\mathfrak{F}$-subnormal) subgroups is a sublattice of the lattice of all subgroups in every group.*

The next result provides a criterion for a saturated formation to be a lattice formation.

Theorem 6.3.3. *Any two of the following assertions about a saturated formation $\mathfrak{F}$ are equivalent:*

1. *$\mathfrak{F}$ is a lattice formation.*
2. *If A and B are $\mathfrak{F}$-subnormal $\mathfrak{F}$-subgroups of a group G, then $\langle A, B \rangle$ is an $\mathfrak{F}$-subgroup of G.*
3. *$\mathfrak{F}$ is a Fitting class and the $\mathfrak{F}$-radical $G_{\mathfrak{F}}$ of a group G contains every $\mathfrak{F}$-subnormal $\mathfrak{F}$-subgroup of G.*

Proof. Assume, arguing by contradiction, that $\mathfrak{F}$ is a lattice formation such that $\mathfrak{F}$ does not satisfy Statement 2. Let G be a group of minimal order among the groups X having two $\mathfrak{F}$-subnormal $\mathfrak{F}$-subgroups H and K such that $\langle H, K \rangle$ is not an $\mathfrak{F}$-group. Among the pairs (H, K) of $\mathfrak{F}$-subnormal $\mathfrak{F}$-subgroups of G such that $\langle H, K \rangle \notin \mathfrak{F}$, we choose a pair (A, B) with $|A|+|B|$ maximal. Because of Lemma 6.1.7 (2) and the choice of G, it must be $G = \langle A, B \rangle$. Moreover if N is a minimal normal subgroup of G, it follows that $G/N \in \mathfrak{F}$ because G/N is generated by the $\mathfrak{F}$-subnormal $\mathfrak{F}$-subgroups AN/N and BN/N. Therefore G is in the boundary of $\mathfrak{F}$. In particular, G is a monolithic primitive group.

Put $N = \mathrm{Soc}(G) = G^{\mathfrak{F}}$. By Lemma 6.1.7 (2) and Proposition 6.1.11, $AN = AF^*(AN)$ and $BN = BF^*(BN)$ are $\mathfrak{F}$-groups. Applying Lemma 6.1.7 (1), we have that AN and BN are $\mathfrak{F}$-subnormal subgroups of G. The choice of the pair (A, B) yields $N \leq A \cap B$.

Let H be a minimal supplement to N in G. By [DH92, A, 9.2(c)], we have $H \cap N \leq \Phi(H)$; since $H/(H \cap N) \cong HN/N = G/N \in \mathfrak{F}$, it follows that $H \in \mathrm{E}_{\Phi}\mathfrak{F} = \mathfrak{F}$. On the other hand, $A = N(A \cap H)$ and $B = N(B \cap H)$. By Lemma 6.1.6 (1) and Lemma 6.1.7 (1), $A \cap H$ is $\mathfrak{F}$-subnormal in G. Hence the normal closure $(A \cap H)^H$ of $A \cap H$ in H is $\mathfrak{F}$-subnormal in G. Note that $N(A \cap H)^H$ is normal in G and A is contained in $N(A \cap H)^H$. Therefore $G = N\big((A \cap H)^H(B \cap H)\big)$. Since $(A \cap H)^H$ and $B \cap H$ are $\mathfrak{F}$-subnormal in G, it follows that $(A \cap H)^H(B \cap H)$ is an $\mathfrak{F}$-subnormal $\mathfrak{F}$-subgroup of G. Applying Proposition 6.1.11, $G \in \mathfrak{F}$ and we have reached the desired contradiction. Therefore $G \in \mathfrak{F}$. We have proved that 1 implies 2.

2 implies 3. Suppose that G is a group such that $G = N_1 N_2$ with $N_i \trianglelefteq G$ and $N_i \in \mathfrak{F}$ for $i = 1, 2$. Then $N_i \in \mathrm{E}\,\mathrm{K}(\mathfrak{F})$, $i = 1, 2$, and so $G \in \mathrm{E}\,\mathrm{K}(\mathfrak{F})$. Applying Proposition 6.1.10, N_i are $\mathfrak{F}$-subnormal in G for $i = 1, 2$. By Statement 2, $G \in \mathfrak{F}$ and we have shown that $\mathfrak{F}$ is N_0-closed. Therefore $\mathfrak{F}$ is a Fitting class because $\mathfrak{F}$ is subgroup-closed.

Let G be a group and $A = \langle X \in \mathfrak{F} : X \text{ is } \mathfrak{F}\text{-subnormal in } G \rangle$. Then A is normal in G and $A \in \mathfrak{F}$ by Statement 2. Hence A is contained in the $\mathfrak{F}$-radical $G_{\mathfrak{F}}$ of G.

3 implies 1. Suppose that $\mathfrak{F}$ is not a lattice formation and derive a contradiction. Let G be a counterexample with least possible order. Then G has two $\mathfrak{F}$-subnormal subgroups U and V such that $\langle U, V \rangle$ is not $\mathfrak{F}$-subnormal. If N is a minimal normal subgroup of G, then $\langle U, V \rangle N/N$ is $\mathfrak{F}$-subnormal in G/N by Lemma 6.1.6 (3). Hence $\langle U, V \rangle N$ is $\mathfrak{F}$-subnormal in G by Lemma 6.1.6 (2). Assume that $\langle U, V \rangle N$ is a proper subgroup of G. Then U and V are $\mathfrak{F}$-subnormal in $\langle U, V \rangle N$ by Lemma 6.1.7 (2). Hence $\langle U, V \rangle$ is $\mathfrak{F}$-subnormal in $\langle U, V \rangle N$ by the minimal choice of G. Therefore $\langle U, V \rangle$ is $\mathfrak{F}$-subnormal in G, contrary to supposition. Hence $G = \langle U, V \rangle N$ for every minimal normal subgroup N of G.

On taking N contained in $\mathrm{Core}_G(\langle U, V \rangle)$, if this is non-trivial, we can conclude $G = \langle U, V \rangle$. This is not possible. Thus $\mathrm{Core}_G(\langle U, V \rangle) = 1$. On the other hand, $U^{\mathfrak{F}}$ and $V^{\mathfrak{F}}$ are subnormal in G by Lemma 6.1.9 (1) and so N normalises $\langle U^{\mathfrak{F}}, V^{\mathfrak{F}} \rangle$ by [DH92, A, 14.3 and 14.4]. Hence $\langle \langle U^{\mathfrak{F}}, V^{\mathfrak{F}} \rangle^G \rangle = \langle \langle U^{\mathfrak{F}}, V^{\mathfrak{F}} \rangle^{\langle U, V \rangle} \rangle \leq \mathrm{Core}_G(\langle U, V \rangle) = 1$. This yields $U \in \mathfrak{F}$ and $V \in \mathfrak{F}$. By Statement 3, U and V are contained in $G_{\mathfrak{F}}$ and so $G = G_{\mathfrak{F}} N$. On taking $N \leq G_{\mathfrak{F}}$, we conclude that $G = G_{\mathfrak{F}} \in \mathfrak{F}$. In particular, $\langle U, V \rangle$ is $\mathfrak{F}$-subnormal in G. This is the final contradiction. $\square$

Corollary 6.3.4. *Let $\mathfrak{F}$ be a saturated lattice formation. If $G \in \mathrm{E}\,\mathrm{K}(\mathfrak{F})$, then $G_{\mathfrak{F}} = \langle X \in \mathfrak{F} : X \text{ is } \mathfrak{F}\text{-subnormal in } G \rangle$.*

Proof. Applying Proposition 6.1.10, every subnormal subgroup of G is $\mathfrak{F}$-subnormal. Hence $G_{\mathfrak{F}} \leq \langle X \in \mathfrak{F} : X \text{ is } \mathfrak{F}\text{-subnormal in } G \rangle$ and the equality holds by Theorem 6.3.3 (3). $\square$

Remark 6.3.5. If $\mathfrak{F} = \mathfrak{S}_p$ for some prime p, there exist groups G such that $1 = \langle X \in \mathfrak{F} : X \text{ is } \mathfrak{F}\text{-subnormal in } G \rangle < G_{\mathfrak{F}} = \mathrm{O}_p(G) < G^{\mathfrak{F}} = \mathrm{O}^p(G)$.

A well-known result of Baer asserts that if p is a prime, then a p-element x of a group G lies in $\mathrm{O}_p(G)$ if, and only if, any two conjugates of x generate a p-subgroup of G. As a consequence a subgroup H of a group G is contained in the Hall π-subgroup of $\mathrm{F}(G)$, π a set of primes, if, and only if, $\langle H, H^g \rangle$ is a nilpotent π-group for every $g \in G$ ([DH92, A, 14.11]). This result does not hold for saturated Fitting formations. For instance, if $\mathfrak{F}$ is the class of all groups with nilpotent length at most 2 and $G = \mathrm{Sym}(4)$, then $\langle H, H^g \rangle \in \mathfrak{F}$ for every subgroup H generated by a transposition and every $g \in G$. However H is not contained in $\mathrm{Alt}(4) = G_{\mathfrak{N}^2}$.

Our next theorem shows that lattice formations $\mathfrak{F}$ do enjoy the above property in groups with soluble residual. This result was proved in the soluble universe in [BBDPR92].

Theorem 6.3.6. *Let $\mathfrak{F}$ be a lattice formation of characteristic π. For a subgroup H of a π-group $G \in \mathfrak{S}\mathfrak{F}$, the following statements are equivalent:*

1. *H is contained in the $\mathfrak{F}$-radical $G_{\mathfrak{F}}$ of G;*
2. *$\langle H, H^g \rangle$ is an $\mathfrak{F}$-group for every $g \in G$.*

Proof. 1 implies 2. If H is contained in $G_{\mathfrak{F}}$, then $\langle H, H^g \rangle \leq G_{\mathfrak{F}}$ for all $g \in G$. Hence $\langle H, H^g \rangle$ is an $\mathfrak{F}$-group for all $g \in G$.

2 implies 1. By Lemma 6.1.7 (1), the subgroup H is $\mathfrak{F}$-subnormal in $\langle H, H^g \rangle$ for all $g \in G$. By Theorem 6.2.17, H is $\mathfrak{F}$-subnormal in G. Since $H \in \mathfrak{F}$, it follows that $H \leq G_{\mathfrak{F}}$ by Theorem 6.3.3 (3).   □

Lemma 6.3.7. *Let $\mathfrak{F}$ be a K-lattice formation. Then $\mathfrak{F}$ is a lattice formation.*

Proof. Assume the result is false and let G be a group of minimal order among the groups X for which $s_{n\mathfrak{F}}(X)$ is not a sublattice of the subgroup lattice of X. Then G has two $\mathfrak{F}$-subnormal subgroups U and V such that $\langle U, V \rangle$ is not $\mathfrak{F}$-subnormal in G. Let N be a minimal normal subgroup of G. Then $\langle U, V \rangle N$ is $\mathfrak{F}$-subnormal in G by Lemma 6.1.6 (3) and Lemma 6.1.6 (2). Put $\mathfrak{H} = \mathrm{E}\,\mathrm{K}(\mathfrak{F})$. Applying Lemma 6.1.9 (2), $U^{\mathfrak{H}} = V^{\mathfrak{H}} = G^{\mathfrak{H}}$. If $G \notin \mathfrak{H}$, then $N \leq G^{\mathfrak{H}}$ and so $\langle U, V \rangle = \langle U, V \rangle N$. This contradiction yields $G \in \mathfrak{H}$ and so $s_{n\mathfrak{F}}(G) = s_{n\mathrm{K}\text{-}\mathfrak{F}}(G)$ by Proposition 6.1.10. We have reached a contradiction. Therefore $\mathfrak{F}$ is a lattice formation.   □

Lemma 6.3.8. *Let $\mathfrak{F}$ be a saturated K-lattice formation. Then every K-$\mathfrak{F}$-subnormal $\mathfrak{F}$-subgroup of a group G is contained in the $\mathfrak{F}$-radical of G.*

Proof. We proceed by induction on $|G|$; we may clearly suppose that $G \notin \mathfrak{F}$. Let $1 \neq H$ be a K-$\mathfrak{F}$-subnormal subgroup of G such that $H \in \mathfrak{F}$. Let N be a minimal normal subgroup of G. By Lemma 6.1.6 (3), HN/N is K-$\mathfrak{F}$-subnormal in G/N. Applying induction $HN/N \leq (G/N)_{\mathfrak{F}} = A/N$. If A is a proper subgroup of G, then H is contained in $A_{\mathfrak{F}}$ because H is K-$\mathfrak{F}$-subnormal in A by Lemma 6.1.7 (2). Since $A_{\mathfrak{F}}$ is a normal $\mathfrak{F}$-subgroup of G, it follows that $A_{\mathfrak{F}} \leq G_{\mathfrak{F}}$. Therefore H is contained in $G_{\mathfrak{F}}$. There remains the possibility that $G/N \in \mathfrak{F}$ for all minimal normal subgroups N of G. Then G is in the boundary of $\mathfrak{F}$ and so G is a primitive group and $N = G^{\mathfrak{F}}$ is the unique minimal normal subgroup of G. Assume that $N \notin \mathfrak{F}$. Then $(HN)_{\mathfrak{F}} \cap N = 1$. This implies that $(HN)_{\mathfrak{F}} \leq C_G(N) \leq N$. On the other hand, H is a proper subgroup of G and therefore H is contained in a proper subgroup M of G such that either $M \trianglelefteq G$ or $G^{\mathfrak{F}} \leq M$. In both cases $N \leq M$ and so HN is a proper subgroup of G. By induction, $H \leq (HN)_{\mathfrak{F}} \leq (HN)_{\mathfrak{F}} \cap N = 1$. This contradiction implies that $N \in \mathfrak{F}$. Therefore $G \in \mathrm{E}\kappa(\mathfrak{F})$ and so H is $\mathfrak{F}$-subnormal in G by Proposition 6.1.10. In this case, H is contained in $G_{\mathfrak{F}}$ by Lemma 6.3.7 and Theorem 6.3.3 (3). This is the final contradiction. $\square$

Theorem 6.3.9. *Let $\mathfrak{F}$ be a saturated formation. Then $\mathfrak{F}$ is a lattice formation if and only if $\mathfrak{F}$ is a K-lattice formation.*

Proof. Only the necessity of the condition is in doubt. Assume, arguing by contradiction, that $\mathfrak{F}$ is a lattice formation and there exists a group G for which $s_{n\,\mathrm{K}\text{-}\mathfrak{F}}(G)$ is not a sublattice of the subgroup lattice of G. Furthermore let G be a group of smallest order with this property. Then G has two K-$\mathfrak{F}$-subnormal subgroups U and V such that $\langle U, V \rangle$ is not K-$\mathfrak{F}$-subnormal in G. Let N be a minimal normal subgroup of G. Since UN/N and VN/N are K-$\mathfrak{F}$-subnormal in G by Lemma 6.1.6 (3), it follows that $\langle U, V \rangle N/N$ is K-$\mathfrak{F}$-subnormal in G/N by the minimal choice of G. Hence $\langle U, V \rangle N$ is K-$\mathfrak{F}$-subnormal in G by Lemma 6.1.6 (2). If $\langle U, V \rangle N$ were a proper subgroup of G, then $\langle U, V \rangle$ would be K-$\mathfrak{F}$-subnormal in $\langle U, V \rangle N$ by minimality of G (note that U and V are K-$\mathfrak{F}$-subnormal in $\langle U, V \rangle$ by Lemma 6.1.7 (2)). Applying Lemma 6.1.6 (1), $\langle U, V \rangle$ is K-$\mathfrak{F}$-subnormal in G. This contradiction yields $G = \langle U, V \rangle N$ for every minimal normal subgroup N of G. In particular, $\mathrm{Core}_G(\langle U, V \rangle) = 1$. By Lemma 6.1.9 (1), $U^{\mathfrak{F}}$ and $V^{\mathfrak{F}}$ are subnormal subgroups of G. Therefore $\langle U^{\mathfrak{F}}, V^{\mathfrak{F}} \rangle = D$ is subnormal in G and $\mathrm{Soc}(G) \leq \mathrm{N}_G(D)$ by [DH92, A, 14.3 and 14.4]. Hence $D^G = D^{\langle U, V \rangle N} = D^{\langle U, V \rangle} \leq \langle U, V \rangle$. This means that $D^G \leq \mathrm{Core}_G(\langle U, V \rangle) = 1$. Hence U and V belong to $\mathfrak{F}$. Applying Lemma 6.3.8, $\langle U, V \rangle$ is contained in $G_{\mathfrak{F}}$. Hence $G = G_{\mathfrak{F}} N$ for every minimal normal subgroup N of G. In particular, $G = G_{\mathfrak{F}}$ and $\langle U, V \rangle$ is K-$\mathfrak{F}$-subnormal in G by Lemma 6.1.7 (1). This is the desired contradiction. $\square$

Lemma 6.3.10. *Let $\{\mathfrak{F}_i : i \in \mathcal{I}\}$ be a family of saturated lattice formations. Then $\mathfrak{F} = \bigcap_{i \in \mathcal{I}} \mathfrak{F}_i$ is a saturated lattice formation.*

Proof. It is sufficient to see that $\mathfrak{F}$ satisfies Statement 3 of Theorem 6.3.3. It is clear that $\mathfrak{F}$ is a saturated Fitting formation. Moreover $X^{\mathfrak{F}_i}$ is contained in $X^{\mathfrak{F}}$ for every group X, $i \in \mathcal{I}$. Hence every $\mathfrak{F}$-subnormal subgroup is $\mathfrak{F}_i$-subnormal for all $i \in \mathcal{I}$ by Lemma 6.1.7 (1).

Let G be a group and let H be an $\mathfrak{F}$-subnormal $\mathfrak{F}$-subgroup of G. Then H is an $\mathfrak{F}_i$-subnormal $\mathfrak{F}_i$-subgroup of G for every $i \in \mathcal{I}$. By Theorem 6.3.3 (3),

H is contained in $\bigcap_{i \in \mathcal{I}} G_{\mathfrak{F}_i}$, which is a normal $\mathfrak{F}$-subgroup of G because $\mathfrak{F}_i$ is subgroup-closed for every $i \in \mathcal{I}$. Therefore H is contained in $G_{\mathfrak{F}}$ and $\mathfrak{F}$ is a lattice formation. $\qquad \square$

Lemma 6.3.11. *Let $\mathcal{I}$ be a non-empty set. For each $i \in \mathcal{I}$, let $\mathfrak{F}_i$ be a subgroup-closed saturated lattice formation. Assume that $\pi(\mathfrak{F}_i) \cap \pi(\mathfrak{F}_j) = \emptyset$ for all $i, j \in \mathcal{I}$, $i \neq j$. Then $\mathfrak{F} = \mathsf{X}_{i \in \mathcal{I}} \mathfrak{F}_i$ is a subgroup-closed saturated lattice formation.*

Proof. By Remark 2.2.13, $\mathfrak{F}$ is a subgroup-closed saturated formation.

Assume that $\mathfrak{F}$ does not satisfy Statement 2 of Theorem 6.3.3 and derive a contradiction. Let G be a counterexample of minimal order. Then G has two $\mathfrak{F}$-subnormal $\mathfrak{F}$-subgroups A and B such that $\langle A, B \rangle$ is not an $\mathfrak{F}$-group. Then obviously $A \neq 1$ and $B \neq 1$. Observe that $\langle A, B \rangle$ and any epimorphic image of G inherits the conditions of G. Therefore $G = \langle A, B \rangle$ and $G/N \in \mathfrak{F}$ for every minimal normal subgroup N of G. Since $G \notin \mathfrak{F}$, it follows that $N = G^{\mathfrak{F}}$ is the unique minimal normal subgroup of G and $\mathrm{C}_G(N) \leq N$. Since A is $\mathfrak{F}$-subnormal in AN by Lemma 6.1.7 (2) and N is a quasinilpotent normal subgroup of G, it follows that AN belongs to $\mathfrak{F}$ by Proposition 6.1.11. Hence there exists $i \in \mathcal{I}$ such that $N \in \mathfrak{F}_i$. Moreover, $\mathrm{C}_G(N) \leq N$ forces $AN \in \mathfrak{F}_i$. The same arguments can be applied to B. We then conclude that AN, $BN \in \mathfrak{F}_i$. Since $G/N \in \mathfrak{F}$ and $\mathfrak{F} = \mathsf{X}_{i \in \mathcal{I}} \mathfrak{F}_i$, it follows that G/N has a normal $\pi(\mathfrak{F}_i)$-Hall subgroup. Since AN/N and BN/N are $\pi(\mathfrak{F}_i)$-groups, we have that G/N is a $\pi(\mathfrak{F}_i)$-group. In particular, G is a $\pi(\mathfrak{F}_i)$-group and so A and B are $\mathfrak{F}_i$-subnormal $\mathfrak{F}_i$-subgroups of G. Therefore $G = \langle A, B \rangle \in \mathfrak{F}_i \subseteq \mathfrak{F}$ by Theorem 6.3.3 (2). This contradiction confirms that $\mathfrak{F}$ is a lattice formation. $\qquad \square$

Let $\mathfrak{Z}$ be a class of groups. A group G is called s-*critical for* $\mathfrak{Z}$, or simply $\mathfrak{Z}$-*critical*, if G is not in $\mathfrak{Z}$ but all proper subgroups of G are in $\mathfrak{Z}$. Critical groups associated with some classes of groups will play a central role in Section 6.4.

Lemma 6.3.12. *Let $\mathfrak{F}$ be a saturated Fitting formation. Assume that each of the following conditions holds:*

1. *$\mathfrak{F} = \mathfrak{S}_p \mathfrak{F}$ for all $p \in \mathrm{char}\, \mathfrak{F}$.*
2. *$\mathfrak{F}$ is an $\mathfrak{F}^2$-normal Fitting class.*
3. *Every $\mathfrak{F}$-critical group G with $\Phi(G) = 1$ is either cyclic or G is monolithic such that $\mathrm{Soc}(G)$ is non-abelian and $G/\mathrm{Soc}(G)$ is a cyclic group of prime power order.*

Then $\mathfrak{F}$ is a lattice formation.

Proof. It will be established that every $\mathfrak{F}$-subnormal $\mathfrak{F}$-subgroup H of a group G is contained in the $\mathfrak{F}$-radical of G. This will be accomplished by induction on $|G|$, which we suppose greater than 1. Obviously we may suppose $G \notin \mathfrak{F}$ and $1 \neq H < G$. Let N be a minimal normal subgroup of G. Then HN/N is an

$\mathfrak{F}$-subnormal $\mathfrak{F}$-subgroup of G/N by Lemma 6.1.6 (3). By induction, HN/N is contained in $(G/N)_{\mathfrak{F}} = T/N$. If T is a proper subgroup of G, then H is $\mathfrak{F}$-subnormal in T by Lemma 6.1.7 (2) and H is contained in $T_{\mathfrak{F}}$ by induction. Since T is normal in G, it follows that $T_{\mathfrak{F}} \leq G_{\mathfrak{F}}$ and H is contained in $G_{\mathfrak{F}}$. Hence we may assume that $G/N \in \mathfrak{F}$ for every minimal normal subgroup N of G. This implies that G is a monolithic primitive group and $N = G^{\mathfrak{S}}$ is the unique minimal normal subgroup of G and $\mathrm{C}_G(N) \leq N$. By Lemma 6.1.7 (2) and Proposition 6.1.11, $HN = HF^*(HN)$ is an $\mathfrak{F}$-group. Hence $N \in \mathfrak{F}$ and $G \in \mathfrak{F}^2$. By Statement 2, $G_{\mathfrak{F}}$ is the $\mathfrak{F}$-injector of G.

If N were abelian, then N would be a p-group for some prime $p \in \mathrm{char}\,\mathfrak{F}$. Then $G \in \mathfrak{S}_p\mathfrak{F} = \mathfrak{F}$, contrary to supposition. Hence N is non-abelian. If $HG_{\mathfrak{F}}$ were a proper subgroup of G, then H would be contained in $(HG_{\mathfrak{F}})_{\mathfrak{F}}$. Thus $HG_{\mathfrak{F}} \in \mathfrak{F}$ and $HG_{\mathfrak{F}} = G_{\mathfrak{F}}$ by the $\mathfrak{F}$-maximality of $G_{\mathfrak{F}}$ in G. Consequently we may assume that $G = HG_{\mathfrak{F}}$. Let M be a maximal subgroup of G containing $G_{\mathfrak{F}}$. Then $M = (H \cap M)G_{\mathfrak{F}}$ and $H \cap M$ is $\mathfrak{F}$-subnormal in M by Lemma 6.1.7 (2). Since $H \cap M \in \mathfrak{F}$, it follows that H is contained in $M_{\mathfrak{F}}$ by induction. This forces $M \in \mathfrak{F}$ and so $M = G_{\mathfrak{F}}$ by the $\mathfrak{F}$-maximality of $G_{\mathfrak{F}}$ in G. Hence $G/G_{\mathfrak{F}}$ is a cyclic group of order p, for a prime number $p \in \mathrm{char}\,\mathfrak{F}$. Let H_p and J be Sylow p-subgroups of H and $G_{\mathfrak{F}}$, respectively, such that $P = H_pJ$ is a Sylow p-subgroup of G ([Hup67, VI, 4.7]). Then $G = PG_{\mathfrak{F}}$. Consider the subgroup PN of G. Since $N = G^{\mathfrak{S}}$, it follows that PN is $\mathfrak{F}$-subnormal in G by Lemma 6.1.9 (1). Moreover, PN is the product of its $\mathfrak{F}$-subnormal $\mathfrak{F}$-subgroups H_pN and JN. If $G = PN$, then H_pN is subnormal in G. Since $H_pN \in \mathfrak{F}$, it follows that $H_pN \leq G_{\mathfrak{F}}$. Consequently $G = G_{\mathfrak{F}}$, contrary to the choice of G. Hence we may assume that PN is a proper subgroup of G. By induction $PN \in \mathfrak{F}$. This implies that P is $\mathfrak{F}$-subnormal in G.

Let A be a maximal subgroup of G such that $A \neq G_{\mathfrak{F}}$. Then $G = AG_{\mathfrak{F}}$ and $A = A_p(A \cap G_{\mathfrak{F}})$ for some Sylow p-subgroup A_p of G. Without loss of generality we may assume that A_p is contained in P. Then, by Lemma 6.1.6 (1), A_p is $\mathfrak{F}$-subnormal in G because A_p is $\mathfrak{F}$-subnormal in P by Lemma 6.1.9 (4). Since A_p and $A \cap G_{\mathfrak{F}}$ are two $\mathfrak{F}$-subnormal $\mathfrak{F}$-subgroups of A, it follows that $A \in \mathfrak{F}$ by induction. Therefore G is an $\mathfrak{F}$-critical group. By Statement 3, G/N is a cyclic group of order p^{α} for some $\alpha \geq 1$. But then $G = PN$. This contradicts our supposition. Therefore G satisfies Statement 3 of Theorem 6.3.3 and $\mathfrak{F}$ is a lattice formation. $\square$

Example 6.3.13. Let $\mathfrak{F} = \mathfrak{S}_{\pi}$ be the class of all soluble π-groups for a set of primes π. Then $\mathfrak{F}$ is a lattice formation as $\mathfrak{F}$ satisfies Statements 1–3 of Lemma 6.3.12.

There exist non-soluble saturated lattice formations as the next example due to A. F. Vasil'ev, S. F. Kamornikov, and V. N. Semenchuk [VKS93] shows:

Example 6.3.14. Let S be a non-abelian simple group with the property that if $T < S$, then T is soluble (e. g., $G = \mathrm{Alt}(5)$). Let $\mathfrak{F} = \mathfrak{S}_{\pi}\,\mathrm{D}_0(1, S)$, for $\pi = \pi(S)$. By Proposition 2.2.11, $\mathfrak{F}$ is a formation. Moreover, by [DH92, II,

1.9], $\mathfrak{F}$ is S_n-closed. It is not difficult to prove that $\mathfrak{F}$ is also N_0-closed and saturated. Hence $\mathfrak{F}$ is a saturated Fitting formation contained in $\mathfrak{E}_\pi$.

Suppose, by way of contradiction, that $\mathfrak{F}$ is not subgroup-closed, and choose a group G of minimal order such that $G \in \mathfrak{F}$ and G has a subgroup $H \notin \mathfrak{F}$. Let N be a normal soluble π-subgroup of G such that $G/N \in \mathrm{D}_0(1, S) \subseteq \mathfrak{F}$. If $N \neq 1$, then $HN/N \in \mathfrak{F}$ by the minimal choice of G. Hence $H \in \mathrm{S}_\pi\mathfrak{F} = \mathfrak{F}$, contrary to supposition. Therefore $N = 1$ and $G = S_1 \times \cdots \times S_n$ is a direct product of copies of S. For $i = 1, \ldots, n$, let π_i denote the projection of G onto the ith component of the direct product. Let $\mathcal{A}$ denote the subset of $\{1, \ldots, n\}$ defined by

$$i \in \mathcal{A} \quad \text{if and only if} \quad \pi_i(H) = S_i.$$

Set $K = \bigcap_{i \in \mathcal{A}} \mathrm{Ker}\big((\pi_i)_H\big)$ and $K^* = \bigcap_{i \notin \mathcal{A}} \mathrm{Ker}\big((\pi_i)_H\big)$. Then $H/K \in \mathrm{D}_0(1, S)$ and H/K^* is soluble. Since $H/KK^* \in \mathrm{D}_0(1, S)$ and H/KK^* is soluble, it follows that $H = KK^* = K \times K^*$ as $K \cap K^* = 1$. Hence $H \in \mathfrak{F}$. This contradiction shows that $\mathfrak{F}$ is subgroup-closed.

Assume that $\mathfrak{F}$ is not a lattice formation and choose a group G of minimal order having an $\mathfrak{F}$-subnormal $\mathfrak{F}$-subgroup H which is not contained in $G_\mathfrak{F}$. Clearly $H \neq 1$. By familiar arguments, $G \in \mathrm{b}(\mathfrak{F})$ and so G is a monolithic primitive group. Let N be the unique minimal normal subgroup of G. If N is abelian, then $G \in \mathrm{S}_\pi\mathfrak{F} = \mathfrak{F}$, which contradicts our assumption. Hence N is non-abelian and $\mathrm{C}_G(N) = 1$. By Lemma 6.1.7 (2) and Proposition 6.1.11, $HN = HF^*(HN)$ is an $\mathfrak{F}$-group. Since $\mathrm{C}_G(N) = 1$, it follows that HN has no normal soluble π-subgroups. Thus $HN \in \mathrm{D}_0(1, S)$ and $HN = N \leq G_\mathfrak{F}$. This is the final contradiction. Applying Theorem 6.3.3 (3), $\mathfrak{F}$ is a lattice formation. $\qquad\square$

We have now arrived at our first main objective, namely the classification of the subgroup-closed saturated lattice formations.

Theorem 6.3.15. *Let* $\mathfrak{F} = \mathrm{LF}(F)$ *be a saturated formation. Then* $\mathfrak{F}$ *is a lattice formation if and only if* $\mathfrak{F} = \mathfrak{M} \times \mathfrak{G}$ *for some subgroup-closed saturated formations* $\mathfrak{M}$ *and* $\mathfrak{G}$ *satisfying the following conditions:*

1. *$\pi(\mathfrak{M}) \cap \pi(\mathfrak{G}) = \emptyset$.*
2. *There exists a set of prime numbers π^* and a partition $\{\pi_i : i \in \mathcal{I}\}$ of π^* such that $\mathfrak{G} = \mathsf{X}_{i \in \mathcal{I}} \, \mathfrak{S}_{\pi_i}$.*
3. *$\mathfrak{M} = \mathfrak{S}_p\mathfrak{M}$ for all $p \in \pi(\mathfrak{M})$ and $\mathfrak{M}$ is an $\mathfrak{M}^2$-normal Fitting class.*
4. *Every non-cyclic $\mathfrak{M}$-critical group G with $\Phi(G) = 1$ is a primitive group of type 2 such that $G/\mathrm{Soc}(G)$ is a cyclic group of prime power order.*

Proof. First of all, applying Proposition 3.1.40, $F(p)$ is a subgroup-closed formation for every prime $p \in \pi = \mathrm{char}\,\mathfrak{F}$.

Assume that $\mathfrak{F}$ is a lattice formation. For the ease of reading we break the argument into separately-stated steps.

1. *For each $p \in \pi$, every primitive group G of type 1 in $\mathfrak{F} \cap \mathrm{b}\big(F(p)\big)$ is cyclic.*

It is clear that $N = \mathrm{Soc}(G)$, the unique minimal normal subgroup of G, is a q-group for some prime $q \neq p$. By [DH92, B, 10.9], G has an irreducible and faithful G-module V over $\mathrm{GF}(p)$. We claim that G has a unique core-free maximal subgroup, which provides the result.

Suppose that M_1 and M_2 are maximal subgroups of G such that $M_1 \neq M_2$ and $\mathrm{Core}_G(M_i) = 1$ for $i = 1, 2$ and derive a contradiction. Then $M_i \in F(p)$, $i = 1, 2$. Consider the semidirect product $H = [V]G$, with respect to the action of G on V. Clearly $H \notin \mathfrak{F}$ because $G \notin F(p)$. Hence $H^{\mathfrak{F}} = V$ and G is not $\mathfrak{F}$-subnormal in H. But for $i = 1, 2$, VM_i is an $\mathfrak{F}$-normal maximal subgroup of H, and M_i is $\mathfrak{F}$-subnormal in VM_i because $VM_i \in \mathfrak{S}_p F(p) = F(p) \subseteq \mathfrak{F}$, that is, M_i is $\mathfrak{F}$-subnormal in H (Lemma 6.1.6 (1) and Lemma 6.1.7 (1)). Since $\mathfrak{F}$ is a lattice formation, it follows that $G = \langle M_1, M_2 \rangle$ is $\mathfrak{F}$-subnormal in H, contrary to supposition.

2. *If p and q belong to π, and $q \in \mathrm{char}\, F(p)$, then $p \in \mathrm{char}\, F(q)$.*

Assume that $C_p \notin F(q)$ and consider an irreducible and faithful C_q-module V over $\mathrm{GF}(p)$ ([DH92, B, 10.9]). Then the semidirect product $[V]C_q$, with respect to the action of C_q on V, is a non-cyclic primitive group of type 1 in $\mathfrak{F} \cap \mathrm{b}\big(F(p)\big)$. This contradicts Step 1. Therefore $C_p \in F(q)$ and $p \in \mathrm{char}\, F(q)$.

3. *If p, $q \in \pi$ and $p \in \mathrm{char}\, F(q)$, then $\mathrm{char}\, F(p) = \mathrm{char}\, F(q)$.*

If $r \in \mathrm{char}\, F(q) \setminus \mathrm{char}\, F(p)$, then $r \neq q$ and $C_q \in F(r)$, because of Step 2. Consider an irreducible and faithful C_q-module V over $\mathrm{GF}(r)$. Then $[V]C_q \in \mathfrak{F} \cap \mathrm{b}\big(F(p)\big)$ and $[V]C_q$ is non-cyclic primitive group of type 1. This contradicts Step 1. Therefore $\mathrm{char}\, F(q) \subseteq \mathrm{char}\, F(p)$ and analogously $\mathrm{char}\, F(p) \subseteq \mathrm{char}\, F(q)$.

4. *If p, $q \in \pi$ and $p \in \mathrm{char}\, F(q)$, then $\mathfrak{S}_p \subseteq F(q)$.*

Since $F(q)$ is subgroup-closed, and a p-group of order p^n is isomorphic with a subgroup of the n-fold iterated wreath product $H_n = \big(\ldots (C_p \wr C_p) \ldots\big) \wr C_p$, it is enough to prove that $H_n \in F(q)$ for all $n \in \mathbb{N}$. Denote inductively $H_1 = C_p$ and $H_n = H_{n-1} \wr C_p$ for $n \geq 2$. We can assume that $p \neq q$. Since $\mathrm{Z}(H_n)$ is cyclic, H_n has a unique minimal normal subgroup, and consequently there exists an irreducible and faithful H_n-module V over $\mathrm{GF}(q)$ by [DH92, B, 10.9]. Consider the semidirect product $G = [V]H_n$, with respect to the action of H_n on V. If $(H_{n-1})^{\natural}$ denotes the base group of H_n, then $H_n = (H_{n-1})^{\natural} C_p$. Since $(H_{n-1})^{\natural}$ and C_p are $F(q)$-groups, it follows that $V(H_{n-1})^{\natural}$ and VC_p belong to $F(q)$. Moreover they are $\mathfrak{F}$-subnormal in G. Hence $G \in \mathfrak{F}$ by Theorem 6.3.3 (2) and so $H_n \in F(q)$.

5. *If p, $q \in \pi$ and $q \in \mathrm{char}\, F(q)$, then $\mathfrak{S}_p F(q) = F(q)$.*

Assume that $F(q) \neq \mathfrak{S}_p F(q)$ and derive a contradiction. Let G be a group of minimal order in the supposed non-empty class $\mathfrak{S}_p F(q) \setminus F(q)$. Then, since $F(q)$ is a subgroup-closed formation, G has a unique minimal normal subgroup, M say, and $G/M \in F(q)$. Moreover M is a p-group and every maximal subgroup of G belongs to $F(q)$. If $M \leq \Phi(G)$, then $G \in \mathfrak{F}$ and we may argue as in Step 1 to obtain that G is cyclic. Consequently $G \in F(q)$ by Step 4.

This contradiction implies that M is not contained in $\Phi(G)$. Let R be a maximal subgroup of G such that $G = MR$. Then $M \cap R = 1$, $R \in F(q)$ and $M = G^{\mathfrak{F}}$. Clearly we may assume that $p \neq q$. Hence considering a faithful and irreducible G-module over $\mathrm{GF}(q)$, it is rather clear that R must be a cyclic r-group for some prime r, $r \in \pi$. From Step 3 and Step 4, it follows that $G \in \mathfrak{S}_p\mathfrak{S}_r \subseteq \mathfrak{S}_pF(p) = F(p) \subseteq \mathfrak{F}$. This is the desired contradiction. Therefore $F(q) = \mathfrak{S}_pF(q)$.

Calling two elements p, $q \in \pi$ *equivalent* if and only if $\operatorname{char} F(p) = \operatorname{char} F(q)$, we obtain an equivalence relation on π whose equivalence classes $\{\pi_i : i \in \mathcal{I}\}$ form a partition of π.

Let $p \in \pi_i$, $i \in \mathcal{I}$. Since $F(p)$ is a subgroup-closed formation, it follows that $F(p) \subseteq \mathfrak{E}_{\pi_i}$. If $2 \notin \pi_i$, then every group in $\mathfrak{E}_{\pi_i}$ is soluble by the Odd Order Theorem [FT63]. Therefore $F(p) \subseteq \mathfrak{S}_{\pi_i}$. In fact, we have:

6. *If* $p \in \pi_i$, $i \in \mathcal{I}$, *and* $2 \notin \pi_i$, *then* $F(p) = \mathfrak{S}_{\pi_i}$.

Assume that $F(p) \neq \mathfrak{S}_{\pi_i}$ and choose a group $G \in \mathfrak{S}_{\pi_i} \setminus F(p)$ of minimal order. Then G has a unique minimal normal subgroup N, N is a q-group for some prime $q \in \pi_i$, and $G/N \in F(p)$. By Step 5, $G \in \mathfrak{S}_qF(p) = F(p)$. This contradiction forces $F(p) = \mathfrak{S}_{\pi_i}$.

Put $\mathfrak{M} = (1)$ if $2 \notin \pi$ and $\mathfrak{M} = \mathfrak{F} \cap \mathfrak{E}_{\pi_{i_0}}$ if $2 \in \pi_{i_0}$ for some $i_0 \in I$.

Assume that $2 \in \pi$. Then $\{\pi_i : i \neq i_0\}$ is a partition of $\pi^* = \pi \setminus \pi_{i_0}$. Let $\mathfrak{G} = \bigtimes_{i \in \mathcal{I} \setminus \{i_0\}} \mathfrak{S}_{\pi_i}$. Then $\mathfrak{G}$ is a subgroup-closed lattice-formation by Lemma 6.3.11 and Example 6.3.13.

7. $\mathfrak{F} = \mathfrak{M} \times \mathfrak{G}$, $\pi(\mathfrak{M}) \cap \pi(\mathfrak{G}) = \emptyset$.

It is clear that $\mathfrak{M} \times \mathfrak{G}$ is contained in $\mathfrak{F}$. Suppose, for a contradiction, that this inclusion is proper, and choose a group G of minimal order in $\mathfrak{F} \setminus (\mathfrak{M} \times \mathfrak{G})$. Then G is a monolithic primitive group because $\mathfrak{M} \times \mathfrak{G}$ is a saturated formation. Let N be the unique minimal normal subgroup of G. Suppose that N is non-abelian. Then 2 divides $|N|$ and $G \in F(2) \subseteq \mathfrak{F} \cap \mathfrak{E}_{\pi_{i_0}} = \mathfrak{M}$, contrary to our choice of G. Therefore N is abelian. Let p be the prime dividing $|N|$. Then $G \in F(p)$. If $p \in \pi_{i_0}$, it follows that $G \in \mathfrak{M}$. If $p \in \pi_i$ for some $i \in \mathcal{I} \setminus \{i_0\}$, we have $G \in \mathfrak{G}$. In both cases, $G \in \mathfrak{M} \times \mathfrak{G}$, another contradiction.

Evidently, $\pi_{i_0} = \pi(\mathfrak{M})$ and $\pi^* = \pi \setminus \pi_{i_0} = \pi(\mathfrak{G})$. Hence $\pi(\mathfrak{M}) \cap \pi(\mathfrak{G}) = \emptyset$.

8. $\mathfrak{M} = \mathfrak{S}_p\mathfrak{M}$ *for all* $p \in \pi(\mathfrak{M})$.

Let $p \in \pi(\mathfrak{M})$. Assume that $\mathfrak{S}_p\mathfrak{M}$ is not contained in $\mathfrak{M}$ and derive a contradiction. Let $G \in \mathfrak{S}_p\mathfrak{M} \setminus \mathfrak{M}$ be a group of minimal order. By familiar reasoning, G is a primitive group of type 1 and $N = \operatorname{Soc}(G)$ is an abelian p-group. Our eventual goal is to show that every core-free maximal subgroup M of G is cyclic. Suppose that M_1 and M_2 are maximal subgroups of M such that $M_1 \neq M_2$. Since $\mathfrak{S}_p\mathfrak{M}$ is subgroup-closed, it follows that $NM_i \in \mathfrak{M} \subseteq \mathfrak{F}$ for $i = 1$, 2. Moreover, NM_i are $\mathfrak{F}$-subnormal in G because $N = G^{\mathfrak{F}}$ (Lemma 6.1.7 (1)). By Theorem 6.3.3 (2), $G = N\langle M_1, M_2\rangle \in \mathfrak{F}$, which contradicts the assumption that $G \notin \mathfrak{M}$. Therefore M has a unique maximal subgroup and so M is a cyclic group of prime power order. By Step 4, $M \in F(q)$, where $q \in \pi(M)$. Therefore $G \in \mathfrak{S}_pF(q) = F(q)$ by Step 5. We conclude then that $G \in \mathfrak{F} \cap \mathfrak{E}_{\pi_{i_0}}$. This final contradiction completes the proof.

9. $\mathfrak{M}$ *is an* $\mathfrak{M}^2$-*normal Fitting class.*

It is clear that $\mathfrak{M}$ is a Fitting class. Let $G \in \mathfrak{M}^2$ be a group and let J be an $\mathfrak{M}$-maximal subgroup of G containing $G_{\mathfrak{M}}$. Since $G/G_{\mathfrak{M}} \in \mathfrak{M} \subseteq \mathfrak{F}$, it follows that $G^{\mathfrak{F}} \leq G_{\mathfrak{M}}$. Consequently J is $\mathfrak{F}$-subnormal in G by Lemma 6.1.7 (1). Applying Theorem 6.3.3 (3), $J \leq G_{\mathfrak{F}} = G_{\mathfrak{M}}$ because $G \in \mathfrak{C}_{\pi_{i_0}}$. Therefore $G_{\mathfrak{F}}$ is $\mathfrak{F}$-maximal in G. Let H be a subnormal subgroup of G. Then H is actually $\mathfrak{F}$-subnormal because $G \in \text{E}\,\text{K}(\mathfrak{F})$ (Proposition 6.1.10). In addition, $H \cap G_{\mathfrak{F}}$ is contained in $H_{\mathfrak{F}}$ as $H \cap G_{\mathfrak{F}}$ is an $\mathfrak{F}$-subnormal $\mathfrak{F}$-subgroup of H (Lemma 6.1.7 (2) and Theorem 6.3.3 (3)). Consequently $H \cap G_{\mathfrak{F}} = H_{\mathfrak{F}} = H_{\mathfrak{M}}$ and $H_{\mathfrak{M}} = H \cap G_{\mathfrak{M}}$ is $\mathfrak{M}$-maximal in H. This means that $G_{\mathfrak{M}}$ is an $\mathfrak{M}$-injector of G.

10. *If* G *is a non-cyclic* $\mathfrak{M}$-*critical group and* $\Phi(G) = 1$, *then* G *is a primitive group of type 2 such that* $\text{Soc}(G)$ *is non-abelian and* $G/\text{Soc}(G)$ *is a cyclic group of prime power order.*

Let G be a non-cyclic $\mathfrak{M}$-critical group such that $\Phi(G) = 1$. Then G is a monolithic primitive group because $\mathfrak{M}$ is saturated. Assume that $N = \text{Soc}(G)$ is abelian. Then $N < G$ because G is non-cyclic, and so N is a p-group for some prime $p \in \pi(\mathfrak{M})$. Hence $G \in \mathfrak{S}_p\mathfrak{M} = \mathfrak{M}$ by Step 8. This contradiction implies that N is non-abelian. Suppose that $N < G$. Let M_1 and M_2 be two different maximal subgroups of G containing N. Then $M_i \in \mathfrak{M}$ and M_i is $\mathfrak{F}$-subnormal in G, $i = 1, 2$, as $N = G^{\mathfrak{F}}$ (Lemma 6.1.7 (1)). Applying Theorem 6.3.3 (2), $G \in \mathfrak{F}$. Since $G \in \mathfrak{C}_{\pi_{i_0}}$, it follows that $G \in \mathfrak{M}$, contrary to the assumption that G is $\mathfrak{M}$-critical. This contradiction proves that G/N has a unique maximal subgroup and so G/N is cyclic of prime power order.

Applying Lemma 6.3.12, $\mathfrak{M}$ is a lattice formation.

Conversely, assume that $\mathfrak{F} = \mathfrak{M} \times \mathfrak{G}$ for subgroup-closed saturated formations $\mathfrak{M}$ and $\mathfrak{G}$ satisfying Statements 1 to 4. By Example 6.3.13, Lemma 6.3.11, and Lemma 6.3.12, $\mathfrak{F}$ is a lattice formation. □

Corollary 6.3.16. *Let* $\mathfrak{F}$ *be a saturated formation of soluble groups of characteristic* π. *Then* $\mathfrak{F}$ *is a lattice formation if and only if there exists a partition* $\{\pi_i : i \in I\}$ *of* π *such that* $\mathfrak{F} = \bigtimes_{i \in I} \mathfrak{S}_{\pi_i}$.

Corollary 6.3.16 holds not only for subgroup-closed saturated formations but also for s_n-closed saturated ones. This was proved in [VKS93].

Lockett [Loc71] described the $\mathfrak{F}$-injectors of soluble π-groups, here $\mathfrak{F}$ is a lattice formation of soluble groups of characteristic π. It turns out that if G is a soluble π-group, the $\mathfrak{F}$-injectors of G are exactly the $\mathfrak{F}$-maximal subgroups of G containing $G_{\mathfrak{F}}$, that is, $\mathfrak{F}$ is a dominant Fitting class in $\mathfrak{S}_{\pi}$.

Theorem 6.3.17 ([Loc71]). *Let* π *be a non-empy set of primes and let* G *be a soluble* π-*group. Assume that* $\{\pi_i : i \in I\}$ *is a partition of* π *and* $\mathfrak{F} = \bigtimes_{i \in \mathcal{I}} \mathfrak{S}_{\pi_i}$. *For each* $i \in \mathcal{I}$, *let* V_i *be a Hall* π_i-*subgroup of* $C_i = C_G\left(\bigtimes_{j \neq i} O_{\pi_j}(G)\right)$. *Then:*

1. $[V_i, V_j] = 1$ if $i \neq j$,

2. the subgroup $\langle V_i : i \in \mathcal{I} \rangle = \mathsf{X}_{i \in \mathcal{I}} V_i$ is an $\mathfrak{F}$-subgroup of G containing $G_{\mathfrak{F}} = \mathsf{X}_{i \in \mathcal{I}} \mathrm{O}_{\pi_i}(G)$.

Let $\mathrm{I}(G)$ be the set of all such subgroups $\langle V_i : i \in \mathcal{I} \rangle$ obtained from the various choices of $V_i \in \mathrm{Hall}_{\pi_i}(C_i)$. Then

3. if W is an $\mathfrak{F}$-subgroup of G containing $G_{\mathfrak{F}}$, then $W \leq V$ for some $V \in \mathrm{I}(G)$.

4. $\mathrm{I}(G) = \mathrm{Inj}_{\mathfrak{F}}(G)$.

Proof. It is clear that $\mathrm{F}(G)$ is contained in $G_{\mathfrak{F}} = \mathsf{X}_{i \in \mathcal{I}} \mathrm{O}_{\pi_i}(G)$ and $\mathrm{C}_G\big(\mathrm{F}(G)\big)$ is contained in $\mathrm{F}(G)$ because G is soluble.

1. Take $i, j \in I$, $i \neq j$. Then $[V_i, C_j] \leq C_i \cap C_j \leq \mathrm{C}_G(G_{\mathfrak{F}}) \leq \mathrm{C}_G\big(\mathrm{F}(G)\big) \leq \mathrm{F}(G) \leq G_{\mathfrak{F}}$. Therefore C_j normalises $V_i G_{\mathfrak{F}} = V_i \times \mathsf{X}_{i \neq j} \mathrm{O}_{\pi_j}(G)$. Since $V_i = \mathrm{O}_{\pi_i}(V_i G_{\mathfrak{F}})$, it follows that C_j normalises V_i. In particular, V_j normalises V_i. By a similar argument V_i normalises V_j. Hence $[V_i, V_j] \leq V_i \cap V_j = 1$.

2. We deduce at once from Statement 1 that $\langle V_i : i \in \mathcal{I} \rangle$ is the direct product of its Hall π_i-subgroups and also that $G_{\mathfrak{F}} \leq \langle V_i : i \in \mathcal{I} \rangle \in \mathfrak{F}$.

3. Let $i \in \mathcal{I}$. Since $W \in \mathfrak{F}$, the Hall π_i-subgroup W_i of W centralises $\mathrm{O}_{\pi_j}(W)$, which contains $\mathrm{O}_{\pi_j}(G)$ by assumption, $i \neq j$. Therefore W_i is contained in a Hall π_i-subgroup, V_i say, of $\mathrm{C}_G\big(\mathsf{X}_{j \neq i} \mathrm{O}_{\pi_i}(G)\big)$. Hence $W = \mathsf{X}_{i \in \mathcal{I}} W_i \leq \mathsf{X}_{i \in \mathcal{I}} V_i \in \mathrm{I}(G)$.

4. It is enough to prove that $\mathrm{I}(G)$ is a conjugacy class of G.

Let $V = \mathsf{X}_{i \in \mathcal{I}} V_i$ and $\bar{V} = \mathsf{X}_{i \in \mathcal{I}} \bar{V}_i$ be two typical elements of $\mathrm{I}(G)$. For each $i \in \mathcal{I}$, there exists an element $x_i \in \mathrm{C}_G\big(\mathsf{X}_{j \neq i} \mathrm{O}_{\pi_j}(G)\big)$ such that $\bar{V}_i = V_i^{x_i}$. Let $x = \prod_{i \in \mathcal{I}} x_i$, where the product may be taken in any order. If $i \neq j$, the element x_j normalises each conjugate of V_i, and therefore $V^x = \mathsf{X}_{i \in \mathcal{I}} V_i^x = \mathsf{X}_{i \in \mathcal{I}} V_i^{x_i} = \bar{V}$.   □

The next result, due to A. Ballester-Bolinches, K. Doerk, and M. D. Pérez-Ramos [BBDPR92], shows that these injectors have a good behaviour with respect to $\mathfrak{F}$-subnormal subgroups.

Theorem 6.3.18. *Let $\mathfrak{F}$ be a lattice formation of soluble groups of characteristic π. If G is a soluble π-group and V is an $\mathfrak{F}$-injector of G and H is an $\mathfrak{F}$-subnormal subgroup of G, then $V \cap H$ is an $\mathfrak{F}$-injector of H.*

Proof. Assume that the result is not true and let G be a counterexample of minimal order. Clearly we may suppose that H is an $\mathfrak{F}$-normal maximal subgroup of G. Hence $G/\mathrm{Core}_G(H)$ is a π_i-group for some member π_i of $\{\pi_i : i \in \mathcal{I}\}$, where $\{\pi_i : i \in \mathcal{I}\}$ is the partition of π such that $\mathfrak{F} = \mathsf{X}_{i \in \mathcal{I}} \mathfrak{S}_{\pi_i}$. Write $\pi = \pi_i$. Then $\mathrm{Core}_G(H)$ contains every Hall π'-subgroup of G.

Note that $H_{\mathfrak{F}}$ is contained in $G_{\mathfrak{F}}$ because $H_{\mathfrak{F}}$ is an $\mathfrak{F}$-subnormal $\mathfrak{F}$-subgroup of G (Lemma 6.1.6 (1) and Theorem 6.3.3 (3)). Let V be an $\mathfrak{F}$-injector of G such that $V \cap H$ is not an $\mathfrak{F}$-injector of H. Since $H_{\mathfrak{F}}$ is contained in $V \cap H$, it follows that $V \cap H$ is not $\mathfrak{F}$-maximal in H. Let R be an $\mathfrak{F}$-maximal subgroup of H containing $V \cap H$. It is clear that R is an $\mathfrak{F}$-injector of H. Since the

Hall π'-subgroup $V_{\pi'}$ of V is contained in the Hall π'-subgroup $R_{\pi'}$ of R and $R_{\pi'}$ is contained in $\mathrm{Core}_G(H)$, it follows that $V_{\pi'} = R_{\pi'}$ as $V \cap \mathrm{Core}_G(H)$ is an $\mathfrak{F}$-injector of $\mathrm{Core}_G(H)$. On the other hand, according to Lockett's result, the Hall π-subgroup V_π of V is a Hall π-subgroup of $C = \mathrm{C}_G\left(\mathsf{X}_{j \neq i}\, \mathrm{O}_{\pi_j}(G)\right)$ and the Hall π-subgroup R_π of R is a Hall π-subgroup of $\mathrm{C}_H\left(\mathsf{X}_{j \neq i}\, \mathrm{O}_{\pi_j}(H)\right)$. Moreover $V_\pi \cap H \leq R_\pi$. Since $G/\mathrm{Core}_G(H)$ is a π-group, it follows that $\mathsf{X}_{j \neq i}\, \mathrm{O}_{\pi_j}(G) = \mathsf{X}_{j \neq i}\, \mathrm{O}_{\pi_j}(H)$ and so there exists an element $g \in C$ such that $V_\pi^g \cap H = R_\pi$. If $C = G$, then V_π is a Hall π-subgroup of G. Thus $G = \mathrm{Core}_G(H)V_\pi$ and $V_\pi^g \cap H = V_\pi^h \cap H$ for some $h \in \mathrm{Core}_G(H)$. This implies that $|V_\pi \cap H| = |R_\pi|$ and $R_\pi = V_\pi \cap H$, contrary to our supposition. Consequently C is a proper subgroup of G. Since C is normal in G, it follows that $V \cap C$ is an $\mathfrak{F}$-injector of C. Moreover $H \cap C$ is $\mathfrak{F}$-subnormal in C by Lemma 6.1.7 (2). The minimality of G yields $V \cap H \cap C$ is an $\mathfrak{F}$-injector of $H \cap C$. In particular $V \cap H \cap C = R \cap C$. Since R_π is a Hall π-subgroup of $R \cap C$, we have that $V_\pi \cap H = R_\pi$. This contradiction proves the result. $\qquad\square$

The following result is a characterisation of saturated lattice formations of soluble groups by means of properties of Fitting type. Most of the work is already contained in the above theorem.

Theorem 6.3.19. *Let $\mathfrak{F}$ be a saturated formation of soluble groups of characteristic π. The following statements are pairwise equivalent:*

1. $\mathfrak{F}$ *is a lattice formation.*
2. $\mathfrak{F}$ *is a Fitting class satisfying that if G is a soluble π-group, V is an $\mathfrak{F}$-injector of G and H is an $\mathfrak{F}$-subnormal subgroup of G, then $V \cap H$ is an $\mathfrak{F}$-injector of H.*
3. $\mathfrak{F}$ *is a Fitting class and if H is an $\mathfrak{F}$-subnormal $\mathfrak{F}$-subgroup of a soluble π-group G, then $\langle H, H^g \rangle \in \mathfrak{F}$ for every $g \in G$.*

Proof. Applying Theorem 6.3.3 (3) and Theorem 6.3.18, we have that 1 implies 2.

Assume that Statement 2 holds. Let H be an $\mathfrak{F}$-subnormal $\mathfrak{F}$-subgroup of a soluble π-group G. If $g \in G$, then H^g is contained in every $\mathfrak{F}$-injector of G. Therefore $\langle H, H^g \rangle \in \mathfrak{F}$.

Suppose, arguing by contradiction, that the statement 3 is true but $\mathfrak{F}$ is not a lattice formation. On this supposition, by Theorem 6.3.3 (3), there exists a group G of minimal order having an $\mathfrak{F}$-subnormal $\mathfrak{F}$-subgroup $1 \neq H$ which is not contained in the $\mathfrak{F}$-radical of G. If N is a minimal normal subgroup of G, then HN/N is contained in the $\mathfrak{F}$-radical K/N of G/N by minimality of G. Since H is $\mathfrak{F}$-subnormal in K by Lemma 6.1.7 (2), and $K_\mathfrak{F}$ is contained in $G_\mathfrak{F}$, it follows that $K = G$. Hence $G/N \in \mathfrak{F}$ for every minimal normal subgroup of G. Thus G is a monolithic primitive group. Let $A = G^\mathfrak{F}$ be the unique minimal normal subgroup of G. By Lemma 6.1.7 (2) and Proposition 6.1.11, $T = HA = HF^*(T)$ is an $\mathfrak{F}$-group. Note that $\langle T, T^g \rangle$ is $\mathfrak{F}$-subnormal in G for all $g \in G$ because A is contained in T (Lemma 6.1.7 (1)). Applying

Statement 3, it follows that $\langle T^G \rangle \in \mathfrak{F}$. Since $\langle T^G \rangle$ is normal in G and $\mathfrak{F}$ is a Fitting class, T^G is contained in the $\mathfrak{F}$-radical of G. In particular, H is contained in $G_{\mathfrak{F}}$, which contradicts our assumption. Hence $\mathfrak{F}$ is a lattice formation. $\square$

Remark 6.3.20. In [BBMPPR00], it is proved that an s_n-closed saturated formation of soluble groups of full characteristic satisfying Statement 3 of Theorem 6.3.19 is actually subgroup-closed. Therefore Theorem 6.3.19 hold not only for subgroup-closed saturated formations of soluble groups, but also for s_n-closed ones.

We round this section off with a characterisation of lattice formations of soluble groups.

It is not always true in general that a lattice formation of soluble groups is saturated. It is enough to consider the formation of all abelian groups. In the sequel we shall take a closer look at this family of formations, following ideas of A. F. Vasil'ev and S. F. Kamornikov [VK02].

Therefore until further notice we make the following general assumption.

Hypothesis 6.3.21. $\mathfrak{F}$ *is not only a subgroup-closed formation but also soluble.*

Let $\mathfrak{Z}_{\mathfrak{F}}$ be the class of all groups G such that every subgroup of G is $\mathfrak{F}$-subnormal in G. The basic properties of $\mathfrak{F}$-subnormal subgroups imply that $\mathfrak{Z}_{\mathfrak{F}}$ is an homomorph containing $\mathfrak{F}$.

The formation of all abelian groups shows that the inclusion could be proper. Moreover it is rather easy to see that $\mathfrak{Z}_{\mathfrak{F}} = \mathfrak{F}$ if $\mathfrak{F}$ is saturated.

We gather together in a convenient "portmanteau" lemma some relevant properties of $\mathfrak{Z}_{\mathfrak{F}}$, when $\mathfrak{F}$ is a lattice formation.

Lemma 6.3.22. *Let $\mathfrak{F}$ be a lattice formation. Then:*

1. $\mathfrak{Z}_{\mathfrak{F}}$ *is a subgroup-closed formation of soluble groups.*
2. $\pi(\mathfrak{F}) = \pi(\mathfrak{Z}_{\mathfrak{F}})$ *and $\mathfrak{Z}_{\mathfrak{F}}$ contains all nilpotent $\pi(\mathfrak{F})$-groups.*
3. $\mathfrak{Z}_{\mathfrak{F}}$ *is a Fitting class.*

Proof. 1. First we prove that $\mathfrak{Z}_{\mathfrak{F}}$ is a soluble class. Suppose, by way of contradiction, that $\mathfrak{Z}_{\mathfrak{F}}$ is not contained in $\mathfrak{S}$. Then $\mathfrak{Z}_{\mathfrak{F}} \setminus \mathfrak{S}$ is not empty. Let G be a group of minimal order in $\mathfrak{Z}_{\mathfrak{F}} \setminus \mathfrak{S}$. By familiar reasoning, G is a non-abelian simple group such that every subgroup of G is $\mathfrak{F}$-subnormal in G. Let M be a maximal subgroup of G. Then $1 \neq M$ and $G^{\mathfrak{F}}$ is contained in M because M is $\mathfrak{F}$-subnormal in G. But then $G^{\mathfrak{F}} = 1$ because G is simple. This means that $G \in \mathfrak{F}$ and so G is soluble, contrary to supposition. Hence $\mathfrak{Z}_{\mathfrak{F}}$ is composed of soluble groups.

It is clear that $\mathfrak{Z}_{\mathfrak{F}}$ is a homomorph. Let N_1 and N_2 be minimal normal subgroups of a group G such that $N_1 \cap N_2 = 1$ and $G/N_i \in \mathfrak{Z}_{\mathfrak{F}}$ for $i = 1, 2$. Let P be a Sylow subgroup of G. Then our assumption implies that PN_i/N_i is

$\mathfrak{F}$-subnormal in G/N_i for $i = 1, 2$. By Lemma 6.1.7 (3), $PN_1 \cap PN_2 = P$ is $\mathfrak{F}$-subnormal in G. Moreover since N_i/N_i is $\mathfrak{F}$-subnormal in G/N_i, it follows that N_i is $\mathfrak{F}$-subnormal in G for $i = 1, 2$ by Lemma 6.1.6 (2). Hence $1 = N_1 \cap N_2$ is $\mathfrak{F}$-subnormal in G. By Lemma 6.1.7 (2), 1 is $\mathfrak{F}$-subnormal in P. Therefore every subgroup of P is $\mathfrak{F}$-subnormal in G by Lemma 6.1.9 (3) and Lemma 6.1.6 (2). Since every subgroup of G is generated by its subgroups of prime power order, and $\mathfrak{F}$ is a lattice formation, it follows that $G \in 3_{\mathfrak{F}}$. Applying [DH92, II, 2.6], $3_{\mathfrak{F}}$ is R_0-closed and so $3_{\mathfrak{F}}$ is a formation.

Let $G \in 3_{\mathfrak{F}}$ and let H be a subgroup of G. Since every subgroup of H is $\mathfrak{F}$-subnormal in G, it follows by Lemma 6.1.7 (2) that $H \in 3_{\mathfrak{F}}$. Hence $3_{\mathfrak{F}}$ is subgroup-closed.

2. It is clear that $\pi(\mathfrak{F}) \subseteq \pi(3_{\mathfrak{F}})$ because $\mathfrak{F} \subseteq 3_{\mathfrak{F}}$. Let $p \in \pi(3_{\mathfrak{F}})$ and let G be a group in $3_{\mathfrak{F}}$ such that p divides $|G|$. Then $C_p \in 3_{\mathfrak{F}}$ because $3_{\mathfrak{F}}$ is subgroup-closed. Hence 1 is $\mathfrak{F}$-subnormal in C_p and so $C_p \in \mathfrak{F}$. This shows that $p \in \pi(\mathfrak{F})$.

If P is a p-group for some prime $p \in \pi(\mathfrak{F}) = \pi(3_{\mathfrak{F}})$, then $C_p \in 3_{\mathfrak{F}}$ because $3_{\mathfrak{F}}$ is subgroup-closed. By Lemma 6.1.9 (4), every subgroup of P is $\mathfrak{F}$-subnormal in P. Hence $P \in 3_{\mathfrak{F}}$. This implies that every nilpotent $\pi(\mathfrak{F})$-group is a $3_{\mathfrak{F}}$-group.

3. It is clear that only the N_0-closure of $3_{\mathfrak{F}}$ needs checking. Let A and B be normal subgroups of a group G such that $G = AB$ and A and B belong to $3_{\mathfrak{F}}$. We prove that $G \in 3_{\mathfrak{F}}$ by induction on the order of G. If G is nilpotent, then $G \in 3_{\mathfrak{F}}$ because $\pi(G) \subseteq \pi(3_{\mathfrak{F}})$ and $3_{\mathfrak{F}}$ contains all nilpotent $\pi(3_{\mathfrak{F}})$-groups. Hence we may suppose that G is not nilpotent. Let P be a Sylow subgroup of G. Then P is the product of $P \cap A$ and $P \cap B$, which are obviously normal subgroups of P. Moreover, $P \cap A$ and $P \cap B$ are $3_{\mathfrak{F}}$-subgroups of P because $3_{\mathfrak{F}}$ is subgroup-closed. Since P is a proper subgroup of G, the induction hypothesis leads to the conclusion that $P \in 3_{\mathfrak{F}}$. Furthermore, G is soluble because $3_{\mathfrak{F}}$ is composed of soluble groups. Therefore $G \in \mathrm{K}\,\mathrm{E}(\mathfrak{F})$ as $\pi(\mathfrak{F}) = \pi(3_{\mathfrak{F}}) = \mathrm{char}\, 3_{\mathfrak{F}}$. This implies that A and B are $\mathfrak{F}$-subnormal in G by Proposition 6.1.10. Since A and B belong to $3_{\mathfrak{F}}$, it follows that $P \cap A$ and $P \cap B$ are $\mathfrak{F}$-subnormal subgroups of G by Lemma 6.1.6 (1). Therefore $P = (P \cap A)(P \cap B)$ is $\mathfrak{F}$-subnormal in G because $\mathfrak{F}$ is a lattice formation. Hence every subgroup of P is $\mathfrak{F}$-subnormal in G Lemma 6.1.6 (1). Since every subgroup of G is generated by its subgroups of prime power order, it follows that $G \in 3_{\mathfrak{F}}$. Consequently $3_{\mathfrak{F}}$ is N_0-closed and so $3_{\mathfrak{F}}$ is a Fitting class. $\qquad\square$

Combining Theorem 2.5.2 and Lemma 6.3.22, we have:

Proposition 6.3.23. *Let $\mathfrak{F}$ be a lattice formation. Then $3_{\mathfrak{F}}$ is a saturated formation.*

Theorem 6.3.24. *Let $\mathfrak{F}$ be a lattice formation. Then $3_{\mathfrak{F}}$ is a lattice formation.*

Proof. By Theorem 6.3.3 (3), it is sufficient to prove that every $3_{\mathfrak{F}}$-subnormal $3_{\mathfrak{F}}$-subgroup H of a group G is contained in the $3_{\mathfrak{F}}$-radical $G_{3_{\mathfrak{F}}}$ of G.

Suppose, by way of contradiction, that there exists a group G of minimal order having a $3_{\mathfrak{F}}$-subnormal $3_{\mathfrak{F}}$-subgroup H such that H is not contained in $G_{3_{\mathfrak{F}}}$. Among the $3_{\mathfrak{F}}$-subnormal $3_{\mathfrak{F}}$-subgroups of G that are not contained in $G_{3_{\mathfrak{F}}}$, let H be one of maximal order. Let N be a minimal normal subgroup of G. Then HN/N is a $3_{\mathfrak{F}}$-subnormal $3_{\mathfrak{F}}$-subgroup of G/N by Lemma 6.1.6 (3). The choice of G implies that $HN/N \leq (G/N)_{3_{\mathfrak{F}}} = L/N$. Assume that L is a proper subgroup of G. The minimality of G forces the conclusion that H is contained in $L_{3_{\mathfrak{F}}}$ as H is $3_{\mathfrak{F}}$-subnormal $3_{\mathfrak{F}}$-subgroup of L. Since L is normal in G, it follows that $L_{3_{\mathfrak{F}}}$ is contained in $G_{3_{\mathfrak{F}}}$. This contradiction shows that $L = G$ and so $G/N \in 3_{\mathfrak{F}}$ for every minimal normal subgroup of G. Consequently G is a monolithic primitive group and $N = G^{3_{\mathfrak{F}}}$ is the unique minimal normal subgroup of G. Moreover $C_G(N) \leq N$. By Lemma 6.1.7 (2) and Proposition 6.1.11, $HN = HF^*(HN)$ is an $3_{\mathfrak{F}}$-group. Since HN is $3_{\mathfrak{F}}$-subnormal in G by Lemma 6.1.7 (1), it follows that $N \leq H$ by the choice of the pair (G, H). Therefore $H = N(H \cap M)$, where M is a core-free maximal subgroup of G complementing N in G. On the other hand, H/N is $\mathfrak{F}$-subnormal in G/N. Hence H is $\mathfrak{F}$-subnormal in G by Lemma 6.1.6 (2). Since $H \in 3_{\mathfrak{F}}$, it follows that $H \cap M = X$ is also $\mathfrak{F}$-subnormal in G Lemma 6.1.6 (1). Consequently X^M, the normal closure of X in M, is $\mathfrak{F}$-subnormal in G because $\mathfrak{F}$ is a lattice formation. Furthermore $X^M \in 3_{\mathfrak{F}}$ because $M \in 3_{\mathfrak{F}}$ and $3_{\mathfrak{F}}$ is subgroup-closed. Since X^M is $3_{\mathfrak{F}}$-subnormal in G, it follows that X^M is $3_{\mathfrak{F}}$-subnormal in NX^M by Lemma 6.1.7 (2), and NX^M belongs to $3_{\mathfrak{F}}$ by Proposition 6.1.11. Since NX^M is a normal subgroup of G and $3_{\mathfrak{F}}$ is a Fitting class, it follows that $H \leq NX^M \leq G_{3_{\mathfrak{F}}}$. This contradiction proves the theorem. $\square$

We are now in a position to state and prove Vasil'ev and Kamornikov's characterisation of lattice formations of soluble groups.

Theorem 6.3.25 ([VK02]). *Let $\mathfrak{F}$ be a formation of soluble groups. The following statements are pairwise equivalent:*

1. *The set of all K-$\mathfrak{F}$-subnormal subgroups is a sublattice of the subgroup lattice of every group.*
2. *The set of all $\mathfrak{F}$-subnormal subgroups is a sublattice of the subgroup lattice of every group.*
3. *There exists a partition $\{\pi_i : i \in \mathcal{I}\}$ of the set $\pi(\mathfrak{F})$ such that $\mathfrak{F} = \mathsf{X}_{i \in \mathcal{I}} \, \mathfrak{F}_{\pi_i}$, where $\mathfrak{F}_{\pi_i} = \mathfrak{F} \cap \mathfrak{S}_{\pi_i}$. Moreover, $\mathfrak{F}_{\pi_i} = \mathfrak{S}_{\pi_i}$ for all $i \in \mathcal{I}$ such that $|\pi_i| > 1$.*

Proof. Of the three statements in the theorem, it follows that 1 implies 2.

Assume that Statement 2 holds. Then the preceding results show that $3_{\mathfrak{F}}$ is a saturated lattice formation of soluble groups. Hence, by Theorem 6.3.15, there exists a partition $\{\pi_i : i \in \mathcal{I}\}$ of $\pi := \text{char } 3_{\mathfrak{F}} = \pi(3_{\mathfrak{F}}) = \pi(\mathfrak{F})$ such that $3_{\mathfrak{F}} = \mathsf{X}_{i \in \mathcal{I}} \, \mathfrak{S}_{\pi_i}$. Hence $\mathfrak{F} = \mathsf{X}_{i \in \mathcal{I}} \, \mathfrak{F}_{\pi_i}$, where $\mathfrak{F}_{\pi_i} = \mathfrak{F} \cap \mathfrak{S}_{\pi_i}$ for all $i \in \mathcal{I}$.

Let G be a soluble primitive π_i-group for some $i \in \mathcal{I}$. Then $G \in 3_{\mathfrak{F}}$ and so every subgroup of G is $\mathfrak{F}$-subnormal in G. In particular, $G \in \mathfrak{F}$ and thus $G \in \mathfrak{F}_{\pi_i}$. Suppose, in addition, that $|\pi_i| \geq 2$. If A is a π_i-group such that $|\pi(A)| \geq 2$, then by a theorem of Hawkes [Haw75, Theorem 1], A is isomorphic to a subgroup of a multiprimitive $\pi(A)$-group G, that is, every epimorphic image of G is primitive. Since G is primitive and G is a π_i-group, it follows that $G \in \mathfrak{F}_{\pi_i}$. Hence $A \in \mathfrak{F}_{\pi_i}$ because $\mathfrak{F}_{\pi_i}$ is subgroup-closed. If A is a π_i-group, then A is isomorphic to a subgroup of a π_i-group B such that $|\pi(B)| \geq 2$ as $|\pi_i| \geq 2$. Hence $A \in \mathfrak{F}_{\pi_i}$. Consequently $\mathfrak{F}_{\pi_i} = \mathfrak{S}_{\pi_i}$ for all $i \in \mathcal{I}$ such that $|\pi_i| \geq 2$ and Statement 3 is true.

To complete the proof we now show that 3 implies 1. Suppose that $\mathfrak{F}$ is a formation such that $\mathfrak{F} = \bigtimes_{i \in \mathcal{I}} \mathfrak{F}_{\pi_i}$ for a partition $\{\pi_i : i \in \mathcal{I}\}$ of $\pi = \pi(\mathfrak{F})$. Assume, in addition, that $\mathfrak{F}_{\pi_i} = \mathfrak{S}_{\pi_i}$ if $|\pi_i| \geq 2$. Consider the subgroup-closed formation $\mathfrak{H} = \bigtimes_{i \in \mathcal{I}} \mathfrak{S}_{\pi_i}$. By Lemma 6.3.11, $\mathfrak{H}$ is a saturated lattice formation and $\operatorname{char} \mathfrak{H} = \pi$. We aim to show that $\mathrm{s}_{n\mathrm{K}\text{-}\mathfrak{H}}(G) = \mathrm{s}_{n\mathrm{K}\text{-}\mathfrak{F}}(G)$ for every group G. Assume, arguing by contradiction, there exists a group G of minimal order such that $\mathrm{s}_{n\mathrm{K}\text{-}\mathfrak{H}}(G) \neq \mathrm{s}_{n\mathrm{K}\text{-}\mathfrak{F}}(G)$. Clearly $\mathrm{s}_{n\mathrm{K}\text{-}\mathfrak{F}}(G) \subseteq \mathrm{s}_{n\mathrm{K}\text{-}\mathfrak{H}}(G)$ because $\mathfrak{F} \subseteq \mathfrak{H}$. Hence there exists a subgroup $H \in \mathrm{s}_{n\mathrm{K}\text{-}\mathfrak{H}}(G) \setminus \mathrm{s}_{n\mathrm{K}\text{-}\mathfrak{F}}(G)$. Then H is a proper subgroup of G and thus there exists a subgroup M of G such that either M is normal in G or M is an $\mathfrak{H}$-normal maximal subgroup of G. Since H is K-$\mathfrak{H}$-subnormal in M, it follows that H is K-$\mathfrak{F}$-subnormal in M by minimality of G. If M were normal in G, we would have that H would be K-$\mathfrak{F}$-subnormal in G. This would contradict our choice of H. Hence M is an $\mathfrak{H}$-normal maximal subgroup of G and so $G^{\mathfrak{H}}$ is contained in M. Then $G/\operatorname{Core}_G(M)$ is a π_i-group for some $\pi_i \subseteq \pi$ as $G/\operatorname{Core}_G(M)$ is a soluble primitive $\mathfrak{H}$-group. Note that $|\pi_i| > 1$ because M is not normal in G. Therefore $G/\operatorname{Core}_G(M) \in \mathfrak{F}_{\pi_i}$. This means that M is $\mathfrak{F}$-normal in G and H is K-$\mathfrak{F}$-normal in G, contrary to our initial supposition. Therefore $\mathrm{s}_{n\mathrm{K}\text{-}\mathfrak{F}}(X) = \mathrm{s}_{n\mathrm{K}\text{-}\mathfrak{H}}(X)$ for all groups X. Applying Corollary 6.3.16 and Theorem 6.3.9, the set $\mathrm{s}_{n\mathrm{K}\text{-}\mathfrak{F}}(X)$ is a sublattice of the subgroup lattice of X for all groups X. $\qquad\square$

Example 6.3.26. Let $\mathfrak{F}$ be the formation of all abelian groups. Then $\mathfrak{F}$ is a lattice formation of soluble groups such that $\mathfrak{Z}_{\mathfrak{F}} = \mathfrak{N}$, the class of all nilpotent groups. It is clear that $\mathfrak{F}_p \neq \mathfrak{S}_p$ for all $p \in \pi(\mathfrak{F}) = \mathbb{P}$.

In [VK01], A. F. Vasil'ev and S. F. Kamornikov consider w-inherited subgroup functors f enjoying the following property:

If G is a group and $H, K \in \mathrm{f}(G)$, then $H \cap K \in \mathrm{f}(G)$ and $\langle H, K \rangle \in \mathrm{f}(G)$.

They called them *subgroup NTL-functors*. The techniques employed in this section allow them to prove the following nice result in the universe of all soluble groups:

Theorem 6.3.27. *Let* f *be a subgroup NTL-functor. Then:*

1. *The class* $\chi_{\mathrm{f}} = \{G \mid \mathrm{f}(G) = \mathrm{s}(G)\}$ *is a subgroup-closed saturated formation,*

2. *there exists a partition $\{\pi_i \mid i \in \mathcal{I}\}$ of $\pi(\chi_{\mathrm{f}})$ such that $\chi_{\mathrm{f}} = \mathsf{X}_{i \in \mathcal{I}} \, \mathfrak{S}_{\pi_i}$,*
3. *For every group G, $\mathrm{f}(G) = \mathrm{s}_{n \chi_{\mathrm{f}}}(G)$.*

Consequently the subgroup NTL-functors in the soluble universe are exactly $\mathrm{f} = \mathrm{s}_{n \mathfrak{F}}$, for some subgroup-closed saturated lattice formation $\mathfrak{F}$.

The authors also consider the problem in the general finite universe. The best they were able to prove is the following:

Proposition 6.3.28. *Let f be a subgroup NTL-functor. Then:*

1. *The class $\chi_{\mathrm{f}} = \{G \mid \mathrm{f}(G) = \mathrm{s}(G)\}$ is a subgroup-closed solubly saturated formation,*
2. *For every group G, $\mathrm{s}_{n \chi_{\mathrm{f}}}(G)$ is contained in $\mathrm{f}(G)$.*

Consequently the following question remains open.

Open question 6.3.29 ([VK01]). *Let f be a subgroup NTL-functor. Is there a solubly saturated formation $\mathfrak{F}$ such that $\mathrm{f} = \mathrm{s}_{n \mathfrak{F}}$?.*

The reader is referred to [KS03] for more information about subgroup functors and classes of groups.

Postscript

Lattice formations have been also involved in the study of $\mathfrak{F}$-normality associated with subgroup-closed saturated formations $\mathfrak{F}$ in the soluble universe. As it is known, this notion was primarily associated with maximal subgroups.

A first attempt to give a definition valid for arbitrary subgroups was made by A. Ballester-Bolinches, K. Doerk, and M. D. Pérez-Ramos in [BBDPR95]. In the case $\mathfrak{F} = \mathfrak{N}$, the class of all nilpotent groups, the $\mathfrak{F}$-normality coincides with the classical normality and, for a general subgroup-closed saturated formation $\mathfrak{F}$, the $\mathfrak{F}$-subnormality turns out to be associated naturally with the $\mathfrak{F}$-normality in the obvious way. However, the results concerning lattice properties of $\mathfrak{F}$-normal subgroups differ from the corresponding ones for $\mathfrak{F}$-subnormal subgroups.

More recently, M. Arroyo-Jordá and M. D. Pérez-Ramos [AJPR01] study an alternative definition of $\mathfrak{F}$-normality, the $\mathfrak{F}$-Dnormality. It was suggested by K. Doerk. This new definition satisfies all the desired properties. Moreover, in this case, lattice formations turn out to be the subgroup-closed saturated formations for which the set of all $\mathfrak{F}$-Dnormal subgroups is a sublattice of the subgroup lattice in every soluble group.

The same authors [AJPR04a], [AJPR04b], studied Fitting classes with stronger closure properties involving $\mathfrak{F}$-subnormal subgroups, for a lattice formation $\mathfrak{F}$ of full characteristic.

Definition 6.3.30. *1. Let $\mathfrak{F}$ be a lattice formation containing the class $\mathfrak{N}$ of nilpotent groups. A class $\mathfrak{X}$ of groups is said to be an $\mathfrak{F}$-Fitting class if: a) for every $G \in \mathfrak{X}$ and every $\mathfrak{F}$-subnormal subgroup H of G we have $H \in \mathfrak{X}$; and b) for $G = \langle H, K \rangle$ with H, K $\mathfrak{F}$-subnormal in G, if H, $K \in \mathfrak{X}$, then $G \in \mathfrak{X}$.*

2. A subgroup of a group G is said to be an $(\mathfrak{X}, \mathfrak{F})$-injector if, for every $\mathfrak{F}$-subnormal subgroup K of G, $V \cap K$ is $\mathfrak{X}$-maximal in K.

Every $\mathfrak{F}$-Fitting class is also a Fitting class. They proved in [AJPR04b] the following nice result (see Theorem 2.4.26):

Theorem 6.3.31. *Let $\mathfrak{F}$ be a lattice formation containing $\mathfrak{N}$, and $\mathfrak{X}$ an $\mathfrak{F}$-Fitting class. Then for every group G, a subgroup V of G is an $(\mathfrak{X}, \mathfrak{F})$-injector if and only if it is an $\mathfrak{X}$-injector.*

6.4 $\mathfrak{F}$-subnormal subgroups and $\mathfrak{F}$-critical groups

We saw in Section 6.3 that if $\mathfrak{F}$ is a saturated formation, then $\mathfrak{F}$ is a lattice formation if and only if $\mathfrak{F}$ contains all groups generated by two $\mathfrak{F}$-subnormal $\mathfrak{F}$-subgroups (Theorem 6.3.3 (2)). As a consequence, a saturated lattice formation $\mathfrak{F}$ enjoys the following property:

If A and B are $\mathfrak{F}$-subnormal $\mathfrak{F}$-subgroups of a group G and $G = AB$, then $G \in \mathfrak{F}$.
$\qquad\qquad\qquad\qquad\qquad\qquad\qquad\qquad\qquad\qquad\qquad\qquad$ (6.1)

It turns out that Condition (6.1) is not sufficient for a subgroup-closed saturated formation to be a lattice formation: the formation of all p-nilpotent groups, p a prime, satisfies Condition (6.1), but it is not a lattice formation (see Example 6.3.1). Moreover, the formation of all groups with nilpotent length at most two does not satisfy Condition (6.1). Consequently the question of determining the subgroup-closed saturated formations which are closed under taking products of $\mathfrak{F}$-subnormal subgroups arises (see [MK99, Problem 14.99]). This problem has already been settled and solved in the soluble universe by A. Ballester-Bolinches in [BB92] for subgroup-closed saturated formations of full characteristic (see also [Sem92]).

The first result of this section puts a rich source of subgroup-closed saturated formations satisfying Condition 6.1 at our disposal.

Proposition 6.4.1. *Let $\mathfrak{F}$ be a saturated formation. Suppose that, for every $p \in \pi = \operatorname{char} \mathfrak{F}$, there exists a set of primes $\pi(p)$ with $p \in \pi(p)$ such that $\mathfrak{F}$ is locally defined by the formation function f given by $f(p) = \mathfrak{E}_{\pi(p)}$ if $p \in \pi$ and $f(q) = \emptyset$ if $q \notin \pi$. Then $\mathfrak{F}$ is closed under taking products of $\mathfrak{F}$-subnormal subgroups.*

Proof. Assume that the result is false and derive a contradiction. Then there exists a group G of minimal order with two $\mathfrak{F}$-subnormal $\mathfrak{F}$-subgroups A and B

such that $G = AB$ and $G \notin \mathfrak{F}$. If N is a minimal normal subgroup of G, then it is clear that G/N is the product of the $\mathfrak{F}$-subnormal $\mathfrak{F}$-subgroups AN/N and BN/N by Lemma 6.1.6 (3). The choice of G implies that $G/N \in \mathfrak{F}$. Therefore G is in the boundary of $\mathfrak{F}$ and so G is a monolithic primitive group. Then $N = G^{\mathfrak{F}}$ is the unique minimal normal subgroup of G and $\mathrm{C}_G(N) \leq N$. By Lemma 6.1.7 (2) and Proposition 6.1.11, $AN = AF^*(AN)$ is an $\mathfrak{F}$-group. Analogously $BN \in \mathfrak{F}$. Since $\mathrm{C}_G(N) \subseteq N$, it follows that AN and BN belong to $\mathfrak{E}_{\pi(p)}$ for all $p \in \pi(N)$. Therefore $G \in \mathfrak{E}_{\pi(p)}$ for each prime p dividing $|N|$. This implies that $G \in \mathfrak{F}$, contrary to our supposition. Consequently $\mathfrak{F}$ is closed under taking products of $\mathfrak{F}$-subnormal products.   □

Note that the above result also holds if we replace $\mathfrak{E}_{\pi(p)}$ by $\mathfrak{S}_{\pi(p)}$, for all $p \in \operatorname{char} \mathfrak{F}$.

Unfortunately, the converse of Proposition 6.4.1 is not true in general, as the following example shows.

Example 6.4.2. Let S be a non-abelian simple group, and consider the saturated formation $\mathfrak{H} = \big(G : S \notin \mathrm{Q}(G)\big)$. Let $\mathfrak{F}$ be the largest subgroup-closed formation contained in $\mathfrak{H}$. By Theorem 3.1.42, $\mathfrak{F} = \big(G : \mathrm{s}(G) \subseteq \mathfrak{H}\big)$ is saturated. In addition, $\mathfrak{F}$ cannot be locally defined by a formation function as in Proposition 6.4.1.

We assert that $\mathfrak{F}$ is closed under taking products of $\mathfrak{F}$-subnormal subgroups. Suppose, for a contradiction, that this is not true and let G be a counterexample of least order. Then G has two proper $\mathfrak{F}$-subnormal $\mathfrak{F}$-subgroups A and B such that $G = AB$ and $G \notin \mathfrak{F}$. Let N be a minimal normal subgroup of G. Since G/N is a product of the $\mathfrak{F}$-subnormal $\mathfrak{F}$-subgroups AN/N and BN/N, the choice of G implies that $G/N \in \mathfrak{F}$. Therefore G is in the boundary of $\mathfrak{F}$ and so G is a monolithic primitive group. In particular, $N = G^{\mathfrak{F}}$ is the unique minimal normal subgroup of G. Assume $G \notin \mathfrak{H}$. Then $G \in \mathrm{b}(\mathfrak{H}) = (S)$. Hence G is non-abelian and simple. This implies that $N = G$ and therefore $G = A = B$, contrary to supposition. Consequently $G \in \mathfrak{H}$. Since $G \notin \mathfrak{F}$, it follows that $\mathrm{s}(G)$ is not contained in $\mathfrak{H}$. Among the proper subgroups X of G not belonging to $\mathfrak{H}$, we choose H of minimal order. Then every proper subgroup of H belongs to $\mathfrak{H}$. Applying [DH92, III, 2.2(c)], there exists a normal subgroup K of H such that $H/K \in \mathrm{b}(\mathfrak{H})$. Hence H/K is a non-abelian simple group. Since $H/H \cap N$ belongs to $\mathfrak{F}$, it follows that $H = (H \cap N)K$. It $H \cap N$ were a proper subgroup of H, we would have $H \cap N \in \mathfrak{F}$ and so $H/K \in \mathfrak{F} \subseteq \mathfrak{H}$, contrary to supposition. Hence $H \cap N = H$ and H is a subgroup of N. By Lemma 6.1.7 (2) and Proposition 6.1.11, $AN = HF^*(AN)$ is an $\mathfrak{F}$-group. Consequently $N \in \mathfrak{F}$ and so $H \in \mathfrak{F}$. This final contradiction proves that $\mathfrak{F}$ is closed under taking $\mathfrak{F}$-subnormal subgroups.

At the time of writing no useful characterisation of subgroup-closed saturated formations satisfying (6.1) is known. The picture improves, however, if attention is confined just to subgroup-closed saturated formations of soluble groups. The following result supports this view.

Theorem 6.4.3. *Let $\mathfrak{F}$ be a saturated formation of soluble groups of characteristic π. The following statements are equivalent:*

1. *For each prime $p \in \pi$, there exists a set of primes $\pi(p)$, with $p \in \pi(p)$, such that $\mathfrak{F}$ is locally defined by the formation function f given by $f(p) = \mathfrak{S}_{\pi(p)}$ and $f(q) = \emptyset$ if $q \notin \pi$.*
2. *$\mathfrak{F}$ satisfies Condition (6.1).*

Proof. It follows at once from Proposition 6.4.1 that 1 implies 2.

2 implies 1. We are assuming in this chapter that $\mathfrak{F}$ is subgroup-closed. Hence, for every $p \in \pi$, $F(p)$ is a subgroup-closed formation by Proposition 3.1.40. Therefore $F(p)$ is contained in $\mathfrak{F} \cap \mathfrak{E}_{\pi(p)} \subseteq \mathfrak{S}_{\pi(p)}$, where $\pi(p) = \operatorname{char} F(p)$. Suppose, for a contradiction, that the inclusion is proper and choose a group G of minimal order in $(\mathfrak{F} \cap \mathfrak{S}_{\pi(p)}) \setminus F(p)$. Then every proper subgroup of G belongs to $F(p)$ and G is a soluble monolithic group. Assume that G contains two inconjugate maximal subgroups, L and M say. Then $G = ML$ by [DH92, A, 16.2]. Moreover M and L belong to $F(p)$. Let W be a faithful G-module over $\operatorname{GF}(p)$ and denote by $Z = [W]G$ the corresponding semidirect product. Then $Z^{\mathfrak{F}}$ is contained in W and therefore WM and WL are two $\mathfrak{F}$-subnormal subgroups of Z by Lemma 6.1.7 (1). Moreover WM an WL belong to $\mathfrak{S}_p F(p) = F(p)$. Hence WM and WL are $\mathfrak{F}$-groups. Since $Z = (WM)(WL)$ and $\mathfrak{F}$ satisfies (6.1), it follows that $Z \in \mathfrak{F}$. This implies that $G \in F(p)$, which is clearly not the case. Hence G has a single conjugacy class of maximal subgroups. This implies that G is a cyclic group whose order is a power of a prime, q say. Moreover, $q \in \pi(p)$. On the other hand, it is rather easy to see that $\mathfrak{F}$ is clearly a Fitting class as $\mathfrak{F}$ is closed under taking products of $\mathfrak{F}$-subnormal subgroups. Hence $F(p)$ is also a Fitting class by Proposition 3.1.40. Since $q \in \pi(p)$, it follows that $F(p)$ contains $\mathfrak{S}_q$ by [DH92, IX, 1.9]. Hence $G \in F(p)$ and we have reached a contradiction. Consequently $F(p) = \mathfrak{F} \cap \mathfrak{S}_{\pi(p)}$ for all $p \in \pi$. It remains to prove that $\mathfrak{F} = \operatorname{LF}(f)$, where f is the formation function defined by $f(p) = \mathfrak{S}_{\pi(p)}$, $p \in \pi$, and $f(q) = \emptyset$ if $q \notin \pi$. To this end assume, by way of contradiction, that $\mathfrak{M} = \operatorname{LF}(f)$ is not contained in $\mathfrak{F}$ and let G be a group of minimal order in $\mathfrak{M} \setminus \mathfrak{F}$. Then G is a soluble primitive group and $N = \operatorname{Soc}(G) = G^{\mathfrak{F}}$ is the unique minimal normal subgroup of G. Let q be the prime dividing $|N|$. Then $q \in \pi$ and $G/N \in f(q) = \mathfrak{S}_{\pi(q)}$. Hence $G/N \in \mathfrak{F} \cap \mathfrak{S}_{\pi(q)} = F(q)$. By Remark 3.1.7 (2), $G \in \mathfrak{F}$. It follows that our supposition is wrong and hence $\mathfrak{M}$ is contained in $\mathfrak{F}$. Since $\mathfrak{F}$ is obviously contained in $\mathfrak{M}$, we have $\mathfrak{F} = \mathfrak{M}$ and the proof of the theorem is complete. $\qquad\square$

From now on we focus our attention on formations whose associated critical groups have special properties. In order to carry out our task we shall need some definitions.

Recall that if $\mathfrak{Z}$ be a class of groups, a group G is called s-*critical for $\mathfrak{Z}$*, or simply $\mathfrak{Z}$-*critical*, if G is not in $\mathfrak{Z}$ but all proper subgroups of G are in $\mathfrak{Z}$. Following [DH92, VII, 6.1], we denote $\operatorname{Crit}_{\operatorname{S}}(\mathfrak{Z})$ the class of all $\mathfrak{Z}$-critical groups. The motivation for investigating such minimal classes is that detailed knowledge of groups that just fail to have a group theoretic property is likely to give some insight into just what makes a group have the property. The minimal classes have been investigated for a number of classes of groups. For instance, O. J. Schmidt (see [Hup67, III, 5.2]) studied the 𝔑-critical groups. These groups are also called *Schmidt groups*. They have a very restricted structure and they are useful in proving a known result of H. Wielandt about groups with nilpotent Hall π-subgroups (see [Hup67, III, 5.8]). K. Doerk [Doe66] studied the critical groups with respect to the class of all supersoluble groups and R. W. Carter, B. Fischer, and T. O. Hawkes (see [DH92, VII, Section 6]) used a method of extreme classes to study the soluble ℑ-critical groups in the case when ℑ is the formation of all soluble groups with nilpotent length less than or equal to r. K. Doerk and T. O. Hawkes [DH92, VII, 6.18] gave a complete description of a soluble group G which is not in ℑ but all maximal subgroups are in ℑ, where ℑ is an arbitrary (not necessarily subgroup-closed) saturated formation of soluble groups (note that such a group G is ℑ-critical if ℑ is a subgroup-closed formation). This result was extended by A. Ballester-Bolinches and M. C. Pedraza-Aguilera to the general universe of all finite groups in [BBPA96].

The reader is referred to [Rob02], [BBERR05], and [BBERss] for further information about critical groups associated with some interesting classes of groups.

A useful property for a formation ℑ in this connection is that of having ℑ-critical groups with a well-known structure. For instance, if ℑ is either a soluble saturated lattice formation or the formation of all p-nilpotent groups for a prime p, then every ℑ-critical group is either a Schmidt group or a cyclic group of prime order. Therefore a subgroup-closed class 𝔷 is contained in ℑ if and only if ℑ contains every Schmidt group and every cyclic group of prime order in 𝔷.

This raises the following question.

Which are the saturated formations ℑ such that every ℑ-critical group is either a Schmidt group or a cyclic group of prime order?

This question was proposed by L. A. Shemetkov in [MK92, Problem 9.74]. Hence we shall say that a formation ℑ has the *Shemetkov property* or ℑ is a *Š-formation* if every ℑ-critical group is a Schmidt group or a cyclic group of prime order.

The first investigation of Š-formations was taken up by V. N. Semenchuk and A. F. Vasil'ev [SV84] in the soluble realm. A. Ballester-Bolinches and M. D. Pérez-Ramos [BBPR95] determined necessary and sufficient conditions for a subgroup-closed saturated formation to be a Š-formation. This result can be used to give examples of subgroup-closed saturated Š-formations of different nature.

On the other hand, L. A. Shemetkov [She92, Problem 10.22] proposes the following question:

Let ℑ be a subgroup-closed Š-formation. Is ℑ saturated?

A. N. Skiba [Ski90] answered this question affirmatively in the soluble universe. However his result does not remain true in the general case as A. Ballester-Bolinches and M. D. Pérez-Ramos showed in [BBPR96b]. In this paper, they gave a criterion for a subgroup-closed Š-formation to be saturated from which Skiba's result emerges. An alternative approach to Shemetkov's question is due to S. F. Kamornikov [Kam94]. There he proved that a subgroup-closed Š-formation is a Baer-local formation.

We shall begin our treatment of this material with a general result concerning formations $\mathfrak{F}$ whose $\mathfrak{F}$-critical groups have the composition factors of their $\mathfrak{F}$-residual in a class of simple groups $\mathfrak{X}$. We shall then specialise $\mathfrak{X}$ to $\mathfrak{J}$ and in this class aim to give a detailed account of the present state of knowledge.

Theorem 6.4.4 ([BB05]). *Let $\emptyset \neq \mathfrak{X}$ be a class of simple groups satisfying $\pi(\mathfrak{X}) = \operatorname{char} \mathfrak{X}$. Denote $\mathfrak{Y} = \mathfrak{X} \cap \mathbb{P}$, the abelian groups in $\mathfrak{X}$.*

For a formation $\mathfrak{F}$, the following statements are equivalent:

1. *Every $\mathfrak{F}$-critical group G such that $G^{\mathfrak{F}}$ is contained in $O_{\mathfrak{X}}(G)$ is either a Schmidt group or a cyclic group of prime order.*

2. *Every $\mathfrak{F}$-critical group G whose $\mathfrak{F}$-residual is contained in $O_{\mathfrak{X}}(G)$ is soluble, $\mathfrak{F}$ is a $\mathfrak{Y}$-local formation, and for each prime $p \in \mathfrak{Y} \cap \operatorname{char} \mathfrak{F}$ there exists a set of primes $\pi(p)$ with $p \in \pi(p)$ such that $\mathfrak{F}$ is $\mathfrak{Y}$-locally defined by the $\mathfrak{Y}$-formation function f given by*

$$f(S) = \begin{cases} \mathfrak{E}_{\pi(p)} & \text{if } S \cong C_p,\ p \in \mathfrak{Y} \cap \operatorname{char} \mathfrak{F}, \\ \emptyset & \text{if } S \cong C_p,\ p \in \mathfrak{Y} \setminus \operatorname{char} \mathfrak{F}, \\ \mathfrak{F} & \text{if } S \in \mathfrak{X}' \cup (\mathfrak{X} \setminus \mathfrak{Y}). \end{cases}$$

Proof. 1 implies 2. It is clear that every $\mathfrak{F}$-critical group G with $G^{\mathfrak{F}} \leq O_{\mathfrak{X}}(G)$ is soluble by Statement 1. The next stage of the proof is to show that $\mathfrak{F}$ is a $\mathfrak{Y}$-local formation. Applying Lemma 3.1.21, it is enough to prove that $\mathfrak{F}$ is (C_p)-local for all primes $p \in \mathfrak{Y}$.

Let p be a prime in $\mathfrak{Y}$. By Theorem 3.1.11, the smallest (C_p)-local formation $\mathfrak{F}_1$ containing $\mathfrak{F}$ is (C_p)-locally defined by the (C_p)-local formation function f given by

$$\underline{f}(p) = {}_{\mathrm{Q}}{}_{\mathrm{R}_0}\big(A/\operatorname{C}_A(H/K) : A \in \mathfrak{F}$$

and H/K is an abelian p-chief factor of A), and

$$\underline{f}(S) = {}_{\mathrm{Q}}{}_{\mathrm{R}_0}\big(A/L : A/L \text{ is monolithic and if } \operatorname{Soc}(A/L) \in {}_{\mathrm{E}}(S)\big) \quad \text{if } S \in (C_p)'.$$

Denote $\pi(p) = \pi\big(\mathfrak{S}_p\underline{f}(p)\big)$, and consider the (C_p)-local formation $\mathfrak{M} = \operatorname{LF}_{(C_p)}(g)$, where

$$g(p) = \mathfrak{E}_{\pi(p)},$$
$$g(S) = \mathfrak{F} \qquad\qquad \text{if } S \in (C_p)'.$$

It is clear that $\mathfrak{F}$ is contained in $\mathfrak{M}$. Assume that $\mathfrak{M}$ is not contained in $\mathfrak{F}$ and derive a contradiction. Let G be a group of minimal order in the non-empty class $\mathfrak{M} \setminus \mathfrak{F}$. Since G is monolithic and $G \notin \mathfrak{F}$, it follows that $N := \mathrm{Soc}(G)$ is a p-group. Then $p \in \mathrm{char}\, \mathfrak{F}$. Moreover there exists a subgroup H of G such that $H \in \mathrm{Crit}_S(\mathfrak{F})$. By Statement 1, H is a Schmidt group as $\pi(G) \subseteq \mathrm{char}\, \mathfrak{F}$.

Suppose that H is a proper subgroup of G. Since $HN/N \in \mathfrak{F}$ and $H \notin \mathfrak{F}$, it follows that $H \cap N \neq 1$. In particular, $|H| = p^a q^b$ for some prime $q \neq p$ and one of the non-trivial Sylow subgroups of H is normal in H by [Hup67, III, 5.2]. Assume that a Sylow q-subgroup Q of H is normal in H. Then $H/Q \in \mathfrak{F}$ because $p \in \mathrm{char}\, \mathfrak{F}$ and so $\mathfrak{M}$ contains $\mathfrak{S}_p$. Since $H/(H \cap N)$ belongs to $\mathfrak{F}$, it follows that $H \in \mathrm{R}_0 \mathfrak{F} = \mathfrak{F}$. This would contradict the choice of G. Therefore H has a normal Sylow p-subgroup and a Sylow q-subgroup of H is cyclic by [Hup67, III, 5.2].

Suppose that q does not belong to $\pi(p)$. Then $H \cap N$ is a Sylow p-subgroup of H. Assume not, and let P be a Sylow p-subgroup of H containing $H \cap N$. Since $[P, Q]=P$ (see the proof of [Hup67, III, 5.2(c)]), it follows that $Q(H \cap N)/(H \cap N)$ is not contained in $\mathrm{O}_{p',p}(H/H \cap N)$. This implies that $q \in \pi(p)$, contrary to our supposition. Hence $H \cap N$ is a Sylow p-subgroup of G. Let $C = \mathrm{C}_G(N)$. If H is a subgroup of C, then H is nilpotent. This is not possible. Hence H is not contained in C. Since $H \cap N$ is contained in C, it follows that q divides $|G/C|$. Denote $A = [N](G/C)$. By Corollary 2.2.5, $A \in \mathfrak{M}$. Hence $q \in \pi(p)$. This contradiction proves that $q \in \pi(p)$.

The definition of $\pi(p)$ implies the existence of a group $B \in \mathfrak{F}$ and an abelian p-chief factor L/M of B satisfying that q divides $\left|B/ \mathrm{C}_B(L/M)\right|$. By Corollary 2.2.5, $C = [L/M]\left(B/ \mathrm{C}_B(L/M)\right) \in \mathfrak{F}$. Denote $V = L/M$ and $B^* = B/ \mathrm{C}_B(L/M)$, and $E = \langle g\, \mathrm{C}_B(L/M) \rangle$ for some element $g \in B$ such that $\mathrm{o}\!\left(g\, \mathrm{C}_B(L/M)\right) = q$. It is clear that V is a faithful and irreducible B^*-module over $\mathrm{GF}(p)$. Moreover V, regarded as E-module, is completely reducible by Maschke's theorem [DH92, B, 4.5]. Since V is faithful for B^* and E is a cyclic group of order q, we can find an irreducible E-submodule W of V such that W is a faithful E-module. Let $F = WE$ be the corresponding semidirect product. Then F is isomorphic to $\mathrm{E}(q|p)$, the unique Schmidt primitive group defined in [DH92, B, 12.5]. Then $F \in \mathfrak{F}$ because $\mathfrak{F}$ is subgroup-closed. If $\Phi\!\left(\mathrm{O}_p(H)\right) \neq 1$, then $H/\Phi\!\left(\mathrm{O}_p(H)\right)$ is isomorphic to $\mathrm{E}(q|p)$ constructed above. This implies that $H/\Phi\!\left(\mathrm{O}_p(H)\right) \in \mathfrak{F} \subseteq \mathfrak{M}$ and so $H \in \mathfrak{M}$ because $\mathfrak{M}$ is (C_p)-saturated by Theorem 3.2.14. The minimality of G implies that $H \in \mathfrak{F}$, and this contradicts the fact that $H \in \mathrm{Crit}_S(\mathfrak{F})$. Thus our supposition is false and $\Phi\!\left(\mathrm{O}_p(H)\right) = 1$. But then H is isomorphic to $\mathrm{E}(q|p) \in \mathfrak{F}$ and we have reached the contradiction that $H \in \mathfrak{F}$.

Consequently $H = G$ and G is a Schmidt group with a normal Sylow p-subgroup, P say. Let Q be a Sylow q-subgroup of G. Since N is the unique minimal normal subgroup of G and $\Phi(Q)$ is normal in G, we have that $\Phi(Q) = 1$ and Q is a cyclic group of order q. Note that $\Phi(P)$ is elementary abelian by [DH92, VII, 6.18] and $G/\Phi(P) \cong \mathrm{E}(q|p)$ and $\Phi(P) = \Phi(G)$. Hence G is an epimorphic image of the maximal Frattini extension E of $\mathrm{E}(q|p)$ with p-elementary abelian kernel (see [DH92, Appendix β]). Denote by A the kernel of the above extension. Then E/A is isomorphic to $\mathrm{E}(q|p)$ and $A \leq \varPhi(E)$. Moreover $\mathrm{C}_E\big(\mathrm{Soc}(A)\big)$ is a p-group by [GS78, Theorem 1]. Let Q^* be a Sylow q-subgroup of E. If AQ^* belongs to $\mathfrak{F}$, then E is $\mathfrak{F}$-critical because $G \notin \mathfrak{F}$ and A is contained in each maximal subgroup of E (note that every Sylow subgroup of E belongs to $\mathfrak{F}$). This implies that E is a Schmidt group. In particular, AQ^* is nilpotent and then $1 \neq Q^* \leq \mathrm{C}_E\big(\mathrm{Soc}(G)\big)$. This contradiction yields $AQ^* \notin \mathfrak{F}$. In this case, we can find a subgroup J of AQ^* such that $J \in \mathrm{Crit}_S(\mathfrak{F})$. By Statement 1, J should be a Schmidt group with an elementary abelian Sylow p-subgroup. This implies that J is isomorphic to $\mathrm{E}(q|p) \in \mathfrak{F}$, and we have a contradiction. Therefore $\mathfrak{F} = \mathfrak{M}$ is a $\mathfrak{Y}$-local formation, and we have completed the proof of the implication.

2 implies 1. Suppose that $\mathfrak{F}$ is a $\mathfrak{Y}$-local formation and there exists a set of primes $\pi(p)$ with $p \in \pi(p)$, for each $p \in \pi = \mathrm{char}\,\mathfrak{F}$, such that $\mathfrak{F}$ is $\mathfrak{Y}$-locally defined by the $\mathfrak{Y}$-formation function f given by $f(p) = \mathfrak{E}_{\pi(p)}$ if $p \in \pi \cap \mathfrak{Y}$, $f(q) = \emptyset$, if $p \in \mathfrak{Y} \setminus \pi$, and $f(E) = \mathfrak{F}$ for every simple group $E \in \mathfrak{X}' \cup (\mathfrak{X} \setminus \mathfrak{Y})$. We shall prove that every group in $\mathrm{Crit}_S(\mathfrak{F})$ whose $\mathfrak{F}$-residual is an $\mathfrak{X}$-group is a Schmidt group or a cyclic group of prime order.

Let G be a group in $\mathrm{Crit}_S(\mathfrak{F})$ such that $G^{\mathfrak{F}} \leq \mathrm{O}_{\mathfrak{X}}(G)$. By Condition 2, G is soluble. We prove by induction on $|G|$ that G is a Schmidt group or a cyclic group of prime order.

If G is a p-group for some prime p and G has not order p, then $p \in \pi \cap \mathfrak{Y}$ and so G is an $\mathfrak{E}_{\pi(p)}$-group. In particular, G is an $\mathfrak{F}$-group. This contradicts our choice of G. Hence G is cyclic group of prime order.

Assume that G has not prime power order and there exists a minimal normal subgroup B of G such that G/B is not an $\mathfrak{F}$-group. Then B has to be contained in $\varPhi(G)$ because G is $\mathfrak{F}$-critical. Therefore $G/B \in \mathrm{Crit}_S(\mathfrak{F})$. By induction, G/B is either a Schmidt group or a cyclic group of prime order. If G/B is a cyclic group of prime order, then so is G. This contradiction shows that G/B is a Schmidt group. Let p be the prime dividing $|B|$. Then G is a $\{p, q\}$-group, for some prime $q \neq p$ and either G has a normal Sylow p-subgroup or G has a normal Sylow q-subgroup. Suppose that G has a normal Sylow p-subgroup, P say. Then $G^{\mathfrak{F}}$ is a p- group and so $p \in \pi \cap \mathfrak{Y}$. Since G is not nilpotent because it is $\mathfrak{F}$-critical, then there exists a q-element $g \in G$ such that g does not centralise P. Let us choose g of minimal order. Then every proper subgroup of $N = \langle g \rangle$ centralises P. Consequently PN is a Schmidt group. Suppose that PN is a proper subgroup of G. Then $PN \in \mathfrak{F}$. Hence $PN/\mathrm{O}_{p',p}(PN)$ belongs to $\mathfrak{E}_{\pi(p)}$. It follows that $q \in \mathfrak{E}_{\pi(p)}$, $G \in f(p) = \mathfrak{E}_{\pi(p)}$ and G is an $\mathfrak{F}$-group. This contradiction yields $G = PN$ and G is a Schmidt group. If G has a normal Sylow q-subgroup, a similar argument can be used to conclude that G is a Schmidt group.

Consequently we may assume that $G/B \in \mathfrak{F}$ for every minimal normal subgroup B of G. Then $N = G^{\mathfrak{F}} = \mathrm{Soc}(G)$ is a minimal normal subgroup of G and it is a p-group for some prime $p \in \pi \cap \mathfrak{Y}$. If N is contained in

$\Phi(G)$, then $O_{p',p}(G/N) = O_{p',p}(G)/N = O_p(G)/N$ and so $G/O_p(G) \in \mathfrak{E}_{\pi(p)}$. Hence $G \in f(p) = \mathfrak{E}_{\pi(p)}$ and G is an $\mathfrak{F}$-group. This contradiction implies that $\Phi(G) = 1$ and G is a monolithic primitive group. If $N = G$, then G is a cyclic group of prime order. Thus we may assume that N is a proper subgroup of G. Since $G \notin \mathfrak{F}$, it follows that $G \notin f(p) = \mathfrak{E}_{\pi(p)}$. Hence there exists an element $g \in G$ whose order is a prime $q \notin \pi(p)$. Denote $A = \langle g \rangle$. If NA were a proper subgroup of G, then $NA \in \mathfrak{F}$. Hence $NA/O_{p',p}(NA) \cong A$ belongs to $\mathfrak{E}_{\pi(p)}$. This contradiction yields $G = AN$ and then every maximal subgroup of G is nilpotent. Consequently G is a Schmidt group and the Statement 1 of the theorem is now clear.   □

If $\mathfrak{X} = \mathfrak{J}$, the class of all simple groups, then $\mathfrak{Y}$ is the class of all abelian simple groups. Therefore we have:

Corollary 6.4.5. *Let $\mathfrak{F}$ be a formation. The following statements are equivalent:*

1. *Every $\mathfrak{F}$-critical group is either a Schmidt group or a cyclic group of prime order.*
2. *Every $\mathfrak{F}$-critical group is soluble, $\mathfrak{F}$ is solubly saturated and, for each prime $p \in \operatorname{char} \mathfrak{F}$, there exists a set of primes $\pi(p)$ with $p \in \pi(p)$ such that $\mathfrak{F}$ is $\mathbb{P}$-locally defined by the $\mathbb{P}$-formation function f given by*

$$f(S) := \begin{cases} \mathfrak{E}_{\pi(p)} & \text{if } S \cong C_p,\ p \in \operatorname{char} \mathfrak{F}, \\ \emptyset & \text{if } S \cong C_p,\ p \notin \operatorname{char} \mathfrak{F}, \\ \mathfrak{F} & \text{if } S \in \mathfrak{J} \setminus \mathbb{P}. \end{cases}$$

In particular, every subgroup closed Š-formation is solubly saturated (see [Kam94]). It is clear that every soluble formation is saturated if and only if it is solubly saturated. Hence combining Theorem 6.4.4 and Theorem 6.4.3 we have

Corollary 6.4.6. *Let $\mathfrak{F}$ be a soluble Š-formation. Then $\mathfrak{F}$ is saturated and it is closed under taking products of $\mathfrak{F}$-subnormal subgroups.*

The saturated formation $\mathfrak{S}$ of all soluble groups shows that the converse of the above result does not hold.

There also exist non-saturated Š-formations.

Example 6.4.7 ([BBPR96b]). Let

$$\mathfrak{F} = (G : \text{every } \{3,5\}\text{-subgroup of } G \text{ is nilpotent}).$$

By [DH92, VII, 6.5], $\mathfrak{F}$ is a subgroup-closed formation. Let G be an $\mathfrak{F}$-critical group. Then G has a $\{3,5\}$-subgroup H such that H is not nilpotent. The choice of G implies that $H = G$. Especially, G is a $\{3,5\}$-group which is not nilpotent but all its subgroups are nilpotent. Therefore G is a Schmidt group.

Take $G = \mathrm{Alt}(5)$, the alternating group of degree 5. Then $G \in \mathfrak{F}$. Let E be the maximal Frattini extension of G with 5-elementary abelian kernel. Then $E/\Phi(E)$ is isomorphic to G and $C_G(\Phi(E)) = O_{5'}(G) = 1$ by [GS78, Proposition 5]. If $\mathfrak{F}$ were saturated, it would be true that $E \in \mathfrak{F}$. But this is not true because $\Phi(E)P$, for a Sylow 3-subgroup P of E, is not nilpotent inasmuch as P does not centralise $\Phi(E)$.

The following result provides a criterion for a $\check{\mathrm{S}}$-formation to be saturated.

Theorem 6.4.8 ([BBPR96b]). *Let $\mathfrak{F}$ be a $\check{\mathrm{S}}$-formation. The following statements are equivalent:*

1. *$\mathfrak{F}$ is a saturated formation.*
2. *Let G be a primitive group of type 2 such that $G \in \mathfrak{F}$. If p is a prime dividing $|\mathrm{Soc}(G)|$ and V is an irreducible and faithful G-module over $\mathrm{GF}(p)$, then every Schmidt subgroup isomorphic to $\mathrm{E}(q|p)$ of $[V]G$ belongs to $\mathfrak{F}$.*

Proof. If $\mathfrak{F}$ is a saturated formation, then the statement 2 is always true: Let $G \in \mathfrak{F}$ be a primitive group of type 2; then $G \in F(p)$ for every prime $p \in \pi(\mathrm{Soc}(G))$, where F is the canonical local definition of $\mathfrak{F}$. The semidirect product $[V]G \in \mathfrak{S}_p F(p) = F(p) \subseteq \mathfrak{F}$, for each irreducible and faithful G-module V over $\mathrm{GF}(p)$ and p dividing the order of $\mathrm{Soc}(G)$. Now the result is clear because $\mathfrak{F}$ is subgroup-closed.

To complete the proof we now show that 2 implies 1. By Corollary 6.4.5, $\mathfrak{F}$ is solubly saturated and, for each prime $p \in \mathrm{char}\,\mathfrak{F}$, there exists a set of primes $\pi(p)$ with $p \in \pi(p)$ such that $\mathfrak{F} = \mathrm{LF}_{\mathbb{P}}(f)$, where f is the $\mathbb{P}$-formation function given by

$$
f(S) := \begin{cases} \mathfrak{E}_{\pi(p)} & \text{if } S \cong C_p,\ p \in \mathrm{char}\,\mathfrak{F}, \\ \emptyset & \text{if } S \cong C_p,\ p \notin \mathrm{char}\,\mathfrak{F}, \\ \mathfrak{F} & \text{if } S \in \mathfrak{J} \setminus \mathbb{P}. \end{cases}
$$

Applying Theorem 3.4.5, the formation $\mathfrak{H} = \mathrm{LF}(f)$ is the largest saturated formation contained in $\mathfrak{F}$. Suppose, by way of contradiction, that the class $\mathfrak{F} \setminus \mathfrak{H}$ is non empty, and let G be a group of minimal order in this class. Then G is a primitive group of type 2 and, since $G \notin \mathfrak{H}$, there exists a prime $p \in \pi(\mathrm{Soc}(G)) \subseteq \mathrm{char}\,\mathfrak{F}$ such that $G \notin f(p) = \mathfrak{E}_{\pi(p)}$. Consequently there exists an element $g \in G$ of order q, for some prime $q \notin \pi(p)$. Furthermore, by [DH92, B, 10.9], G has an irreducible and faithful module V over $\mathrm{GF}(p)$. Let $X = [V]G$ be the corresponding semidirect product. Denote $A = \langle g \rangle$ and consider the subgroup VA of X. V, regarded as an A-module, is semisimple by [DH92, B, 4.5]. Moreover, since V is faithful and A is a cyclic group of order q, we can find an irreducible A-submodule W of V such that W is a faithful A-module. Let $B = WA$ be the corresponding semidirect product. It is clear that B is a Schmidt group which is isomorphic to $\mathrm{E}(q|p)$. By Condition 2, $B \in \mathfrak{F}$. It yields $B/C_B(W) \in f(p)$. Hence $q \in \pi(p)$. This contradiction shows that $\mathfrak{H} = \mathfrak{F}$ and $\mathfrak{F}$ is saturated. $\qquad\square$

Remark 6.4.9. Let $\mathfrak{F}$ be a saturated Š-formation. According to Corollary 6.4.5, $\mathfrak{F} = \mathrm{LF}_{\mathbb{P}}(f)$, where f is the $\mathbb{P}$-formation function given by $f(p) = \mathfrak{E}_{\pi(p)}$, $p \in \pi(p)$, if $p \in \mathrm{char}\,\mathfrak{F}$, $f(p) = \emptyset$ if $p \notin \mathrm{char}\,\mathfrak{F}$ and $f(S) = \mathfrak{F}$ if $S \in \mathfrak{J} \setminus \mathbb{P}$. By Theorem 3.1.17, the canonical $\mathbb{P}$-local definition of $\mathfrak{F}$, F say, is given by

$$F(S) = \begin{cases} \mathfrak{F} \cap \mathfrak{E}_{\pi(p)} & \text{if } p \in \mathrm{char}\,\mathfrak{F}, \\ \emptyset & \text{if } p \notin \mathrm{char}\,\mathfrak{F}, \\ \mathfrak{F} & \text{if } S \in \mathfrak{J} \setminus \mathbb{P}. \end{cases}$$

Furthermore, by Corollary 3.1.18, the canonical local definition of $\mathfrak{F}$ is $F(p) = \mathfrak{F} \cap \mathfrak{E}_{\pi(p)}$ if $p \in \mathrm{char}\,\mathfrak{F}$ and $F(p) = \emptyset$ otherwise. Using familiar arguments it can be proved that $\mathfrak{F} = \mathrm{LF}(f)$.

Unfortunately, not every saturated formation which is locally defined as above is a Š-formation.

Example 6.4.10 ([BBPR95]). Consider $\mathfrak{F} = \mathrm{LF}(f)$ which is locally defined by the formation function given by $f(2) = f(3) = \mathfrak{E}_{\{2,3\}}$, $f(5) = \mathfrak{E}_{\{2,5\}}$, and $f(q) = \emptyset$ if $q \neq 2, 3, 5$. Then $\mathfrak{F}$ is subgroup-closed and $\mathrm{char}\,\mathfrak{F} = \{2,3,5\}$; $\mathrm{Alt}(5)$ is $\mathfrak{F}$-critical but it is neither a Schmidt group nor a cyclic group of prime order.

For saturated formations of soluble groups, the following characterisation holds.

Theorem 6.4.11. *Let $\mathfrak{F}$ be a saturated formation of soluble groups. Then every soluble $\mathfrak{F}$-critical group is either a Schmidt group or a cyclic group of prime order if and only if $\mathfrak{F}$ satisfies the following condition: there exists a formation function f, defined by $f(p) = \mathfrak{S}_{\pi(p)}$ for a set of primes $\pi(p)$ such that $p \in \pi(p)$ if $p \in \mathrm{char}\,\mathfrak{F}$ and $f(p) = \emptyset$ otherwise, such that $\mathfrak{F} = \mathrm{LF}(f)$.*

Proof. Assume that every soluble $\mathfrak{F}$-critical group is either a Schmidt group or a cyclic group of prime order. Let $\mathfrak{F}$ be the canonical local definition of $\mathfrak{F} = \mathrm{LF}(F)$. Since $\mathfrak{F}$ is subgroup-closed, it follows that $F(p)$ is subgroup-closed for each $p \in \pi = \mathrm{char}\,\mathfrak{F}$ by Proposition 3.1.40. Hence $F(p)$ is contained in $\mathfrak{F} \cap \mathfrak{S}_{\pi(p)}$, where $\pi(p) = \mathrm{char}\,F(p)$ for every $p \in \pi$. Assume that there exists a prime $p \in \pi$ such that $F(p) \neq \mathfrak{S}_{\pi(p)} \cap \mathfrak{F}$ and let G be a group of minimal order in the non-empty class $(\mathfrak{F} \cap \mathfrak{S}_{\pi(p)}) \setminus F(p)$. Then $1 \neq \mathrm{Soc}(G)$ is the unique minimal normal subgroup of G which is not a p-group. By [DH92, B, 10.9] there exists an irreducible and faithful G-module V over $\mathrm{GF}(p)$. Let $X = [V]G$ be the corresponding semidirect product. It is clear that X is a primitive group and $V = \mathrm{Soc}(X)$ is the unique minimal normal subgroup of X. Since $G \notin F(p)$, we have that X is not an $\mathfrak{F}$-group. Let M be a maximal subgroup of X. If $\mathrm{Core}_X(M) = 1$, then M is isomorphic to G. Hence $M \in \mathfrak{F}$. Assume that $\mathrm{Core}_X(M) \neq 1$. Then $V \leq M$ and $M \cap G$ is a maximal subgroup of G. In this case $M \cap G \in F(p)$ and so $M \in \mathfrak{S}_p F(p) = F(p) \subseteq \mathfrak{F}$. Hence X is an $\mathfrak{F}$-critical soluble group. By hypothesis, X is a Schmidt group (clearly X cannot be a cyclic group of prime order). In particular G is a nilpotent group. Assume that $\mathrm{Soc}(G)$ is a proper subgroup of G. Let A be a maximal subgroup of G containing $\mathrm{Soc}(G)$. Then VA is a maximal subgroup of X and so VA is nilpotent. Let q be the prime dividing $|\mathrm{Soc}(G)|$. Then the Sylow q-subgroup A_q of A is non-trivial and $A_q \leq \mathrm{C}_{VA}(V) = V$. This contradiction yields $\mathrm{Soc}(G) = G$ and hence G is a cyclic group of order $q \in \pi(p) = \mathrm{char}\, F(p)$. This means that $G \in F(p)$, contrary to our supposition. Consequently, for each prime $p \in \pi$, we have that $F(p) = \mathfrak{F} \cap \mathfrak{S}_{\pi(p)}$, where $\pi(p) = \mathrm{char}\, F(p)$.

We are now close to completing the proof of the implication. Let f be the formation function given by $f(p) = \mathfrak{S}_{\pi(p)}$ if $p \in \pi$ and $f(q) = \emptyset$, if $q \neq \pi$ and $F(q) = \emptyset$. It is clear that $\mathfrak{F}$ is contained in $\mathrm{LF}(f)$. Assume that the equality is not true and take a group $G \in \mathrm{LF}(f) \setminus \mathfrak{F}$ of minimal order. Since $\mathrm{LF}(f)$ is composed of soluble groups, it follows that G is a soluble primitive group. Let p be the prime dividing $|Soc(G)|$. Then $G/N \in \mathfrak{S}_{\pi(p)} \cap \mathfrak{F} = F(p)$. Consequently $G \in \mathfrak{S}_p F(p) \subseteq \mathfrak{F}$, and we have reached a contradiction. Hence $\mathfrak{F} = \mathrm{LF}(f)$.

Suppose now that there exists a formation function f, defined by $f(p) = \mathfrak{S}_{\pi(p)}$ for a set of primes $\pi(p)$ such that $p \in \pi(p)$ if $p \in \mathrm{char}\, \mathfrak{F}$ and $f(p) = \emptyset$ otherwise, such that $\mathfrak{F} = \mathrm{LF}(f)$. Let G a soluble $\mathfrak{F}$-critical group. Assume that $\Phi(G) = 1$. Then G is a primitive group. Let p be the prime dividing $\mathrm{Soc}(G)$. If $q \neq p$ is a prime dividing the order of G, and $g \in G$ is an element of G of order q, then g does not centralise N. Denote $A = \langle g \rangle$. If NA were a proper subgroup of G, then $NA \in \mathfrak{F}$. Hence $NA/\mathrm{O}_{p',p}(NA) \cong A$ belongs to $\mathfrak{S}_{\pi(p)}$ and $q \in \pi(p)$. Since G does not belong to $\mathfrak{S}_{\pi(p)}$, it follows that $G = AN$ for some subgroup A of G of prime order. This means that G is a Schmidt group. Hence, in this case, G is either a Schmidt group or a cyclic group of prime order.

Assume that $\Phi(G) \neq 1$. The group $G^* = G/\Phi(G)$ is an $\mathfrak{F}$-critical group and $\Phi(G^*) \neq 1$. The above argument implies that G^* is either a Schmidt group or a cyclic group of prime order. Consequently, G is a Schmidt group and the other implication of the theorem is now clear. $\square$

We now present a set of necessary and sufficient conditions for a saturated formation to be a Š-formation.

Theorem 6.4.12 ([BBPR95]). *Let $\mathfrak{F}$ be a saturated formation. Then $\mathfrak{F}$ is a Š-formation if and only if $\mathfrak{F}$ satisfies the following two conditions:*

1. *There exists a formation function f, defined by $f(p) = \mathfrak{E}_{\pi(p)}$ for a set of primes $\pi(p)$ such that $p \in \pi(p)$ if $p \in \mathrm{char}\, \mathfrak{F}$ and $f(p) = \emptyset$ otherwise, such that $\mathfrak{F} = \mathrm{LF}(f)$; this formation function f satisfies the following property: If $G \in \mathrm{Crit}_S(\mathfrak{F}) \cap \mathrm{b}(\mathfrak{F})$ and G is an almost simple group such that $G \notin f(p)$ for some prime $p \in \pi(\mathrm{Soc}(G))$, then $G \notin f(q)$ for each prime $q \in \pi(\mathrm{Soc}(G))$.* (6.2)
2. *$\mathrm{Crit}_S(\mathfrak{F}) \cap \mathrm{b}(\mathfrak{F})$ does not contain non-abelian simple groups.*

Proof. Denote $\pi := \operatorname{char} \mathfrak{F}$. If $\mathfrak{F}$ is a Š-formation, every group $G \in \operatorname{Crit}_S(\mathfrak{F}) \cap$ b $(\mathfrak{F})$ has abelian socle. Bearing in mind Remark 6.4.9, only the sufficiency of the conditions is in doubt.

Assume that there exists a set of primes $\pi(p)$ with $p \in \pi(p)$, for each $p \in \pi$, such that $\mathfrak{F}$ is locally defined by the formation function f given by $f(p) = \mathfrak{E}_{\pi(p)}$ if $p \in \pi$, and $f(q) = \emptyset$ if $q \notin \pi$. Then $\mathfrak{F} = \operatorname{LF}_{\mathbb{P}}(\hat{f})$, where $\hat{f}$ is the $\mathbb{P}$-formation function defined by $\hat{f}(p) = f(p)$ for all $p \in \mathbb{P}$ and $\hat{f}(S) = \mathfrak{F}$ for all $S \in \mathfrak{J} \setminus \mathbb{P}$ (see Corollary 3.1.13 and Corollary 3.1.18). By Corollary 6.4.5, it will be sufficient to show that every $\mathfrak{F}$-critical group is soluble to conclude that $\mathfrak{F}$ is a Š-formation. Suppose that $\operatorname{Crit}_S(\mathfrak{F}) \setminus \mathfrak{S}$ is not empty and derive a contradiction. Let G be a group of minimal order in $\operatorname{Crit}_S(\mathfrak{F}) \setminus \mathfrak{S}$. Then $\Phi(G) = 1$ and G is a monolithic primitive group in b$(\mathfrak{F})$. Let $N = \operatorname{Soc}(G)$ be the unique minimal normal subgroup of G. If $N = G$, then G is simple. Since this contradicts 2, we must have $N < G$, so that $N \in \mathfrak{F}$. Assume that N is non-abelian. Then $C_G(N) = 1$ and $\pi(N) \subseteq \pi(p)$, for every $p \in \pi(N)$. Now since $G \notin \mathfrak{F}$, there exists a prime $q \in \pi(G)$ such that $q \notin \pi(p)$ for some prime $p \in \pi(N)$; in particular, $q \notin \pi(N)$. Let g be an element of G of order q. Denote $A = \langle g \rangle$. The group A operates by conjugation on N and $(|N|, |A|) = 1$. By [DH92, I, 1.3], there exists an A-invariant Sylow p-subgroup N_p of N. Since G is not soluble, it follows that $N_p A$ is a proper subgroup of G. Hence $N_p A \in \mathfrak{F}$. Since N_p is normal in $N_p A$ and $q \notin \pi(p)$, it follows that $A \leq C_G(N_p)$. On the other hand, $N = N_1 \times \cdots \times N_r$ is a direct product of non-abelian simple groups N_i, $1 \leq i \leq r$, which are pairwise isomorphic. Since $N_p = (N_1)_p \times \cdots \times (N_r)_p$ for some Sylow p-subgroup $(N_i)_p$ of N_i, $1 \leq i \leq r$, and $(N_i)_p \leq N_i \cap N_i^g$, it follows that $N_i = N_i^g$, $1 \leq i \leq r$. Hence A normalises N_i for all $i \in \{1, \ldots, r\}$. Suppose that $N_i A \neq G$ for every $i \in \{1, \ldots, r\}$. Then $N_i A \in \mathfrak{F}$. Consequently $A \leq C_G(N_i)$ for every $i \in \{1, \ldots, r\}$ because $q \notin \pi(p)$. This implies that $A \leq C_G(N) = 1$, which is impossible. We conclude for this contradiction that $G = N_i A$ for some $i \in \{1, \ldots, r\}$ and $N = N_i$ is a non-abelian simple group. In particular, G is an almost simple group. We may apply now Condition 1 and deduce that $G \notin f(r)$ for each $r \in \pi(N)$. But the above argument shows that A centralises a Sylow r-subgroup of N for each $r \in \pi(N)$. Hence $A \leq C_G(N) = 1$, and again we have a contradiction. Therefore N must be abelian. Let p be the prime dividing $|N|$. Then $p \in \pi$ and $G \notin f(p)$ because $G \notin \mathfrak{F}$. Let g be an element of G whose order is a prime $q \notin \pi(p)$. Denote again $A = \langle g \rangle$. If $NA = G$, then G is soluble. This contradiction implies that NA is a proper subgroup of G. But in this case $q \in \pi(p)$ because N is self-centralising in G, contrary to our initial supposition that $q \notin \pi(p)$.

Thus we are forced to the conclusion that every $\mathfrak{F}$-critical group is soluble and $\mathfrak{F}$ is a Š-formation. $\square$

Remark 6.4.13. None of the conditions 1 and 2 can be dispensed with in Theorem 6.4.12 (see [BBPR95, Examples]).

With the help of the preceding theorem we can now give examples of subgroup-closed saturated Š-formations of different nature. The simplest example is the formation $\mathfrak{F}$ of the p-nilpotent groups, p a prime number. It is clear that $\mathfrak{F} = \mathrm{LF}(f)$, where $f(p) = \mathfrak{S}_p$ and $f(q) = \mathfrak{E}$ for every prime $q \neq p$. Hence $\mathfrak{F}$ belongs to the family of saturated formations described in Theorem 6.4.12. Let $G \in \mathrm{Crit}_S(\mathfrak{F}) \cap b(\mathfrak{F})$. Then G is not a p'-group. Thus $p \in \pi(G)$. Since G is not p-nilpotent, we can apply the p-nilpotence criterion of Frobenius [Hup67, IV, 5.8] to conclude that $G = \mathrm{N}_G(P)$ for some p-subgroup $1 \neq P$ of G. Hence $\mathrm{Soc}(G)$ is abelian and $\mathfrak{F}$ satisfies Conditions 1 and 2 of Theorem 6.4.12. Consequently $\mathfrak{F}$ is a Š-formation. This is a classical result due to Itô ([Hup67, IV, 5.4]).

Less trivial is the following result.

Theorem 6.4.14 ([BBPR95]). *Let $\{\pi_i : i \in \mathcal{I}\}$ be a family of pairwise disjoint sets of primes and put $\pi = \bigcup\{\pi_i : i \in \mathcal{I}\}$. Let $\mathfrak{F}$ be the saturated formation locally defined by the formation function f given by $f(p) = \mathfrak{E}_{\pi_i}$ if $p \in \pi_i$, $i \in \mathcal{I}$, and $f(q) = \emptyset$ if $q \notin \pi$. Then $\mathfrak{F}$ is a subgroup-closed saturated Š-formation.*

Proof. By Proposition [DH92, IV, 3.14], $\mathfrak{F}$ is a subgroup-closed saturated formation. It is clear that $\pi = \mathrm{char}\,\mathfrak{F}$. Note that a group G belongs to $\mathfrak{F}$ if and only if G has a normal Hall π_i-subgroup for every $i \in \mathcal{I}$.

We claim that $\mathfrak{F}$ satisfies Conditions 1 and 2 of Theorem 6.4.12. On one hand, the formation function defined above satisfies Condition 1. On the other hand, assume that G is a non-abelian simple group in $\mathrm{Crit}_S(\mathfrak{F}) \cap b(\mathfrak{F})$ and derive a contradiction. Then $2 \in \mathrm{char}\,\mathfrak{F}$, by the Odd Order Theorem [FT63], and so there exists an element $i \in \mathcal{I}$ such that $2 \in \pi_i$. Denote $\pi_1 = \pi \setminus \pi_i$ and $\pi_2 = \pi_i$. If X is a group in $\mathfrak{F}$, we denote by X_1 the normal Hall π_1-subgroup of X. The normal Hall π_2-subgroup of X is denoted by X_2. We reach a contradiction after the following steps:

Step 1. Let M be a maximal subgroup of G such that $M_1 \neq 1$ and $M_2 \neq 1$. Then $\mathrm{Syl}_p(M) \subseteq \mathrm{Syl}_p(G)$, for every prime p dividing the order of M.

Let p be a prime dividing $|M|$ and let $M_p \in \mathrm{Syl}_p(M)$. There exists a Sylow p-subgroup G_p of G such that $M_p \subseteq G_p$. Assume, arguing by contradiction, that M_p is a proper subgroup of G_p. Then M_p is a proper subgroup of $T_p = \mathrm{N}_{G_p}(M_p)$. Suppose that $p \in \pi_2$ (similar arguments can be used if $p \in \pi_1$). In this case, we have that $M_1 \leq \mathrm{N}_G(M_p)$ and so $\langle M_1, T_p \rangle \leq \mathrm{N}_G(M_p)$, which is a proper subgroup of G because G is a non-abelian simple group. Let L be a maximal subgroup of G such that $\mathrm{N}_G(M_p) \leq L$. Then $L = L_1 \times L_2$ because $L \in \mathfrak{F}$. Furthermore, $L_1 \neq 1$ and $L_2 \neq 1$ as $p \in \pi(L)$ and $M_1 \leq L_1$. Hence $\langle M_2, L_2 \rangle \leq \mathrm{N}_G(M_1) = M$ and so M_p is a Sylow p-subgroup of L_2. But then T_p is a Sylow p-subgroup of L_2 containing properly a Sylow p-subgroup of L_2. This contradiction yields $M_p = G_p$ and M_p is a Sylow p-subgroup of G.

Step 2. Let p be a prime in π_1 and let $1 \neq P$ be a p-subgroup of G. Then $\mathrm{N}_G(P)$ is of Glauberman type with respect to the prime p (cf. [Gor80, 4.1, page 281]).

It is clear that $N_G(P)$ is a proper subgroup of G. Hence $N_G(P) \in \mathfrak{F}$ and $N_G(P) = N_G(P)_1 \times N_G(P)_2$. In particular, $N_G(P)$ is p-soluble. Suppose that $p = 3$, then $\mathrm{SL}(2,3)$ is not involved in G because the Sylow 2-subgroup of $\mathrm{SL}(2,3)$ is not centralised by a Sylow 3-subgroup of $\mathrm{SL}(2,3)$. Hence $N_G(P)$ is strongly p-soluble in the sense of [Gor80, page 234]. Moreover, $O_p\big(N_G(P)\big) \neq 1$ and p is an odd prime. Therefore we can apply [Gor80, pages 268–269] to conclude that $N_G(P)$ is p-constrained and p-stable. By [Gor80, Theorem 8.2.11, page 279] we have that $N_G(P)$ is of Glauberman type with respect to the prime p.

Step 3. Let p be a prime in π_1 and let $1 \neq P$ be a Sylow p-subgroup of G. Then $P \leq N'$, where $N = N_G\big(\mathrm{ZJ}(P)\big)$ and $\mathrm{ZJ}(P)$ is the centre of the Thompson subgroup of P.

By Step 2, the normaliser of every nonidentity p-subgroup of G is of Glauberman type with respect to the prime p. Applying [Gor80, Theorem 8.4.3, page 282], we conclude that $P \cap G' = P \cap G = P \cap N'$.

Step 4. Let M be a maximal subgroup of G. Then M is either a π_1-group or a π_2-group.

Since G is $\mathfrak{F}$-critical, we have that every maximal subgroup of G belongs to $\mathfrak{F}$. Assume that the above statement is not true. Then the set

$$\Sigma := \{M : M \text{ is a maximal subgroup of } G, \ M_1 \neq 1, \text{ and } M_2 \neq 1\}$$

is non-empty. We define a binary relation $\mathcal{R}$ in Σ by $M \mathcal{R} L$ if and only if $M_2 \leq L_2$. Clearly $\mathcal{R}$ is reflexive and transitive. Moreover, if $M \mathcal{R} L$ and $L \mathcal{R} M$, then $M_2 = L_2$ and so $M = N_G(M_2) = N_G(L_2) = L$. Hence $(\Sigma, \mathcal{R})$ is a partially ordered set. Let M be a maximal element of $(\Sigma, \mathcal{R})$. Since $M_1 \neq 1$ and M_1 is soluble, by the Feit-Thompson theorem, we have that $(M_1)'$ is a proper subgroup of M_1. Let p be a prime dividing $|M_1 : (M_1)'|$ and let P be a Sylow p-subgroup of M. Then, by Step 1, P is a Sylow p-subgroup of G and moreover $P \leq N'$, where $N = N_G\big(\mathrm{ZJ}(P)\big)$, by Step 3. Clearly N is a proper subgroup of G and $M_2 \leq N$. Let L be a maximal subgroup of G containing N. Then $M_2 \leq L_2$. Moreover $L_1 \neq 1$ because $p \in \pi_1$. Therefore $L \in \Sigma$ and $M \mathcal{R} L$. By the maximality of M, we have that $L = M$. In particular, P is contained in $(M_1)'$ because $N' \leq (M_1)' \times (M_2)'$. Hence $|M_1 : (M_1)'|$ is a p'-number, contrary to our supposition.

Step 5. G has a maximal subgroup of odd order.

If every maximal subgroup of G were of even order, then G would be an $\mathfrak{E}_{\pi_2}$-group by Step 4. This would imply that $G \in \mathfrak{F}$, and we would have a contradiction. Hence we conclude that G has a maximal subgroup of odd order.

Applying [LS91, Theorem 2], we have that G is one of the following groups: $\mathrm{Alt}(p)$, p a prime number, $p \equiv 3 \pmod 4$ and $p \neq 7, 11, 23$; $\mathrm{L}_2(q)$, $q \equiv 3 \pmod 4$, $\mathrm{L}_p^\varepsilon(q)$, $\varepsilon = \pm 1$, p odd prime, and $G \neq \mathrm{U}_3(3)$ or $\mathrm{U}_5(2)$; M_{23}, Th, F_2, or F_1.

In the remaining steps we rule out the above possibilities for the non-abelian simple group G.

Step 6. G is not of the type $\mathrm{Alt}(p)$, p *a prime number,* $p \equiv 3 \pmod 4$ *and* $p \neq 7, 11, 23$.

Suppose that $G = \mathrm{Alt}(p)$ for some prime p, $p \equiv 3 \pmod 4$. It is clear that $\mathrm{Alt}(p-1)$ is a maximal subgroup of G, $\mathrm{Alt}(p-1) \in \mathfrak{F}$ and $\mathrm{Alt}(p-1) \in \mathfrak{E}_{\pi_2}$. Let P be a Sylow p-subgroup of G and let M be a maximal subgroup of G such that $\mathrm{N}_G(P) \leq M$. By Step 4, M is either a π_1-group or a π_2-group. If $M \in \mathfrak{E}_{\pi_2}$, then $G \in \mathfrak{E}_{\pi_2}$ and so $G \in \mathfrak{F}$. This contradiction yields $M \in \mathfrak{E}_{\pi_1}$. Hence $M = P = \mathrm{N}_G(P)$ and we have a contradiction.

Step 7. $G \neq \mathrm{L}_2(q)$, $q \equiv 3 \pmod 4$.

Assume that $G = \mathrm{L}_2(q)$, for some $q \equiv 3 \pmod 4$. Then by [LS91, Theorem 2], if M is a maximal subgroup of odd order, then M is isomorphic to a semidirect product of an elementary abelian group of order q and a cyclic group of order $(q-1)/2$. On the other hand, by the theorem of Dickson [Hup67, II, 8.27], G has a subgroup H which is isomorphic to the dihedral group of order $2\big((q-1)/2\big)$. Then $H \in \mathfrak{E}_{\pi_2}$ by Step 4, and therefore $M \in \mathfrak{E}_{\pi_2}$. It means that $G \in \mathfrak{E}_{\pi_2}$. This contradiction confirms Step 7.

Step 8. $G \neq \mathrm{L}_p^{\varepsilon}(q)$, $\varepsilon \in \{\pm 1\}$, p *odd prime.*

Assume that $G = \mathrm{L}_p^{\varepsilon}(q)$ for some odd prime p, $\varepsilon \in \{\pm 1\}$ and $G \neq \mathrm{U}_3(3)$ or $\mathrm{U}_5(2)$. Again, by [LS91, Theorem 2], if M is a maximal subgroup of G of odd order, then the order of M is $p\big((q^p - \varepsilon)/(q - \varepsilon)(q - \varepsilon, p)\big)$. From Tables 3.5A and 3.5B and the corresponding results of Chapter 4 of [KL90], it follows that there exists a proper subgroup M of G of even order such that $p \in \pi(M)$. Hence $M \in \mathfrak{E}_{\pi_2}$ and then $G \in \mathfrak{E}_{\pi_2}$. This contradiction proves Step 8.

Step 9. G *is not of type* M_{23}, Th, F_2, *or* F_1.

Using the *Atlas* [CCN$^+$85] as reference for the list of maximal subgroups of G, we see that in this case G should be a π_2-group. This final contradiction proves the theorem. $\qquad\square$

We now turn our attention to an application of Theorem 6.4.12 leading to a characterisation of the subgroup-closed saturated Š-formations.

Theorem 6.4.15 ([BBPR96b]). *Let* $\mathfrak{F} = \mathrm{LF}(F)$ *be a subgroup-closed saturated formation. Denote* $\pi = \mathrm{char}\,\mathfrak{F}$ *and* $\pi(p) = \mathrm{char}\,F(p)$, *for every* $p \in \pi$. *Any two of the following statements are equivalent:*

1. $\mathfrak{F}$ *is a Š-formation.*
2. *A* π-group G *belongs to* $\mathfrak{F}$ *if and only if* $\mathrm{N}_G(Q)/\,\mathrm{C}_G(Q)$ *belongs to* $\mathfrak{E}_{\pi(p)}$ *for each* p-subgroup Q *of* G *and each prime* $p \in \pi$.
3. *A* π-group G *belongs to* $\mathfrak{F}$ *if and only if* $\mathrm{N}_G(Q) \in \mathfrak{E}_{\pi \setminus \{p\}}\mathfrak{E}_{\pi(p)}$ *for each non-trivial* p-subgroup Q *of* G *and each prime* $p \in \pi$.

Proof. 1 implies 2. Assume that $\mathfrak{F}$ is a Š-formation. According to Theorem 6.4.12, we have that $F(p) = \mathfrak{E}_{\pi(p)} \cap \mathfrak{F}$, for every $p \in \pi$. Let G be a π-group in $\mathfrak{F}$. Suppose that a prime $p \in \pi$ is fixed and let Q be a p-subgroup of G. Then $\mathrm{N}_G(Q) \in \mathfrak{F}$ because $\mathfrak{F}$ is subgroup-closed. In particular $\mathrm{N}_G(Q)/\,\mathrm{O}_{p'}\big(\mathrm{N}_G(Q)\big) \in F(p) \subseteq \mathfrak{E}_{\pi(p)}$. Since Q is a normal p-subgroup of $\mathrm{N}_G(Q)$, it follows that $\mathrm{O}_{p'}\big(\mathrm{N}_G(Q)\big) \leq \mathrm{C}_G(Q)$. This means that

$N_G(Q)/C_G(Q) \in \mathfrak{E}_{\pi(p)}$. Conversely, assume that G is a π-group such that $N_G(Q)/C_G(Q)$ belongs to $\mathfrak{E}_{\pi(p)}$ for each p-subgroup Q of G and each $p \in \pi$, but G is not an $\mathfrak{F}$-group. If we choose G of minimal order among the groups $X \notin \mathfrak{F}$ satisfying the above property, we have that G is an $\mathfrak{F}$-critical group because this property holds in every subgroup of G. Since $\mathfrak{F}$ is a Š-formation, it follows that G is a Schmidt group. In particular $\pi(G) = \{p, q\}$ for two distinct primes p and q in π and G has a normal Sylow p-subgroup, P say. By hypothesis, we have that $G/C_G(P) \in \mathfrak{E}_{\pi(p)}$. If q were not in $\pi(p)$, it would be true that $Q \leq C_G(P)$. This is not possible. Hence $q \in \pi(p)$ and then $Q \in \mathfrak{E}_{\pi(p)} \cap \mathfrak{F} = F(p)$. Therefore $G \in \mathfrak{S}_p F(p) = F(p) \subseteq \mathfrak{F}$. This contradiction yields $G \in \mathfrak{F}$.

2 implies 1. We see that $\mathfrak{F}$ satisfies the Statements 1 and 2 of Theorem 6.4.12.

(a) *For each prime $p \in \pi$, we have that $F(p) = \mathfrak{E}_{\pi(p)} \cap \mathfrak{F}$.*

Let p be a prime in π. Since $F(p)$ is subgroup-closed by Proposition 3.1.40, it follows that $F(p)$ is contained in $\mathfrak{E}_{\pi(p)} \cap \mathfrak{F}$. Assume, by way of contradiction, that $F(p) \neq \mathfrak{E}_{\pi(p)} \cap \mathfrak{F}$ and let G be a group of minimal order in the non-empty class $(\mathfrak{E}_{\pi(p)} \cap \mathfrak{F}) \setminus F(p)$. Then $1 \neq \operatorname{Soc}(G)$ is the unique minimal normal subgroup of G and it is not a p-group. By [DH92, B, 10.9], there exists an irreducible and faithful G-module over $\operatorname{GF}(p)$. Let $X = [V]G$ be the corresponding semidirect product. X is a primitive group and $X \notin \mathfrak{F}$ because $G \notin F(p)$. Let q be a prime in π and let Q be a non-trivial q-subgroup of G. Suppose that $p \neq q$. Then $N_X(Q)$ is a proper subgroup of G because V is the unique minimal normal subgroup of X. Let L be a maximal subgroup of G containing $N_X(Q)$. If $\operatorname{Core}_X(L) = 1$, then L is isomorphic to G and if $\operatorname{Core}_G(L) \neq 1$, then $V \leq L$ and $L = V(G \cap L)$. Since $G \cap L \in F(p)$, it follows that $L \in \mathfrak{S}_p F(p) = F(p) \subseteq \mathfrak{F}$. In both cases, $L \in \mathfrak{F}$. Therefore $N_X(Q) \in \mathfrak{F}$ and so $N_X(Q)/C_X(Q) \in \mathfrak{E}_{\pi(q)}$. Now, if $p = q$ and $N_X(Q)$ is a proper subgroup of G, we can argue as above to conclude that $N_X(Q)/C_X(Q) \in \mathfrak{E}_{\pi(p)}$. If Q is a normal subgroup of X, then $V = Q$ and $X/C_X(Q) \in \mathfrak{E}_{\pi(p)}$ because it is isomorphic to G. Since X is a π-group, we can apply Statement 2 to conclude that $X \in \mathfrak{F}$. This contradiction shows that $F(p) = \mathfrak{E}_{\pi(p)} \cap \mathfrak{F}$.

(b) $\operatorname{Crit}_S(\mathfrak{F}) \cap b(\mathfrak{F})$ *does not contain primitive groups of type 2.*

Assume that G is a primitive group of type 2 in $\operatorname{Crit}_S(\mathfrak{F}) \cap b(\mathfrak{F})$. Since G is $\mathfrak{F}$-critical, it follows that G is a π-group. On the other hand, applying Statement 2, we can determine a prime $p \in \pi$ and a p-subgroup Q of G such that $N_G(Q)/C_G(Q) \notin \mathfrak{E}_{\pi(p)}$. Then Q is non-trivial. Suppose that $N_G(Q)$ is a proper subgroup of G. Then $N_G(Q) \in \mathfrak{F} \subseteq \mathfrak{E}_{\pi \setminus \{p\}} \mathfrak{E}_{\pi(p)}$. This means that $N_G(Q)/C_G(Q) \in \mathfrak{E}_{\pi(p)}$, contrary to our supposition. Hence Q is a normal subgroup of G. But then $\operatorname{Soc}(G) \leq Q$ and $\operatorname{Soc}(G)$ is abelian. This contradiction confirms Statement b.

From Statements a and b we deduce that $\mathfrak{F}$ enjoys the properties given in Theorem 6.4.12. This means that $\mathfrak{F}$ is a Š-formation.

Assume now that G is a π-group. Let p be a prime in π and let Q be a non-trivial p-subgroup of G. If $N_G(Q) \in \mathfrak{E}_{\pi \setminus \{p\}} \mathfrak{E}_{\pi(p)}$, then $N_G(Q)/C_G(Q) \in \mathfrak{E}_{\pi(p)}$.

This elementary remark proves that 2 implies 3. Now, if Statement 3 holds, we can repeat the arguments used in the proof of 2 implies 1 to conclude that $\mathfrak{F}$ is a Š-formation. Consequently 3 implies 1 and the circle of implications is complete. □

Illustration 6.4.16. Let $\mathfrak{F}$ be the saturated formation of p-nilpotent groups, p a prime number. It is clear that $\mathfrak{F} = \mathrm{LF}(F)$, where $F(p) = \mathfrak{S}_p$ and $F(q) = \mathfrak{F}$ for every prime $q \neq p$. We have seen above that $\mathfrak{F}$ is a Š-formation. Therefore $\mathfrak{F}$ satisfies Condition 2 of Theorem 6.4.15. Hence a group G is p-nilpotent if and only if $N_G(Q)/C_G(Q)$ is a p-group for every p-subgroup Q of G. The statement 3 of this theorem says that a group G is p-nilpotent if and only if $N_G(Q)$ is p-nilpotent for every p-subgroup Q of G. These statements are two equivalent forms of the well known p-nilpotence criterion due to Frobenius.

The next topic we broach concerns the relation between the $\mathfrak{F}$-residual of a group and the subgroup generated by the $\mathfrak{F}$-residuals of some of its $\mathfrak{F}$-critical subgroups. The springboard for these results was a theorem of Berkovich [Ber99] stating that the nilpotent residual of a group G is the subgroup generated by the nilpotent residuals of the subgroups A of G such that $A/\Phi(A)$ is a Schmidt group.

Berkovich's result is a particular case of a more general theorem as we shall see below.

Let $\mathfrak{F}$ be a formation. Denote by $\mathfrak{B}_{\mathfrak{F}}$ the class of all groups G such that $G/\Phi(G)$ is an $\mathfrak{F}$-critical group. Note that if $\mathfrak{F} = \mathfrak{N}$, the class of all nilpotent groups, $\mathfrak{B}_{\mathfrak{N}}$ is the class of all groups such that $G/\Phi(G)$ is a Schmidt group (see [Ber99]).

Let G be a group and let $\mathrm{T}(G) = \langle A^{\mathfrak{F}} : A \leq G; A \in \mathfrak{B}_{\mathfrak{F}} \rangle$ if $\mathfrak{B}_{\mathfrak{F}} \cap \mathrm{s}(G) \neq \emptyset$; otherwise, we let $\mathrm{T}(G) = 1$.

Theorem 6.4.17 ([ABB02]). *Let $\mathfrak{F}$ be a saturated formation, and let G be a group. Then $\mathrm{T}(G) = G^{\mathfrak{F}}$.*

Proof. Clearly $X^{\mathfrak{F}} \leq G^{\mathfrak{F}}$ for every subgroup X of G because $\mathfrak{F}$ is subgroup-closed. Hence $T = \mathrm{T}(G)$ is contained in $G^{\mathfrak{F}}$.

Assume, arguing by contradiction, that $G/T \notin \mathfrak{F}$. Then G/T has an $\mathfrak{F}$-critical subgroup, A/T say. Choose now a minimal supplement A_0 of T in A. Then $A_0 \cap T$ is contained in $\Phi(A_0)$. Since A/T is isomorphic to $A_0/A_0 \cap T$, it follows that $A_0/A_0 \cap T$ is $\mathfrak{F}$-critical. Therefore the factor group $(A_0/A_0 \cap T)/\Phi(A_0/A_0 \cap T)$ is also $\mathfrak{F}$-critical because $\mathfrak{F}$ is saturated. It means that $A_0 \in \mathfrak{B}_{\mathfrak{F}}$ and so $A_0^{\mathfrak{F}}$ is contained in T. Hence $A_0^{\mathfrak{F}} \leq A_0 \cap T \leq \Phi(A_0)$. It follows that $A_0/\Phi(A_0) \in \mathfrak{F}$. Now since $\mathfrak{F}$ is saturated, we conclude that $A_0 \in \mathfrak{F}$. This contradiction completes the proof. □

We continue the section with an application of Theorem 6.4.17 leading to a characterisation of the Š-formations in the soluble universe among the subgroup-closed saturated formations. It rests on the following result.

Theorem 6.4.18 ([ABB02]). *Let $\mathfrak{F}$ be a saturated formation of soluble groups of full characteristic such that every soluble group in $\mathrm{Crit_S}(\mathfrak{F})$ is a Schmidt group. If A is a group in $\mathfrak{B}_{\mathfrak{F}}$, then $A^{\mathfrak{N}} = A^{\mathfrak{F}}$.*

Proof. We shall argue by induction on $|A|$. Firstly, if $\Phi(A) = 1$, then A is an $\mathfrak{F}$-critical group. Assume that A is not soluble. Then A is a non-abelian simple group. In this case $A^{\mathfrak{F}} = A = A^{\mathfrak{N}}$. Thus we may suppose that A is soluble and then A is a Schmidt group. Hence there exists a normal abelian Sylow p-subgroup of A, P say, for some prime p. It is rather clear that P coincides with both the nilpotent residual and the $\mathfrak{F}$-residual of A. Hence we can assume that $\Phi(A) \neq 1$. Let N be a minimal normal subgroup of A contained in $\Phi(A)$. Then $A/N \in \mathfrak{B}_{\mathfrak{F}}$. Hence the induction hypothesis implies that $(A/N)^{\mathfrak{F}} = (A/N)^{\mathfrak{N}}$. This yields $A^{\mathfrak{F}} N = A^{\mathfrak{N}} N$.

If $A^{\mathfrak{N}} \cap N = 1$, then $A^{\mathfrak{N}} = A^{\mathfrak{N}} \cap A^{\mathfrak{F}} N = A^{\mathfrak{F}}(A^{\mathfrak{N}} \cap N) = A^{\mathfrak{F}}$. Thus we can suppose that N is contained in $A^{\mathfrak{N}}$ and $A^{\mathfrak{N}} = A^{\mathfrak{F}} N$. If N is contained in $A^{\mathfrak{F}}$, then $A^{\mathfrak{N}} = A^{\mathfrak{F}}$ and the theorem is true. Consequently we shall assume that $N \cap A^{\mathfrak{F}} = 1$, and hence $\Phi(A) \cap A^{\mathfrak{F}} = 1$. Note that we can suppose that $A/\Phi(A)$ is an extension of a p-group by a q-group for some primes p and q. Since this class is a saturated formation, we have that A is also an extension of a p-group by a q-group. Therefore $A^{\mathfrak{N}}$ is a p-group, and consequently N is a p-group, too.

Let us have a look now at the structure of the $\mathfrak{F}$-group $A/A^{\mathfrak{F}}$. Given a subgroup H of A, denote by $\bar{H}$ the corresponding subgroup $HA^{\mathfrak{F}}/A^{\mathfrak{F}}$ of $A/A^{\mathfrak{F}} = \bar{A}$. By Theorem 6.4.11, we have that the class $\mathfrak{F}$ is locally defined by a formation function f given by $f(r) = \mathfrak{S}_{\pi(r)}$, where $\pi(r)$ is a set of primes such that $r \in \pi(r)$, for all primes r. Now note that $\bar{N}$ is a minimal normal subgroup of $\bar{A}$. Therefore, $\bar{A}/\mathrm{C}_{\bar{A}}(\bar{N}) \in \mathfrak{S}_{\pi(p)}$. We can conclude that $A/\mathrm{C}_A(N) \cong \bar{A}/\mathrm{C}_{\bar{A}}(\bar{N}) \in \mathfrak{S}_{\pi(p)}$. If $q \in \pi(p)$, then $A \in \mathfrak{S}_{\pi(p)} = f(p)$ and so $A \in \mathfrak{F}$, against the supposition that $A/\Phi(A)$ is $\mathfrak{F}$-critical. Therefore $q \notin \pi(p)$ and we have that $A/\mathrm{C}_A(N)$ is a p-group. Since the normal Sylow p-subgroup of A centralises N, it follows that N is central in G. On the other hand, $A/\Phi(A)$ is an $\mathfrak{F}$-critical group with trivial Frattini subgroup. Since $A^{\mathfrak{F}} \cap \Phi(A) = 1$ and $A^{\mathfrak{F}}\Phi(A)/\Phi(A) = \big(A/\Phi(A)\big)^{\mathfrak{F}}$, it follows that $A^{\mathfrak{F}}$ is abelian. But the equality $A^{\mathfrak{N}} = A^{\mathfrak{F}} \times N$ yields that $A^{\mathfrak{N}}$ is complemented in A by a Carter subgroup of A by Theorem 4.2.17. We can conclude that there exists a Carter subgroup C of A such that $A = A^{\mathfrak{N}}C$ and $A^{\mathfrak{N}} \cap C = 1$. Now N is central in A. Hence $N \leq \mathrm{N}_G(C) = C$. Consequently $N \leq A^{\mathfrak{N}} \cap C = 1$, contrary to supposition. This final contradiction proves the result. $\square$

Theorem 6.4.19 ([ABB02]). *Let $\mathfrak{F}$ be a saturated Fitting formation of soluble groups of full characteristic. The following statements are equivalent:*

1. *Every soluble group in $\mathrm{Crit_S}(\mathfrak{F})$ is a Schmidt group.*
2. *$G^{\mathfrak{F}} = \langle A^{\mathfrak{N}} : A \leq G; A \in \mathfrak{B}_{\mathfrak{F}} \rangle$ for every group G.*

Proof. By Theorem 6.4.17, we have that $G^{\mathfrak{F}} = \langle A^{\mathfrak{F}} : A \leq G, A \in \mathfrak{B}_{\mathfrak{F}} \rangle$. Hence if every soluble group in $\mathrm{Crit_S}(\mathfrak{F})$ is a Schmidt group, we apply Theorem 6.4.18

to conclude that $G^{\mathfrak{F}} = \langle A^{\mathfrak{N}} : A \leq G; A \in \mathfrak{B}_{\mathfrak{F}} \rangle$ for every group G. Therefore 1 implies 2. Therefore, only the sufficiency of the Condition 2 is in doubt. To prove that every soluble group in $\text{Crit}_{\text{S}}(\mathfrak{F})$ is a Schmidt group, we shall use Theorem 6.4.11. Write $\mathfrak{F} = \text{LF}(F)$, where F denotes the canonical local defin- ition of $\mathfrak{F}$. Consider any prime p. We prove that $F(p) = \mathfrak{S}_{\pi(p)} \cap \mathfrak{F}$, where $\pi(p) = \text{char } F(p)$. Since $\mathfrak{F}$ is a subgroup-closed Fitting formation, we have that $F(p)$ is subgroup-closed Fitting formation by Proposition 3.1.40. Since F is integrated, we have that $F(p) \subseteq \mathfrak{S}_{\pi(p)} \cap \mathfrak{F}$. Assume that $\mathfrak{S}_{\pi(p)} \cap \mathfrak{F} \neq F(p)$ and take a group G in $(\mathfrak{S}_{\pi(p)} \cap \mathfrak{F}) \setminus F(p)$ of minimal order. By familiar reas- oning, $1 \neq \text{Soc}(G)$ is the unique minimal normal subgroup of G. Moreover, $\text{Soc}(G)$ cannot be a p-group since, being F full, it holds that $F(p) = \mathfrak{S}_p F(p)$. Note that, in fact, $O_p(G) = 1$. By [DH92, B, 10.9], there exists an irreducible and faithful G-module V over $\text{GF}(p)$. Consider now the corresponding semi- direct product $X = [V]G$. Note that if $X \in \mathfrak{F}$, then $X/\text{C}_X(V) \in F(p)$ and thus $X/V \in F(p)$. This is impossible because $G \cong X/V$. Therefore $X \notin \mathfrak{F}$ and X is in fact an $\mathfrak{F}$-critical group.

We are ready at this point to reach our final contradiction. Since $X^{\mathfrak{F}} = \langle A^{\mathfrak{N}} : A \leq X; A \in \mathfrak{B}_{\mathfrak{F}} \rangle$, and $X \in \mathfrak{B}_{\mathfrak{F}}$, we have that $X^{\mathfrak{F}} = X^{\mathfrak{N}} = V$. Therefore $G \cong X/V = X/X^{\mathfrak{N}}$ is nilpotent. Then G is a q-group for some prime $q \in \text{char } F(p)$. Since $F(p)$ is a Fitting class of soluble groups, it follows that $\mathfrak{S}_q$ is contained in $F(p)$ by [DH92, IX, 1.9] and then $G \in F(p)$. This contradiction yields $F(p) = \mathfrak{S}_{\pi(p)} \cap \mathfrak{F}$. It follows then that $\mathfrak{F} = \text{LF}(f)$, where f is the formation function defined by $f(p) = \mathfrak{S}_{\pi(p)}$ if $p \in \text{char } \mathfrak{F}$ and $f(p) = \emptyset$ otherwise. Applying Theorem 6.4.11, every soluble group in $\text{Crit}_{\text{S}}(\mathfrak{F})$ is a Schmidt group. □

Remark 6.4.20. The formation in the above theorem is not a Š-formation in general (see Example 6.4.10).

We close our extended treatment of Š-formations with a survey describing another context where this family of saturated formations appears.

In [Keg65] O. H. Kegel introduced the notion of a triple factorisation. This is a factorisation of a group G involving three subgroups A, B, and C of the type $G = AB = AC = BC$. The evidence is that the existence of a triple factorisation can have greater consequences for the group structure than does a single factorisation. For example, Kegel shows that a group which has a triple factorisation by nilpotent groups is nilpotent. Consequently, it seems natural to wonder which are the saturated formations $\mathfrak{F}$ which are closed under taking triple factorisations. The first contribution to the solution of this problem was made by Vasil'ev [Vas87, Vas92] in the soluble universe. The following three results are proved in that universe.

Theorem 6.4.21 (Vasil'ev). *Let $\mathfrak{F}$ be an s_n-closed saturated formation. Then the following statements are equivalent:*

1. $\mathfrak{F}$ *is a Š-formation.*
2. *(Kegel's property)* $\mathfrak{F}$ *contains each group* $G = AB = AC = BC$ *where* A, B, $C \in \mathfrak{F}$.
3. $\mathfrak{F}$ *contains each group having three pairwise non-conjugate maximal subgroups belonging to* $\mathfrak{F}$.

The above result has been improved in [BBPAMP00].

Theorem 6.4.22. *Let* $\mathfrak{F}$ *be an* s_n-*closed saturated formation of full characteristic. The following statements are equivalent.*

1. $\mathfrak{F}$ *is a Š-formation.*
2. $\mathfrak{F}$ *satisfies the property:*
   If G *is a group of the form* $G = AB = AC = BC$, *where* A *and* B *are* $\mathfrak{F}$-*subgroups of* G *and* C *is an* $\mathfrak{F}$-*subnormal* $\mathfrak{F}$-*subgroup of* G, *then* G *is an* $\mathfrak{F}$-*group.*

In the study of factorised groups, the case of a triply factorised group $G = AB = AC = BC$ where C is a normal subgroup of G is of particular interest. For instance, the factoriser of a normal subgroup of a factorised group always has this form. Hence the following characterisation of the above formations is also of interest ([BBPAMP00]).

Theorem 6.4.23. *Let* $\mathfrak{F}$ *be an* s_n-*closed saturated formation of full characteristic. The following statements are equivalent:*

1. $\mathfrak{F}$ *is a Š-formation.*
2. $\mathfrak{F}$ *satisfies the property:*
   If G *is a group of the form* $G = AB = AC = BC$, *where* A *and* B *are* $\mathfrak{F}$-*subgroups of* G *and* C *is a normal subgroup of* G, *then* $G^{\mathfrak{F}} = C^{\mathfrak{F}}$.

Bearing in mind the above results, a natural question arises:

Let $\mathfrak{F}$ *be a Fitting formation, non-necessarily subgroup-closed, with the Kegel property. Is* $\mathfrak{F}$ *saturated?*

This question, proposed by Vasil'ev in the *Kourovka Notebook* [MK99] for formations of soluble groups, was partially answered in [BBE05].

Theorem 6.4.24. *Let* $\mathfrak{F}$ *be a Fitting formation with the following property:*

for every prime $p \in$ char $\mathfrak{F}$, *whenever* G *is a primitive* $\mathfrak{F}$-*group whose socle is a* p-*group, all groups* $\mathrm{E}(q|p)$ *are in* $\mathfrak{F}$ *for all primes* $q \neq p$ *such that* q *divides* $|G/\operatorname{Soc}(G)|$. $\qquad (6.3)$

Then $\mathfrak{F}$ *satisfies the Kegel property if and only if* $\mathfrak{F}$ *is a subgroup-closed Š-formation.*

Note that if $\mathfrak{F}$ is saturated, then $\mathfrak{F}$ satisfies (6.3).

Let $\mathfrak{F}$ be a Fitting formation. If for some primes p, q, the group $\mathrm{E}(q|p) \in \mathfrak{F}$, then $\mathfrak{S}_p(C_q) \subseteq \mathfrak{F}$ by [DH92, XI, 2.5]. Since $\mathfrak{S}_p(C_q) \subseteq \mathfrak{F} \cap \mathfrak{N}^2$ and $\mathfrak{F} \cap \mathfrak{N}^2$ is a Fitting formation of metanilpotent groups, it follows that $\mathfrak{S}_p\mathfrak{S}_q \subseteq \mathfrak{F} \cap \mathfrak{N}^2 \subseteq \mathfrak{F}$ by [DH92, XI, 2.4]. Hence $\mathrm{E}(q|p) \in \mathfrak{F}$ if and only if $\mathfrak{S}_p\mathfrak{S}_q \subseteq \mathfrak{F}$.

6.5 Wielandt operators

One of the significant properties of subnormal subgroups is that the nilpotent residual of the subgroup generated by two subnormal subgroups of a group is the subgroup generated by the nilpotent residuals of the subgroups. This is a consequence of an elegant theory of operators created by H. Wielandt for proving results on permutability of subnormal subgroups.

For a group G and the lattice $s_n(G)$ of all subnormal subgroups of G, a map $\omega\colon s_n(G) \longrightarrow s_n(G)$ is called a *Wielandt operator* in G if, for any H, $K \in s_n(G)$, the following conditions are satisfied:

$$\langle H, K \rangle^\omega = \langle H^\omega, K^\omega \rangle, \tag{6.4}$$

$$\text{if } H \trianglelefteq K, \text{ then } H^\omega \trianglelefteq K. \tag{6.5}$$

Here, of course, H^ω denotes the image of H under the map ω. Note that Condition 6.5 implies that H^ω is a normal subgroup of H.

The importance of the theory of operators is suggested by the following result of H. Wielandt.

Theorem 6.5.1 ([Wie57]). *Let φ and ψ be two Wielandt operators in a group G. Assume that two subnormal subgroups H and K of G are permutable if $H = H^\varphi = H^\psi$. Then $A^\varphi B^\psi = B^\psi A^\varphi$ for any pair (A, B) of subnormal subgroups of G.*

It is a consequence of the above result that each new operator leads to the discovering of a new case of permutability of subnormal subgroups and gives new insights on the construction of subnormal subgroup generation. Wielandt's theory of operators is clearly of interest in relation to the theory of classes of groups and may repay further study.

Suppose that a Wielandt operator ω is defined in all groups G. If ω satisfies $(X^\omega)^\alpha = (X^\alpha)^\omega$ for any homomorphism α of a group X, then the class $\mathfrak{F} := (X : X^\omega = 1)$ is a Fitting formation and G^ω is the $\mathfrak{F}$-residual of G for every group G. Conversely if $\mathfrak{F}$ is a Fitting formation, then the map $\delta\colon s_n(G) \longrightarrow s_n(G)$, $H^\delta = H^\mathfrak{F}$ for all $H \in s_n(G)$, defines a Wielandt operator in every group G, permuting with all homomorphisms provided that δ satisfies Condition 6.4.

Consequently, the problem of finding Wielandt operators which are permutable with homomorphisms is reduced to the description of Fitting formations $\mathfrak{F}$ satisfying the following property:

If U and V are subnormal subgroups of a group G, then $\langle U, V \rangle^\mathfrak{F} = \langle U^\mathfrak{F}, V^\mathfrak{F} \rangle$. (6.6)

Let us state this property in a formal definition.

Definition 6.5.2. *Let $\mathfrak{F}$ be a formation. We say that $\mathfrak{F}$ satisfies the Wielandt property for residuals if whenever U and V are subnormal subgroups of $\langle U, V \rangle$ in a group G, then $\langle U, V \rangle^\mathfrak{F} = \langle U^\mathfrak{F}, V^\mathfrak{F} \rangle$.*

The formations appearing in this section are not subgroup-closed in general.

Not all formations have the Wielandt property for residuals. For instance, let $\mathfrak{F}$ be the saturated formation composed of all groups with no epimorphic image isomorphic to $\mathrm{Alt}(5)$. Then if G is the symmetric group of degree 5, it follows that $G^{\mathfrak{F}} = 1 \neq \langle \mathrm{Alt}(5), 1 \rangle^{\mathfrak{F}} = \mathrm{Alt}(5)$. In fact, we have:

Proposition 6.5.3. *Let $\mathfrak{F}$ be a formation. If $\mathfrak{F}$ satisfies the Wielandt property for residuals, then $\mathfrak{F}$ is a Fitting formation.*

Proof. Let G be a group in $\mathfrak{F}$, and N a subnormal subgroup of G. Then $N^{\mathfrak{F}} \leq \langle N^{\mathfrak{F}}, G^{\mathfrak{F}} \rangle = \langle N, G \rangle^{\mathfrak{F}} = G^{\mathfrak{F}} = 1$. Hence $N \in \mathfrak{F}$ and $\mathfrak{F}$ is s_n-closed.

Suppose that $G = N_1 N_2$ for normal subgroups N_1 and N_2 such that $N_i \in \mathfrak{F}$, $i = 1, 2$. Then $G^{\mathfrak{F}} = N_1^{\mathfrak{F}} N_2^{\mathfrak{F}} = 1$. This means that $G \in \mathfrak{F}$ and $\mathfrak{F}$ is N_0-closed.

Consequently, $\mathfrak{F}$ is a Fitting class. $\square$

The validity of the converse is not known at the time of writing and seems to be quite difficult.

Our aim in the first part of this section is to show that many of the known Fitting formations have the Wielandt property for residuals.

The procedure we describe here is based on the papers [KS95] and [BBCE01].

The basic strategy is the following: first we prove a reduction theorem for a minimal counterexample. This allows us to reduce the problem in many cases to considering a restricted class of groups in the boundary of the formation. As an application, we deduce that many known Fitting formations have the Wielandt property for residuals.

The main obstacle in giving the complete answer for the problem is in understanding the restriction of an irreducible module to a subnormal subgroup. Although a certain amount of information can be derived from repeated application of the Clifford theorems, the closed relation between the components of the restriction is lost. In particular, for a subnormal subgroup, it is difficult to find the relationship between the kernels of the action of the subnormal subgroup on each component of the restriction.

We begin by describing two ways to obtain new formations with the Wielandt property from some old ones.

Proposition 6.5.4. *If $\mathfrak{F}_1$, $\mathfrak{F}_2$, and $\mathfrak{F}_i$, $i \in \mathcal{I}$, are formations satisfying the Wielandt property for residuals, then*

1. the formation $\mathfrak{F}_1 \circ \mathfrak{F}_2$ satisfies the Wielandt property, and
2. the formation $\bigcap_{i \in \mathcal{I}} \mathfrak{F}_i$ satisfies the Wielandt property.

Proof. 1. We have that $X^{\mathfrak{F}_1 \circ \mathfrak{F}_2} = (X^{\mathfrak{F}_2})^{\mathfrak{F}_1}$ by Proposition 2.2.11 (4) for any group X. Let G be a group and U and V subgroups of G such that U and V are subnormal subgroups of $H = \langle U, V \rangle$. Then $H^{\mathfrak{F}_1 \circ \mathfrak{F}_2} = (H^{\mathfrak{F}_2})^{\mathfrak{F}_1} = \langle U^{\mathfrak{F}_2}, V^{\mathfrak{F}_2} \rangle^{\mathfrak{F}_1} = \langle U^{\mathfrak{F}_1 \circ \mathfrak{F}_2}, V^{\mathfrak{F}_1 \circ \mathfrak{F}_2} \rangle$.

2. We have that $X^{\cap_{i\in\mathcal{I}}\mathfrak{F}_i} = \prod_{i\in\mathcal{I}} X^{\mathfrak{F}_i}$ for any group X, where in the product only a finite set of residuals appear since X is finite. Consider a group G and U and V subgroups of G such that U and V are subnormal subgroups of $H = \langle U, V \rangle$. Then $H^{\cap_{i\in\mathcal{I}}\mathfrak{F}_i} = \prod_{i\in\mathcal{I}} H^{\mathfrak{F}_i} = \prod_{i\in\mathcal{I}}\langle U^{\mathfrak{F}_i}, V^{\mathfrak{F}_i}\rangle = \langle \prod_{i\in\mathcal{I}} U^{\mathfrak{F}_i}, \prod_{i\in\mathcal{I}} V^{\mathfrak{F}_i}\rangle = \langle U^{\cap_{i\in\mathcal{I}}\mathfrak{F}_i}, V^{\cap_{i\in\mathcal{I}}\mathfrak{F}_i}\rangle$.   □

Note that if $\mathfrak{F}$ is a Fitting formation, then $U^{\mathfrak{F}}$ is contained in $G^{\mathfrak{F}}$ for every subnormal subgroup U of G. Therefore it is always true that $\langle U^{\mathfrak{F}}, V^{\mathfrak{F}}\rangle$ is contained in $\langle U, V\rangle^{\mathfrak{F}}$ provided that U and V are subnormal in $\langle U, V\rangle$. If U and V permute, the equality holds as the next result shows.

Proposition 6.5.5. *Let $\mathfrak{F}$ be a Fitting formation. If U and V are subgroups of G such that $UV = VU$ and U and V are subnormal in UV, then $(UV)^{\mathfrak{F}} = U^{\mathfrak{F}}V^{\mathfrak{F}}$.*

Proof. Assume that the result is false and let G be a counterexample of least order. Let U and V be subnormal subgroups of $UV = VU$ such that $|U| + |V|$ is maximal doing false the result. Clearly U and V are proper subgroups of G and $G = UV$. Let N be a proper normal subgroup of G such that $U \le N$. Then $N = U(V \cap N)$. The minimality of G yields $N^{\mathfrak{F}} = U^{\mathfrak{F}}(V \cap N)^{\mathfrak{F}}$. If U is a proper subgroup of N, then $G^{\mathfrak{F}} = N^{\mathfrak{F}}V^{\mathfrak{F}}$ by the maximality of the pair (U, V). Hence $G^{\mathfrak{F}} = U^{\mathfrak{F}}V^{\mathfrak{F}}$. This contradiction shows that U and V are maximal normal subgroups of G. Thus $U^{\mathfrak{F}}$ and $V^{\mathfrak{F}}$ are normal in G. Assume that one of them, $U^{\mathfrak{F}}$ say, is not trivial, and let N be a minimal normal subgroup of G such that $N \le U^{\mathfrak{F}}$. It follows that G/N is a group generated by the subnormal subgroups UN/N and VN/N of G/N. Then, by minimality of G, we have that $G^{\mathfrak{F}} = U^{\mathfrak{F}}(V^{\mathfrak{F}}N) = U^{\mathfrak{F}}V^{\mathfrak{F}}$, contrary to our initial supposition. Hence $U^{\mathfrak{F}} = 1 = V^{\mathfrak{F}}$ or, equivalently, U and V are in $\mathfrak{F}$. Since $\mathfrak{F}$ is a Fitting class, we deduce that $G \in \mathfrak{F}$, i.e. $G^{\mathfrak{F}} = 1$. This final contradiction proves the proposition.   □

Corollary 6.5.6. *Let $\mathfrak{F}$ be a Fitting formation. If U and V are subgroups of a group G such that U and V are subnormal in $\langle U, V\rangle$ and $U \in \mathfrak{F}$, then $\langle U, V\rangle^{\mathfrak{F}} = V^{\mathfrak{F}}$.*

Proof. Since U is a subnormal subgroup of $\langle U, V\rangle$ and $U \in \mathfrak{F}$, we have that U is contained in the $\mathfrak{F}$-radical $\langle U, V\rangle_{\mathfrak{F}}$ of $\langle U, V\rangle$. Hence $\langle U, V\rangle = \langle U, V\rangle_{\mathfrak{F}}V$. By Proposition 6.5.5, we deduce that $\langle U, V\rangle^{\mathfrak{F}} = (\langle U, V\rangle_{\mathfrak{F}})^{\mathfrak{F}}V^{\mathfrak{F}} = V^{\mathfrak{F}}$.   □

A well-known result of H. Wielandt (see [Wie94b]) asserts that the Fitting subgroup of a group G normalises the nilpotent residual of each subnormal subgroup of G. The next corollary extends this result to an arbitrary Fitting formation.

Corollary 6.5.7. *Let $\mathfrak{F}$ be a Fitting formation. If U and V are subgroups of a group G such that U and V are subnormal in $\langle U, V\rangle$, it follows that $U_{\mathfrak{F}}$ normalises $V^{\mathfrak{F}}$. In particular, $G_{\mathfrak{F}}$ normalises the $\mathfrak{F}$-residual $H^{\mathfrak{F}}$ of each subnormal subgroup H of G.*

Proof. Consider the subgroup $K = \langle U_{\mathfrak{F}}, V \rangle$ generated by its subnormal subgroups $U_{\mathfrak{F}}$ and V. Then $K^{\mathfrak{F}} = V^{\mathfrak{F}}$ by Corollary 6.5.6 and $K^{\mathfrak{F}}$ is normal in K. Hence $U_{\mathfrak{F}}$ normalises $V^{\mathfrak{F}}$.   □

Let $\mathfrak{F}$ be a Fitting formation. Given a group X, we denote by $\mathcal{W}(X, \mathfrak{F})$ the set of all pairs (A, B) such that A and B are subnormal subgroups of $\langle A, B \rangle \leq X$ and $\langle A^{\mathfrak{F}}, B^{\mathfrak{F}} \rangle < \langle A, B \rangle^{\mathfrak{F}}$. Let $\mathcal{B}(\mathfrak{F})$ denote the class of all groups X such that $\mathcal{W}(X, \mathfrak{F}) \neq \emptyset$. If $\mathfrak{F}$ does not satisfy the Wielandt property for residuals, then the class $\mathcal{B}(\mathfrak{F})$ is non-empty. In the following we analyse the structure of a group G of minimal order in $\mathcal{B}(\mathfrak{F})$. We consider a pair (U, V) in $\mathcal{W}(G, \mathfrak{F})$ such that $|U| + |V|$ is maximal. Denote $H = \langle U^{\mathfrak{F}}, V^{\mathfrak{F}} \rangle$ and $A = U \cap V$. By Proposition 6.5.5, U and V do not permute. In particular, neither U nor V is normal in G. Moreover, $U^{\mathfrak{F}} \neq 1 \neq V^{\mathfrak{F}}$ by Corollary 6.5.6.

Statement 6.5.8. $G = \langle U, V \rangle$. *Moreover,* $U^{\mathfrak{F}} \neq V^{\mathfrak{F}}$.

Proof. By minimality of G, it is clear that $G = \langle U, V \rangle$ and $1 \neq G^{\mathfrak{F}}$. If $N = U^{\mathfrak{F}} = V^{\mathfrak{F}}$, then N is normal in G. The minimal choice of G implies that $G^{\mathfrak{F}}/N = \langle U^{\mathfrak{F}}/N, V^{\mathfrak{F}}/N \rangle = 1$. Then $N = G^{\mathfrak{F}}$. This contradiction yields $U^{\mathfrak{F}} \neq V^{\mathfrak{F}}$.   □

Statement 6.5.9. $G^{\mathfrak{F}} = HN$ *for every minimal normal subgroup* N *of* G. *In particular,* H *is core-free in* G. *Moreover,* H *is normal in* $G^{\mathfrak{F}}$.

Proof. Let N be a minimal normal subgroup of G. We consider $G/N = \langle UN/N, VN/N \rangle$. By minimality of G, we deduce that $G^{\mathfrak{F}}N = HN$. If N is not contained in $G^{\mathfrak{F}}$, then $N \cap G^{\mathfrak{F}} = 1$. This means that $G^{\mathfrak{F}}N = G^{\mathfrak{F}} \times N$. Since $H \leq G^{\mathfrak{F}}$, it follows that $H \cap N = 1$. But $G^{\mathfrak{F}}N = HN$ implies that $|G^{\mathfrak{F}}| = |H|$ and then $G^{\mathfrak{F}} = H$, contrary to our supposition. Hence $\mathrm{Soc}(G) \leq G^{\mathfrak{F}}$ and $G^{\mathfrak{F}} = HN$ for any minimal normal subgroup N of G.

By [DH92, A, 14.3], $\mathrm{Soc}(G)$ normalises H because H is subnormal in G. Hence H is normal in $G^{\mathfrak{F}}$.

Assume that H is not core-free in G. Then H contains a minimal normal subgroup of G, N say. Hence $G^{\mathfrak{F}} = G^{\mathfrak{F}}N = HN = H$, against to our choice of G. Therefore H is core-free in G.   □

Statement 6.5.10. *If* $\mathrm{Soc}(G)$ *is non-abelian, then* $\mathrm{Soc}(G)$ *is a minimal normal subgroup of* G *and* G *is in the boundary of* $\mathfrak{F}$. *In this case,* $G^{\mathfrak{F}}$ *is the unique minimal normal subgroup of* G.

Proof. First, note that for every minimal normal subgroup N of G, since $H \cap N$ is normal in N, we have that $N = (H \cap N) \times N^*$ and $G^{\mathfrak{F}} = H \times N^*$ with $N^* \neq 1$. This implies that H centralises N^*. If there exist two minimal normal subgroups N_1 and N_2 of G, then $G^{\mathfrak{F}} = H \times N_i^* \leq \mathrm{C}_G(N_{3-i}^*)$, for $i = 1$, 2. Therefore $N_i^* \leq \mathrm{Z}(G^{\mathfrak{F}})$ and both N_1 and N_2 are abelian. In other words, if $\mathrm{Soc}(G)$ is not a minimal normal subgroup of G, then $\mathrm{Soc}(G)$ is abelian.

Assume that $N = \mathrm{Soc}(G)$ is non-abelian. Then N is a minimal normal subgroup of G and $\mathrm{C}_G(N) = 1$. It is clear that N is a direct product of copies of a non-abelian simple group, E say. This means that $N \in \mathrm{D}_0(1, E) = \mathfrak{X}$. Then $G^{\mathfrak{F}}/H \in \mathfrak{X}$ and $(G^{\mathfrak{F}})^{\mathfrak{X}} \leq H$. Since $(G^{\mathfrak{F}})^{\mathfrak{X}}$ is normal in G, it follows that $(G^{\mathfrak{F}})^{\mathfrak{X}} = 1$ by Statement 6.5.9. Hence $G^{\mathfrak{F}} \in \mathfrak{X}$. Assume that N is a proper subgroup of $G^{\mathfrak{F}}$. Then there exists a copy of E centralising N. This is a contradiction. Hence $N = G^{\mathfrak{F}}$. In particular, G is in the boundary of $\mathfrak{F}$. ☐

Statement 6.5.11. *If* $\mathrm{Soc}(G)$ *is abelian, then* $G^{\mathfrak{F}}$ *is an elementary abelian p-group for some prime p.*

Proof. Let N be a minimal normal subgroup of G. By Statement 6.5.9, $G^{\mathfrak{F}} = HN$. Since $\mathrm{Soc}(G)$ is abelian, N is an elementary abelian p-group for some prime p. In particular, $\mathrm{O}^p(G^{\mathfrak{F}})$ and $(G^{\mathfrak{F}})'$ are normal subgroups of G contained in H. Since H is core-free in G, $\mathrm{O}^p(G^{\mathfrak{F}}) = (G^{\mathfrak{F}})' = 1$, and $G^{\mathfrak{F}}$ is an abelian p-group.

If, on the other hand, $\Phi(G^{\mathfrak{F}}) = 1$, then we can take N to be contained in $\Phi(G^{\mathfrak{F}})$. In this case, $G^{\mathfrak{F}} = HN = H$. This contradiction leads to $\Phi(G^{\mathfrak{F}}) = 1$, and $G^{\mathfrak{F}}$ is an elementary abelian p-group. ☐

Statement 6.5.12. $H = U^{\mathfrak{F}}V^{\mathfrak{F}}$. *Furthermore,* $U^{\mathfrak{F}}$ *and* $V^{\mathfrak{F}}$ *are proper subgroups of* U *and* V, *respectively.*

Proof. Whether or not $\mathrm{Soc}(G)$ is abelian, every subnormal subgroup of $G^{\mathfrak{F}}$ is a normal subgroup of $G^{\mathfrak{F}}$. In particular, $U^{\mathfrak{F}}$ and $V^{\mathfrak{F}}$ are normal in $G^{\mathfrak{F}}$. Therefore $H = U^{\mathfrak{F}}V^{\mathfrak{F}}$. Assume that $U^{\mathfrak{F}} = U$. Then U normalises $V^{\mathfrak{F}}$. This would imply that $V^{\mathfrak{F}}$ is normal in G, contrary to Statement 6.5.9. Therefore $U^{\mathfrak{F}} < U$ and $V^{\mathfrak{F}} < V$. ☐

Statement 6.5.13. $A = G_{\mathfrak{F}}$ *and* $G^{\mathfrak{F}}$ *is contained in* A. *Moreover,*

1. A *is a maximal normal subgroup of* U *and* V, *and* G/A *is a q-group for some prime* $q \in \mathrm{char}\,\mathfrak{F}$;
2. *if* $\mathrm{Soc}(G)$ *is a p-group, then* $p \in \mathrm{char}\,\mathfrak{F}$; *and*
3. $G^{\mathfrak{N}} = \mathrm{O}^q(G)$ *is contained in* A.

Proof. Let M be a proper subnormal subgroup of G such that $U \leq M$ and consider the subgroup $Y = \langle U, M \cap V \rangle$. The lattice properties of the subnormal subgroups imply that Y is subnormal in G. Furthermore Y is contained in M. Assume that U is a proper subgroup of Y. Then, by maximality of the pair (U, V), we have that $G^{\mathfrak{F}} = \langle Y^{\mathfrak{F}}, V^{\mathfrak{F}} \rangle$. By minimality of G, it follows that $Y^{\mathfrak{F}} = \langle U^{\mathfrak{F}}, (M \cap V)^{\mathfrak{F}} \rangle$. Therefore $G^{\mathfrak{F}} = \langle U^{\mathfrak{F}}, (M \cap V)^{\mathfrak{F}}, V^{\mathfrak{F}} \rangle = \langle U^{\mathfrak{F}}, V^{\mathfrak{F}} \rangle = H$ and we have reached a contradiction. Hence $U = Y$ and so $M \cap V \leq U$. In particular $M \cap V = U \cap V = A$. The arguments for a proper subnormal subgroup of G containing V are analogous.

Let M be a maximal normal subgroup of G such that $U \leq M$. By the foregoing arguments, we have that $M \cap V = A$. Therefore, A is a normal subgroup of V. Moreover, $V/A = V/(V \cap U) = V/(M \cap V) \cong VM/M = G/M$ is a simple group (note that V is not contained in M). Analogously, we deduce that A is normal in U and that U/A is a simple group. This implies that A is a normal subgroup of G and that A is a maximal normal subgroup of U and V. Since $A^{\mathfrak{F}}$ is a normal subgroup of G contained in H, it follows that $A^{\mathfrak{F}} = 1$ by Statement 6.5.9. This means that $A \in \mathfrak{F}$ and A is contained in $G_{\mathfrak{F}}$.

Since $G_{\mathfrak{F}}$ is normal in G, we have that $(UG_{\mathfrak{F}})^{\mathfrak{F}} = U^{\mathfrak{F}}$ by Corollary 6.5.6. Therefore $UG_{\mathfrak{F}}$ is a proper subnormal subgroup of G. Assume that $U < UG_{\mathfrak{F}}$. Since $G = \langle UG_{\mathfrak{F}}, V \rangle$, we deduce that $G^{\mathfrak{F}} = \langle (UG^{\mathfrak{F}})^{\mathfrak{F}}, V^{\mathfrak{F}} \rangle = \langle U^{\mathfrak{F}}, V^{\mathfrak{F}} \rangle$ by maximality of the pair (U, V), contrary to supposition. Hence $G_{\mathfrak{F}}$ is a subgroup of U. Analogously, $G_{\mathfrak{F}}$ is contained in V, and we have the equality.

Assume that U/A is a non-abelian simple group. Then U/A normalises V/A by Theorem 2.2.19, and V is normal in G. This contradiction yields that U/A is a cyclic group of prime order, q say. The same argument for V proves that V/A is a cyclic group of prime order, r say. If $r \neq q$, then $[U/A, V/A] \leq [\mathrm{O}_q(G/A), \mathrm{O}_r(G/A)] = 1$. Then $G/A = U/A \times V/A$ is abelian, and U and V are normal subgroups of G. This possibility cannot happen. Therefore, $r = q$ and G/A is a group generated by two subnormal q-subgroups, U/A and V/A, and so G/A is a q-group. Suppose that $A = 1$. Then G is a q-group. If $q \notin \operatorname{char} \mathfrak{F}$, then $U^{\mathfrak{F}} = U$ and $V^{\mathfrak{F}} = V$, contrary to Statement 6.5.12. Therefore we must have $q \in \operatorname{char} \mathfrak{F}$ and so $G \in \mathfrak{F}$ by [DH92, IX, 1.9]. This contradiction yields $A \neq 1$ and then A contains a minimal normal subgroup of G. On the other hand, we can assume that either $\operatorname{Soc}(G)$ is an elementary abelian p-group for some prime p, or $\operatorname{Soc}(G)$ is a non-abelian minimal normal subgroup of G by Statements 6.5.10 and 6.5.11. In both cases, we have that $\operatorname{Soc}(G)$ and $G^{\mathfrak{F}}$ are subgroups of A and Statement 2 holds. Since $G^{\mathfrak{F}} \leq A$, we have that the q-group G/A belongs to $\mathfrak{F}$. Therefore $q \in \operatorname{char} \mathfrak{F}$ and Statement 1 holds.

Now we prove that if M is a maximal normal subgroup of G, then A is contained in M. Assume that there exists a maximal normal subgroup M of G such that A is not contained in M. Then $G = AM$, $U = A(U \cap M)$, and $V = A(V \cap M)$. The subnormal subgroup $T = \langle U \cap M, V \cap M \rangle$ is a supplement of A in G and $A \cap M \leq T \leq M$. Hence $M = G \cap M = TA \cap M = T(A \cap M) = T$. By minimality of G, $M^{\mathfrak{F}} = \langle (U \cap M)^{\mathfrak{F}}, (V \cap M)^{\mathfrak{F}} \rangle$. On the other hand, since $G = MU$, by Proposition 6.5.5, it follows that $G^{\mathfrak{F}} = M^{\mathfrak{F}} U^{\mathfrak{F}} = \langle (U \cap M)^{\mathfrak{F}}, (V \cap M)^{\mathfrak{F}}, U^{\mathfrak{F}} \rangle = \langle (V \cap M)^{\mathfrak{F}}, U^{\mathfrak{F}} \rangle = \langle U^{\mathfrak{F}}, V^{\mathfrak{F}} \rangle$, contrary to our initial supposition. Therefore, every maximal normal subgroup of G contains A.

Clearly, $G^{\mathfrak{N}}$ is contained in A. Since every maximal subgroup of $G/G^{\mathfrak{N}}$ is normal in $G/G^{\mathfrak{N}}$, it follows that $A/G^{\mathfrak{N}} \leq \Phi(G/G^{\mathfrak{N}})$. Since G/A is a q-group, we deduce that $G/G^{\mathfrak{N}}$ is a q-group. Then $G^{\mathfrak{N}} = \mathrm{O}^q(G)$ and 3 holds. $\qquad \square$

Next, we assume that $\operatorname{Soc}(G)$ is abelian. This implies that $G^{\mathfrak{F}}$ is an elementary abelian p-group for some prime $p \in \operatorname{char} \mathfrak{F}$ by Statements 6.5.11 and 6.5.13. Denote $B = G^{\mathfrak{F}}$. Then B is a G-module over the field $\mathrm{GF}(p)$.

Statement 6.5.14. *B is a completely reducible A-module over* GF(p).

Proof. We denote by $\mathrm{J}(B_A)$ the intersection of all maximal A-submodules of B_A. Since A is normal in G, the action of G permutes these maximal submodules, and thus $\mathrm{J}(B_A)$ is a normal subgroup of G.

Suppose that $\mathrm{J}(B_A) \neq 0$, and let N be a minimal normal subgroup of G such that $N \leq \mathrm{J}(B_A)$. By Statement 6.5.9, we have, in additive notation, that $B = H + N$. Since B, H, and N are A-submodules and $N \leq \mathrm{J}(B_A)$, we have that $B = H$ by Nakayama's lemma ([HB82a, VII, 6.4]). This is a contradiction. Therefore, $\mathrm{J}(B_A) = 0$, and B is a completely reducible A-module over GF(p) by [HB82a, VII, 1.6]. $\qquad\square$

It is clear that H is an A-submodule of B.

Statement 6.5.15. *Let Z be an arbitrary irreducible A-submodule of H. Then if Z_1 is an irreducible A-submodule of B, then there exists $g \in G$ such that Z_1 is A-isomorphic to Z^g.*

Proof. Let Z be an irreducible A-submodule of H and consider the normal closure $\langle Z^G \rangle = \sum_{g \in G} Z^g$. Then $\langle Z^G \rangle_A$ is a completely reducible A-module and is a direct sum of its irreducible submodules which are isomorphic to some conjugate of Z. Let N be a minimal normal subgroup of G such that $N \leq \langle Z^G \rangle$. Hence N_A is again a completely reducible A-module and is a direct sum of its irreducible submodules, which are isomorphic to some conjugate of Z. On the other hand, $B = H + N$ by Statement 6.5.9. Therefore, every A-composition factor of B/H is isomorphic to a conjugate of Z.

Let Z_1 be an irreducible A-submodule of B. The normal closure $N_1 = \langle Z_1^G \rangle$ is not contained in H, and every A-composition factor of N_1 is isomorphic to a conjugate of Z_1. Again by Statement 6.5.9, $B = H + N_1$ and so every A-composition factor of B/H is isomorphic to a conjugate of Z_1. This implies that Z_1 is A-isomorphic to a conjugate of Z. $\qquad\square$

The following lemma is needed in the proof of our next statement.

Lemma 6.5.16. *Let K be a field of characteristic p, and let G be a group with a normal subgroup N such that G/N is a p-group. If W is an irreducible KN-module, then the induced KG-module W^G has all of its composition factors isomorphic.*

Proof. Let T be the inertia subgroup of W in G. First note that $(W^T)_N = \bigoplus_g Wg$, where g runs over a transversal of N in T. This is a particular case of Mackey's theorem ([DH92, B, 6.21]). Since T is the inertia subgroup of W in G, we have that $Wg \cong W$ for all $g \in T$. Therefore $(W^T)_N$ is homogeneous, and all of its composition factors are isomorphic to W. In particular, if U/V is a composition factor of W^T, then $(U/V)_N$ is homogeneous, and all its composition factors are isomorphic to W.

If U/V is a composition factor of W^T, then the G-module $(U/V)^G \cong U^G/V^G$ is irreducible by [DH92, B, 7.4]. It is thus sufficient to prove that all composition factors of W^T are isomorphic. Let U/V be a composition factor of W^T. Then, by [DH92, B, 8.3], $(U/V)_N$ is an irreducible N-module. Hence $(U/V)_N$ is isomorphic to W. By [DH92, B, 5.17], all composition factors of W^T are isomorphic. □

Statement 6.5.17. *If $p = q$, then all composition factors of B are isomorphic.*

Proof. Suppose that $p = q$. Let Z be an irreducible A-submodule of B. By Lemma 6.5.16, the induced module Z^G has all its composition factors isomorphic. Let M be a composition factor of B, and let Z_1 be an irreducible A-submodule of M_A. By Statement 6.5.15, Z_1 is A-isomorphic to Z^g for some $g \in G$. Then $Z_1^{g^{-1}}$ is an irreducible A-submodule of M which is isomorphic to Z. In other words, M_A has an irreducible submodule isomorphic to Z, that is, $0 \neq \operatorname{Hom}_{KA}(Z, M_A)$. By Nakayama's reciprocity theorem [DH92, B, 6.5], it follows that $0 \neq \operatorname{Hom}_{KG}(Z^G, M)$. Therefore a composition factor of Z^G is isomorphic to M, and then all composition factors of Z^G are isomorphic to M. □

Statement 6.5.18. *If $p \neq q$, then $B = \operatorname{Soc}(G)$.*

Proof. Let Z be an irreducible A-submodule of B_A. Since $p \neq q$, it follows that Z^G is a completely reducible G-module by [HB82a, VII, 9.4].

Denote by α the inclusion of Z in B_A. Applying [HB82a, VII, 4.4], there exists a KG-homomorphism $\alpha' \colon Z^G \longrightarrow B$ such that $(z \otimes g)\alpha' = z^g$ for all $g \in G$ and all $z \in Z$. Hence $\operatorname{Im}(\alpha') = \langle Z^G \rangle$, the normal closure of Z in G. Therefore $\langle Z^G \rangle$ is a completely reducible G-module and $\langle Z^G \rangle \leq \operatorname{Soc}(G)$. In particular, Z is contained in $\operatorname{Soc}(G)$. Since, by Statement 6.5.14, B_A is a completely reducible A-module, it follows that B is contained in $\operatorname{Soc}(G)$ and the equality holds by Statement 6.5.9. □

The most important examples of Fitting formations are as follows:

1. The solubly saturated Fitting formations (see Chapter 3).

2. The Fitting formations constructed by Fitting families of modules ([DH92, Chapter IX, 2, Construction F]). Fix a prime r. Let K be an extension field of $\operatorname{GF}(r)$. For any r-soluble group G, denote $\mathfrak{T}_K(G)$ the class of all irreducible KG-modules V such that V is a composition factor of the module $W^K = W \otimes K$, where W is an r-chief factor of G.

Suppose that, for every group G, a class of irreducible KG-modules $\mathfrak{M}(G)$ is defined. Then the class $\mathfrak{M} := \bigcup_G \mathfrak{M}(G)$ is called a *Fitting family* if it satisfies the four properties listed in Definition 2.5.5. Applying Theorem 2.5.6, the class

$$\mathfrak{T}(1, \mathfrak{M}) = \big(G : G \text{ is } r\text{-soluble and } \mathfrak{T}_K(G) \subseteq \mathfrak{M}(G) \big)$$

is a Fitting formation provided that $\mathfrak{M}$ is a Fitting family.

In both cases, we have a way to distinguish between the abelian r-chief factors of any group X in the following sense:

1. If $\mathfrak{F}$ is a solubly saturated formation defined by the canonical $\mathbb{P}$-local formation function F, then an abelian r-chief factor M of X can be $\mathfrak{F}$-central if $X/C_X(M) \in F(r)$ or $\mathfrak{F}$-eccentric otherwise.
2. If $\mathfrak{F} = \mathfrak{T}(1, \mathfrak{M})$ is a Fitting formation constructed by a Fitting family of modules $\mathfrak{M}$, then an abelian r-chief factor M of X can be such that all composition factors of M^K are in $\mathfrak{M}(X)$ or not.

Let X be an arbitrary group, and let M be an X-module over $GF(r)$. Denote by $\mathrm{Irr}(M)$ the class of all irreducible X-modules occurring as composition factors of M.

Suppose that $\mathfrak{F}$ is either a solubly saturated formation or a Fitting formation defined by a Fitting family of modules, and let $\mathrm{Mod}_{\mathfrak{F}}(U)$ denote the class of all irreducible U-modules occurring as

1. $\mathfrak{F}$-central chief factors of U below B, if $\mathfrak{F}$ is a solubly saturated Fitting formation, or

2. abelian chief factors M of U below B such that every composition factor of M^K is in $\mathfrak{M}(U)$, if $\mathfrak{F} = \mathfrak{T}(1, \mathfrak{M})$ is a Fitting formation constructed by a Fitting family of modules $\mathfrak{M}$.

Analogously, let $\mathrm{Mod}_{\mathfrak{F}}(V)$ denote the corresponding set for V.

Statement 6.5.19. *If $\mathfrak{F}$ is either a solubly saturated Fitting formation or a Fitting formation defined by a Fitting family of modules, then G is in the boundary of $\mathfrak{F}$.*

Proof. Assume first that $p = q$. In this case, all composition factors of B are isomorphic G-modules by Statement 6.5.17. We consider a G-composition series of B, $0 = B_0 \leq B_1 \leq \cdots \leq B_r = B$ say.

The composition factor B_1 is a minimal normal subgroup of G and so $B = HB_1$ by Statement 6.5.9. Since H is core-free in G, it follows that $B_1 \neq U^{\mathfrak{F}}$. Moreover, B_1 is a completely reducible U-module by Clifford's theorems [DH92, B, 7.3]. It then decomposes as $B_1 = B_1' \oplus B_1^*$, where $B_1' = B_1 \cap U^{\mathfrak{F}}$. Since B_1 is not contained in $U^{\mathfrak{F}}$, we have that $B_1^* \neq 0$.

Let M be a U-composition factor of B_1^*. Then M is isomorphic to a U-composition factor of B_1/B_1', which is a section of $U/U^{\mathfrak{F}} \in \mathfrak{F}$. This implies that $M \in \mathrm{Mod}_{\mathfrak{F}}(U)$ and $\mathrm{Irr}(B_1^*)$ is contained in $\mathrm{Mod}_{\mathfrak{F}}(U)$.

Assume now that $B_1' \neq 0$ and let M be an irreducible U-submodule of B_1'. Then $U^{\mathfrak{F}} = M \oplus M_1$, for some U-submodule M_1 of $U^{\mathfrak{F}}$. Since $U/M_1 \notin \mathfrak{F}$ and $U/U^{\mathfrak{F}} \in \mathfrak{F}$, it follows that M is not in $\mathrm{Mod}_{\mathfrak{F}}(U)$. Consequently $\mathrm{Irr}(B_1') \cap \mathrm{Mod}_{\mathfrak{F}}(U) = \emptyset$ if $B_1' \neq 0$.

The arguments for V are completely analogous. Hence B_1, considered as V-module, decomposes as $B_1 = B_1'' \oplus B_1^{**}$ where $B_1'' \leq B_1 \cap V^{\mathfrak{F}}$ and B_1^{**} is a non-trivial V-submodule of B_1 such that $\mathrm{Irr}(B_1^{**})$ is contained in $\mathrm{Mod}_{\mathfrak{F}}(V)$. Moreover, $\mathrm{Irr}(B_1'') \cap \mathrm{Mod}_{\mathfrak{F}}(V) = \emptyset$ provided that $B_1'' \neq 0$.

Let $B' = U^{\mathfrak{F}} + B_{r-1}$. Assume that $B'/B_{r-1} \neq 0$ and let M/B_{r-1} be an irreducible U-submodule of B'/B_{r-1}. Since B'/B_{r-1} is completely reducible as U-module, it follows that $B'/B_{r-1} = M/B_{r-1} \oplus M_1/B_{r-1}$. Note that

$M_1 = M_1 \cap (U^{\mathfrak{F}} + B_{r-1}) = (M_1 \cap U^{\mathfrak{F}}) + B_{r-1}$ and $U^{\mathfrak{F}} + M_1 = U^{\mathfrak{F}} + B_{r-1} = B'$. Therefore $U^{\mathfrak{F}}/(M_1 \cap U^{\mathfrak{F}})$ is U-isomorphic to $(U^{\mathfrak{F}} + M_1)/M_1 \cong B'/M_1$ and B'/M_1 is U-isomorphic to M/B_{r-1}. Since $M/B_{r-1} \neq 0$, we have that $U^{\mathfrak{F}}$ is not contained in M_1 and so $U/(M_1 \cap U^{\mathfrak{F}})$ is not in $\mathfrak{F}$. Hence $M/B_{r-1} \notin \mathrm{Mod}_{\mathfrak{F}}(U)$. Consequently $\mathrm{Irr}(B'/B_{r-1}) \cap \mathrm{Mod}_{\mathfrak{F}}(U) = \emptyset$.

The same argument holds for V, that is, if $B'' = V^{\mathfrak{F}} + B_{r-1}$, then $\mathrm{Irr}(B''/B_{r-1}) \cap \mathrm{Mod}_{\mathfrak{F}}(V) = \emptyset$ provided that $B''/B_{r-1} \neq 0$.

Suppose that $r \geq 2$. Then the composition factors B_1 and B/B_{r-1} are different. Furthermore $B = B' + B''$.

Assume that $B'/B_{r-1} = 0$. Then $B = B''$ and $\mathrm{Irr}(B/B_{r-1}) \cap \mathrm{Mod}_{\mathfrak{F}}(V) = \emptyset$. Let $\varphi \colon B/B_{r-1} \longrightarrow B_1$ be a G-isomorphism. Since φ is a V-isomorphism, it follows that $\mathrm{Irr}(B_1) \cap \mathrm{Mod}_{\mathfrak{F}}(V) = \emptyset$. This is a contradiction because B_1^{**} is a non-trivial V-submodule of B_1 such that $\mathrm{Irr}(B_1^{**}) \subseteq \mathrm{Mod}_{\mathfrak{F}}(V)$. Consequently $B'/B_{r-1} \neq 0$ and $B''/B_{r-1} \neq 0$. Moreover $\varphi(B'/B_{r-1})$ is contained in B_1' and $\varphi(B''/B_{r-1})$ is contained in B_1''. Hence $B_1 = \varphi(B/B_{r-1}) = \varphi(B'/B_{r-1} + B''/B_{r-1}) = \varphi(B'/B_{r-1}) + \varphi(B''/B_{r-1}) = B_1' + B_1'' \leq (B_1 \cap U^{\mathfrak{F}}) + (B_1 \cap V^{\mathfrak{F}}) \leq H$. This is a contradiction. Therefore $r = 1$ and B is an irreducible G-module.

Consider now the case where $p \neq q$. Then, by Statement 6.5.18, B is a completely reducible G-module. By Clifford's theorem, $U^{\mathfrak{F}}$ is a completely reducible U-module. If M is an irreducible U-submodule of $U^{\mathfrak{F}}$, then there exists a U-submodule M_0 of $U^{\mathfrak{F}}$ such that $U^{\mathfrak{F}} = M \oplus M_0$. Since U/M_0 is not in $\mathfrak{F}$, we have that $M \notin \mathrm{Mod}_{\mathfrak{F}}(U)$. That is, $\mathrm{Irr}(U^{\mathfrak{F}}) \cap \mathrm{Mod}_{\mathfrak{F}}(U) = \emptyset$. On the other hand, since $U/U^{\mathfrak{F}} \in \mathfrak{F}$, it follows that $M \in \mathrm{Mod}_{\mathfrak{F}}(U)$, for every chief factor M of U between $U^{\mathfrak{F}}$ and U. That is, $\mathrm{Irr}(B/U^{\mathfrak{F}})$ is contained in $\mathrm{Mod}_{\mathfrak{F}}(U)$. With a similar argument, we deduce the corresponding result for V. Hence

$$\mathrm{Irr}(B_U/U^{\mathfrak{F}}) \cap \mathrm{Irr}(U^{\mathfrak{F}}) = \emptyset = \mathrm{Irr}(V^{\mathfrak{F}}) \cap \mathrm{Irr}(B_V/V^{\mathfrak{F}}).$$

Now suppose that $B = N_1 \times \cdots \times N_r$, where N_i is a minimal normal subgroup of G, $i = 1, \ldots, r$, and $r \geq 2$. Each N_i can be decomposed as $N_i = N_i^* \oplus (N_i \cap U^{\mathfrak{F}})$, where N_i^* is a complement of $N_i \cap U^{\mathfrak{F}}$ in N_i as U-modules. Then $B_U = (N_1^* \oplus \cdots \oplus N_r^*) \oplus ((N_1 \cap U^{\mathfrak{F}}) \oplus \cdots \oplus (N_r \cap U^{\mathfrak{F}}))$. Denote $B^* = N_1^* \oplus \cdots \oplus N_r^*$. Then $U^{\mathfrak{F}} \cap B^* = 0$ because $\mathrm{Irr}(U^{\mathfrak{F}}) \cap \mathrm{Irr}(B_U/U^{\mathfrak{F}}) = \emptyset$. Hence $U^{\mathfrak{F}} = (N_1 \cap U^{\mathfrak{F}}) \oplus \cdots \oplus (N_r \cap U^{\mathfrak{F}})$. The same arguments hold for V: $V^{\mathfrak{F}} = (N_1 \cap V^{\mathfrak{F}}) \oplus \cdots \oplus (N_r \cap V^{\mathfrak{F}})$. Comparing the two decomposition as vector spaces, $B = N_1 + H = N_1 + U^{\mathfrak{F}} + V^{\mathfrak{F}} = N_1 + ((N_1 \cap U^{\mathfrak{F}}) \oplus \cdots \oplus (N_r \cap U^{\mathfrak{F}})) + ((N_1 \cap V^{\mathfrak{F}}) \oplus \cdots \oplus (N_r \cap V^{\mathfrak{F}})) = N_1 \oplus ((N_2 \cap U^{\mathfrak{F}}) + (N_2 \cap V^{\mathfrak{F}})) \oplus \cdots \oplus ((N_r \cap U^{\mathfrak{F}}) + (N_r \cap V^{\mathfrak{F}})) = N_1 \oplus \cdots \oplus N_r$, we deduce that $N_i = (N_i \cap U^{\mathfrak{F}}) + (N_i \cap V^{\mathfrak{F}})$ for each $i \geq 2$. Therefore $N_i \leq U^{\mathfrak{F}} + V^{\mathfrak{F}} = H$ for each $i \geq 2$. This contradiction leads to $r = 1$ and B is a minimal normal subgroup of G.

Consequently, in both cases, we have that G is a monolithic group in the boundary of $\mathfrak{F}$. $\square$

For convenience, we incorporate the class of all groups satisfying the above statements in a formal definition.

Definition 6.5.20. *Let* $\mathfrak{F}$ *be a Fitting formation. Define* $b_3(\mathfrak{F})$ *as the class of all triples* (G, U, V) *such that*

1. $G \in b(\mathfrak{F})$ *and* U *and* V *are subnormal subgroups of* G;
2. $G = \langle U, V \rangle$;
3. $U \cap V = G_{\mathfrak{F}} \neq 1$;
4. $U/G_{\mathfrak{F}}$ *and* $V/G_{\mathfrak{F}}$ *are cyclic groups of order* q, *a prime.*

Note that if $(G, U, V) \in b_3(\mathfrak{F})$, then $G^{\mathfrak{F}}$ is contained in $G_{\mathfrak{F}}$ and $G/G_{\mathfrak{F}}$ is a q-group, $q \in \operatorname{char} \mathfrak{F}$.

The above statements lead to the following result.

Theorem 6.5.21. *Let* $\mathfrak{F}$ *be a Fitting formation. Suppose that either*

1. $\mathfrak{F}$ *is a solubly saturated Fitting formation, or*
2. $\mathfrak{F} = \mathfrak{T}(1, \mathfrak{M})$ *is a Fitting formation defined by a Fitting family of modules* $\mathfrak{M}$ *constructed over an extension field* K *of* $\operatorname{GF}(r)$.

Then the following statements are equivalent:

1. $\mathfrak{F}$ *satisfies the Wielandt property for residuals.*
2. *For every triple* $(G, U, V) \in b_3(\mathfrak{F})$, *we have that* $G^{\mathfrak{F}} = \langle U^{\mathfrak{F}}, V^{\mathfrak{F}} \rangle$.

Applying Theorem 6.5.21, a large number of Fitting formations satisfying the Wielandt property for residuals appear.

Corollary 6.5.22. *Let* $\mathfrak{F}$ *be a Fitting formation. Then* $\mathfrak{F}$ *satisfies the Fitting property for residuals provided that one of the following conditions hold:*

1. $\mathfrak{S}_p\mathfrak{F} = \mathfrak{F}$, *for all primes* $p \in \operatorname{char} \mathfrak{F}$.
2. $\mathfrak{F}\mathfrak{S}_p = \mathfrak{F}$, *for all primes* $p \in \operatorname{char} \mathfrak{F}$.
3. $\mathfrak{F}$ *is solubly saturated, and its boundary is composed of non-abelian simple groups.*
4. $\operatorname{char} \mathfrak{F} = \emptyset$.
5. $\mathfrak{F} = {}_{\mathrm{E}}\mathfrak{X}$ *for some class* $\mathfrak{X}$ *of simple groups.*
6. $\mathfrak{F} = {}_{\mathrm{D}_0}(1, \mathfrak{X}_1)$, *where* $\mathfrak{X}_1$ *is a class of non-abelian simple groups.*

Let p be a prime, and let $\mathfrak{M}_p$ be the class of all groups whose abelian p-chief factors are central. It is rather clear that $\mathfrak{M}_p$ is a Fitting formation. Moreover, $\mathfrak{M}_p$ is solubly saturated by Lemma 3.2.15 and $\mathfrak{M}_p \cap \mathfrak{S}$ is the class of all soluble p-nilpotent groups. The $\mathfrak{M}_p$-radical of a group G is the intersection of the centralisers of the abelian p-chief factors of G. This subgroup also appears when a $\mathbb{P}$-local definition of a solubly saturated formation is considered (see Section 3.2).

Corollary 6.5.23. *Let* p *be a prime. Then* $\mathfrak{M}_p$ *satisfies the Wielandt property for residuals.*

Proof. Applying Theorem 6.5.21, we need only consider triples in $\text{b}_3(\mathfrak{M}_p)$.

Suppose that (G, U, V) is a triple in $\text{b}_3(\mathfrak{M}_p)$. Then $G = \langle U, V \rangle$ is a monolithic group in $\text{b}(\mathfrak{M}_p)$, U and V are subnormal subgroups of G, $U \cap V$ is the $\mathfrak{M}_p$-radical of G and $G/(U \cap V)$ is a q-group for some prime $q \in \text{char } \mathfrak{M}_p = \mathbb{P}$. Denote by N the $\mathfrak{M}_p$-residual of G. Then N is an abelian p-group contained in $U \cap V = A$. Since N is a completely reducible A-module, it follows that $A \leq C_G(N)$. Consequently $G = QA = Q\,C_G(N)$ for every Sylow q-subgroup Q of G. Let $B = NQ$. We have that B is soluble and N is a minimal normal subgroup of Q. It is clear that Q does not centralise N because $G \notin \mathfrak{M}_p$. This implies that $B^{\mathfrak{M}_p} = N$. On the other hand, $U = A(Q \cap U)$ and $V = A(Q \cap V)$. Hence $G = A\langle Q \cap U, Q \cap V \rangle$ and $Q = \langle Q \cap U, Q \cap V \rangle$. It means that $B = N\langle Q \cap U, Q \cap V \rangle = \langle N(Q \cap U), N(Q \cap V) \rangle = \langle U \cap B, V \cap B \rangle$. Note that $B^{\mathfrak{M}_p} = B^{\mathfrak{F}}$, where $\mathfrak{F}$ is the saturated formation of all p'-nilpotent groups. Combining Proposition 6.5.4 (1) and Corollary 6.5.22 (1), it follows that $\mathfrak{F}$ satisfies the Wielandt property for residuals. Therefore $B^{\mathfrak{M}_p} = \langle (U \cap B)^{\mathfrak{M}_p}, (V \cap B)^{\mathfrak{M}_p} \rangle \leq \langle U^{\mathfrak{M}_p}, V^{\mathfrak{M}_p} \rangle$ and so $N = \langle U^{\mathfrak{M}_p}, V^{\mathfrak{M}_p} \rangle$. $\square$

Let $\mathfrak{F}$ be a solubly saturated formation. Then, applying Theorem 3.2.14, there exists a Baer function f such that $\mathfrak{F} = \text{LF}_{\mathbb{P}}(f)$. Denote $\text{Supp}(f) = \{p \in \mathbb{P} : f(p) \neq \emptyset\} \cup \{S \in \mathfrak{J} \setminus \mathbb{P} : f(S) \neq \emptyset\}$. Then it rather clear that $\mathfrak{F} = \bigcap_{p \in \text{Supp}(f)} \mathfrak{M}_p \circ f(p) \cap \bigcap_{S \in \text{Supp}(f) \setminus \mathbb{P}} \text{E}\big((S)'\big) \circ f(S)$ by Remarks 3.1.2 and Remark 3.1.9.

Therefore, applying Proposition 6.5.4 and Corollary 6.5.23, we have:

Theorem 6.5.24 ([KS95]). *Let $\mathfrak{F}$ be a solubly saturated formation and let f be a Baer function $\mathbb{P}$-locally defining $\mathfrak{F}$. If for all $S \in \text{Supp}(f)$, $f(S)$ satisfies the Wielandt property for residuals, then $\mathfrak{F}$ satisfies the Wielandt property for residuals.*

Corollary 6.5.25. *Let $\mathfrak{F}$ be a saturated formation locally defined by a formation function f. If for all primes p, the formations $f(p)$ satisfy the Wielandt property for residuals, then $\mathfrak{F}$ satisfies the Wielandt property for residuals.*

Proof. Set

$$g(J) = \begin{cases} f(p) & \text{when } J \cong C_p,\ p \in \mathbb{P} \text{ and} \\ \bigcap_{p \,||\, J} f(p) & \text{when } J \in \mathfrak{J} \setminus \mathbb{P}, \end{cases}$$

then it is clear that $\mathfrak{F} = \text{LF}_{\mathbb{P}}(g)$. Applying Proposition 6.5.4 (2), $g(J)$ satisfies the Wielandt property for residuals if $J \in \mathfrak{J} \setminus \mathbb{P}$. By Theorem 6.5.24, $\mathfrak{F}$ satisfies the Wielandt property for residuals. $\square$

Corollary 6.5.26. *Any soluble subgroup-closed Fitting formation satisfies the Wielandt property for residuals.*

Proof. Any soluble subgroup-closed Fitting formation $\mathfrak{F}$ is a primitive saturated formation. Therefore, $\mathfrak{F}$ has a local definition f such that $f(p)$ satisfies the Wielandt property for residuals for all prime numbers p (see [DH92, page 497]). $\square$

Example 6.5.27. (see Example 2.2.17) Let $\mathfrak{Q}$ be the Fitting formation of all quasinilpotent groups. Then $\mathfrak{Q}$ is a solubly saturated formation $\mathbb{P}$-locally defined by the $\mathbb{P}$-local formation function f given by

$$f(S) = \begin{cases} (1) & \text{when } S \cong C_p, \text{ and} \\ D_0(1, S) & \text{when } S \in \mathfrak{J} \setminus \mathbb{P}. \end{cases}$$

Since $f(S)$ satisfies the Wielandt property for residuals for all S, it follows that $\mathfrak{Q}$ satisfies the Wielandt property for residuals by Theorem 6.5.24.

In the next examples, we work in the universe of all soluble groups.

Let $\mathfrak{X}_i$ be Fitting formations, $i = 1, 2$. For every group G, denote by $\mathfrak{M}(G)$ the class of all irreducible KG-modules V such that $V = U \otimes W$ with U π-special, W π'-special, and $G/\operatorname{Ker}(G \text{ on } U) \in \mathfrak{X}_1$ and $G/\operatorname{Ker}(G \text{ on } W) \in \mathfrak{X}_2$. Applying Theorem 2.5.10, $\mathfrak{M} = \mathfrak{M}(K, \mathcal{P}, \mathfrak{X}_1, \mathfrak{X}_2) = \bigcup_G \mathfrak{M}(G)$ is a Fitting family. Let $\mathfrak{T}(1, \mathfrak{M}) = \mathfrak{T}(1, r, \mathcal{P}, \mathfrak{X}_1, \mathfrak{X}_2)$ be the Fitting formation defined by $\mathfrak{M}$.

Theorem 6.5.28. *Let π be a set of primes and consider the partition $\mathcal{P} = \{\pi, \pi'\}$ of the set of all prime numbers. The Fitting formation $\mathfrak{F} = \mathfrak{T}(1, \mathfrak{M}) = \mathfrak{T}(1, r, \mathcal{P}, \mathfrak{X}_1, \mathfrak{X}_2)$ satisfies the Wielandt property for residuals in the following case: $\mathfrak{X}_1 = \mathfrak{S}_\rho$ and $\mathfrak{X}_2 = \mathfrak{S}_\sigma$ for some sets of primes ρ and σ (not both empty).*

The following result is used in the proof of Theorem 6.5.28. It can be proved by using similar arguments to those used in the proof of [HB82a, VII, 9.13].

Lemma 6.5.29. *Let N be a normal subgroup of G, and let V_1 and V_2 be two KG-modules such that*

1. *$(V_1)_N$ is absolutely irreducible, and*
2. *V_2 is absolutely irreducible and $(V_2)_N$ is homogeneous, and all of its constituents are isomorphic to $(V_1)_N$. Write $(V_2)_N \cong s(V_1)_N$.*

Then there exists an irreducible $K(G/N)$-module W with $\dim W = s$ such that $V_2 \cong V_1 \otimes W$.

Proof (of Theorem 6.5.28). We use only the restriction on the $\mathfrak{X}_i$ at one point, and so have written the proof as far as possible to be independent of that hypothesis. Applying Theorem 6.5.21, we need only consider groups in $\mathrm{b}_3(\mathfrak{F})$. Hence we suppose that G is in the boundary of $\mathfrak{F}$ and, moreover, that U and V are subnormal subgroups of G satisfying $G = \langle U, V \rangle$, $A = U \cap V = G_\mathfrak{F} \neq 1$, and U/A and V/A are of prime order q, $q \in \operatorname{char} \mathfrak{F}$. Note that G/A is a q-group and $O^q(G) = G^\mathfrak{N} = O^q(U) = O^q(V) = O^q(A)$. Furthermore, G has a unique minimal normal subgroup $B = G^\mathfrak{F}$ which is a p-group for some prime p. First, we observe that $p = r$ (the characteristic of K), since otherwise all r-chief factors would come from $G/B \in \mathfrak{F}$, and so G would be in $\mathfrak{F}$. We are working with a field K which is algebraically closed. However, when dealing with dimensions of KX-modules for a subgroup X of G, we can assume that K is a splitting field for G and all its subgroups. In fact, by Brauer's theorem [HB82a, VII, 2.6], we can assume that K is a finite Galois extension of $k = \mathrm{GF}(p)$. We are interested in the behaviour of the irreducible components of B^K. By [HB82a, VII, 1.15], the KG-module B^K is completely reducible. Let N be an irreducible component of B^K. Applying [HB82a, VII, 1.18 (b)], every irreducible KG-submodule of B^K is G-isomorphic to N^η for some $\eta \in \mathrm{G}(K/k)$.

We collect some properties we need. First, if L is a normal subgroup of G, then a KL-module Q is π-special if and only if all of its G-conjugates are π-special, and $L/\mathrm{Ker}(L \,\mathrm{on}\, Q) \in \mathfrak{X}_i$ if and only if the same is true for all of the G-conjugates of Q. Further, Q is π-special if and only if all of its Galois conjugates are special and $L/\mathrm{Ker}(L \,\mathrm{on}\, Q) \in \mathfrak{X}_i$ if and only if the same is true for all of the Galois conjugates of Q.

Clearly we may assume that $q \in \pi$. Suppose, by way of contradiction, that $B \neq \langle U^{\mathfrak{F}}, V^{\mathfrak{F}} \rangle$. Then $\mathrm{Core}_G(\langle U^{\mathfrak{F}}, V^{\mathfrak{F}} \rangle) = 1$. We have that B_U is completely reducible as U-module and so $B = U^{\mathfrak{F}} \oplus B_0$, with $B \neq B_0 \neq 0$. It follows that $(U^{\mathfrak{F}})^K$ can contain no components in $\mathfrak{M}(U)$ and $(B_0)^K$ must have all its components in $\mathfrak{M}(U)$. Let N be an irreducible component of B^K. If no component of $(B_U)^K$ is in $\mathfrak{M}(U)$, then no component of $(B_U)^K$ is in $\mathfrak{M}(U)$ and thus $B_0 = 0$. This is a contradiction. If every component of N_U is in $\mathfrak{M}(U)$, then every component of $(B_U)^K$ is in $\mathfrak{M}(U)$. This implies that $U^{\mathfrak{F}} = 1$ (or, equivalently, $B = B_0$). It is also a contradiction. Hence, if we denote by $D(N_U)$ the sum of all irreducible KU-submodules of N_U which do not lie in $\mathfrak{M}(U)$, then $0 \neq D(N_U) \neq N$. In particular, N_U is not a homogeneous module. Similar remarks apply to V.

Furthermore, $N_A = N_1 \oplus \cdots \oplus N_t$, where the N_i are irreducible KA-modules, all conjugate by elements of G. Since $A \in \mathfrak{F}$, for each i we have that $N_i = Z_i \otimes X_i$, where Z_i is a π-special irreducible KA-module with $A/\mathrm{Ker}(A \,\mathrm{on}\, Z_i) \in \mathfrak{X}_1$ and X_i is a π'-special irreducible KA-module with $A/\mathrm{Ker}(A \,\mathrm{on}\, X_i) \in \mathfrak{X}_2$. Note that since all N_i are G-conjugates, so are the Z_i and the X_i, because if $N_i^g \cong N_1$ for some $G \in G$, then $Z_i^g \otimes X_i^g \cong (Z_i \otimes X_i)^g = N_i^g \cong N_1 = Z_1 \otimes X_1$, and thus $Z_i^g \cong Z_1$ and $X_i^g \cong X_1$, by [CK87, 2.4].

We break the proof into a number of cases.

Case 1. Suppose that all of the Z_i, as well as all of the X_i, are isomorphic. This is equivalent to saying that N_A is homogeneous. If $p = q$, then N_A is irreducible by [DH92, B, 8.3]. This implies that N_U is irreducible and either $N \in \mathfrak{M}(U)$ or $N \notin \mathfrak{M}(U)$. This contradiction yields $p \neq q$. Since N_U is a completely irreducible U-module, we can write $N_U = L_1 \oplus \cdots \oplus L_u$, where L_i are irreducible KU-modules. Analogously, $N_V = P_1 \oplus \cdots \oplus P_v$, where P_i are irreducible KV-modules.

If L_j is an irreducible component of N_U such that N_i is a component of $(L_j)_A$, then $(L_j)_A \cong t_j N_i$ for some t_j. Since q divides $|K| - 1$ by [HB82a, VII, 2.6], we have that t_j is either 1 or q by [DH92, B, 8.5].

Analogously, if P_k is an irreducible component of N_V such that N_i is a component of $(P_k)_A$, then either $(P_k)_A = N_i$ or $(P_k)_A \cong qN_i$. We have that

$N_A = N_1 \oplus \cdots \oplus N_r$, and each irreducible component N_i is $N_i = Z \otimes X$, with Z a π-special KA-module and X a π'-special KA-module. Applying [CK87, 2.3], there is a unique π'-special KU-module Y contained in X^U such that $X = Y_A$. Moreover, Z^U is completely reducible by [HB82a, VII, 9.4]. Let W be an irreducible component of Z^U. By the Nakayama's reciprocity theorem, $0 \neq \mathrm{Hom}_U(Z^U, W) \cong \mathrm{Hom}_A(Z, W_A)$ ([DH92, B, 6.5]). Therefore Z is an irreducible component of W_A. Since Z is π-special, then so is W by [CK87, 2.3]. It is clear that the inertia subgroup of Z in U is the whole U. Then W_A is homogeneous, i.e. $W_A \cong tZ$. Again, by [DH92, B, 8.5], either $t = 1$ or $t = q$. Assume that $t = q$. Therefore, we have that $\dim W = \dim Z^U = q \dim Z$. This implies that $W \cong Z^U$ and Z^U is a π-special KU-module. Let L be any irreducible KU-module such that $Z \otimes X$ is a component of L_A. It follows that $Z^U \otimes Y$ is irreducible by [CK87, 2.4]. By [HB82a, VII, 4.5 (a)], we have that $(Z \otimes X)^U = (Z \otimes Y_A)^U \cong Z^U \otimes Y$. Applying Nakayama's reciprocity theorem ([DH92, B, 6.5]), it follows that $0 \neq \mathrm{Hom}_A(Z \otimes X, L_A) \cong \mathrm{Hom}_U((Z \otimes X)^U, L)$. Consequently $Z^U \otimes Y \cong L$. This implies that $L_i \cong Z^U \otimes Y$ for all $i \in \{1, \ldots, u\}$ and N_U is homogeneous, contrary to $0 \neq D(N_U) \neq N$. Hence $t = 1$, and W has the same dimension as Z. Consequently, $W \otimes Y$ is an irreducible KU-module with $(W \otimes Y)_A = Z \otimes X$.

For any irreducible component L_j of N_U, it follows from Lemma 6.5.29 that $L_j = (W \otimes Y) \otimes J_j$, where J_j is an irreducible $K(U/A)$-module (regarded as KU-module) and $\dim J_j = 1$ or q. Since U/A is cyclic, it follows that $\dim J_j = 1$ by [DH92, B, 9.2]. Hence $(L_j)_A = N_i$.

Arguing with V, we have that if $N_V = P_1 \oplus \cdots \oplus P_v$, with the P_i irreducible V-modules, and P_k is an irreducible component of N_V such that N_i is a component of $(P_k)_A$, then $(P_k)_A = N_i$ is irreducible.

It implies that N_i is in fact U-module and V-module. Therefore $N_i = N$ is an irreducible G-module. This is a contradiction.

Case 2. Suppose that not all of the X_i are isomorphic. We let T denote the inertia subgroup of X_1 and note that $A \leq T \neq G$. Since G/A is a q-group generated by U/A and V/A, we have that there is a maximal normal subgroup M of G satisfying $T \leq M$ and so either U or V is not contained in M. We may suppose that U is not contained in M. Recall that all X_i are isomorphic to G-conjugates of X_1, and so the inertia subgroups are conjugate in G. It then follows that U is not contained in the inertia subgroup of any X_i. Now let L be a component of N_U and suppose that N_1 is a component of L_A. If L is π-factorable, then $L = D \otimes E$ with D π-special and E π'-special. Note that $L_A = D_A \otimes E_A$; if $D_A = D_1 \oplus \cdots \oplus D_m$ with all D_i irreducible A-modules, then D_i is π-special for all $i \in \{1, \ldots, m\}$ by [CK87, 2.2]. Suppose that E_A is irreducible. Then $L_A = (D_1 \otimes E_A) \oplus \cdots \oplus (D_m \otimes E_A)$. Therefore E_A is isomorphic to X_1 by [CK87, 2.4], and then U is contained in the inertia subgroup of X_1, contrary to supposition. Hence we cannot have E_A irreducible. By Clifford's theorem, since the inertia subgroup of X_1 in U is A, we have that E is the direct sum of $q = |U/A|$ irreducible modules conjugate to X_1. But then the dimension of E is not a π'-number. This contradiction yields that L cannot be a π-factorable module. It follows that no component of N_U can be π-factorable, and so no component of N_U can be in $\mathfrak{M}(U)$, i.e. $N_U = D(N_U)$, and we have reached a contradiction.

Case 3. Suppose that all of the X_i are isomorphic. By Case 1, we may assume that not all the Z_i are isomorphic and let T be the inertia subgroup of Z_1. As before, it follows that we may suppose that U is not contained in the inertia subgroup of any Z_i.

Now let L be any irreducible KU-module such that $Z_1 \otimes X_1$ is a component of L_A. We then have that X_1 has a unique extension to a π'-special KU-module, $(X_1)^*$ say by [CK87, 2.3]. Also, since Z_1 is not U-invariant, we have that $(Z_1)^U$ is irreducible by [DH92, B, 7.8] and π-special by [CK87, 2.3]. It follows that $(Z_1)^U \otimes (X_1)^*$ is irreducible by [CK87, 2.4]. By [HB82a, VII, 4.5], we have that $(Z_1 \otimes X_1)^U = \left(Z_1 \otimes ((X_1)^*)_A \right)^U \cong Z_1^U \otimes (X_1)^*$. Now $0 \neq \operatorname{Hom}_A(Z_1 \otimes X_1, L_A) \cong \operatorname{Hom}_U\left((Z_1 \otimes X_1)^U, L \right)$ by the Nakayama's reciprocity theorem ([DH92, B, 6.5]). Then L is isomorphic to $(Z_1)^U \otimes (X_1)^*$.

It follows that if $N_U = L_1 \oplus \cdots \oplus L_U$ with the L_i irreducible, then $L_i = (Z_i)^* \oplus (X_i)^*$ with $(Z_i)^*$ π-special and $(X_i)^*$ π'-special, $1 \leq i \leq u$. In each case the π'-special factor is isomorphic to $(X_1)^*$, and thus if $U/\operatorname{Ker}\left(U \text{ on}(X_1)^* \right)$ is not in $\mathfrak{X}_2$, then no component of N_U is in $\mathfrak{M}_U$, i.e. $N_U = D(N_U)$. This contradiction proves that $U/\operatorname{Ker}\left(U \text{ on}(X_1)^* \right) \in \mathfrak{X}_2$. Some of the L_j is in $\mathfrak{M}(U)$. Suppose $L_i \in \mathfrak{M}(U)$. Then the group $U/\operatorname{Ker}\left(U \text{ on}(Z_i)^* \right) \in \mathfrak{X}_1$. On the other hand, since $A \in \mathfrak{F}$, the group $A/\operatorname{Ker}(A \text{ on } Z_j)$ belongs to $\mathfrak{X}_1$ for all j. Recall that all Z_j are conjugate and then so are the $\operatorname{Ker}(A \text{ on } Z_j)$. Since $(Z_j)^* = (Z_j)^U$, we have that $\operatorname{Ker}\left(U \text{ on}(Z_j)^* \right) = \operatorname{Core}_U\left(\operatorname{Ker}(A \text{ on } Z_j) \right)$. Thus $A/\operatorname{Ker}\left(U \text{ on}(Z_j)^* \right) \in \mathfrak{X}_1$.

At this point, we must invoke the special form of $\mathfrak{X}_i$, $i = 1, 2$. Since $U/\operatorname{Ker}\left(U \text{ on}(Z_i)^* \right) \in \mathfrak{S}_\rho$ and $\operatorname{Ker}\left(U \text{ on}(Z_i)^* \right)$ is contained in A, we must have $Q \in \rho$. Then $U/\operatorname{Ker}\left(U \text{ on}(Z_j)^* \right)$ is a ρ-group and hence is in $\mathfrak{X}_1$ for all j. Thus $L_j \in \mathfrak{M}(U)$ for all j. In other words, $D(N_U) = 0$. This final contradiction proves $G^{\mathfrak{F}} = \langle U^{\mathfrak{F}}, V^{\mathfrak{F}} \rangle$ and then $\mathfrak{F}$ has the Wielandt property for residuals. $\qquad \square$

Examples 6.5.30. 1. Set $\mathcal{P} = \{\pi = \{p\}, \pi' = \{p\}'\}$, p a prime, $\mathfrak{X}_1 = (1)$ and $\mathfrak{X}_2 = \mathfrak{S}$, then $\mathfrak{T}(1, p, \mathcal{P}, \mathfrak{X}_1, \mathfrak{X}_2) = \mathfrak{T}(1, \mathfrak{M}^p)$ are the Fitting classes introduced by T. O. Hawkes in [Haw70]. Applying Theorem 6.5.28, $\mathfrak{T}(1, \mathfrak{M}^p)$ satisfies the Wielandt property for residuals.

2. The Fitting formations studied by K. L. Haberl and H. Heineken in [HH84] can be seen as Fitting formations constructed by the Cossey-Kanes method with $\mathfrak{X}_1 = \mathfrak{S}$ and $\mathfrak{X}_2 = (1)$. Hence, by Theorem 6.5.28, these classes also satisfy the Wielandt property for residuals.

Let $\mathfrak{F}$ be a Fitting formation satisfying the Wielandt property for residuals. In general, the $\mathfrak{F}$-residual of a group generated by two $\mathfrak{F}$-subnormal subgroups is not the subgroup generated by their $\mathfrak{F}$-residuals, as the following example shows.

Example 6.5.31. Let $\mathfrak{F}$ be the saturated Fitting formation of all groups of nilpotent length at most 2. Then $\mathfrak{F}$ is a subgroup-closed formation of soluble groups. Applying Corollary 6.5.26, $\mathfrak{F}$ has the Wielandt property for residuals.

Let G be the symmetric group of degree 4. If A is the alternating group of degree 4 and B is a Sylow 2-subgroup of G, then A and B are both $\mathfrak{F}$-subnormal in $G = \langle A, B \rangle$, A and B belong to $\mathfrak{F}$, but $G \notin \mathfrak{F}$.

From this example the following problem arises:

Find a precise description of those formations $\mathfrak{F}$ for which the $\mathfrak{F}$-residual of a group generated by two $\mathfrak{F}$-subnormal subgroups is the subgroup generated by their $\mathfrak{F}$-residuals.

We will be mainly concerned with this problem from now on. Our treatment of the question closely follows the approaches developed in the papers of S. F. Kamornikov [Kam96] and A. Ballester-Bolinches, M. C. Pedraza-Aguilera, and M. D. Pérez-Ramos [BBPAPR96], and A. Ballester-Bolinches [BB05].

For the purposes of this discussion, let $\mathfrak{F}$ be a fixed, but arbitrary, formation.

Definition 6.5.32. *1. We say that* $\mathfrak{F}$ *has the generalised Wielandt property for residuals,* $\mathfrak{F}$ *is a GWP-formation for short, if* $\mathfrak{F}$ *enjoys the following property:*

If G is a group generated by two $\mathfrak{F}$-subnormal subgroups A and B, then $G^{\mathfrak{F}} = \langle A^{\mathfrak{F}}, B^{\mathfrak{F}} \rangle$.

2. $\mathfrak{F}$ satisfies the Kegel-Wielandt property for residuals, $\mathfrak{F}$ is a KW-formation for short, if $\mathfrak{F}$ has the following property:

Let $G = \langle A, B \rangle$ be a group generated by two K-$\mathfrak{F}$-subnormal subgroups A and B. Then $G^{\mathfrak{F}} = \langle A^{\mathfrak{F}}, B^{\mathfrak{F}} \rangle$.

Obviously, every KW-formation is a GWP-formation. We show in the following that that the converse holds for saturated formations, and the soluble GWP-formations are exactly the soluble subgroup-closed saturated lattice formations. We need a couple of preliminary results.

To be a subgroup-closed Fitting formation is a necessary condition for a formation $\mathfrak{F}$ to have the generalised Wielandt property for residuals.

Lemma 6.5.33. *If $\mathfrak{F}$ is a GWP-formation, then $\mathfrak{F}$ is a subgroup-closed Fitting formation.*

Proof. First suppose, by way of contradiction, that $\mathfrak{F}$ is not subgroup-closed. Let G be an $\mathfrak{F}$-group of minimal order having a subgroup not in $\mathfrak{F}$ and, among subgroups of G not in $\mathfrak{F}$, let M be one of maximal order. Then M is a maximal subgroup of G. Since $G^{\mathfrak{F}} = 1$, it follows that M is $\mathfrak{F}$-subnormal in G. Since $\mathfrak{F}$ is a GWP-formation, we have that $M^{\mathfrak{F}} = \langle M^{\mathfrak{F}}, 1 \rangle = \langle M^{\mathfrak{F}}, G^{\mathfrak{F}} \rangle = G^{\mathfrak{F}} = 1$. This contradicts the choice of G. Consequently $\mathfrak{F}$ is subgroup-closed. In particular, $\mathfrak{F}$ is s_n-closed. To complete the proof we now show that $\mathfrak{F}$ is N_0-closed. Suppose that this is not true and derive a contradiction. Let G be a group of minimal order having two normal subgroups N_1 and N_2 such that $G = N_1 N_2$ and $N_i \in \mathfrak{F}$ for $i = 1$, 2. If N is a minimal normal subgroup of G, it follows that $G/N \in \mathfrak{F}$. Therefore G is in the boundary of $\mathfrak{F}$ and $N = G^{\mathfrak{F}}$ is the unique minimal normal subgroup of G. It is clear that $N_i \neq 1$, $i = 1$, 2. Hence N is contained in $N_1 \cap N_2$ and thus N_i is $\mathfrak{F}$-subnormal in G, $i = 1$, 2 by Lemma 6.1.7 (1). Since $\mathfrak{F}$ is a GWP-formation, it follows that $G^{\mathfrak{F}} = N_1^{\mathfrak{F}} N_2^{\mathfrak{F}} = 1$, contrary to supposition. Therefore $\mathfrak{F}$ is N_0-closed. The proof of the lemma is now complete.   □

The following result is another step to attain our objectives.

Theorem 6.5.34. *Let $\mathfrak{F}$ be a GWP-formation. Then $\mathfrak{F}$ is a lattice formation.*

Proof. By Lemma 6.5.33, $\mathfrak{F}$ is subgroup-closed. Hence the intersection of $\mathfrak{F}$-subnormal subgroups of a group is $\mathfrak{F}$-subnormal by Lemma 6.1.7 (3).

Suppose that $\mathfrak{F}$ is not a lattice formation and derive a contradiction. By this supposition, there exists a group G of minimal order such that $s_{n\mathfrak{F}}(G)$ is not a sublattice of the subgroup lattice of G. In particular, G has two $\mathfrak{F}$-subnormal subgroups A and B such that $\langle A, B \rangle$ is not $\mathfrak{F}$-subnormal in G.

Let N be a minimal normal subgroup of G. Then AN/N and BN/N are $\mathfrak{F}$-subnormal in G/N by Lemma 6.1.6 (3). Hence $\langle AN/N, BN/N \rangle = \langle A, B \rangle N/N$ is $\mathfrak{F}$-subnormal in G/N by minimality of G. Therefore $X = \langle A, B \rangle N$ is $\mathfrak{F}$-subnormal in G by Lemma 6.1.6 (2). Since A and B are $\mathfrak{F}$-subnormal in X by Lemma 6.1.7 (2), it follows that $\langle A, B \rangle$ is $\mathfrak{F}$-subnormal in X provided that X is a proper subgroup of G. This would imply the $\mathfrak{F}$-subnormality of $\langle A, B \rangle$ in G by Lemma 6.1.6 (1). Consequently $G = \langle A, B \rangle N$ for every minimal normal subgroup N of G. Hence either $G = \langle A, B \rangle$ or $\text{Core}_G(\langle A, B \rangle) = 1$. If $G = \langle A, B \rangle$, then $\langle A, B \rangle$ is $\mathfrak{F}$-subnormal in G, contrary to supposition. Hence $\text{Core}_G(\langle A, B \rangle) = 1$. On the other hand, $A^{\mathfrak{F}}$ and $B^{\mathfrak{F}}$ are subnormal subgroups of G by Lemma 6.1.9 (1). Hence $\langle A^{\mathfrak{F}}, B^{\mathfrak{F}} \rangle$ is subnormal in G and so N normalises $\langle A^{\mathfrak{F}}, B^{\mathfrak{F}} \rangle$ ([DH92, A, 14.3 and 14.4]). Since $\mathfrak{F}$ is a GWP-formation, we have that $\langle A^{\mathfrak{F}}, B^{\mathfrak{F}} \rangle = \langle A, B \rangle^{\mathfrak{F}}$. This implies that $\langle A, B \rangle^{\mathfrak{F}}$ is normal in G. Hence $\langle A, B \rangle^{\mathfrak{F}} \leq \text{Core}_G(\langle A, B \rangle) = 1$ and $\langle A, B \rangle$ is an $\mathfrak{F}$-group. Let us consider the subgroup AN. Clearly N is not contained in A. If $N^{\mathfrak{F}} = N$, then no simple component of N belongs to $\mathfrak{F}$ and thus $(AN)^{\mathfrak{F}} = N$. This contradicts the fact that A is $\mathfrak{F}$-subnormal in AN (Lemma 6.1.6 (2)). Therefore $N \in \mathfrak{F}$. This implies that $G \in \text{E K}(\mathfrak{F})$ and so N is $\mathfrak{F}$-subnormal in G by Proposition 6.1.10. Then N is an $\mathfrak{F}$-subnormal subgroup of AN by Lemma 6.1.6 (2). In particular, $(AN)^{\mathfrak{F}} = A^{\mathfrak{F}} N^{\mathfrak{F}} = 1 = (BN)^{\mathfrak{F}}$ because the property of $\mathfrak{F}$. Since $G = \langle AN, BN \rangle$ and $\mathfrak{F}$ is a GWP-formation, it follows that $G^{\mathfrak{F}} = \langle (AN)^{\mathfrak{F}}, (BN)^{\mathfrak{F}} \rangle = 1$. This final contradiction proves that $\mathfrak{F}$ is a lattice formation.   □

A challenging unsolved problem in the theory of formations is whether the converse of Theorem 6.5.34 is true. The chance of finding the answer seems remote. With our present knowledge even the saturated case remains unanswered.

We shall prove a result that provides a test for a subgroup-closed saturated lattice formation to be a GWP-formation in terms of its boundary. This allows us to present the complete answer to the problem in the soluble universe and give interesting examples.

As in the case of groups generated by subnormal subgroups, we thought it would be desirable to collect the arguments common to our next results. Let $\mathfrak{F}$ be a subgroup-closed Fitting formation. Given a group Z, we denote by $\mathcal{R}(Z, \mathfrak{F})$ the set of all pairs (H, K) such that H and K are $\mathfrak{F}$-subnormal subgroups of $\langle H, K \rangle$ and $\langle H^{\mathfrak{F}}, K^{\mathfrak{F}} \rangle < \langle H, K \rangle^{\mathfrak{F}}$. Let $\mathcal{W}(\mathfrak{F})$ denote the class of all groups Z such that $\mathcal{R}(Z, \mathfrak{F}) \neq \emptyset$.

If $\mathfrak{F}$ is not a GWP-formation, then the class $\mathcal{W}(\mathfrak{F})$ is not empty.

In the following we analyse the structure of a group G of minimal order in $\mathcal{W}(\mathfrak{F})$. Then G has two $\mathfrak{F}$-subnormal subgroups A and B such that $\langle A, B \rangle^{\mathfrak{F}} \neq \langle A^{\mathfrak{F}}, B^{\mathfrak{F}} \rangle$. Choose A and B with $|A| + |B|$ maximal.

Arguing as in the subnormal case, we have:

Result 6.5.35. $G = \langle A, B \rangle$, and

Result 6.5.36. $\operatorname{Soc}(G) \leq G^{\mathfrak{F}}$ and $G^{\mathfrak{F}} = \langle A^{\mathfrak{F}}, B^{\mathfrak{F}} \rangle N$ for any minimal normal subgroup of G. In particular, $\operatorname{Core}_G(\langle A^{\mathfrak{F}}, B^{\mathfrak{F}} \rangle) = 1$.

Result 6.5.37. $\langle A^{\mathfrak{F}}, B^{\mathfrak{F}} \rangle$ is normal in $G^{\mathfrak{F}}$.

Proof. Applying Lemma 6.1.9 (1), $A^{\mathfrak{F}}$ and $B^{\mathfrak{F}}$ are subnormal subgroups of G. Hence $\operatorname{Soc}(G) \leq \operatorname{N}_G(\langle A^{\mathfrak{F}}, B^{\mathfrak{F}} \rangle)$ by [DH92, A, 14.3 and 14.4]. This implies that $\langle A^{\mathfrak{F}}, B^{\mathfrak{F}} \rangle$ is normal in $G^{\mathfrak{F}}$. □

Result 6.5.38. $G^{\mathfrak{F}} \in \operatorname{Q} \operatorname{R}_0(N)$ for any minimal normal subgroup N of G.

Proof. Let N be a minimal normal subgroup of G. Then $G^{\mathfrak{F}} = \langle A^{\mathfrak{F}}, B^{\mathfrak{F}} \rangle N$ and $\langle A^{\mathfrak{F}}, B^{\mathfrak{F}} \rangle \trianglelefteq G^{\mathfrak{F}}$ by Results 6.5.36 and 6.5.37. Hence $(G^{\mathfrak{F}})^{\operatorname{Q} \operatorname{R}_0(N)}$ is a normal subgroup of G contained in $\langle A^{\mathfrak{F}}, B^{\mathfrak{F}} \rangle$. Since $\operatorname{Core}_G(\langle A^{\mathfrak{F}}, B^{\mathfrak{F}} \rangle) = 1$ by Result 6.5.36, we have that $(G^{\mathfrak{F}})^{\operatorname{Q} \operatorname{R}_0(N)} = 1$ and $G^{\mathfrak{F}} \in \operatorname{Q} \operatorname{R}_0(N)$. □

Result 6.5.39. $N \in \mathfrak{F}$ for any minimal normal subgroup N of G.

Proof. Since $N^{\mathfrak{F}}$ is normal in G, we have that either $N^{\mathfrak{F}} = 1$ or $N^{\mathfrak{F}} = N$. Assume that $N^{\mathfrak{F}} = N$. By Lemma 6.1.6 (2), A is $\mathfrak{F}$-subnormal in AN. Hence $A^{\mathfrak{F}}$ is normal in AN by Lemma 6.1.9 (1) and [DH92, A, 14.3]. This implies that $AN/A^{\mathfrak{F}}N \in \mathfrak{F}$ and so $(AN)^{\mathfrak{F}} = A^{\mathfrak{F}}N$. Hence $AN = A(AN)^{\mathfrak{F}}$. The $\mathfrak{F}$-subnormality of A in AN yields $AN = A$. Since $\mathfrak{F}$ is subgroup-closed, it follows that $NA^{\mathfrak{F}}/A^{\mathfrak{F}} \in \mathfrak{F}$, whence $N/(N \cap A^{\mathfrak{F}}) \in \mathfrak{F}$. Therefore $N = N \cap A^{\mathfrak{F}}$ and $G^{\mathfrak{F}} = \langle A^{\mathfrak{F}}, B^{\mathfrak{F}} \rangle$, contrary to our initial supposition. Consequently $N \in \mathfrak{F}$. □

Result 6.5.40. $G^{\mathfrak{F}} \in \mathfrak{F}$.

Proof. Let N be a minimal normal subgroup of G contained in $G^{\mathfrak{F}}$. Then $N \in \mathfrak{F}$ by Result 6.5.39 and $G^{\mathfrak{F}} \in {}_{\mathrm{Q}}\mathrm{R}_0(N)$ by Result 6.5.38. Hence $G^{\mathfrak{F}} \in \mathfrak{F}$. $\square$

Result 6.5.41. $G^{\mathfrak{F}}$ *is contained in* $A \cap B$. *In particular,* $A^{\mathfrak{F}} B^{\mathfrak{F}}$ *is a subgroup of* G.

Proof. Clearly $AG^{\mathfrak{F}}$ is a proper subgroup of G. Hence $AG^{\mathfrak{F}}$ is contained in a maximal $\mathfrak{F}$-normal subgroup of G. The minimality of G yields $(AG^{\mathfrak{F}})^{\mathfrak{F}} = A^{\mathfrak{F}}$. Assume that A is a proper subgroup of $AG^{\mathfrak{F}}$. Since $AG^{\mathfrak{F}}$ is $\mathfrak{F}$-subnormal in G, it follows that $G^{\mathfrak{F}} = \langle (AG^{\mathfrak{F}})^{\mathfrak{F}}, B^{\mathfrak{F}} \rangle = \langle A^{\mathfrak{F}}, B^{\mathfrak{F}} \rangle$ by the choice of the pair (A, B), contrary to our initial supposition. Hence $A = AG^{\mathfrak{F}}$ and $G^{\mathfrak{F}}$ is contained in A. Analogously $G^{\mathfrak{F}}$ is contained in B. $\square$

With the same arguments to those used in Statement 6.5.10, we have:

Result 6.5.42. *If* $G^{\mathfrak{F}}$ *is non-abelian, then* G *is in the boundary of* $\mathfrak{F}$.

Suppose now that there exists a family of subgroup-closed formations $\{\mathfrak{F}_i\}_{i \in \mathcal{I}}$ such that $\pi(\mathfrak{F}_i) \cap \pi(\mathfrak{F}_j) = \emptyset$, $i \neq j$, and $\mathfrak{F} = \bigtimes_{i \in \mathcal{I}} \mathfrak{F}_i$.

Result 6.5.43. *There exist* i, $j \in I$ *such that* $G/G^{\mathfrak{F}} \in \mathfrak{F}_i$ *and* $G^{\mathfrak{F}} \in \mathfrak{F}_j$. *Moreover if either* $G \notin \mathrm{b}(\mathfrak{F})$ *or* $G^{\mathfrak{F}}$ *is non-abelian, then* $i = j$.

Proof. By Result 6.5.40, we have that $G^{\mathfrak{F}} \in \mathfrak{F}$ and, by Result 6.5.38, $G^{\mathfrak{F}}$ is a direct product of copies of a simple group. Hence there exists $j \in I$ such that $G^{\mathfrak{F}} \in \mathfrak{F}_j$. On the other hand, $G/G^{\mathfrak{F}} = X_{i_1}/G^{\mathfrak{F}} \times \cdots \times X_{i_t}/G^{\mathfrak{F}}$, where $X_{i_k}/G^{\mathfrak{F}} \in \mathfrak{F}_{i_k}$ is a Hall $\pi(\mathfrak{F}_{i_k})$-subgroup of $G/G^{\mathfrak{F}}$, $1 \leq k \leq t$, for some set $\{i_1, \ldots, i_t\} \subseteq I$. Let $k \in \{1, \ldots, t\}$. Then $X_{i_k}/G^{\mathfrak{F}} = \langle (A \cap X_{i_k})/G^{\mathfrak{F}}, (B \cap X_{i_k})/G^{\mathfrak{F}} \rangle = \langle A \cap X_{i_k}, B \cap X_{i_k} \rangle/G^{\mathfrak{F}}$ and $X_{i_k} = \langle A \cap X_{i_k}, B \cap X_{i_k} \rangle$. Applying Lemma 6.1.7 (2), $A \cap X_{i_k}$ and $B \cap X_{i_k}$ are $\mathfrak{F}$-subnormal subgroups of X_{i_k}. Assume that X_{i_k} is a proper subgroup of G for all $k \in \{1, \ldots, t\}$. Then $X_{i_k}^{\mathfrak{F}} = \langle (A \cap X_{i_k})^{\mathfrak{F}}, (B \cap X_{i_k})^{\mathfrak{F}} \rangle$ by the minimal choice of G, leading to $X_{i_k}^{\mathfrak{F}} = 1$. This is due to the fact that $X_{i_k}^{\mathfrak{F}}$ is a normal subgroup of G contained in $\langle A^{\mathfrak{F}}, B^{\mathfrak{F}} \rangle$ and $\mathrm{Core}_G(\langle A^{\mathfrak{F}}, B^{\mathfrak{F}} \rangle) = 1$ by Result 6.5.36. Hence $G \in \mathrm{N}_0 \mathfrak{F} = \mathfrak{F}$, contrary to hypothesis. Therefore there exists an index $i = i_k \in \{i_1, \ldots, i_t\}$ such that $X_i = G$. This means that $G/G^{\mathfrak{F}} \in \mathfrak{F}_i$.

Suppose that $i \neq j$. Then $G^{\mathfrak{F}}$ is a Hall $\pi(\mathfrak{F}_j)$-subgroup of G and there exists a Hall $\pi(\mathfrak{F}_i)$-subgroup C of G such that $G = G^{\mathfrak{F}}C$ and $G^{\mathfrak{F}} \cap C = 1$. It follows that $A/A^{\mathfrak{F}} = G^{\mathfrak{F}}/A^{\mathfrak{F}} \times (C \cap A)A^{\mathfrak{F}}/A^{\mathfrak{F}}$ and so A normalises $A^{\mathfrak{F}} B^{\mathfrak{F}}$. Analogously B normalises $A^{\mathfrak{F}} B^{\mathfrak{F}}$. It implies that $A^{\mathfrak{F}} B^{\mathfrak{F}} = 1$ by Result 6.5.36 and $G^{\mathfrak{F}}$ is a minimal normal subgroup of G. Hence $G \in \mathrm{b}(\mathfrak{F})$.

If $G^{\mathfrak{F}}$ is non-abelian, $\mathrm{C}_G(G^{\mathfrak{F}}) = 1$. Since $A = G^{\mathfrak{F}} \times (C \cap A)$ and $B = G^{\mathfrak{F}} \times (C \cap B)$, it follows that $A = B = G^{\mathfrak{F}}$. Then, by Results 6.5.35 and 6.5.41, $A = B = G$, and this contradicts our initial hypothesis.

Consequently if either $G \notin \mathrm{b}(\mathfrak{F})$ or $G^{\mathfrak{F}}$ is non-abelian, we have that $i = j$. $\square$

Assume now that $\mathfrak{F}$ is a saturated GWP-formation. Then $\mathfrak{F}$ is a lattice formation by Theorem 6.5.34 and, since $\mathfrak{F}$ is saturated, it follows that $\mathfrak{F}$ is a K-lattice formation by Theorem 6.3.9.

Suppose that $\mathfrak{F}$ is not a KW-formation. Then there exists a group G and a pair (A, B) of K-$\mathfrak{F}$-subnormal subgroups of G such that $G = \langle A, B \rangle$ and $G^{\mathfrak{F}} \neq \langle A^{\mathfrak{F}}, B^{\mathfrak{F}} \rangle$. Let us take (A, B) satisfying $|A| + |B|$ maximal. Then, as in the above reductions, G enjoys the properties stated in Results 6.5.35, 6.5.36, 6.5.37, and 6.5.38. G also has the following property.

Result 6.5.44. $G^{\mathfrak{F}} \in \mathfrak{F}$.

Proof. Consider the subgroup $M = \langle A, B^{\mathfrak{F}} \rangle$. Suppose that $M = G$. Then $G = AG^{\mathfrak{F}}$. Since by Lemma 6.1.9 (1), $A^{\mathfrak{F}}$ is subnormal in G and $G^{\mathfrak{F}}$ is a direct product of isomorphic simple groups by Result 6.5.38, it follows that $G^{\mathfrak{F}}$ normalises $A^{\mathfrak{F}}$ and so $A^{\mathfrak{F}}$ is a normal subgroup of G. By Result 6.5.36, we have that $A \in \mathfrak{F}$. By virtue of Lemma 6.3.8, it follows that $A \leq G_{\mathfrak{F}}$. If $G_{\mathfrak{F}} \cap G^{\mathfrak{F}} \neq 1$, then $G^{\mathfrak{F}} \in \mathfrak{F}$ and the result follows. Hence $G_{\mathfrak{F}} \cap G^{\mathfrak{F}} = 1$ and $G = G_{\mathfrak{F}} \times G^{\mathfrak{F}}$. By Result 6.5.36, we have that $\mathrm{Soc}(G) \leq G^{\mathfrak{F}}$. It implies that $G_{\mathfrak{F}} = 1$ and so $A = 1$ and $G = B$, giving a contradiction. Therefore we may assume that M is a proper subgroup of G. The choice of G, Lemma 6.1.6 (1) and Lemma 6.1.7 (2) imply that $M^{\mathfrak{F}} = \langle A, B^{\mathfrak{F}} \rangle^{\mathfrak{F}} = \langle A^{\mathfrak{F}}, (B^{\mathfrak{F}})^{\mathfrak{F}} \rangle$.

Arguing in a similar way with B, we have $\langle A^{\mathfrak{F}}, B \rangle^{\mathfrak{F}} = \langle (A^{\mathfrak{F}})^{\mathfrak{F}}, B^{\mathfrak{F}} \rangle$. If either $A < \langle A, B^{\mathfrak{F}} \rangle$ or $B < \langle A^{\mathfrak{F}}, B \rangle$, it follows that $G^{\mathfrak{F}} = \langle \langle A, B^{\mathfrak{F}} \rangle^{\mathfrak{F}}, \langle A^{\mathfrak{F}}, B \rangle^{\mathfrak{F}} \rangle = \langle A^{\mathfrak{F}}, B^{\mathfrak{F}} \rangle$ by the choice of G (note that $\mathfrak{F}$ is subgroup-closed). This contradiction yields $A = \langle A, B^{\mathfrak{F}} \rangle$ and $B = \langle A^{\mathfrak{F}}, B \rangle$. Then $B^{\mathfrak{F}}$ is contained in A and $A^{\mathfrak{F}}$ is a normal subgroup of $G^{\mathfrak{F}}$. Hence $(A \cap G^{\mathfrak{F}})/A^{\mathfrak{F}}$ is a K-$\mathfrak{F}$-subnormal $\mathfrak{F}$-subgroup of $G^{\mathfrak{F}}/A^{\mathfrak{F}}$ by Lemma 6.1.7 (2) and Lemma 6.1.6 (3). Applying Lemma 6.3.8, $(A \cap G^{\mathfrak{F}})/A^{\mathfrak{F}}$ is contained in $(G^{\mathfrak{F}}/A^{\mathfrak{F}})_{\mathfrak{F}}$. If $A^{\mathfrak{F}} \neq A \cap G^{\mathfrak{F}}$, then $(G^{\mathfrak{F}}/A^{\mathfrak{F}})_{\mathfrak{F}} \neq 1$ and $G^{\mathfrak{F}} \in \mathfrak{F}$ because $G^{\mathfrak{F}}$ is a direct product of simple groups. Therefore we may assume $A^{\mathfrak{F}} = A \cap G^{\mathfrak{F}}$. In this case, $B^{\mathfrak{F}}$ is contained in $A^{\mathfrak{F}}$. Arguing in a similar way with B, we conclude that $A^{\mathfrak{F}}$ is contained in $B^{\mathfrak{F}}$. Consequently $A^{\mathfrak{F}} = B^{\mathfrak{F}}$ is a normal subgroup of G. By Result 6.5.36, A and B are $\mathfrak{F}$-groups. By Lemma 6.3.8, $G \in \mathfrak{F}$ and $G^{\mathfrak{F}} = 1$. $\qquad\square$

This completes our preparations, and we can now deduce the main results.

Theorem 6.5.45. *Let $\mathfrak{F}$ be a saturated formation. Then:*
$\mathfrak{F}$ is a GWP-formation if and only if $\mathfrak{F}$ is a KW-formation.

Proof. Only the necessity of the condition is in doubt. Assume that $\mathfrak{F}$ is a GWP-formation which is not a KW-formation. Then there exists a group G and a pair (A, B) of K-$\mathfrak{F}$-subnormal subgroups of G such that $G = \langle A, B \rangle$ and $G^{\mathfrak{F}} \neq \langle A^{\mathfrak{F}}, B^{\mathfrak{F}} \rangle$. If $|A| + |B|$ maximal, then $G^{\mathfrak{F}} \in \mathfrak{F}$ by Result 6.5.44. Then $s_{n\mathfrak{F}}(G) = s_{n\mathrm{K}\text{-}\mathfrak{F}}(G)$ by Proposition 6.1.10. This contradiction yields that $\mathfrak{F}$ is a KW-formation. $\qquad\square$

Our next main result provides a test for a subgroup-closed saturated lattice formation to have the generalised Wielandt property for residuals in terms of its boundary.

If $\mathfrak{F}$ is a subgroup-closed Fitting formation, let $\mathrm{b}_n(\mathfrak{F})$ denote the class of all groups $G \in \mathrm{b}(\mathfrak{F})$ such that $\mathrm{Soc}(G)$ is not abelian and G has the properties stated in Results 6.5.35–6.5.43.

Theorem 6.5.46. *Let $\mathfrak{F}$ be a subgroup-closed saturated lattice formation. Then $\mathfrak{F}$ is a GWP-formation if and only if the following condition is fulfilled by all groups $G \in \mathrm{b}_n(\mathfrak{F})$:*

If $G = \langle A, B \rangle$ with A and B $\mathfrak{F}$-subnormal subgroups of G, then $G^{\mathfrak{F}} = \langle A^{\mathfrak{F}}, B^{\mathfrak{F}} \rangle$. $\qquad\qquad$ (6.7)

Proof. It is clear that only the sufficiency of the condition is in doubt.

Assume that Property (6.7) holds. We suppose that $\mathfrak{F}$ is not a GWP-formation and derive a contradiction. Since $\mathcal{W}(\mathfrak{F})$ is not empty, a group G of minimal order in $\mathcal{W}(\mathfrak{F})$ satisfies the properties stated in Results 6.5.35–6.5.43 for a pair of $\mathfrak{F}$-subnormal subgroups A and B of G with $|A| + |B|$ maximal.

Applying Theorem 6.3.15, $\mathfrak{F} = \mathfrak{M} \times \mathfrak{H}$, where $\mathfrak{M}$ is a subgroup-closed saturated Fitting formation such that $\mathfrak{S}_{\pi(\mathfrak{M})}\mathfrak{M} = \mathfrak{M}$ and $\mathfrak{H} = \bigtimes_{i \in I} \mathfrak{S}_{\pi_i}$, with $\pi_l \cap \pi_k = \emptyset$ for all $k \neq l$ in I. Moreover $\pi(\mathfrak{M}) \cap \pi(\mathfrak{H}) = \emptyset$. Since $G \notin \mathrm{b}_n(\mathfrak{F})$, $G^{\mathfrak{F}}$ is an elementary abelian p-group for some prime p by Results 6.5.38 and 6.5.42. Therefore $G^{\mathfrak{F}} \in \mathfrak{M}$ or $G^{\mathfrak{F}} \in \mathfrak{S}_{\pi_i}$ for some $i \in \mathcal{I}$. In addition, by Result 6.5.43, $G/G^{\mathfrak{F}} \in \mathfrak{M}$ or $G/G^{\mathfrak{F}} \in \mathfrak{S}_{\pi_j}$ for some $j \in I$. If $G^{\mathfrak{F}} \in \mathfrak{M}$ and $G/G^{\mathfrak{F}} \in \mathfrak{M}$, then $p \in \pi(\mathfrak{M})$ and $G \in \mathfrak{S}_{\pi(\mathfrak{M})}\mathfrak{M} = \mathfrak{M} \subseteq \mathfrak{F}$, contradicting $G \in \mathcal{W}(\mathfrak{F})$. Assume now that $G/G^{\mathfrak{F}} \in \mathfrak{S}_{\pi_j}$ for some $j \in I$. Then $G^{\mathfrak{F}}$ is a Hall $\pi(\mathfrak{M})$-subgroup of G and there exists a Hall π_j-subgroup C of G such that $G = G^{\mathfrak{F}}C$ and $G^{\mathfrak{F}} \cap C = 1$. Then $A/A^{\mathfrak{F}} = G^{\mathfrak{F}}/A \times (C \cap A)A^{\mathfrak{F}}/A^{\mathfrak{F}}$. It follows that A normalises $A^{\mathfrak{F}}B^{\mathfrak{F}}$. Analogously B normalises $A^{\mathfrak{F}}B^{\mathfrak{F}}$. Consequently $A^{\mathfrak{F}}B^{\mathfrak{F}} = 1$ and A and B are $\mathfrak{F}$-groups. Since $\mathfrak{F}$ is a lattice formation and A and B are $\mathfrak{F}$-subnormal in G, we have that $G \in \mathfrak{F}$ by Theorem 6.3.3 (3). It contradicts our supposition. Suppose that $G^{\mathfrak{F}} \in \mathfrak{S}_{\pi_i}$. If $G/G^{\mathfrak{F}} \in \mathfrak{M}$ or $G/G^{\mathfrak{F}} \in \mathfrak{S}_{\pi_j}$ for some $j \in I$, $i \neq j$, we can argue as above and obtain a contradiction. Hence $G/G^{\mathfrak{F}} \in \mathfrak{S}_{\pi_i}$ and so $G \in \mathfrak{S}_{\pi_i} \subseteq \mathfrak{F}$, contradicting $G \in \mathcal{W}(\mathfrak{F})$. It follows that our supposition is wrong and hence $\mathfrak{F}$ is a GWP-formation. $\qquad\square$

If $\mathfrak{F}$ is a soluble subgroup-closed saturated lattice formation, then $\mathrm{b}_n(\mathfrak{F}) = \emptyset$. Moreover, if $\mathfrak{F}$ is a soluble GWP-formation, then $\mathfrak{F}$ is a subgroup-closed Fitting formation by Lemma 6.5.33, and hence saturated by Theorem 2.5.2. Therefore we have:

Corollary 6.5.47 (see [Kam96, BBPAPR96]). *Let $\mathfrak{F}$ be a soluble formation. Then $\mathfrak{F}$ is a GWP-formation if and only if $\mathfrak{F}$ is a subgroup-closed saturated lattice formation.*

Another interesting examples of GWP-formations follow from the following result.

Corollary 6.5.48 ([Kam96]). *Let $\mathfrak{F}$ be a saturated formation representable as $\mathfrak{F} = \mathfrak{M} \times \mathfrak{H}$, where $\pi(\mathfrak{M}) \cap \pi(\mathfrak{H}) = \emptyset$, $\mathfrak{M} = \mathfrak{M}^2$ is a subgroup-closed saturated Fitting formation, $\mathfrak{H} = \bigtimes_{i \in I} \mathfrak{S}_{\pi_i}$, and moreover $\pi_l \cap \pi_k = \emptyset$ for all $k \neq l$ in I. Then $\mathfrak{F}$ is a GWP-formation.*

Proof. Applying Theorem 6.3.15, $\mathfrak{F}$ is a subgroup-closed saturated lattice formation. Hence, by Theorem 6.5.46, it is enough to check the property in groups in $b_n(\mathfrak{F})$ generated by two $\mathfrak{F}$-subnormal subgroups. Let G be one of them. Then $G \in b(\mathfrak{F})$ and $\mathrm{Soc}(G) = G^{\mathfrak{F}}$ is non-abelian. Moreover, by Result 6.5.43, $G^{\mathfrak{F}} \in \mathfrak{M}$ and $G/G^{\mathfrak{F}} \in \mathfrak{M}$ (note that $\mathfrak{H}$ is a soluble formation). Hence $G \in \mathfrak{M}^2 = \mathfrak{M} \subseteq \mathfrak{F}$. This contradiction proves that $\mathfrak{F}$ is a GWP-formation. $\square$

This completes our discussion about GWP-formations. We can turn this situation on its head and ask the following.

Let $\mathfrak{F}$ be a subgroup-closed formation and let G be a group generated by two $\mathfrak{F}$-subnormal subgroups A and B of G. When do we have $G^{\mathfrak{F}} = \langle A^{\mathfrak{F}}, B^{\mathfrak{F}} \rangle$?

The question is answered in [BBEPA02] for subgroup-closed saturated formations. It is proved there that if G is a group whose derived subgroup is nilpotent, then $G^{\mathfrak{F}} = \langle A^{\mathfrak{F}}, B^{\mathfrak{F}} \rangle$ provided that A and B are $\mathfrak{F}$-subnormal in $G = \langle A, B \rangle$. Furthermore the class $\mathfrak{N}\mathfrak{A}$ of all groups whose derived subgroup is nilpotent is characterised as the largest subgroup-closed saturated formation enjoying that property.

Let $\mathfrak{F}$ be a GWP-formation. Then $\mathfrak{F}$ has the following property:

If A and B are K-$\mathfrak{F}$-subnormal ($\mathfrak{F}$-subnormal) subgroups of a group G and $G = AB$, then $G^{\mathfrak{F}} = A^{\mathfrak{F}} B^{\mathfrak{F}}$. $\qquad\qquad$ (6.8)

In general, Property 6.8 does not characterise the GWP-formations as the class of all 2-nilpotent groups shows. Hence the question of how one subgroup-closed formation satisfying Property 6.8 can be characterised arises. This question is closely related to the characterisation of the subgroup-closed formations satisfying Property 6.1.

The above question has a nice answer in the soluble universe for subgroup-closed saturated formations of full characteristic.

Theorem 6.5.49 ([BBPAPR96]). *Let $\mathfrak{F}$ be a subgroup-closed saturated formation of soluble groups of full characteristic. The following statements are pairwise equivalent:*

1. *$\mathfrak{F}$ satisfies the property:*
    If A and B are two $\mathfrak{F}$-subnormal subgroups of a soluble group G and $G = AB$, then $G^{\mathfrak{F}} = A^{\mathfrak{F}} B^{\mathfrak{F}}$.
2. *For each prime number p, there exists a set of primes $\pi(p)$ with $p \in \pi(p)$ such that $\mathfrak{F}$ is locally defined by the formation function f given by $f(p) = \mathfrak{S}_{\pi(p)}$.*
    These sets of primes satisfy the following property:

If $q \in \pi(p)$, then $\pi(q) \subseteq \pi(p)$ for every pair of primes p and q.

Let $\mathfrak{F}$ be a subgroup-closed saturated formation of full characteristic satisfying the conditions of the above theorem. Then a soluble group G is an $\mathfrak{F}$-group if and only if G has a normal $\pi(p)$-complement for every prime p, where $\pi(p)$ is the set of primes such that $p \in \pi(p)$.

7

Fitting classes and injectors

7.1 A non-injective Fitting class

After B. Fischer, W. Gaschütz, and B. Hartley's result about the injective character of the Fitting classes of soluble groups (Theorem 2.4.26), and bearing in mind the extension of the projective theory to the general universe of finite groups, it seemed to be reasonable to think about the validity of Theorem 2.4.26 outside the soluble realm. It was conjectured then that if $\mathfrak{F}$ is an arbitrary Fitting class and G is a finite group, then $\mathrm{Inj}_{\mathfrak{F}}(G) \neq \emptyset$. In the eighties of the last century, a big effort of some mathematicians was addressed to find methods to obtain injectors for Fitting classes in all finite groups. These efforts were successful for a big number of Fitting classes and they will be presented in Section 7.2. In this atmosphere, the construction of E. Salomon [Sal] of an example of a non-injective Fitting class caused a deep shock.

Salomon's construction, never published, is based in a pull-back construction of induced extensions due to F. Gross and L. G. Kovács (see Section 1.1). The aim of this section is to present the Salomon's example in full detail.

We begin with a quick insight to the group $A = \mathrm{Aut}\big(\mathrm{Alt}(6)\big)$. Let D denote the normal subgroup of inner automorphisms $D \cong \mathrm{Alt}(6)$ of A. It is well-known that the quotient group A/D is isomorphic to an elementary abelian 2-group of order 4 and A does not split over D, i.e. there is no complement of D in A (see [Suz82]).

If u is an involution of $\mathrm{Sym}(6)$, the symmetric group of degree 6, then $\langle u \rangle$ is a complement of $\mathrm{Alt}(6)$ in $\mathrm{Sym}(6)$ and the element u acts on $\mathrm{Alt}(6)$ as an outer automorphism.

Likewise, $\mathrm{Alt}(6) \cong \mathrm{PSL}(2,9)$ but $\mathrm{Sym}(6) \not\cong \mathrm{PGL}(2,9)$ (see [Hup67, pages 183 and 184]). There exist elements of order 2 in $\mathrm{PGL}(2,9)$ which are not in $\mathrm{PSL}(2,9)$ (for instance the coclass of the matrix $\begin{pmatrix} 1 & \\ & -1 \end{pmatrix}$ in the quotient group $\mathrm{GL}(2,9)/\mathrm{Z}\big(\mathrm{GL}(2,9)\big) \cong \mathrm{PGL}(2,9)$). If v is one of these involutions, then $\langle v \rangle$ is a complement of $\mathrm{PSL}(2,9)$ in $\mathrm{PGL}(2,9)$ and the element v acts on $\mathrm{Alt}(6) \cong \mathrm{PSL}(2,9)$ as an outer automorphism.

The subgroup $B = D\langle u \rangle \cong \mathrm{Sym}(6)$ and the subgroup $C = D\langle v \rangle \cong \mathrm{PGL}(2,9)$ are normal subgroups of A of index 2. Clearly $A = BC$ and $B \cap C = D$.

Let S be a non-abelian simple group. If x is an involution in S, define the group homomorphism

$$\alpha_1 \colon B \longrightarrow S \quad \text{such that } \mathrm{Ker}(\alpha_1) = D, \ B^{\alpha_1} = \langle x \rangle,$$

Put $|S \colon \mathrm{Im}(\alpha_1)| = |S|/2 = n_1$, and consider the right transversal

$$T_1 = \{s_1 = 1, s_2, \ldots, s_{n_1}\},$$

of $\mathrm{Im}(\alpha_1)$ in S and the transitive action

$$\rho_1 \colon S \longrightarrow \mathrm{Sym}(n_1)$$

on the set of indices $\mathcal{I}_1 = \{1, \ldots, n_1\}$. For each $i \in \mathcal{I}_1$ and each $s \in S$, $s_i s = x_{i,s} s_j$, for some $x_{i,s} \in \mathrm{Im}(\alpha_1)$ and $i^{s^{\rho_1}} = j$. Write $P_S = S^{\rho_1} \leq \mathrm{Sym}(n_1)$ and consider the monomorphism (see Lemma 1.1.26)

$$\lambda_1 = \lambda_{T_1} \colon S \longrightarrow \mathrm{Im}(\alpha_1) \wr_{\rho_1} P_S,$$

defined by $s^{\lambda_1} = (x_{1,s}, \ldots, x_{n_1,s}) s^{\rho_1}$, for any $x \in S$, and the epimorphism

$$\bar{\alpha}_1 \colon W_1 = B \wr_{\rho_1} P_S \longrightarrow \mathrm{Im}(\alpha_1) \wr_{\rho_1} P_S$$

defined by $\big((b_1, \ldots, b_{n_1})\tau\big)^{\bar{\alpha}_1} = (b_1^{\alpha_1}, \ldots, b_{n_1}^{\alpha_1})\tau$, for $b_1, \ldots, b_{n_1} \in B$ and $\tau \in P_S$. Write $M_1 = \mathrm{Ker}(\bar{\alpha}_1) = D^{n_1} \cong \mathrm{Alt}(6)^{n_1}$.

Construct the induced extension G_1, defined by α_1 (see Definition 1.1.27),

$$E\lambda_1 \colon 1 \longrightarrow M_1 \longrightarrow G_1 \overset{\sigma_1}{\longrightarrow} S \longrightarrow 1$$

Recall that

$$G_1 = \{w \in W_1 : w^{\bar{\alpha}_1} = s^{\lambda_1} \quad \text{for some } s \in S\},$$

and

$$\sigma_1 \colon G_1 \longrightarrow S \quad \text{defined by } w^{\sigma_1} = s, \text{ where } w^{\bar{\alpha}_1} = s^{\lambda_1}.$$

The following diagram is commutative:

$$
\begin{array}{ccccccccc}
E\lambda_1 \colon 1 & \longrightarrow & M_1 & \longrightarrow & G_1 & \overset{\sigma_1}{\longrightarrow} & S & \longrightarrow & 1 \\
 & & \downarrow{\scriptstyle \mathrm{id}} & & \downarrow & & \downarrow{\scriptstyle \lambda_1} & & \\
E \colon 1 & \longrightarrow & M_1 & \longrightarrow & W_1 & \overset{\bar{\alpha}_1}{\longrightarrow} & \mathrm{Im}(\alpha_1) \wr_{\rho_1} P_s & \longrightarrow & 1
\end{array}
$$

Then, applying Theorem 1.1.35, G_1 splits over M_1, since B splits over D.

For the group C we repeat the previous arguments to construct a similar group G_2. Let T be a non-abelian simple group. If y is an involution in T, define the group homomorphism

$$\alpha_2 \colon C \longrightarrow T \quad \text{such that } \mathrm{Ker}(\alpha_2) = D,\ C^{\alpha_2} = \langle y \rangle.$$

Put $|T : \mathrm{Im}(\alpha_2)| = |T|/2 = n_2$, and consider the right transversal

$$\mathcal{T}_2 = \{t_1 = 1, t_2, \ldots, t_{n_2}\}$$

of $\mathrm{Im}(\alpha_2)$ in T and the transitive action

$$\rho_2 \colon T \longrightarrow \mathrm{Sym}(n_2)$$

on the set of indices $\mathcal{I}_2 = \{1, \ldots, n_2\}$. For each $i \in \mathcal{I}_2$ and each $t \in T$, $t_i t = y_{i,t} t_j$, for some $y_{i,t} \in \mathrm{Im}(\alpha_2)$ and $i^{t^{\rho_2}} = j$.

With the obvious changes of notation, construct the induced extension defined by α_2 as in Definition 1.1.27. Then, for $G_2 = \{w \in W_2 = C \wr_{\rho_2} P_T : w^{\bar{\alpha}_2} = t^{\lambda_2} \text{ for some } t \in T\}$ and $\sigma_2 \colon G_2 \longrightarrow T$ defined as above, we also have that the following diagram is commutative

$$
\begin{array}{ccccccccc}
E\lambda_2 \colon 1 & \longrightarrow & M_2 & \longrightarrow & G_2 & \xrightarrow{\sigma_2} & T & \longrightarrow & 1 \\
& & \downarrow{\scriptstyle \mathrm{id}} & & \downarrow & & \downarrow{\scriptstyle \lambda_2} & & \\
E_2 \colon 1 & \longrightarrow & M_2 & \longrightarrow & W_2 & \xrightarrow{\bar{\alpha}_2} & \mathrm{Im}(\alpha_2) \wr_{\rho_2} P_T & \longrightarrow & 1
\end{array}
$$

Then, again by Theorem 1.1.35, G_2 splits over M_2 since C splits over D.

Finally, consider the homomorphism $\alpha \colon A \longrightarrow S \times T$ such that $b^\alpha = (b^{\alpha_1}, 1)$, $c^\alpha = (1, c^{\alpha_2})$ for any $b \in B$, $c \in C$. Then, $\mathrm{Ker}(\alpha) = D$ and $\mathrm{Im}(\alpha) = \mathrm{Im}(\alpha_1) \times \mathrm{Im}(\alpha_2)$. Put $|S \times T : \mathrm{Im}(\alpha)| = \frac{|S|}{2} \frac{|T|}{2} = n_1 n_2$, and consider the right transversal of $\mathrm{Im}(\alpha)$ in $S \times T$

$$
\begin{aligned}
\mathcal{T} &= \mathcal{T}_1 \times \mathcal{T}_2 \\
&= \{(s_1, t_1) = (1,1), (s_1, t_2), \ldots, (s_1, t_{n_2}), (s_2, t_1), (s_2, t_2), \ldots, (s_{n_1}, t_{n_2})\}.
\end{aligned}
$$

The transitive action $\rho \colon S \times T \longrightarrow \mathrm{Sym}(n_1 n_2)$ on the set of indices $\mathcal{I} = \mathcal{I}_1 \times \mathcal{I}_2 = \{(1,1), \ldots, (n_1, n_2)\}$ (lexicographically ordered) gives $P = (S \times T)^\rho = P_S \times P_T$.

Consider the monomorphism

$$\lambda = \lambda_{\mathcal{T}} \colon S \times T \longrightarrow \mathrm{Im}(\alpha) \wr_\rho P,$$

defined by

$$(s,t)^\lambda = \big((x_{1,s}, y_{1,t}), (x_{1,s}, y_{2,t}), \ldots, (x_{n_1,s}, y_{n_2,t})\big)(s,t)^\rho$$

for any $s \in S$, $t \in T$, the epimorphism

$$\bar{\alpha} \colon W = A \wr_\rho P \longrightarrow \operatorname{Im}(\alpha) \wr_\rho P$$

defined by

$$\left((a_{(1,1)}, a_{(1,2)}, \ldots, a_{(n_1,n_2)}) \tau \right)^{\bar{\alpha}} = (a_{(1,1)}^\alpha, a_{(1,2)}^\alpha, \ldots, a_{(n_1,n_2)}^\alpha) \tau$$

for $a_{(1,1)}, a_{(1,2)}, \ldots, a_{(n_1,n_2)} \in A$ and $\tau \in P$, and write $M = \operatorname{Ker}(\bar{\alpha}) = D^\natural = D^{n_1 n_2} \cong \operatorname{Alt}(6)^{n_1 n_2}$.

Construct the induced extension defined by the homomorphism $\alpha \colon A \longrightarrow S \times T$:

$$
\begin{array}{ccccccccc}
E\lambda \colon 1 & \longrightarrow & M & \longrightarrow & G & \xrightarrow{\ \sigma\ } & S \times T & \longrightarrow & 1 \\
& & \downarrow{\scriptstyle \mathrm{id}} & & \downarrow & & \downarrow{\scriptstyle \lambda} & & \\
E \colon 1 & \longrightarrow & M & \longrightarrow & W & \xrightarrow{\ \bar{\alpha}\ } & \operatorname{Im}(\alpha) \wr_\rho (P_S \times P_T) & \longrightarrow & 1
\end{array}
$$

Then,

$$G = \{ w \in W = A \wr_\rho P : w^{\bar{\alpha}} = (s,t)^\lambda \quad \text{for some } (s,t) \in S \times T \}$$

and $\sigma \colon G \longrightarrow S \times T$ defined by $w^\sigma = (s,t)$ such that $w^{\bar{\alpha}} = (s,t)^\lambda$, for all $w \in G$. Now applying Theorem 1.1.35, the group G does not split over M, since A does not split over D.

Every element $w \in W$ can be written uniquely as

$$w = (a_{(1,1)}, \ldots, a_{(n_1,n_2)})(\tau_1, \tau_2)$$

where $a_{(1,1)}, a_{(1,2)}, \ldots, a_{(n_1,n_2)} \in A$ for all $(i,j) \in \mathcal{I}$, $\tau_1 \in P_S$ and $\tau_2 \in P_T$. If $w \in G$, and $w^{\bar{\alpha}} = (s,t)^\lambda$, then

$$
\begin{aligned}
w^{\bar{\alpha}} &= (a_{(1,1)}^\alpha, \ldots, a_{(n_1,n_2)}^\alpha)(\tau_1, \tau_2) \\
&= w^{\sigma\lambda} \\
&= \left((x_{1,s}, y_{1,t}), (x_{1,s}, y_{2,t}), \ldots, (x_{n_1,s}, y_{n_2,t}) \right)(s,t)^\rho
\end{aligned}
$$

and $a_{(i,j)}^\alpha = (x_{i,s}, y_{j,t})$, for all $(i,j) \in \mathcal{I}$, $s^{\rho_1} = \tau_1$ and $t^{\rho_2} = \tau_2$.

Proposition 7.1.1. *The group W possesses subgroups $W_{(1)}$ and $W_{(2)}$ which are isomorphic to W_1 and W_2, respectively.*

Proof. Let $W_{(1)}$ be the subset of all elements w in W such that

1. $a_{(i,1)} = a_{(i,2)} = \cdots = a_{(i,n_2)}$, for all $i = 1, \ldots, n_1$,
2. $a_{(i,j)} \in B$, for all $(i,j) \in \mathcal{I}$, and
3. $\tau_2 = 1$.

Then $W_{(1)}$ is a subgroup of W and the map $\psi_1\colon W_1 \longrightarrow W_{(1)}$ such that $\big((b_1,\ldots,b_{n_1})\tau\big)^{\psi_1}$ is the element $w \in W_{(1)}$ such that

1. $a_{(i,1)} = a_{(i,2)} = \cdots = a_{(i,n_2)} = b_i$, for all $i = 1,\ldots,n_1$,
2. $\tau_1 = \tau$ and $\tau_2 = 1$,

is a group isomorphism. Put $M_{(1)} = M_1^{\psi_1}$.

A similar argument and construction holds for W_2. □

Proposition 7.1.2. *The group G possesses two subgroups which are isomorphic to G_1 and G_2, respectively.*

Proof. Consider the subgroup $G_{(1)} = W_{(1)} \cap G$ and note that

$$G_{(1)} = \{x \in W_{(1)} : x^{\bar{\alpha}} = (s,1)^{\lambda} \text{ for some } s \in S\}.$$

Note that the kernel of the group epimorphism

$$\sigma_{(1)} = \sigma\pi_1\colon G_{(1)} \longrightarrow S,$$

where $\pi_1\colon S \times T \longrightarrow S$ is the canonical projection, is $M_{(1)} = M_1^{\psi_1}$, as in Proposition 7.1.1. Define the group homomorphism

$$\beta_1 = \iota_{(1)}\psi_1^{-1}\colon G_{(1)} \longrightarrow W_1,$$

where $\iota_{(1)}\colon G_{(1)} \longrightarrow W_{(1)}$ is the canonical inclusion and ψ_1 as in Proposition 7.1.1.

Consider an element $x = (a_{(1,1)},\ldots,a_{(n_1,n_2)})(\tau_1,1) \in G_{(1)}$. Then, if $x^{\bar{\alpha}} = (s,1)^{\lambda}$, we have that $s^{\rho_1} = \tau_1$ and $a_{(i,j)}^{\alpha} = (x_{i,s},1) \in S\times 1$, for all $i = 1,\ldots,n_1$, i.e. $a_{(i,j)} \in B$ and $a_{(i,j)}^{\alpha_1} = x_{i,s}$, for all $i = 1,\ldots,n_1$. Observe that

$$x^{\bar{\alpha}} = (s,1)^{\lambda} = \big((x_{1,s},1),(x_{1,s},1)\ldots,(x_{n_1,s},1)\big)(s^{\rho_1},1),$$

and

$$\begin{aligned}
x^{\beta_1\bar{\alpha}_1} = x^{\iota_{(1)}\psi_1^{-1}\bar{\alpha}_1} &= x^{\psi_1^{-1}\bar{\alpha}_1} = \big((a_{(1,1)},\ldots,a_{(n_1,1)})\tau_1\big)^{\bar{\alpha}_1} = \\
&= (a_{(1,1)}^{\alpha_1},\ldots,a_{(n_1,1)}^{\alpha_1})\tau_1 = (x_{1,s},\ldots,x_{n_1,s})s^{\rho_1} = \\
&= s^{\lambda_1} = (s,1)^{\pi_1\lambda_1} = x^{\sigma\pi_1\lambda_1} = x^{\sigma_{(1)}\lambda_1}.
\end{aligned}$$

Then the diagram

$$
\begin{array}{ccccccccc}
1 & \longrightarrow & M_1 & \longrightarrow & G_{(1)} & \overset{\sigma_{(1)}}{\longrightarrow} & S & \longrightarrow & 1 \\
& & \downarrow{\scriptstyle \text{id}} & & \downarrow{\scriptstyle \beta_1} & & \downarrow{\scriptstyle \lambda_1} & & \\
1 & \longrightarrow & M_1 & \longrightarrow & W_1 & \overset{\bar{\alpha}_1}{\longrightarrow} & \mathrm{Im}(\alpha_1) \wr_{\rho_1} P_s & \longrightarrow & 1
\end{array}
$$

is commutative.

By the universal property, Theorem 1.1.23 (2), we have that $G_{(1)}$ is isomorphic to G_1.

Analogously we can proceed with G_2 and it appears a subgroup $G_{(2)}$ in $W_{(2)}$ which is isomorphic to G_2.   □

Let S and T be two non-abelian simple groups. Recall that the class $\mathfrak{F} = \mathrm{D}_0(S, T, 1)$ composed by the trivial group and all groups which are direct products of the form

$$S_1 \times \cdots \times S_n \times T_1 \times \cdots \times T_m,$$

where $S_i \cong S$, $T_j \cong T$, $1 \leq i \leq n$, $1 \leq j \leq m$, for some positive integers n and m, is a Fitting formation (see Lemma 2.2.3).

Theorem 7.1.3. *Let S and T be two non-abelian simple groups. Suppose that S and T satisfy the three following conditions:*

1. *no subgroup of S is isomorphic to T,*
2. *no subgroup of T is isomorphic to S, and*
3. *either S or T are isomorphic to no subgroup of a direct product of copies of the alternating group $\mathrm{Alt}(6)$ of degree 6.*

Consider the Fitting formation $\mathfrak{F} = \mathrm{D}_0(S, T, 1)$. Then the group G, constructed above, has no $\mathfrak{F}$-injectors.

Proof. The group G possesses two subgroups, $\tilde{S}$ and $\tilde{T}$, which are isomorphic to S and T, respectively. Write $G/M = (H_1/M) \times (H_2/M)$, with $H_1/M \cong S$ and $H_2/M \cong T$. Observe that $\tilde{S}M/M \cong \tilde{S}/(\tilde{S} \cap M) = \tilde{S}$, since $\tilde{S} \cap M = 1$, by condition 3. If $(H_1/M) \cap (\tilde{S}M/M) = 1$, then the group $G/H_1 \cong T$ would have a subgroup isomorphic to S, and this is not possible by Condition 2. Hence $H_1 = \tilde{S}M$. A similar argument with $\tilde{T}$ and H_2 leads to $H_2 = \tilde{T}M$. Both H_1 and H_2 are maximal normal subgroups of G.

We observe that $\mathrm{Max}_{\mathfrak{F}}(\tilde{S}M) = \{U : UM = \tilde{S}M, U \cong S\}$. If $U \in \mathrm{Max}_{\mathfrak{F}}(\tilde{S}M)$, then $U \cap M = 1$ by condition 3. Since $U \in \mathfrak{F}$ and $UM \leq \tilde{S}M$, we have that $U \cong S$ and $UM = \tilde{S}M$.

Similarly $\mathrm{Max}_{\mathfrak{F}}(\tilde{T}M) = \{V : VM = \tilde{T}M, V \cong T\}$.

Suppose that X is an $\mathfrak{F}$-injector of G. Then, the subgroup $X \cap \tilde{S}M = R_1$ is $\mathfrak{F}$-maximal in $\tilde{S}M$. Hence $R_1 \cong S$. Likewise, $X \cap \tilde{T}M = R_2 \cong T$. Hence $R_1 \times R_2$ is a normal subgroup of X and $R_1 \times R_2 \cong S \times T$. Moreover, $(R_1 \times R_2) \cap M = 1$. Since $|G| = |M||S \times T| = |M||R_1 \times R_2|$, we conclude that $R_1 \times R_2$ is a complement of M in G, i.e. G splits over M. But this is not true. Therefore the group G has no $\mathfrak{F}$-injectors and $\mathfrak{F}$ is a non-injective Fitting class.   □

Remark 7.1.4. The simple groups $S = \mathrm{Alt}(7)$ and $T = \mathrm{PSL}(2, 11)$ satisfy the above conditions 1, 2, and 3.

7.2 Injective Fitting classes

We have proved in Corollary 2.4.28 that every Fitting class $\mathfrak{F}$ is injective in the universe $\mathfrak{FS}$. In fact, in the attempt of investigating classes of groups, larger than the soluble one, in which there exist $\mathfrak{F}$-injectors for a particular Fitting class $\mathfrak{F}$, the first remarkable contribution comes from A. Mann in [Man71]. There, following some ideas due to B. Fischer and E. C. Dade (see [DH92, page 623]), it is proved that in every $\mathfrak{N}$-constrained group G, there exists a single conjugacy class of $\mathfrak{N}$-injectors and each $\mathfrak{N}$-injector is an $\mathfrak{N}$-maximal subgroup containing the Fitting subgroup. A group G is said to be $\mathfrak{N}$-*constrained* if $C_G(F(G)) \leq F(G)$. It is well-known that every soluble group is $\mathfrak{N}$-constrained (see [DH92, A, 10.6]).

In [BL79] D. Blessenohl and H. Laue proved that the class $\mathfrak{Q}$ of all quasin-ilpotent groups is an injective Fitting class in $\mathfrak{E}$. In fact they prove something more (see [DH92, IX, 4.15]).

Theorem 7.2.1 (D. Blessenohl and H. Laue). *Every finite group G has a single conjugacy class of $\mathfrak{Q}$-injectors, and this consists of those $\mathfrak{Q}$-maximal subgroups of G containing $F^*(G)$.*

In the decade of the eighties of the last century there was a considerable amount of contributions to obtain more injective Fitting classes. P. Förster proved the existence of a certain non-empty characteristic conjugacy class of $\mathfrak{N}$-injectors in every finite group in [För85a]. Later M. J. Iranzo and F. Pérez-Monasor obtained the existence of injectors in all finite groups with respect to various Fitting classes, including a new type of $\mathfrak{N}$-injectors. Their investigations, together with M. Torres, gave light to a "test" to prove the injectivity of a number of Fitting classes. Some of the most interesting results obtained from this test have been published recently by M. J. Iranzo, J. Lafuente, and F. Pérez-Monasor. Their achievements illuminate the validity of a L. A. Shemetkov conjecture saying that any Fitting class composed of soluble groups is injective.

We present here some of the fruits of these investigations.

Proposition 7.2.2. *Let $\mathfrak{F}$ be a Fitting class and G be a group.*

1. *A perfect comonolithic subnormal subgroup E of G is an $\mathfrak{F}$-component of G if and only of $EG_{\mathfrak{F}}/G_{\mathfrak{F}}$ is a component of $G/G_{\mathfrak{F}}$.*
2. *If E is an $\mathfrak{F}$-component of G, the $\mathfrak{F}$-maximal subgroups of E containing $E_{\mathfrak{F}}$ are $\mathfrak{F}$-injectors of E.*

Proof.   1. Let E be a perfect comonolithic subnormal subgroup of a group G. Suppose that E is an $\mathfrak{F}$-component of G. Then $N(E)$ is a subnormal $\mathfrak{F}$-subgroup of G, i.e. $N(E) \leq G_{\mathfrak{F}}$. Therefore $EG_{\mathfrak{F}}/G_{\mathfrak{F}}$ is isomorphic to a quotient group of $E/N(E)$, and then $EG_{\mathfrak{F}}/G_{\mathfrak{F}}$ is a quasisimple subnormal subgroup of $G/G_{\mathfrak{F}}$. Conversely, if $EG_{\mathfrak{F}}/G_{\mathfrak{F}}$ is a component of $G/G_{\mathfrak{F}}$, then $E/(E \cap G_{\mathfrak{F}})$ is a quasisimple group. Since E is subnormal in G, $E_{\mathfrak{F}} = E \cap G_{\mathfrak{F}}$

by Remark 2.4.4. If $E \in \mathfrak{F}$, then E is contained in $G_{\mathfrak{F}}$, contrary to supposition. Hence $E_{\mathfrak{F}} \leq \mathrm{Cosoc}(E)$. Moreover, $\mathrm{Cosoc}(E)/E_{\mathfrak{F}} = Z(E/E_{\mathfrak{F}})$. Therefore $\mathrm{N}(E) = [E, \mathrm{Cosoc}(E)] \leq E_{\mathfrak{F}}$. Hence $\mathrm{N}(E) \in \mathfrak{F}$.

2. Suppose E is an $\mathfrak{F}$-component of G and V is an $\mathfrak{F}$-maximal subgroup of E such that $E_{\mathfrak{F}} \leq V$. Since $\mathrm{N}(E) \leq E_{\mathfrak{F}} \leq \mathrm{Cosoc}(E)$ and $\mathrm{Cosoc}(E)/\mathrm{N}(E)$ is abelian, $E_{\mathfrak{F}}$ is the $\mathfrak{F}$-injector of $\mathrm{Cosoc}(G)$. Moreover, $V \cap \mathrm{Cosoc}(E)$ is normal in $\mathrm{Cosoc}(E)$ and then is a subnormal $\mathfrak{F}$-subgroup of E. Hence $V \cap \mathrm{Cosoc}(E) = E_{\mathfrak{F}}$ and V is an $\mathfrak{F}$-injector of E. □

Proposition 7.2.3. *Let K be a subnormal subgroup of a group G. If E is an $\mathfrak{F}$-component of G such that E is not contained in K, we have that $[K, E] \leq \mathrm{N}(E)$.*

Proof. Denote $M = \mathrm{Cosoc}(E)$. By Theorem 2.2.19, the subgroup K normalises E. Therefore K normalises M. Clearly K is subnormal in KE and KM is normal in KE. Since $K \cap E$ is subnormal in the comonolithic group E and $E \not\leq K$, we have that $K \cap E \leq M$. Therefore

$$[K, E] \leq [KM, E] \leq KM \cap E = M(K \cap E) \leq M.$$

Hence

$$[K, E, E] = [E, K, E] \leq [M, E] = \mathrm{N}(E)$$

and the Three-Subgroups Lemma (see [KS04, 1.5.6]) yields that $[E, K] = [E, E, K] \leq \mathrm{N}(E)$. □

Now we are ready to state and prove the result of Iranzo, Pérez-Monasor, and Torres.

Theorem 7.2.4 ([IPMT90]). *Let $\mathfrak{F}$ be a Fitting class and G a group. Let $\{E_1, \dots, E_n\}$ be a set of $\mathfrak{F}$-components of G which is invariant by conjugation of the elements of G. For each $i = 1, \dots, n$, let J_i be an $\mathfrak{F}$-injector of E_i. Consider the subgroup $J = \langle J_1, \dots, J_n \rangle$.*
Then $\mathrm{Inj}_{\mathfrak{F}}\big(\mathrm{N}_G(J)\big) \subseteq \mathrm{Inj}_{\mathfrak{F}}(G)$.

Proof. Note that, by Proposition 7.2.2 (2) and Proposition 7.2.3, J is a normal product $J = J_1 \cdots J_n$, and therefore $J \in \mathfrak{F}$. Let H be an $\mathfrak{F}$-injector of $\mathrm{N}_G(J)$. We have to prove that for any subnormal subgroup S of G, the subgroup $H \cap S$ is $\mathfrak{F}$-maximal in S. To do that we consider an $\mathfrak{F}$-subgroup K of S such that $H \cap S \leq K$ and argue that $H \cap S = K$.

We may assume without loss of generality that the $\mathfrak{F}$-components $E_1, \dots, E_m$ are those contained in S, for $m \leq n$, and the other ones are not in S. This implies that $\{E_1, \dots, E_m\}$ is a set of $\mathfrak{F}$-components of S which is invariant by conjugation of the elements of S.

Observe that $J \leq \mathrm{N}_G(J)_{\mathfrak{F}} \leq H$. Therefore, for any $i = 1, \dots, m$, we have that

$$J_i \leq J \cap E_i \leq H \cap E_i \leq H \cap S \cap E_i \leq K \cap E_i \in \mathfrak{F},$$

since $K \cap E_i$ is subnormal in K. Therefore

$$J_i = J \cap E_i = H \cap E_i = K \cap E_i,$$

since $J_i \in \mathrm{Max}_{\mathfrak{F}}(E_i)$, $i = 1, \ldots, m$.

Observe that if $x \in K$, for every $i \in \{1, \ldots, m\}$, there exists an index $j \in \{1, \ldots, m\}$ such that

$$J_i^x = (J \cap E_i)^x = K \cap E_i^x = K \cap E_j = J_j.$$

Choose now $j \in \{m+1, \ldots, n\}$. Applying Proposition 7.2.3, it can be deduced that $[J_j, S] \leq [E_j, S] \leq \mathrm{N}(E_j) \leq J_j$. This is to say that S normalises J_j for every $j \in \{m + 1, \ldots, n\}$. Therefore

$$K \leq \mathrm{N}_S(J_1 \ldots J_m) \leq \mathrm{N}_S(J).$$

Hence $H \cap S \leq K \leq \mathrm{N}_S(J)$ and then $H \cap S = H \cap \mathrm{N}_S(J)$.

The subgroup $\mathrm{N}_S(J)$ is subnormal in $\mathrm{N}_G(J)$. Since $H \in \mathrm{Inj}_{\mathfrak{F}}\big(\mathrm{N}_G(J)\big)$, we have that $H \cap S \in \mathrm{Max}_{\mathfrak{F}}(\mathrm{N}_S(J))$. This implies that $H \cap S = K$, as desired. $\qquad\square$

Theorem 7.2.4 is a crucial result when proving the injectivity of a Fitting class by inductive arguments: with the above notation, if $\mathrm{Inj}_{\mathfrak{F}}\big(\mathrm{N}_G(J)\big) \neq \emptyset$, then the group G possesses $\mathfrak{F}$-injectors. Equipped with this theorem we can obtain several results of M. J. Iranzo, J. Lafuente, and F. Pérez-Monasor in [ILPM03] and [ILPM04], which go much further on the theorems about the existence of injectors.

Lemma 7.2.5 (see [ILPM03]). *Let G be a group and $\mathfrak{m}$ a preboundary of perfect groups. Set $\mathfrak{B} = \mathrm{Fit}\big(\mathrm{Cosoc}(Z) : Z \in \mathfrak{m}\big)$.*

1. *If $X, Y \in \mathrm{b}_{\mathfrak{m}}(G)$, then*
   a) $\mathrm{Cosoc}(X) = X_{\mathfrak{B}}$, $[X, Y] \leq X \cap Y$ and $(XY)_{\mathfrak{B}} = X_{\mathfrak{B}} Y_{\mathfrak{B}}$,
   b) $X \neq Y$ if and only if $X G_{\mathfrak{B}}/G_{\mathfrak{B}} \neq Y G_{\mathfrak{B}}/G_{\mathfrak{B}}$.
2. *Suppose that $\mathrm{b}_{\mathfrak{m}}(G) = \{X_1, \ldots, X_n\} \neq \emptyset$ and write $E = \mathrm{E}_{\mathfrak{m}}(G)$; then*
   a) $E = X_1 \ldots X_n$ and $E_{\mathfrak{B}} = (X_1)_{\mathfrak{B}} \ldots (X_n)_{\mathfrak{B}}$,
   b) $E/E_{\mathfrak{B}} \cong X_1/(X_1)_{\mathfrak{B}} \times \cdots \times X_n/(X_n)_{\mathfrak{B}}$ is a direct product of non-abelian simple groups.

Proof. 1a. By definition of $\mathfrak{B}$, we have that $\mathrm{Cosoc}(X) \in \mathfrak{B}$. Assume that $X \in \mathfrak{B}$. Then $X \in \mathrm{s}_n\big(\mathrm{Cosoc}(Z) : Z \in \mathfrak{m}\big)$, by [DH92, XI, 4.14]. But this is not possible since $\mathfrak{m}$ is subnormally independent. Therefore $\mathrm{Cosoc}(X) = X_{\mathfrak{B}}$.

Trivially, if $X = Y$, then $[X, Y] \leq X \cap Y$. Suppose that $X \neq Y$. Observe that, since $\mathfrak{m}$ is subnormally independent, we have that $X \not\leq Y$ and $Y \not\leq X$. By Theorem 2.2.19, Y normalises X and X normalises Y. Hence $[X, Y] \leq X \cap Y$.

If $X \neq Y$, then $X \cap Y \leq \mathrm{Cosoc}(X) \cap \mathrm{Cosoc}(Y) = X_{\mathfrak{B}} \cap Y_{\mathfrak{B}}$. Moreover,

$$XY_{\mathfrak{B}} \cap YX_{\mathfrak{B}} = (X \cap YX_{\mathfrak{B}})Y_{\mathfrak{B}} = (X \cap Y)X_{\mathfrak{B}}Y_{\mathfrak{B}} = X_{\mathfrak{B}}Y_{\mathfrak{B}}$$

and then
$$XY/X_\mathfrak{B} Y_\mathfrak{B} = XY_\mathfrak{B}/X_\mathfrak{B} Y_\mathfrak{B} \times YX_\mathfrak{B}/X_\mathfrak{B} Y_\mathfrak{B}$$

is a direct product of non-abelian simple groups. Since $(XY)_\mathfrak{B}/X_\mathfrak{B} Y_\mathfrak{B} \leq Z(XY/X_\mathfrak{B} Y_\mathfrak{B})$ by [DH92, IX, 1.1], we conclude that $(XY)_\mathfrak{B} = X_\mathfrak{B} Y_\mathfrak{B}$.

1b. Observe that $XG_\mathfrak{B}/G_\mathfrak{B} \cong X/(X \cap G_\mathfrak{B}) = X/X_\mathfrak{B}$ is a non-abelian simple group. Suppose that $X \neq Y$ and $XG_\mathfrak{B}/G_\mathfrak{B} = YG_\mathfrak{B}/G_\mathfrak{B}$. Notice that $[X, Y] \leq X \cap Y \in \mathfrak{B}$, and then, $XG_\mathfrak{B}/G_\mathfrak{B} = (XG_\mathfrak{B}/G_\mathfrak{B})' = [XG_\mathfrak{B}/G_\mathfrak{B}, YG_\mathfrak{B}/G_\mathfrak{B}] = [X, Y]G_\mathfrak{B}/G_\mathfrak{B} = 1$. This is a contradiction.

Part 2 follows immediately from 1. □

Lemma 7.2.6 (M. J. Iranzo, J. Lafuente, and F. Pérez-Monasor, unpublished). *Let $\mathfrak{F}$ be a Fitting class and $\mathfrak{n}$ a subclass of $\bar{\mathrm{b}}(\mathfrak{F})$. Then*

$$\mathrm{Fit}(\mathfrak{F}, \mathfrak{n}) = \mathfrak{F} \cdot \mathrm{Fit}\,\mathfrak{n} = \big(G \in \mathfrak{E} : G = G_\mathfrak{F} \, \mathrm{E}_\mathfrak{n}(G)\big).$$

Proof. Let G be a group. If $X \in b_\mathfrak{n}(G)$, then clearly $\mathrm{Cosoc}(X) = X_\mathfrak{F}$.

Write $\mathfrak{X} = \big(G \in \mathfrak{E} : G = G_\mathfrak{F} \, \mathrm{E}_\mathfrak{n}(G)\big)$ and $\mathfrak{Y} = \mathrm{Fit}\,\mathfrak{n}$. For each group G, the subgroup $\mathrm{E}_\mathfrak{n}(G)$ is in $\mathrm{Fit}\,\mathfrak{n}$, i.e. $\mathrm{E}_\mathfrak{n}(G) \leq G_\mathfrak{Y}$. Therefore $\mathfrak{X} \subseteq \mathfrak{F} \cdot \mathrm{Fit}\,\mathfrak{n} \subseteq \mathrm{Fit}(\mathfrak{F}, \mathfrak{n})$. Let us prove that $\mathfrak{X}$ is a Fitting class.

If $G \in \mathfrak{X}$, then $G/G_\mathfrak{F} \cong \mathrm{E}_\mathfrak{n}(G)/\mathrm{E}_\mathfrak{n}(G)_\mathfrak{F}$ is a direct product of non-abelian simple groups by Lemma 7.2.5 (2b). Let N be a normal subgroup of G. Then $b_\mathfrak{n}(N) \subseteq b_\mathfrak{n}(G)$. Thus, if $b_\mathfrak{n}(N) = \{X_1, \ldots, X_r\}$, then

$$NG_\mathfrak{F}/G_\mathfrak{F} = X_1 G_\mathfrak{F}/G_\mathfrak{F} \times \cdots \times X_r G_\mathfrak{F}/G_\mathfrak{F}$$

and then $N = N \cap NG_\mathfrak{F} = N \cap X_1 \ldots X_r G_\mathfrak{F} = N \cap \mathrm{E}_\mathfrak{n}(N)G_\mathfrak{F} = \mathrm{E}_\mathfrak{n}(N)N_\mathfrak{F} \in \mathfrak{X}$.

If N and M are normal subgroups of a group $G = NM$ and $N, M \in \mathfrak{X}$, then $G = NM = N_\mathfrak{F} \, \mathrm{E}_\mathfrak{n}(N)M_\mathfrak{F} \, \mathrm{E}_\mathfrak{n}(M) \leq G_\mathfrak{F} \, \mathrm{E}_\mathfrak{n}(G)$. Hence $G \in \mathfrak{X}$.

Therefore $\mathfrak{X}$ is a Fitting class. It is clear that $\mathfrak{F}$ and $\mathfrak{n}$ are contained in $\mathfrak{X}$. Hence $\mathfrak{X} = \mathrm{Fit}(\mathfrak{F}, \mathfrak{n})$. □

Lemma 7.2.7. *Let $\mathfrak{T}$ be a Fitting class such that $\mathfrak{T} = \mathfrak{T}\mathfrak{S}$. Consider $\mathfrak{F} = \mathfrak{T}^b = \mathrm{Fit}\big(\mathrm{Cosoc}(X) : X \in \mathrm{b}(\mathfrak{T})\big)$. Then $\mathrm{b}(\mathfrak{T}) = \bar{\mathrm{b}}(\mathfrak{T}) \subseteq \bar{\mathrm{b}}(\mathfrak{F})$.*

Proof. Let G be a group in $\mathrm{b}(\mathfrak{T})$. Then G is a comonolithic perfect group and $\mathrm{Cosoc}(G) \in \mathfrak{F}$. If $G \in \mathfrak{F}$, then $G \in s_n\big(\mathrm{Cosoc}(X) : X \in \mathrm{b}(\mathfrak{T})\big)$ by [DH92, XI, 4.14]. This is to say that there exists a group $X \in \mathrm{b}(\mathfrak{T})$ such that G is a proper subnormal subgroup of X. In particular $G \in \mathfrak{T}$, and this contradicts our assumption. Hence $G \in \bar{\mathrm{b}}(\mathfrak{F})$. □

Theorem 7.2.8. *Let $\mathfrak{T}$ be a class of groups. The following statements are equivalent:*

1. *$\mathfrak{T}$ is a Fitting class such that $\mathfrak{T} = \mathfrak{T}\mathfrak{S}$.*
2. *$\mathfrak{T} = (G \in \mathfrak{E} : G_\mathfrak{X} \in \mathfrak{F})$ for a pair of Fitting classes $\mathfrak{X}$ and $\mathfrak{F}$ such that $\mathfrak{F} = \mathfrak{X} \cap \mathfrak{F}\mathfrak{A}$.*

In this case, for each group G, we have $G_\mathfrak{T} = \mathrm{C}_G(G_\mathfrak{X}/G_\mathfrak{F})$.

Proof. 1 implies 2. Set $\mathfrak{m} = \mathrm{b}(\mathfrak{T})$, and consider the Fitting classes $\mathfrak{F} = \mathfrak{T}^{\mathrm{b}}$ and $\mathfrak{X} = \mathrm{Fit}\,\mathfrak{m}$. Clearly $\mathfrak{F} \subseteq \mathfrak{X} \cap \mathfrak{T}$. Since $\mathfrak{T} = \mathfrak{T}\mathfrak{S}$, we have that $\mathfrak{m} = \bar{\mathrm{b}}(\mathfrak{T}) \subseteq \bar{\mathrm{b}}(\mathfrak{F})$, by the above lemma. Then we can apply Lemma 7.2.6 and conclude that

$$\mathfrak{X} = \mathrm{Fit}(\mathfrak{F}, \mathfrak{m}) = \big(G \in \mathfrak{E} : G = G_{\mathfrak{F}}\,\mathrm{E}_{\mathfrak{m}}(G)\big).$$

If $G \in \mathfrak{X} \cap \mathfrak{F}\mathfrak{A}$, then $G/G_{\mathfrak{F}} \cong \mathrm{E}_{\mathfrak{m}}(G)/\big(\mathrm{E}_{\mathfrak{m}}(G) \cap G_{\mathfrak{F}}\big)$ and this group is abelian and a direct product of non-abelian simple groups, by Lemma 7.2.5 (2b). Hence $G \in \mathfrak{F}$, and then $\mathfrak{F} = \mathfrak{X} \cap \mathfrak{F}\mathfrak{A}$.

Set $\mathfrak{H} = (G \in \mathfrak{E} : G_{\mathfrak{X}} \in \mathfrak{F})$. If a group $G \in \mathfrak{H} \setminus \mathfrak{T}$, there exists a subnormal subgroup N of G such that $N \in \mathfrak{m}$. Thus $N \leq G_{\mathfrak{X}} \in \mathfrak{F} \subseteq \mathfrak{T}$, and this is a contradiction. Hence $\mathfrak{H} \subseteq \mathfrak{T}$. Conversely if G is a group in $\mathfrak{T}$ and $N = G_{\mathfrak{X}}$, then $N = N_{\mathfrak{F}}\mathrm{E}_{\mathfrak{m}}(N)$. But since $\mathfrak{T}$ is a Fitting class, $\mathrm{E}_{\mathfrak{m}}(G) = 1 = \mathrm{E}_{\mathfrak{m}}(N)$. Then $N \in \mathfrak{F}$. Therefore $G \in \mathfrak{H}$. Hence $\mathfrak{H} = \mathfrak{T}$.

2 implies 1. We see that, under these hypotheses, the class $\mathfrak{T}$ is a Fitting class. Let N be a normal subgroup of a $\mathfrak{T}$-group G. Clearly $N_{\mathfrak{X}} \leq G_{\mathfrak{X}} \in \mathfrak{F}$, and then $N \in \mathfrak{T}$. Consider now a group $G = NM$ such that N and M are normal $\mathfrak{T}$-subgroups of G. Then $N_{\mathfrak{X}}$, $M_{\mathfrak{X}} \in \mathfrak{F}$ and the subgroup $F = N_{\mathfrak{X}}M_{\mathfrak{X}} \in \mathfrak{F}$. By [DH92, IX, 1.1], we have that $G_{\mathfrak{X}}/F \leq \mathrm{Z}(G/F)$, and then $G_{\mathfrak{X}} \in \mathfrak{X} \cap \mathfrak{F}\mathfrak{A} = \mathfrak{F}$. Therefore $G \in \mathfrak{T}$. Thus, $\mathfrak{T}$ is a Fitting class.

Suppose that N is a normal $\mathfrak{T}$-subgroup of a group G, such that $G/N \in \mathfrak{A}$. Then $N_{\mathfrak{X}} \in \mathfrak{F}$. Since $G_{\mathfrak{X}}/N_{\mathfrak{X}} = G_{\mathfrak{X}}/(N \cap G_{\mathfrak{X}}) \cong NG_{\mathfrak{X}}/N \in \mathfrak{A}$, we have that $G_{\mathfrak{X}} \in \mathfrak{X} \cap \mathfrak{F}\mathfrak{A} = \mathfrak{F}$. Therefore $G \in \mathfrak{T}$. This implies that $\mathfrak{T} = \mathfrak{T}\mathfrak{S}$.

Finally, observe that in this situation $\mathfrak{F} = \mathfrak{X} \cap \mathfrak{T}$. Therefore $G_{\mathfrak{F}} = G_{\mathfrak{T}} \cap G_{\mathfrak{X}}$. Thus $G_{\mathfrak{T}} \leq \mathrm{C}_G(G_{\mathfrak{X}}/G_{\mathfrak{F}}) = C$. Obviously $(C \cap G_{\mathfrak{X}})/G_{\mathfrak{F}}$ is an abelian group and then $C_{\mathfrak{X}} = C \cap G_{\mathfrak{X}} \in \mathfrak{F}$, since $\mathfrak{F} = \mathfrak{X} \cap \mathfrak{F}\mathfrak{A}$. Therefore $C \in \mathfrak{T}$ and $C = G_{\mathfrak{T}}$. $\square$

Corollary 7.2.9. *Let $\mathfrak{T}$ be a Fitting class such that $\mathfrak{T} = \mathfrak{T}\mathfrak{S}$. Then*

$$\mathrm{Fit}\big(\mathrm{b}(\mathfrak{T})\big) \cap \mathfrak{T} = \mathfrak{T}^{\mathrm{b}}.$$

Proof. Set $\mathfrak{m} = \mathrm{b}(\mathfrak{T})$ and consider again the Fitting classes $\mathfrak{F} = \mathfrak{T}^{\mathrm{b}}$ and $\mathfrak{X} = \mathrm{Fit}\,\mathfrak{m}$. By the above arguments, if a group G is in $\mathfrak{X} \cap \mathfrak{T}$, then $G = G_{\mathfrak{F}}\,\mathrm{E}_{\mathfrak{m}}(G) \in \mathfrak{T}$. Hence $\mathrm{E}_{\mathfrak{m}}(G) \in \mathfrak{T}$, and this implies that $\mathrm{E}_{\mathfrak{m}}(G) = 1$. Thus $G \in \mathfrak{F}$. Therefore $\mathfrak{X} \cap \mathfrak{T} = \mathfrak{F}$. $\square$

The following proposition is motivated by a result due to W. Gaschütz (see [DH92, X, 3.14]).

Proposition 7.2.10. *Let $\mathfrak{F}$ and $\mathfrak{G}$ be two Fitting classes in the same Lockett section such that $\mathfrak{F} \subseteq \mathfrak{G}$. For each group G denote*

$$\psi \colon G_{\mathfrak{G}}/G_{\mathfrak{F}} \longrightarrow (G_{\mathfrak{G}}G')/(G_{\mathfrak{F}}G')$$

the natural epimorphism. If p is a prime divisor of $|\mathrm{Ker}(\psi)|$, then $\mathfrak{G}\mathfrak{S}_p \neq \mathfrak{G}$.

Proof. Observe that $\mathrm{Ker}(\psi) = (G_{\mathfrak{G}}/G_{\mathfrak{F}}) \cap (G/G_{\mathfrak{F}})'$. Let p be a prime divisor of $|\mathrm{Ker}(\psi)|$ and suppose that $\mathfrak{G}\mathfrak{S}_p = \mathfrak{G}$. If $P/G_{\mathfrak{F}}$ is a Sylow p-subgroup of $G/G_{\mathfrak{F}}$, then $P \in \mathfrak{F}\mathfrak{S}_p \subseteq \mathfrak{G}\mathfrak{S}_p = \mathfrak{G}$. Since $\mathfrak{F}$ and $\mathfrak{G}$ are in the same Lockett section and $\mathfrak{F} \subseteq \mathfrak{G}$, the groups $P/P_{\mathfrak{F}}$ and $G_{\mathfrak{G}}/G_{\mathfrak{F}}$ are abelian, by [DH92, X, 1.21]. Thus $P' \le P_{\mathfrak{F}}$ and $P \cap G_{\mathfrak{G}}$ is a normal subgroup of $G_{\mathfrak{G}}$. Hence $P' \cap G_{\mathfrak{G}} \in \mathfrak{F}$ and $P' \cap G_{\mathfrak{G}}$ is subnormal in $G_{\mathfrak{G}}$. Therefore $P' \cap G_{\mathfrak{G}} \le (G_{\mathfrak{G}})_{\mathfrak{F}} = G_{\mathfrak{F}}$. Then $(P/G_{\mathfrak{F}})' \cap (G_{\mathfrak{G}}/G_{\mathfrak{F}}) = 1$. By [DH92, X, 1.21] again, $G_{\mathfrak{G}}/G_{\mathfrak{F}} \le \mathrm{Z}(G/G_{\mathfrak{F}})$ and then

$$(P/G_{\mathfrak{F}}) \cap (G/G_{\mathfrak{F}})' \cap (G_{\mathfrak{G}}/G_{\mathfrak{F}}) \le (P/G_{\mathfrak{F}}) \cap (G/G_{\mathfrak{F}})' \cap \mathrm{Z}(G/G_{\mathfrak{F}}) \le (P/G_{\mathfrak{F}})'$$

by [Hup67, IV, 2.2]. Thus, $(P/G_{\mathfrak{F}}) \cap (G/G_{\mathfrak{F}})' \cap (G_{\mathfrak{G}}/G_{\mathfrak{F}}) = 1$ and this contradicts the choice of P.   □

Lemma 7.2.11. *Let $\mathfrak{T}$ be a Fitting class such that $\mathfrak{T}\mathfrak{S} = \mathfrak{T}$. Then*

$$\mathfrak{T}^b \subseteq \mathfrak{T}_* \subseteq \mathfrak{T} = \mathfrak{T}^*.$$

Proof. By [DH92, X, 1.8], we have that $\mathfrak{T} = \mathfrak{T}^*$. If $X \in \mathrm{b}(\mathfrak{T})$, then X is perfect. By Proposition 7.2.10, $X_{\mathfrak{T}} = X_{\mathfrak{T}_*}$. Then $\mathrm{Cosoc}(X) \in \mathfrak{T}_*$ and $\mathfrak{T}^b \subseteq \mathfrak{T}_*$.   □

Theorem 7.2.12 (see [ILPM04]). *Let $\mathfrak{T}$ be a Fitting class such that $\mathfrak{T}\mathfrak{S} = \mathfrak{T}$. The correspondence $\mathfrak{F} \longrightarrow \mathfrak{F} \cdot \mathrm{Fit}\big(\mathrm{b}(\mathfrak{T})\big)$, for every Fitting class $\mathfrak{F} \in \mathrm{Sec}(\mathfrak{T}^b, \mathfrak{T})$, defines a bijection*

$$\mathrm{Sec}(\mathfrak{T}^b, \mathfrak{T}) \longrightarrow \mathrm{Sec}\big(\mathrm{Fit}\big(\mathrm{b}(\mathfrak{T})\big), \mathfrak{T} \cdot \mathrm{Fit}\big(\mathrm{b}(\mathfrak{T})\big)\big)$$

whose inverse is defined by $\mathfrak{G} \longrightarrow \mathfrak{G} \cap \mathfrak{T}$, for every $\mathfrak{G} \in \mathrm{Sec}\big(\mathrm{Fit}\big(\mathrm{b}(\mathfrak{T})\big), \mathfrak{T} \cdot \mathrm{Fit}\big(\mathrm{b}(\mathfrak{T})\big)\big)$.

Moreover, the restriction of this bijection to the Lockett section $\mathrm{Locksec}(\mathfrak{T})$ gives a bijection

$$\mathrm{Locksec}(\mathfrak{T}) \longrightarrow \mathrm{Locksec}\big(\mathfrak{T} \cdot \mathrm{Fit}\big(\mathrm{b}(\mathfrak{T})\big)\big).$$

Proof. Set $\mathfrak{m} = \mathrm{b}(\mathfrak{T})$, $\mathfrak{M} = \mathrm{Fit}\,\mathfrak{m}$, $\mathfrak{B} = \mathfrak{T}^b$ and $\mathfrak{R} = \mathfrak{T} \cdot \mathfrak{M}$.

If $\mathfrak{F} \in \mathrm{Sec}(\mathfrak{B}, \mathfrak{T})$, then $\mathfrak{F} \cdot \mathfrak{M}$ is a Fitting class by [DH92, XI, 4.7] and Lemma 7.2.6. Obviously $\mathfrak{F} \cdot \mathfrak{M} \in \mathrm{Sec}(\mathfrak{M}, \mathfrak{R})$ and $\mathfrak{F} \subseteq \mathfrak{F} \cdot \mathfrak{M} \cap \mathfrak{T}$. Let G be a group in $\mathfrak{F} \cdot \mathfrak{M} \cap \mathfrak{T}$. Then $G_{\mathfrak{M}} \in \mathfrak{M} \cap \mathfrak{T} = \mathfrak{B}$, by Corollary 7.2.9. Hence $G = G_{\mathfrak{F}}G_{\mathfrak{M}} \in \mathfrak{F}$. Thus, $\mathfrak{F} = \mathfrak{F} \cdot \mathfrak{M} \cap \mathfrak{T}$.

On the other hand, if $\mathfrak{G} \in \mathrm{Sec}(\mathfrak{M}, \mathfrak{R})$, then $\mathfrak{T} \cap \mathfrak{G} \in \mathrm{Sec}(\mathfrak{B}, \mathfrak{T})$ by Corollary 7.2.9 and $(\mathfrak{T} \cap \mathfrak{G}) \cdot \mathfrak{M} \subseteq \mathfrak{G}$. Let G be a group in $\mathfrak{G}$. Then $G_{\mathfrak{T}} = G_{\mathfrak{T} \cap \mathfrak{G}}$ and, since $\mathfrak{G} \subseteq \mathfrak{R}$, we have that

$$G = G_{\mathfrak{T}}G_{\mathfrak{M}} = G_{\mathfrak{T} \cap \mathfrak{G}}G_{\mathfrak{M}} \in (\mathfrak{T} \cap \mathfrak{G}) \cdot \mathfrak{M}$$

and then $\mathfrak{G} = (\mathfrak{T} \cap \mathfrak{G}) \cdot \mathfrak{M}$.

Hence it only remains to prove the properties of the second bijection. We have to prove that $\mathfrak{R}$ is a Lockett class and $\mathfrak{R}_* = \mathfrak{T}_* \cdot \mathfrak{M}$.

If G and H are groups, then it is clear that $\mathrm{E_m}(G \times H) = \mathrm{E_m}(G) \times \mathrm{E_m}(H)$. Since $\mathfrak{T}$ is a Lockett class, by Theorem 7.2.11, we also have that $(G \times H)_\mathfrak{T} = G_\mathfrak{T} \times H_\mathfrak{T}$. Hence

$$(G \times H)_\mathfrak{R} = (G \times H)_\mathfrak{T}\,\mathrm{E_m}(G \times H) = G_\mathfrak{T}\,\mathrm{E_m}(G) \times H_\mathfrak{T}\,\mathrm{E_m}(H) = G_\mathfrak{R} \times H_\mathfrak{R},$$

and $\mathfrak{R}$ is a Lockett class.

Let $\mathfrak{s}(\mathfrak{R})$ denote the largest Fitting subclass of $\mathfrak{R}$ which has a generating system of perfect groups. Then $\mathfrak{M} \subseteq \mathfrak{s}(\mathfrak{R}) \subseteq \mathfrak{R}_*$. Hence $\mathfrak{T}_* \cdot \mathfrak{M} \subseteq \mathfrak{R}_*$. On the other hand, for an arbitrary group G, we have that

$$[G_\mathfrak{R}, G] = [G_\mathfrak{T} G_\mathfrak{M}, G] = [G_\mathfrak{T}, G][G_\mathfrak{M}, G] \le G_{\mathfrak{T}_*} G_\mathfrak{M},$$

by [DH92, X, 1.3]. Hence $\mathfrak{T}_* \cdot \mathfrak{M} \in \mathrm{Locksec}(\mathfrak{R})$ by [DH92, X, 1.21]. Therefore $\mathfrak{T}_* \cdot \mathfrak{M} = \mathfrak{R}_*$ and we conclude the proof. $\square$

Lemma 7.2.13. *Let $\mathfrak{T}$ be a Fitting class such that $\mathfrak{T} = \mathfrak{T}\mathfrak{S}$.*

1. *Set $\mathfrak{M} = \mathrm{Fit}\big(\mathrm{b}(\mathfrak{T})\big)$. If U is an $\mathfrak{M}$-subgroup of a group G containing $G_\mathfrak{M}$, then U is a subgroup of $G_\mathfrak{M} G_\mathfrak{T}$.*
2. *The class $\mathfrak{T} \cdot \mathrm{Fit}\big(\mathrm{b}(\mathfrak{T})\big)$ is a normal Fitting class.*

Proof. Denote $\mathfrak{m} = \mathrm{b}(\mathfrak{T})$ and $\mathfrak{B} = \mathfrak{T}^\mathrm{b}$.

1. We can assume that $G \notin \mathfrak{T}$ and then $\mathrm{b_m}(G) = \{X_1, \ldots, X_n\}$ is a non-empty set and $\mathrm{E_m}(G) = X_1 \cdots X_n \le G_\mathfrak{M} \le U$. Hence $\mathrm{b_m}(U) = \{X_1, \ldots, X_n, \ldots, X_t\}$, for $n \le t$, and $\mathrm{E_m}(U) = \mathrm{E_m}(G)L$, for $L = X_{n+1} \cdots X_t$. As in the proof of Theorem 7.2.8, $G_\mathfrak{M} = G_\mathfrak{B}\,\mathrm{E_m}(G)$ and $U = U_\mathfrak{B}\,\mathrm{E_m}(U)$.

Since $X_i \not\le U_\mathfrak{B}$ for each index i, we have that $[U_\mathfrak{B}, X_i] \le U_\mathfrak{B} \cap X_i \le (X_i)_\mathfrak{B}$. Thus

$$[\mathrm{E_m}(G), U_\mathfrak{B}] = [X_1, U_\mathfrak{B}] \cdots [X_n, U_\mathfrak{B}] \le (X_1)_\mathfrak{B} \cdots (X_n)_\mathfrak{B} = \mathrm{E_m}(G)_\mathfrak{B},$$

by Lemma 7.2.5 (2a). Analogously, by Lemma 7.2.5 (1a), $[X_i, L] \le X_i \cap L \le (X_i)_\mathfrak{B}$, for each i. Hence $[\mathrm{E_m}(G), L] \le \mathrm{E_m}(G)_\mathfrak{B}$. Therefore

$$[G_\mathfrak{M}, U_\mathfrak{B} L] = [G_\mathfrak{B}\,\mathrm{E_m}(G), U_\mathfrak{B} L] \le G_\mathfrak{B}[\mathrm{E_m}(G), U_\mathfrak{B}][\mathrm{E_m}(G), L] \le G_\mathfrak{B}.$$

By Theorem 7.2.8, $U_\mathfrak{B} L \le G_\mathfrak{T}$ and $U = U_\mathfrak{B}\,\mathrm{E_m}(G)L \le \mathrm{E_m}(G)G_\mathfrak{T} = G_\mathfrak{M} G_\mathfrak{T}$.

2. To see that the class $\mathfrak{R} = \mathfrak{T} \cdot \mathfrak{M}$ is a normal Fitting class consider a group G and suppose that U is an $\mathfrak{R}$-subgroup such that $G_\mathfrak{R} \le U \le G$. By Statement 1, $U_\mathfrak{M} \le G_\mathfrak{M} G_\mathfrak{T} = G_\mathfrak{R}$. On the other hand, using the arguments of the proof of Statement 1, $[\mathrm{E_m}(G), U_\mathfrak{T}] \le U_\mathfrak{T} \cap \mathrm{E_m}(G) \le \mathrm{E_m}(G)_\mathfrak{B}$. Then

$$[G_\mathfrak{M}, U_\mathfrak{T}] = [G_\mathfrak{B}\,\mathrm{E_m}(G), U_\mathfrak{T}] \le G_\mathfrak{B}[\mathrm{E_m}(G), U_\mathfrak{T}] \le G_\mathfrak{B}\,\mathrm{E_m}(G)_\mathfrak{B} \le G_\mathfrak{B}.$$

Hence $U_\mathfrak{T} \le \mathrm{C}_G(G_\mathfrak{M}/G_\mathfrak{B}) = G_\mathfrak{T}$, by Theorem 7.2.8. Thus, $U = U_\mathfrak{M} U_\mathfrak{T} \le G_\mathfrak{R}$ and $U = G_\mathfrak{R}$. $\square$

Lemma 7.2.14. *If $\mathfrak{T}$ is a Fitting class such that $\mathfrak{T} = \mathfrak{T}\mathfrak{S}$, X is a group in $\mathrm{b}(\mathfrak{T})$ and $\mathfrak{F} \in \mathrm{Locksec}(\mathfrak{T})$, then $X_{\mathfrak{F}}$ is not $\mathfrak{F}$-maximal in X.*

Proof. If $\mathfrak{F} \in \mathrm{Locksec}(\mathfrak{T})$, then, in particular, $\mathfrak{T}^{\mathrm{b}} \subseteq \mathfrak{F} \subseteq \mathfrak{T}$ by Lemma 7.2.11. Moreover $\mathrm{b}(\mathfrak{T}) \subseteq \mathrm{b}(\mathfrak{F})$ by [DH92, XI, 4.7]. Since $X \in \mathrm{b}(\mathfrak{T})$, then $\mathrm{Cosoc}(X) = X_{\mathfrak{F}}$. Suppose that $X_{\mathfrak{F}}$ is $\mathfrak{F}$-maximal in X. Consider a soluble subgroup $Y/X_{\mathfrak{F}}$ of $X/X_{\mathfrak{F}}$. Then $Y \in \mathfrak{T}\mathfrak{S} = \mathfrak{T}$, and by maximality of $X_{\mathfrak{F}}$ in X, we have that $X_{\mathfrak{F}} = Y_{\mathfrak{F}}$. Since $\mathfrak{F} \in \mathrm{Locksec}(\mathfrak{T})$, the quotient $Y/X_{\mathfrak{F}}$ is abelian, by [DH92, X, 1.21]. Then $X/X_{\mathfrak{F}}$ is soluble, and this is a contradiction. $\qquad\square$

Theorem 7.2.15 (see [ILPM04]). *Let $\mathfrak{T}$ be a Fitting class such that $\mathfrak{T} = \mathfrak{T}\mathfrak{S}$. If $\mathfrak{H} \in \mathrm{Sec}\big(\mathfrak{T}_*, \mathfrak{T} \cdot \mathrm{Fit}\big(\mathrm{b}(\mathfrak{T})\big)\big)$, then*

1. $\mathfrak{H}$ *is an injective Fitting class;*
2. $\mathfrak{H}$ *is a normal Fitting class if and only if $\mathfrak{H} \in \mathrm{Locksec}\big(\mathfrak{T} \cdot \mathrm{Fit}\big(\mathrm{b}(\mathfrak{T})\big)\big)$.*

Proof. 1. Write $\mathfrak{m} = \mathrm{b}(\mathfrak{T})$, $\mathfrak{F} = \mathfrak{T} \cap \mathfrak{H}$ and $\mathfrak{G} = \mathfrak{F} \cdot \mathrm{Fit}\,\mathfrak{m}$. If $H \in \mathfrak{H}$, then $H = H_{\mathfrak{T}}\,\mathrm{E}_{\mathfrak{m}}(H)$, by Lemma 7.2.6, since $\mathfrak{H} \subseteq \mathfrak{T} \cdot \mathrm{Fit}\,\mathfrak{m}$. Thus, $H_{\mathfrak{T}} \in \mathfrak{H} \cap \mathfrak{T} = \mathfrak{F}$. Hence $H = H_{\mathfrak{F}}\,\mathrm{E}_{\mathfrak{m}}(H) \in \mathfrak{F} \cdot \mathrm{Fit}\,\mathfrak{m} = \mathfrak{G}$. Hence $\mathfrak{H} \subseteq \mathfrak{G}$.

To see that $\mathfrak{H}$ is injective, let G be a group and let us prove that G possesses $\mathfrak{H}$-injectors. If $\mathrm{b}_{\mathfrak{m}}(G) = \emptyset$, then $G \in \mathfrak{T}$. Hence $G_{\mathfrak{F}} = G_{\mathfrak{H}}$. Since $\mathfrak{F} \in \mathrm{Locksec}(\mathfrak{T})$ by Theorem 7.2.12, the quotient $G/G_{\mathfrak{H}}$ is abelian. Therefore $G_{\mathfrak{H}}$ is a normal $\mathfrak{H}$-injector of G.

Assume that $\mathrm{b}_{\mathfrak{m}}(G) \neq \emptyset$. Since $G_{\mathfrak{H}}$ is a normal subgroup of G we can assume that $\mathrm{b}_{\mathfrak{m}}(G_{\mathfrak{H}}) = \{X_1, \ldots, X_r\}$ and $\mathrm{b}_{\mathfrak{m}}(G) = \{X_1, \ldots, X_n\}$, for $r \leq n$. If $r = n$, then $G_{\mathfrak{H}} = G_{\mathfrak{F}}\,\mathrm{E}_{\mathfrak{m}}(G_{\mathfrak{H}}) = G_{\mathfrak{F}}\,\mathrm{E}_{\mathfrak{m}}(G) = G_{\mathfrak{G}}$. By Theorem 7.2.12, $\mathfrak{G} \in \mathrm{Locksec}\big(\mathfrak{T} \cdot \mathrm{Fit}\big(\mathrm{b}(\mathfrak{T})\big)\big)$. Since, by Lemma 7.2.13, $\mathfrak{T} \cdot \mathrm{Fit}\big(\mathrm{b}(\mathfrak{T})\big)$ is a normal Fitting class, we deduce that so is $\mathfrak{G}$, by [DH92, X, 3.3]. Therefore $G_{\mathfrak{G}}$ is $\mathfrak{G}$-injector of G and $G_{\mathfrak{H}}$ is $\mathfrak{H}$-injector of G.

Now assume that $r < n$. Fix an index $i \in \{r+1, \ldots, n\}$. Clearly, X_i is a perfect comonolithic group such that $X_i \notin \mathfrak{H}$. In addition, $\mathrm{Cosoc}(X_i) \in \mathfrak{H}$, by virtue of Lemma 7.2.11. In particular, X_i is an $\mathfrak{H}$-component of G, By Proposition 7.2.2, X_i possesses $\mathfrak{H}$-injectors. Consider $H = H_{r+1} \cdots H_n$, with $H_i \in \mathrm{Inj}_{\mathfrak{H}}(X_i)$ (note that H_i normalises H_j, $i, j \in \{r+1, \ldots, n\}$, by Lemma 7.2.3). By induction on the order of G, if $\mathrm{N}_G(H)$ is a proper subgroup of G, then $\mathrm{N}_G(H)$ possesses $\mathfrak{H}$-injectors. Then G possesses $\mathfrak{H}$-injectors by Theorem 7.2.4. Therefore we can suppose that H is a normal subgroup of G. Then H_i is a normal subgroup of X_i and then $H_i = \mathrm{Cosoc}(X_i) = (X_i)_{\mathfrak{H}}$. Thus $(X_i)_{\mathfrak{H}}$ is an $\mathfrak{F}$-maximal subgroup of X_i, which contradicts Lemma 7.2.14.

2. It is shown in Theorem 7.2.12 that $\mathfrak{T} \cdot \mathrm{Fit}\,\mathfrak{m}$ is a Lockett class. Moreover, by Lemma 7.2.13, it is a normal Fitting class. If $\mathfrak{H} \in \mathrm{Locksec}(\mathfrak{T} \cdot \mathrm{Fit}\,\mathfrak{m})$, then $\mathfrak{H}$ is also a normal Fitting class by [DH92, X, 3.3]. For the converse, consider $\mathfrak{H} \notin \mathrm{Locksec}(\mathfrak{T} \cdot \mathrm{Fit}\,\mathfrak{m})$. Observe that $(\mathfrak{T} \cdot \mathrm{Fit}\,\mathfrak{m})_* = \mathfrak{T}_* \cdot \mathrm{Fit}\,\mathfrak{m}$, by Theorem 7.2.12 and then $\mathrm{Fit}\,\mathfrak{m} \not\subseteq \mathfrak{H}$. Let X be a group in $\mathfrak{m} \setminus \mathfrak{H}$. Then X is a perfect and comonolithic group and $\mathrm{Cosoc}(X) \in \mathfrak{H} \cap \mathfrak{T} = \mathfrak{F}$. Hence $X_{\mathfrak{F}} = \mathrm{Cosoc}(X)$. Since

$\mathfrak{T}_*$ is contained in $\mathfrak{F}$, it follows that $\mathfrak{F} \in \mathrm{Locksec}(\mathfrak{T})$. By Lemma 7.2.14, $X_{\mathfrak{F}}$ is not $\mathfrak{F}$-maximal in X. Therefore $\mathfrak{H}$ is not a normal Fitting class.   □

Corollary 7.2.16 (see [ILPM04]). *If $\mathfrak{F}$ is a Fitting class in* $\mathrm{Locksec}(\mathfrak{S})$, *then $\mathfrak{F}$ is injective.*

Proof. If $\mathfrak{F} \in \mathrm{Sec}\big(\mathfrak{S}_*, \mathfrak{S} \cdot \mathrm{Fit}(b(\mathfrak{S}))\big) = \mathrm{Sec}\big(\mathfrak{S}_*, \mathfrak{S}^* \cdot \mathrm{Fit}(b(\mathfrak{S}))\big)$, then $\mathfrak{F}$ is an injective Fitting class. In particular if $\mathfrak{F} \in \mathrm{Locksec}(\mathfrak{S}) = \{\mathfrak{F} : \mathfrak{S}_* \subseteq \mathfrak{F} \subseteq \mathfrak{S} = \mathfrak{S}^*\}$, then $\mathfrak{F}$ is injective.   □

Remarks 7.2.17. The example of a non-injective Fitting class in Section 7.1 affords counterexamples to possible extensions of Theorem 7.2.15:

1. Fitting classes $\mathfrak{H} \in \mathrm{Sec}\big(\mathfrak{T}^b, \mathrm{Fit}(b(\mathfrak{T}))\big)$ need not be injective;
2. if $\mathfrak{T} = \mathfrak{T}\mathfrak{S}$, then $\mathrm{Fit}(b(\mathfrak{T}))$ need not be injective;
3. Fitting classes $\mathfrak{H} \in \mathrm{Sec}\big(\mathfrak{T}^b, \mathrm{Fit}(b(\mathfrak{T}))\big)$ need not be normal. There are normal Fitting classes which does not belong to $\mathrm{Sec}\big(\mathfrak{T}^b, \mathrm{Fit}(b(\mathfrak{T}))\big)$.

Proof. Let S and T be non-abelian simple groups such that $\mathrm{D}_0(S, T, 1)$ is a non-injective Fitting class.

1. Let R be a non-abelian simple group and consider the regular wreath product $W = (S \times T) \wr R$. Then W is a perfect comonolithic group (see [DH92, A, 18.8]). Hence $\mathfrak{m} = (W)$ is a preboundary and $\mathfrak{T} = \mathrm{h}(\mathfrak{m})$ is a Fitting class such that $\mathfrak{T} = \mathfrak{T}\mathfrak{S}$ by Theorem 2.4.12 (3). Note that $\mathfrak{T}^b = \mathrm{Fit}\big(\mathrm{Cosoc}(W)\big) = \mathrm{D}_0(S, T)$ is not injective.

2. If $\mathfrak{m} = (S, T, 1)$ and $\mathfrak{T} = \mathrm{h}(\mathfrak{m})$, then $\mathfrak{T} = \mathfrak{T}\mathfrak{S}$ and $\mathrm{Fit}(b(\mathfrak{T})) = \mathrm{D}_0(S, T, 1)$ is a non-injective Fitting class.

3. Let $\mathfrak{D}$ denote the class of all direct products of non-abelian simple groups. Let E and F be any two non-abelian simple groups. The regular wreath product $W = E \wr F$ is a perfect comonolithic group. Set $\mathfrak{m} = (W)$, $\mathfrak{T} = \mathrm{h}(\mathfrak{m})$ and $\mathfrak{H} = \mathfrak{S}_*\mathfrak{D}$. Then $\mathfrak{T}^b = \mathrm{D}_0(E, 1) \subseteq \mathfrak{H}$. Moreover, $\mathfrak{H}$ is the smallest normal Fitting class, by [DH92, X, 3.27], and then $\mathfrak{H} \subseteq \mathfrak{T} \cdot \mathrm{Fit}(b(\mathfrak{T}))$ by Lemma 7.2.13. If R is a non-abelian simple group, $R \not\cong F$, then the regular wreath product $G = E \wr R \in \mathfrak{T}$. The base subgroup is $E^\natural = G_{\mathfrak{H}}$ and $G/G_{\mathfrak{H}} \cong R$ is non-abelian. Therefore $\mathfrak{T}_* \not\subseteq \mathfrak{H}$, by [DH92, X, 1.2]. Clearly $\mathrm{Fit}(b(\mathfrak{T})) = \mathrm{Fit}(W) \not\subseteq \mathfrak{H}$. Note that $\mathfrak{T}^b$ is not normal.   □

Corollary 7.2.18. *If $\mathfrak{F}$ is a Fitting class such that $\mathfrak{F}\mathfrak{S} = \mathfrak{F}$, then $\mathfrak{F}$ is injective. In particular, the class $\mathfrak{S}$ of all soluble groups is injective.*

Corollary 7.2.19. *A group G possesses a single conjugacy class of $\mathfrak{S}$-injectors if and only if G is soluble.*

Proof. Applying Theorem 2.4.26, only the necessity of the condition is in doubt. Assume that a group G possesses a single conjugacy class of $\mathfrak{S}$-injectors. Let p and q be two different primes dividing the order of $E_{\mathfrak{S}}(G)$

and let P and Q be a Sylow p-subgroup and a Sylow q-subgroup of $E_{\mathfrak{S}}(G)$ respectively. Applying Proposition 7.2.2 (2) and Theorem 7.2.4, there exist $\mathfrak{S}$-injectors V and W of G such that $P \leq V$ and $Q \leq W$. Since V and W are conjugate in G and $E_{\mathfrak{S}}(G)$ is normal in G, it follows that $V \cap E_{\mathfrak{S}}(G)$ contains a Sylow q-subgroup of $E_{\mathfrak{S}}(G)$ for each prime q dividing $|E_{\mathfrak{S}}(G)|$. Therefore $E_{\mathfrak{S}}(G)$ is contained in V and so $E_{\mathfrak{S}}(G) = 1$. This yields that G is soluble. $\square$

Theorem 7.2.20. *Let $\mathfrak{X}$ be a class of quasisimple groups and consider the class*

$$\mathfrak{K}(\mathfrak{X}) = (G : \text{every component of } G \text{ is in } \mathfrak{X}).$$

Then $\mathfrak{K}(\mathfrak{X})$ is an injective Fitting class.

Proof. Let $\mathfrak{X}$ be a class of quasisimple groups and denote $\mathfrak{K} = \mathfrak{K}(\mathfrak{X})$. We first prove that $\mathfrak{K}$ is a Fitting class.

If $G \in \mathfrak{K}$ and N is a normal subgroup of G, then every component of N is a component of G. Hence every component of N is in $\mathfrak{X}$ and then $N \in \mathfrak{K}$.

Suppose that a group G is product $G = NM$, where N and M are normal $\mathfrak{K}$-subgroups of G. Let E be a component of G. Assume that E is not contained in M and E is not contained in N. Applying Proposition 7.2.3, it follows that E centralises MN. Hence E is central in G. This is a contradiction. Therefore either E is contained in M or E is contained in N. Hence E belongs to $\mathfrak{X}$. It implies that $G \in \mathfrak{K}$.

Let E be a component of a group $G \in \mathfrak{K}\mathfrak{S}$. Then $E \in \mathfrak{K}\mathfrak{S}$. Since E is perfect, it follows that $E \in \mathfrak{K}$. Hence $\mathfrak{K} = \mathfrak{K}\mathfrak{S}$ and therefore $\mathfrak{K}$ is injective by Corollary 7.2.18. $\square$

Let $\mathfrak{K}$ be a Fitting class as in Theorem 7.2.20. By Proposition 2.4.6 (5) and Proposition 2.4.6 (2), $\mathfrak{F} \diamond \mathfrak{K} \diamond \mathfrak{S} = \mathfrak{F} \diamond \mathfrak{K}$ for each Fitting class $\mathfrak{F}$. Hence we have the following:

Corollary 7.2.21. *Let $\mathfrak{X}$ be a class of quasisimple groups and consider the class $\mathfrak{K} = \mathfrak{K}(\mathfrak{X})$ as in Theorem 7.2.20. Then $\mathfrak{F} \diamond \mathfrak{K}$ is an injective Fitting class for any Fitting class $\mathfrak{F}$.*

Note that [För87, 2.5(b)] is a consequence of the above corollary.

In the following, we describe another injective Fitting class, the class of all $\mathfrak{F}$-constrained groups.

Proposition 7.2.22. *Let $\mathfrak{F}$ be a Fitting class. In a group G, the following statements are equivalent:*

1. $C_G(G_{\mathfrak{F}}) \leq G_{\mathfrak{F}}$,
2. $F^*(G) \in \mathfrak{F}$.

Proof. 1 implies 2. Suppose that E is a component of G such that $E \nleq G_{\mathfrak{F}}$. Then $[G_{\mathfrak{F}}, E] = 1$, by Proposition 7.2.3. Therefore $E \leq C_G(G_{\mathfrak{F}}) \leq G_{\mathfrak{F}}$. This contradiction yields $E(G) \leq G_{\mathfrak{F}}$.

Denote $\pi = \operatorname{char} \mathfrak{F}$. Applying Proposition 2.2.22 (2) we have that $\mathrm{F}^*(G) = \mathrm{F}(G)\,\mathrm{E}(G) = \mathrm{O}_{\pi'}(\mathrm{F}(G))\,\mathrm{O}_{\pi}(\mathrm{F}(G))\,\mathrm{E}(G)$. On the other hand, the normal $\mathfrak{F}$-subgroup $\mathrm{O}_{\pi'}(\mathrm{F}(G)) \cap G_{\mathfrak{F}}$ is a nilpotent π'-group. Hence $\mathrm{O}_{\pi'}(\mathrm{F}(G)) \cap G_{\mathfrak{F}} = 1$ and then $\mathrm{O}_{\pi'}(\mathrm{F}(G)) \leq \mathrm{C}_G(G_{\mathfrak{F}}) \leq G_{\mathfrak{F}}$. Therefore $\mathrm{O}_{\pi'}(\mathrm{F}(G)) = 1$ and $\mathrm{F}(G) = \mathrm{O}_{\pi}(\mathrm{F}(G)) \in \mathfrak{F}$. Then $\mathrm{F}^*(G) \in \mathfrak{F}$.

2 implies 1. Since $\mathrm{F}^*(G) \in \mathfrak{F}$, it follows that $\mathrm{F}^*(G) \leq G_{\mathfrak{F}}$. Thus, by Proposition 2.2.22 (4),

$$C_G(G_{\mathfrak{F}}) \leq C_G\big(\mathrm{F}^*(G)\big) \leq \mathrm{F}^*(G) \leq G_{\mathfrak{F}}. \qquad \square$$

Corollary 7.2.23. *Let $\mathfrak{F}$ be a Fitting class. Let G be a group such that $C_G(G_{\mathfrak{F}}) \leq G_{\mathfrak{F}}$. Then for any subnormal subgroup S of G, we have that $C_S(S_{\mathfrak{F}}) \leq S_{\mathfrak{F}}$.*

Corollary 7.2.24 ([IPM86]). *Let $\mathfrak{F}$ be a Fitting class and $\pi = \operatorname{char} \mathfrak{F}$. For any group G, write $\bar{G} = G/\mathrm{O}_{\pi'}(G)$ and adopt the "bar convention:" if $H \leq G$, then $\bar{H} = H\,\mathrm{O}_{\pi'}(G)/\mathrm{O}_{\pi'}(G)$.*

The following statements are pairwise equivalent:

1. $C_{\bar{G}}(\bar{G}_{\mathfrak{F}}) \leq \bar{G}_{\mathfrak{F}}$,
2. $\mathrm{E}(\bar{G}) \in \mathfrak{F}$,
3. $\mathrm{F}^*(\bar{G}) \in \mathfrak{F}$.

Definition 7.2.25. *For a Fitting class $\mathfrak{F}$, a group G is said to be $\mathfrak{F}$-constrained if G satisfies one condition of Corollary 7.2.24.*

Note that every group is $\mathfrak{Q}$-constrained by Proposition 2.2.22 (4) and a group G is $\mathfrak{N}$-constrained if $C_G(\mathrm{F}(G)) \leq \mathrm{F}(G)$.

Corollary 7.2.26. *Let $\mathfrak{F}$ be a Fitting class. The class of all $\mathfrak{F}$-constrained groups is an injective Fitting class.*

Proof. Let $\mathfrak{X}$ be the class of all quasisimple $\mathfrak{F}$-groups and consider the Fitting class $\mathfrak{K} = \mathfrak{K}(\mathfrak{X})$. A group G is $\mathfrak{F}$-constrained if and only if $\mathrm{E}\big(G/\mathrm{O}_{\pi'}(G)\big) \in \mathfrak{F}$. This is equivalent to say that every component of the group $G/\mathrm{O}_{\pi'}(G) \in \mathfrak{X}$. This happens if and only if $G/\mathrm{O}_{\pi}(G) \in \mathfrak{K}$, or, in other words, if and only if $G \in \mathfrak{E}_{\pi'} \diamond \mathfrak{K}$. Therefore the class of all $\mathfrak{F}$-constrained groups is the Fitting class $\mathfrak{E}_{\pi'} \diamond \mathfrak{K}$. By Corollary 7.2.21, is an injective Fitting class. $\qquad \square$

Recall that the first result of existence and conjugacy of $\mathfrak{N}$-injectors in a universe larger that the soluble groups is due to Mann working on $\mathfrak{N}$-constrained groups [Man71]. Theorem 7.2.1 proves that every group, i.e. every $\mathfrak{Q}$-constrained group, possesses a unique conjugacy class of $\mathfrak{Q}$-injectors. Thus it seems that for every Fitting class $\mathfrak{F}$, the property of being an $\mathfrak{F}$-constrained group is closely related to the conjugacy of $\mathfrak{F}$-injectors. In general the equivalence does not hold as we observed in Corollary 7.2.19 inasmuch as the class $\mathfrak{S}$ of all soluble groups is properly contained in the class of all $\mathfrak{S}$-constrained groups (which is the same as the class of all $\mathfrak{N}$-constrained groups). For Fitting classes $\mathfrak{F}$ such that $\mathfrak{N} \subseteq \mathfrak{F} \subseteq \mathfrak{Q}$, we have the following result.

Proposition 7.2.27 ([IPM86]). *Let $\mathfrak{F}$ be a Fitting class such that $\mathfrak{N} \subseteq \mathfrak{F} \subseteq \mathfrak{Q}$.*

If G is an $\mathfrak{F}$-constrained group, then

1. *G possesses a single conjugacy class of $\mathfrak{F}$-injectors, and*
2. *the $\mathfrak{F}$-injectors and the $\mathfrak{Q}$-injectors of G coincide.*

Conversely, if G is a group such that the $\mathfrak{Q}$-injectors are in $\mathfrak{F}$, then G is an $\mathfrak{F}$-constrained group.

Proof. Let G be an $\mathfrak{F}$-constrained group. Then, since char $\mathfrak{F} = \mathbb{P}$, we have that $\mathrm{F}^*(G) = G_{\mathfrak{F}}$, by Corollary 7.2.24. Let V be an $\mathfrak{Q}$-injector of G. Then V is an $\mathfrak{Q}$-maximal subgroup containing $\mathrm{F}^*(G)$ [BL79]. Observe that, since $\mathrm{F}^*(G) \leq V_{\mathfrak{F}}$, we have that

$$\mathrm{C}_V(V_{\mathfrak{F}}) \leq \mathrm{C}_V\big(\mathrm{F}^*(G)\big) \leq \mathrm{F}^*(G) \leq V_{\mathfrak{F}},$$

and V is an $\mathfrak{F}$-constrained group. Thus $V = \mathrm{F}^*(V) = V_{\mathfrak{F}}$ and V is an $\mathfrak{F}$-maximal subgroup of G.

If S is a subnormal subgroup of G, then $V \cap S$ is an $\mathfrak{Q}$-injector of S. Since $\mathfrak{F}$ is contained in $\mathfrak{Q}$, we have that $V \cap S$ is $\mathfrak{F}$-maximal in S.

In order to obtain the conjugacy of all $\mathfrak{F}$-injectors of G, it is enough to prove that each $\mathfrak{F}$-injector of G is an $\mathfrak{Q}$-injector of G. Let H be an $\mathfrak{F}$-injector of G, then H is an $\mathfrak{F}$-maximal subgroup of G containing $G_{\mathfrak{F}} = \mathrm{F}^*(G)$. Hence H is an $\mathfrak{Q}$-subgroup of G containing $\mathrm{F}^*(G)$ and there exists a $\mathfrak{Q}$-injector V of G such that $H \leq V$. By the previous arguments, $V = H$.

The converse is obvious.   □

Lemma 7.2.28. *Let $\mathfrak{H}$ and $\mathfrak{F}$ be Fitting classes and let G be a group such that*

$$\mathrm{C}_G(G_{\mathfrak{H} \diamond \mathfrak{F}}/G_{\mathfrak{H}}) \leq G_{\mathfrak{H} \diamond \mathfrak{F}}.$$

Let J be subgroup of G containing $G_{\mathfrak{H} \diamond \mathfrak{F}}$. Then

1. *$J \in \mathrm{Max}_{\mathfrak{H} \diamond \mathfrak{F}}(G)$ if and only if $J/G_{\mathfrak{H}} \in \mathrm{Max}_{\mathfrak{F}}(G/G_{\mathfrak{H}})$.*
2. *$J \in \mathrm{Inj}_{\mathfrak{H} \diamond \mathfrak{F}}(G)$ if and only if $J/G_{\mathfrak{H}} \in \mathrm{Inj}_{\mathfrak{F}}(G/G_{\mathfrak{H}})$.*

Proof. The condition $\mathrm{C}_G(G_{\mathfrak{H} \diamond \mathfrak{F}}/G_{\mathfrak{H}}) \leq G_{\mathfrak{H} \diamond \mathfrak{F}}$ is equivalent to $\mathrm{C}_{\bar{G}}(\bar{G}_{\mathfrak{F}}) \leq \bar{G}_{\mathfrak{F}}$ for the quotient group $\bar{G} = G/G_{\mathfrak{H}}$. Let S be a subnormal subgroup of G. By Corollary 7.2.23 we have that $\mathrm{C}_{\bar{S}}(\bar{S}_{\mathfrak{F}}) \leq \bar{S}_{\mathfrak{F}}$, for $\bar{S} = SG_{\mathfrak{H}}/G_{\mathfrak{H}}$. But, since $S_{\mathfrak{H}} = G_{\mathfrak{H}} \cap S$, we have that $\bar{S} \cong S/S_{\mathfrak{H}}$. Therefore, for any subnormal subgroup S of G, $\mathrm{C}_S(S_{\mathfrak{H} \diamond \mathfrak{F}}/S_{\mathfrak{H}}) \leq S_{\mathfrak{H} \diamond \mathfrak{F}}$.

Let K be a subgroup of G such that $G_{\mathfrak{H} \diamond \mathfrak{F}} \leq K$. Observe that $G_{\mathfrak{H}} \leq K$ implies that $G_{\mathfrak{H}} \leq K_{\mathfrak{H}} \cap G_{\mathfrak{H} \diamond \mathfrak{F}}$. On the other hand $K_{\mathfrak{H}} \cap G_{\mathfrak{H} \diamond \mathfrak{F}}$ is a normal $\mathfrak{H}$-subgroup of K and then of $G_{\mathfrak{H} \diamond \mathfrak{F}}$, i.e.

$$K_{\mathfrak{H}} \cap G_{\mathfrak{H} \diamond \mathfrak{F}} \leq (G_{\mathfrak{H} \diamond \mathfrak{F}})_{\mathfrak{H}} \leq G_{\mathfrak{H}}$$

and therefore $G_{\mathfrak{H}} = K_{\mathfrak{H}} \cap G_{\mathfrak{H} \diamond \mathfrak{F}}$. Thus $[K_{\mathfrak{H}}, G_{\mathfrak{H} \diamond \mathfrak{F}}] \leq G_{\mathfrak{H}}$. This implies that

$$K_{\mathfrak{H}} \leq C_G(G_{\mathfrak{H} \diamond \mathfrak{F}})/G_{\mathfrak{H}} \leq G_{\mathfrak{H} \diamond \mathfrak{F}}$$

and then $G_{\mathfrak{H}} = K_{\mathfrak{H}}$.

Using this fact, the proof is a routine checking. $\qquad\Box$

Corollary 7.2.29. *Let $\mathfrak{F}$ be a Fitting class containing the class of all nilpotent groups $\mathfrak{N}$. Assume that every $\mathfrak{F}$-constrained group possesses $\mathfrak{F}$-injectors.*

Then, for every Fitting class $\mathfrak{H}$, the class $\mathfrak{H} \diamond \mathfrak{F}$ is injective.

Proof. We have to prove that $\mathrm{Inj}_{\mathfrak{H} \diamond \mathfrak{F}}(G) \neq \emptyset$ for every group G. Let G be a minimal counterexample. First we notice that a subgroup E is an $\mathfrak{H} \diamond \mathfrak{F}$-component of G such that $\mathrm{N}(E) \in \mathfrak{H}$ if and only if $EG_{\mathfrak{H}}/G_{\mathfrak{H}}$ is a component of $G/G_{\mathfrak{H}}$ such that $EG_{\mathfrak{H}}/G_{\mathfrak{H}} \notin \mathfrak{F}$.

Let $\mathcal{E} = \{E_1, \ldots, E_n\}$ be the set of all $\mathfrak{H} \diamond \mathfrak{F}$-components of G such that $\mathrm{N}(E_i) \in \mathfrak{H}$ and suppose that $\mathcal{E} \neq \emptyset$. For $J_i \in \mathrm{Inj}_{\mathfrak{H} \diamond \mathfrak{F}}(E_i)$, $i = 1, \ldots, n$, construct the product $J = J_1 \cdots J_n$. If $\mathrm{N}_G(J)$ is a proper subgroup of G, then $\mathrm{Inj}_{\mathfrak{H} \diamond \mathfrak{F}}(\mathrm{N}_G(J)) \neq \emptyset$, by minimality of G. Since the set $\mathcal{E}$ is invariant by conjugation of the elements of G, we can apply Theorem 7.2.4 and then $\mathrm{Inj}_{\mathfrak{H} \diamond \mathfrak{F}}(G) \neq \emptyset$. This contradicts our assumption. Therefore J is a normal subgroup of G and then each J_i is normal in E_i, for $i = 1, \ldots, n$. This implies that $J_i \leq \mathrm{Cosoc}(E_i)$.

Let $P/(E_i)_{\mathfrak{H}}$ be a Sylow subgroup of $E_i/(E_i)_{\mathfrak{H}}$. Then $P \in \mathfrak{H} \diamond \mathfrak{F}$. Observe that, since $J_i/\mathrm{N}(E_i) \leq \mathrm{Z}(E_i/\mathrm{N}(E_i))$, the subgroup P is normal in PJ_i. Then $PJ_i \in \mathfrak{H} \diamond \mathfrak{F}$. By maximality of J_i, we have that $P \leq J_i$. Since this happens for any Sylow subgroup of E_i, we have that $E_i \leq J_i$, which is a contradiction. Hence $\mathcal{E} = \emptyset$ and every component of $G/G_{\mathfrak{H}}$ is in $\mathfrak{F}$. Therefore $\mathrm{E}(G/G_{\mathfrak{H}}) \in \mathfrak{F}$. This implies that $G/G_{\mathfrak{H}}$ is $\mathfrak{F}$-constrained, i.e. $C_G(G_{\mathfrak{H} \diamond \mathfrak{F}})/G_{\mathfrak{H}} \leq G_{\mathfrak{H} \diamond \mathfrak{F}}$ by Corollary 7.2.24 . By hypothesis, the group $G/G_{\mathfrak{H}}$, possesses $\mathfrak{F}$-injectors. By Lemma 7.2.28, the group G possesses $\mathfrak{H} \diamond \mathfrak{F}$-injectors. This is the final contradiction. $\qquad\Box$

Corollary 7.2.30 (M. J. Iranzo and F. Pérez-Monasor). *Let $\mathfrak{F}$ be a Fitting class such that $\mathfrak{N} \subseteq \mathfrak{F} \subseteq \mathfrak{Q}$. Then, for every Fitting class $\mathfrak{H}$, the class $\mathfrak{H} \diamond \mathfrak{F}$ is injective.*

In particular, the class $\mathfrak{N}$ of all nilpotent groups is injective (P. Förster [För85a]).

Observe that $\mathfrak{E}_{\pi'}\mathfrak{N}_{\pi} = \mathfrak{E}_{\pi'}\mathfrak{N}$. This leads us to the following.

Corollary 7.2.31. *Let π be a set of prime numbers. The Fitting class $\mathfrak{E}_{\pi'}\mathfrak{N}_{\pi}$ is injective.*

In particular, for any prime p, the Fitting class $\mathfrak{E}_{p'}\mathfrak{S}_p$ of all p-nilpotent groups is injective.

Remark 7.2.32. Let p be a prime. We say that a group G is *p-constrained* if G is $\mathfrak{S}_p$-constrained group. M. J. Iranzo and M. Torres proved in [IT89] that a group G possesses a unique conjugacy class of p-nilpotent injectors if and only if G is p-constrained. Moreover, in this case,

$$\mathrm{Inj}_{\mathfrak{E}_{p'}\mathfrak{S}_p}(G) = \{O_{p',p}(G)P : P \in \mathrm{Syl}_p(G)\},$$

and the p-nilpotent injectors of G are the p-nilpotent maximal subgroups of G containing $O_{p',p}(G)$.

Theorem 7.2.33 ([IPM88]). *Every extensible saturated Fitting formation is injective.*

Proof. Assume the result is false and let G be counterexample of least order. Clearly $\pi = \mathrm{char}\,\mathfrak{F} = \pi(\mathfrak{F})$ and $\mathfrak{N}_\pi \subseteq \mathfrak{F} \subseteq \mathfrak{E}_\pi$ since $\mathfrak{F}$ is saturated.

Assume the result is false and let G be counterexample of least order. Since G possesses $\mathfrak{F}$-injectors if and only if $G/O_{\pi'}(G)$ possesses $\mathfrak{F}$-injectors, it follows that $O_{\pi'}(G) = 1$. Also, since $\mathfrak{F}$ is an extensible homomorph, G has $\mathfrak{F}$-injectors if and only if $G/G_\mathfrak{F}$ possesses $\mathfrak{F}$-injectors. Therefore $G_\mathfrak{F} = 1$.

Consider, as in Theorem 7.2.4, the set $\mathcal{E} = \{E_1, \ldots, E_n\}$ of all $\mathfrak{F}$-components of G and suppose that $\mathcal{E} \neq \emptyset$. Observe that, since $G_\mathfrak{F} = 1$, the $\mathfrak{F}$-components of G are just the components. Let $i = 1, \ldots, n$. Then every $\mathfrak{F}$-maximal subgroup J_i of E_i containing the $\mathfrak{F}$-radical of E_i is an $\mathfrak{F}$-injector of E_i by Proposition 7.2.2 (2). Consider the subgroup $J = \langle J_1, \ldots, J_n \rangle$. By Theorem 7.2.4, we have that J is normal in G. Moreover, J is an $\mathfrak{F}$-group. Hence J is contained in $G_\mathfrak{F}$ and then $J_i = 1$. This implies that $E_i \in \mathfrak{E}_{\pi'}$ and, since E_i is subnormal in G, we obtain that $E_i = 1$. Then $\mathrm{E}(G) = 1$ and $\mathrm{F}^*(G) = \mathrm{F}(G) = O_\pi(\mathrm{F}(G)) \times O_{\pi'}(\mathrm{F}(G))$. But $O_\pi(\mathrm{F}(G)) \leq G_\mathfrak{F} = 1$ and $O_{\pi'}(\mathrm{F}(G)) \leq O_{\pi'}(G) = 1$. Hence $\mathrm{F}^*(G) = 1$. This contradiction proves the theorem. $\qquad\square$

It is not difficult to prove that every extensible saturated Fitting formation $\mathfrak{F}$ is of the form

$$\mathfrak{F} = \big(G : \text{all composition factors of } G \text{ belong to } \mathfrak{F} \cap \mathfrak{J}\big).$$

The most popular extensible saturated Fitting formations are the class $\mathfrak{E}_\pi$, π a set of primes, and the class $\mathfrak{S}$ of all soluble groups.

Applying the above result, every finite group possesses $\mathfrak{E}_\pi$-injectors. In general, if V is an $\mathfrak{E}_\pi$-injector of a group G, then V is a maximal π-subgroup of G containing $O_\pi(G)$; but $|G : V|$ need not to be a π'-number. If G possesses Hall π-subgroups, in particular if G is soluble, then the $\mathfrak{E}_\pi$-injectors of G are the Hall π-subgroups of G.

Concluding Remarks 7.2.34. There are many other injective Fitting classes closely related to the ones presented in the section. For instance, for each prime p, let us consider the class $\mathfrak{E}_{p^*p}$, the *p^*p-groups*, defined by H. Bender (see [HB82b]). This is the class composed by all groups G factorising as $G = N\,\mathrm{C}_G^*(P)$ for any normal subgroup N and any $P \in \mathrm{Syl}_p(N)$, where $\mathrm{C}_G^*(P)$

is the largest normal subgroup of $N_G(P)$ acting nilpotently on P. A group $G \in \mathfrak{E}_{p^*p}$ such that $O^p(G) = G$ is said to be a p^*-*group* and the class of all p^*-groups is denoted by $\mathfrak{E}_{p^*}$. The class $\mathfrak{E}_{p^*p}$ is an injective Fitting class and, in fact, any Fitting class $\mathfrak{F}$ such that $\mathfrak{E}_{p^*p} \subseteq \mathfrak{F} \subseteq \mathfrak{E}_{p^*}\mathfrak{S}_p$ is injective (see [IT89]).

Other examples of injective Fitting classes are the class $\mathfrak{E}_{p'}\mathfrak{Q}$ of all p-quasinilpotent groups and the class $\mathfrak{D}^p = \big(G : G/\operatorname{C}_G(O_p(G)) \in \mathfrak{S}_p\big)$ (see [MP92]). These classes satisfy the following chain

$$\mathfrak{E}_{p'}\mathfrak{Q} \subset \mathfrak{E}_{p^*p} \subset \mathfrak{E}_{p^*}\mathfrak{S}_p \subset \mathfrak{D}^p$$

where all containments are strict.

Finally let us mention the contribution of M. J. Iranzo, J. Medina, and F. Pérez-Monasor in [IMPM01] that, using that the class $\mathfrak{E}_\pi$ is injective, proves that the class of all p-decomposable groups is an injective Fitting class.

Bearing in mind Salomon's example in Section 7.1 and the results of the present section, the following question arises:

Open question 7.2.35. *Is it possible to characterise the injective Fitting classes?*

7.3 Supersoluble Fitting classes

It is well-known that the product of two supersoluble normal subgroups of a group need not to be supersoluble. In other words, the class $\mathfrak{U}$ of all supersoluble groups is not a Fitting class, although $\mathfrak{U}$ is closed for subnormal subgroups. This failure is the starting point of two fruitful lines of research.

1. Obviously the direct product of supersoluble subgroups is always supersoluble; hence the study of different types of products, with extra conditions, such that those special products of supersoluble subgroups give a new supersoluble subgroup makes sense; following these ideas a considerable amount of papers has been published in the last years dealing with totally permutable products, mutually permutable products, ... (see, for instance, [AS89], [BBPR96a])

2. On the other hand we can analyse the properties of supersoluble Fitting classes, i.e. those Fitting classes contained in the class $\mathfrak{U}$ of all supersoluble groups. This investigation was encouraged by the excellent results obtained in metanilpotent Fitting classes due to T. O. Hawkes, T. R. Berger, R. A. Bryce, and J. Cossey (see [DH92, XI, Section 2]).

The question of the existence of Fitting classes composed of supersoluble groups was settled by M. Menth in [Men95b]. In this paper he presented a family of supersoluble non-nilpotent Fitting classes. These Fitting classes are constructed via Dark's method (see [DH92, IX, Section 5]). Terminology and notation are mainly taken from [DH92, IX, Sections 5 and 6] and the papers of Menth [Men94, Men95b, Men95a, Men96].

Following Dark's strategy, we start with a identification of the universe of groups to consider. Let p be a prime such that $p \equiv 1 \pmod 3$, and n a primitive 3rd root of unity in the field $\mathrm{GF}(p)$. The universe to consider will be the class $\mathfrak{S}_p \mathfrak{S}_3$.

Now the ingredients are:

1. *The key section $\kappa(G)$ of a group $G \in \mathfrak{S}_p \mathfrak{S}_3$ is $\kappa(G) = O^p(G)$.*
2. *The associated class $\mathfrak{X}$.* Consider the groups

$$T = \langle a, b : a^p = b^p = [a, b, a, a] = [a, b, a, b] = [a, b, b, b] = 1 \rangle$$

and

$$V = \langle T, s : s^3 = 1, a^s = a^n, b^s = b^n \rangle.$$

These groups have the following properties:
   a) $|T| = p^5$, $T' = Z_2(T)$ and the factors of the central series are $T/T' \cong C_p \times C_p$, $T'/Z(T) \cong C_p$, and $Z(T) \cong C_p \times C_p$;
   b) $Z(V) = Z(T)$ and the conjugation by s induces on T/T' the power automorphism $x \longmapsto x^n$, on $T'/Z(T)$ the power automorphism $x \longmapsto x^{n^2}$, and centralises $Z(T)$;
   c) every extension of T by an elementary abelian 3-group is supersoluble; in particular V is supersoluble.

Let $\mathfrak{V}_0$ be the class of all finite groups G which can be factorised as $G = XY$ where
   a) $X = O_p(G)$ is a central product of copies T_i of T (the empty product, i.e. the case $O_p(G) = 1$, is admitted);
   b) $Y \in \mathrm{Syl}_3(G)$ and for every index i, we have that $Y/C_Y(T_i) \cong C_3$ and $[T_i](Y/C_Y(T_i)) \cong V$.
3. *The class $\mathfrak{V} = D^p(\mathfrak{V}_0) = \big(G \in \mathfrak{S}_p \mathfrak{S}_3 : \kappa(G) \in \mathfrak{V}_0 \big)$.*

The following result is due to Menth. We quote it here without proof.

Theorem 7.3.1 ([Men95b, 4.2]). *The class $\mathfrak{V} = \mathrm{Fit}(V)$ is the Fitting class generated by V. If $G \in \mathfrak{V}$ and write $P = O_p(G)$, $V_0 = O^p(G)$, and $C = O_3\big(Z_\infty(V_0)\big)$, then*

   1. *G is supersoluble;*
   2. *$F(G) = PC$ and $G/F(G)$ is an elementary abelian 3-group;*
   3. *$G = C_P(Y)V_0$ for every Sylow 3-subgroup Y of G;*
   4. *$\mathrm{Soc}(G) \leq Z(G)$.*
   Moreover, $\mathfrak{V}$ is a Lockett class ([Men94, 2.2]).

This supersoluble Fitting class is contained in $\mathfrak{S}_p \mathfrak{S}_3$. The above construction can be generalised to include examples of supersoluble Fitting classes in $\mathfrak{S}_p \mathfrak{S}_q$ for other odd primes q. In [Tra98], G. Traustason gives an example of a supersoluble Fitting class in $\mathfrak{S}_p \mathfrak{S}_2$. This class is also constructed following Dark's strategy.

In contrast with metanilpotent Fitting classes, supersoluble Fitting classes are extremely restricted in additional closure properties. This is also proved by M. Menth in [Men95a]. In this section we will present the most relevant results of this paper.

Lemma 7.3.2. *Let G be a supersoluble group. Then, $\mathrm{Fit}(G)$ is supersoluble if and only if $\mathrm{Fit}(G) \subseteq \mathrm{lform}(G)$.*

Proof. Denote $\mathfrak{G} = \mathrm{lform}(G)$. Since G is supersoluble, $\mathfrak{G} \subseteq \mathfrak{U}$. Hence $\mathrm{Fit}(G)$ is a supersoluble Fitting class.

For the converse, observe that since G is supersoluble, the quotient group $G/\mathrm{O}_{p',p}(G)$ is an abelian group of exponent $e(p)$ dividing $p-1$ for each prime p by [DH92, IV, 3.4 (f)]. Applying Theorem 3.1.11, the saturated formation $\mathfrak{G}$ is locally defined by the formation function f, where $f(p) = \mathrm{form}\big(G/\mathrm{O}_{p',p}(G)\big)$, if p divides $|G|$, and $f(p) = \emptyset$ if p does not divide $|G|$. It is rather easy to see that $f(p) = \mathfrak{A}\big(e(p)\big)$, where $\mathfrak{A}(m)$ denotes the class of all abelian groups of exponent dividing m. Since $f(p)$ is subgroup-closed for all primes p, the formation $\mathfrak{G} = \mathrm{LF}(f)$ is subgroup-closed by [DH92, IV, 3.14]. Hence the class $\mathrm{Fit}(G) \cap \mathfrak{G}$ is s_n-closed.

Let X be a group which is the product of two normal subgroups N_1, N_2 of X such that $N_1, N_2 \in \mathrm{Fit}(G) \cap \mathfrak{G}$. For each prime p, we have that $X/\mathrm{O}_{p',p}(X)$ is the normal product of $N_1\,\mathrm{O}_{p',p}(X)/\mathrm{O}_{p',p}(X)$ and $N_2\,\mathrm{O}_{p',p}(X)/\mathrm{O}_{p',p}(X)$. Since $X \in \mathrm{Fit}(G)$, then X is supersoluble and so $X/\mathrm{O}_{p',p}(X)$ is abelian by [DH92, IV, 3.4 (f)]. Moreover, for $i = 1, 2$, we have that

$$N_i\,\mathrm{O}_{p',p}(X)/\mathrm{O}_{p',p}(X) \cong N_i/\mathrm{O}_{p',p}(N_i) \in \mathfrak{A}\big(e(p)\big),$$

since $N_i \in \mathrm{LF}(f)$. Hence $X/\mathrm{O}_{p',p}(X) \in \mathfrak{A}\big(e(p)\big)$. Hence $X \in \mathfrak{G}$. This is to say that the class $\mathrm{Fit}(G) \cap \mathfrak{G}$ is N_0-closed.

Therefore $\mathrm{Fit}(G) \cap \mathfrak{G}$ is a Fitting class containing G. Thus, $\mathrm{Fit}(G) \subseteq \mathrm{lform}(G)$. $\qquad\Box$

Lemma 7.3.3. *Let X be a group such that the regular wreath product $W = X \wr C$ is a supersoluble group for some non-trivial group C. Then X is nilpotent.*

Proof. Suppose that the result is false and let X be a counterexample of minimal order. Then X is a non-nilpotent group and the regular wreath product $W = X \wr C$ is a supersoluble group for some non-trivial group C. Denote by $X^\natural$ the base of group of W. If Y is a subgroup of X, denote by $Y^\natural$ the corresponding subgroup of $X^\natural$. Let N be a minimal normal subgroup of X. Then $(X/N) \wr C \cong W/N^\natural$ by [DH92, A, 18.2(d)]. Moreover $(X/N) \wr C$ is supersoluble. By minimality of X, we have that X/N is nilpotent. Since X is non-nilpotent, it follows that $X \in \mathrm{b}(\mathfrak{N})$ and so X is a primitive group. Since X is a supersoluble non-nilpotent primitive group, then X possesses a unique minimal normal subgroup Y which is a cyclic group of prime order, q say, and $\mathrm{Z}(X) = 1$. Then $Y^\natural$ is a minimal normal subgroup of W by [DH92, A, 18.5(a)]), and W is primitive by [DH92, A, 18.5(b)]. In particular, the order of the minimal normal subgroup of W is a prime. Note that the order of $Y^\natural$ is $q^{|C|}$. This contradiction proves the lemma. $\square$

Theorem 7.3.4. *Let $\mathfrak{F}$ be a supersoluble Fitting class. Assume that X is a group and p is a prime such that the regular wreath product $X \wr C_p \in \mathfrak{F}$. Then X is a p-group.*

Proof. Set $G = X \wr C_p \in \mathfrak{F}$. We can assume, without loss of generality that $\mathfrak{F} = \mathrm{Fit}(G)$. By Lemma 7.3.2, $\mathfrak{F} \subseteq \mathrm{lform}(G)$. We can apply now some results due to P. Hauck (see [DH92, X, 2.9 and 2.10]) to deduce that $X \wr P \in \mathfrak{F} \subseteq \mathrm{lform}(G)$, for every p-group P.

Suppose further that X is not a p-group. Then there exists a prime divisor $q \neq p$ of $|X|$. Since $X \wr C_p$ is supersoluble, it follows that X is nilpotent by Lemma 7.3.3. Therefore $X = \mathrm{O}_{q',q}(X)$.

Applying Theorem 3.1.11, $\mathrm{lform}(G) = \mathrm{LF}(f)$ is locally defined by the formation function f, where $f(r) = \mathrm{form}(G/\mathrm{O}_{r',r}(G))$, if r divides $|G|$, and $f(r) = \emptyset$ if r does not divide $|G|$. Then $P \in f(q)$ for all p-groups P. Hence $\mathfrak{S}_p \subseteq f(q) = \mathrm{form}(G/\mathrm{O}_{q',q}(G))$.

Observe that for every natural number e, the class $\mathfrak{S}_p^{(e)} = (G \in \mathfrak{S}_p : \exp(G) \leq p^e)$ is a subformation of $\mathfrak{S}_p$. Hence $\mathrm{form}(G/\mathrm{O}_{q',q}(G))$ has infinitely many subformations, and this contradicts the theorem of R. M. Bryant, R. A. Bryce, and B. Hartley ([DH92, VII, 1.6]). $\square$

Fitting classes with the property of Theorem 7.3.4 are called *abstoßend* by P. Hauck. This term is translated into English as *repellent* (see [DH92, X, 2, Exercise 4]).

Proposition 7.3.5. *Let $\mathfrak{F}$ be a Fitting class of soluble groups. Suppose that the group G is a semidirect non-direct product $G = [N]A$ of the normal subgroup N by a q-subgroup A, q a prime. Suppose that A induces the automorphism group A^* on N and consider the semidirect product $G^* = [N]A^*$. Then $G \in \mathfrak{F}$ if and only if $G^* \in \mathfrak{F}$.*

Proof. First observe that $A^* \cong A/\mathrm{C}_A(N)$ and $C = \mathrm{C}_A(N)$ is a normal subgroup of G. Thus, the group $G^* \cong G/C$ is an epimorphic image of G. Moreover, since the semidirect product is non-direct, $C \neq A$.

Suppose that $G \in \mathfrak{F}$. Then $\mathfrak{S}_q \subseteq \mathfrak{F}$, by [DH92, IX, 1.9], and $G/N \cong A \in \mathfrak{F}$. Moreover $N \cap C = 1$ and $G/NC \cong A^*$ is nilpotent. By Lemma 2.4.2, the $G^* \cong G/C \in \mathfrak{F}$.

The same arguments show that G is in $\mathfrak{F}$ if $G^* \cong G/C \in \mathfrak{F}$. $\square$

Proposition 7.3.6. *Let $\mathfrak{F}$ be a Fitting class and suppose that G is an $\mathfrak{F}$-group such that G is the semidirect product $G = [N]\langle s \rangle$ where $N = N_1 \times \cdots \times N_n$, N_i normal in G, $1 \leq i \leq n$. Let σ_i be the automorphism of N_i induced by conjugation of s. For each $i = 1, \ldots, n$, consider a copy $\overline{N_i} \cong N_i$ and construct the semidirect product $H_i = [\overline{N_i} \times N_i]\langle s \rangle$, where s induces on $\overline{N_i}$ the automorphism σ_i^{-1}. Then $H_i \in \mathfrak{F}$.*

Proof. Without loss of generality, we can argue with the normal subgroup N_1. Consider the direct product $N^* = \overline{N_1} \times N_1 \times \cdots \times N_n$ and a cyclic group $\langle t \rangle$ such that $\langle s \rangle \cong \langle t \rangle$. Construct the semidirect product $G^* = [N^*](\langle s \rangle \times \langle t \rangle)$, where $\overline{N_1}$ and all factors N_i are normal in G^* and the operation of s and t on the N_i is as follows: s centralises $\overline{N_1}$ and acts on N_i in the same way as σ_i; t centralises N_1, operates on N_i in the same way as σ_i for $2 \leq i \leq n$ and on $\overline{N_1}$ as σ_1. Since $N_1 \in \mathfrak{F}$, we have that $\overline{N_1} \in \mathfrak{F}$. Therefore $\langle N^*, s \rangle \cong \langle N^*, t \rangle \cong \overline{N_1} \times G \in \mathfrak{F}$. Then $G^* \in \mathfrak{F}$. The normal subgroup $\langle N^*, st^{-1} \rangle$ of G^* is an $\mathfrak{F}$-group. Finally, observe that $H_1 \cong \langle \overline{N_1} \times N_1, st^{-1} \rangle$ and this is normal in $\langle N^*, st^{-1} \rangle$. Hence $H_1 \in \mathfrak{F}$. □

Remarks and notation 7.3.7. Let p and q be different primes, p odd, such that q divides $p - 1$. Let e and r be natural numbers.

1. Recall that $\mathrm{Aut}(C_{p^e}) \cong C_{p^{e-1}(p-1)}$ (see [DH92, A, 21.1]). Each natural number m, with $\gcd(m,p) = 1$ and $1 \leq m \leq p^e$ can be uniquely written in the form $m = tp + k$, for $0 \leq t \leq p^{e-1} - 1$ and $1 \leq k \leq p - 1$. The pair (t,k) uniquely determines the automorphism $\sigma(t,k)$ of the cyclic group $C_{p^e} = \langle x \rangle$ of order p^e, defined by $x^{\sigma(t,k)} = x^{tp+k} = x^m$.

2. Therefore there exists an automorphism $\alpha = \sigma(t,k)$ of C_{p^e} of order q. This means that $n = tp + k \neq 1$ is an integer such that $n^q \equiv 1 \pmod{p^e}$. Moreover any automorphism of C_{p^e} of order q is of the form α^t for $1 \leq t \leq q - 1$. If x is a generator of the cyclic group C_{p^e}, then $x^{\alpha^t} = x^{n^t}$.

3. Let X_r be the direct product of r copies of the cyclic group of order p^e. Construct the semidirect product $G_r = [X_r]C$ of X_r and a cyclic group $C = \langle s \rangle \cong C_q$ where s raises all elements of X_r to the same n-th power. If $\{x_1, \ldots, x_r\}$ is a set of r generators (a basis) of X_r, observe that all subgroups of the form $\langle x_i, s \rangle$, for $i = 1, \ldots, r$, are isomorphic to $\mathrm{E}(q|p^e)$ (see [DH92, B, 12.5]).

Lemma 7.3.8. *Consider the Fitting class, $\mathrm{Fit}(G_r)$, generated by the group G_r. For any natural number k, let $H_k = [X_k]C$ denote a group which is a semidirect product of the homocyclic abelian group X_k of exponent p^e and rank $k \geq 1$ by a cyclic group $C = \langle \alpha \rangle$ such that α is an automorphism of X_k of order q and $\det(\alpha) = 1$. Then $H_k \in \mathrm{Fit}(G_r)$.*

Proof. The prime q is a divisor of $p - 1$ and then $\gcd(q,p) = 1$. By [DH92, A, 11.6], X_k has a direct decomposition $X_k = X_{k(1)} \times \cdots \times X_{k(s)}$ into $\langle \alpha \rangle$-admissible subgroups $X_{k(i)}$ with the following properties for each $i = 1, \ldots, s$:

1. $X_{k(i)}$ is indecomposable as a $\langle \alpha \rangle$-module;
2. $Y_{k(i)} = X_{k(i)}/\Phi(X_{k(i)})$ is an irreducible $\mathrm{GF}(p)\langle \alpha \rangle$-module.

The finite field $\mathrm{GF}(p)$ contains a primitive q-th root of unity n. This implies that every irreducible representation of the cyclic group C_q over the field $\mathrm{GF}(p)$ is linear ([DH92, B, 8.9 (d)]). Therefore $Y_{k(i)} \cong C_p$ for each $i = 1, \ldots, s$. Therefore $X_{k(i)} \cong C_{p^e}$ for each $i = 1, \ldots, s$. This is to say that there exists a basis of X_k such that the action of α on X_k, according to this basis, can be written as a diagonal matrix $\mathrm{diag}(n^{\lambda_1},\ldots,n^{\lambda_{k-1}},n^\lambda)$, where $\lambda = -(\lambda_1 + \cdots + \lambda_{k-1})$.

Consider the homocyclic group X_{k+r-1} of exponent p^e and rank $k+r-1$ and fix a basis $\{x_1,\ldots,x_k,y_1,\ldots,y_{r-1}\}$ of X_{k+r-1}. For each $j = 1,\ldots,k-1$, consider the extension $L_j = [X_{k+r-1}]\langle\alpha_j\rangle$ of X_{k+r-1} such that $x_j^{\alpha_j} = x_j^{n^{\lambda_j}}$, $x_l^{\alpha_j} = x_l$, if $l \in \{1,\ldots,k\}\setminus\{j\}$, and $y_s^{\alpha_j} = y_s^{n^{\lambda_j}}$, for $s = 1,\ldots,r-1$. Consider also the extension $L_k = [X_{k+r-1}]\langle\alpha_k\rangle$ of X_{k+r-1} such that $x_k^{\alpha_k} = x_k^{n^\lambda}$, $x_l^{\alpha_k} = x_l$, if $l \in \{1,\ldots,k-1\}$, and $y_s^{\alpha_k} = y_s^{n^\lambda}$, for $s = 1,\ldots,r-1$.

In other words, the action of the automorphism α_j on X_{k+r-1}, in the fixed basis, can be written as a diagonal matrix

$$\alpha_j = \mathrm{diag}(\underbrace{1,\ldots,1}_{j-1},n^{\lambda_j},\underbrace{1,\ldots,1}_{k-j},\underbrace{n^{\lambda_j},\ldots,n^{\lambda_j}}_{r-1}), \qquad \text{if } 1 \le j \le k-1,$$

and

$$\alpha_k = \mathrm{diag}(\underbrace{1,\ldots,1}_{k-1},\underbrace{n^\lambda,\ldots,n^\lambda}_{r}).$$

Hence, for all $j = 1,\ldots,k$, we have that $L_j \cong G_r \times X_{k-1}$ and therefore $L_j \in \mathrm{Fit}(G_r)$.

Set $L = [X_{k+r-1}]\langle\alpha_1,\ldots,\alpha_k\rangle$. Clearly L is a normal product of $L_1,\ldots,L_k$. Hence $L \in \mathrm{Fit}(G_r)$. Consider the product

$$\alpha = \prod_{j=1}^{k}\alpha_j = \mathrm{diag}(n^{\lambda_1},n^{\lambda_2},\ldots,n^{\lambda_{k-1}},n^\lambda,\underbrace{1,\ldots,1}_{r-1})$$

and the normal subgroup $L_0 = [X_{k+r-1}]\langle\alpha\rangle$ of L. Identify $X_k = \langle x_1,\ldots,x_k\rangle$ and observe that the subgroup $\langle X_k,\alpha\rangle$ is isomorphic to H_k and $L_0 \cong H_k \times X_{r-1}$. Therefore H_k is isomorphic to a subnormal subgroup of L. Hence $H_k \in \mathrm{Fit}(G_r)$. $\square$

Lemma 7.3.9. *Let α be any nontrivial automorphism of X_r of order a power of q and write $G = [X_r]\langle\alpha\rangle$. Then $G_q \in \mathrm{Fit}(G)$.*

Proof. If the order of α is q^m and $m > 1$, then the order of $\alpha^{q^{m-1}}$ is q. Since $\langle X_r,\alpha^{q^{m-1}}\rangle$ is normal in G, then $\langle X_r,\alpha^{q^{m-1}}\rangle \in \mathrm{Fit}(G)$. Therefore we can assume that the order of α is q. As in Lemma 7.3.8, there exists a basis $\{x_1,\ldots,x_r\}$ of X_r such that the matrix of α with respect to this basis is diagonal and $\alpha = \mathrm{diag}(n^{\lambda_1},\ldots,n^{\lambda_r})$. Since $\alpha \ne \mathrm{id}$, not all λ_i are equal to 0. Without loss of generality we can assume that $\lambda_1 = 1$. As a consequence of Proposition 7.3.6, the class $\mathrm{Fit}(G)$ contains the group $E_1 = [X_q]\langle\beta_1\rangle$ which is an extension of X_q by the automorphism β_1 such that in a fixed basis of X_q has a diagonal matrix expression as follows: $\beta_1 = \mathrm{diag}(n,n^{-1},1,\ldots,1)$. Clearly, this group is isomorphic to $E_2 = [X_q]\langle\beta_2\rangle$, where the automorphism β_2 in the fixed basis of X_q has a diagonal matrix expression $\beta_2 = \mathrm{diag}(1, n^2, n^{-2}, 1, \ldots, 1)$. Hence E_2 belongs to $\mathrm{Fit}(G)$. Therefore the class $\mathrm{Fit}(G)$ contains the extensions of X_q by the automorphisms β_j, for $j = 1, \ldots, q-1$ such that in the fixed basis have diagonal matrix expressions as follows:

$$\beta_1 = \mathrm{diag}(n, n^{-1}, 1, \ldots, 1)$$
$$\beta_2 = \mathrm{diag}(1, n^2, n^{-2}, 1, \ldots, 1)$$
$$\ldots$$
$$\beta_{q-1} = \mathrm{diag}(1, \ldots, 1, n^{q-1}, n)$$

Thus $\mathrm{Fit}(G)$ contains the extension of X_q by the automorphism

$$\beta = \prod_{i=1}^{q-1} \beta_i = \mathrm{diag}(n, \ldots, n)$$

and then $G_q = [X_q]\langle\beta\rangle \in \mathrm{Fit}(G)$. $\square$

Lemma 7.3.10. *Let X be a homocyclic group of exponent p^e and let $G = [X]Q$ be a semidirect non-direct product of X and a q-group Q.*

1. If $q \geq 3$, then $C_{p^e} \wr C_q \in \mathrm{Fit}(G)$.

2. If $q = 2$, then $\mathrm{Fit}(G)$ contains the extension of X_4 by $\langle\alpha, \beta\rangle$, where α and β are automorphisms of X_4, i.e. members of the group $\mathrm{GL}(4, \mathbb{Z}/p^e\mathbb{Z})$, such that in a fixed basis $\{x_1, x_2, x_3, x_4\}$ of X_4 have matrix expressions

$$\alpha = \begin{pmatrix} 0 & 0 & 1 & 0 \\ 0 & 0 & 0 & 1 \\ 1 & 0 & 0 & 0 \\ 0 & 1 & 0 & 0 \end{pmatrix}, \qquad \beta = \begin{pmatrix} -1 & & & \\ & -1 & & \\ & & 1 & \\ & & & 1 \end{pmatrix}$$

3. In both cases 1 and 2 the Fitting class $\mathrm{Fit}(G)$ is not supersoluble.

Proof. By Proposition 7.3.5, we can assume that Q is a group of automorphisms of X. Since the semidirect product is non-direct, there exists an element $s \in Q$ which is a non-trivial automorphism of X of order a power of q. It is clear that $[X]\langle s\rangle$ is subnormal in G and then $H = [X]\langle s\rangle \in \mathrm{Fit}(G)$.

By Lemma 7.3.9, we have that $\mathrm{Fit}(G_q) \subseteq \mathrm{Fit}([X]\langle s\rangle) \subseteq \mathrm{Fit}(G)$. By Lemma 7.3.8, the class $\mathrm{Fit}(G)$ contains all extensions of a homocyclic group X of exponent p^e by $\alpha \in \mathrm{Aut}(X)$ of order q such that $\det\alpha = 1$.

1. Suppose that q is odd. Observe that the regular wreath product $C_{p^e} \wr C_q$ is isomorphic to a extension of the homocyclic group X_q of exponent p^e and rank q by an automorphism α of order q whose action on X_q has matrix

$$\begin{pmatrix} 0\,1\,0\,0 \ldots 0\,0 \\ 0\,0\,1\,0 \ldots 0\,0 \\ 0\,0\,0\,1 \ldots 0\,0 \\ 0\,0\,0\,0 \ldots 0\,0 \\ \vdots\,\vdots\,\vdots\,\vdots\quad\ \vdots\,\vdots \\ 0\,0\,0\,0 \ldots 0\,1 \\ 1\,0\,0\,0 \ldots 0\,0 \end{pmatrix}$$

whose determinant is $(-1)^{q-1} = 1$. Hence $C_{p^e} \wr C_q \in \mathrm{Fit}(G)$.

2. Since α and β have both order 2 and determinant 1, the extensions $\langle X_4, \alpha \rangle$ and $\langle X_4, \beta \rangle$ are in $\mathrm{Fit}(G)$. The group $\langle \alpha, \beta \rangle$ is isomorphic to a dihedral group of order 8. Therefore the extension $H = [X_4]\langle \alpha, \beta \rangle$ is a subnormal product of $\langle X_4, \alpha \rangle$ and $\langle X_4, \beta \rangle$ and then $H \in \mathrm{Fit}(G)$.

3. In Case 1, the Fitting class $\mathrm{Fit}(G)$ is not supersoluble by Theorem 7.3.4.

In Case 2, suppose that the group H is supersoluble and consider the Frattini quotient $Y_4 = X_4/\Phi(X_4)$. The group $H^* = [Y_4]\langle \alpha, \beta \rangle$ is an epimorphic image of H and then H^* is supersoluble. Denote $Y_4 = \langle y_1, y_2, y_3, y_4 \rangle$, where $y_i = x_i \Phi(X_4)$, for $i = 1, 2, 3, 4$. Now the respective actions of α and β on the 4-dimensional $\mathrm{GF}(p)$-vector space Y_4 have the same matrix representation, but now considered in $\mathrm{GL}(4, p)$. Let N be a minimal normal subgroup of H^* contained in Y_4. Since H^* is supersoluble, the group N is cyclic, $N = \langle y \rangle$ say. This is to say that y is an eigenvector for α and for β. Since y is an eigenvector for β, then either $y = x_1^{n_1} x_2^{n_2}$ or $y = x_3^{n_3} x_4^{n_4}$. But then y is not an eigenvector for α. Hence H is not supersoluble and $\mathrm{Fit}(G)$ is not a supersoluble Fitting class. □

Theorem 7.3.11. *If $\mathfrak{F}$ is a supersoluble Fitting class, then every metabelian $\mathfrak{F}$-group is nilpotent.*

Proof. Assume that the result is not true and let G be a metabelian non-nilpotent $\mathfrak{F}$-group of minimal order. Note that $N = G'$ is abelian. For every element $x \notin N$, $N\langle x \rangle$ is a metanilpotent normal subgroup of G. If $N\langle x \rangle$ were a proper subgroup of G for each element $x \in G$, then G would be nilpotent. This would contradict the choice of G. Therefore $G = N\langle x \rangle$, for some element $x \notin N$. By the same argument, we can assume that x is a q-element for some prime q. Clearly N is not a q-group and $G = O_{q'}(N)Q$ for some $Q \in \mathrm{Syl}_q(G)$ such that $x \in Q$. The subgroup $G_0 = O_{q'}(N)\langle x \rangle$ is subnormal in G. Hence $G_0 \in \mathfrak{F}$. If G_0 were nilpotent, then $G = NG_0$ would be a product of two subnormal nilpotent subgroups and therefore G would be nilpotent, contrary to supposition. The minimal choice of G implies that $G = G_0$, i. e., we can assume that N is a q'-group. We also may suppose that x is of order q. For a prime p with $p \neq q$, the subgroup $O_p(N)$ is normal in G. If $O_p(N)\langle x \rangle$ is nilpotent, then x centralises $O_p(N)$. In this case $G = N^*\langle x \rangle \times O_p(N)$, where N^* is the Hall p'-subgroup of N. By minimality of G, $N^*\langle x \rangle$ is nilpotent. Thus G is nilpotent, and this contradicts our choice of G. Hence, we can assume that $N = N_1 \times \cdots \times N_n$, where $N_i \in \mathrm{Syl}_{p_i}(N)$, for all primes p_i dividing $|N|$, and x induces on each N_i a non-trivial automorphism σ_i. Since x does not centralise N_1, it follows that x does not centralise some chief factor of G below N_1. This implies that q divides $p_1 - 1$ since G is supersoluble.

Consider the semidirect product $H = [P]C$, where $P = N_0 \times N_1$, with $N_0 \cong N_1$, and $C = \langle x \rangle$. Suppose that x induces on N_1 the automorphism σ_1 and on N_0 the automorphism σ_1^{-1}. By Proposition 7.3.6, we have that $H \in \mathfrak{F}$ and H is non-nilpotent.

By [DH92, A, 11.6], we have that N_0 has a direct decomposition $N_0 = A_{1(0)} \times \cdots \times A_{k(0)}$ with the following properties for each $i = 1, \ldots, k$:

1. $A_{i(0)}$ is indecomposable as a C-module;
2. $A_{i(0)}/\Phi(A_{i(0)})$ is an irreducible $\mathrm{GF}(p_1)C$-module;
3. $A_{i(0)}$ is homocyclic.

Note that $A_{i(0)}/\Phi(A_{i(0)})$ is a faithful C-module and so its dimension is 1 because q divides $p_1 - 1$ ([DH92, B, 8.9 (d)]). Therefore $A_{i(0)} \cong C_{p_1^e}$ for each $i = 1, \ldots, k$. Moreover x induces on each $A_{i(0)}$ an automorphism σ_1^{-1}. Analogously $N_1 = A_{1(1)} \times \cdots \times A_{k(1)}$, $A_{i(1)} \cong C_{p_1^e}$ for each $i = 1, \ldots, k$ and x induces on each $A_{i(1)}$ the automorphism σ_1. By Lemma 7.3.6, we have that $[A_{1(0)} \times A_{1(1)}]C \in \mathfrak{F}$. Hence Lemma 7.3.10 implies that $\mathfrak{F}$ is not supersoluble. This contradiction proves the theorem. $\qquad\square$

Theorem 7.3.12. *Let $\mathfrak{F}$ be a supersoluble non-nilpotent Fitting class. Then $\mathfrak{F}$ is not closed with respect to any of the operators* Q, s, *and* E$_\Phi$.

Proof. Assume that $\mathfrak{F}$ is a Q-closed non-nilpotent supersoluble Fitting class. Let H be a supersoluble non-nilpotent $\mathfrak{F}$-group of minimal order. Then H/N is nilpotent $\mathfrak{F}$-group for every minimal normal subgroup N of H. Consequently $H \in \mathrm{b}(\mathfrak{N})$ and so H is a primitive group. Then, by Theorem 1, $N = \mathrm{Soc}(H)$ is a minimal normal subgroup of H and $N = \mathrm{C}_H(N)$ and N is cyclic of prime order. In particular, H is metabelian. This contradicts Theorem 7.3.11. Therefore the class $\mathfrak{F}$ is not Q-closed.

Suppose that $\mathfrak{F}$ is an E$_\Phi$-closed supersoluble non-nilpotent Fitting class. Since $\mathfrak{F}$ is composed of metanilpotent groups we can apply the theorem [DH92, XI, 2.16] to conclude that $\mathfrak{F}$ is s-closed. Applying Theorem 2.5.2, $\mathfrak{F}$ is a saturated formation. In particular, $\mathfrak{F}$ is Q-closed. This contradiction proves that $\mathfrak{F}$ is not E$_\Phi$-closed. Note that $\mathfrak{F}$ cannot be subgroup-closed either. $\qquad\square$

Recall that a *Fischer class* is a Fitting class $\mathfrak{F}$ satisfying the following property: if G is a group in $\mathfrak{F}$ and H/K is a normal nilpotent subgroup of G/K for some normal subgroup K of G, it follows that $H \in \mathfrak{F}$. These classes were originally introduced by Fischer in the soluble universe. If $\mathfrak{F}$ is a Fischer class of soluble groups, then the $\mathfrak{F}$-injectors of a soluble group are exactly the *Fischer $\mathfrak{F}$-subgroups*, which are the natural duals of Gaschütz's covering subgroups (see [DH92, IX, Section 3]).

Corollary 7.3.13. *Let $\mathfrak{F}$ be a supersoluble non-nilpotent Fitting class. Then $\mathfrak{F}$ is not a Fischer class.*

Proof. Assume that $\mathfrak{F}$ is a Fischer class. We shall prove that $\mathfrak{F}$ is subgroup-closed. Suppose that this is not true and let G be a group of minimal order such that $G \in \mathfrak{F}$ but $M \notin \mathfrak{F}$ for some subgroup M of G. Among the subgroups of G which are not in $\mathfrak{F}$, we choose M of maximal order. Clearly M is a maximal subgroup of G. If G' is contained in M, then M is normal and so $M \in \mathfrak{F}$, contrary to supposition. Consequently, $G = MG'$. Since, by [DH92, VII, 2.2], M has prime index, it follows that $M/M \cap G'$ is a cyclic group of prime order. Note that G' is nilpotent and $M \cap G'$ has prime index in G'. This implies that $M \cap G'$ is normal in G'. Therefore $M \cap G'$ is normal in G. Since $\mathfrak{F}$ is a Fischer class, we have that $M \in \mathfrak{F}$, contrary to the choice of M. Then $\mathfrak{F}$ is subgroup-closed. This contradicts Theorem 7.3.12. Consequently, $\mathfrak{F}$ is not a Fischer class.

Since metanilpotent R_0-closed Fitting classes need not be Q-closed, the exclusion of the R_0-closure cannot be argued in the same way. What Menth shows is that the supersoluble Fitting class $\mathfrak{V}$ introduced at the beginning of the section is not R_0-closed.

Theorem 7.3.14. *The class $\mathfrak{V}$ is not R_0-closed.*

Proof. We will use the notation introduced at the beginning of the section. Let us consider the direct product $W = V \times V^\varphi$ of two copies of V. The diagonal subgroup $D = \{(x, x^\varphi) : x \in V\}$ of W is isomorphic to V. The subgroups $A = \{(x, 1) : x \in T'\}$ and $B = \{(1, x^\varphi) : x \in T'\}$ are normal in W and $A \cap B = (1, 1)$. Observe that the subgroup $G = \langle A, D \rangle$ is a semidirect product $G = [A]D = [B]D$ and $G/A \cong G/B \cong V \in \mathfrak{V}$. Next we see that $G \notin \mathfrak{V}$.

The element (s, s^φ) is a 3-element and then $(s, s^\varphi) \in O^p(G)$. Hence the commutator $[(a, a^\varphi), (s, s^\varphi)] = (a, a^\varphi)^{n-1} \in O^p(G)$ and also $(b, b^\varphi)^{n-1} \in O^p(G)$ for the generators a, b of T. Therefore D is contained in $O^p(G)$. There exists an element $t \in T' \setminus Z(T)$ such that $t^s = t^{n^2}$. Hence $[(t, 1), (s, s^\varphi)] = ([t, s], 1) = (t^{n^2-1}, 1) \in O^p(G)$. Since n is a primitive cube root of unity in $\mathrm{GF}(p)$, we have that p divides $n^3 - 1$ but $\gcd(p, n^2 - 1) = 1$. Therefore $(t, 1) \in O^p(G)$. Then $[(t, 1), (a, a^\varphi)] = ([t, a], 1)$ and $[(t, 1), (b, b^\varphi)] = ([t, b], 1)$ are in $O^p(G)$. Then $A \leq O^p(G)$. Therefore $G = O^p(G)$ and the group G is p-perfect.

Observe that the subgroup $Z(T) \times Z(T)^\varphi$ is a subgroup of $Z(O_p(G))$ of order p^4. If we suppose that $G \in \mathfrak{V}$, then $G \in \mathfrak{V}_0$ and then $O_p(G)$ is a central product of copies of T. Since $|O_p(G)| = p^8$, we need exactly two copies of T, T_1, T_2 say, such that $|T_1 \cap T_2| = p^2$. Therefore $Z(T_1) = Z(T_2) = Z(O_p(G))$ has order p^2. This contradicts the previous observation. Hence $G \notin \mathfrak{V}$. We conclude then that $\mathfrak{V}$ is not R_0-closed.   □

Let $\mathfrak{F}$ be a Fitting class of soluble groups. If π is a set of primes, $\mathfrak{F}$ is said to be *Hall-π-closed* provided that whenever H is a Hall π-subgroup of G and $G \in \mathfrak{F}$, then $H \in \mathfrak{F}$. The class $\mathfrak{F}$ is said to be *Hall-closed* if it is Hall-π-closed for all sets of primes π.

Theorem 7.3.15. *Every metanilpotent Lockett class is Hall-closed.*

Proof. Assume that the result is false and let $\mathfrak{F}$ be a metanilpotent Fitting class that is not Hall-closed. There exists a set π of primes and a group $G \in \mathfrak{F}$ such that G has a Hall π-subgroup $H \notin \mathfrak{F}$. Set $F = \mathrm{F}(G)$, and let $p_1, \ldots,$ p_n be the prime divisors of $|F|$. Then F is the direct product of its Sylow p_i-subgroups P_i, $1 \leq i \leq n$, and G/F is nilpotent. Having numbered the primes suitably, there is an integer k $(1 \leq k \leq n)$ such that $p_1, \ldots, p_k$ are elements of π. Note that $k < n$ because otherwise H would be subnormal in G. Then $P = H \cap F = P_1 \cdots P_k$. The quotient H/P is isomorphic to a subgroup of G/F and therefore nilpotent. Hence H/P is generated by cyclic subgroups $\langle x_i P \rangle$. At least one of the subgroups $\langle P, x_i \rangle$ is not an $\mathfrak{F}$-group. Let us choose $H^* = \langle P, x \rangle$ such that $|H^*|$ is of minimal order. Then $H^*_{\mathfrak{F}}$ is a normal maximal subgroup of H^*. Now we replace G by $G^* = \langle F, x \rangle$, because $G^* \in \mathfrak{F}$ and H^* is a Hall π-subgroup of G^*. Set $Q = P_{k+1} \cdots P_n$. We define a direct product $D = \langle P, x_1 \rangle \times \langle Q, x_2 \rangle$, where $\langle P, x_1 \rangle$ is a copy of H^* and $\langle Q, x_2 \rangle$ is a copy of $Q\langle x \rangle$. Then $K = PQ\langle x_1 x_2 \rangle$ is a normal subgroup of D isomorphic to G^*. Hence K is contained in $D_{\mathfrak{F}} = \langle P, x_1 \rangle_{\mathfrak{F}} \times \langle Q, x_2 \rangle_{\mathfrak{F}}$. Since $|\langle P, x_1 \rangle : \langle P, x_1 \rangle_{\mathfrak{F}}| = P$ and $|\langle Q, x_2 \rangle : \langle Q, x_2 \rangle_{\mathfrak{F}}| = p$, it follows that $|D : D_{\mathfrak{F}}| = p^2$. However $|D : K| = p$. This contradiction proves the theorem. $\qquad\square$

Not every supersoluble Fitting class is a Lockett class ([Men96, Example 1]). In the following we shall prove that every supersoluble Fitting class is contained in a supersoluble Lockett class.

Theorem 7.3.16. *Every supersoluble Fitting class is contained in a supersoluble Lockett class.*

Proof. Assume that $\mathfrak{F}$ is a supersoluble Fitting class. If $G \in \mathfrak{F}^*$, then $D = \{(g, g^{-1}) : g \in G\}$ is a subgroup of $(G \times G)_{\mathfrak{F}}$ by [DH92, X, 1.5, 1.9]. Therefore D is supersoluble. Since G is an epimorphic image of D, it follows that G is supersoluble. Therefore $\mathfrak{F}^*$ is a supersoluble Lockett class. $\qquad\square$

7.4 Fitting sets, Fitting sets pairs, and outer Fitting sets pairs

This section has two main themes. The first is connected with Fitting sets and injectors. The second subject under investigation is the localised theory of Fitting pairs and outer Fitting pairs developed in [AJBBPR00].

As mentioned in Section 2.4, the theory of Fitting classes has been enriched by the introduction of Fitting sets by W. Anderson in [And75]. Recall that a subgroup H of a group G is an injector of G if H is an $\mathcal{F}$-injector of G for some Fitting set $\mathcal{F}$ of G. One the most important motivating questions in the theory of Fitting sets is to determine which subgroups are injectors. Some results in this direction are presented in [DH92, VIII, Section 3]. There Doerk and Hawkes proposed the problem of describing injectors of soluble groups without explicit use of the concept of a Fitting set.

This problem is complicated by the general nature of injectors: there are likely to be many Fitting sets for a given group, often leading to different sets of injectors. For example, the set of injectors of a soluble group includes all its normal subgroups, all its Hall subgroups, and all its maximal subgroups [DH92, VIII, 3.5]. An injector A of a finite soluble group B must have rather strong properties that can be described without direct reference to Fitting sets: $A \cap K$ must be a CAP subgroup of K and pronormal (see [DH92, Section I, 6]) in B for each normal subgroup K of B [DH92, VIII, 2.14]. However, these properties are inadequate to characterise injectors [DH92, Exercise 2, p. 553]. We present here the best attempt to accomplish that task. This characterisation, unpublished at the moment of writing this, was communicated privately by its authors, R. Dark and A. Feldman ([DF]), to us.

If G is a group, denote by $\mathrm{Inj}(G)$ the set of all injectors of G. The following result is a very useful characterisation of this set. Recall that if H is a subgroup of G then

$$\mathrm{s}_n\, H^G = \{S \le G : S \text{ is a subnormal subgroup of } H^g, \text{ for some } g \in G\}.$$

Lemma 7.4.1 ([DH92, VIII, 3.3]). *Let G be a soluble group and H a subgroup of G. Then any two of the following statements are equivalent*

1. *$H \in \mathrm{Inj}(G)$*
2. *$\mathrm{s}_n\, H^G$ is a Fitting set of G.*
3. *$\mathrm{s}_n\, H^G$ is the smallest Fitting set of G which contains H.*

Lemma 7.4.2. *Suppose S and T are pronormal subgroups of a soluble group G and x, $y \in G$. If S and T are subnormal in $\langle S, T \rangle$ and S^x and T^y are subnormal in $\langle S^x, T^y \rangle$, then there exists $z \in G$ with $S^x = S^z$ and $T^y = T^z$.*

Proof. Let Σ be a Hall system of G which reduces into $\langle S, T \rangle$. Applying [DH92, I, 6.3], S and T are normal in $\langle S, T \rangle = ST$. By [DH92, I, 4.21], Σ reduces into both S and T. Analogously, S^x and T^y are normal in $\langle S^x, T^y \rangle = S^x T^y$. Then by [DH92, I, 6.11], $S^x T^y = (ST)^z$ for some $z \in G$. This implies that Σ^z, which reduces into $(ST)^z$, reduces into the subnormal subgroups S^x and S^z and T^y and T^z of that group. But the pronormality of S and T then implies, by [DH92, I, 6.6], that $S^x = S^z$ and $T^y = T^z$, as claimed.   □

Now we prove a result that will supply the inductive step in our eventual characterisation of injectors:

Theorem 7.4.3. *Let G be a soluble group and suppose H is a subgroup of G and M is a normal subgroup of G. Assume that the following condition holds:*

> *Whenever S is a subnormal subgroup of H, $g \in G$, $S^g \leq HM$ and $S_1 = H \cap S^g M$ is subnormal in H, then S_1 and S^g are conjugate in $J = \langle S_1, S^g \rangle$.* (7.1)

Then

1. *if S is a subnormal subgroup of H, then S is pronormal in $\mathrm{N}_G(SM)$ and*
2. *if $HM \in \mathrm{Inj}(G)$, then $H \in \mathrm{Inj}(G)$.*

Proof. 1. Let g be an element of $\mathrm{N}_G(SM)$, so that $S^g M = (SM)^g = SM$. Note that if S is subnormal in H, then SM is subnormal in HM, and therefore $S_1 = H \cap S^g M = H \cap SM = S(H \cap M)$ is subnormal in H. Applying (7.1) with $g = 1$ yields S and $S_1 = S(H \cap M)$ are conjugate. Now, by order considerations, $S = S_1$. By (7.1) then, S and S^g are conjugate in $\langle S, S^g \rangle$; i.e. S is pronormal in $\mathrm{N}_G(SM)$.

2. Suppose that S and T are subnormal subgroups of H and $a, b \in G$ with S^a and T^b normal in $S^a T^b$. By Lemma 7.4.1, it suffices to find an element w such that $S^a T^b$ is subnormal in H^w. Now SM and TM are subnormal subgroups of HM and $S^a M$ and $T^b M$ are normal in $Y = S^a T^b M = S^a M T^b M$, and because $HM \in \mathrm{Inj}(G)$, there exists $c \in G$ such that Y is subnormal in $(HM)^c = H^c M$. Let $H_0 = H^c$ and $S_0 = S^c$. Note that condition (7.1) still holds when H is replaced by the conjugate H_0. Replacing S and g by S_0 and $c^{-1}a$ we have S_0 is subnormal in H_0, $S_0^g = S^a \leq H_0 M$, and $S_0^g M = S^a M$ is normal in Y which is subnormal in $H_0 M$. Hence $S_0^g M$ is subnormal in $H_0 M$, and $S_1 = H_0 \cap S_0^g M$ is subnormal in H_0. Then by (7.1), S_1 and S^a are conjugate in $\langle S_1, S^a \rangle \leq S^a M \leq Y$. Similarly, $T_1 = H_0 \cap T^b M$ is subnormal in H_0, and T^b is conjugate in Y to T_1; hence there are elements $x, y \in Y$ such that $S_1^x = S^a$ and $T_1^y = T^b$.

Now $S^a M = H_0 M \cap S_0^g M = (H_0 \cap S_0^g M)M = S_1 M$ and then $Y \leq \mathrm{N}_G(S^a M) = \mathrm{N}_G(S_1 M)$, and if follows from Assertion 1 that S_1 is pronormal in Y. Similarly, T_1 is pronormal in Y. We also have that S_1 and T_1 are subnormal in $\langle S_1, T_1 \rangle$ and S_1^x, T_1^y normal in $S_1^x T_1^y$. By Lemma 7.4.2, there exists $z \in Y$ with $S_1^x = S_1^z$ and $T_1^y = T_1^z$. Hence $S^a T^b = S_1^x T_1^y = (S_1 T_1)^z$ is subnormal in $H_0^z = H^{cz}$, so setting $w = cz$ yields our result. $\square$

Now we are ready to prove that two properties that do not involve Fitting sets are equivalent to that of being an injector. Not surprisingly, conjugation, which is crucial to the definition of Fitting set and normality (and therefore indirectly, subnormality) play an important role in these properties. In particular, for convenience we introduce the following definition:

Definition 7.4.4. *If H and X are subgroups of a soluble group G and $g \in G$, we say H is (X, g)-pronormal if $H \cap X$ and $H^g \cap X$ are conjugate in $J = \langle H \cap X, H^g \cap X \rangle$.*

Note that H is a pronormal subgroup of G if and only if H is (G, g)-pronormal for all $g \in G$.

We now can prove:

Theorem 7.4.5 (R. Dark and A. Feldman). *Let G be a soluble group, and suppose that H is a subgroup of G. Then any two of the following conditions are equivalent:*

1. *H is an injector of G;*
2. *whenever $H \leq K \leq G$, $g \in G$, and X and $X^{g^{-1}}$ are subnormal subgroups of K, then H is (X, g)-pronormal;*
3. *whenever M/N is a chief factor of G which is not covered by H, S is a subnormal subgroup of H such that $H \cap N \leq S$, $g \in G$, and $S^g \leq HM$ with $S_1 = H \cap S^g M$ subnormal in H, then S_1 and S^g are conjugate in $J = \langle S_1, S^g \rangle$.*

Proof. 1 implies 2. Suppose that H is an $\mathcal{F}$-injector of G for some Fitting set $\mathcal{F}$ of G. Then, with K and X as in 2 and J as in the definition of (X, g)-pronormal, H is an $\mathcal{F}_K$-injector of K by [DH92, VIII, 2.13], and then $H \cap X$ is an $\mathcal{F}_X$-injector of X by [DH92, VIII, 2.6], and hence $H \cap X$ is an $\mathcal{F}_J$-injector of J by [DH92, VIII, 2.13] again. Similarly, H^g is an $\mathcal{F}_{K^g}$-injector of K^g, and X is subnormal in K^g by hypothesis, and then $H^g \cap X$ is an $\mathcal{F}_X$-injector of X, and $H^g \cap X$ is an $\mathcal{F}_J$-injector of J. Thus by Theorem 2.4.26, $H \cap X$ and $H^g \cap X$ are conjugate in J, establishing 2.

2 implies 3. First observe that, in these hypotheses, we certainly have that H avoids M/N. With $X = M$, we see that $H \cap M$ and $H^g \cap M$ are conjugate in $J = \langle H \cap M, H^g \cap M \rangle$, and then $(H \cap M)N$ and $(H^g \cap M)N$ are conjugate in JN. But $JN/N \leq M/N$, which is abelian, and it follows that $(H \cap M)N = (H^g \cap M)N$. This holds for all $g \in G$ because $X = M$ is normal in G, and then $(H \cap M)N$ is normal in G. Since H does not cover the chief factor M/N of G, we have that $(H \cap M)N < M$. Then $(H \cap M)N = N$, establishing the result.

Assume the hypotheses of 3 and take $X = S^g M$. Then X is subnormal in $H^g M$ and $X^{g^{-1}}$ is subnormal in HM. Also, $X \leq HM$, and $X = HM \cap S^g M = (H \cap S^g M)M = S_1 M$ is subnormal in HM. Moreover, $H \cap X = S_1$ by definition, and $H^g \cap X = H^g \cap S^g M = S^g(H^g \cap M)$, which equals $S^g(H^g \cap N)$ inasmuch as H^g avoids M/N. But $H \cap N \leq S$ by hypothesis, and then $H^g \cap N \leq S^g$, and $H^g \cap X = S^g$. Thus 2 yields that S_1 and S^g are conjugate in $\langle S_1, S^g \rangle$, as claimed.

To see that 3 implies 1, we pass through an intermediate Step 4.

4. *Whenever M/N is a chief factor of G which is not covered by H, and such that $\mathrm{Core}_G(H) \leq N < M \leq \langle H^G \rangle$, and S is a subnormal subgroup of H such that $H \cap N \leq S$, $g \in G$, and $S^g \leq HM$ with $S_1 = H \cap S^g M$ subnormal in H, then S_1 and S^g are conjugate in $J = \langle S_1, S^g \rangle$.*

It is clear that 3 implies 4. Hence we have to prove that 4 implies 1.

Note first that if $C = \mathrm{Core}_G(H)$, it is easy to see that if 4 holds for $H \leq G$, then 4 also holds for $H/C \leq G/C$. Moreover, if $H/C \in \mathrm{Inj}(G/C)$, then $H \in \mathrm{Inj}(G)$ by [DH92, VIII, 2.17]. Thus it suffices to prove that if 4 holds for H/C in G/C, then $H/C \in \mathrm{Inj}(G/C)$, and we may assume that $C = 1$, i.e. H is core-free in G.

We proceed by induction on the index $|\langle H^G \rangle : H|$. If $|\langle H^G \rangle : H| = 1$, then $H = 1$ inasmuch as H is core-free. In this case H is obviously an injector of G. Hence we may assume that $|\langle H^G \rangle : H| > 1$. Let M_1 be a minimal normal subgroup of G such that $M_1 \leq \langle H^G \rangle$. Since H is core-free, H does not cover M_1. We see next that because 4 holds for H, it also holds for HM_1.

Suppose that M/N is a chief factor of G which $1 < M_1 \leq \mathrm{Core}_G(HM_1) \leq N < M \leq \langle (HM_1)^G \rangle = \langle H^G \rangle$ and M/N is not covered by HM_1.

Now suppose that $g \in G$ and $\bar{S}$ is a subnormal subgroup of HM_1 such that $HM_1 \cap N \leq \bar{S}$, $g \in G$, and $\bar{S}^g \leq (HM_1)M$ with $\bar{S}_1 = HM_1 \cap \bar{S}^g M$ subnormal in HM_1,

Consider $S = H \cap \bar{S}$. Then S is subnormal in H. Since $M_1 \leq HM_1 \cap N \leq \bar{S}$, then $\bar{S} = HM_1 \cap \bar{S} = (H \cap \bar{S})M_1 = SM_1$, and then $\bar{S}^g M = S^g M$. Observe also that $H \cap N = H \cap HM_1 \cap N \leq H \cap \bar{S} = S$ and $S^g \leq \bar{S}^g \leq HM$. Finally, it is clear that $S_1 = H \cap S^g M = H \cap (HM_1 \cap \bar{S}^g M) = H \cap \bar{S}_1$ is subnormal in H.

Thus the hypotheses of 4 hold, implying S_1 and S^g are conjugate in $J = \langle S_1, S^g \rangle$. Moreover, $\bar{S} = SM_1$, and $\bar{S}^g = S^g M_1$, and $\bar{S}_1 = HM_1 \cap \bar{S}^g M = (H \cap S^g M)M_1 = S_1 M_1$, and $\bar{J} = \langle \bar{S}_1, \bar{S}^g \rangle = JM_1$. Hence $\bar{S}_1$ and $\bar{S}^g$ are conjugate in $\bar{J}$.

Observe that $|\langle H^G \rangle| = |\langle (HM_1)^G \rangle : HM_1| < |\langle H^G \rangle : H|$. Thus the induction hypothesis implies that $HM_1 \in \mathrm{Inj}(G)$. To complete the proof, we apply Theorem 7.4.3 (2) with $M = M_1$. With $N = 1$, and by 4 applied to the chief factor M_1/N, Condition (7.1) of Theorem 7.4.3 holds. Thus, Theorem 7.4.3 (2) shows that $H \in \mathrm{Inj}(G)$. □

Corollary 7.4.6. *Let G be a soluble group. Suppose that H is an injector of G and M a normal subgroup of G. Then $H \cap M$ is pronormal in G.*

Applying [DH92, VIII, 3.5], a maximal subgroup of a group is always an injector. Hence, in particular, in a soluble group the intersection of a maximal subgroup and a normal subgroup is pronormal in the group.

By [DH92, VIII, 3.8] every normally embedded subgroup of a soluble group is an injector. In the following we give a proof of this fact using Theorem 7.4.5.

Corollary 7.4.7. *Suppose H is a normally embedded subgroup of a soluble group G. Then $H \in \mathrm{Inj}(G)$.*

Proof. Assume that H is normally embedded in G, $H \leq K \leq G$, and $X, X^{g^{-1}}$ are subnormal in K for some $g \in G$. We shall show that H is (X, g)-pronormal.

First we show that $H \cap X$ and $H^g \cap X$ are locally conjugate in X. For an arbitrary prime, p, let $P \in \mathrm{Syl}_p(H)$; let $P_1 \in \mathrm{Syl}_p(K)$ such that $P_1 \cap H = P$. Because X is subnormal in K, $P_1 \cap X \in \mathrm{Syl}_p(X)$, by [DH92, I, 4.21]. Also, $H \cap X$ is subnormal in H, and $P \cap X = P \cap (H \cap X) \in \mathrm{Syl}_p(H \cap X)$. Now H normally embedded in G implies $P \in \mathrm{Syl}_p(\langle P^G \rangle)$, and $P \leq \langle P^G \rangle \cap P_1 \leq \langle P^G \rangle$, and then $P = \langle P^G \rangle \cap P_1$. Because $\langle P^G \rangle \cap X$ is normal in X, $(P_1 \cap X) \cap (\langle P^G \rangle \cap X) \in \mathrm{Syl}_p(\langle P^G \rangle \cap X)$. But $(\langle P^G \rangle \cap X) \cap (P_1 \cap X) = (\langle P^G \rangle \cap P_1) \cap X = P \cap X \in \mathrm{Syl}_p(H \cap X)$. Hence any Sylow p-subgroup of $H \cap X$ is a Sylow p-subgroup of $\langle P^G \rangle \cap X$. By similar arguments, $P^g \in \mathrm{Syl}_p(H^g)$ implies $P^g \cap X \in \mathrm{Syl}_p(\langle (P^g)^G \rangle \cap X) = \mathrm{Syl}_p(\langle P^G \rangle \cap X)$ and $P^g \cap X \in \mathrm{Syl}_p(H^g \cap X)$. Thus we have Sylow p-subgroups of $H \cap X$ and $H^g \cap X$ that are Sylow p-subgroups of the same subgroup of X, and they are conjugate in X, as desired.

Now note that $\langle P^G \rangle \cap X$ is normal in X, and since this works for all primes p, $H \cap X$ and $H^g \cap X$ are normally embedded in X. Thus $H \cap X$ and $H^g \cap X$ are locally pronormal [DH92, I, 7.13] and therefore pronormal [DH92, I, 6.14] in X. Thus $H \cap X$ and $H^g \cap X$ are locally conjugate and locally pronormal subgroups in X, and they are conjugate in X [DH92, I, 6.16]. Finally, the pronormality of $H \cap X$ in X implies that $H \cap X$ and $H^g \cap X$ are conjugate in their join; i.e. H is (X, g)-pronormal, establishing the result. $\qquad\square$

Let $\mathfrak{F}$ be a Fitting class. Blessenohl and Gaschütz [BG70] introduced the notion of $\mathfrak{F}$-Fitting pair which turns out to be useful for the construction of normal Fitting classes in the Lockett section of $\mathfrak{F}$.

We need to deal with arbitrary (possibly infinite) groups. Hence if we denote a group by G, we are assuming that the group G is finite. Otherwise, we put **G**.

Definition 7.4.8. *If N and M are groups, an* embedding *is a group monomorphism $\nu: N \longrightarrow M$.*

If N^ν is a normal subgroup of M, then ν is said to be a normal embedding.

Definition 7.4.9 ([BG70]). *Let $\mathfrak{F}$ be a Fitting class. An $\mathfrak{F}$-Fitting pair is a pair $(d, \mathbf{A})$ which consists of a group $\mathbf{A}$ and a family $\big(d_U \in \mathrm{Hom}(U, \mathbf{A})$: $U \in \mathfrak{F}\big)$ such that for each normal embedding $\nu: U \longrightarrow V \in \mathfrak{F}$, the assertion $d_U = \nu d_V$ holds.*

It can be proved that in this case $\{(g)^{d_G} : g \in G, G \in \mathfrak{F}\}$ is an abelian subgroup of $\mathbf{A}$ ([DH92, IX, 2.12 (b)]). Hence, without loss of generality, we may assume that $\mathbf{A}$ is abelian.

In the same paper, Blessenohl and Gaschütz gave examples of Fitting pairs and proved the following result, which remains valid in the general finite universe.

Proposition 7.4.10 (see [DH92, IX, 2.11]). *Let $\mathfrak{F}$ be a Fitting class and let $(d, \mathbf{A})$ be an $\mathfrak{F}$-Fitting pair. Then the class $\mathfrak{R} = \mathrm{Ker}(d, \mathbf{A})$ of all groups $G \in \mathfrak{F}$ such that $G^{d_G} = 1$ is a normal Fitting class such that $\mathfrak{F}_* \subseteq \mathfrak{R} \subseteq \mathfrak{F}$.*

Lausch [Lau73] showed that every non-trivial normal Fitting class in the soluble universe can be described as the kernel of a Fitting pair. He also described a universal $\mathfrak{F}$-Fitting pair, leading to the so-called Lausch group. He carried out the construction for the case $\mathfrak{F} = \mathfrak{S}$, but as Bryce and Cossey pointed out in [BC75], Lausch's method applies to an arbitrary Fitting class (see [DH92, X, Section 4] for details).

J. Pense, in his Dissertation [Pen87], generalised the concept of an $\mathfrak{F}$-Fitting pair to that of outer $\mathfrak{F}$-Fitting pair.

Definition 7.4.11 (see [Pen88]). *Let $\mathfrak{F}$ be a Fitting class. An* outer $\mathfrak{F}$-Fitting pair *is a pair $(d, \mathbf{A})$ which consists of a group $\mathbf{A}$ and a family $\big(d_U \in \mathrm{Hom}(U, \mathbf{A}) : U \in \mathfrak{F}\big)$ such that for each normal embedding $\nu \colon U \longrightarrow V \in \mathfrak{F}$, there exists an inner automorphism α of $\mathbf{A}$ such that $d_U \alpha = \nu d_V$.*

Obviously, if $\mathbf{A}$ is an abelian group, then an outer $\mathfrak{F}$-Fitting pair is just an $\mathfrak{F}$-Fitting pair.

Pense extended the definition of a Fitting set to an infinite group by requiring it to mean a set of *finite* subgroups closed under conjugation and under the usual operations of taking normal subgroups and forming finite normal products. He also introduced the concept of $\mathcal{F}$-Fitting sets pair $(d, \mathbf{A})$, where $\mathbf{A}$ is an abelian group, to develop a local version of the Lausch group in certain type of groups ([Pen87]).

Definition 7.4.12. *If N and M are finite subgroups of $\mathbf{G}$, a $\mathbf{G}$-embedding is a group monomorphism $\nu \colon N \longrightarrow M$ which is the restriction to N of an inner automorphism of $\mathbf{G}$.*

If N^ν is a normal subgroup of M, then ν is said to be a normal $\mathbf{G}$-embedding.

Definition 7.4.13. *Let $\mathcal{F}$ be a Fitting set of a group $\mathbf{G}$. An $\mathcal{F}$-Fitting sets pair relative to $\mathbf{G}$ is a pair $(d, \mathbf{A})$ which consists of a group $\mathbf{A}$ and a family $\big(d_U \in \mathrm{Hom}(U, \mathbf{A}) : U \in \mathcal{F}\big)$ such that for each normal $\mathbf{G}$-embedding $\nu \colon U \longrightarrow V \in \mathcal{F}$, the assertion $d_U = \nu d_V$ holds.*

Note that, in our definition of $\mathcal{F}$-Fitting sets pair, we do not require that $\mathbf{A}$ is an abelian group. An outer $\mathcal{F}$-Fitting sets pair is defined as follows:

Definition 7.4.14 ([AJBBPR00]). *Let $\mathcal{F}$ be a Fitting set of a group $\mathbf{G}$. An* outer $\mathcal{F}$-Fitting sets pair relative to $\mathbf{G}$ *is a pair $(d, \mathbf{A})$ which consists of a group $\mathbf{A}$ and a family $\big(d_U \in \mathrm{Hom}(U, \mathbf{A}) : U \in \mathcal{F}\big)$ such that for each normal $\mathbf{G}$-embedding $\nu \colon U \longrightarrow V \in \mathcal{F}$, there exists an inner automorphism α of $\mathbf{A}$ such that $d_U \alpha = \nu d_V$.*

If $\mathfrak{F}$ is a Fitting class, then $\mathrm{Tr}_{\mathfrak{F}}(\mathbf{G})$ is a Fitting set of the group $\mathbf{G}$, and if $(d, \mathbf{A})$ is an (outer) $\mathfrak{F}$-Fitting pair, then the pair $(d, \mathbf{A})$, for $\big(d_U \in \mathrm{Hom}(U, \mathbf{A}) : U \in \mathrm{Tr}_{\mathfrak{F}}(\mathbf{G})\big)$, is an (outer) $\mathrm{Tr}_{\mathfrak{F}}(\mathbf{G})$-Fitting sets pair relative to $\mathbf{G}$.

Definition 7.4.15. *Two outer $\mathcal{F}$-Fitting sets pairs $(d_i, \mathbf{A}_i)$, $i = 1, 2$, are equivalent if there exists an isomorphism $\sigma \colon \mathbf{A}_1 \longrightarrow \mathbf{A}_2$, such that for each $U \in \mathcal{F}$, there exists $\alpha_U \in \mathrm{Inn}(\mathbf{A}_2)$ such that $d_{2U} = d_{1U}\sigma\alpha_U$.*

In [AJBBPR00], P. Arroyo-Jordá, A. Ballester-Bolinches, and M. D. Pérez-Ramos made a complete study of outer Fitting sets pairs. In the sequel, we will present the main results of this paper.

To begin with, we point out that there are some differences between Fitting pairs and Fitting sets pairs. We shall show two of them.

Remarks 7.4.16.   1. In Definition 7.4.9 of Fitting pair, the group $\mathbf{A}$ can be assumed abelian without loss of generality. This is not true for Fitting sets pairs in general.

Let G be the alternating group of degree 5, $G = \mathrm{Alt}(5)$, and $\mathcal{F}$ the trace in G of the Fitting class $\mathfrak{F} = \mathfrak{S}_3\mathfrak{S}_5\mathfrak{S}_2$. In other words, the Fitting set $\mathcal{F}$ is composed of all subgroups of G of prime-power order, and the normalisers of the Sylow 5- and 3-subgroups. Consider the symmetric group $S = \mathrm{Sym}(3)$ of degree 3. If X is a subgroup of prime-power order of G, then put $d_X \colon X \longrightarrow S$ to be the trivial homomorphism: $x^{d_X} = 1$ for all $x \in X$. If $P \in \mathrm{Syl}_3(G)$ and $N_3 = N_G(P)$, then put $d_{N_3} \colon N_3 \longrightarrow S$ to be a homomorphism such that $P = \mathrm{Ker}(d_{N_3})$ and $\mathrm{Im}(d_{N_3}) = \langle (12) \rangle$. If $Q \in \mathrm{Syl}_5(G)$ and $N_5 = N_G(Q)$, then put $d_{N_5} \colon N_5 \longrightarrow S$ to be a homomorphism such that $Q = \mathrm{Ker}(d_{N_5})$ and $\mathrm{Im}(d_{N_5}) = \langle (23) \rangle$.

The pair $(\{d_H : H \in \mathcal{F}\}, S)$ is an $\mathcal{F}$-Fitting sets pair relative to G.

Observe that S is not abelian and $S = \langle h^{d_H} : H \in \mathcal{F}, h \in H \rangle$.

2. Pense [Pen87, Kollollar 3.30] shows that if (d, A) is a outer Fitting pair with A finite, then it is equivalent to a Fitting pair. This is not true for outer Fitting sets pairs.

Let $Q = \langle x, y : x^4 = 1, x^2 = y^2, x^y = x^{-1} \rangle$ be a quaternion group of order 8 and fix a subgroup $C = \langle x \rangle$ of order 4 of Q. The set of all subgroups of C is a Fitting set $\mathcal{F}$ of Q. The inclusion $\iota \colon C \longrightarrow Q$ induces a family of monomorphisms between the members of $\mathcal{F}$ and Q. The pair (ι, Q) is an outer $\mathcal{F}$-Fitting sets pair relative to Q. The inner automorphism α_y of Q induced by y gives a normal Q-embedding of $\nu \colon C \longrightarrow C$ such that $x^\nu = x^{-1}$ and $\iota\alpha_y = \nu\iota$.

If (ι, Q) were equivalent to a $\mathcal{F}$-Fitting sets pair (d, A), there would exist an isomorphism $\psi \colon Q \longrightarrow A$ such that for each subgroup T of C there would exist $\alpha_T \in \mathrm{Inn}(A)$ such that $d_T = \iota_T\psi\alpha_T$. Since $d_C = \nu d_C$, we have that $x^2 \in \mathrm{Ker}(d_C)$. But $\iota_C\psi\alpha_C$ is a monomorphism and therefore $d_C \neq \iota_C\psi\alpha_C$. Thus (ι, Q) cannot be equivalent to an $\mathcal{F}$-Fitting sets pair (d, A).

The following result is the "Fitting sets" version of [Pen87, Satz 3.2].

Theorem 7.4.17. *Let $(d, \mathbf{A})$ be an outer $\mathcal{F}$-Fitting sets pair relative to $\mathbf{G}$ and let $\mathcal{H}$ be a Fitting set of $\mathbf{A}$.*

1. *The collection $\mathcal{H}d^{-1} = \{U \in \mathcal{F} : U^{d_U} \in \mathcal{H}\}$ of finite subgroups of $\mathbf{G}$ is a Fitting set of $\mathbf{G}$.*

2. *If $U \in \mathcal{F}$, then $U_{\mathcal{H}d^{-1}} = \left((U^{d_U})_{\mathcal{H}}\right)^{d_U^{-1}}$.*

Proof.   1. If N is a normal subgroup of $U \in \mathcal{H}d^{-1}$, then N^{d_N} is conjugate in $\mathbf{A}$ to the normal subgroup N^{d_U} of $U^{d_U} \in \mathcal{H}$. Thus $N \in \mathcal{H}d^{-1}$.

Assume that N_1 and N_2 are subgroups of $\mathbf{G}$ which are normal in $T = N_1 N_2$ and $N_i \in \mathcal{H}d^{-1}$, for $i = 1, 2$. Then $T^{d_T} = N_1^{d_T} N_2^{d_T}$ and $N_i^{d_T}$ is normal in T^{d_T}, for $i = 1, 2$. Moreover, $N_i^{d_T}$ is conjugate in $\mathbf{A}$ to $N_i^{d_{N_i}}$, for $i = 1, 2$. Therefore $T \in \mathcal{H}d^{-1}$.

2. Let $C = \left((U^{d_U})_{\mathcal{H}}\right)^{d_U^{-1}}$. By Statement 1, C is a normal $\mathcal{H}d^{-1}$-subgroup of U. If M is a normal subgroup of U, with $M \in \mathcal{H}d^{-1}$, then $M^{d_M} \in \mathcal{H}$ and it is conjugate in $\mathbf{A}$ to M^{d_U}. Hence $M^{d_U} \leq (U^{d_U})_{\mathcal{H}}$ and then $M \leq C$.   □

Definition 7.4.18. *For an outer $\mathcal{F}$-Fitting sets pair relative to $\mathbf{G}$, $(d, \mathbf{A})$, and a homomorphism $\varphi \colon \mathbf{A} \longrightarrow \mathbf{B}$, we define the induced outer $\mathcal{F}$-Fitting pair relative to $\mathbf{G}$, $(d\varphi, \mathbf{B})$, by $(d\varphi)_T = d_T \varphi$, for every $T \in \mathcal{F}$.*

The next theorem provides a criterion for the Fitting sets constructed by means of outer Fitting sets pairs to be injective.

Theorem 7.4.19 ([AJBBPR00]). *Let $\mathbf{G}$ be a group and denote by $\mathcal{E}_{\mathbf{G}}$ the Fitting set composed of all finite subgroups of $\mathbf{G}$. Let $(d, \mathbf{A})$ be an outer $\mathcal{E}_{\mathbf{G}}$-Fitting sets pair relative to $\mathbf{G}$. Suppose that $\mathcal{F}$ is a Fitting set of $\mathbf{A}$ and the pair $(d, \mathbf{A})$ satisfies the following condition:*

For each $\mathbf{G}$-embedding $\nu \colon V \longrightarrow U$, for $U, V \in \mathcal{E}_{\mathbf{G}}$ such that $U_{\mathcal{F}d^{-1}} \leq V^\nu$, there exists $\eta \in \mathrm{Inn}(A)$ such that $\nu d_U = d_V \eta$.   (7.2)

Let $X \in \mathcal{E}_{\mathbf{G}}$. If the group X^{d_X} possesses a single conjugacy class of $\mathcal{F}$-injectors, then X also possesses a single conjugacy class of $\mathcal{F}d^{-1}$-injectors.

Proof. Let X be a subgroup of $\mathbf{G}$ and assume that T is an $\mathcal{F}$-injector of X^{d_X}. Denote by $U = T^{d_X^{-1}}$. We shall see that U is an $\mathcal{F}d^{-1}$-injector of X. Since T is an $\mathcal{F}$-injector of X^{d_X}, it follows that $(X^{d_X})_{\mathcal{F}}$ is a subgroup of T. Hence $X_{\mathcal{F}d^{-1}} = ((X^{d_X})_{\mathfrak{F}})^{d_X^{-1}}$ by Theorem 7.4.17 (2) and it is contained in U. By property (7.2) there exists $a \in \mathbf{A}$ such that $(U^{d_U})^a = U^{d_X}$. Since $T = U^{d_X} \in \mathcal{F}$ it follows that $U \in \mathcal{F}d^{-1}$.

Let N be a subnormal subgroup of X and suppose that $U \cap N \leq W \leq N$, where $W \in \mathcal{F}d^{-1}$. Since N is a subnormal subgroup of X, it holds that $N_{\mathcal{F}d^{-1}} = N \cap X_{\mathcal{F}d^{-1}} \leq N \cap U \leq W$. By (7.2), the subgroup W^{d_N} is conjugate in $\mathbf{A}$ to W^{d_W} which is in $\mathcal{F}$. On the other hand, since $(d, \mathbf{A})$ is an outer $\mathcal{E}_{\mathbf{G}}$-Fitting sets pair relative to $\mathbf{G}$, there exists $\theta \in \mathrm{Inn}(A)$ such that d_N is $d_X \theta$ restricted to N. Hence W^{d_N} is conjugate in $\mathbf{A}$ to W^{d_X}. Consequently $W^{d_X} \in \mathcal{F}$. Now $\mathrm{Ker}(d_X) \leq X_{\mathfrak{F}d^{-1}} \leq U$. Hence $(U \cap N)^{d_X} = T \cap N^{d_X}$ which is contained in $W^{d_X} \leq N^{d_X}$. Since T is an $\mathcal{F}$-injector of X^{d_X} and $W^{d_X} \in \mathcal{F}$, it follows that $T \cap N^{d_X} = W^{d_X}$. Therefore $W \leq U$ and $U \cap N = W$. This means that U is an $\mathcal{F}d^{-1}$-injector of X.

Suppose now that X^{d_X} has a single conjugacy class of $\mathcal{F}$-injectors. Let U and $\tilde{U}$ be two $\mathcal{F}d^{-1}$-injectors of X. A straightforward proof using analogous arguments provides that U^{d_X} and $\tilde{U}^{d_X}$ are $\mathcal{F}$-injectors of X^{d_X}. By hypothesis, there exists $x \in X$ such that $U^{d_X} = (\tilde{U}^x)^{d_X}$. Since $\mathrm{Ker}(d_X) \leq U \cap \tilde{U}$, it follows that $U = \tilde{U}^x$.   □

The rest of the section is devoted to construct injective Fitting sets using outer Fitting sets pairs. We shall give some examples of outer Fitting sets pairs which are local versions of the outer Fitting pairs constructed in [Pen88, Sections 4 and 5]. These local constructions provide further information and show that Fitting sets pairs are worth investigating.

Our first example leads to a p-supersoluble Fitting set, p a prime, in every group. This Fitting set is dominant in the set of all p-constrained groups (see Definition 2.4.29).

Example 7.4.20. Let G be a group and let J be a simple group. Suppose that n_G is the largest natural number such that $|J|^{n_G}$ divides $|G|$. Denote by $\mathrm{D}_J(n_G)$ the direct product of n_G copies of J. If $n_G = 0$, we agree that $\mathrm{D}_J(n_G) = 1$. Let $\mathrm{A}_J(n_G) = \mathrm{Aut}\big(\mathrm{D}_J(n_G)\big)$ and $\mathrm{O}_J(n_G) = \mathrm{Out}\big(\mathrm{D}_J(n_G)\big)$. It is known that

1. if J is non-abelian, then $\mathrm{A}_J(n_G)$ is isomorphic to the natural wreath product

$$\mathrm{A}_J(n_G) \cong \mathrm{Aut}(J) \wr_{\mathrm{nat}} \mathrm{Sym}(n_G) \quad \text{and} \quad \mathrm{O}_J(n_G) \cong \mathrm{Out}(J) \wr_{\mathrm{nat}} \mathrm{Sym}(n_G).$$

2. if $J \cong C_p$, for a prime p, then

$$\mathrm{A}_J(n_G) \cong \mathrm{GL}(p, n_G).$$

Also let $\mathbf{D}_J$ be the restricted direct product of countably infinitely many copies of J and let $\mathbf{A}_J = \mathrm{Aut}^0(\mathbf{D}_J)$ be the group of all automorphisms of $\mathbf{D}_J$ with finite support Denote $\mathbf{O}_J$ the group of outer automorphisms of $\mathbf{D}_J$ with finite support.

Let $\mathfrak{F}$ and $\mathfrak{G}$ be two Fitting classes such that $\mathfrak{G} \subseteq \mathfrak{F}$.

1. ([Pen88, Theorem II]) For any group G and any chief series Γ of G through $G_{\mathfrak{F}}$ and $G_{\mathfrak{G}}$, let $\mathbf{D}_J(\Gamma, \mathfrak{F}/\mathfrak{G})$ be the direct product of all the J-chief factors of Γ between $G_{\mathfrak{F}}$ and $G_{\mathfrak{G}}$, taken in the order of occurrence in Γ. We consider this group as the subgroup of $\mathbf{D}_J$ consisting of the first direct components of $\mathbf{D}_J$. The group G operates on every such $\mathbf{D}_J(\Gamma, G_{\mathfrak{F}}/G_{\mathfrak{G}})$ and by identical continuation also on $\mathbf{D}_J$. This action defines a homomorphism

$$d_G^{J, \mathfrak{F}/\mathfrak{G}} : G \longrightarrow \mathbf{A}_J.$$

Then the pair $(d^{J, \mathfrak{F}/\mathfrak{G}}, \mathbf{A}_J)$ is an outer $\mathfrak{E}$-Fitting pair. This is called the *chief factor product Fitting pair*.

The construction is dependent on the inherent choices only within equivalence of outer Fitting pairs.

2. ([AJBBPR00, Ex. IV]) Let G be a group. Let $\mathcal{E}_G$ denote the Fitting set composed of all subgroups of G. For each $T \in \mathcal{E}_G$, i.e. for each subgroup T of G, we consider a chief series Γ_T of T through $T_{\mathfrak{F}}$ and $T_{\mathfrak{G}}$. Let $\mathrm{D}_J(\Gamma_T)$ be the direct product of all the J-chief factors of T taken in the order of occurrence in Γ_T. We consider this group as the subgroup of $\mathrm{D}_J(n_G)$ consisting of the first direct components of $\mathrm{D}_J(n_G)$. T acts by conjugacy on $\mathrm{D}_J(\Gamma_T)$ and in trivial way on the rest of components of $\mathrm{D}_J(n_G)$. This action defines a homomorphism

$$d_T^{J,\mathfrak{F}/\mathfrak{G}} : T \longrightarrow \mathrm{A}_J(n_G).$$

Then the pair $\left(d^{J,\mathfrak{F}/\mathfrak{G}}, \mathrm{A}_J(n_G)\right)$ is an outer $\mathcal{E}_G$-Fitting sets pair relative to G. This is called the *chief factor product Fitting sets pair* relative to G.

The construction is dependent on the inherent choices only within equivalence of outer Fitting sets pairs.

Remark 7.4.21. With the above notation, if $\mathcal{F}$ is a Fitting set of $\mathbf{A}_J$, then $\mathfrak{F} = \mathcal{F}d^{-1}$ is a Fitting class defined by the chief factor product Fitting pair by [Pen87, Satz 3.2]. Then $\mathrm{Tr}_{\mathfrak{F}}(G)$ is the Fitting set of G defined by the chief factor product Fitting sets pair relative to G (see Theorem 7.4.17).

There exist Fitting sets associated with chief factor product Fitting sets pairs which cannot be obtained in this way.

Let G be a group and p a prime dividing $|G|$. Following the notation the above example, we take $J = C_p$, the cyclic group of order p, $\mathfrak{F} = \mathfrak{E}$ the class of all finite groups, and $\mathfrak{G} = (1)$, the trivial class. Let n_G be the natural number such that p^{n_G} is the order of a Sylow p-subgroup of G. Then $\mathrm{D}_J(n_G)$ is an elementary abelian p-group of order p^{n_G} and $\mathrm{A}_J(n_G) = \mathrm{GL}(n_G, p)$. Denote by $\left(d, \mathrm{GL}(n_G, p)\right)$ the chief factor product Fitting sets pair relative to G of Example 7.4.20 (2), that is $d = d^{C_p, \mathfrak{E}/(1)}$.

Let $\mathcal{F} = \{U \leq \mathrm{GL}(n_G, p) \colon U \leq \mathrm{Z}(\mathrm{GL}(n_G, p))\}$. Since $\mathrm{Z}(\mathrm{GL}(n_G, p))$ is a normal subgroup of $\mathrm{GL}(n_G, p)$, it is clear that $\mathcal{F}$ is a Fitting set of $\mathrm{GL}(n_G, p)$. By Theorem 7.4.17 we have that $\mathcal{F}_Z = \mathcal{F}d^{-1}$ is a Fitting set of G.

It is proved in [AJBBPR00, Ex. VI]) that there exist groups G for which $\mathcal{F}_Z$ is not the trace in G of any Fitting class. In particular, $\mathcal{F}_Z$ is not the trace in G of the Fitting class obtained by the inverse image of a Fitting set of $\mathbf{A}_{C_p}$ through the chief factor product Fitting pair.

We study the Fitting set $\mathcal{F}_Z$ in a group G. We assume that $n_G \neq 0$. For any subgroup $B \leq G$, write p^{n_B} the order of a Sylow p-subgroup of B. If $x \in B$, then

$$x^{d_B} = \begin{pmatrix} M(x) & 0 \\ 0 & I_{n_G - n_B} \end{pmatrix},$$

where $M(x) \in \mathrm{GL}(n_B, p)$ is the matrix of the action of x on the p-chief factors of a fixed chief series of B.

If $B \in \mathcal{F}_Z$, then $x^{d_B} = \lambda I_{n_G}$, for some non-zero scalar λ of $\mathrm{GF}(p)$. Hence, the p-chief factors of B are simple and all of them are B-isomorphic. In particular, $\mathrm{Ker}(d_B) = \mathrm{O}_{p',p}(B)$. Moreover, $B/\mathrm{Ker}(d_B)$ is a subgroup of $\mathrm{Z}\left(\mathrm{GL}(n_G, p)\right)$

and then it is isomorphic to a cyclic group of order dividing $p - 1$. If B does not contain any Sylow p-subgroup of G, then B is p-nilpotent; that is, $B = \mathrm{Ker}(d_B)$.

Note that all p-nilpotent subgroups of G are in $\mathcal{F}_Z$. If H is a subgroup of G, the order of a Sylow p-subgroup of H is denoted by $|H|_p$.

Lemma 7.4.22. *Let H be a subgroup of G. Assume that H is a p-soluble group of p-length at most 1. Then:*

1. *$H_{\mathcal{F}_Z}$ is the unique $\mathcal{F}_Z$-maximal subgroup of H containing $O_{p',p}(H)$; in particular $H_{\mathcal{F}_Z}$ is the unique $\mathcal{F}_Z$-injector of H.*
2. *If $|H|_p < p^{n_G}$, then $H_{\mathcal{F}_Z} = O_{p',p}(H)$.*
3. *If $|H|_p = p^{n_G}$, then $H_{\mathcal{F}_Z}$ is the set of all $m \in H$ such that m has scalar action on the direct product of the p-chief factors of H in a chief series of H.*

Proof.   1. Let M be an $\mathcal{F}_Z$-subgroup of H containing $O_{p',p}(H)$. We claim that M is normal in H, so that the conclusion is clear.

Since the p-length of H is smaller than or equal to 1, then $M/O_{p',p}(H)$ is a p'-group. Consequently the p-chief factors of H are completely reducible $\mathrm{GF}(p)M$-modules. Hence the direct product of the p-chief factors of H in a chief series of H, viewed as a $\mathrm{GF}(p)M$-module in the natural way, is $\mathrm{GF}(p)$ M-isomorphic to the direct product of the p-chief factors of M in a chief series of M. Since $M \in \mathcal{F}_Z$, then M has scalar action on the above mentioned direct product of the p-chief factors of H. Therefore $[M, H] \leq O_{p',p}(H) \leq M$. In particular M is normal in H.

2. If $|H|_p < p^{n_G}$, it is clear that $H_{\mathcal{F}_Z}$ is p-nilpotent and then $H_{\mathcal{F}_Z} = O_{p',p}(H)$.

3. Assume now that $|H|_p = p^{n_G}$. Denote by S the set of all $m \in H$ such that m has scalar action on the direct product of the p-chief factors of H in a chief series of H. It is clear that S is a normal subgroup of H containing $O_{p',p}(H)$. Note that the p-chief factors of H are completely reducible as $\mathrm{GF}(p)H_{\mathcal{F}_Z}$-modules and also as $\mathrm{GF}(p)S$-modules because $H_{\mathcal{F}_Z}$ and S are normal subgroups of H. Moreover, since $|H|_p = p^{n_G}$ we can easily deduce that $S \in \mathcal{F}_Z$ and also that $S = H_{\mathcal{F}_Z}$. $\square$

Recall that the class $\mathfrak{E}_{p'}\mathfrak{S}_p$ of all p-nilpotent groups is injective, and a group G possesses a unique conjugacy class of $\mathfrak{E}_{p'}\mathfrak{S}_p$-injectors if and only if G is p-constrained (see Corollary 7.2.31 and Remark 7.2.32). Moreover, in this case,

$$\mathrm{Inj}_{\mathfrak{E}_{p'}\mathfrak{S}_p}(G) = \{O_{p',p}(G)P : P \in \mathrm{Syl}_p(G)\},$$

and the p-nilpotent injectors of G are the p-nilpotent maximal subgroups of G containing $O_{p',p}(G)$.

Lemma 7.4.23. *Let H be a p-constrained subgroup of G such that $|H|_p = p^{n_G}$. Suppose that M is an $\mathcal{F}_Z$-maximal subgroup of H containing $O_{p',p}(H)$.*

1. *There exists a p-nilpotent injector I of H such that $I = O_{p',p}(M)$.*
2. *Moreover, M is the $\mathcal{F}_Z$-radical of $N_H(I)$ and is the set of all elements $m \in N_H(I)$ such that m has scalar action on the direct product of the p-chief factors of $N_H(I)$ in a chief series of $N_H(I)$.*

Proof. Suppose that $|M|_p < p^{n_G}$. In this case since $M \in \mathcal{F}_Z$, we have that M is a p-nilpotent group and then M is contained in a p-nilpotent injector, X say, of H, because $O_{p',p}(H) \leq M$. But clearly $X \in \mathcal{F}_Z$, which implies $X = M$. In particular $M = O_{p'}(H)H_p$ for some $H_p \in \mathrm{Syl}_p(H)$, which is a contradiction. Consequently there exists a Sylow p-subgroup H_p of H such that $O_{p',p}(H)H_p \leq O_{p',p}(M)$. But $I = O_{p',p}(H)H_p$ is a p-nilpotent injector of H, which implies that $I = O_{p',p}(M)$.

Observe that $I \leq M \leq N_H(I)$ and $I = O_{p',p}(N_H(I))$. Since $N_H(I)$ is a p-soluble group of p-length at most 1, the conclusion follows from Lemma 7.4.22. □

Theorem 7.4.24. *Let H be a p-constrained subgroup of G. Then H has a unique conjugacy class of $\mathcal{F}_Z$-injectors. Moreover, the $\mathcal{F}_Z$-injectors of H are exactly the $\mathcal{F}_Z$-maximal subgroups of H containing $O_{p',p}(H)$, or equivalently, the $\mathcal{F}_Z$-radical of H.*

Moreover, we have:

1. *If $|H|_p < p^{n_G}$, then the $\mathcal{F}_Z$-injectors of H are exactly the p-nilpotent injectors of H.*
2. *If $|H|_p = p^{n_G}$, then the set of $\mathcal{F}_Z$-injectors of H is exactly*

$$\mathrm{Inj}_{\mathcal{F}_Z}(G) = \left\{ \left(N_H(I)\right)_{\mathcal{F}_Z} : I \in \mathrm{Inj}_{\mathfrak{E}_{p'}\mathfrak{S}_p}(H) \right\}$$

In particular, the $\mathcal{F}_Z$-injectors of H are the subgroups composed of all elements $m \in N_H(I)$ such that m has scalar action on the direct product of the p-chief factors of $N_H(I)$ in a chief series of $N_H(I)$, where I is a p-nilpotent injector of H.

Proof. Note that if $|H|_p < p^{n_G}$, then the $\mathcal{F}_Z$-subgroups of H are exactly the p-nilpotent subgroups. On the other hand, if $|H|_p = p^{n_G}$, it is clear by Lemma 7.4.23 that the set of $\mathcal{F}_Z$-maximal subgroups of H containing $O_{p',p}(H)$ is exactly the set $\left\{ \left(N_H(I)\right)_{\mathcal{F}_Z} : I \in \mathrm{Inj}_{\mathfrak{E}_{p'}\mathfrak{S}_p}(H) \right\}$ which is a conjugacy class of subgroups of H. Since $O_{p',p}(H) \leq H_{\mathcal{F}_Z}$, we deduce that this set also coincides with the set of all $\mathcal{F}_Z$-maximal subgroups of H containing $H_{\mathcal{F}_Z}$.

Therefore the Fitting set $\mathcal{F}_Z$ is dominant in the set $\mathcal{X} = \{H \leq G : H$ is p-constrained$\}$. □

J. Pense ([Pen87, 4.14]) presented a type of Fitting classes, constructed by means of Fitting pairs, with respect to which every finite group has a unique conjugacy class of injectors. An improved version of this result is presented in [Pen90c]. We shall show in the sequel that Pense's result is actually a particular case of a more general one.

Definition 7.4.25. *Let G be a group and let S be a perfect comonolithic group whose head is isomorphic to a simple group J. Let L be the subgroup generated by all subnormal subgroups of G isomorphic to S*

$$L = \langle T : T \text{ is subnormal in } G \text{ and } T \cong S \rangle$$

and let

$$M = \langle \operatorname{Cosoc}(T) : T \text{ is subnormal in } G \text{ and } T \cong S \rangle$$

(which is a normal subgroup of L by Theorem 2.2.19). The factor group L/M is called the S-head-section of G.

By Theorem 2.2.19, $L = T_1 \cdots T_m$, where all T_i are normal subgroups of L and $T_i \cong S$. Note that if S a perfect comonolithic subnormal subgroup of a group which is the join of two subnormal subgroups S_1 and S_2, then either S is contained in S_1 or S is contained in S_2 ([Wie39]). This implies that $T_i \cap M = \operatorname{Cosoc}(T_i)$ and then $T_i M/M \cong J$. Hence L/M is a group in the Fitting class $\operatorname{Fit}(J)$ generated by J, i.e. L/M is isomorphic to a direct product of copies of J, by Example 2.2.3 (1).

Example 7.4.26. Let G be a group and let S be a perfect comonolithic group whose head is isomorphic to a simple group J. Let $\mathbf{D}_J(n_G)$, $\mathbf{A}_J(n_G)$, $\mathfrak{F}$, and $\mathfrak{G}$ be as in Example 7.4.20.

1. ([Pen88, Theorem III]) For any group G fix an embedding of the S-head-section of $G_{\mathfrak{F}}/G_{\mathfrak{G}}$ as the first components of $\mathbf{D}_J$. Then G operates on $\mathbf{D}_J$ via this embedding, and therefore we have a homomorphism

$$H_G^{S,\mathfrak{F}/\mathfrak{G}} : G \longrightarrow \mathbf{A}_J.$$

The pair $(H^{S,\mathfrak{F}/\mathfrak{G}}, \mathbf{A}_J)$ is an outer $\mathfrak{E}$-Fitting pair.

2. ([AJBBPR00, Ex. V])

For each subgroup T of the group G, we fix an embedding of the S-head-section of $T_{\mathfrak{F}}/T_{\mathfrak{G}}$ as the first components of $D_J(n_G)$. Then T operates on $D_J(n_G)$ via this embedding, and therefore we have a homomorphism

$$h_T^{S,\mathfrak{F}/\mathfrak{G}} : T \longrightarrow \mathbf{A}_J(n_G).$$

Denote by $\mathcal{E}_G$ the Fitting set of all subgroups of G. Thus the pair

$$\left(h^{S,\mathfrak{F}/\mathfrak{G}}, \mathbf{A}_J(n_G) \right)$$

is an outer $\mathcal{E}_G$-Fitting sets pair relative to G.

Let S be a perfect comonolithic group whose head is isomorphic to a non-abelian simple group J. Consider the Fitting classes $\mathfrak{F} = \mathfrak{E}$, the class of all finite groups, and $\mathfrak{G} = (1)$, the trivial class. Write $H^{S,\mathfrak{F}/\mathfrak{G}} = H^S$. Then it appears the outer $\mathfrak{E}$-Fitting pair, $(H^S, \mathbf{A}_J)$ say. Consider the projection from

$\mathbf{A}_J$ to $\mathbf{O}_J$ and let $(\tilde{H}^S, \mathbf{O}_J)$ be the induced outer Fitting pair from the pair $(H^S, \mathbf{A}_J)$.

Analogously, if we consider the projection from $\mathbf{A}_J(n_G)$ onto $\mathbf{O}_J(n_G) = \mathrm{Out}(\mathbf{D}_J(n_G))$ and let $(\tilde{h}^S, \mathbf{O}_J(n_G))$ be the induced outer Fitting sets pair relative to G from the pair $(h^S, \mathbf{A}_J(n_G))$.

Theorem 7.4.27. *With the notation introduced above, let $\mathcal{F}$ be a Fitting set of $\mathbf{O}_J(n_G)$ all whose elements are subgroups of the base group of $\mathbf{O}_J(n_G)$ and let $\mathcal{T} = \mathcal{F}(\tilde{h}^S)^{-1}$ be the Fitting set corresponding to the pair $(\tilde{h}^S, \mathbf{O}_J(n_G))$.*

If $\mathrm{Out}(J)$ is soluble, then each subgroup of G has exactly a conjugacy class of $\mathcal{T}$-injectors.

Proof. Note that for every subgroup B of $\mathbf{O}_J(n_G)$, the $\mathcal{F}$-injectors of $B \cap \mathrm{Out}(J)^\natural$, where $\mathrm{Out}(J)^\natural$ is the base group of $\mathbf{O}_J(n_G)$, are exactly the $\mathcal{F}$-injectors of B. Therefore each subgroup of $\mathbf{O}_J(n_G)$ possesses a single conjugacy class of $\mathcal{F}$-injectors by Theorem 2.4.26. Then it is enough to show that the pair $(\tilde{h}^S, \mathbf{O}_J(n_G))$ satisfies the property (7.2) of Theorem 7.4.19.

Write $f = \tilde{h}^S$. Let $\nu\colon V \longrightarrow U$ be a G-embedding between subgroups U and V of G such that $U_\mathcal{T} \leq V^\nu$. We consider L_U/M_U and L_{V^ν}/M_{V^ν} the S-head-section of U and V^ν respectively. It is clear that L_U/M_U is the S-head-section of L_U and so $L_U^{f_{L_U}} = 1 \in \mathcal{F}$. Then $L_U \in \mathcal{F}f^{-1} = \mathcal{T}$ and $L_U \leq U_\mathcal{T}$ and so also $L_U \leq V^\nu$. This implies that $L_U \leq L_{V^\nu}$. Now suppose that there exists a subnormal subgroup X of V^ν such that $X \cong S$ and X is not subnormal in U. Then, for any subnormal subgroup T of U such that $T \cong S$, we have that X and T are normal in XT, by Theorem 2.2.19, and then $[X, T] \leq \mathrm{Cosoc}(T)$. Hence $[X, L_U] \leq M_U$. Therefore $X \leq \mathrm{C}_{V^\nu}(L_U/M_U) \leq \mathrm{C}_U(L_U/M_U)$. Since $\mathrm{C}_U(L_U/M_U) \leq \mathrm{Ker}(f_U) \leq U_\mathcal{T} \leq V^\nu$, it follows that X is subnormal in $\mathrm{C}_U(L_U/M_U)$ and also is in U, contrary to supposition.

Therefore the S-head-section of V^ν coincides with the S-head-section of U and then it is conjugate to the S-section of V. By construction of the Fitting sets pair, it follows that there exists $\eta \in \mathrm{Inn}(\mathbf{O}_J(n_G))$, such that $\nu f_U = f_V \eta$. $\qquad \square$

Now we deduce the aforesaid result of J. Pense.

Theorem 7.4.28 ([Pen90c]). *Let S be a perfect comonolithic group with head J. Consider the outer Fitting pair $(\tilde{H}^S, \mathbf{O}_J)$. Let $\mathcal{F}$ be a Fitting set in the base group of $\mathbf{O}_J$ and let $\mathfrak{F} = \mathcal{F}(\tilde{H}^S)^{-1}$ be the corresponding Fitting class. If the outer automorphism group of J is soluble, then every finite group has exactly a conjugacy class of $\mathfrak{F}$-injectors.*

Proof. First of all, note that $\mathbf{A}_J = \lim_{n\to\infty}\big(\mathrm{Aut}(\overbrace{J \times \cdots \times J}^{(n\ \mathrm{copies})})\big)$ and so $\mathbf{A}_J$ is the (restricted, natural) wreath product $\lim_{n\to\infty}\big(\mathrm{Aut}(J) \wr_{\mathrm{nat}} S_n\big)$ with base group $\mathrm{Aut}(J)^\natural$. Then $\mathbf{O}_J$ is $\mathbf{A}_J/\mathrm{Inn}(J)^\natural$ with base group $\mathrm{Out}(J)^\natural$.

For each group G we consider $O_J(n_G)$ as a subgroup of $\mathbf{O}_J$. With respect to the outer $\mathcal{E}_G$-Fitting sets pair relative to G, $\left(\tilde{h}^S, O_J(n_G)\right)$ and for each subgroup T of G, we have

$$(t)^{\tilde{h}_T^S} = (t)^{\tilde{H}_T^S} \in O_J(n_G) \leq \mathbf{O}_J \qquad \text{for every } t \in T.$$

Therefore it follows that $\mathrm{Tr}_G(\mathfrak{F}) = \left(\mathrm{Tr}_{O_J(n_G)}(\mathcal{F})\right)(\tilde{H}^S)^{-1}$. Applying Theorem 7.4.27, G has a conjugacy class of $\mathfrak{F}$-injectors. $\qquad\square$

Recall finally Schreier's conjecture, whose validity has been proved using the classification of finite simple groups, which states that the group $\mathrm{Out}(J)$, of all outer automorphisms of a non-abelian simple group J, is always soluble (see [KS04, page 151]).

References

[ABB02]   M. J. Alejandre and A. Ballester-Bolinches. On a theorem of Berkovich. *Israel J. Math.*, 131:149–156, 2002.

[AJBBPR00] P. Arroyo-Jordá, A. Ballester-Bolinches, and M. D. Pérez-Ramos. Fitting sets pairs. *J. Algebra*, 231:574–588, 2000.

[AJPR01]   M. Arroyo-Jordá and M. D. Pérez-Ramos. On the lattice of $\mathfrak{F}$-Dnormal subgroups in finite soluble groups. *J. Algebra*, 242:198–212, 2001.

[AJPR04a] M. Arroyo-Jordá and M. D. Pérez-Ramos. Fitting classes and lattice formations. I. *J. Aust. Math. Soc.*, 76(1):93–108, 2004.

[AJPR04b] M. Arroyo-Jordá and M. D. Pérez-Ramos. Fitting classes and lattice formations. II. *J. Aust. Math. Soc.*, 76(2):175–188, 2004.

[And75]   W. Anderson. Injectors in finite solvable groups. *J. Algebra*, 36:333–338, 1975.

[AS85]   M. Aschbacher and L. Scott. Maximal subgroups of finite groups. *J. Algebra*, 92:44–80, 1985.

[AS89]   M. Asaad and A. Shaalan. On the supersolvability of finite groups. *Arch. Math. (Basel)*, 53(4):318–326, 1989.

[Bae57]   R. Baer. Classes of finite groups and their properties. *Illinois J. Math.*, 1:115–187, 1957.

[Bar72]   D. W. Barnes. On complemented chief factors of finite soluble groups. *Bull. Austral. Math. Soc.*, 7:101–104, 1972.

[Bar77]   D. Bartels. Subnormality and invariant relations on conjugacy classes in finite groups. *Math. Z.*, 157:13–17, 1977.

[BB74]   J. Beidleman and B. Brewster. $\mathfrak{F}$-normalizers in finite π-solvable groups. *Boll. Un. Mat. Ital. (4)*, 10:14–27, 1974.

[BB76]   D. Blessenohl and B. Brewster. Über Formationen und komplementierbare Hauptfaktoren. *Arch. Math.*, 27:347–351, 1976.

[BB89a]   A. Ballester-Bolinches. $\mathfrak{H}$-normalizers ahd local definitions of saturated formations of finite groups. *Israel J. Math.*, 67:312–326, 1989.

[BB89b]   A. Ballester-Bolinches. *Normalizadores y subgrupos de prefrattini en grupos finitos*. PhD thesis, Facultat de Matemàtiques, Universitat de València, 1989.

[BB91]   A. Ballester-Bolinches. Remarks on formations. *Israel J. Math.*, 73(1):97–106, 1991.

[BB92]   A. Ballester-Bolinches. A note on saturated formations. *Arch. Math. (Basel)*, 58(2):110–113, 1992.

[BB05]   A. Ballester-Bolinches. $\mathfrak{F}$-critical groups, $\mathfrak{F}$-subnormal subgroups and the generalised Wielandt property for residuals. Preprint, 2005.

[BBCE01]   A. Ballester-Bolinches, J. Cossey, and L. M. Ezquerro. On formations of finite groups with the Wielandt property for residuals. *J. Algebra*, 243(2):717–737, 2001.

[BBCER03]   A. Ballester-Bolinches, C. Calvo, and R. Esteban-Romero. A question from the Kourovka Notebook on formation products. *Bull. Austral. Math. Soc.*, 68(3):461–470, 2003.

[BBCER05]   A. Ballester-Bolinches, C. Calvo, and R. Esteban-Romero. $\mathfrak{X}$-saturated formations of finite groups. *Comm. Algebra*, 33(4):1053–1064, 2005.

[BBCER06]   A. Ballester-Bolinches, C. Calvo, and R. Esteban-Romero. Products of formations of finite groups. To appear in *J. Algebra*, 2006.

[BBCS05]   A. Ballester-Bolinches, C. Calvo, and L. A. Shemetkov. On partially saturated formations. Preprint, 2005.

[BBDPR92]   A. Ballester-Bolinches, K. Doerk, and M. D. Pérez-Ramos. On the lattice of $\mathfrak{F}$-subnormal subgroups. *J. Algebra*, 148(1):42–52, 1992.

[BBDPR95]   A. Ballester-Bolinches, K. Doerk, and M. D. Pérez-Ramos. On $\mathfrak{F}$-normal subgroups of finite soluble groups. *J. Algebra*, 171(1):189–203, 1995.

[BBE91]   A. Ballester-Bolinches and L. M. Ezquerro. On maximal subgroups of finite groups. *Comm. Algebra*, 19(8):2373–2394, 1991.

[BBE95]   A. Ballester-Bolinches and L. M. Ezquerro. The Jordan-Hölder theorem and pre-Frattini subgroups of finite groups. *Glasgow Math. J.*, 37:265–277, 1995.

[BBE98]   A. Ballester-Bolinches and L. M. Ezquerro. On a theorem of Bryce and Cossey. *J. Austral. Math. Soc. Ser. A*, 57:455–460, 1998.

[BBE05]   A. Ballester-Bolinches and L. M. Ezquerro. On formations with the Kegel property. *J. Group Theory*, 8(5):605–611, 2005.

[BBEPA02]   A. Ballester-Bolinches, L. M. Ezquerro, and M. C. Pedraza-Aguilera. A characterization of the class of finite groups with nilpotent derived subgroup. *Math. Nachr.*, 239–240:5–10, 2002.

[BBERR05]   A. Ballester-Bolinches, R. Esteban-Romero, and D. J. S. Robinson. On finite minimal non-nilpotent groups. *Proc. Amer. Math. Soc.*, 133(12):3455–3462, 2005.

[BBERss]   A. Ballester-Bolinches and R. Esteban-Romero. On minimal non-supersoluble groups. *Rev. Mat. Iberoamericana*, in press.

[BBH70]   R. M. Bryant, R. A. Bryce, and B. Hartley. The formation generated by a finite group. *Bull. Austral. Math. Soc.*, 2:347–357, 1970.

[BBMPPR00] A. Ballester-Bolinches, A. Martínez-Pastor, and M. D. Pérez-Ramos. Nilpotent-like Fitting formations of finite soluble groups. *Bull. Austral. Math. Soc.*, 62(3):427–433, 2000.

[BBPA96]   A. Ballester-Bolinches and M. C. Pedraza-Aguilera. On minimal subgroups of finite groups. *Acta Math. Hungar.*, 73(4):335–342, 1996.

[BBPAMP00] A. Ballester-Bolinches, M. C. Pedraza-Aguilera, and A. Martínez-Pastor. Finite trifactorized groups and formations. *J. Algebra*, 226:990–1000, 2000.

[BBPAPR96] A. Ballester-Bolinches, M. C. Pedraza-Aguilera, and M. D. Pérez-Ramos. On 𝔉-subnormal subgroups and 𝔉-residuals of finite groups. *J. Algebra*, 186(1):314–322, 1996.

[BBPR90] A. Ballester-Bolinches and M. D. Pérez-Ramos. On 𝔉-subnormal subgroups. *Supl. Rend. Circ. Mat. Palermo (2)*, 23:25–28, 1990.

[BBPR91] A. Ballester-Bolinches and M. D. Pérez-Ramos. 𝔉-subnormal closure. *J. Algebra*, 138(1):91–98, 1991.

[BBPR94a] A. Ballester-Bolinches and M. D. Pérez-Ramos. A note on the 𝔉-length of maximal subgroups in finite soluble groups. *Math. Nachr.*, 166:67–70, 1994.

[BBPR94b] A. Ballester-Bolinches and M. D. Pérez-Ramos. On 𝔉-subnormal subgroups and Frattini-like subgroups of a finite group. *Glasgow Math. J.*, 36(2):241–247, 1994.

[BBPR95] A. Ballester-Bolinches and M. D. Pérez-Ramos. On 𝔉-critical groups. *J. Algebra*, 174(3):948–958, 1995.

[BBPR96a] A. Ballester-Bolinches and M. D. Pérez-Ramos. A question of R. Maier concerning formations. *J. Algebra*, 182(3):738–747, 1996.

[BBPR96b] A. Ballester-Bolinches and M. D. Pérez-Ramos. Two questions of L. A. Shemetkov on critical groups. *J. Algebra*, 179(3):905–917, 1996.

[BBPR98] A. Ballester-Bolinches and M. D. Pérez-Ramos. Some questions of the Kourovka notebook concerning formation products. *Comm. Algebra*, 26(5):1581–1587, 1998.

[BBS97] A. Ballester-Bolinches and L. A. Shemetkov. On lattices of p-local formations of finite groups. *Math. Nachr.*, 186:57–65, 1997.

[BC72] R. A. Bryce and J. Cossey. Fitting formations of finite soluble groups. *Math. Z.*, 127:217–223, 1972.

[BC75] R. A. Bryce and John Cossey. A problem in the theory of normal Fitting classes. *Math. Z.*, 141:99–110, 1975.

[BC78] T. R. Berger and J. Cossey. More Fitting formations. *J. Algebra*, 51: 573–578, 1978.

[BC82] R. A. Bryce and J. Cossey. Subgroup-closed Fitting classes are formations. *Math. Proc. Cambridge Philos. Soc.*, 91:225–258, 1982.

[BCMV84] A. Bolado-Caballero and J. R. Martínez-Verduch. The Fitting class 𝔉𝔖. *Arch. Math. (Basel)*, 42:307–310, 1984.

[Bec64] H. Bechtell. Pseudo-Frattini subgroups. *Pacific J. Math.*, 14:1129–1136, 1964.

[Ben70] H. Bender. On groups with abelian sylow 2-subgroups. *Math. Z.*, 117:164–176, 1970.

[Ber99] Y. Berkovich. Some corollaries to Frobenius' normal p-complement theorem. *Proc. Amer. Math. Soc.*, 127:2505–2509, 1999.

[BG70] D. Blessenohl and W. Gaschütz. Über normale Schunck- und Fittingklassen. *Math. Z.*, 118:1–8, 1970.

[BH03] J. C. Beidleman and H. Heineken. On the Fitting core of a formation. *Bull. Austral. Math. Soc.*, 68(1):107–112, 2003.

[Bir69] G. Birkhoff. *Lattice Theory*, volume 25 of *Amer. Math. Soc. Colloquium Pub.* Amer. Math. Soc. Providence, RI, USA, 1969.

[BK66] D. W. Barnes and O. H. Kegel. Gaschütz functors on finite soluble groups. *Math. Z.*, 94:134–142, 1966.

358   References

[BL79]   D. Blessenohl and H. Laue. Fittingklassen endlicher Gruppen, in
            denen gewisse Hauptfaktoren einfach sind. *J. Algebra*, 56:516–532,
            1979.

[Bra88]   A. Brandis. Moduln und verschränkte Homomorphismen endlicher
            Gruppen. *J. Reine Angew. Math.*, 385:102–116, 1988.

[Cam81]   P. J. Cameron. Finite permutation groups and finite simple groups.
            Bull. London Math. Soc., 13:1–22, 1981.

[Car61]   R. Carter. Nilpotent self-normalizing subgroups of soluble groups.
            Math. Z., 75:136–139, 1960/1961.

[CCN⁺85]   J. H. Conway, R. T. Curtis, S. P. Norton, R. A. Parker, and R. A.
            Wilson. *Atlas of Finite Groups*. Oxford Univ. Press, London, 1985.

[CFH68]   R. W. Carter, B. Fischer, and T. O. Hawkes. Extreme classes of finite
            soluble groups. *J. Algebra*, 9:285–313, 1968.

[CH67]   R. Carter and T. Hawkes. The $\mathcal{F}$-normalizers of a finite soluble group.
            J. Algebra, 5:175–202, 1967.

[Cha72]   G. A. Chambers. On f-prefrattini subgroups. *Canad. Math. Bull.*,
            15(3):345–348, 1972.

[CK87]   J. Cossey and C. Kanes. A construction for Fitting formations.
            J. Algebra, 107:117–133, 1987.

[CM98]   A. Carocca and M. Maier. Hypercentral embedding and pronormality.
            Arch. Math. (Basel), 71:433–436, 1998.

[CO87]   J. Cossey and E. A. Ormerod. A construction for Fitting-Schunck
            classes. *J. Austral. Math. Soc. Ser. A*, 43:91–94, 1987.

[COM71]   J. Cossey and S. Oates-MacDonald. On the definition of saturated
            formations of groups. *Bull. Austral. Math. Soc.*, 4:9–15, 1971.

[Cos89]   J. Cossey. A construction for Fitting formations II. *J. Austral. Math.
            Soc. Ser. A*, 47:95–102, 1989.

[Dar72]   R. S. Dark. Some examples in the theory of injectors of finite soluble
            groups. *Math. Z.*, 127:145–156, 1972.

[Des59]   W. E. Deskins. On maximal subgroups. *Proc. Symp. in Pure Math.
            Amer. Math. Soc.*, 1:100–104, 1959.

[DF]   R. Dark and A. Feldman. A characterization of injectors in finite
            groups. Private communication.

[DH78]   K. Doerk and T. O. Hawkes. On the residual of a direct product.
            Arch. Math., 30:458–468, 1978.

[DH92]   K. Doerk and T. Hawkes. *Finite Soluble Groups*. Number 4 in De
            Gruyter Expositions in Mathematics. Walter de Gruyter, Berlin, New
            York, 1992.

[Doe66]   K. Doerk. Minimal nicht überauflösbare, endliche Gruppen. *Math.
            Z.*, 91:198–205, 1966.

[Doe71]   K. Doerk. *Über Homomorphe und Formationen endlicher auflösbarer
            Gruppen*. Habilitationsschrift, Johannes Gutenberg-Universität Mainz,
            Mainz, 1971.

[Doe73]   K. Doerk. Die maximale lokale Erklärung einer gesättigten Formation.
            Math. Z., 133:133–135, 1973.

[Doe74]   K. Doerk. Über Homomorphe endlicher auflösbarer Gruppen. *J.
            Algebra*, 30:12–30, 1974.

[Eri82]   R. P. Erickson. Projectors of finite groups. *Comm. Alg.*, 10:1919–1938,
            1982.

[ESE]   L. M. Ezquerro and X. Soler-Escrivà. On certain distributive lattices
            of subgroups of finite soluble groups. To appear in *Acta Math. Sinica.*
[ESE05]   L. M. Ezquerro and X. Soler-Escrivà. Some new permutability prop-
            erties of hypercentrally embedded groups. *J. Austral. Math. Soc. Ser.
            A*, 79(2):243–255, 2005.
[Ezq86]   L. M. Ezquerro. On generalized covering subgroups and normalizers
            of finite soluble groups. *Arch. Math.*, 47:385–394, 1986.
[FGH67]   B. Fischer, W. Gaschütz, and B. Hartley. Injektoren endlichen
            auflösbarer Gruppen. *Math. Z.*, 102:337–339, 1967.
[Fis66]   B. Fischer. *Klassen konjugierter und Untergurppen in endlichen
            auflösbarer Gruppen.* Habilitationsschrift, Universität Frankfurt am
            Mainz, Frankfurt, 1966.
[För78]   P. Förster. Charakterisierungen einiger Schunckklassen endlicher
            auflösbarer Gruppen. *J. Algebra*, 55:155–187, 1978.
[För79]   P. Förster. Closure operations for Schunck classes and formations of
            finite solvable groups. *Math. Proc. Cambridge Philos. Soc.*, 85(2):253–
            259, 1979.
[För82]   P. Förster. Homomorphs and wreath product extensions. *Math. Proc.
            Cambridge Philos. Soc.*, 92(1):93–99, 1982.
[För83]   P. Förster. Prefrattini subgroups. *J. Austral. Math. Soc. (Ser. A)*,
            34:234–247, 1983.
[För84a]   P. Förster. A note on primitive groups with small maximal subgroups.
            Publ. Sec. Mat. Univ. Autònoma Barcelona, 28(2-3):19–27, 1984.
[För84b]   P. Förster. Projektive Klassen endlicher Gruppen. I: Schunck- und
            Gaschützklassen. *Math. Z.*, 186:149–178, 1984.
[För85a]   P. Förster. Nilpotent injectors in finite groups. *Bull. Austral. Math.
            Soc.*, 32:293–298, 1985.
[För85b]   P. Förster. Projektive Klassen endlicher Gruppen. IIa. Gesättigte
            Formationen: ein allgemeiner Satz von Gaschütz-Lubeseder-Baer-
            Typ. *Publ. Sec. Mat. Univ. Autònoma Barcelona*, 29(2-3):39–76, 1985.
[För85c]   P. Förster. Projektive Klassen endlicher Gruppen. IIb. Gesättigte
            Formationen: Projektoren. *Arch. Math. (Basel)*, 44(3):193–209, 1985.
[För87]   P. Förster. Maximal quasinilpotent subgroups and injectors for Fitting
            classes in finite groups. *Southeast Asian Bull. Math.*, 11:1–11, 1987.
[För88]   P. Förster. Chief factors, crowns, and the generalised Jordan-Hölder
            theorem. *Comm. Algebra*, 16(8):1627–1638, 1988.
[För89]   P. Förster. An elementary proof of Lubeseder's theorem. *Arch. Math.
            (Basel)*, 52(5):417–419, 1989.
[Föra]   P. Förster. Projectors of Soluble type in Finite Groups. Preprint.
[Förb]   P. Förster. Salomon Subgroups in Finite Groups. Preprint.
[FS85]   P. Förster and E. Salomon. Local definitions of local homomorphs
            and formations of finite groups. *Bull. Austral. Math. Soc.*, 31(1):5–34,
            1985.
[FT63]   W. Feit and J. G. Thompson. Solvability of groups of odd order.
            Pacific J. Math., 13:775–1029, 1963.
[Gaj79]   D. Gajendragadkar. A characteristic class of characters of finite
            π-separable groups. *J. Algebra*, 59:237–259, 1979.
[Gas62]   W. Gaschütz. Praefrattinigruppen. *Arch. Math.*, 13:418–426, 1962.
[Gas63]   W. Gaschütz. Zur Theorie der endlichen auflösbaren Gruppen. *Math.
            Z.*, 80:300–305, 1963.

360   References

[Gas69]   W. Gaschütz. Selected topics in the theory of soluble groups. Canberra, 1969. Lectures given at the 9th Summer Research Institute of the Australian Math. Soc. Notes by J. Looker.

[GK84]   F. Gross and L. G. Kovács. On normal subgroups which are direct products. *J. Algebra*, 90:133–168, 1984.

[GL63]   W. Gaschütz and U. Lubeseder. Kennzeichnung gesättigter Formationen. *Math. Z.*, 82:198–199, 1963.

[Gla66]   G. Glauberman. On the automorphism groups of a finite group having no non-identity normal subgroups of odd order. *Math. Z.*, 93:154–160, 1966.

[Gor80]   D. Gorenstein. *Finite Groups*. Chelsea Pub. Co., New York, 1980.

[GS78]   R. L. Griess and P. Schmid. The Frattini module. *Arch. Math.*, 30: 256–266, 1978.

[Hal28]   P. Hall. A note of soluble groups. *J. London Math. Soc.*, 3:98–105, 1928.

[Hal37]   P. Hall. On the system normalizers of a soluble group. *Proc. London Math. Soc.*, 43:507–525, 1937.

[Hal59]   P. Hall. On the finiteness of certain soluble groups. *Proc. London Math. Soc. (3)*, 9:595–622, 1959.

[Hal63]   P. Hall. On non-strictly simple groups. *Proc. Cambridge Philos. Soc.*, 59:531–553, 1963.

[Har72]   M. E. Harris. On normal subgroups of *p*-solvable groups. *Math. Z.*, 129:55, 1972.

[Haw67]   T. Hawkes. Analogues of Prefrattini subgroups. In *Proc. Internat. Conf. Theory of Groups (Canberra, 1965)*, pages 145–150. Gordon and Breach, New York, 1967.

[Haw69]   T. Hawkes. On formation subgroups of a finite soluble group. *J. London Math. Soc.*, 44:243–250, 1969.

[Haw70]   T. O. Hawkes. On Fitting formations. *Math. Z.*, 117:177–182, 1970.

[Haw73]   T. Hawkes. Closure operations for Schunck classes. *J. Austral. Math. Soc. Ser. A*, 16:316–318, 1973.

[Haw75]   T. O. Hawkes. Two applications of twisted wreath products to finite soluble groups. *Trans. Amer. Math. Soc.*, 214:325–335, 1975.

[Haw98]   I. Hawthorn. The existence and uniqueness of injectors for Fitting sets of solvable groups. *Proc. Amer. Math. Soc.*, 126:2229–2230, 1998.

[HB82a]   B. Huppert and N. Blackburn. *Finite Groups II*, volume 242 of *Grundlehren der Mathematischen Wissenschaften*. Springer-Verlag, Berlin-Heidelberg-New York, 1982.

[HB82b]   B. Huppert and N. Blackburn. *Finite groups. III*, volume 243 of *Grundlehren der Mathematischen Wissenschaften [Fundamental Principles of Mathematical Sciences]*. Springer-Verlag, Berlin, 1982.

[Hei94]   H. Heineken. Fitting classes of certain metanilpotent groups. *Glasgow Math. J.*, 36(2):185–195, 1994.

[Hei97]   H. Heineken. More metanilpotent Fitting classes with bounded chief factor ranks. *Rend. Sem. Mat. Univ. Padova*, 98:241–251, 1997.

[HH84]   K. L. Haberl and H. Heineken. Fitting classes defined by chief factor ranks. *J. London Math. Soc.*, 29:34–40, 1984.

[Höl89]   O. Hölder. Zurückführung einer beliebigen algebraischen Gleichung auf eine Kette von Gleichungen. *Math. Ann.*, pages 26–56, 1889.

[Hun80]   T. W. Hungerford. *Algebra*, volume 73 of *Graduate Texts in Mathematics*. Springer-Verlag, New York, 1980. Reprint of the 1974 original.

[Hup67]   B. Huppert. *Endliche Gruppen I*. Springer-Verlag, Berlin, Heidelberg, New-York, 1967.

[ILPM03]   M. J. Iranzo, Julio P. Lafuente, and F. Pérez-Monasor. Preboundaries of perfect groups. *J. Group Theory*, 6(1):57–68, 2003.

[ILPM04]   M. J. Iranzo, J. P. Lafuente, and F. Pérez-Monasor. Preboundaries of perfect groups II. *J. Group Theory*, 7:113–125, 2004.

[IMPM01]   M. J. Iranzo, J. Medina, and F. Pérez-Monasor. On p-decomposable groups. *Siberian Math. J.*, 42:59–63, 2001.

[IPM86]   M. J. Iranzo and F. Pérez-Monasor. Fitting classes $\mathfrak{F}$ such that all finite groups have $\mathfrak{F}$-injectors. *Arch. Math. (Basel)*, 46:205–210, 1986.

[IPM88]   M. J. Iranzo and F. Pérez-Monasor. Existencia de inyectores en grupos finitos respecto de ciertas clases de Fitting. *Publ. Mat. Univ. Autònoma Barcelona*, 32:57–59, 1988.

[IPMT90]   M. J. Iranzo, F. Pérez-Monasor, and M. Torres. A criterion for the existence of injectors in finite groups. *Supl. Rend. Circ. Mat. Palermo (2)*, 23:193–196, 1990.

[Isa84]   I. M. Isaacs. Characters of π-separable groups. *J. Algebra*, 86:98–128, 1984.

[IT89]   M. J. Iranzo and M. Torres. The p^*p-injectors of a finite group. *Rend. Sem. Mat. Uni. Padova*, 82:233–237, 1989.

[Jor70]   C. Jordan. *Traité des substitutions et des équations algébriques*. Gauthier-Villars, Paris, 1870.

[JS96]   P. Jiménez Seral. El teorema de O'Nan-Scott. Ph. D. course, 1996–97, Universidad de Zaragoza, 1996.

[Kam92]   S. F. Kamornikov. On a problem of Kegel. *Mat. Zametki*, 51(5):51–56, 157, 1992.

[Kam93]   S. F. Kamornikov. On some properties of the formation of quasinilpotent groups. *Mat. Zametki*, 53(2):71–77, 1993.

[Kam94]   S. F. Kamornikov. On two problems of L. A. Shemetkov. *Sibirsk. Mat. Zh.*, 35(4):801–812, ii, 1994. Russian. Translation in *Siberian Math. J.*, **35**, no. 4 (1994), pages 713–721.

[Kam96]   S. F. Kamornikov. Permutability of subgroups and $\mathfrak{F}$-subnormality. *Siberian Math. J.*, 37(5):936–949, 1996.

[Kat77]   U. Kattwinkel. Die größte untergruppenabgeschlossene Teilklasse einer Schunckklasse endlicher auflösbarer Gruppen. *Arch. Math.*, 29:337–343, 1977.

[Keg65]   O. H. Kegel. Zur Struktur mehrfach faktorisierter endlicher Gruppen. *Math. Z.*, 87:42–48, 1965.

[Keg78]   O. H. Kegel. Untergruppenverbände endlicher Gruppen, die den Subnormalteilerverband echt enthalten. *Arch. Math. (Basel)*, 30(3):225–228, 1978.

[KL90]   P. B. Kleidman and M. W. Liebeck. *The subgroup structure of the finite classical groups*, volume 129 of *London Math. Soc. Lecture Notes Series*. Cambridge Univ. Press, Cambridge, UK, 1990.

[Kli77]   A. A. Klimowicz. $\mathfrak{X}$-prefrattini subgroups of π-soluble groups. *Arch. Math. (Basel)*, 28:572–576, 1977.

[Kov86]   L. G. Kovács. Maximal subgroups in composite finite groups. *J. Algebra*, 99(1):114–131, 1986.

362   References

[Kov88]   L. G. Kovács. Primitive permutation groups of simple diagonal type.
            Israel J. Math., 63:119–127, 1988.
[Kov89]   L. G. Kovács. Primitive subgroups of wreath products in product
            action. *Proc. London Math. Soc. (3)*, 58:306–332, 1989.
[KS95]   S. F. Kamornikov and L. A. Shemetkov. On coradicals of subnormal
            subgroups. *Algebra i Logika*, 34(5):493–513, 608, 1995.
[KS03]   S. F. Kamornikov and M. V. Sel'kin. *Subgroup functors and classes
            of finite groups.* Belaruskaya Nauka, Minsk, 2003.
[KS04]   H. Kurzweil and B. Stellmacher. *The theory of finite groups. An
            introduction.* Springer-Verlag, Berlin-Heidelberg-New York, 2004.
[Kur89]   H. Kurzweil. Die Praefrattinigruppe im Intervall eines Untergruppen-
            verbandes. *Arch. Math. (Basel)*, 53(3):235–244, 1989.
[Laf78]   J. Lafuente. Homomorphs and formations of a given derived class.
            Math. Proc. Cambridge Philos. Soc., 84:437–441, 1978.
[Laf84a]   J. Lafuente. Nonabelian crowns and Schunck classes of finite groups.
            Arch. Math. (Basel), 42(1):32–39, 1984.
[Laf84b]   J. Lafuente. On restricted twisted wreath products of groups. *Arch.
            Math. (Basel)*, 43(3):208–209, 1984.
[Laf89]   J. Lafuente. Maximal subgroups and the Jordan-Hölder theorem.
            J. Austral. Math. Soc. Ser. A, 46(3):356–364, 1989.
[Lau73]   H. Lausch. On normal Fitting classes. *Math. Z.*, 130:67–72, 1973.
[Loc71]   P. Lockett. *On the theory of Fitting classes of finite soluble groups.*
            PhD thesis, University of Warwick, 1971.
[LPS88]   M. W. Liebeck, C. E. Praeger, and J. Saxl. On the O'Nan-Scott
            theorem for finite primitive permutation groups. *J. Austral. Math.
            Soc. (Ser. A)*, 44:389–396, 1988.
[LS87]   J. C. Lennox and S. E. Stonehewer. *Subnormal Subgroups of Groups.*
            Clarendon Press, Oxford, 1987.
[LS91]   M. W. Liebeck and J. Saxl. On point stabilizers in primitive permuta-
            tion groups. *Comm. Algebra*, 19(10):2777–2789, 1991.
[Lub63]   U. Lubeseder. *Formationsbildungen in endlichen auflösbaren Grup-
            pen.* Dissertation, Universität Kiel, Kiel, 1963.
[Mak70]   A. Makan. Another characteristic conjugacy class of subgroups of
            finite soluble groups. *J. Austral. Math. Soc. Ser. A*, 11:395–400, 1970.
[Mak73]   A. Makan. On certain sublattices of the lattice of subgroups gen-
            erated by the prefrattini subgroups, the injectos and the formation
            subgroups. *Canad. J. Math. Soc.*, 25:862–869, 1973.
[Man70]   A. Mann. $\mathfrak{H}$-normalizers of a finite soluble group. *J. Algebra*, 14:312–
            325, 1970.
[Man71]   A. Mann. Injectors and normal subgroups of finite groups. *Israel
            J. Math.*, 56:554–558, 1971.
[Men94]   M. Menth. Examples of supersoluble Lockett sections. *Bull. Austral.
            Math. Soc.*, 94:325–332, 1994.
[Men95a]   M. Menth. Closure properties of supersoluble Fitting classes. In
            Groups '93 Galway/St. Andrews, Vol. 2, volume 212 of *London Math.
            Soc. Lecture Note Ser.*, pages 418–425. Cambridge Univ. Press, Cam-
            bridge, 1995.
[Men95b]   M. Menth. A family of Fitting classes of supersoluble groups. *Math.
            Proc. Cambridge Philos. Soc.*, 118(1):49–57, 1995.

[Men96]   M. Menth. A note on Hall closure of metanilpotent Fitting classes. *Bull. Austral. Math. Soc.*, 53(2):209–212, 1996.

[MK84]   V. D. Mazurov and E. I. Khukhro, editors. *Unsolved problems in Group Theory: The Kourovka Notebook.* Institute of Mathematics, Sov. Akad., Nauk SSSR, Siberian Branch, Novosibirsk, SSSR, 9 edition, 1984.

[MK90]   V. D. Mazurov and E. I. Khukhro, editors. *Unsolved problems in Group Theory: The Kourovka Notebook.* Institute of Mathematics, Sov. Akad., Nauk SSSR, Siberian Branch, Novosibirsk, SSSR, 11 edition, 1990.

[MK92]   V. D. Mazurov and E. I. Khukhro, editors. *Unsolved problems in Group Theory: The Kourovka Notebook.* Institute of Mathematics, Sov. Akad., Nauk SSSR, Siberian Branch, Novosibirsk, SSSR, 12 edition, 1992.

[MK99]   V. D. Mazurov and E. I. Khukhro, editors. *Unsolved problems in Group Theory: The Kourovka Notebook.* Institute of Mathematics, Sov. Akad., Nauk SSSR, Siberian Branch, Novosibirsk, SSSR, 14 edition, 1999.

[MP92]   A. Martínez-Pastor. *Classes inyectivas de grupos finitos.* PhD thesis, Facultat de Matemàtiques, Universitat de València, València, 1992.

[Pen87]   J. Pense. *Äußere Fittingpaare.* Dissertation, Johannes Gutenberg-Universität Mainz, Mainz, 1987.

[Pen88]   J. Pense. Outer Fitting pairs. *J. Algebra*, 119(1):34–50, 1988.

[Pen90a]   J. Pense. Allgemeines über äußere Fittingpaare. *J. Austral. Math. Soc. Ser. A*, 49(2):241–249, 1990.

[Pen90b]   J. Pense. Fittingmengen und Lockettabschnitte. *J. Algebra*, 133(1): 168–181, 1990.

[Pen90c]   J. Pense. Notiz über Injektoren. *Arch. Math. (Basel)*, 54(5):422–426, 1990.

[Pen92]   J. Pense. Ränder und Erzeugendensysteme von Fittingklassen. *Math. Nachr.*, 156:117–127, 1992.

[Plo58]   B. I. Plotkin. Generalized soluble and nilpotent groups. *Uspehi Mat. Nauk.*, 13:89–172, 1958. Translation in *Amer. Math. Soc. Translations* (2), 17, 29–115 (1961).

[Rob02]   D. J. S. Robinson. Minimality and Sylow-permutability in locally finite groups. *Ukr. Math. J.*, 54(6):1038–1049, 2002.

[Sal]   E. Salomon. A non-injective Fitting class. Private communication.

[Sal85]   E. Salomon. Über lokale und Baerlokale Formationen endlicher Gruppen. Master's thesis, Johannes Gutenberg-Universität, Mainz, 1983.

[Sal87]   E. Salomon. *Strukturerhaltende untergruppen, Schunkklassen und extreme klassen endlicher Gruppen.* Dissertation, Johannes Gutenberg-Universität, Mainz, 1987.

[Sch66]   H. Schunck. *Zur Konstruktion von Systemen konjugierter Untergruppen in endlichen auflösbaren Gruppen.* PhD thesis, Christian-Albrechts-Universität zu Kiel, 1966.

[Sch67]   H. Schunck. 𝔥-Untergruppen in endichen auflösbaren Gruppen. *Math. Z.*, 97:326–330, 1967.

[Sch74]   P. Schmid. Lokale Formationen endlicher Gruppen. *Math. Z.*, 137:31–48, 1974.

364   References

[Sch77]   K.-U. Schaller. Über die maximale Formation in einem gesättigten Homomorph. *J. Algebra*, 45(453–464), 1977.

[Sch78]   P. Schmid. Every saturated formation is a local formation. *J. Algebra*, 51:144–148, 1978.

[Sco80]   L. Scott. Representations in characteristic *p*. In *Proc. Symp. Pure Math. The Santa Cruz Conf. on finite groups*, volume 37, page 327. AMS, 1980.

[SE02]   X. Soler-Escrivà. *On certain lattices of subgroups of finite groups. Factorizations*. PhD thesis, Nafarroako Unibertsitate publikoa-Universidad Pública de Navarra, 2002.

[Sem92]   V. N. Semenchuk. A characterization of š-formations. *Problems in Algebra*, 7:103–107, 1992. Russian.

[She72]   L. A. Shemetkov. Formation properties of finite groups. *Dokl. Akad. Nauk. SSSR*, 204(6):851–855, 1972.

[She74a]   L. A. Shemetkov. The complementability of the *F*-coradical and the properties of the *F*-hypercenter of a finite group. *Dokl. Akad. Nauk BSSR*, 18:204–206, 282, 1974.

[She74b]   L. A. Shemetkov. Graduated formations of groups. *Mat. Sb. (N.S.)*, 94(136):628–648, 656, 1974.

[She75]   L. A. Shemetkov. Two trends in the development of the theory of nonsimple finite groups. *Uspehi Mat. Nauk*, 30(2(182)):179–198, 1975.

[She76]   L. A. Shemetkov. Factorizaton of nonsimple finite groups. *Algebra i Logika*, 15(6):684–715, 744, 1976.

[She78]   L. A. Shemetkov. *Formations of finite groups*. Nauka, Moscow, 1978. Russian.

[She84]   L. A. Shemetkov. The product of formations. *Dokl. Akad. Nauk BSSR*, 28(2):101–103, 1984.

[She92]   L. A. Shemetkov. Some ideas and results in the theory of formations of finite groups. *Problems in Algebra*, 7:3–38, 1992.

[She97]   L. A. Shemetkov. Frattini extensions of finite groups and formations. *Comm. Algebra*, 25(3):955–964, 1997.

[She00]   L. A. Shemetkov. Radical and residual classes of finite groups. In *Proceedings of the International Algebraic Conference on the Occasion of the 90th. Birthday of A. G. Kurosh, Moscow, Russia, May 25–30, 1998*, pages 331–344, Berlin-New York, 2000. Yuri Bahturin, Walter de Gruyter.

[She01]   L. A. Shemetkov. On partially saturated formations and residuals of finite groups. *Comm. Algebra*, 29(9):4125–4137, 2001. Special issue dedicated to Alexei Ivanovich Kostrikin.

[Ski90]   A. N. Skiba. On a class of formations of finite groups. *Dokl. Akad. Nauk Belarus*, 34(11):982–985, 1990. Russian.

[Ski97]   A. N. Skiba. *Algebra formatsii*. Izdatel'stvo Belaruskaya Navuka, Minsk, 1997.

[Ski99]   A. N. Skiba. On factorizations of compositional formations. *Mat. Zametki*, 65(3):389–395, 1999.

[SR97]   A. N. Skiba and V. N. Ryzhik. Factorizations of *p*-local formations. In *Problems in algebra, No. 11 (Russian)*, pages 76–89. Gomel. Gos. Univ., Gomel', 1997.

[SS89]   L. A. Shemetkov and A. N. Skiba. On inherently non-decomposable formations. *Dokl. Akad. Nauk BSSR*, 37(7):581–583, 1989.

[SS95]   A. N. Skiba and L. A. Shemetkov. On partially local formations. *Dokl. Akad. Nauk Belarusi*, 39(3):9–11, 123, 1995.

[SS99]   A. N. Skiba and L. A. Shemetkov. Partially compositional formations of finite groups. *Dokl. Nats. Akad. Nauk Belarusi*, 43(4):5–8, 123, 1999.

[SS00a]   L. A. Shemetkov and A. N. Skiba. Multiply ω-local formations and Fitting classes of finite groups [translation of Mat. Tr. **2** (1999), no. 2, 114–147;]. *Siberian Adv. Math.*, 10(2):112–141, 2000.

[SS00b]   A. N. Skiba and L. A. Shemetkov. Multiply $\mathfrak{L}$-composition formations of finite groups. *Ukr. Math. J.*, 52(6):898–913, 2000.

[Suz82]   M. Suzuki. *Group theory I*, volume 247 of *Grundlehren der Mathematischen Wischenschaften*. Springer-Verlag, Berlin-Heidelberg-New York, 1982.

[Suz86]   M. Suzuki. *Group theory. II*, volume 248 of *Grundlehren der Mathematischen Wissenschaften (Fundamental Principles of Mathematical Sciences)*. Springer-Verlag, New York, 1986.

[SV84]   V. N. Semenchuk and A. F. Vasil′ev. Characterization of local formations $\mathfrak{F}$ by given properties of minimal non-$\mathfrak{F}$-groups. In V. I. Sergienko, editor, *Investigation of the normal and subgroup structure of finite groups. Proceedings of the Gomel's seminar, Minsk 1984*, volume 224, pages 175–181, Minsk, 1984. Nauka i Tekhnika. Russian.

[SW70]   G. M. Seitz and C. R. B. Wright. On complements of $\mathfrak{F}$-residuals in finite solvable groups. *Arch. Math. (Basel)*, 21:139–150, 1970.

[Tom75]   M. J. Tomkinson. Prefrattini subgroups and cover-avoidance properties in $\mathfrak{U}$-groups. *Canadian J. Math.*, 27:837–851, 1975.

[Tra98]   G. Traustason. A note of supersoluble Fitting classes. *Arch. Math. (Basel)*, 70:1–8, 1998.

[Vas87]   A. F. Vasil′ev. On the problem of the enumeration of local formations with a given property. *Problems in algebra, No. 3*, 126:3–11, 1987. Russian.

[Vas92]   A. F. Vasil′ev. On the enumeration of local formations with the Kegel condition. *Problems in algebra, No. 7*, pages 86–93, 1992. Russian.

[Ved88]   V. A. Vedernikov. On some classes of finite groups. *Dokl. Akad. Nauk BSSR*, 32(10):872–875, 1988.

[VK01]   A. F. Vasil′ev and S. F. Kamornikov. On the functor method for studying lattices of subgroups of finite groups. *Sibirsk. Mat. Zh.*, 42(1):30–40, i, 2001.

[VK02]   A. F. Vasil′ev and S. F. Kamornikov. On the Kegel-Shemetkov problem on lattices of generalized subnormal subgroups of finite groups. *Algebra Logika*, 41(4):411–428, 510, 2002.

[VKS93]   A. F. Vasil′ev, S. F. Kamornikov, and V. N. Semenchuk. On lattices of subgroups of finite groups. In N. S. Chernikov, editor, *Infinite groups and related algebraic structures*, pages 27–54, Kiev, 1993. Institut Matematiki AN Ukrainy. Russian.

[Vor93]   N. T. Vorob'ev. On factorizations of nonlocal formations of finite groups. *Problems in Algebra, No. 6*, pages 21–24, 1993.

[Wie39]   H. Wielandt. Eine Verallgemeinerung der invarianten Untergruppen. *Math. Z.*, 45:209–244, 1939.

[Wie57]   H. Wielandt. Vertauschbare nachinvariante Untergruppen. *Abh. Math. Sem. Univ. Hamburg*, 21(1-2):55–62, 1957.

[Wie58]   H. Wielandt. Über die Existenz von Normalteilern in endlichen Gruppen. *Math. Nachr.*, 18:274–280, 1958.

[Wie74]   H. Wielandt.   Kriterien für Subnormalität in endlichen Gruppen.
Math. Z., 138:199–203, 1974.

[Wie94a]   H. Wielandt. *Mathematische Werke-Mathematical Works*, volume 1:
Group Theory. Walter de Gruyter, 1994. Edited by B. Huppert and
H. Schneider.

[Wie94b]   H. Wielandt. Subnormale Untergruppen endlicher Gruppen. In *Mathematical Works* [Wie94a], pages 413–479. Vorlesung Univ. Tübingen,
1971.

[Yen70]   T. Yen. On $\mathfrak{F}$-normalizers. *Proc. Amer. Math. Soc.*, 26:49–56, 1970.

List of symbols

$[N]_\alpha H$ semidirect product of the
  H-group N with H via the action
  α

$[a, b]$ commutator of a and b, $a^{-1}b^{-1}ab$

$\mathrm{AGL}(n, p)$ 5

$\mathrm{A}_J(n_G)$ 348

$\mathrm{A}_p(G)$ 5

$\mathfrak{A}$ 87

$\mathrm{Alt}(n)$ 3

$\mathrm{Aut}(G)$ group of automorphisms of the
  group G

$\mathrm{Aut}_G(H/K)$ 41

$\mathfrak{B}_{\mathfrak{F}}$ 281

$\mathrm{BLF}(f)$ 97

$\mathcal{B}(\mathfrak{F})$ 288

$\mathrm{C}^{\mathfrak{X}_p}(G)$ 128

char $\mathfrak{X}$ 88

$\mathrm{Core}_G(H)$ 2

$\mathrm{Cosoc}(G)$ 98

$\mathrm{Cov}_{\mathfrak{H}^d(G)}$ 226

$\mathrm{Cov}_{\mathfrak{H}}(G)$ 101

$\mathrm{Crit}_S(3)$ 267

$\mathrm{D}_J(\Gamma_T)$ 349

$\mathrm{D}_J(n_G)$ 348

$\Delta_K(G)$ 123

$\mathfrak{E}$ 87

$\mathfrak{E}(\pi)$ 231

$\mathfrak{E}(n)$ 91

$\mathfrak{E}(p)$ 135

$\mathrm{E}(G)$ 98

$\mathrm{E}(q|p^e)$ 333

$\mathrm{E}_{\mathfrak{F}}(G)$ 113

$\mathfrak{F} \circ \mathfrak{G}$ 95

$\mathfrak{F} \times \mathfrak{G}$ 95

$\mathrm{F}'(G)$ 78

$\mathrm{F}(G)$ Fitting subgroup of the group G

$\mathrm{F}^*(G)$ 97

$\mathrm{Fit}\, 3$ 110

$\mathrm{GF}(q)$ finite field of q elements

$\mathrm{GL}(n, q)$ general linear group of
  dimension n over $\mathrm{GF}(q)$

$\mathrm{Hall}_\pi(G)$ 99

$\mathrm{Hom}(U, A)$ set of homomorphisms
  between U and A

$\mathrm{Hom}_{KG}(V, W)$ set of KG-
  homomorphisms from V to
  W

$\mathrm{I}(G)$ 258

$\mathrm{I}_K(E)$ 223

$\mathrm{Inj}_{\mathfrak{F}}(G)$ 114

$\mathrm{Inj}_{\mathcal{F}}(G)$ 114

$\mathrm{Inn}(G)$ group of inner automorphisms
  of the group G

$\mathrm{Irr}(M)$ 293

$\mathfrak{J}$ 87

$\mathrm{J}(KG)$ 5

$\mathrm{K}_G(X)$ 239

$\mathrm{K}_n(G)$ 93

$\mathfrak{K}(\mathfrak{X})$ 324

$\mathrm{Ker}(d, \mathbf{A})$ 344

$\mathrm{Ker}(f)$ kernel of the homomorphism f

$\mathrm{Ker}(x \text{ on } U)$ kernel of the action of the
  element $x \in G$ on the KG-module
  U

$\mathrm{LF}(f)$ 97

$\mathrm{LF}_{\mathfrak{X}}(f)$ 126

$\mathfrak{L}(\mathfrak{X})$ 142

$\mathrm{L}(G)$ 197

$\mathrm{L}_{\mathfrak{H}}(G)$ 197

$\mathrm{Locksec}(\mathfrak{X})$ 112

$\mathfrak{M}(G)$ 120

$\mathfrak{M}_p$ 295

$\mathrm{Max}(G)$ 52

$\mathrm{Max}(G)_{\mathfrak{N}}^n$ 197

$\mathrm{Max}^*(G)$ 53

$\mathrm{Max}^*(G)_{\mathfrak{H}}^a$ 192

$\mathrm{Max}^*(G)_{\mathfrak{N}}^a$ 197

$\mathrm{Max}_{\mathfrak{H}^d}(G)$ 225

$\mathrm{Max}_{\mathfrak{H}}(G)$ 101

$\mathrm{Mod}_{\mathfrak{F}}(U)$ 293

$\mathbb{N}$, set of all natural numbers

$\mathfrak{N}$ 87

$\mathfrak{N}_c$ 91

$\mathrm{Nor}_{\mathfrak{H}}(G)$ 171

$\mathrm{N}_G(\Sigma)$ 183

$\mathrm{N}_G(\mathbf{X}(\Sigma))$ 85

$\mathrm{O}^{p'}(G)$ smallest normal subgroup of G
  of p'-index

$\mathrm{O}^p(G)$ smallest normal subgroup of G
  of p-index

$\mathrm{O}_J(n_G)$ 348

$\mathrm{O}_{\mathfrak{Y}}(G)$ 126

$\mathrm{O}_\pi(G)$ largest normal π-subgroup of G

$\mathrm{O}_{p',p}(G)$ largest p-nilpotent normal
  subgroup of the group G

$\mathrm{O}_p(G)$ largest normal p-subgroup of G

Index of authors

Index